ANNE-MARIE [illegible]AT

W9-ARJ-234

#3[illegible]

tel: 741-2536

adresse: 727 Ludgate Court.

#9 p137

VECTOR CALCULUS

VECTOR CALCULUS

SECOND EDITION

JERROLD E. MARSDEN
UNIVERSITY OF CALIFORNIA, BERKELEY

ANTHONY J. TROMBA
UNIVERSITY OF CALIFORNIA, SANTA CRUZ

with the assistance of
Michael Hoffman and Joanne Seitz

W. H. FREEMAN AND COMPANY
SAN FRANCISCO

Project Editor: Larry Olsen
Copy Editor: Rebecca Stein
Designer: Sharon H. Smith
Production Coordinator: Bill Murdock
Illustration Coordinator: Audre Loverde
Artist: Evan Gillespie
Compositor: Syntax International
Printer and Binder: The Maple-Vail Book Manufacturing Group

The motion of waves is described by the equations of fluid mechanics, which are discussed on page 453. The notions of vector calculus can give insight into this and other complex phenomena. Photograph of Duxbury reef, Bolinas, California, by Nancy Williams.

Library of Congress Cataloging in Publication Data

Marsden, Jerrold E
Vector calculus.

Includes index.
1. Calculus. 2. Vector analysis. I. Tromba, Anthony, joint author. II. Title.
QA303.M338 1981 515'.63 80-24663
ISBN 0-7167-1244-X

Copyright © 1976, 1981 by W. H. Freeman and Company

No part of this book may be reproduced by any mechanical, photographic, or electronic process, or in the form of a phonographic recording, nor may it be stored in a retrieval system, transmitted, or otherwise copied for public or private use, without written permission from the publisher.

AMS 1970 subject classifications: 26A60, 35-01

Printed in the United States of America

9 8 7 6 5 4 3 2 1

To Ralph Abraham,
teacher and friend

Some calculus tricks are quite easy. Some are enormously difficult. The fools who write the textbooks of advanced mathematics seldom take the trouble to show you how easy the easy calculations are.

Silvanus P. Thompson
Calculus Made Easy
Macmillan (1910)

CONTENTS

PREFACE

This text is intended for a one-semester course in the calculus of functions of several variables and vector analysis at the sophomore level. Normally the course is preceded by a beginning course in linear algebra, but this is not an essential prerequisite. We require only the bare rudiments of matrix algebra, and the necessary concepts are developed in the text. However, we do assume a knowledge of the fundamentals of one-variable calculus—differentiation and integration of the standard functions.

The text includes most of the basic theory as well as many concrete examples and problems. Teaching experience at this level shows that it is desirable to omit many of the technical proofs; they are difficult for beginning students and are included in the text mainly for reference or supplementary reading. In particular, some of the technical proofs for theorems in Chapters 2 and 5 are given in the appendices. Section 2.2, on limits and continuity, is designed to be treated lightly and is deliberately brief. More sophisticated theoretical topics, such as compactness and delicate proofs in integration theory, have been omitted since they usually belong to and are better treated in a more advanced course.

Computational skills and intuitive understanding are important at this level, and we have endeavored to meet this need by making the book as concrete and student-oriented as possible. For example, although we formulate the definition of the derivative correctly, it is done by using matrices of partial derivatives rather than linear transformations. This device alone can save one or two weeks of teaching time and can save endless headaches for those students whose linear algebra is not in top form. Also, we include quite a large number of physical illustrations. Specifically, we have included examples from such sciences as fluid mechanics, gravitation, electromagnetic theory, and economics, although prior knowledge of these subjects is not assumed.

In this second edition we have added more optional sections (set off by heavy horizontal rules) on such theoretical topics as the implicit function theorem, properties of the integral, and the geometric meaning of divergence and curl. Furthermore, additional applications in the text and the exercises have been included to give greater versatility. We have expanded the section on Lagrange multipliers and their applications to include applications in economics and have added new material on fluid mechanics. The concept of area zero has been eliminated from the integration chapters, although more proofs are given in the optional Appendix B. Cylindrical and spherical coordinates have been moved to Chapter 1 and are given a more prominent role throughout the text. In this edition we have also added numerous historical notes to give the student a greater sense of the history of vector calculus.

Another special feature of the text is the early introduction of vector fields, divergence, and curl in Chapter 3, before integration. Vector analysis usually suffers in a course of this type, and the present arrangement is designed to offset this tendency. To go even further, one might consider teaching Chapter 4 (Taylor's theorem, maxima and minima, Lagrange multipliers) after Chapter 7 (vector analysis).

Many colleagues and students in the mathematical community have made valuable contributions and suggestions since this book was begun. An early draft of the book was written in collaboration with Ralph Abraham. We thank him for allowing us to draw upon his work. It is impossible to list all those who assisted with this book, but we wish especially to thank our assistants Michael Hoffman and Joanne Seitz; we also received valuable comments from Mary Anderson, John Ball, Frank Gerrish, Richard Koch, Andrew Lenard, David Merriell, Jeanette Nelson, Dan Norman, Keith Phillips, Anne Perleman, Kenneth Ross, Diane Sauvageot, Joel Smoller, Ralph and Bob Tromba, Steve Wan, Alan Weinstein, and John Wilker.

A final word of thanks goes to those who helped in the preparation of the manuscript and the production of the book. We especially thank Connie Calica, Nora Lee, Marnie McElhiney, Rosemarie Stampful, Ruth Suzuki, and Ikuko Workman for their excellent typing of the manuscript; Herb

Holden of Stanford Research Institute and Jerry Kazdan of the University of Pennsylvania for suggesting and preparing the computer-generated figures; and Peter Renz, Larry Olsen, and Rebecca Stein, who have given valuable advice on mathematical and editorial aspects of the project.

November 1980

Jerrold E. Marsden
Anthony J. Tromba

NOTATION

The student is assumed to have studied the calculus of functions of a real variable, including analytic geometry in the plane. Some students may have had some exposure to matrices as well, although what we shall need is given in Sections 1.3 and 1.5.

The student is also assumed to be familiar with functions of elementary calculus, such as $\sin x$, $\cos x$, e^x, and $\log x$ (we write $\log x$ for the natural logarithm, which is sometimes denoted $\ln x$ or $\log_e x$). The student is expected to know, or to review as the course proceeds, the basic rules of differentiation and integration for functions of one variable, such as the Chain Rule, the quotient rule, integration by parts, and so forth.

We shall review here the notations to be used later, often without explicit mention. This can be read through quickly now, then referred to later if the need arises.

The collection of all real numbers is denoted $\mathbb{R}$. Thus $\mathbb{R}$ includes the *integers*, . . . , -3, -2, -1, 0, 1, 2, 3, . . . ; the *rational numbers*, p/q, where p and q are integers ($q \neq 0$); and the *irrational numbers*, like $\sqrt{2}$, π, e, etc. Members of $\mathbb{R}$ may be visualized as points on the real-number line as shown in Figure 0.1.

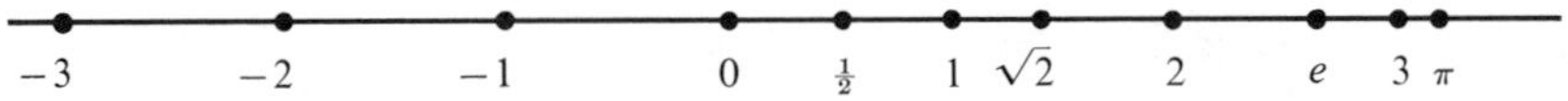

FIGURE 0.1

The geometric representation of points on the real number line.

When we write $a \in \mathbb{R}$ we mean that a is a member of the set $\mathbb{R}$; in other words, that a is a real number. Given two real numbers a and b with $a < b$ (that is, a is less than b), we can form the *closed interval* $[a, b]$, consisting of all x such that $a \leq x \leq b$ and the *open interval* $]a, b[$, consisting of all x such that $a < x < b$. (In other books the open interval is often denoted (a, b).) Similarly, we may form half-open intervals $]a, b]$ and $[a, b[$ (Figure 0.2).

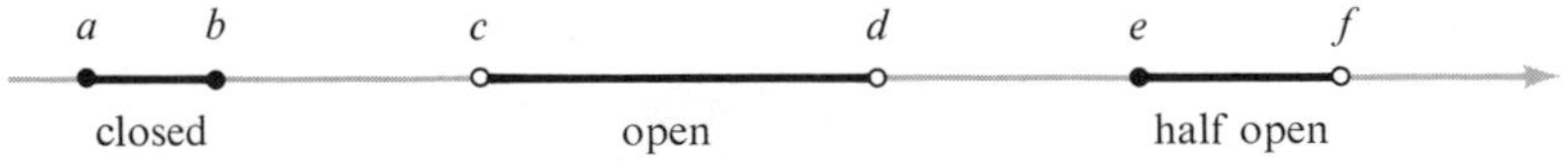

FIGURE 0.2

The geometric representation of the intervals $[a,b]$, $]c,d[$ *and* $[e,f[$.

The *absolute value* of a number $a \in \mathbb{R}$ is written $|a|$ and is defined as

$$|a| = \begin{cases} a & \text{if } a \geq 0 \\ -a & \text{if } a < 0 \end{cases}$$

For example, $|3| = 3$, $|-3| = 3$, $|0| = 0$, and $|-6| = 6$. The inequality $|a + b| \leq |a| + |b|$ always holds. The *distance from a to b* is given by $|a - b|$. Thus, the distance from 6 to 10 is 4 and from -6 to 3 is 9.

If we write $A \subset \mathbb{R}$, we mean A is a *subset* of $\mathbb{R}$. For example, A could equal the set of integers $\{\ldots, -3, -2, -1, 0, 1, 2, 3, \ldots\}$. Another example of a subset of $\mathbb{R}$ is the set $\mathbb{Q}$ of rational numbers. Generally, for two collections of objects (i.e., sets) A and B, $A \subset B$ means A is a subset of B; that is, every member of A is also a member of B.

The symbol $A \cup B$ means the *union* of A and B, the collection whose members are members of either A or B. Thus

$$\{\ldots, -3, -2, -1, 0\} \cup \{-1, 0, 1, 2, \ldots\} = \{\ldots, -3, -2, -1, 0, 1, 2, \ldots\}.$$

Similarly $A \cap B$ means the *intersection* of A and B; that is, this set consists of those members of A and B that are in both A and B. Thus the intersection of the two sets above is $\{-1, 0\}$.

We shall write $A \backslash B$ for those members of A that are not in B. Thus

$$\{\ldots, -3, -2, -1, 0\} \backslash \{-1, 0, 1, 2, \ldots\} = \{\ldots, -3, -2\}.$$

We can also specify sets as in the following examples:

$$\{a \in \mathbb{R} \mid a \text{ is an integer}\} = \{\ldots, -3, -2, -1, 0, 1, 2, \ldots\}$$

$$\{a \in \mathbb{R} \mid a \text{ is an even integer}\} = \{\ldots, -2, 0, 2, 4, \ldots\}$$

$$\{x \in \mathbb{R} \mid a \leq x \leq b\} = [a, b]$$

A *function* $f: A \to B$ is a rule that assigns to each $a \in A$ one specific member $f(a)$ of B. The fact that the function f sends a to $f(a)$ is denoted symbolically by $a \mapsto f(a)$. For example, $f(x) = x^3/(1 - x)$ assigns the number $x^3/(1 - x)$ to each $x \neq 1$ in $\mathbb{R}$. We can specify a function f by giving the rule for $f(x)$. Thus, the above function f can be defined by the rule $x \mapsto x^3/(1 - x)$.

If $A \subset \mathbb{R}$, $f: A \subset \mathbb{R} \to \mathbb{R}$ means that f assigns a value in $\mathbb{R}$, $f(x)$, to each $x \in A$. The set A is called the *domain* of f, and we say f has *range* $\mathbb{R}$, since that is where the values of f are taken. The *graph* of f consists of all the points $(x, f(x))$ in the plane (Figure 0.3). Generally, a *mapping* (=function = transformation = map) $f: A \to B$, where A and B are sets, is a rule that assigns to each $x \in A$ a specific point $f(x) \in B$.

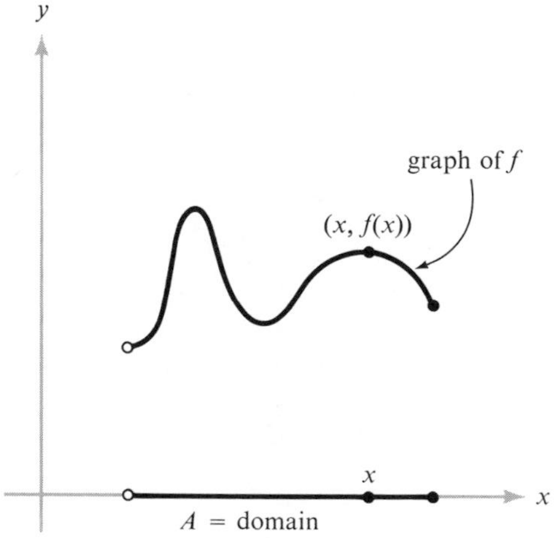

FIGURE 0.3

The graph of a function with the half-open interval A as domain.

The notation $\sum_{i=1}^{n} a_i$ means $a_1 + \cdots + a_n$, where $a_1, \ldots, a_n$ are given numbers. The sum of the first n integers is

$$1 + 2 + \cdots + n = \sum_{i=1}^{n} i = \frac{n(n+1)}{2}$$

The *derivative* of a function $f(x)$ is denoted $f'(x)$ or

$$\frac{df}{dx}$$

and the *definite integral* is written

$$\int_a^b f(x)\,dx$$

If we set $y = f(x)$, the derivative is also denoted by

$$\frac{dy}{dx}.$$

Readers are assumed to be familiar with the Chain Rule, integration by parts, and other rules that govern the calculus of functions of one variable. In particular, they should know how to differentiate and integrate exponential, logarithmic, and trigonometric functions. Appendix C offers a short table of derivatives and integrals, which is adequate for the needs of this text.

The following notations are used synonymously: $e^x = \exp(x)$, $\ln x = \log x$, and $\sin^{-1} x = \arcsin x$.

VECTOR CALCULUS

CHAPTER 1

THE GEOMETRY OF EUCLIDEAN SPACE

Quaternions came from Hamilton . . . and have been an unmixed evil to those who have touched them in any way. Vector is a useless survival . . . and has never been of the slightest use to any creature.

LORD KELVIN

In this chapter we shall consider some basic operations on vectors in three-dimensional space: vector addition, scalar multiplication, and the dot and cross products. In Section 1.5 we shall generalize some of these notions to n-space, and briefly review some properties of matrices that will be needed in Chapters 2 and 3.

1.1 VECTORS IN THREE-DIMENSIONAL SPACE

We recall that points P in the plane may be represented by ordered pairs of real numbers (a, b) called Cartesian coordinates. If we draw two perpendicular lines and label them the x- and y- axes, we may drop perpendiculars from P to these axes as in Figure 1.1.1; a is called the x-component of P and b is called the y-component.

Points in space may be similarly represented as ordered triples of real numbers. To construct such a representation, we choose three mutually perpendicular lines that meet at a point in space. These lines are called the x-axis, y-axis, and z-axis, and the point at which they meet is called the *origin*

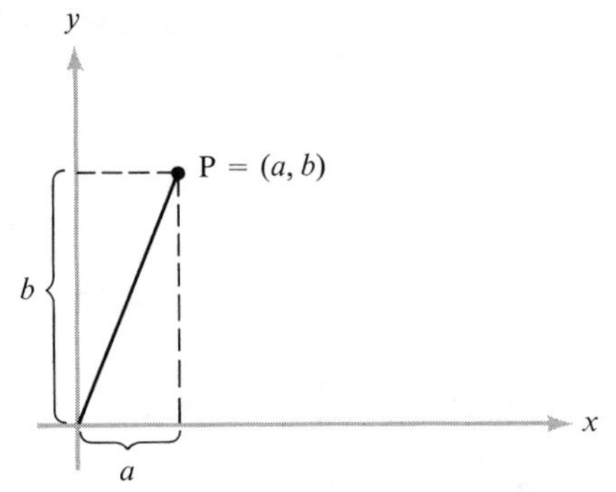

FIGURE 1.1.1
Cartesian coordinates in the plane.

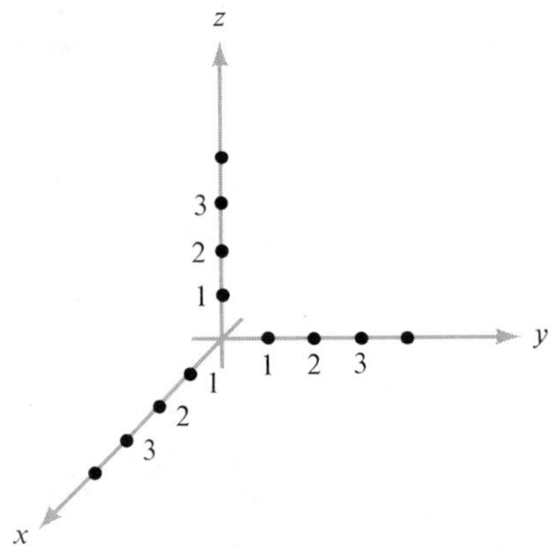

FIGURE 1.1.2
Cartesian coordinates in space.

(this is our reference point). The set of axes is often referred to as a *coordinate system*, and it is drawn as shown in Figure 1.1.2.

If we identify a real number with each point on these axis lines (as we did with the real number line), then we may assign to each point P in space a unique ordered triple of real numbers (a, b, c) and, conversely, to each triple we may assign a unique point in space.

Let the triple $(0, 0, 0)$ correspond to the origin of the coordinate system and let the arrows on the axes indicate the positive directions. Then, for example, the triple $(2, 4, 4)$ represents a point 2 units from the origin in the positive direction along the x-axis, 4 units in the positive direction along the y-axis, and 4 units in the positive direction along the z-axis. This can be done in any order (Figure 1.1.3).

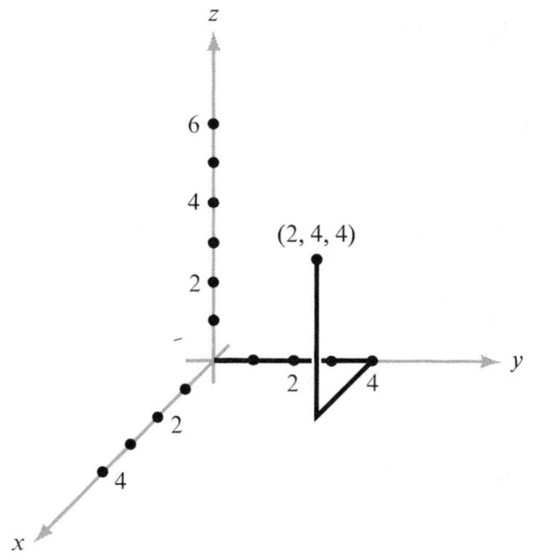

FIGURE 1.1.3
Geometric representation of the point (2, 4, 4) *in Cartesian coordinates.*

Because we can associate points in space and ordered triples in this way, we often use the expression "the point (a, b, c)" instead of the longer phrase "the point P that corresponds to the triple (a, b, c)." If the triple (a, b, c) corresponds to P, we say that a is the x-coordinate (or first coordinate), b is the y-coordinate (or second coordinate), and c is the z-coordinate (or third coordinate) of P. With this method of representing points in mind, we see that the x-axis consists of the points of the form $(a, 0, 0)$, where a is any real number, the y-axis consists of the points $(0, a, 0)$, and the z-axis consists of the points $(0, 0, a)$. It is also common to denote points in space with the letters x, y, and z in place of a, b, and c. Thus the triple (x, y, z) represents a point whose first coordinate is x, second coordinate is y, and third coordinate is z.

What we have just constructed is a model of three-dimensional space. In the next few paragraphs we shall examine the mathematical properties of this model, and then consider its geometric interpretation.

We employ the following notation for the line, the plane, and three-dimensional space.

(*i*) The real line is denoted $\mathbb{R}^1$ (thus, $\mathbb{R}$ and $\mathbb{R}^1$ are identical).

(*ii*) The set of all ordered pairs (x, y) of real numbers is denoted $\mathbb{R}^2$.

(*iii*) The set of all ordered triples (x, y, z) of real numbers is denoted $\mathbb{R}^3$.

When speaking of $\mathbb{R}^1$, $\mathbb{R}^2$, and $\mathbb{R}^3$ collectively, we write $\mathbb{R}^n$, $n = 1, 2,$ or 3; or $\mathbb{R}^m$, $m = 1, 2, 3$.

The operation of addition can be extended from $\mathbb{R}$ to $\mathbb{R}^2$ and $\mathbb{R}^3$. For $\mathbb{R}^3$, this proceeds as follows. Given the two triples (x, y, z) and (x', y', z'), we define their *sum* by

$$(x, y, z) + (x', y', z') = (x + x', y + y', z + z')$$

EXAMPLE 1.

$$\begin{aligned}(1, 1, 1) + (2, -3, 4) &= (3, -2, 5)\\ (x, y, z) + (0, 0, 0) &= (x, y, z)\\ (1, 7, 3) + (2, 0, 6) &= (3, 7, 9)\end{aligned}$$

The element $(0, 0, 0)$ is called the *zero element* of $\mathbb{R}^3$. The element $(-x, -y, -z)$ is called the *additive inverse* of (x, y, z), and we write $(x, y, z) - (x', y', z')$ for $(x, y, z) + (-x', -y', -z')$.

There are some very important product operations in $\mathbb{R}^3$. One of these, called the inner product, assigns a real number to each pair of elements of $\mathbb{R}^3$. We shall discuss the inner product in detail in Section 1.2. Another important product operation for $\mathbb{R}^3$ is called scalar multiplication (the word "scalar" is here a synonym for "real number"). This product combines scalars (real numbers) and elements of $\mathbb{R}^3$ (ordered triples) to yield elements of $\mathbb{R}^3$ as follows: given a scalar α and a triple (x, y, z), we define the scalar multiple *or scalar product* by

$$\alpha(x, y, z) = (\alpha x, \alpha y, \alpha z)$$

EXAMPLE 2.

$$\begin{aligned}2(4, e, 1) &= (2 \cdot 4, 2 \cdot e, 2 \cdot 1) = (8, 2e, 2)\\ 6(1, 1, 1) &= (6, 6, 6)\\ 1(x, y, z) &= (x, y, z)\\ 0(x, y, z) &= (0, 0, 0)\\ (\alpha + \beta)(x, y, z) &= ((\alpha + \beta)x, (\alpha + \beta)y, (\alpha + \beta)z)\\ &= (\alpha x + \beta x, \alpha y + \beta y, \alpha z + \beta z)\\ &= \alpha(x, y, z) + \beta(x, y, z)\end{aligned}$$

It is an immediate result of the definitions that addition and scalar multiplication for $\mathbb{R}^3$ satisfy the following identities:

(*i*) $(\alpha\beta)(x, y, z) = \alpha(\beta(x, y, z))$ (associativity)

(*ii*) $(\alpha + \beta)(x, y, z) = \alpha(x, y, z) + \beta(x, y, z)$ (distributivity)

(*iii*) $\alpha((x, y, z) + (x', y', z')) = \alpha(x, y, z) + \alpha(x', y', z')$ (distributivity) (*iv*) $\alpha(0, 0, 0) = (0, 0, 0)$ (properties of zero elements)

(*v*) $0(x, y, z) = (0, 0\ 0)$

(*vi*) $1(x, y, z) = (x, y, z)$ (property of the identity element)

For $\mathbb{R}^2$, addition is defined just as in $\mathbb{R}^3$, by

$$(x, y) + (x', y') = (x + x', y + y')$$

and scalar multiplication is defined by

$$\alpha(x, y) = (\alpha x, \alpha y)$$

We often identify $\mathbb{R}^2$ with the set of triples $(x, y, 0)$ and speak of $\mathbb{R}^2$ geometrically as the plane (or more precisely, the xy-plane).

Now let us turn to the geometry of our model. One of the more fruitful tools of mathematics and its applications has been the notion of a *vector*. We define a (geometric) vector to be a directed line segment beginning at the origin, that is, a line segment with specified magnitude and direction, with initial point at the origin. Have you heard pilots say: "We are now vectoring in on the landing strip"? They are referring to the vector giving the direction and distance of the airplane from the landing strip. Needless to say, both direction and distance are important here.

Figure 1.1.4 shows several vectors. Vectors may thus be thought of as arrows beginning at the origin. They are generally printed thus: **v**.

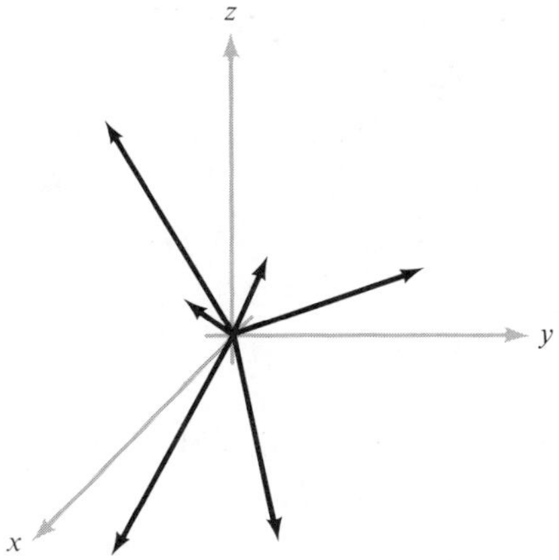

FIGURE 1.1.4
Geometrically, vectors are thought of as arrows emanating from the origin.

Using this definition of a vector, we may associate with each vector **v** the point in space (x, y, z) where **v** terminates, and conversely with each point (x, y, z) in space we can associate a vector **v**. Thus we shall identify **v** with

$(x\ y, z)$ and write $\mathbf{v} = (x, y, z)$. For this reason, the elements of $\mathbb{R}^3$ not only are ordered triples of real numbers, but are also called vectors. The triple $(0, 0, 0)$ is denoted $\mathbf{0}$.

We say that two vectors are *equal* if and only if they have the same direction and the same magnitude. This condition may be expressed algebraically by saying that if $\mathbf{v}_1 = (x, y, z)$ and $\mathbf{v}_2 = (x', y', z')$, then

$$\mathbf{v}_1 = \mathbf{v}_2 \quad \text{if and only if} \quad x = x', y = y', z = z'$$

Geometrically, we define vector addition as follows. In the plane containing the vectors $\mathbf{v}_1$ and $\mathbf{v}_2$ (see Figure 1.1.5), let us form the parallelogram having $\mathbf{v}_1$ as one side and $\mathbf{v}_2$ as its adjacent side. Then the sum $\mathbf{v}_1 + \mathbf{v}_2$ is the directed line segment along the diagonal of the parallelogram. This geometric view of vector addition is useful in many physical situations, as we shall see later. For an easily visualized example, consider a bird or an airplane flying through the air with velocity $\mathbf{v}_1$, in a wind with velocity $\mathbf{v}_2$. The resultant velocity, $\mathbf{v}_1 + \mathbf{v}_2$, is what you see.

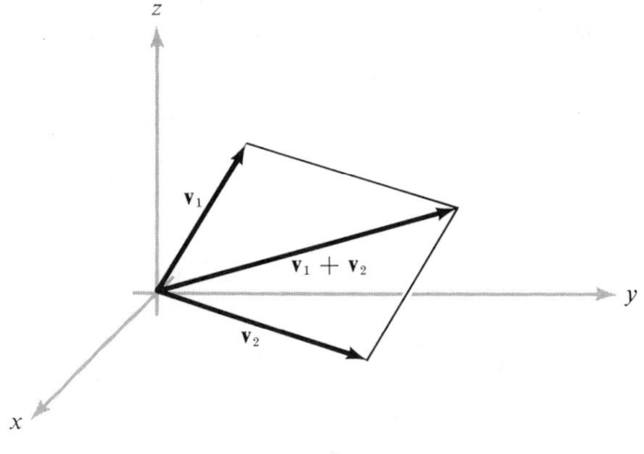

FIGURE 1.1.5
The geometry of vector addition.

To show that our geometric definition of addition is consistent with our algebraic definition, we must demonstrate that $\mathbf{v}_1 + \mathbf{v}_2 = (x + x', y + y', z + z')$. We shall prove this result in the plane, and leave the reader to formulate the proposition for three-dimensional space. Thus we wish to show that if $\mathbf{v}_1 = (x, y)$ and $\mathbf{v}_2 = (x', y')$, then $\mathbf{v}_1 + \mathbf{v}_2 = (x + x', y + y')$.

In Figure 1.1.6 let $\mathbf{v}_1 = (x, y)$ be the vector ending at the point A, and let $\mathbf{v}_2 = (x', y')$ be the vector ending at point B. By definition, the vector $\mathbf{v}_1 + \mathbf{v}_2$ ends at the vertex C of parallelogram OBCA. Hence, to verify that $\mathbf{v}_1 + \mathbf{v}_2 = (x + x', y + y')$, it is sufficient to show that the coordinates of C are $(x + x', y + y')$.

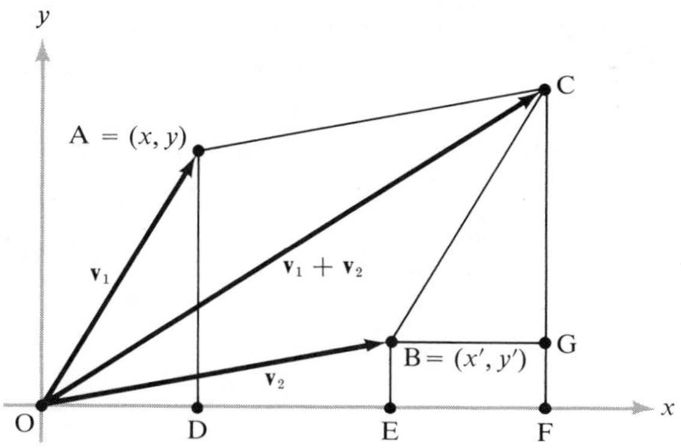

FIGURE 1.1.6
Construction for the proof that $(x, y) + (x', y') = (x + x', y + y')$.

From Figure 1.1.6, it is clear that triangle OAD is congruent to triangle BCG. By the congruence relation, BG = OD; and since BGFE is a rectangle, we have EF = BG. Furthermore, OD = x and OE = x'. Hence EF = BG = OD = x. Since OF = EF + OE, it follows that OF = $x + x'$. This shows that the x-coordinate of C is $x + x'$. The proof for the y-coordinate is analogous. With a similar argument for the other quadrants, we see that the geometric definition of vector addition is equivalent to the algebraic definition in terms of coordinates.

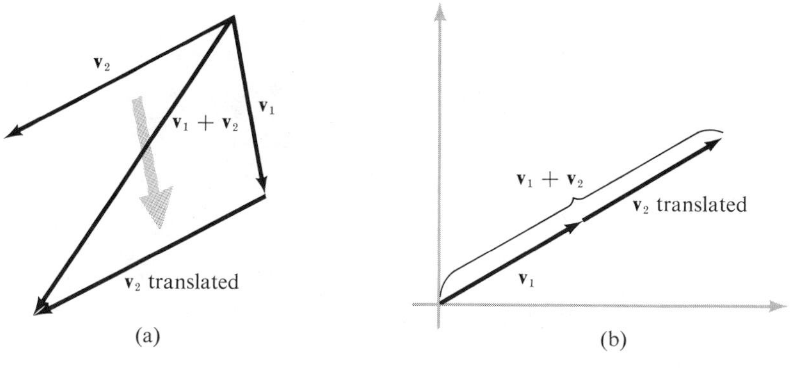

FIGURE 1.1.7
Vector addition may be visualized in terms of triangles as well as parallelograms, as demonstrated in (a). *However, the triangle collapses when* $\mathbf{v}_1$, *and* $\mathbf{v}_2$ *are collinear, as can be seen in* (b).

Figure 1.1.7(a) illustrates another way of looking at vector addition; that is, we translate (without rotation) the directed line segment representing the vector $\mathbf{v}_2$ so that it begins at the end of the vector $\mathbf{v}_1$. The endpoint of the resulting directed segment is the endpoint of the vector $\mathbf{v}_1 + \mathbf{v}_2$. We note that when $\mathbf{v}_1$ and $\mathbf{v}_2$ are collinear, the triangle collapses. This situation is illustrated in Figure 1.1.7(b)

Scalar multiples of vectors have similar geometric interpretations. If α is a scalar and $\mathbf{v}$ a vector, we define $\alpha\mathbf{v}$ to be the vector which is α times as long as $\mathbf{v}$ with the same direction as $\mathbf{v}$ if $\alpha > 0$, but with the opposite direction if $\alpha < 0$. Figure 1.1.8 illustrates several examples.

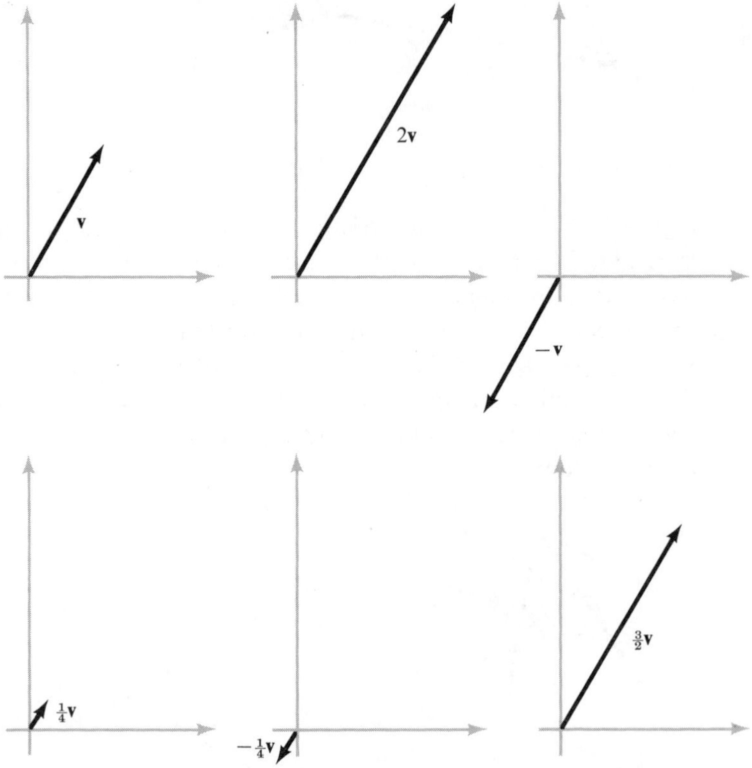

FIGURE 1.1.8
Some scalar multiples of a vector $\mathbf{v}$.

Using an argument that depends on similar triangles we can prove that if $\mathbf{v} = (x, y, z)$, then

$$\alpha\mathbf{v} = (\alpha x, \alpha y, \alpha z) \tag{1}$$

that is, the geometric definition coincides with the algebraic one.

How do we represent the vector $\mathbf{b} - \mathbf{a}$ geometrically? Since $\mathbf{a} + (\mathbf{b} - \mathbf{a}) = \mathbf{b}$, $\mathbf{b} - \mathbf{a}$ is that vector which when added to a $\mathbf{a}$ gives $\mathbf{b}$. In view of this, we may conclude that $\mathbf{b} - \mathbf{a}$ is the vector parallel to, and with the same magnitude as, the directed line segment beginning at the endpoint of $\mathbf{a}$ and terminating at the endpoint of $\mathbf{b}$ (see Figure 1.1.9).

$$\frac{86400}{1,4 \times 10^{-5}} = 6,17 \times 10^{9}$$

$8,64 \times 10^{4}$

1s → $1,4 \times 10^{-5}$

$\times 10^{-4}$

$186400 \times 10^{+19}$

$0,61714,286 \times 10^{-5} \times 10 \times 10^{4}$

$6,2 \times 10^{9} \times 10^{5}$

86400×10^{5}

$$\frac{1,400 \times 10^{-5}}{864}$$

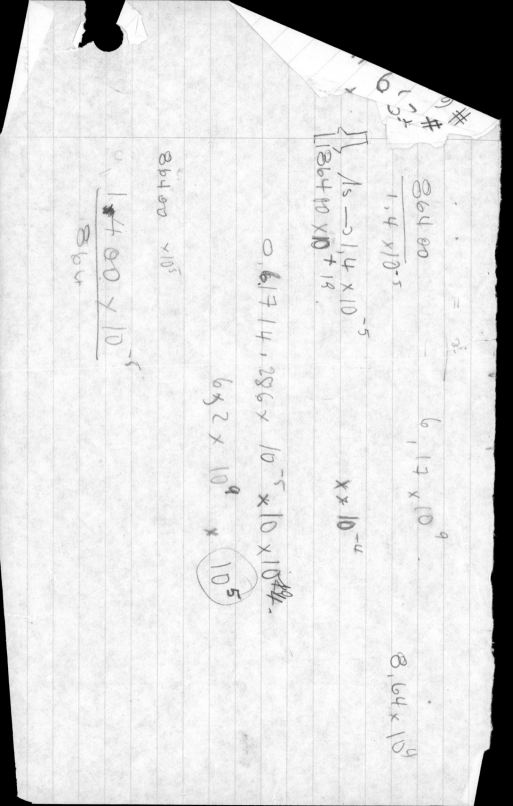

chap onde. optique en Physique. + devoir de physique.

Lire chap en informatique. + pgm.

Dessin.

Lab de chimie.

Lab de physique.

CSI 1590 -
DESSIN

p146.

p161.

#12 p197

#18 p198

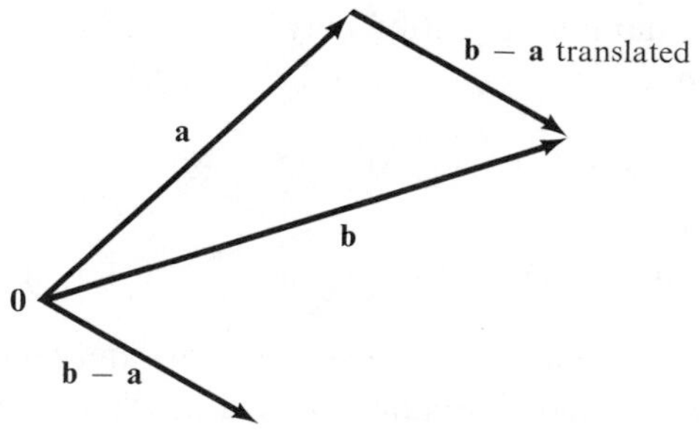

FIGURE 1.1.9
The geometry of vector subtraction.

Let us denote by **i** the vector that ends at (1, 0, 0), **j** the vector that ends at (0, 1, 0), and **k** the vector that ends at (0, 0, 1). Then using (1) above we find that if $\mathbf{v} = (x, y, z)$,

$$\mathbf{v} = x(1, 0, 0) + y(0, 1, 0) + z(0, 0, 1) = x\mathbf{i} + y\mathbf{j} + z\mathbf{k}$$

Hence we can represent any vector in three-dimensional space in terms of the vectors, **i**, **j**, and **k**. For this reason the vectors **i**, **j**, and **k** are called the *standard basis vectors* for $\mathbb{R}^3$.

EXAMPLE 3. The vector ending at (2, 3, 2) is $2\mathbf{i} + 3\mathbf{j} + 2\mathbf{k}$, and the vector ending at $(0, -1, 4)$ is $-\mathbf{j} + 4\mathbf{k}$ (see Figure 1.1.10).

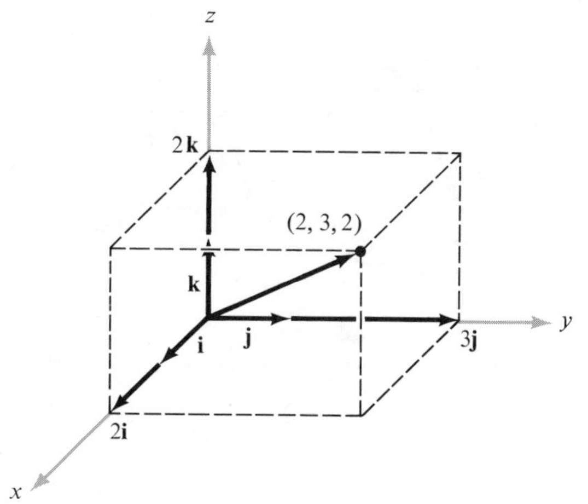

FIGURE 1.1.10
Representation of (2, 3, 2) *in terms of the standard basis* **i**, **j**, *and* **k**.

Addition and scalar multiplication may be written in terms of the standard basis vectors as follows:

$$(x\mathbf{i} + y\mathbf{j} + z\mathbf{k}) + (x'\mathbf{i} + y'\mathbf{j} + z'\mathbf{k}) = (x + x')\mathbf{i} + (y + y')\mathbf{j} + (z + z')\mathbf{k}$$

and

$$\alpha(x\mathbf{i} + y\mathbf{j} + z\mathbf{k}) = (\alpha x)\mathbf{i} + (\alpha y)\mathbf{j} + (\alpha z)\mathbf{k}$$

Because of the correspondence between vectors and points, we may sometimes refer to a *point* $\mathbf{a}$ under circumstances in which $\mathbf{a}$ has been defined to be a vector. The reader should realize that by this statement we mean the *endpoint* of the vector $\mathbf{a}$.

EXAMPLE 4. As an example of the use of vectors, let us describe the points that lie in the parallelogram whose adjacent sides are the vectors $\mathbf{a}$ and $\mathbf{b}$.

Consider Figure 1.1.11. If P is any point in the given parallelogram and we construct lines l_1 and l_2 through P parallel to the vectors $\mathbf{a}$ and $\mathbf{b}$, respectively, we see that l_1 intersects the side of the parallelogram determined by the vector $\mathbf{b}$ at some point $t\mathbf{b}$, where $0 \leq t \leq 1$. Likewise, l_2 intersects the side determined by the vector $\mathbf{a}$ at some point $s\mathbf{a}$, where $0 \leq s \leq 1$.

Since P is now the endpoint of the diagonal of a parallelogram having adjacent sides $s\mathbf{a}$ and $t\mathbf{b}$, if $\mathbf{v}$ denotes the vector ending at P, we see that $\mathbf{v} = s\mathbf{a} + t\mathbf{b}$. Thus, all the points in the given parallelogram are endpoints of vectors of the form $s\mathbf{a} + t\mathbf{b}$ for $0 \leq s \leq 1$ and $0 \leq t \leq 1$. By reversing our steps it is easy to see that all vectors of this form end within the parallelogram.

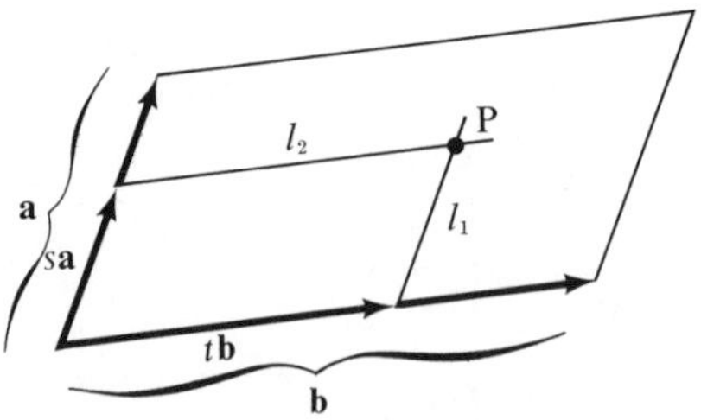

FIGURE 1.1.11
Describing points within the parallelogram formed by vectors $\mathbf{a}$ *and* $\mathbf{b}$.

Since two lines through the origin determine a plane, two nonparallel vectors also determine a plane. If we apply the same reasoning as in the example above we see that the plane formed by two nonparallel vectors

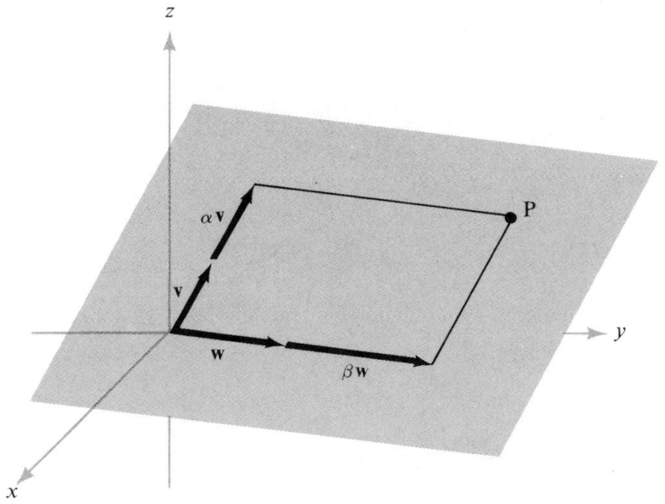

FIGURE 1.1.12
Describing points P *in the plane formed from vectors* **v** *and* **w**.

v and **w** consists of all points of the form $\alpha\mathbf{v} + \beta\mathbf{w}$ where α and β vary over the real numbers. Notice that any point P in the plane formed by the two vectors will be the opposite vertex of the parallelogram determined by $\alpha\mathbf{v}$ and $\beta\mathbf{w}$, where α and β are some scalars (see Figure 1.1.12).

The plane determined by **v** and **w** is called the plane *spanned by* **v** and **w**. When $\mathbf{v} = \gamma\mathbf{w}$, a multiple of $\mathbf{w}\,(\mathbf{w} \neq \mathbf{0})$, then **v** and **w** are parallel and the plane degenerates to a straight line. When $\mathbf{v} = \mathbf{w} = \mathbf{0}$ (zero vectors) we obtain a single point.

There are three particular planes that arise naturally in a coordinate system and which will be of use to us later. We call the plane spanned by vectors **i** and **j** the xy-plane, the plane spanned by **j** and **k** the yz-plane, and the plane spanned by **i** and **k** the xz-plane. These planes are illustrated in Figure 1.1.13.

Planes and lines are geometric objects that can be represented by equations. We shall defer until Section 1.3 a study of equations representing planes. However, using the geometric interpretation of vector addition and scalar multiplication, we may find the equation of a line l that passes through the endpoint of the vector **a**, with the direction of a vector **v** (see Figure 1.1.14). As t varies through all real values, the points of the form $t\mathbf{v}$ are all scalar multiples of the vector **v**, and therefore exhaust the points of the line passing through the origin in the direction of **v**. Since every point on l is the endpoint of the diagonal of a parallelogram with sides **a** and $t\mathbf{v}$, for some suitable value of t, we see that all the points on l are of the form $\mathbf{a} + t\mathbf{v}$. Thus, the line l

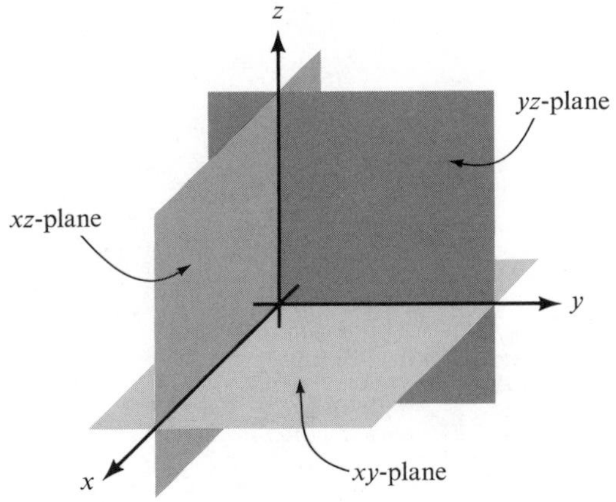

FIGURE 1.1.13
The three coordinate planes.

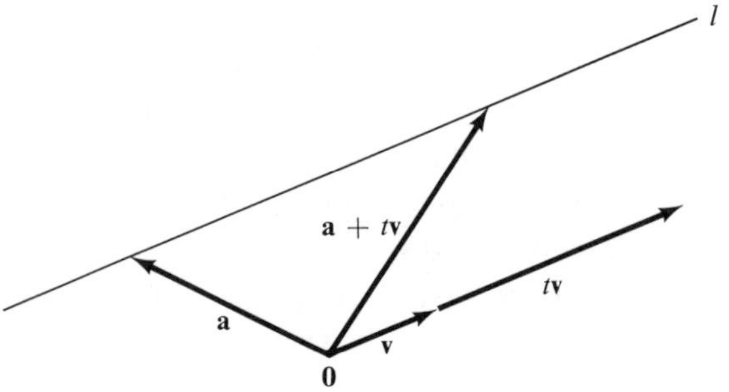

FIGURE 1.1.14
The line l, parametrically given by $\mathbf{l}(t) = \mathbf{a} + t\mathbf{v}$, *lies in the direction* $\mathbf{v}$ *and passes through the tip of* $\mathbf{a}$.

may be expressed by the equation $\mathbf{l}(t) = \mathbf{a} + t\mathbf{v}$. We say that l is expressed *parametrically*, with t the parameter. At $t = 0$, $\mathbf{l}(t) = \mathbf{a}$. As t increases, the point $\mathbf{l}(t)$ moves away from $\mathbf{a}$ in the direction of $\mathbf{v}$. As t decreases from $t = 0$ through negative values, $\mathbf{l}(t)$ moves away from $\mathbf{a}$ in the direction of $-\mathbf{v}$.

There can be many parametrizations of the same line. These may be obtained by choosing instead of $\mathbf{a}$ a different point on the given line, and forming the parametric equation of the line beginning at that point and in the direction of $\mathbf{v}$. For example, the endpoint of $\mathbf{a} + \mathbf{v}$ is on the line $\mathbf{l}(t) = \mathbf{a} + t\mathbf{v}$, and thus, $\mathbf{l}_1(t) = (\mathbf{a} + \mathbf{v}) + t\mathbf{v}$ represents the same line. Still other parametrizations may be obtained by observing that if $\alpha \neq 0$, the vector $\alpha\mathbf{v}$ has the same (or opposite) direction as $\mathbf{v}$. Thus $\mathbf{l}_2(t) = \mathbf{a} + t\alpha\mathbf{v}$ is another parametrization of $\mathbf{l}(t) = \mathbf{a} + t\mathbf{v}$.

EXAMPLE 5. Determine the equation of the line passing through (1, 0, 0) in the direction of $\mathbf{j}$.

The desired line can be expressed parametrically as $\mathbf{l}(t) = \mathbf{i} + t\mathbf{j}$ (Figure 1.1.15). In terms of coordinates we have

$$\mathbf{l}(t) = (1, 0, 0) + t(0, 1, 0) = (1, t, 0)$$

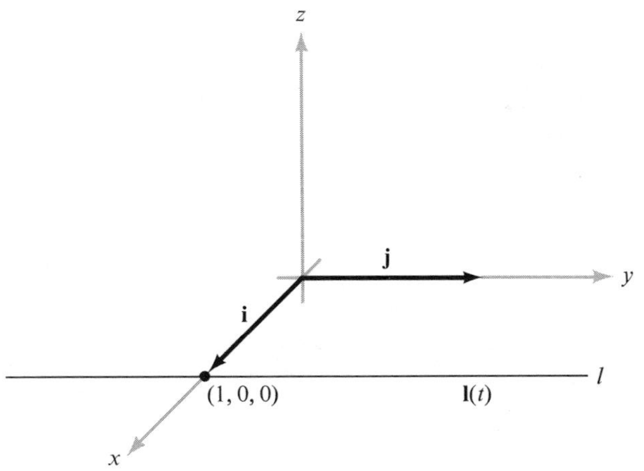

FIGURE 1.1.15
The line l, parametrically given by $\mathbf{l}(t) = \mathbf{i} + t\mathbf{j}$, passes through the tip of $\mathbf{i}$ in the direction $\mathbf{j}$.

We may also derive the equation of a line passing through the endpoints of two given vectors $\mathbf{a}$ and $\mathbf{b}$.

Since the vector $\mathbf{b} - \mathbf{a}$ is parallel to the directed line segment from $\mathbf{a}$ to $\mathbf{b}$, what we really wish to do here is calculate the parametric equations of the line passing through $\mathbf{a}$ in the direction of $\mathbf{b} - \mathbf{a}$ (Figure 1.1.16). Thus $\mathbf{l}(t) = \mathbf{a} + t(\mathbf{b} - \mathbf{a})$, that is, $\mathbf{l}(t) = (1 - t)\mathbf{a} + t\mathbf{b}$. As t increases from 0 to 1, $t(\mathbf{b} - \mathbf{a})$ starts as the zero vector and increases in length (remaining in the direction of $\mathbf{b} - \mathbf{a}$) until at $t = 1$ it *is* the vector $\mathbf{b} - \mathbf{a}$. Thus for $\mathbf{l}(t) = \mathbf{a} + t(\mathbf{b} - \mathbf{a})$, as t increases from 0 to 1, the vector $\mathbf{l}(t)$ moves from the endpoint of $\mathbf{a}$ to the endpoint of $\mathbf{b}$ along the directed line segment from $\mathbf{a}$ to $\mathbf{b}$.

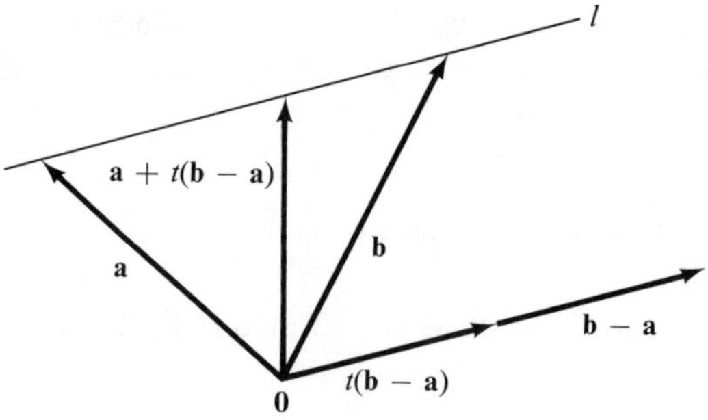

FIGURE 1.1.16
The line l, parametrically given by $\mathbf{l}(t) = \mathbf{a} + t(\mathbf{b} - \mathbf{a})$, *passes through the tips of* $\mathbf{a}$ *and* $\mathbf{b}$.

EXAMPLE 6. Find the equation of the line passing through $(-1, 1, 0)$ and $(0, 0, 1)$ (see Figure 1.1.17).

Letting $\mathbf{a} = -\mathbf{i} + \mathbf{j}$, $\mathbf{b} = \mathbf{k}$, we have

$$\begin{aligned}\mathbf{l}(t) &= (1 - t)(-\mathbf{i} + \mathbf{j}) + t\mathbf{k}\\ &= -(1 - t)\mathbf{i} + (1 - t)\mathbf{j} + t\mathbf{k}\end{aligned}$$

The equation of this line may thus be written as

$$\mathbf{l}(t) = (t - 1)\mathbf{i} + (1 - t)\mathbf{j} + t\mathbf{k}$$

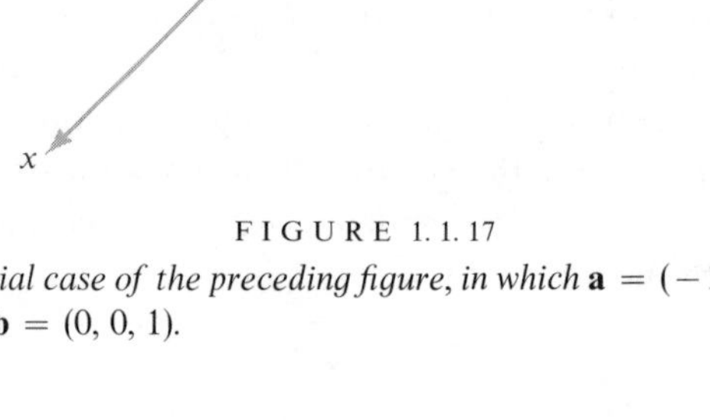

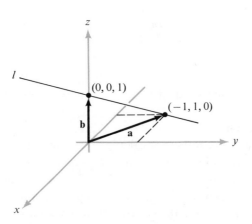

FIGURE 1.1.17
Special case of the preceding figure, in which $\mathbf{a} = (-1, 1, 0)$ *and* $\mathbf{b} = (0, 0, 1)$.

or, equivalently,

$$x = t - 1, \quad y = 1 - t, \quad z = t$$

We note that any vector of the form $\mathbf{c} = \lambda\mathbf{a} + \mu\mathbf{b}$, where $\lambda + \mu = 1$, is on the line passing through the endpoints of **a** and **b**. To see this, observe that $\mathbf{c} = (1 - \mu)\mathbf{a} + \mu\mathbf{b} = \mathbf{a} + \mu(\mathbf{b} - \mathbf{a})$.

In terms of components, the equation of the line through the two points (x_1, y_1, z_1) and (x_2, y_2, z_2) is

$$x = x_1 + t(x_2 - x_1) \qquad y = y_1 + t(y_2 - y_1) \qquad z = z_1 + t(z_2 - z_1)$$

These are sometimes written as

$$\frac{x - x_1}{x_2 - x_1} = \frac{y - y_1}{y_2 - y_1} = \frac{z - z_1}{z_2 - z_1}$$

(by eliminating t).

EXAMPLE 7. As an example of the power of the vector concept, let us give a simple proof that the diagonals of a parallelogram bisect each other.

Let the adjacent sides of the parallelogram be represented by the vectors **a** and **b**, as shown in Figure 1.1.18. We first calculate the vector to the midpoint of PQ. Since $\mathbf{b} - \mathbf{a}$ is parallel and equal in length to the directed segment from P to Q, $(\mathbf{b} - \mathbf{a})/2$ is parallel and equal in length to the directed line segment from P to the midpoint of PQ. Thus, the vector $\mathbf{a} + (\mathbf{b} - \mathbf{a})/2 = (\mathbf{a} + \mathbf{b})/2$ ends at the midpoint of PQ.

Next, we calculate the vector to the midpoint of OR. We know $\mathbf{a} + \mathbf{b}$ ends at R; thus $(\mathbf{a} + \mathbf{b})/2$ ends at the midpoint of OR. Since we have shown that the vector $(\mathbf{a} + \mathbf{b})/2$ ends at both the midpoint of OR and the midpoint of PQ, it follows that OR and PQ bisect each other.

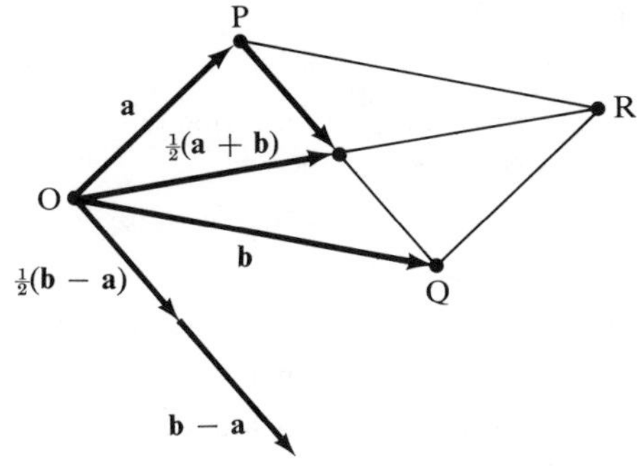

FIGURE 1.1.18

Constructions used in the proof that the diagonals of a parallelogram bisect each other.

HISTORICAL NOTE

Many scientists resisted the use of vectors in favor of the more complicated theory of quaternions until around 1900. The book that popularized vector methods was *Vector Analysis*, by E. B. Wilson (reprinted by Dover in 1960), which was based on lectures delivered by J. W. Gibbs at Yale in 1899–1900. Wilson was reluctant to take Gibbs' course since he had just completed a full-year course in quaternions at Harvard under J. M. Pierce, a champion of quaternionic methods, but was forced by a dean to add the course to his program. (For more details, see M. J. Crowe, *A History of Vector Analysis*, University of Notre Dame Press, Notre Dame, 1967.)

EXERCISES

1. What restrictions must be made on x, y, and z, so that the triple (x, y, z) will represent a point on the y-axis? On the z-axis? In the xz-plane? In the yz-plane?
2. Sketch the vectors $\mathbf{v} = (2, 3, -6)$ and $\mathbf{w} = (-1, 1, 1)$. On your sketch, draw in $-\mathbf{v}$, $\mathbf{v} + \mathbf{w}$, $2\mathbf{v}$, and $\mathbf{v} - \mathbf{w}$.
3. (a) Generalize the geometric construction in Figure 1.1.6 to show that if $\mathbf{v}_1 = (x, y, z)$ and $\mathbf{v}_2 = (x', y', z')$ then $\mathbf{v}_1 + \mathbf{v}_2 = (x + x', y + y', z + z')$.
   (b) Using an argument based on similar triangles, prove that $\alpha\mathbf{v} = (\alpha x, \alpha y, \alpha z)$ when $\mathbf{v} = (x, y, z)$.

Complete the following computations:

4. $(3, 4, 5) + (6, 2, -6) = (?, ?, ?)$
5. $(-21, 23) - (?, 6) = (-25, ?)$
6. $3(133, -0.33, 0) + (-399, 0.99, 0) = (?, ?, ?)$
7. $(8a, -2b, 13c) = (52, 12, 11) + \frac{1}{2}(?, ?, ?)$
8. $(2, 3, 5) - 4\mathbf{i} + 3\mathbf{j} = (?, ?, ?)$
9. $800(0.03, 0, 0) = ?\mathbf{i} + ?\mathbf{j} + ?\mathbf{k}$

In Exercises 10 to 16 use vector notation to describe the points that lie in the given configurations, as we did in Examples 4, 5, and 6.

10. The parallelogram whose adjacent sides are the vectors $\mathbf{i} + 3\mathbf{k}$ and $-2\mathbf{j}$.
11. The plane spanned by $\mathbf{v}_1 = (2, 7, 0)$ and $\mathbf{v}_2 = (0, 2, 7)$.
12. The line passing through $(0, 2, 1)$ in the direction of $2\mathbf{i} - \mathbf{k}$.
13. The line passing through $(-1, -1, -1)$ and $(1, -1, 2)$.
14. The parallelepiped with sides the vectors $\mathbf{a}$, $\mathbf{b}$, and $\mathbf{c}$. (See Figure 1.3.5 for a picture of the region we have in mind.)
15. The points within the parallelogram with one corner at (x_0, y_0, z_0) whose sides extending from that corner are equal in magnitude and direction to vectors $\mathbf{a}$ and $\mathbf{b}$.

16. The plane determined by the three points (x_0, y_0, z_0), (x_1, y_1, z_1), and (x_2, y_2, z_2).
17. Show that the medians of a triangle intersect at a point, and that this point divides each median in a ratio of 2:1.
18. Find the points of intersection of the line $x = 3 + 2t$, $y = 7 + 8t$, $z = -2 + t$, that is, $\mathbf{l}(t) = (3 + 2t, 7 + 8t, -2 + t)$ with the coordinate planes.
19. Show that there are no points (x, y, z) satisfying $2x - 3y + z - 2 = 0$ and lying on the line $\mathbf{v} = (2, -2, -1) + t(1, 1, 1)$.
20. Show that every point on the line $\mathbf{v} = (1, -1, 2) + t(2, 3, 1)$ satisfies $5x - 3y - z - 6 = 0$.

*21. Find a line that lies in the set defined by the equation $x^2 - y^2 - z^2 = 1$.

1.2 THE INNER PRODUCT

In this section and the next we shall discuss two products of vectors, the inner product and the cross product. These are very useful in physical applications and have interesting geometric interpretations. The first product we shall consider is called the *inner product*. The name *dot product* is often used instead.

Suppose we have two vectors $\mathbf{a}$ and $\mathbf{b}$ (Figure 1.2.1) and we wish to determine the angle between them, that is, the smaller angle subtended by $\mathbf{a}$ and $\mathbf{b}$ in the plane that they span. The inner product enables us to do this. Let us first develop the concept formally and then prove that this product does what we claim.

Let $\mathbf{a} = a_1\mathbf{i} + a_2\mathbf{j} + a_3\mathbf{k}$ and $\mathbf{b} = b_1\mathbf{i} + b_2\mathbf{j} + b_3\mathbf{k}$ be two vectors in $\mathbb{R}^3$. We define the *inner product* of $\mathbf{a}$ and $\mathbf{b}$, written $\mathbf{a} \cdot \mathbf{b}$, to be the real number

$$\mathbf{a} \cdot \mathbf{b} = a_1b_1 + a_2b_2 + a_3b_3$$

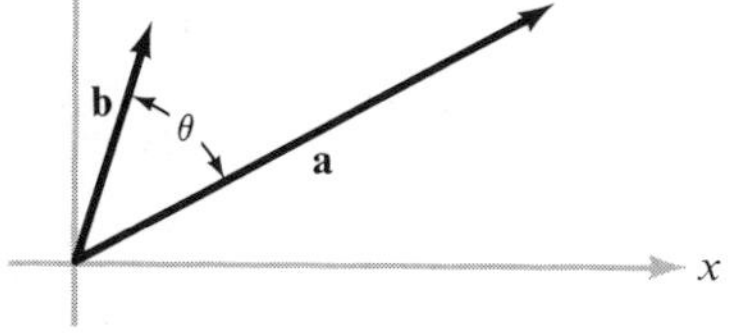

FIGURE 1.2.1
θ is the angle between the vectors **a** *and* **b**.

Note that the inner product of two vectors is a scalar quantity. Sometimes the inner product is denoted $\langle \mathbf{a}, \mathbf{b} \rangle$. This is often done for typographic reasons. *Thus* $\langle \mathbf{a}, \mathbf{b} \rangle$ *and* $\mathbf{a} \cdot \mathbf{b}$ *mean exactly the same thing.*

Certain properties of the inner product follow immediately from the definition. If **a**, **b**, and **c** are vectors in $\mathbb{R}^3$, and α and β are real numbers, then

(*i*) $\mathbf{a} \cdot \mathbf{a} \geq 0$
$\mathbf{a} \cdot \mathbf{a} = 0$ if and only if $\mathbf{a} = \mathbf{0}$

(*ii*) $\alpha\mathbf{a} \cdot \mathbf{b} = \alpha(\mathbf{a} \cdot \mathbf{b})$ and $\mathbf{a} \cdot \beta\mathbf{b} = \beta(\mathbf{a} \cdot \mathbf{b})$

(*iii*) $\mathbf{a} \cdot (\mathbf{b} + \mathbf{c}) = \mathbf{a} \cdot \mathbf{b} + \mathbf{a} \cdot \mathbf{c}$ and $(\mathbf{a} + \mathbf{b}) \cdot \mathbf{c} = \mathbf{a} \cdot \mathbf{c} + \mathbf{b} \cdot \mathbf{c}$

(*iv*) $\mathbf{a} \cdot \mathbf{b} = \mathbf{b} \cdot \mathbf{a}$

To prove (*i*), observe that if $\mathbf{a} = a_1\mathbf{i} + a_2\mathbf{j} + a_3\mathbf{k}$, then $\mathbf{a} \cdot \mathbf{a} = a_1^2 + a_2^2 + a_3^2$. Since a_1, a_2, and a_3 are real numbers we know $a_1^2 \geq 0$, $a_2^2 \geq 0$, $a_3^2 \geq 0$. Thus $\mathbf{a} \cdot \mathbf{a} \geq 0$. Moreover, if $a_1^2 + a_2^2 + a_3^2 = 0$, then $a_1 = a_2 = a_3 = 0$; therefore $\mathbf{a} = \mathbf{0}$ (zero vector). The proofs of the other properties of the inner product are also easily obtained.

It follows from the Pythagorean Theorem that the length of the vector $\mathbf{a} = a_1\mathbf{i} + a_2\mathbf{j} + a_3\mathbf{k}$ is $\sqrt{a_1^2 + a_2^2 + a_3^2}$ (see Figure 1.2.2). The length of the vector **a** is denoted by $\|\mathbf{a}\|$. This quantity is often called the *norm* of **a**. Since $\mathbf{a} \cdot \mathbf{a} = a_1^2 + a_2^2 + a_3^2$, it follows that

$$\|\mathbf{a}\| = (\mathbf{a} \cdot \mathbf{a})^{1/2}$$

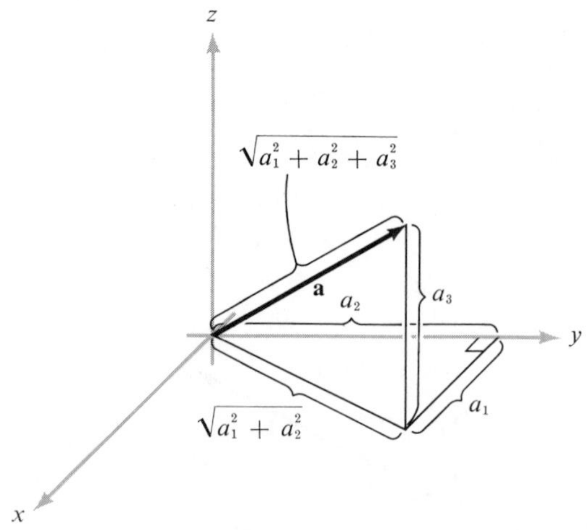

FIGURE 1.2.2
The length of the vector $\mathbf{a} = (a_1, a_2, a_3)$ *is given by the Pythagorean formula:* $\sqrt{a_1^2 + a_2^2 + a_3^2}$.

Vectors with norm 1 are called *unit vectors.* For example, the vectors **i**, **j**, **k** are unit vectors. Observe that for any nonzero vector **a**, $\mathbf{a}/\|\mathbf{a}\|$ is a unit vector and we say that we have *normalized* **a**.

In the plane, the vector $\mathbf{i}_\theta = (\cos\theta)\mathbf{i} + (\sin\theta)\mathbf{j}$ is the unit vector making an angle θ with the x-axis (see Figure 1.2.3). Clearly,

$$\|\mathbf{i}_\theta\| = (\sin^2\theta + \cos^2\theta)^{1/2} = 1.$$

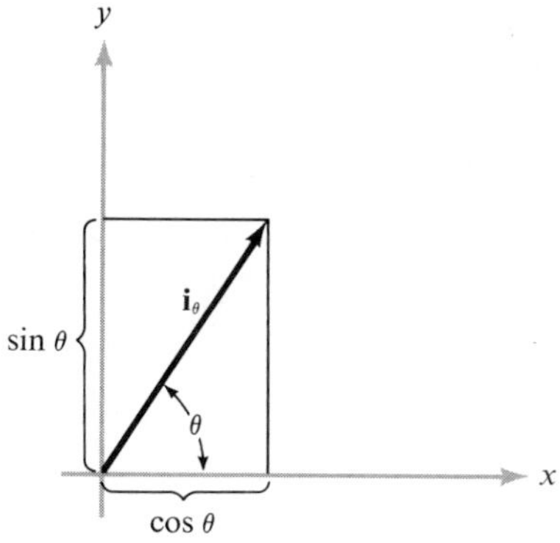

FIGURE 1.2.3
The coordinates of $\mathbf{i}_\theta$ *are* $\cos\theta$ *and* $\sin\theta$.

If **a** and **b** are vectors, we have seen that the vector $\mathbf{b} - \mathbf{a}$ is parallel to and has the same magnitude as the directed line segment from the endpoint of **a** to the endpoint of **b**. It follows that the distance from the endpoint of **a** to the endpoint of **b** is $\|\mathbf{b} - \mathbf{a}\|$ (see Figure 1.2.4).

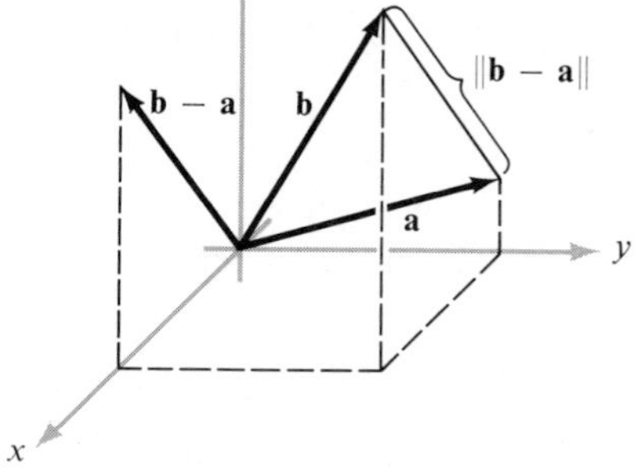

FIGURE 1.2.4
The distance between the tips of **a** *and* **b** *is* $\|\mathbf{b} - \mathbf{a}\|$.

EXAMPLE 1. The distance from the endpoint of the vector **i**, that is, the point (1, 0, 0), to the endpoint of the vector **j**, (0, 1, 0), is

$$\sqrt{(0-1)^2 + (1-0)^2 + (0-0)^2} = \sqrt{2}$$

Let us now show that the inner product does indeed measure the angle between two vectors.

THEOREM 1. *Let* **a** *and* **b** *be two vectors in* $\mathbb{R}^3$ *and let* θ, $0 \leq \theta \leq \pi$, *be the angle between them (Figure 1.2.5). Then*

$$\mathbf{a} \cdot \mathbf{b} = \|\mathbf{a}\| \, \|\mathbf{b}\| \cos \theta$$

Thus we may express the angle between **a** *and* **b** *as*

$$\theta = \cos^{-1} \frac{\mathbf{a} \cdot \mathbf{b}}{\|\mathbf{a}\| \, \|\mathbf{b}\|}$$

if **a** *and* **b** *are nonzero vectors.*

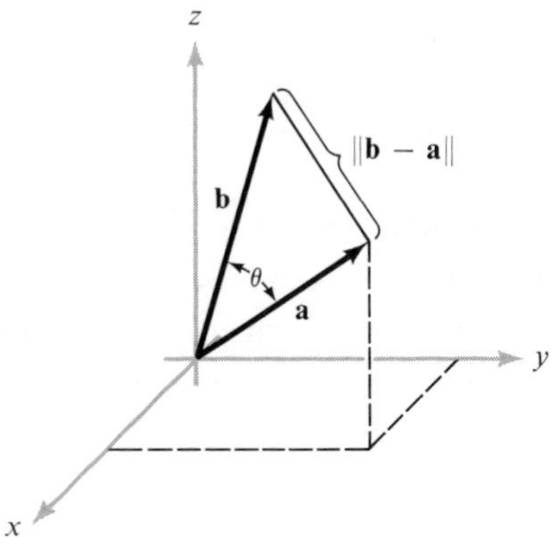

FIGURE 1.2.5
The vectors **a**, **b** *and the angle* θ *between them; the geometry for Theorem 1 and its proof.*

Proof. If we apply the law of cosines from trigonometry to the triangle with one vertex at the origin and adjacent sides determined by the vectors **a** and **b**, it follows that

$$\|\mathbf{b} - \mathbf{a}\|^2 = \|\mathbf{a}\|^2 + \|\mathbf{b}\|^2 - 2\|\mathbf{a}\| \, \|\mathbf{b}\| \cos \theta$$

Since $\|\mathbf{b} - \mathbf{a}\|^2 = (\mathbf{b} - \mathbf{a}) \cdot (\mathbf{b} - \mathbf{a})$, $\|\mathbf{a}\|^2 = \mathbf{a} \cdot \mathbf{a}$, and $\|\mathbf{b}\|^2 = \mathbf{b} \cdot \mathbf{b}$, we can rewrite the above equation as

$$(\mathbf{b} - \mathbf{a}) \cdot (\mathbf{b} - \mathbf{a}) = \mathbf{a} \cdot \mathbf{a} + \mathbf{b} \cdot \mathbf{b} - 2\|\mathbf{a}\| \, \|\mathbf{b}\| \cos \theta$$

Now

$$(\mathbf{b}-\mathbf{a})\cdot(\mathbf{b}-\mathbf{a}) = \mathbf{b}\cdot(\mathbf{b}-\mathbf{a}) - \mathbf{a}\cdot(\mathbf{b}-\mathbf{a})$$
$$= \mathbf{b}\cdot\mathbf{b} - \mathbf{b}\cdot\mathbf{a} - \mathbf{a}\cdot\mathbf{b} + \mathbf{a}\cdot\mathbf{a}$$
$$= \mathbf{a}\cdot\mathbf{a} + \mathbf{b}\cdot\mathbf{b} - 2\mathbf{a}\cdot\mathbf{b}$$

Thus,

$$\mathbf{a}\cdot\mathbf{a} + \mathbf{b}\cdot\mathbf{b} - 2\mathbf{a}\cdot\mathbf{b} = \mathbf{a}\cdot\mathbf{a} + \mathbf{b}\cdot\mathbf{b} - 2\|\mathbf{a}\|\,\|\mathbf{b}\|\cos\theta$$

that is

$$\mathbf{a}\cdot\mathbf{b} = \|\mathbf{a}\|\,\|\mathbf{b}\|\cos\theta. \quad ■$$

This result shows that the inner product of two vectors is the product of their lengths times the cosine of the angle between them. This relationship is often of value in problems of a geometric nature.

COROLLARY (CAUCHY-SCHWARZ INEQUALITY). *For any two vectors* **a** *and* **b**, *we have*

$$|\mathbf{a}\cdot\mathbf{b}| \le \|\mathbf{a}\|\,\|\mathbf{b}\|$$

with equality if and only if **a** *is a scalar multiple of* **b**, *or one of them is* **0**.

Proof. If **a** is not a scalar multiple of **b**, then $|\cos\theta| < 1$ and the inequality holds. When **a** is a scalar multiple of **b** then $\theta = 0$ or π and $|\cos\theta| = 1$. ■

EXAMPLE 2. Find the angle between the vectors $\mathbf{i}+\mathbf{j}+\mathbf{k}$ and $\mathbf{i}+\mathbf{j}-\mathbf{k}$ (Figure 1.2.6).

Using Theorem 1 we have

$$(\mathbf{i}+\mathbf{j}+\mathbf{k})\cdot(\mathbf{i}+\mathbf{j}-\mathbf{k}) = \|\mathbf{i}+\mathbf{j}+\mathbf{k}\|\,\|\mathbf{i}+\mathbf{j}-\mathbf{k}\|\cos\theta$$

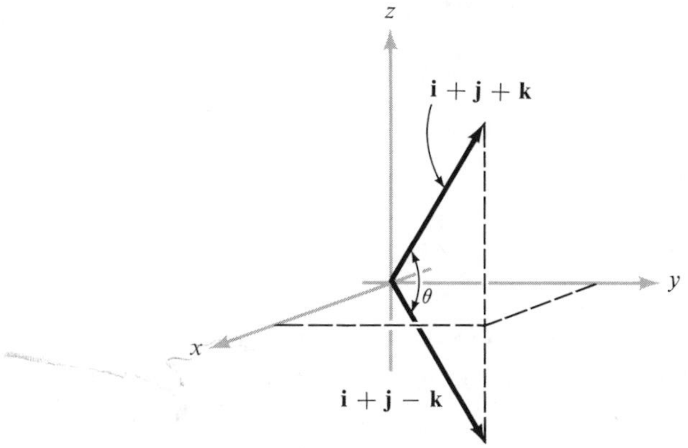

FIGURE 1.2.6
Finding the angle between $\mathbf{a} = \mathbf{i}+\mathbf{j}+\mathbf{k}$ *and* $\mathbf{b} = \mathbf{i}+\mathbf{j}-\mathbf{k}$.

So

$$1 + 1 - 1 = (\sqrt{3})(\sqrt{3})\cos\theta$$

Hence,

$$\cos\theta = \tfrac{1}{3}$$

that is,

$$\theta = \cos^{-1}(\tfrac{1}{3}) \approx 1.23 \text{ radians } (71^\circ)$$

If **a** and **b** are nonzero vectors in $\mathbb{R}^3$ and θ is the angle between them, we see that $\mathbf{a}\cdot\mathbf{b} = 0$ if and only if $\cos\theta = 0$. From this it follows that the inner product of two nonzero vectors is zero if and only if the vectors are perpendicular. Hence, the inner product provides us with a convenient method for determining if two vectors are perpendicular. Often we say that perpendicular vectors are *orthogonal*. The standard basis vectors **i**, **j**, and **k** are mutually orthogonal and normal; any such system is called *orthonormal*. We shall adopt the convention that the zero vector is orthogonal to all vectors.

EXAMPLE 3. The vectors $\mathbf{i}_\theta = (\cos\theta)\mathbf{i} + (\sin\theta)\mathbf{j}$ and $\mathbf{j}_\theta = -(\sin\theta)\mathbf{i} + (\cos\theta)\mathbf{j}$ are orthogonal, since

$$\mathbf{i}_\theta \cdot \mathbf{j}_\theta = -\cos\theta\sin\theta + \sin\theta\cos\theta = 0$$

(see Figure 1.2.7).

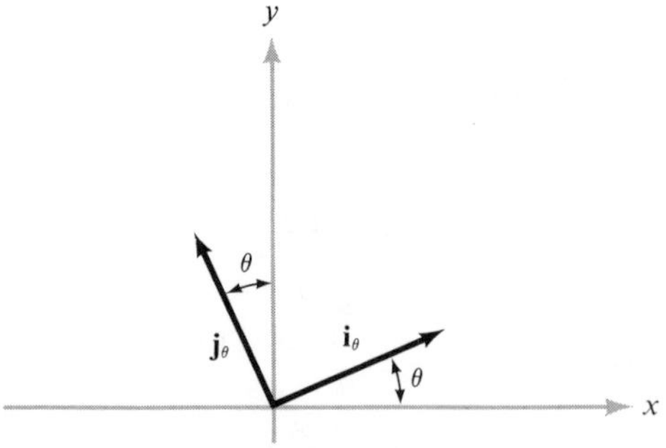

FIGURE 1.2.7
The vectors $\mathbf{i}_\theta$ and $\mathbf{j}_\theta$ are orthogonal.

EXAMPLE 4. Let **a** and **b** be two nonzero orthogonal vectors. Let **c** be another vector in the plane spanned by **a** and **b**. As we have seen, there are scalars α and β, such that $\mathbf{c} = \alpha\mathbf{a} + \beta\mathbf{b}$. The vector $\alpha\mathbf{a}$ is called the *component* of **c** along **a** (or tangent to **a**), and $\beta\mathbf{b}$ the component of **c** along **b**. Thus $\mathbf{c} = \alpha\mathbf{a} + \beta\mathbf{b}$ is a resolution of **c** into two orthogonal vectors. Use the inner product to determine α and β (see Figure 1.2.8).

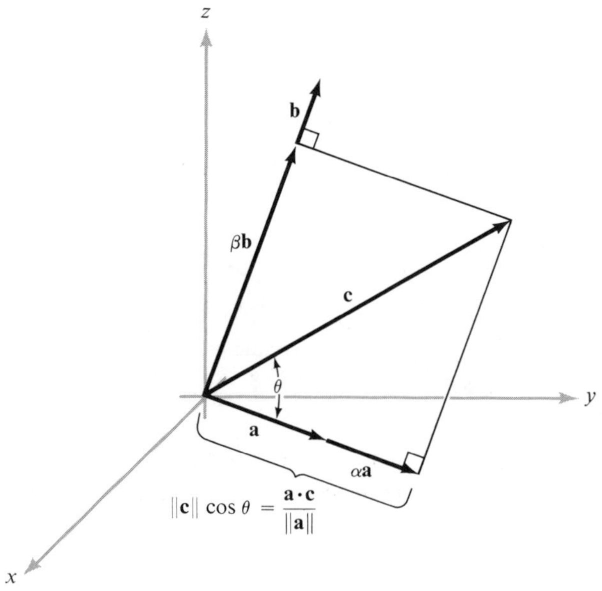

FIGURE 1.2.8
The geometry for finding α and β where $\mathbf{c} = \alpha\mathbf{a} + \beta\mathbf{b}$, as in Example 2.

Taking the inner product of **a** and **c**, we have

$$\mathbf{a}\cdot\mathbf{c} = \mathbf{a}\cdot(\alpha\mathbf{a} + \beta\mathbf{b}) = \alpha\mathbf{a}\cdot\mathbf{a} + \beta\mathbf{a}\cdot\mathbf{b}$$

Since **a** and **b** are orthogonal, $\mathbf{a}\cdot\mathbf{b} = 0$, and so,

$$\alpha = \frac{\mathbf{a}\cdot\mathbf{c}}{\mathbf{a}\cdot\mathbf{a}} = \frac{\mathbf{a}\cdot\mathbf{c}}{\|\mathbf{a}\|^2}$$

Similarly,

$$\beta = \frac{\mathbf{b}\cdot\mathbf{c}}{\mathbf{b}\cdot\mathbf{b}} = \frac{\mathbf{b}\cdot\mathbf{c}}{\|\mathbf{b}\|^2}$$

The result of Example 4 may be obtained using the geometric interpretation of the inner product. Let l be the distance, measured along the line determined by extending **a**, from the origin to the point where the perpendicular from **c** intersects the extension of **a**. It follows that

$$l = \|\mathbf{c}\|\cos\theta$$

where θ is the angle between **a** and **c**. Moreover, $l = \alpha\|\mathbf{a}\|$. Taken together, these results yield

$$\alpha\|\mathbf{a}\| = \|\mathbf{c}\|\cos\theta, \quad \text{or} \quad \alpha = \frac{\|\mathbf{c}\|\cos\theta}{\|\mathbf{a}\|} = \frac{\|\mathbf{c}\|}{\|\mathbf{a}\|}\left(\frac{\mathbf{a}\cdot\mathbf{c}}{\|\mathbf{c}\|\,\|\mathbf{a}\|}\right) = \frac{\mathbf{a}\cdot\mathbf{c}}{\mathbf{a}\cdot\mathbf{a}}$$

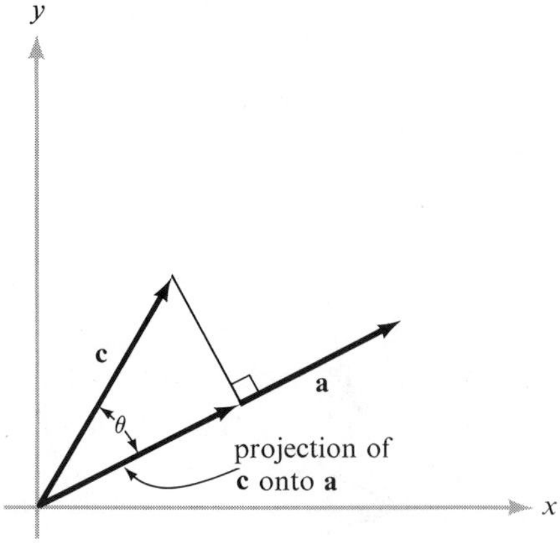

FIGURE 1.2.9
The projection of **c** *on* **a** *is* $(\mathbf{a}\cdot\mathbf{c}/\|\mathbf{a}\|^2)\mathbf{a}$.

In Example 4, the vector $\alpha\mathbf{a}$ is called the *projection* of **c** onto **a**. Similarly, the vector $\beta\mathbf{b}$ is the projection of **c** onto **b**. In general, the length of the projection of a vector **c** onto a vector **a**, where θ is the angle between **a** and **c**, is given by (Figure 1.2.9)

$$\|\mathbf{c}\|\,|\cos\theta| = \frac{|\mathbf{a}\cdot\mathbf{c}|}{\|\mathbf{a}\|}$$

EXERCISES

1. (a) Prove properties (*ii*) and (*iii*) of the inner product.
   (b) Prove that $\mathbf{a}\cdot\mathbf{b} = \mathbf{b}\cdot\mathbf{a}$.
2. Calculate $\mathbf{a}\cdot\mathbf{b}$ where $\mathbf{a} = 2\mathbf{i} + 10\mathbf{j} - 12\mathbf{k}$ and $\mathbf{b} = -3\mathbf{i} + 4\mathbf{k}$.
3. Find the angle between $7\mathbf{j} + 19\mathbf{k}$ and $-2\mathbf{i} - \mathbf{j}$ (to the nearest degree).
4. Compute $\mathbf{u}\cdot\mathbf{v}$, where $\mathbf{u} = \sqrt{3}\mathbf{i} - 315\mathbf{j} + 22\mathbf{k}$ and $\mathbf{v} = \mathbf{u}/\|\mathbf{u}\|$.
5. What is $\|8\mathbf{i} - 12\mathbf{k}\|\cdot\|6\mathbf{j} + \mathbf{k}\| - |(8\mathbf{i} - 12\mathbf{k})\cdot(6\mathbf{j} + \mathbf{k})|$ (to the nearest tenth)?

In Exercises 6 to 11, compute $\|\mathbf{u}\|$, $\|\mathbf{v}\|$, and $\mathbf{u}\cdot\mathbf{v}$ for the given vectors.

6. $\mathbf{u} = 15\mathbf{i} - 2\mathbf{j} + 4\mathbf{k}$, $\mathbf{v} = \pi\mathbf{i} + 3\mathbf{j} - \mathbf{k}$
7. $\mathbf{u} = 2\mathbf{j} - \mathbf{i}$, $\mathbf{v} = -\mathbf{j} + \mathbf{i}$
8. $\mathbf{u} = 5\mathbf{i} - \mathbf{j} + 2\mathbf{k}$, $\mathbf{v} = \mathbf{i} + \mathbf{j} - \mathbf{k}$
9. $\mathbf{u} = -\mathbf{i} + 3\mathbf{j} + \mathbf{k}$, $\mathbf{v} = -2\mathbf{i} - 3\mathbf{j} - 7\mathbf{k}$
10. $\mathbf{u} = -\mathbf{i} + 3\mathbf{k}$, $\mathbf{v} = 4\mathbf{j}$
11. $\mathbf{u} = -\mathbf{i} + 2\mathbf{j} - 3\mathbf{k}$, $\mathbf{v} = -\mathbf{i} - 3\mathbf{j} + 4\mathbf{k}$
12. Normalize the vectors in Exercises 6 to 8.

13. Find two nonparallel vectors both orthogonal to (1, 1, 1).
14. Find the line through (3, 1, −2) that intersects and is perpendicular to the line $x = -1 + t$, $y = -2 + t$, $z = -1 + t$. (HINT: If (x_0, y_0, z_0) is the point of intersection, find its coordinates.)

1.3 THE CROSS PRODUCT

In Section 1.2 we defined a product of vectors that was a scalar. In this section we shall define a product of vectors that is a vector; that is, we shall show how, given two vectors **a** and **b**, we can produce a third vector **a** × **b**, called the *cross product* of **a** and **b**. This new vector will have the pleasing geometric property that it is perpendicular to the plane spanned (determined) by **a** and **b**.

We shall first develop the somewhat lengthy mathematical formalism necessary to state a useful definition of this product. Once this has been accomplished, we can study the geometric implications of the mathematical structure we have built.

We begin by defining a 2 × 2 *matrix* to be an array

$$\begin{bmatrix} a_{11} & a_{12} \\ a_{21} & a_{22} \end{bmatrix}$$

where $a_{11}, a_{12}, a_{21}, a_{22}$, are four scalars. For example,

$$\begin{bmatrix} 2 & 1 \\ 0 & 4 \end{bmatrix}, \quad \begin{bmatrix} -1 & 0 \\ 1 & 1 \end{bmatrix}, \quad \text{and} \quad \begin{bmatrix} 13 & 7 \\ 6 & 11 \end{bmatrix}$$

are 2 × 2 matrices. The *determinant*

$$\begin{vmatrix} a_{11} & a_{12} \\ a_{21} & a_{22} \end{vmatrix}$$

of such a matrix is the real number defined by the equation

$$\begin{vmatrix} a_{11} & a_{12} \\ a_{21} & a_{22} \end{vmatrix} = a_{11}a_{22} - a_{12}a_{21} \tag{1}$$

EXAMPLE 1.

$$\begin{vmatrix} 1 & 1 \\ 1 & 1 \end{vmatrix} = 1 - 1 = 0$$

$$\begin{vmatrix} 1 & 2 \\ 3 & 4 \end{vmatrix} = 4 - 6 = -2$$

$$\begin{vmatrix} 5 & 6 \\ 7 & 8 \end{vmatrix} = 40 - 42 = -2$$

Next we pass to some properties of 3×3 matrices and determinants. A 3×3 matrix is an array

$$\begin{bmatrix} a_{11} & a_{12} & a_{13} \\ a_{21} & a_{22} & a_{23} \\ a_{31} & a_{32} & a_{33} \end{bmatrix}$$

again where each a_{ij} is a scalar; a_{ij} denotes the entry in the array that is in the ith row and the jth column. We define the determinant of a 3×3 matrix by the rule

$$\begin{vmatrix} a_{11} & a_{12} & a_{13} \\ a_{21} & a_{22} & a_{23} \\ a_{31} & a_{32} & a_{33} \end{vmatrix} = a_{11}\begin{vmatrix} a_{22} & a_{23} \\ a_{32} & a_{33} \end{vmatrix} - a_{12}\begin{vmatrix} a_{21} & a_{23} \\ a_{31} & a_{33} \end{vmatrix} + a_{13}\begin{vmatrix} a_{21} & a_{22} \\ a_{31} & a_{32} \end{vmatrix} \qquad (2)$$

Without some mnemonic device, formula (2) would be difficult to memorize. The rule to learn here is that you move along the first row, multiplying a_{1j} by the determinant of the 2×2 matrix obtained by cancelling out the first row and the jth column, and then you add these up, remembering to put a minus in front of the a_{12} term. For example, the matrix involved in the middle term of (2), namely

$$\begin{vmatrix} a_{21} & a_{23} \\ a_{31} & a_{33} \end{vmatrix}$$

is obtained by crossing out the first row and the second column of the given 3×3 matrix, thus:

$$\begin{bmatrix} \not{a}_{11} & \not{a}_{12} & \not{a}_{13} \\ a_{21} & \not{a}_{22} & a_{23} \\ a_{31} & \not{a}_{32} & a_{33} \end{bmatrix}$$

EXAMPLE 2.

$$\begin{vmatrix} 1 & 0 & 0 \\ 0 & 1 & 0 \\ 0 & 0 & 1 \end{vmatrix} = 1\begin{vmatrix} 1 & 0 \\ 0 & 1 \end{vmatrix} - 0\begin{vmatrix} 0 & 0 \\ 0 & 1 \end{vmatrix} + 0\begin{vmatrix} 0 & 1 \\ 0 & 0 \end{vmatrix} = 1$$

$$\begin{vmatrix} 1 & 2 & 3 \\ 4 & 5 & 6 \\ 7 & 8 & 9 \end{vmatrix} = 1\begin{vmatrix} 5 & 6 \\ 8 & 9 \end{vmatrix} - 2\begin{vmatrix} 4 & 6 \\ 7 & 9 \end{vmatrix} + 3\begin{vmatrix} 4 & 5 \\ 7 & 8 \end{vmatrix} = -3 + 12 - 9 = 0$$

An important property of determinants is that interchanging two rows or two columns results in a change of sign. For 2×2 determinants, this is an immediate consequence of the definition. For rows we have

$$\begin{vmatrix} a_{11} & a_{12} \\ a_{21} & a_{22} \end{vmatrix} = a_{11}a_{22} - a_{21}a_{12}$$

$$= -(a_{21}a_{12} - a_{11}a_{22}) = -\begin{vmatrix} a_{21} & a_{22} \\ a_{11} & a_{12} \end{vmatrix}$$

and for columns

$$\begin{vmatrix} a_{11} & a_{12} \\ a_{21} & a_{22} \end{vmatrix} = -(a_{12}a_{21} - a_{11}a_{22}) = -\begin{vmatrix} a_{12} & a_{11} \\ a_{22} & a_{21} \end{vmatrix}$$

We leave it to the reader to verify this property for the 3×3 case. (See Exercise 1 below.)

A second fundamental property of determinants is that we can factor scalars out of any row or column. For 2×2 determinants, this means

$$\begin{vmatrix} \alpha a_{11} & a_{12} \\ \alpha a_{21} & a_{22} \end{vmatrix} = \begin{vmatrix} a_{11} & \alpha a_{12} \\ a_{21} & \alpha a_{22} \end{vmatrix} = \alpha \begin{vmatrix} a_{11} & a_{12} \\ a_{21} & a_{22} \end{vmatrix} = \begin{vmatrix} \alpha a_{11} & \alpha a_{12} \\ a_{21} & a_{22} \end{vmatrix} = \begin{vmatrix} a_{11} & a_{12} \\ \alpha a_{21} & \alpha a_{22} \end{vmatrix}$$

Similarly, for 3×3 determinants we have

$$\begin{vmatrix} \alpha a_{11} & \alpha a_{12} & \alpha a_{13} \\ a_{21} & a_{22} & a_{23} \\ a_{31} & a_{32} & a_{33} \end{vmatrix} = \alpha \begin{vmatrix} a_{11} & a_{12} & a_{13} \\ a_{21} & a_{22} & a_{23} \\ a_{31} & a_{32} & a_{33} \end{vmatrix} = \begin{vmatrix} a_{11} & \alpha a_{12} & a_{13} \\ a_{21} & \alpha a_{22} & a_{23} \\ a_{31} & \alpha a_{32} & a_{33} \end{vmatrix}$$

and so on. These results follow easily from the definitions. In particular, if any row or column consists of zeros, then the value of the determinant is zero.

A third fundamental fact about determinants is the following: If we change a row (respectively, column) by adding another row (respectively, column) to it, the value of the determinant remains the same. For the 2×2 case this means that

$$\begin{vmatrix} a_1 & a_2 \\ b_1 & b_2 \end{vmatrix} = \begin{vmatrix} a_1 + b_1 & a_2 + b_2 \\ b_1 & b_2 \end{vmatrix} = \begin{vmatrix} a_1 & a_2 \\ b_1 + a_1 & b_2 + a_2 \end{vmatrix}$$

$$= \begin{vmatrix} a_1 + a_2 & a_2 \\ b_1 + b_2 & b_2 \end{vmatrix} = \begin{vmatrix} a_1 & a_1 + a_2 \\ b_1 & b_1 + b_2 \end{vmatrix}$$

For the 3×3 case, this means

$$\begin{vmatrix} a_1 & a_2 & a_3 \\ b_1 & b_2 & b_3 \\ c_1 & c_2 & c_3 \end{vmatrix} = \begin{vmatrix} a_1 + b_1 & a_2 + b_2 & a_3 + b_3 \\ b_1 & b_2 & b_3 \\ c_1 & c_2 & c_3 \end{vmatrix} = \begin{vmatrix} a_1 + a_2 & a_2 & a_3 \\ b_1 + b_2 & b_2 & b_3 \\ c_1 + c_2 & c_2 & c_3 \end{vmatrix}$$

and so on. Again, this property can be proved using the definition of determinant.

EXAMPLE 3. Suppose

$$\mathbf{a} = \alpha\mathbf{b} + \beta\mathbf{c}, \quad \text{i.e.,} \quad \mathbf{a} = (a_1, a_2, a_3) = \alpha(b_1, b_2, b_3) + \beta(c_1, c_2, c_3)$$

Let us show that

$$\begin{vmatrix} a_1 & a_2 & a_3 \\ b_1 & b_2 & b_3 \\ c_1 & c_2 & c_3 \end{vmatrix} = 0$$

We shall prove the case $\alpha \neq 0$, $\beta \neq 0$. The case $\alpha = 0 = \beta$ is trivial, and the case where exactly one of α, β is zero is a simple modification of the one we prove. Using the fundamental properties of determinants, the determinant in question is

$$\begin{vmatrix} \alpha b_1 + \beta c_1 & \alpha b_2 + \beta c_2 & \alpha b_3 + \beta c_3 \\ b_1 & b_2 & b_3 \\ c_1 & c_2 & c_3 \end{vmatrix}$$

$$= -\frac{1}{\alpha}\begin{vmatrix} \alpha b_1 + \beta c_1 & \alpha b_2 + \beta c_2 & \alpha b_3 + \beta c_3 \\ -\alpha b_1 & -\alpha b_2 & -\alpha b_3 \\ c_1 & c_2 & c_3 \end{vmatrix}$$

(factoring $-1/\alpha$ out of the second row)

$$= \left(-\frac{1}{\alpha}\right)\left(-\frac{1}{\beta}\right)\begin{vmatrix} \alpha b_1 + \beta c_1 & \alpha b_2 + \beta c_2 & \alpha b_3 + \beta c_3 \\ -\alpha b_1 & -\alpha b_2 & -\alpha b_3 \\ -\beta c_1 & -\beta c_2 & -\beta c_3 \end{vmatrix}$$

(factoring $-1/\beta$ out of the third row)

$$= \frac{1}{\alpha\beta}\begin{vmatrix} \beta c_1 & \beta c_2 & \beta c_3 \\ -\alpha b_1 & -\alpha b_2 & -\alpha b_3 \\ -\beta c_1 & -\beta c_2 & -\beta c_3 \end{vmatrix} \quad \text{(adding the second row to the first row)}$$

$$= \frac{1}{\alpha\beta}\begin{vmatrix} 0 & 0 & 0 \\ -\alpha b_1 & -\alpha b_2 & -\alpha b_3 \\ -\beta c_1 & -\beta c_2 & -\beta c_3 \end{vmatrix} \quad \text{(adding the third row to the first row)}$$

$$= 0$$

HISTORICAL NOTE

Determinants seem to have been invented and first used by Leibniz in 1693, in connection with solutions of linear equations. McLaurin and Cramer developed their properties between 1729 and 1750; in particular, they showed

HISTORICAL NOTE (*Continued*)

that the solution of the system of equations

$$a_{11}x_1 + a_{12}x_2 + a_{13}x_3 = b_1$$
$$a_{21}x_1 + a_{22}x_2 + a_{23}x_3 = b_2$$
$$a_{31}x_1 + a_{32}x_2 + a_{33}x_3 = b_3$$

is

$$x_1 = \frac{1}{\Delta}\begin{vmatrix} b_1 & a_{12} & a_{13} \\ b_2 & a_{22} & a_{23} \\ b_3 & a_{32} & a_{33} \end{vmatrix}, \qquad x_2 = \frac{1}{\Delta}\begin{vmatrix} a_{11} & b_1 & a_{13} \\ a_{21} & b_2 & a_{23} \\ a_{31} & b_3 & a_{33} \end{vmatrix},$$

and

$$x_3 = \frac{1}{\Delta}\begin{vmatrix} a_{11} & a_{12} & b_1 \\ a_{21} & a_{22} & b_2 \\ a_{31} & a_{32} & b_3 \end{vmatrix}$$

where

$$\Delta = \begin{vmatrix} a_{11} & a_{12} & a_{13} \\ a_{21} & a_{22} & a_{23} \\ a_{31} & a_{32} & a_{33} \end{vmatrix}$$

a fact now known as *Cramer's rule*. Later, Vandermonde (1772) and Cauchy (1812), treating determinants as a separate topic worthy of special attention, developed the field more systematically, with contributions by Laplace, Jacobi, and others. Formulas for volumes of parallelpipeds in terms of determinants (see below and Review Exercise 15(a)) are due to Lagrange (1775). Although during the nineteenth century mathematicians studied matrices and determinants, the subjects were considered separate. For the full history up to 1900 see T. Muir, *The Theory of Determinants in the Historical Order of Development*, reprinted by Dover, New York, 1960.

Now that we have established the necessary results about determinants, and discussed their history, we are ready to proceed with the cross product of vectors. Let $\mathbf{a} = a_1\mathbf{i} + a_2\mathbf{j} + a_3\mathbf{k}$ and $\mathbf{b} = b_1\mathbf{i} + b_2\mathbf{j} + b_3\mathbf{k}$ be vectors in $\mathbb{R}^3$. The *cross product* of **a** and **b**, denoted $\mathbf{a} \times \mathbf{b}$, is defined to be the vector

$$\mathbf{a} \times \mathbf{b} = \begin{vmatrix} a_2 & a_3 \\ b_2 & b_3 \end{vmatrix}\mathbf{i} - \begin{vmatrix} a_1 & a_3 \\ b_1 & b_3 \end{vmatrix}\mathbf{j} + \begin{vmatrix} a_1 & a_2 \\ b_1 & b_2 \end{vmatrix}\mathbf{k}$$

or, symbolically,

$$\mathbf{a} \times \mathbf{b} = \begin{vmatrix} \mathbf{i} & \mathbf{j} & \mathbf{k} \\ a_1 & a_2 & a_3 \\ b_1 & b_2 & b_3 \end{vmatrix}$$

EXAMPLE 4.

$$(3\mathbf{i} - \mathbf{j} + \mathbf{k}) \times (\mathbf{i} + 2\mathbf{j} - \mathbf{k}) = \begin{vmatrix} \mathbf{i} & \mathbf{j} & \mathbf{k} \\ 3 & -1 & 1 \\ 1 & 2 & -1 \end{vmatrix} = -\mathbf{i} + 4\mathbf{j} + 7\mathbf{k}$$

Note that the cross product of two vectors is another vector; it is sometimes called the *vector product*.

Certain algebraic properties of the cross product follow immediately from the definition. If **a**, **b**, and **c** are vectors and α, β, and γ are scalars, then

(*i*) $\mathbf{a} \times \mathbf{b} = -(\mathbf{b} \times \mathbf{a})$

(*ii*) $\mathbf{a} \times (\beta\mathbf{b} + \gamma\mathbf{c}) = \beta(\mathbf{a} \times \mathbf{b}) + \gamma(\mathbf{a} \times \mathbf{c})$
$(\alpha\mathbf{a} + \beta\mathbf{b}) \times \mathbf{c} = \alpha(\mathbf{a} \times \mathbf{c}) + \beta(\mathbf{b} \times \mathbf{c})$

Note that $\mathbf{a} \times \mathbf{a} = -(\mathbf{a} \times \mathbf{a})$ by (*i*). Thus, $\mathbf{a} \times \mathbf{a} = \mathbf{0}$. In particular,

$$\mathbf{i} \times \mathbf{i} = \mathbf{0}, \qquad \mathbf{j} \times \mathbf{j} = \mathbf{0}, \qquad \mathbf{k} \times \mathbf{k} = \mathbf{0}$$

Also,

$$\mathbf{i} \times \mathbf{j} = \mathbf{k}, \qquad \mathbf{j} \times \mathbf{k} = \mathbf{i}, \qquad \mathbf{k} \times \mathbf{i} = \mathbf{j}$$

which can be remembered by cyclicly permuting **i**, **j**, **k** like this:

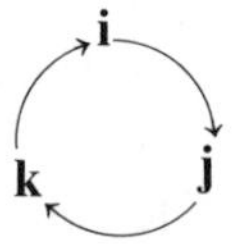

Our next goal is to provide a geometric interpretation of the cross product. To do this, we first introduce the triple product. Given three vectors **a**, **b**, and **c**, the real number

$$\mathbf{a} \cdot (\mathbf{b} \times \mathbf{c})$$

is called the *triple product* of **a**, **b**, and **c** (in that order). Let us obtain a formula for the triple product $\mathbf{a} \cdot (\mathbf{b} \times \mathbf{c})$. If $\mathbf{a} = a_1\mathbf{i} + a_2\mathbf{j} + a_3\mathbf{k}$, $\mathbf{b} = b_1\mathbf{i} + b_2\mathbf{j} + b_3\mathbf{k}$, and $\mathbf{c} = c_1\mathbf{i} + c_2\mathbf{j} + c_3\mathbf{k}$, then

$$\mathbf{a} \cdot (\mathbf{b} \times \mathbf{c}) = (a_1\mathbf{i} + a_2\mathbf{j} + a_3\mathbf{k}) \cdot \left(\begin{vmatrix} b_2 & b_3 \\ c_2 & c_3 \end{vmatrix} \mathbf{i} - \begin{vmatrix} b_1 & b_3 \\ c_1 & c_3 \end{vmatrix} \mathbf{j} + \begin{vmatrix} b_1 & b_2 \\ c_1 & c_2 \end{vmatrix} \mathbf{k} \right)$$

$$= a_1 \begin{vmatrix} b_2 & b_3 \\ c_2 & c_3 \end{vmatrix} - a_2 \begin{vmatrix} b_1 & b_3 \\ c_1 & c_3 \end{vmatrix} + a_3 \begin{vmatrix} b_1 & b_2 \\ c_1 & c_2 \end{vmatrix}$$

This may be written more concisely as

$$\mathbf{a}\cdot(\mathbf{b}\times\mathbf{c}) = \begin{vmatrix} a_1 & a_2 & a_3 \\ b_1 & b_2 & b_3 \\ c_1 & c_2 & c_3 \end{vmatrix}$$

Now suppose that $\mathbf{a}$ is a vector in the plane spanned by the vectors $\mathbf{b}$ and $\mathbf{c}$. This means that the first row in the determinant expression for $\mathbf{a}\cdot(\mathbf{b}\times\mathbf{c})$ is of the form $\mathbf{a} = \alpha\mathbf{b} + \beta\mathbf{c}$, and therefore $\mathbf{a}\cdot(\mathbf{b}\times\mathbf{c}) = 0$ by Example 3. In other words, the vector $\mathbf{b}\times\mathbf{c}$ is orthogonal to any vector in the plane spanned by $\mathbf{b}$ and $\mathbf{c}$, in particular to both $\mathbf{b}$ and $\mathbf{c}$.

Next, we calculate the magnitude of $\mathbf{b}\times\mathbf{c}$. Note that

$$\begin{aligned}\|\mathbf{b}\times\mathbf{c}\|^2 &= \begin{vmatrix} b_2 & b_3 \\ c_2 & c_3 \end{vmatrix}^2 + \begin{vmatrix} b_1 & b_3 \\ c_1 & c_3 \end{vmatrix}^2 + \begin{vmatrix} b_1 & b_2 \\ c_1 & c_2 \end{vmatrix}^2 \\ &= (b_2c_3 - b_3c_2)^2 + (b_1c_3 - c_1b_3)^2 + (b_1c_2 - c_1b_2)^2\end{aligned}$$

Writing out this last expression, we see that it is equal to

$$\begin{aligned}&(b_1^2 + b_2^2 + b_3^2)(c_1^2 + c_2^2 + c_3^2) - (b_1c_1 + b_2c_2 + b_3c_3)^2 \\ &\qquad = \|\mathbf{b}\|^2\|\mathbf{c}\|^2 - (\mathbf{b}\cdot\mathbf{c})^2 = \|\mathbf{b}\|^2\|\mathbf{c}\|^2 - \|\mathbf{b}\|^2\|\mathbf{c}\|^2\cos^2\theta \\ &\qquad = \|\mathbf{b}\|^2\|\mathbf{c}\|^2\sin^2\theta\end{aligned}$$

where θ is the angle between $\mathbf{b}$ and $\mathbf{c}$, $0 \le \theta \le \pi$.

Combining our results, we conclude that $\mathbf{b}\times\mathbf{c}$ *is a vector perpendicular to the plane spanned by* $\mathbf{b}$ *and* $\mathbf{c}$ *with length* $\|\mathbf{b}\|\,\|\mathbf{c}\|\,|\sin\theta|$. However, there are two possible vectors that satisfy these conditions, because there are two choices of direction that are perpendicular (or normal) to the plane P spanned by $\mathbf{b}$ and $\mathbf{c}$. This is clear from Figure 1.3.1 which shows the two choices $\mathbf{n}_1$ and $-\mathbf{n}_1$ perpendicular to P, with $\|\mathbf{n}_1\| = \|-\mathbf{n}_1\| = \|\mathbf{b}\|\,\|\mathbf{c}\|\,|\sin\theta|$.

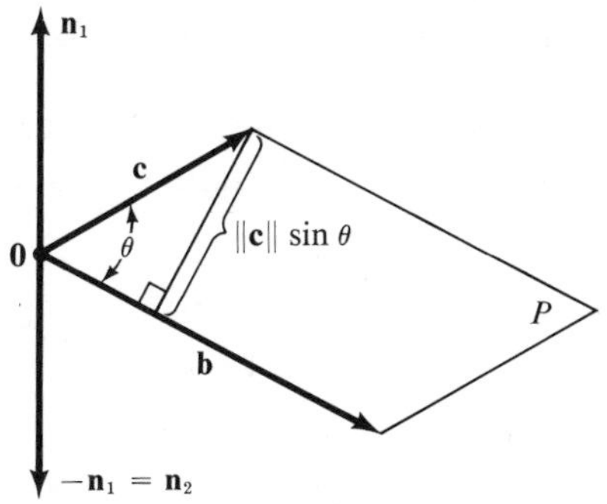

FIGURE 1.3.1

$\mathbf{n}_1$ *and* $\mathbf{n}_2$ *are the two possible vectors orthogonal to both* $\mathbf{b}$ *and* $\mathbf{c}$, *and with norm* $\|b\|\,\|c\|\,|\sin\theta|$.

Which vector represents $\mathbf{b} \times \mathbf{c}$, $\mathbf{n}_1$ or $-\mathbf{n}_1$? The answer is $\mathbf{n}_1 = \mathbf{b} \times \mathbf{c}$. Try a few cases to see, such as $\mathbf{k} = \mathbf{i} \times \mathbf{j}$. The following "right-hand rule" determines the direction of $\mathbf{b} \times \mathbf{c}$.

Take the palm of your right hand and place it in such a way that your fingers curl from $\mathbf{b}$ in the direction of $\mathbf{c}$ through the angle θ. Then your thumb points in the direction of $\mathbf{b} \times \mathbf{c}$ (Figure 1.3.2).

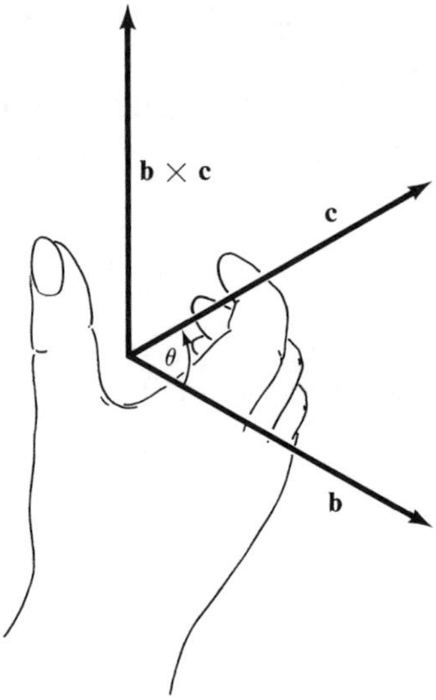

FIGURE 1.3.2
The right-hand rule for determining in which of the two possible directions $\mathbf{b} \times \mathbf{c}$ *points.*

If $\mathbf{b}$ and $\mathbf{c}$ are collinear, $\sin\theta = 0$, and so $\mathbf{b} \times \mathbf{c} = \mathbf{0}$; if $\mathbf{b}$ and $\mathbf{c}$ are not collinear, they then span a plane and $\mathbf{b} \times \mathbf{c}$ is a vector perpendicular to this plane. *The length of* $\mathbf{b} \times \mathbf{c}$, $\|\mathbf{b}\|\,\|\mathbf{c}\|\,|\sin\theta|$, *is just the area of the parallelogram with adjacent sides the vectors* $\mathbf{b}$ *and* $\mathbf{c}$ (Figure 1.3.3).

EXAMPLE 5. Find a unit vector orthogonal to the vectors $\mathbf{i} + \mathbf{j}$ and $\mathbf{j} + \mathbf{k}$.

A vector perpendicular to both $\mathbf{i} + \mathbf{j}$ and $\mathbf{j} + \mathbf{k}$ is the vector

$$(\mathbf{i} + \mathbf{j}) \times (\mathbf{j} + \mathbf{k}) = \begin{vmatrix} \mathbf{i} & \mathbf{j} & \mathbf{k} \\ 1 & 1 & 0 \\ 0 & 1 & 1 \end{vmatrix} = \mathbf{i} - \mathbf{j} + \mathbf{k}$$

Since $\|\mathbf{i} - \mathbf{j} + \mathbf{k}\| = \sqrt{3}$, the vector

$$\frac{1}{\sqrt{3}}(\mathbf{i} - \mathbf{j} + \mathbf{k})$$

is a unit vector perpendicular to $\mathbf{i} + \mathbf{j}$ and $\mathbf{j} + \mathbf{k}$.

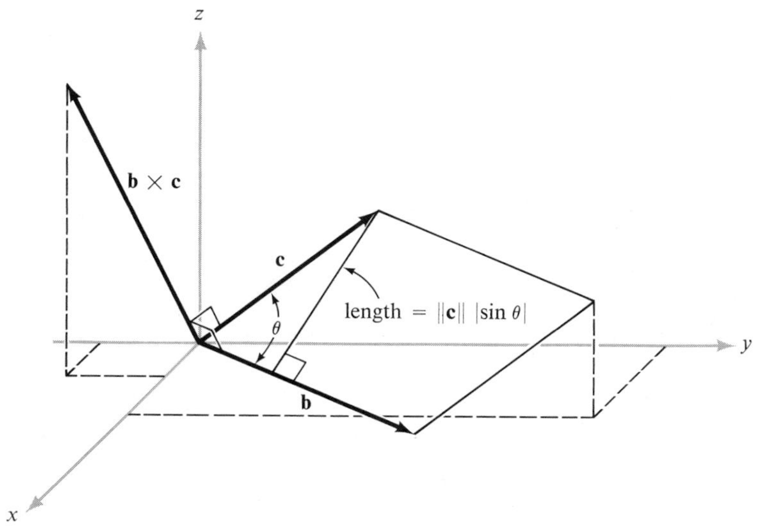

FIGURE 1.3.3
The length of **b** × **c** *equals the area of the parallelogram formed by* **b** *and* **c**.

Using the cross product, we may obtain the basic geometric interpretation of determinants. Let $\mathbf{b} = b_1\mathbf{i} + b_2\mathbf{j}$ and $\mathbf{c} = c_1\mathbf{i} + c_2\mathbf{j}$ be two vectors in the plane. If θ denotes the angle between **b** and **c**, we have seen that $\|\mathbf{b} \times \mathbf{c}\| = \|\mathbf{b}\|\,\|\mathbf{c}\|\,|\sin\theta|$. As noted above, $\|\mathbf{b}\|\,\|\mathbf{c}\|\,|\sin\theta|$ is the area of the parallelogram with adjacent sides **b** and **c** (see Figure 1.3.3). Using the definition of the cross product,

$$\mathbf{b} \times \mathbf{c} = \begin{vmatrix} \mathbf{i} & \mathbf{j} & \mathbf{k} \\ b_1 & b_2 & 0 \\ c_1 & c_2 & 0 \end{vmatrix} = \begin{vmatrix} b_1 & b_2 \\ c_1 & c_2 \end{vmatrix}\mathbf{k}$$

Thus $\|\mathbf{b} \times \mathbf{c}\|$ is the absolute value of the determinant

$$\begin{vmatrix} b_1 & b_2 \\ c_1 & c_2 \end{vmatrix} = b_1c_2 - b_2c_1$$

From this it follows that *the absolute value of the above determinant is the area of the parallelogram with adjacent sides the vectors* $\mathbf{b} = b_1\mathbf{i} + b_2\mathbf{j}$ *and* $\mathbf{c} = c_1\mathbf{i} + c_2\mathbf{j}$.

EXAMPLE 6. Find the area of the triangle with vertices at the points (1, 1), (0, 2), and (3, 2) (Figure 1.3.4).

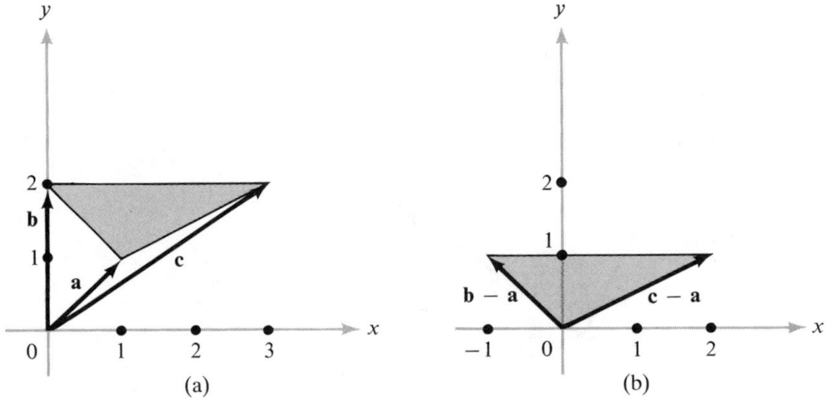

FIGURE 1.3.4
Problem: find the area A of the shaded triangle in (a). *Solution: Express the sides as vector differences* (b) *to get* $A = \frac{1}{2}\|(\mathbf{b} - \mathbf{a}) \times (\mathbf{c} - \mathbf{a})\|$.

Let $\mathbf{a} = \mathbf{i} + \mathbf{j}$, $\mathbf{b} = 2\mathbf{j}$, and $\mathbf{c} = 3\mathbf{i} + 2\mathbf{j}$. It is clear that the triangle whose vertices are the endpoints of the vectors **a**, **b**, and **c** has the same area as the triangle with vertices at **0**, $\mathbf{b} - \mathbf{a}$, and $\mathbf{c} - \mathbf{a}$ (Figure 1.3.4). Indeed, the latter is merely a translation of the former triangle. Since the area of this translated triangle is one-half the area of the parallelogram with adjacent sides $\mathbf{b} - \mathbf{a}$ and $\mathbf{c} - \mathbf{a}$, we find that the area of the triangle with vertices (1, 1), (0, 2), and (3, 2) is the absolute value of

$$\frac{1}{2}\begin{vmatrix} -1 & 1 \\ 2 & 1 \end{vmatrix} = -\frac{3}{2}$$

that is, 3/2.

There is an interpretation of determinants of 3 × 3 matrices as volumes that is analogous to the interpretation of determinants of 2 × 2 matrices as areas. Let $\mathbf{a} = a_1\mathbf{i} + a_2\mathbf{j} + a_3\mathbf{k}$, $\mathbf{b} = b_1\mathbf{i} + b_2\mathbf{j} + b_3\mathbf{k}$, and $\mathbf{c} = c_1\mathbf{i} + c_2\mathbf{j} + c_3\mathbf{k}$ be vectors in $\mathbb{R}^3$. We will show that *the volume of the parallelepiped with adjacent sides* **a**, **b**, *and* **c** (Figure 1.3.5) *is the absolute value of the determinant*

$$D = \begin{vmatrix} a_1 & a_2 & a_3 \\ b_1 & b_2 & b_3 \\ c_1 & c_2 & c_3 \end{vmatrix}$$

We know that $\|\mathbf{a} \times \mathbf{b}\|$ is the area of the parallelogram with adjacent sides **a** and **b**. Moreover, $(\mathbf{a} \times \mathbf{b}) \cdot \mathbf{c} = \|\mathbf{c}\|\,\|\mathbf{a} \times \mathbf{b}\|\cos\psi$, where ψ is the acute angle that **c** makes with the normal to the plane spanned by **a** and **b**. Since

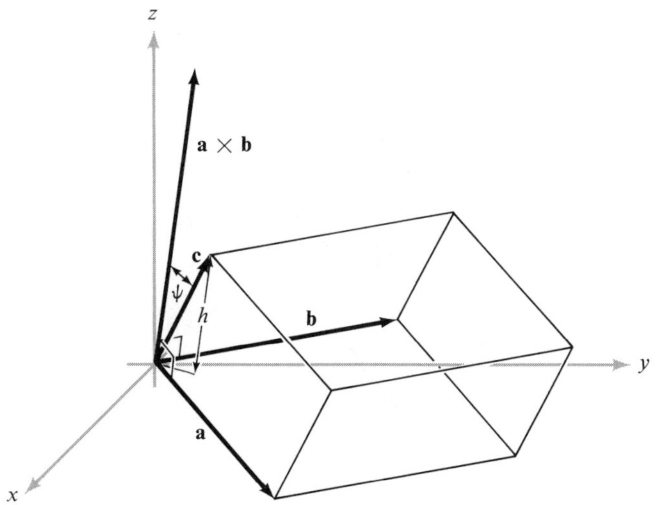

FIGURE 1.3.5
The volume of the parallelepiped formed by **a**, **b**, **c** *is the absolute value of the determinant of the* 3 × 3 *matrix with* **a**, **b**, **c** *as its rows.*

the volume of the parallelepiped with adjacent sides **a**, **b**, and **c** is the product of the area of the base $\|\mathbf{a} \times \mathbf{b}\|$ times the altitude $\|\mathbf{c}\|\cos\psi$, it follows that the volume is merely $|(\mathbf{a} \times \mathbf{b}) \cdot \mathbf{c}|$. We saw on p. 31 that $D = \mathbf{a} \cdot (\mathbf{b} \times \mathbf{c})$. By interchanging rows, $D = -\mathbf{c} \cdot (\mathbf{b} \times \mathbf{a}) = \mathbf{c} \cdot (\mathbf{a} \times \mathbf{b}) = (\mathbf{a} \times \mathbf{b}) \cdot \mathbf{c}$; therefore, the absolute value of D is the volume of the parallelepiped with adjacent sides **a**, **b**, and **c**.

To conclude this section, we shall use vector methods to determine the equation of a plane in space. Let P be a plane in space, **a** a vector ending on the plane, and **n** a vector normal to the plane (see Figure 1.3.6).

If **r** is a vector in $\mathbb{R}^3$, then the endpoint of **r** is on the plane P if and only if $\mathbf{r} - \mathbf{a}$ is parallel to P and, hence, if and only if $(\mathbf{r} - \mathbf{a}) \cdot \mathbf{n} = 0$ (**n** is perpendicular to any vector parallel to P—see Figure 1.3.6). Since the inner product is distributive, this last condition is equivalent to $\mathbf{r} \cdot \mathbf{n} = \mathbf{a} \cdot \mathbf{n}$. Therefore, if we let $\mathbf{a} = a_1\mathbf{i} + a_2\mathbf{j} + a_3\mathbf{k}$, $\mathbf{n} = A\mathbf{i} + B\mathbf{j} + C\mathbf{k}$, and $\mathbf{r} = x\mathbf{i} + y\mathbf{j} + z\mathbf{k}$, it follows that the endpoint of **r** lies on P if and only if

$$Ax + By + Cz = \mathbf{r} \cdot \mathbf{n} = \mathbf{a} \cdot \mathbf{n} = Aa_1 + Ba_2 + Ca_3 \tag{3}$$

Since **n** and **a** are fixed, the right-hand side of equation (3) is a constant, say $-D$. Thus an equation that determines the plane P is

$$Ax + By + Cz + D = 0 \tag{4}$$

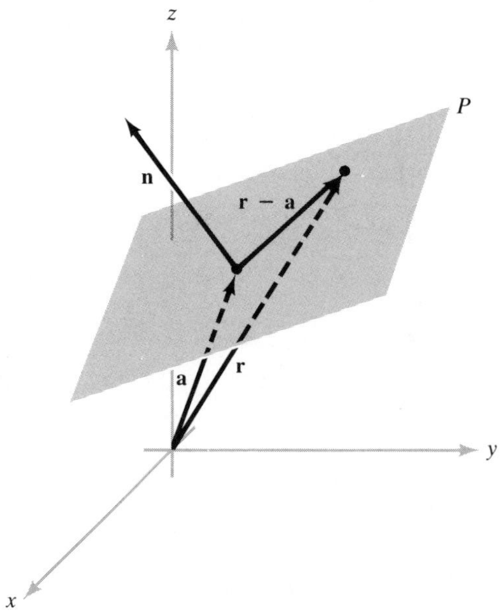

FIGURE 1.3.6
The points **r** *of the plane through* **a** *and perpendicular to* **n** *satisfy the equation* $(\mathbf{r} - \mathbf{a}) \cdot \mathbf{n} = 0$.

where $A\mathbf{i} + B\mathbf{j} + C\mathbf{k}$ is normal to P; conversely, if A, B, and C are not all zero, the set of point (x, y, z) satisfying equation (4) is a plane with normal $A\mathbf{i} + B\mathbf{j} + C\mathbf{k}$. Equation (4) is linear in the three variables x, y, z, which corresponds geometrically to a linear surface, i.e., a plane, in $\mathbb{R}^3$.

The four numbers A, B, C, D are not determined uniquely by P. To see this, note that (x, y, z) satisfies equation (4) if and only if it also satisfies the relation

$$(\lambda A)x + (\lambda B)y + (\lambda C)z + (\lambda D) = 0$$

for $\lambda \neq 0$. If A, B, C, D and A', B', C', D', determine the same plane P then $A = \lambda A'$, $B = \lambda B'$, $C = \lambda C'$, $D = \lambda D'$ for a scalar λ. We say that A, B, C, D are determined by P up to a scalar multiple. Conversely, given A, B, C, D and A', B', C', D', they determine the same plane if $A = \lambda A'$, $B = \lambda B'$, $C = \lambda C'$, $D = \lambda D'$ for some scalar λ. This fact will become more apparent in Example 8 below.

EXAMPLE 7. Determine the equation of the plane perpendicular to the vector $\mathbf{i} + \mathbf{j} + \mathbf{k}$ and containing the point $(1, 0, 0)$.

From the above discussion it follows that the equation of the plane is of the form $x + y + z + D = 0$. Since $(1, 0, 0)$ is on the plane, $1 + 0 + 0 + D = 0$, or $D = -1$. Thus, $x + y + z = 1$ is the equation of the plane.

EXAMPLE 8. Find an equation for the plane containing the points (1, 1, 1), (2, 0, 0), and (1, 1, 0).

Method 1. Any equation of the plane is of the form $Ax+By+Cz+D=0$. Since the points (1, 1, 1), (2, 0, 0), and (1, 1, 0) lie on the plane, we have

$$\begin{aligned} A + B + C + D &= 0 \\ 2A \qquad\qquad + D &= 0 \\ A + B \qquad + D &= 0 \end{aligned}$$

Proceeding by elimination, we reduce the above system to the form

$$\begin{aligned} 2A + D &= 0 && \text{(2nd equation)} \\ 2B + D &= 0 && (2 \times \text{3rd–2nd}) \\ C &= 0 && \text{(1st–3rd)} \end{aligned}$$

Since the numbers A, B, C, and D are determined only up to a scalar multiple, we can fix the value of one of them and then the others will be determined uniquely. If we let $D = -2$, then $A = +1$, $B = +1$, $C = 0$. Thus an equation of the plane that contains the given points is $x + y - 2 = 0$.

Method 2. Let $\mathbf{a} = \mathbf{i} + \mathbf{j} + \mathbf{k}$, $\mathbf{b} = 2\mathbf{i}$, $\mathbf{c} = \mathbf{i} + \mathbf{j}$. Any vector normal to the plane must be orthogonal to the vectors $\mathbf{a} - \mathbf{b}$ and $\mathbf{c} - \mathbf{b}$, which are parallel to the plane since their endpoints lie on the plane. Thus, $\mathbf{n} = (\mathbf{a} - \mathbf{b}) \times (\mathbf{c} - \mathbf{b})$ is normal to the plane. Computing the product,

$$\mathbf{n} = \begin{vmatrix} \mathbf{i} & \mathbf{j} & \mathbf{k} \\ -1 & 1 & 1 \\ -1 & 1 & 0 \end{vmatrix} = -\mathbf{i} - \mathbf{j}$$

Thus, any equation of the plane is of the form $-x - y + D = 0$ (up to a scalar multiple). Since (2, 0, 0) lies on the plane, $D = +2$. After substituting, we obtain $x + y - 2 = 0$.

EXAMPLE 9. Let $A(x - x_0) + B(y - y_0) + C(z - z_0) = 0$ be the equation of a plane P through the point $\mathrm{R} = (x_0, y_0, z_0)$ in $\mathbb{R}^3$. Determine the distance from the point $\mathrm{E} = (x_1, y_1, z_1)$ to the plane (see Figure 1.3.7). What is the distance from E to a plane with the equation $Ax + By + Cz + D = 0$?

Consider the vector

$$\mathbf{n} = \frac{A\mathbf{i} + B\mathbf{j} + C\mathbf{k}}{\sqrt{A^2 + B^2 + C^2}}$$

which is a unit vector normal to the plane. Next, drop a perpendicular from

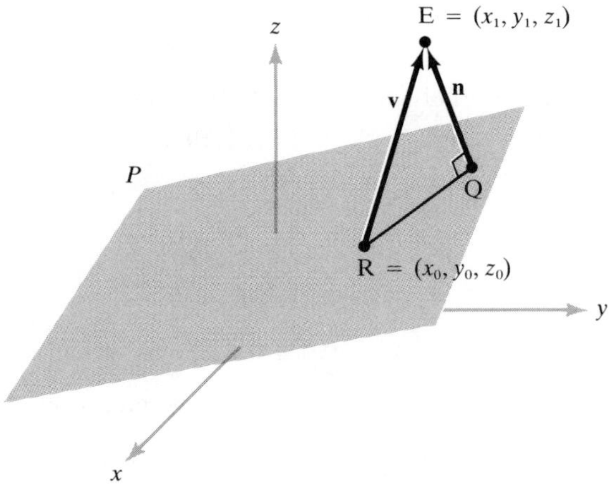

FIGURE 1.3.7

The geometry for determining the distance from the point E *to the plane* P.

E to the plane and construct the triangle REQ shown in Figure 1.3.7. The distance $d = |\mathrm{EQ}|$ is the length of the projection of $\mathbf{v} = \overrightarrow{\mathrm{RE}}$ (the vector from R to E) onto $\mathbf{n}$; thus

$$\begin{aligned} \text{distance} = |\mathbf{v}\cdot\mathbf{n}| &= |[(x_1 - x_0)\mathbf{i} + (y_1 - y_0)\mathbf{j} + (z_1 - z_0)\mathbf{k}]\cdot\mathbf{n}| \\ &= \frac{|A(x_1 - x_0) + B(y_1 - y_0) + C(z_1 - z_0)|}{\sqrt{A^2 + B^2 + C^2}} \end{aligned}$$

If the plane is given in the form $Ax + By + Cz + D = 0$, choose a point (x_0, y_0, z_0) on it and note that $D = -(Ax_0 + By_0 + Cz_0)$. Substitution into the previous formula gives

$$\text{distance} = \frac{|Ax_1 + By_1 + Cz_1 + D|}{\sqrt{A^2 + B^2 + C^2}}$$

EXERCISES

1. Verify that interchanging two rows or two columns of the 3×3 determinant

$$\begin{vmatrix} 1 & 2 & 1 \\ 3 & 0 & 1 \\ 2 & 0 & 2 \end{vmatrix}$$

changes the sign of the determinant (pick any two rows and any two columns).

2. Evaluate

(a) $\begin{vmatrix} 2 & -1 & 0 \\ 4 & 3 & 2 \\ 3 & 0 & 1 \end{vmatrix}$ (b) $\begin{vmatrix} 36 & 18 & 17 \\ 45 & 24 & 20 \\ 3 & 5 & -2 \end{vmatrix}$

(c) $\begin{vmatrix} 1 & 4 & 9 \\ 4 & 9 & 16 \\ 9 & 16 & 25 \end{vmatrix}$ (d) $\begin{vmatrix} 2 & 3 & 5 \\ 7 & 11 & 13 \\ 17 & 19 & 23 \end{vmatrix}$

3. Compute $\mathbf{a} \times \mathbf{b}$, where $\mathbf{a} = \mathbf{i} - 2\mathbf{j} + \mathbf{k}$, $\mathbf{b} = 2\mathbf{i} + \mathbf{j} + \mathbf{k}$.
4. Compute $\mathbf{a} \cdot (\mathbf{b} \times \mathbf{c})$, where $\mathbf{a}$ and $\mathbf{b}$ are as in Exercise 3 and $\mathbf{c} = 3\mathbf{i} - \mathbf{j} + 2\mathbf{k}$.
5. Find the area of the parallelogram with sides the vectors $\mathbf{a}$ and $\mathbf{b}$ given in Exercise 3.
6. What is the volume of the parallelepiped with sides $2\mathbf{i} + \mathbf{j} - \mathbf{k}$, $5\mathbf{i} - 3\mathbf{k}$, and $\mathbf{i} - 2\mathbf{j} + \mathbf{k}$?

In Exercises 7 to 10, describe all unit vectors orthogonal to the given vectors.

7. $\mathbf{i}, \mathbf{j}$
8. $-5\mathbf{i} + 9\mathbf{j} - 4\mathbf{k}$, $7\mathbf{i} + 8\mathbf{j} + 9\mathbf{k}$
9. $-5\mathbf{i} + 9\mathbf{j} - 4\mathbf{k}$, $7\mathbf{i} + 8\mathbf{j} + 9\mathbf{k}$, $\mathbf{0}$
10. $2\mathbf{i} - 4\mathbf{j} + 3\mathbf{k}$, $-4\mathbf{i} + 8\mathbf{j} - 6\mathbf{k}$
11. Determine the distance from the plane $12x + 13y + 5z + 2 = 0$ to the point $(1, 1, -5)$.
12. Find the distance from the plane through the origin that is perpendicular to $\mathbf{i} - 2\mathbf{j} + \mathbf{k}$ to the point $(6, 1, 0)$.
13. Compute $\mathbf{u} + \mathbf{v}$, $\mathbf{u} \cdot \mathbf{v}$, $\|\mathbf{u}\|$, $\|\mathbf{v}\|$, and $\mathbf{u} \times \mathbf{v}$ where $\mathbf{u} = \mathbf{i} - 2\mathbf{j} + \mathbf{k}$, $\mathbf{v} = 2\mathbf{i} - \mathbf{j} + 2\mathbf{k}$.
14. Repeat Exercise 13 for $\mathbf{u} = 3\mathbf{i} + \mathbf{j} - \mathbf{k}$, $\mathbf{v} = -6\mathbf{i} - 2\mathbf{j} - 2\mathbf{k}$.
15. Find an equation for the plane that
    (a) is perpendicular to $\mathbf{v} = (1, 1, 1)$ and passes through $(1, 0, 0)$.
    (b) is perpendicular to $\mathbf{v} = (1, 2, 3)$ and passes through $(1, 1, 1)$.
16. Find an equation for the plane that passes through $(0, 0, 0)$, $(2, 0, -1)$, and $(0, 4, -3)$.
17. (a) Prove the triple-vector-product identities $(\mathbf{A} \times \mathbf{B}) \times \mathbf{C} = (\mathbf{A} \cdot \mathbf{C})\mathbf{B} - (\mathbf{B} \cdot \mathbf{C})\mathbf{A}$ and $\mathbf{A} \times (\mathbf{B} \times \mathbf{C}) = (\mathbf{A} \cdot \mathbf{C})\mathbf{B} - (\mathbf{A} \cdot \mathbf{B})\mathbf{C}$.
    (b) Prove $(\mathbf{u} \times \mathbf{v}) \times \mathbf{w} = \mathbf{u} \times (\mathbf{v} \times \mathbf{w})$ if and only if $(\mathbf{u} \times \mathbf{w}) \times \mathbf{v} = \mathbf{0}$.
    (c) Prove $(\mathbf{u} \times \mathbf{v}) \times \mathbf{w} + (\mathbf{v} \times \mathbf{w}) \times \mathbf{u} + (\mathbf{w} \times \mathbf{u}) \times \mathbf{v} = \mathbf{0}$ (the *Jacobi identity*).
18. (a) Prove, without recourse to geometry, that

$$\begin{aligned} \mathbf{u} \cdot (\mathbf{v} \times \mathbf{w}) &= \mathbf{v} \cdot (\mathbf{w} \times \mathbf{u}) = \mathbf{w} \cdot (\mathbf{u} \times \mathbf{v}) = -\mathbf{u} \cdot (\mathbf{w} \times \mathbf{v}) \\ &= -\mathbf{w} \cdot (\mathbf{v} \times \mathbf{u}) = -\mathbf{v} \cdot (\mathbf{u} \times \mathbf{w}) \end{aligned}$$

    (b) Prove

$$\begin{aligned} (\mathbf{u} \times \mathbf{v}) \cdot (\mathbf{u}' \times \mathbf{v}') &= (\mathbf{u} \cdot \mathbf{u}')(\mathbf{v} \cdot \mathbf{v}') - (\mathbf{u} \cdot \mathbf{v}')(\mathbf{u}' \cdot \mathbf{v}) \\ &= \begin{vmatrix} \mathbf{u} \cdot \mathbf{u}' & \mathbf{u} \cdot \mathbf{v}' \\ \mathbf{u}' \cdot \mathbf{v} & \mathbf{v} \cdot \mathbf{v}' \end{vmatrix} \end{aligned}$$

    (HINT: Use (a) and 17(a).)

19. Verify Cramer's rule, presented in the Historical Note on page 29.
20. Find an equation for the plane that passes through $(2, -1, 3)$ and is perpendicular to $\mathbf{v} = (1, -2, 2) + t(3, -2, 4)$.
21. Find an equation for the plane that passes through $(1, 2, -3)$ and is perpendicular to $\mathbf{v} = (0, -2, 1) + t(1, -2, 3)$.
22. Find the equation of the line that passes through $(1, -2, -3)$ and is perpendicular to the plane $3x - y - 2z + 4 = 0$.
23. Find an equation for the plane containing the two lines

$$\mathbf{v}_1 = (0, 1, -2) + t(2, 3, -1) \quad \text{and} \quad \mathbf{v}_2 = (2, -1, 0) + t(2, 3, -1).$$

24. Find the distance from $(2, 1, -1)$ to the plane $x - 2y + 2z + 5 = 0$.
25. Find an equation for the plane that contains the line $\mathbf{v} = (-1, 1, 2) + t(3, 2, 4)$ and is perpendicular to the plane $2x + y - 3z + 4 = 0$.
26. Find an equation for the plane that passes through $(3, 2, -1)$ and $(1, -1, 2)$ and that is parallel to the line $\mathbf{v} = (1, -1, 0) + t(3, 2, -2)$.
27. Redo Exercises 19 and 20 of Section 1.1 using the dot product and what you know about normals to planes.

1.4 CYLINDRICAL AND SPHERICAL COORDINATES

A standard way to represent a point in the plane $\mathbb{R}^2$ is by means of rectangular coordinates (x, y). However, as the reader has probably learned in elementary calculus, polar coordinates in the plane can be extremely useful. As portrayed in Figure 1.4.1, the coordinates (r, θ) are related to (x, y) by the formulas

$$x = r \cos \theta \quad \text{and} \quad y = r \sin \theta$$

where we usually take $r \geq 0$ and $0 \leq \theta < 2\pi$.

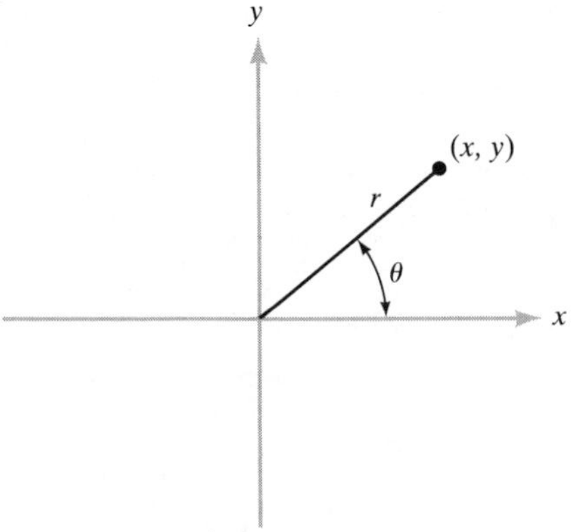

FIGURE 1.4.1
The polar coordinates of (x, y) are (r, θ).

Readers not familiar with polar coordinates are advised to study the relevant section of their calculus texts. We now set forth two ways of representing points in space other than by using rectangular Cartesian coordinates (x, y, z). These alternative coordinate systems are particularly well-suited for certain types of problems, such as the evaluation of integrals (see Section 5.8).

DEFINITION. *(see Figure 1.4.2). The* ***cylindrical coordinates*** (r, θ, z) *of a point* (x, y, z) *are defined by*

$$x = r\cos\theta, \qquad y = r\sin\theta, \qquad z = z \tag{1}$$

or explicitly

$$r = \sqrt{x^2 + y^2}, \qquad z = z, \qquad \theta = \begin{cases} \tan^{-1}(y/x) \text{ if } x > 0 \text{ and } y \geq 0 \\ \pi + \tan^{-1}(y/x) \text{ if } x < 0 \\ 2\pi + \tan^{-1}(y/x) \text{ if } x > 0 \text{ and } y < 0 \end{cases}$$

where $\tan^{-1} u$ *is between* $-\pi/2$ *and* $\pi/2$. *If* $x = 0$, *then* $\theta = \pi/2$ *for* $y > 0$ *and* $3\pi/2$ *for* $y < 0$. *If* $x = y = 0$, θ *is undefined.*

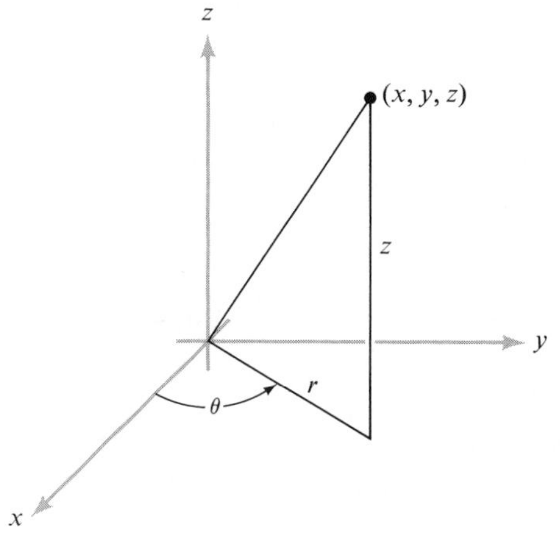

FIGURE 1.4.2
Representing a point (x, y, z) in terms of its cylindrical coordinates r, θ, and z.

In other words, for any point (x, y, z) we represent its first and second coordinates in terms of polar coordinates and leave the third coordinate unchanged. The formula (1) shows that, given (r, θ, z), the triple (x, y, z) is completely determined, and vice versa, if we restrict θ to the interval $[0, 2\pi[$ (sometimes the range $]-\pi, \pi]$ is convenient) and require that $r > 0$.

To see why we use the term "cylindrical coordinates", note that if $0 \leq \theta < 2\pi$, $-\infty < z < \infty$, and $r = a$ is some positive constant, then the locus of these points is a cylinder of radius a (see Figure 1.4.3).

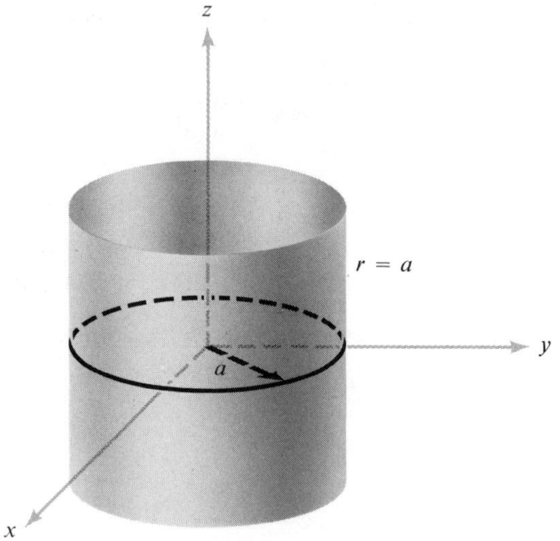

FIGURE 1.4.3
The graph of the points whose cylindrical coordinates satisfy $r = a$ is a cylinder.

EXAMPLE 1. (a) Find and plot the cylindrical coordinates of (6, 6, 8). (b) If a point has cylindrical coordinates $(8, 2\pi/3, -3)$, what are its Cartesian coordinates? Plot.

For part (a), we have $r = \sqrt{6^2 + 6^2} = 6\sqrt{2}$ and $\theta = \tan^{-1}(6/6) = \tan^{-1}(1) = \pi/4$. Thus the cylindrical coordinates are $(6\sqrt{2}, \pi/4, 8)$. This is point P in Figure 1.4.4. For part (b), we have

$$x = r\cos\theta = 8\cos\frac{2\pi}{3} = -\frac{8}{2} = -4$$

and

$$y = r\sin\theta = 8\sin\frac{2\pi}{3} = 8\frac{\sqrt{3}}{2} = 4\sqrt{3}.$$

Thus the Cartesian coordinates are $(-4, 4\sqrt{3}, -3)$. This is point Q in the figure.

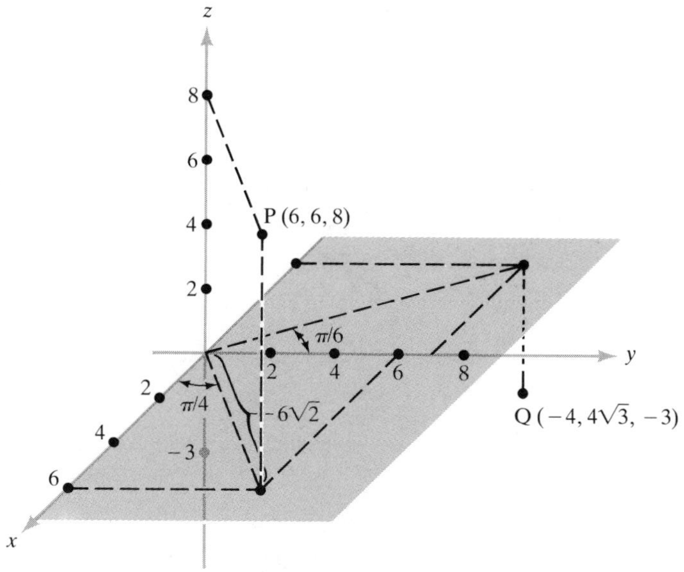

FIGURE 1.4.4
Some examples of the conversion between Cartesian and cylindrical coordinates.

Cylindrical coordinates are not the only possible generalization of polar coordinates to three dimensions. Recall that in two dimensions the magnitude of the vector $x\mathbf{i} + y\mathbf{j}$ (that is, $\sqrt{x^2 + y^2}$) is the r in the polar coordinate system. For cylindrical coordinates, the length of the vector $x\mathbf{i} + y\mathbf{j} + z\mathbf{k}$, namely,

$$\rho = \sqrt{x^2 + y^2 + z^2}$$

is not one of the coordinates of the system—we use only the magnitude $r = \sqrt{x^2 + y^2}$, the angle θ, and the "height" z.

We now modify this by introducing the *spherical coordinate* system, which uses ρ as a coordinate. Spherical coordinates are often useful for problems that possess spherical symmetry (symmetry about a point) while cylindrical coordinates can be applied when cylindrical symmetry (symmetry about a line) is involved.

Given a point $(x, y, z) \in \mathbb{R}^3$, let

$$\rho = \sqrt{x^2 + y^2 + z^2}$$

and represent x and y by polar coordinates in the xy-plane

$$x = r\cos\theta, \qquad y = r\sin\theta \tag{2}$$

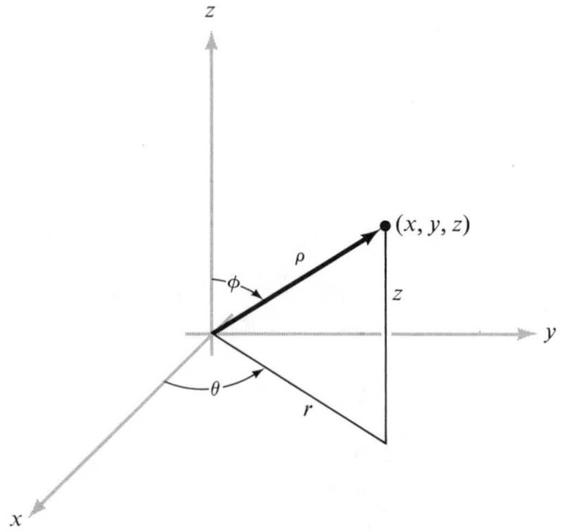

FIGURE 1.4.5
Spherical coordinates (ρ, θ, ϕ); the graph of points satisfying $\rho = a$ is a sphere.

where $r = \sqrt{x^2 + y^2}$ and θ is given by (1). The coordinate z is given by

$$z = \rho \cos \phi$$

where ϕ is the angle (between 0 and π inclusive) the radius vector $\mathbf{v} = x\mathbf{i} + y\mathbf{j} + z\mathbf{k}$ makes with the z-axis, in the plane containing the vector $\mathbf{v}$ and the z-axis (see Figure 1.4.5). Using the dot product we can express ϕ by

$$\cos \phi = \frac{\mathbf{v} \cdot \mathbf{k}}{\|\mathbf{v}\|}, \quad \text{that is,} \quad \phi = \cos^{-1} \frac{\mathbf{v} \cdot \mathbf{k}}{\|\mathbf{v}\|}$$

We take as our coordinates the quantities ρ, θ, ϕ. Since

$$r = \rho \sin \phi$$

we can use (2) to find x, y, z in terms of the spherical coordinates ρ, θ, ϕ.

DEFINITION *The spherical coordinates of (x, y, z) are defined by*

$$x = \rho \sin \phi \cos \theta, \qquad y = \rho \sin \phi \sin \theta, \qquad z = \rho \cos \phi \tag{3}$$

where

$$\rho \geq 0, \qquad 0 \leq \theta < 2\pi, \qquad 0 \leq \phi \leq \pi$$

Note that in spherical coordinates the equation of the sphere of radius a centered at the origin takes on the particularly simple form

$$\rho = a$$

EXAMPLE 2. Express (a) the surface $xz = 1$ and (b) the surface $x^2 + y^2 - z^2 = 1$ in spherical coordinates.

From (3), $x = \rho \sin \phi \cos \theta$, $z = \rho \cos \phi$, so the surface (a) consists of all (ρ, θ, ϕ) such that

$$\rho^2 \sin \phi \cos \theta \cos \phi = 1$$

For part (b) we can write

$$x^2 + y^2 - z^2 = x^2 + y^2 + z^2 - 2z^2 = \rho^2 - 2\rho^2 \cos^2 \phi$$

so the surface is $\rho^2(1 - 2 \cos^2 \phi) = 1$, or $-\rho^2 \cos(2\phi) = 1$.

EXERCISES

1. (a) The following points are given in cylindrical coordinates; express each in rectangular coordinates and spherical coordinates; $(1, 45°, 1)$, $(2, \pi/2, -4)$, $(0, 45°, 10)$, $(3, \pi/6, 4)$.
   (b) Change each of the following points from rectangular coordinates to spherical coordinates and to cylindrical coordinates: $(2, 1, -2)$, $(0, 3, 4)$, $(\sqrt{2}, 1, 1)$, $(-2\sqrt{3}, -2, 3)$.
2. Describe the geometric meaning of the following mappings in cylindrical coordinates.
   (a) $(r, \theta, z) \mapsto (r, \theta, -z)$
   (b) $(r, \theta, z) \mapsto (r, \theta + \pi, -z)$
3. Describe the geometric meaning of the following mappings in spherical coordinates.
   (a) $(\rho, \theta, \phi) \mapsto (\rho, \theta + \pi, \phi)$
   (b) $(\rho, \theta, \phi) \mapsto (\rho, \theta, \pi - \phi)$
4. (a) Describe the surfaces r = constant, θ = constant, and z = constant in the cylindrical coordinate system.
   (b) Describe the surfaces ρ = constant, θ = constant, and ϕ = constant in the spherical coordinate system.
5. Show that in order to represent each point in $\mathbb{R}^3$ by spherical coordinates it is only necessary to take values of θ between 0 and 2π, values of ϕ between 0 and π, and values of $\rho \geq 0$. Are coordinates unique if we allow $\rho \leq 0$?
6. Using cylindrical coordinates and the orthonormal (orthogonal normalized) vectors $\mathbf{e}_r$, $\mathbf{e}_\theta$, and $\mathbf{e}_z$ (see Figure 1.4.6)
   (a) express each of $\mathbf{e}_r$, $\mathbf{e}_\theta$, and $\mathbf{e}_z$ in terms of $\mathbf{i}, \mathbf{j}, \mathbf{k}$ and (x, y, z); and
   (b) calculate $\mathbf{e}_\theta \times \mathbf{j}$ both analytically using (a) and geometrically.

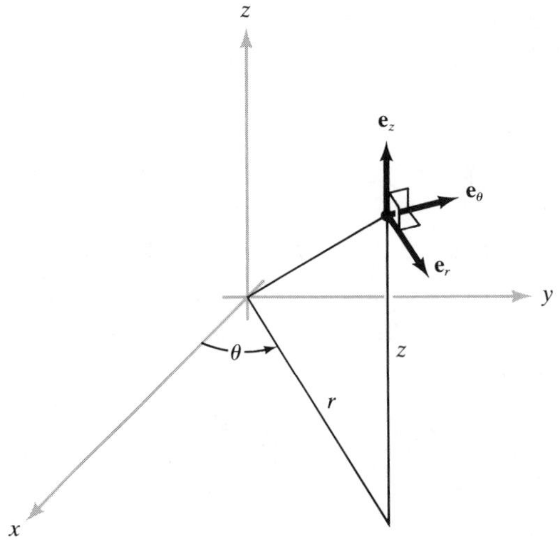

FIGURE 1.4.6
Orthonormal vectors $\mathbf{e}_r$, $\mathbf{e}_\theta$, and $\mathbf{e}_z$ associated with cylindrical coordinates. The vector $\mathbf{e}_r$ is parallel to the line labeled r.

7. Using spherical coordinates and the orthonormal vectors $\hat{\mathbf{e}}_\rho$, $\hat{\mathbf{e}}_\theta$, and $\hat{\mathbf{e}}_\phi$ (see Figure 1.4.7)
   (a) express each of $\hat{\mathbf{e}}_\rho$, $\hat{\mathbf{e}}_\theta$ and $\hat{\mathbf{e}}_\phi$ in terms of **i**, **j**, **k** and (x, y, z); and
   (b) calculate $\hat{\mathbf{e}}_\theta \times \mathbf{j}$ and $\hat{\mathbf{e}}_\phi \times \mathbf{j}$ both analytically and geometrically.

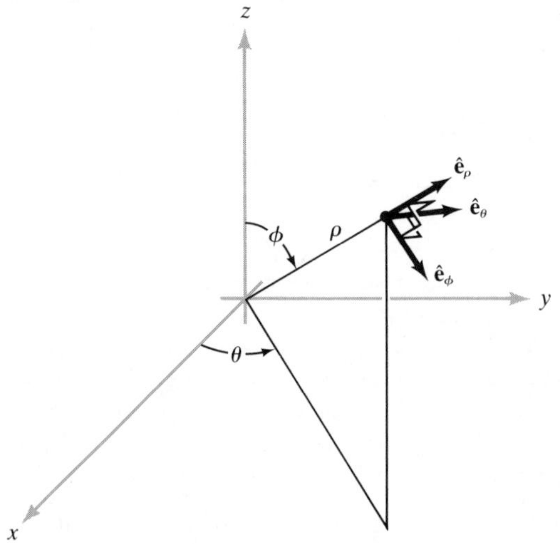

FIGURE 1.4.7
Orthonormal vectors $\hat{\mathbf{e}}_\rho$, $\hat{\mathbf{e}}_\theta$, and $\hat{\mathbf{e}}_\phi$ associated with spherical coordinates.

1.5 *n*-DIMENSIONAL EUCLIDEAN SPACE

In Sections 1.1 and 1.2 we studied the spaces $\mathbb{R}^1(\mathbb{R})$, $\mathbb{R}^2$, and $\mathbb{R}^3$, and gave geometric interpretations to them. For example, a point (x, y, z) in $\mathbb{R}^3$ can be thought of as a geometric object, namely, the directed line segment or vector emanating from the origin and ending at the point (x, y, z). We can therefore think of $\mathbb{R}^3$ in two ways:

(*i*) algebraically, as a set of triples (x, y, z) where x, y, and z are real numbers; or

(*ii*) geometrically, as a set of directed line segments.

These two ways of looking at $\mathbb{R}^3$ are equivalent. For generalization it is easier to use definition (*i*). Specifically, we can define $\mathbb{R}^n$, where n is a positive integer (possibly greater than 3), as the set of all ordered n-tuples $(x_1, x_2, \ldots, x_n)$, where the x_i are real numbers. For instance, $(1, \sqrt{5}, 2, \sqrt{3}) \in \mathbb{R}^4$. It is possible, but more difficult, to formulate a definition of $\mathbb{R}^n$ as a set of directed geometric objects.

The set $\mathbb{R}^n$ defined above is known as *Euclidean n-space* and its elements $\mathbf{x} = (x_1, x_2, \ldots, x_n)$ are known as *vectors* or *n-vectors*. By setting $n = 1, 2,$ or 3, we recover the line, the plane, and three-dimensional space ($\mathbb{R}^3$) respectively.

We shall launch our study of Euclidean n-space by introducing several algebraic operations. These are completely analogous to those introduced in Section 1.1 for $\mathbb{R}^2$ and $\mathbb{R}^3$. The first two, addition and scalar multiplication, are defined by:

(*i*) $(x_1, x_2, \ldots, x_n) + (y_1, y_2, \ldots, y_n) = (x_1 + y_1, x_2 + y_2, \ldots, x_n + y_n)$;

and

(*ii*) for any real number α,

$$\alpha(x_1, x_2, \ldots, x_n) = (\alpha x_1, \alpha x_2, \ldots, \alpha x_n)$$

The geometric significance of these operations for $\mathbb{R}^2$ and $\mathbb{R}^3$ was discussed in Section 1.1.

The n vectors $\mathbf{e}_1 = (1, 0, 0, \ldots, 0)$, $\mathbf{e}_2 = (0, 1, 0, \ldots, 0), \ldots, \mathbf{e}_n = (0, 0, \ldots, 0, 1)$ are called the *standard basis vectors* of $\mathbb{R}^n$, and they generalize the three mutually orthogonal unit vectors $\mathbf{i}, \mathbf{j}, \mathbf{k}$ of $\mathbb{R}^3$. The vector $\mathbf{x} = (x_1, x_2, \ldots, x_n)$ can be written as $\mathbf{x} = x_1\mathbf{e}_1 + x_2\mathbf{e}_2 + \cdots + x_n\mathbf{e}_n$.

For two vectors $\mathbf{x} = (x_1, x_2, x_3)$ and $\mathbf{y} = (y_1, y_2, y_3)$ in $\mathbb{R}^3$, we defined the *dot* or *inner product* $\mathbf{x} \cdot \mathbf{y}$ to be the real number $\mathbf{x} \cdot \mathbf{y} = x_1y_1 + x_2y_2 + x_3y_3$. This definition easily extends to $\mathbb{R}^n$; namely, for $\mathbf{x} = (x_1, x_2, \ldots, x_n)$, $\mathbf{y} = (y_1, y_2, \ldots, y_n)$, define $\mathbf{x} \cdot \mathbf{y} = x_1y_1 + x_2y_2 + \cdots + x_ny_n$. In $\mathbb{R}^n$, the notation $\langle \mathbf{x}, \mathbf{y} \rangle$ is often used in place of $\mathbf{x} \cdot \mathbf{y}$ for the inner product. Continuing the analogy with $\mathbb{R}^3$, we are led to define the abstract notion of the *length* or *norm* of a vector $\mathbf{x}$ by the formula

$$\text{length of } \mathbf{x} = \|\mathbf{x}\| = \sqrt{\mathbf{x} \cdot \mathbf{x}} = \sqrt{x_1^2 + x_2^2 + \cdots + x_n^2}.$$

If $\mathbf{x}$ and $\mathbf{y}$ are two vectors in the plane ($\mathbb{R}^2$) or in space ($\mathbb{R}^3$), then we know that the angle θ between them is given by the formula

$$\cos\theta = \frac{\mathbf{x} \cdot \mathbf{y}}{\|\mathbf{x}\|\,\|\mathbf{y}\|}$$

The right side of this equation can be defined in $\mathbb{R}^n$ as well as in $\mathbb{R}^2$. It still represents the cosine of the angle between $\mathbf{x}$ and $\mathbf{y}$; this angle is well defined since $\mathbf{x}$ and $\mathbf{y}$ lie in a two-dimensional subspace of $\mathbb{R}^n$ (the plane determined by $\mathbf{x}$ and $\mathbf{y}$). The dot product is a powerful mathematical tool, which reflects the geometric notion of the angle between two vectors in $\mathbb{R}^n$.

It will be useful to have available some algebraic properties of the inner product. These are summarized in the next theorem (see formulas (*i*), (*ii*), (*iii*), and (*iv*) of Section 1.2).

THEOREM 2. *For* $\mathbf{x}, \mathbf{y}, \mathbf{z} \in \mathbb{R}^n$ *and* α, β *real numbers, we have*

(*i*) $(\alpha\mathbf{x} + \beta\mathbf{y}) \cdot \mathbf{z} = \alpha(\mathbf{x} \cdot \mathbf{z}) + \beta(\mathbf{y} \cdot \mathbf{z})$

(*ii*) $\mathbf{x} \cdot \mathbf{y} = \mathbf{y} \cdot \mathbf{x}$

(*iii*) $\mathbf{x} \cdot \mathbf{x} \geq 0$

(*iv*) $\mathbf{x} \cdot \mathbf{x} = 0$ *if and only if* $\mathbf{x} = \mathbf{0}$.

Proof. Each assertion above can be proved by a simple computation. For example, to prove (*i*) we write

$$\begin{aligned}(\alpha\mathbf{x} + \beta\mathbf{y}) \cdot \mathbf{z} &= (\alpha x_1 + \beta y_1, \alpha x_2 + \beta y_2, \ldots, \alpha x_n + \beta y_n) \cdot (z_1, z_2, \ldots, z_n)\\ &= (\alpha x_1 + \beta y_1)z_1 + (\alpha x_2 + \beta y_2)z_2 + \cdots + (\alpha x_n + \beta y_n)z_n\\ &= a x_1 z_1 + \beta y_1 z_1 + a x_2 z_2 + \beta y_2 z_2 + \cdots + \alpha x_n z_n + \beta y_n z_n\\ &= \alpha(\mathbf{x} \cdot \mathbf{z}) + \beta(\mathbf{y} \cdot \mathbf{z}).\end{aligned}$$

The other proofs are similar. ■

In Section 1.2 we proved a far more interesting property of dot products, called the Cauchy-Schwarz inequality (sometimes called the Cauchy-Bunyakovskii-Schwarz inequality, or simply CBS inequality, because it was independently discovered in special cases by the French mathematician Cauchy, the Russian mathematician Bunyakovskii, and the German mathematician Schwarz). For $\mathbb{R}^2$ our proof required the use of the law of cosines.

For $\mathbb{R}^n$ we could also use this method, by confining our attention to a plane in $\mathbb{R}^n$. However, we can also give a direct, completely algebraic proof.

THEOREM 3. *Let* $\mathbf{x}, \mathbf{y}$ *be vectors in* $\mathbb{R}^n$. *Then*

$$|\mathbf{x} \cdot \mathbf{y}| \leq \|\mathbf{x}\| \, \|\mathbf{y}\|$$

Proof. Let $a = \mathbf{y} \cdot \mathbf{y}$ and $b = -\mathbf{x} \cdot \mathbf{y}$. If $a = 0$ the theorem is clearly valid, since then $\mathbf{y} = \mathbf{0}$ and both sides of the inequality reduce to 0. Thus we may suppose $a \neq 0$. By Theorem 2 we have

$$\begin{aligned} 0 \leq (a\mathbf{x} + b\mathbf{y}) \cdot (a\mathbf{x} + b\mathbf{y}) &= a^2\mathbf{x} \cdot \mathbf{x} + 2ab\mathbf{x} \cdot \mathbf{y} + b^2\mathbf{y} \cdot \mathbf{y} \\ &= (\mathbf{y} \cdot \mathbf{y})^2\mathbf{x} \cdot \mathbf{x} - (\mathbf{y} \cdot \mathbf{y})(\mathbf{x} \cdot \mathbf{y})^2 \end{aligned}$$

Dividing by $\mathbf{y} \cdot \mathbf{y}$ gives

$$0 \leq (\mathbf{y} \cdot \mathbf{y})(\mathbf{x} \cdot \mathbf{x}) - (\mathbf{x} \cdot \mathbf{y})^2$$

or

$$(\mathbf{x} \cdot \mathbf{y})^2 \leq (\mathbf{x} \cdot \mathbf{x})(\mathbf{y} \cdot \mathbf{y}) = \|\mathbf{x}\|^2 \, \|\mathbf{y}\|^2$$

Taking square roots on both sides of this inequality yields the desired result. ■

There is a very useful consequence of the Cauchy-Schwarz inequality in terms of lengths. The *triangle inequality* is geometrically clear in $\mathbb{R}^3$. In Figure 1.5.1, $\|OQ\| = \|\mathbf{x} + \mathbf{y}\|$, $\|OP\| = \|\mathbf{x}\| = \|RQ\|$, and $\|OR\| = \|\mathbf{y}\|$. Since the sum of the lengths of two sides of a triangle is greater than or equal to the length of the third, we have $\|OQ\| \leq \|OR\| + \|RQ\|$; that is, $\|\mathbf{x} + \mathbf{y}\| \leq \|\mathbf{x}\| + \|\mathbf{y}\|$. The case for $\mathbb{R}^n$ is not as obvious, so we shall provide the analytic proof.

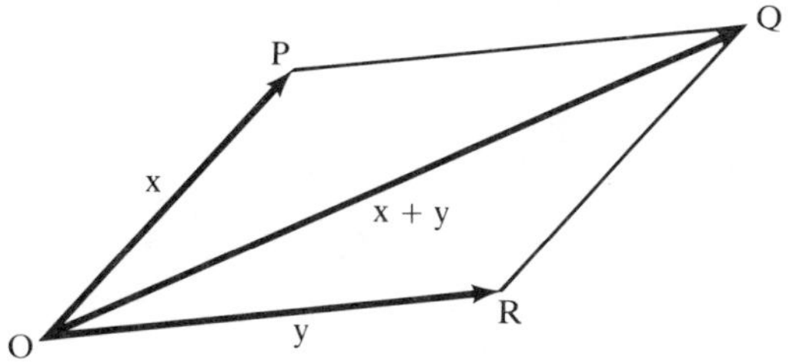

FIGURE 1.5.1
This geometry shows that $\|OQ\| \leq \|OR\| + \|RQ\|$, *in vector notation* $\|\mathbf{x} + \mathbf{y}\| \leq \|\mathbf{x}\| + \|\mathbf{y}\|$, *which is the triangle inequality.*

COROLLARY. *Let* $\mathbf{x}, \mathbf{y}$ *be vectors in* $\mathbb{R}^n$. *Then*

$$\|\mathbf{x} + \mathbf{y}\| \leq \|\mathbf{x}\| + \|\mathbf{y}\|$$

Proof. By Theorem 3, $\mathbf{x} \cdot \mathbf{y} \leq |\mathbf{x} \cdot \mathbf{y}| \leq \|\mathbf{x}\|\,\|\mathbf{y}\|$, so that

$$\|\mathbf{x} + \mathbf{y}\|^2 = \|\mathbf{x}\|^2 + 2\mathbf{x} \cdot \mathbf{y} + \|\mathbf{y}\|^2 \leq \|\mathbf{x}\|^2 + 2\|\mathbf{x}\|\,\|\mathbf{y}\| + \|\mathbf{y}\|^2$$

Hence, we get $\|\mathbf{x} + \mathbf{y}\|^2 \leq (\|\mathbf{x}\| + \|\mathbf{y}\|)^2$; taking square roots gives the result. ■

If Theorem 3 and its corollary are written out algebraically, they become the following rather unobvious inequalities:

$$\left|\sum_{i=1}^{n} x_i y_i\right| \leq \left(\sum_{i=1}^{n} x_i^2\right)^{1/2} \left(\sum_{i=1}^{n} y_i^2\right)^{1/2}$$

$$\left(\sum_{i=1}^{n} (x_i + y_i)^2\right)^{1/2} \leq \left(\sum_{i=1}^{n} x_i^2\right)^{1/2} + \left(\sum_{i=1}^{n} y_i^2\right)^{1/2}$$

These are very useful inequalities; we shall use them often in what follows, but primarily in vector form.

EXAMPLE 1. Let $\mathbf{x} = (1, 2, 0, -1)$, and $\mathbf{y} = (-1, 1, 1, 0)$. Then

$$\|\mathbf{x}\| = \sqrt{1^2 + 2^2 + 0^2 + (-1)^2} = \sqrt{6}$$

$$\|\mathbf{y}\| = \sqrt{(-1)^2 + 1^2 + 1^2 + 0^2} = \sqrt{3}$$

$$\mathbf{x} \cdot \mathbf{y} = 1(-1) + 2 \cdot 1 + 0 \cdot 1 + (-1)0 = 1$$

$$\mathbf{x} + \mathbf{y} = (0, 3, 1, -1)$$

$$\|\mathbf{x} + \mathbf{y}\| = \sqrt{0^2 + 3^2 + 1^2 + (-1)^2} = \sqrt{11}$$

We can directly verify Theorem 3 in this case:

$$\mathbf{x} \cdot \mathbf{y} = 1 \leq 4.24 \approx \sqrt{6}\sqrt{3} = \|\mathbf{x}\|\,\|\mathbf{y}\|$$

Similarly, we can check its corollary:

$$\begin{aligned}\|\mathbf{x} + \mathbf{y}\| &= \sqrt{11} \approx 3.32 \leq 4.18 \\ &= 2.45 + 1.73 \approx \sqrt{6} + \sqrt{3} = \|\mathbf{x}\| + \|\mathbf{y}\|\end{aligned}$$

By analogy with $\mathbb{R}^3$, we can define the notion of distance in $\mathbb{R}^n$; namely, if $\mathbf{x}$ and $\mathbf{y}$ are points in $\mathbb{R}^n$, the distance between $\mathbf{x}$ and $\mathbf{y}$ is defined to be $\|\mathbf{x} - \mathbf{y}\|$, or the length of the vector $\mathbf{x} - \mathbf{y}$. We remark that there is no cross product defined on $\mathbb{R}^n$ except for $n = 3$. It is only the dot product that generalizes.

Generalizing 2×2 and 3×3 matrices (see Section 1.3), we can consider $n \times n$ matrices, arrays of n^2 numbers:

$$A = \begin{bmatrix} a_{11} & a_{12} & \cdots & a_{1n} \\ a_{21} & a_{22} & \cdots & a_{2n} \\ \vdots & \vdots & & \vdots \\ a_{n1} & a_{n2} & \cdots & a_{nn} \end{bmatrix}$$

We shall also write A as (a_{ij}). Let us now define how $n \times n$ matrices are multiplied. If $A = (a_{ij})$, $B = (b_{ij})$ then $AB = C$ has entries given by

$$c_{ij} = \sum_{k=1}^{n} a_{ik}b_{kj}$$

which is the dot product of the ith row of A and the jth column of B:

$$\begin{bmatrix} a_{11} & \cdots & a_{1n} \\ \vdots & & \vdots \\ a_{i1} & \xrightarrow{\cdots} & a_{in} \\ \vdots & & \vdots \\ a_{n1} & \cdots & a_{nn} \end{bmatrix} \begin{bmatrix} b_{11} & \cdots & b_{1j} & \cdots & b_{1n} \\ \vdots & & \vdots\big\downarrow & & \vdots \\ b_{n1} & \cdots & b_{nj} & \cdots & b_{nn} \end{bmatrix}$$

EXAMPLE 2. Let

$$A = \begin{bmatrix} 1 & 0 & 3 \\ 2 & 1 & 0 \\ 1 & 0 & 0 \end{bmatrix} \quad \text{and} \quad B = \begin{bmatrix} 0 & 1 & 0 \\ 1 & 0 & 0 \\ 0 & 1 & 1 \end{bmatrix}$$

Then

$$AB = \begin{bmatrix} 0 & 4 & 3 \\ 1 & 2 & 0 \\ 0 & 1 & 0 \end{bmatrix} \quad \text{and} \quad BA = \begin{bmatrix} 2 & 1 & 0 \\ 1 & 0 & 3 \\ 3 & 1 & 0 \end{bmatrix}$$

Similarly, we can multiply an $n \times m$ matrix (n rows, m columns) by an $m \times p$ matrix (m rows, p columns) to obtain an $n \times p$ matrix (n rows, p columns) by the same rule. Note that for AB to be defined, the number of columns of A must equal the number of rows of B.

EXAMPLE 3. Let

$$A = \begin{bmatrix} 2 & 0 & 1 \\ 1 & 1 & 2 \end{bmatrix} \quad \text{and} \quad B = \begin{bmatrix} 1 & 0 & 2 \\ 0 & 2 & 1 \\ 1 & 1 & 1 \end{bmatrix}$$

Then

$$AB = \begin{bmatrix} 3 & 1 & 5 \\ 3 & 4 & 5 \end{bmatrix}$$

and BA is not defined.

EXAMPLE 4. Let

$$A = \begin{bmatrix} 1 \\ 2 \\ 1 \\ 3 \end{bmatrix} \quad \text{and} \quad B = [2 \quad 2 \quad 1 \quad 2]$$

Then

$$AB = \begin{bmatrix} 2 & 2 & 1 & 2 \\ 4 & 4 & 2 & 4 \\ 2 & 2 & 1 & 2 \\ 6 & 6 & 3 & 6 \end{bmatrix} \quad \text{and} \quad BA = [13]$$

Any $m \times n$ matrix A determines a mapping of $\mathbb{R}^n$ to $\mathbb{R}^m$ as follows: Let $\mathbf{x} = (x_1, \ldots, x_n) \in \mathbb{R}^n$; consider the $n \times 1$ column matrix associated to $\mathbf{x}$ which we shall temporarily denote $\mathbf{x}^T$:

$$\mathbf{x}^T = \begin{bmatrix} x_1 \\ \vdots \\ x_n \end{bmatrix}$$

and multiply A by $\mathbf{x}^T$ (considered as an $n \times 1$ matrix) to get a new $m \times 1$ matrix:

$$A\mathbf{x}^T = \begin{bmatrix} a_{11} & \cdots & a_{1n} \\ \vdots & & \vdots \\ a_{m1} & \cdots & a_{mn} \end{bmatrix} \begin{bmatrix} x_1 \\ \vdots \\ x_n \end{bmatrix} = \begin{bmatrix} y_1 \\ \vdots \\ y_m \end{bmatrix}$$

We then get a vector $\mathbf{y} = (y_1, \ldots, y_m)$. In order to use a matrix A to get a mapping from vectors $\mathbf{x} = (x_1, \ldots, x_n)$ to vectors $\mathbf{y} = (y_1, \ldots, y_m)$ according to the preceding equation, we have to write vectors in column form $\begin{bmatrix} x_1 \\ \vdots \\ x_n \end{bmatrix}$ instead of row form $(x_1, \ldots, x_n)$. This sudden switch from writing $\mathbf{x}$ as a row to writing $\mathbf{x}$ as a column is necessitated by standard conventions on multiplication.* Thus, although it may cause some confusion, we will write $\mathbf{x} = (x_1, \ldots, x_n)$ and $\mathbf{y} = (y_1, \ldots, y_m)$ as column vectors $\mathbf{x} = \begin{bmatrix} x_1 \\ \vdots \\ x_n \end{bmatrix}$, $\mathbf{y} = \begin{bmatrix} y_1 \\ \vdots \\ y_m \end{bmatrix}$

* If mathematicians had evolved the convention to write $\mathbf{x}A$ instead of $A\mathbf{x}$, or had different rules of matrix multiplication, $\mathbf{x}$ could have remained a row.

when dealing with matrix multiplication; that is, we will *identify* these two forms of writing vectors. Thus, we will delete the T on $\mathbf{x}^T$ and view $\mathbf{x}^T$ and $\mathbf{x}$ as the same; that is, $\mathbf{x} = \mathbf{x}^T$.

Thus, $A\mathbf{x} = \mathbf{y}$ will "really" mean the following: Write $\mathbf{x}$ as a column vector, multiply it by A, and let $\mathbf{y}$ be the vector whose components are those of the resulting column vector. The rule $\mathbf{x} \mapsto A\mathbf{x}$ therefore defines a mapping of $\mathbb{R}^n$ to $\mathbb{R}^m$. This mapping is linear; that is, it satisfies

$$A(\mathbf{x} + \mathbf{y}) = A\mathbf{x} + A\mathbf{y}$$

$$A(\alpha\mathbf{x}) = \alpha(A\mathbf{x}), \qquad \alpha \text{ a scalar}$$

as may be easily verified. One learns in a linear-algebra course that, conversely, any linear transformation of $\mathbb{R}^n$ to $\mathbb{R}^m$ is representable in this way by an $m \times n$ matrix.

If $A = (a_{ij})$ is an $m \times n$ matrix and $\mathbf{e}_j$ is the jth standard basis vector of $\mathbb{R}^n$, then $A\mathbf{e}_j$ is a vector in $\mathbb{R}^m$ with components the same as the jth column of A. That is, the ith component of $A\mathbf{e}_j$ is a_{ij}. In symbols, $(A\mathbf{e}_j)_i = a_{ij}$.

EXAMPLE 5. If

$$A = \begin{bmatrix} 1 & 0 & 3 \\ -1 & 0 & 1 \\ 2 & 1 & 2 \\ -1 & 2 & 1 \end{bmatrix}$$

then $\mathbf{x} \mapsto A\mathbf{x}$ of $\mathbb{R}^3$ to $\mathbb{R}^4$ is the mapping defined by

$$\begin{bmatrix} x_1 \\ x_2 \\ x_3 \end{bmatrix} \mapsto \begin{bmatrix} x_1 + 3x_3 \\ -x_1 + x_3 \\ 2x_1 + x_2 + 2x_3 \\ -x_1 + 2x_2 + x_3 \end{bmatrix}$$

EXAMPLE 6. The following illustrates what happens to a specific point when mapped by a 4×3 matrix:

$$A\mathbf{e}_2 = \begin{bmatrix} 4 & 2 & 9 \\ 3 & 5 & 4 \\ 1 & 2 & 3 \\ 0 & 1 & 2 \end{bmatrix} \begin{bmatrix} 0 \\ 1 \\ 0 \end{bmatrix} = \begin{bmatrix} 2 \\ 5 \\ 2 \\ 1 \end{bmatrix} = \text{2nd column of } A$$

Matrix multiplication is not, in general, *commutative:* if A and B are $n \times n$ matrices, then generally

$$AB \neq BA$$

(see Examples 2, 3, and 4). (For both AB and BA to be defined, A and B must be square ($n \times n$) matrices.)

An $n \times n$ matrix is called *invertible* if there is an $n \times n$ matrix B such that

$$AB = BA = I_n$$

where

$$I_n = \begin{bmatrix} 1 & 0 & 0 & \cdots & 0 \\ 0 & 1 & 0 & \cdots & 0 \\ 0 & 0 & 1 & \cdots & 0 \\ \vdots & \vdots & \vdots & & \vdots \\ 0 & 0 & 0 & \cdots & 1 \end{bmatrix}$$

is the $n \times$ n identity matrix: I_n has the property that $I_n C = C I_n = C$ for any $n \times n$ matrix C. We denote B by A^{-1} and call A^{-1} the *inverse* of A. The inverse, when it exists, is unique.

EXAMPLE 7. If

$$A = \begin{bmatrix} 2 & 4 & 0 \\ 0 & 2 & 1 \\ 3 & 0 & 2 \end{bmatrix} \quad \text{then} \quad A^{-1} = \frac{1}{20}\begin{bmatrix} 4 & -8 & 4 \\ 3 & 4 & -2 \\ -6 & 12 & 4 \end{bmatrix}$$

since $AA^{-1} = I_n = A^{-1}A$, as may be checked by matrix multiplication.

Methods of computing inverses are learned in linear algebra; we won't require these methods in this book. If A is invertible, the equation $A\mathbf{x} = \mathbf{y}$ can be solved for the vector $\mathbf{x}$ by multiplying both sides by A^{-1} to obtain $\mathbf{x} = A^{-1}\mathbf{y}$.

In Section 1.3 we defined the determinant of a 3×3 matrix. This can be generalized by induction to $n \times n$ determinants. We illustrate here how to write the determinant of a 4×4 matrix in terms of the determinants of 3×3 matrices:

$$\begin{vmatrix} a_{11} & a_{12} & a_{13} & a_{14} \\ a_{21} & a_{22} & a_{23} & a_{24} \\ a_{31} & a_{32} & a_{33} & a_{34} \\ a_{41} & a_{42} & a_{43} & a_{44} \end{vmatrix} = a_{11}\begin{vmatrix} a_{22} & a_{23} & a_{24} \\ a_{32} & a_{33} & a_{34} \\ a_{42} & a_{43} & a_{44} \end{vmatrix} - a_{12}\begin{vmatrix} a_{21} & a_{23} & a_{24} \\ a_{31} & a_{33} & a_{34} \\ a_{41} & a_{43} & a_{44} \end{vmatrix}$$

$$+ a_{13}\begin{vmatrix} a_{21} & a_{22} & a_{24} \\ a_{31} & a_{32} & a_{34} \\ a_{41} & a_{42} & a_{44} \end{vmatrix} - a_{14}\begin{vmatrix} a_{21} & a_{22} & a_{23} \\ a_{31} & a_{32} & a_{33} \\ a_{41} & a_{42} & a_{43} \end{vmatrix}$$

(see formula (2) of Section 1.3; the signs alternate $+, -, +, -, \ldots$).

The basic properties of 3×3 determinants reviewed in Section 1.3 remain valid for $n \times n$ determinants. In particular, we note the fact that if A is an $n \times n$ matrix and B is the matrix formed by adding a scalar multiple of the kth row (column) of A to the lth row (respectively, column) of A, then the determinant of A (det A) is equal to the determinant of B (det B) (see Example 8 below).

A basic theorem of linear algebra states that an $n \times n$ matrix A is invertible if and only if the determinant of A is not zero. Another basic property is that $\det(AB) = (\det A)(\det B)$. In this text we shall not make use of many details of linear algebra, so we shall leave these assertions unproved.

EXAMPLE 8. Let

$$A = \begin{bmatrix} 1 & 0 & 1 & 0 \\ 1 & 1 & 1 & 1 \\ 2 & 1 & 0 & 1 \\ 1 & 1 & 0 & 2 \end{bmatrix}$$

Then adding $(-1) \times$ first column to the third column, we get

$$\det A = \begin{vmatrix} 1 & 0 & 0 & 0 \\ 1 & 1 & 0 & 1 \\ 2 & 1 & -2 & 1 \\ 1 & 1 & -1 & 2 \end{vmatrix} = 1 \begin{vmatrix} 1 & 0 & 1 \\ 1 & -2 & 1 \\ 1 & -1 & 2 \end{vmatrix}$$

Adding $(-1) \times$ first column to the third column of this 3×3 determinant gives

$$\det A = \begin{vmatrix} 1 & 0 & 0 \\ 1 & -2 & 0 \\ 1 & -1 & 1 \end{vmatrix} = \begin{vmatrix} -2 & 0 \\ -1 & 1 \end{vmatrix} = -2$$

Thus, $\det A = -2 \neq 0$, and so A has an inverse.

If we have three matrices A, B, and C such that the products AB and BC are defined, then the products $(AB)C$ and $A(BC)$ will be defined and equal (that is, matrix multiplication is associative). We call this the *triple product* of matrices and denote it by ABC.

EXAMPLE 9. Let

$$A = \begin{bmatrix} 3 \\ 5 \end{bmatrix}, \qquad B = [1 \quad 1], \quad \text{and} \quad C = \begin{bmatrix} 1 \\ 2 \end{bmatrix}$$

Then

$$ABC = A(BC) = \begin{bmatrix} 3 \\ 5 \end{bmatrix} [3] = \begin{bmatrix} 9 \\ 15 \end{bmatrix}$$

EXAMPLE 10.

$$\begin{bmatrix} 2 & 0 \\ 0 & 1 \end{bmatrix} \begin{bmatrix} 1 & 1 \\ 1 & 1 \end{bmatrix} \begin{bmatrix} 0 & -1 \\ 1 & 1 \end{bmatrix} = \begin{bmatrix} 2 & 0 \\ 0 & 1 \end{bmatrix} \begin{bmatrix} 1 & 0 \\ 1 & 0 \end{bmatrix} = \begin{bmatrix} 2 & 0 \\ 1 & 0 \end{bmatrix}$$

Matrix multiplication is the most important nonelementary algebraic operation on matrices; there are also some elementary operations that deserve mention. Given two $m \times n$ matrices A and B, we can add (subtract) them to obtain a new $m \times n$ matrix $C = A + B$ ($C = A - B$), whose ijth entry c_{ij} is the sum (difference) of a_{ij} and b_{ij}. It is clear that $A + B = B + A$.

EXAMPLE 11.

$$\begin{bmatrix} 2 & 1 & 0 \\ 3 & 4 & 1 \end{bmatrix} + \begin{bmatrix} -1 & 1 & 3 \\ 0 & 0 & 7 \end{bmatrix} = \begin{bmatrix} 1 & 2 & 3 \\ 3 & 4 & 8 \end{bmatrix}$$

EXAMPLE 12.

$$[1 \quad 2] + [0 \quad -1] = [1 \quad 1]$$

EXAMPLE 13.

$$\begin{bmatrix} 2 & 1 \\ 1 & 2 \end{bmatrix} - \begin{bmatrix} 1 & 0 \\ 0 & 1 \end{bmatrix} = \begin{bmatrix} 1 & 1 \\ 1 & 1 \end{bmatrix}$$

Given a scalar λ and an $m \times n$ matrix A, we can multiply A by λ to obtain a new $m \times n$ matrix $\lambda A = C$, whose ijth entry c_{ij} is the product λa_{ij}.

EXAMPLE 14.

$$3\begin{bmatrix} 1 & -1 & 2 \\ 0 & 1 & 5 \\ 1 & 0 & 3 \end{bmatrix} = \begin{bmatrix} 3 & -3 & 6 \\ 0 & 3 & 15 \\ 3 & 0 & 9 \end{bmatrix}$$

EXERCISES

1. Prove (*ii*) to (*iv*) of Theorem 2.
2. In $\mathbb{R}^n$ show that
   (a) $2\|\mathbf{x}\|^2 + 2\|\mathbf{y}\|^2 = \|\mathbf{x} + \mathbf{y}\|^2 + \|\mathbf{x} - \mathbf{y}\|^2$
   (This is known as the parallelogram law.)
   (b) $\|\mathbf{x} + \mathbf{y}\| \, \|\mathbf{x} - \mathbf{y}\| \le \|\mathbf{x}\|^2 + \|\mathbf{y}\|^2$
   (c) $4\langle \mathbf{x}, \mathbf{y} \rangle = \|\mathbf{x} + \mathbf{y}\|^2 - \|\mathbf{x} - \mathbf{y}\|^2$
   (This is called the polarization identity.)

   Interpret these results geometrically in terms of the parallelogram formed by $\mathbf{x}$ and $\mathbf{y}$.
3. Verify the CBS inequality and the triangle inequality for:
   (a) $\mathbf{x} = (2, 0, -1)$, $\mathbf{y} = (4, 0, -2)$
   (b) $\mathbf{x} = (1, 0, 2, 6)$, $\mathbf{y} = (3, 8, 4, 1)$
   (c) $\mathbf{x} = (1, -1, 1, -1, 1)$, $\mathbf{y} = (3, 0, 0, 0, 2)$
4. Verify that if A is an $n \times n$ matrix, the map $\mathbf{x} \mapsto A\mathbf{x}$ of $\mathbb{R}^n$ to $\mathbb{R}^n$ is linear.

5. Assuming the law $\det(AB) = (\det A)(\det B)$, verify that $(\det A)(\det A^{-1}) = 1$ and conclude that if A has an inverse, then $\det A \neq 0$.

6. Compute AB, $\det A$, $\det B$, $\det(AB)$, and $\det(A + B)$ for

$$A = \begin{bmatrix} 3 & 0 & 1 \\ 1 & 2 & -1 \\ 1 & 0 & 1 \end{bmatrix} \qquad B = \begin{bmatrix} 1 & 0 & -1 \\ 2 & 0 & 1 \\ 0 & 1 & 0 \end{bmatrix}$$

7. Verify that the inverse of

$$\begin{bmatrix} a & b \\ c & d \end{bmatrix} \quad \text{is} \quad \frac{1}{ad - bc}\begin{bmatrix} d & -b \\ -c & a \end{bmatrix}$$

8. Use your answer in Exercise 7 to show that the solution of the system

$$ax + by = e$$

$$cx + dy = f$$

is

$$\begin{bmatrix} x \\ y \end{bmatrix} = \frac{1}{ad - bc}\begin{bmatrix} d & -b \\ -c & a \end{bmatrix}\begin{bmatrix} e \\ f \end{bmatrix}$$

9. Use induction on k to prove that if $\mathbf{x}_1, \ldots, \mathbf{x}_k \in \mathbb{R}^n$, then

$$\|\mathbf{x}_1 + \cdots + \mathbf{x}_k\| \leq \|\mathbf{x}_1\| + \cdots + \|\mathbf{x}_k\|$$

10. Prove, using algebra, the *identity of Lagrange*: For real numbers $x_1, \ldots, x_n$ and $y_1, \ldots, y_n$

$$\left(\sum_{i=1}^{n} x_i y_i\right)^2 = \left(\sum_{i=1}^{n} x_i^2\right)\left(\sum_{i=1}^{n} y_i^2\right) - \sum_{i<j} (x_i y_j - x_j y_i)^2$$

Use this to give another proof of the Cauchy-Schwarz inequality in $\mathbb{R}^n$.

*11 Prove by induction that if A is an $n \times n$ matrix, then
(a) $\det(\lambda A) = \lambda^n \det A$; and
(b) if B is a matrix obtained from A by multiplying any row or column by a scalar λ, then $\det B = \lambda \det A$.

In Exercises 12, 13, and 14, A, B, and C denote $n \times n$ matrices.

12. Is $\det(A + B) = \det A + \det B$? Give a proof or counterexample.

13. Does $(A + B)(A - B) = A^2 - B^2 = (A)(A) - (B)(B)$?

14. Prove that $\det(ABC) = (\det A)(\det B)(\det C)$.

*15. (This exercise assumes a knowledge of integration of continuous functions of one variable.) Note that the proof of the CBS inequality (Theorem 3) depends only on the properties of the inner product listed in Theorem 2. Use this observation to establish the following inequality for continuous functions $f, g\colon [0, 1] \to \mathbb{R}$

$$\left|\int_0^1 f(x)g(x)\,dx\right| \leq \sqrt{\int_0^1 [f(x)]^2\,dx}\,\sqrt{\int_0^1 [g(x)]^2\,dx}$$

Do this by

(a) verifying that the space of continuous functions from $[0, 1]$ to $\mathbb{R}$ forms a vector space; that is, we may think of functions f, g abstractly as "vectors" that can be added to each other and multiplied by scalars.

(b) introducing the inner product of functions

$$f \cdot g = \int_0^1 f(x)g(x)\,dx$$

and verifying that it satisfies conditions (*i*) to (*iv*) of Theorem 2.

*16. Define the transpose A^T of an $n \times n$ matrix A as follows: the ij^{th} element of A^T is a_{ji} where a_{ij} is the ij^{th} entry of A. Show that A^T is characterized by the following property: For all $\mathbf{x}$, $\mathbf{y}$ in $\mathbb{R}^n$,

$$(A^T\mathbf{x}) \cdot \mathbf{y} = \mathbf{x} \cdot (A\mathbf{y})$$

REVIEW EXERCISES FOR CHAPTER 1

1. Let $\mathbf{v} = 3\mathbf{i} + 4\mathbf{j} + 5\mathbf{k}$ and $\mathbf{w} = \mathbf{i} - \mathbf{j} + \mathbf{k}$. Compute $\mathbf{v} + \mathbf{w}$, $3\mathbf{v}$, $6\mathbf{v} + 8\mathbf{w}$, $-2\mathbf{v}$, $\mathbf{v} \cdot \mathbf{w}$, $\mathbf{v} \times \mathbf{w}$. Interpret each operation geometrically by graphing the vectors.

2. (a) Find the equation of the line through (0, 1, 0) in the direction of $3\mathbf{i} + \mathbf{k}$.
   (b) Find the equation of the line passing through (0, 1, 1) and (0, 1, 0).
   (c) Find an equation for the plane perpendicular to $(-1, 1, -1)$ and passing through (1, 1, 1).

3. Use vector notation to describe the triangle in space whose vertices are the origin and the endpoints of vectors $\mathbf{a}$ and $\mathbf{b}$.

4. Show that three vectors $\mathbf{a}$, $\mathbf{b}$, $\mathbf{c}$ lie in the same plane through the origin if and only if there are three scalars α, β, γ, not all zero, such that $\alpha\mathbf{a} + \beta\mathbf{b} + \gamma\mathbf{c} = \mathbf{0}$.

5. For real numbers $a_1, a_2, a_3, b_1, b_2, b_3$ show that

$$(a_1b_1 + a_2b_2 + a_3b_3)^2 \le (a_1^2 + a_2^2 + a_3^2)(b_1^2 + b_2^2 + b_3^2)$$

6. Let $\mathbf{u}$, $\mathbf{v}$, $\mathbf{w}$ be unit vectors that are orthogonal to each other. If $\mathbf{a} = \alpha\mathbf{u} + \beta\mathbf{v} + \gamma\mathbf{w}$, show that

$$\alpha = \mathbf{a} \cdot \mathbf{u}, \qquad \beta = \mathbf{a} \cdot \mathbf{v}, \qquad \gamma = \mathbf{a} \cdot \mathbf{w}$$

   Interpret the results geometrically.

7. Let $\mathbf{a}$, $\mathbf{b}$ be two vectors in the plane, $\mathbf{a} = (a_1, a_2)$, $\mathbf{b} = (b_1, b_2)$, and let λ be a real number. Show that the area of the parallelogram determined by $\mathbf{a}$ and $\mathbf{b} + \lambda\mathbf{a}$ is the same as that determined by $\mathbf{a}$ and $\mathbf{b}$. Sketch. Relate this result to a known property of determinants.

8. Find the volume of the parallelepiped determined by the vertices (0, 1, 0), (1, 1, 1), (0, 2, 0), (3, 1, 2).

9. Given nonzero vectors $\mathbf{a}$ and $\mathbf{b}$ in $\mathbb{R}^3$, show that the vector

$$\mathbf{v} = \|\mathbf{a}\|\mathbf{b} + \|\mathbf{b}\|\mathbf{a}$$

   bisects the angle between $\mathbf{a}$ and $\mathbf{b}$.

10. Use vector methods to prove that the distance from the point (x_1, y_1) to the line $ax + by = c$ is

$$\frac{|ax_1 + by_1 - c|}{\sqrt{a^2 + b^2}}$$

11. Verify that the direction of $\mathbf{b} \times \mathbf{c}$ is given by the right-hand rule, by considering $\mathbf{b}, \mathbf{c}$ to be the vectors $\mathbf{i}, \mathbf{j}$, or $\mathbf{k}$.

12. (a) Suppose $\mathbf{a} \cdot \mathbf{b} = \mathbf{a}' \cdot \mathbf{b}$ for all $\mathbf{b}$. Show that $\mathbf{a} = \mathbf{a}'$.
(b) Suppose $\mathbf{a} \times \mathbf{b} = \mathbf{a}' \times \mathbf{b}$ for all $\mathbf{b}$. Is it true that $\mathbf{a} = \mathbf{a}'$?

13. (a) Using vector methods, show that the distance between two nonparallel lines l_1 and l_2 is given by

$$d = \frac{|(\mathbf{v}_2 - \mathbf{v}_1) \cdot (\mathbf{a}_1 \times \mathbf{a}_2)|}{\|\mathbf{a}_1 \times \mathbf{a}_2\|}$$

where $\mathbf{v}_1, \mathbf{v}_2$ are any points on l_1 and l_2, respectively, and $\mathbf{a}_1$ and $\mathbf{a}_2$ are the directions of l_1 and l_2.
(HINT: Consider the plane through l_2 which is parallel to l_1. Show that $(\mathbf{a}_1 \times \mathbf{a}_2)/\|\mathbf{a}_1 \times \mathbf{a}_2\|$ is a unit normal for this plane; now project $\mathbf{v}_2 - \mathbf{v}_1$ onto this normal direction.)
(b) Find the distance between the line l_1 determined by the points $(-1, -1, 1)$ and $(0, 0, 0)$ and the line l_2 determined by the points $(0, -2, 0)$ and $(2, 0, 5)$.

14. Show that two planes given by the equations $Ax + By + Cz + D_1 = 0$ and $Ax + By + Cz + D_2 = 0$ are parallel, and that the distance between two such planes is

$$\frac{|D_1 - D_2|}{\sqrt{A^2 + B^2 + C^2}}$$

15. (a) Prove that the area of the triangle in the plane with vertices (x_1, y_1), (x_2, y_2), (x_3, y_3) is the absolute value of

$$\frac{1}{2}\begin{vmatrix} 1 & 1 & 1 \\ x_1 & x_2 & x_3 \\ y_1 & y_2 & y_3 \end{vmatrix}$$

(b) Find the area of the triangle with vertices $(1, 2)$, $(0, 1)$, $(-1, 1)$.

16. Convert the following points from Cartesian to cylindrical and spherical coordinates and plot:
(a) $(0, 3, 4)$
(b) $(-\sqrt{2}, 1, 0)$
(c) $(0, 0, 0)$
(d) $(-1, 0, 1)$
(e) $(-2\sqrt{3}, -2, 3)$

17. Convert the following points from cylindrical to Cartesian and spherical coordinates and plot:
(a) $(1, \pi/4, 1)$
(b) $(3, \pi/6, -4)$
(c) $(0, \pi/4, 1)$
(d) $(2, -\pi/2, 1)$
(e) $(-2, -\pi/2, 1)$

18. Convert the following points from spherical to Cartesian and cylindrical coordinates and plot:
(a) $(1, \pi/2, \pi)$
(b) $(2, -\pi/2, \pi/6)$
(c) $(0, \pi/8, \pi/35)$
(d) $(2, -\pi/2, -\pi)$
(e) $(-1, \pi, \pi/6)$

19. Rewrite the equation $z = x^2 - y^2$ using cylindrical and spherical coordinates.

20. Using spherical coordinates, show that

$$\phi = \cos^{-1}(\mathbf{u} \cdot \mathbf{k}/\|\mathbf{u}\|)$$

where $\mathbf{u} = x\mathbf{i} + y\mathbf{j} + z\mathbf{k}$. Interpret geometrically.

21. Verify the Cauchy-Schwarz and triangle inequalities for

$$\mathbf{x} = (3, 2, 1, 0) \quad \text{and} \quad \mathbf{y} = (1, 1, 1, 2)$$

22. Multiply the matrices

$$A = \begin{bmatrix} 3 & 0 & 1 \\ 2 & 0 & 1 \\ 1 & 0 & 1 \end{bmatrix} \quad \text{and} \quad B = \begin{bmatrix} 1 & 0 & 1 \\ 1 & 1 & 1 \\ 0 & 0 & 1 \end{bmatrix}$$

Does $AB = BA$?

23. (a) Show that for two $n \times n$ matrices A and B, and $\mathbf{x} \in \mathbb{R}^n$,

$$(AB)\mathbf{x} = A(B\mathbf{x})$$

(b) What does (a) imply about the relationship between the composition of the mappings $\mathbf{x} \mapsto Bx$, $\mathbf{y} \mapsto A\mathbf{y}$ and matrix multiplication?

24. Is

$$\begin{bmatrix} 1 & 0 & 1 \\ 1 & 1 & 1 \\ 0 & 0 & 1 \end{bmatrix} \quad \text{invertible?}$$

25. Verify that any linear mapping of $\mathbb{R}^n$ to $\mathbb{R}^n$ is determined by an $n \times n$ matrix; that is, it comes from an $n \times n$ matrix in the manner explained on p. 53.

26. Find an equation for the plane that contains $(3, -1, 2)$ and the line $\mathbf{v} = (2, -1, 0) + t(2, 3, 0)$.

CHAPTER 2

DIFFERENTIATION

I turn away with fright and horror from the lamentable evil of functions which do not have derivatives.

CHARLES HERMITE, in a letter to Thomas Jan Stieljes

In this chapter we shall extend the differential calculus from functions of one variable to functions of several variables. We shall begin with a section on the geometry of real-valued functions. The study of their graphs is useful for visualizing these functions.

Section 2.2 gives some basic definitions relating to limits and continuity. This subject is treated briefly for the reason that it requires time and mathematical maturity to develop fully, and is therefore best left to a more advanced course*. Fortunately, a complete understanding of all the subtleties of the limit concept is not necessary; the student who has difficulty with this section should bear that in mind. However, we hasten to add that the notion of limit is central to the definition of the derivative, but not to the computation of derivatives in specific problems, as we already know from one-variable calculus.

Section 2.3 and 2.4 deal with the definition of the derivative, and establish some basic rules of calculus; namely, how to differentiate a sum, product, quotient, or composition. (Some more technical facts related to the basic

* See, for example, J. Marsden, *Elementary Classical Analysis*, W. H. Freeman and Company, San Francisco, 1974.

theory of limits and differential calculus are presented in Appendix A.) In Section 2.5 we study directional derivatives and tangent planes, relating these ideas to those in Section 2.1. Finally, in Section 2.6 we consider some properties of higher-order derivatives.

In generalizing calculus from one to several dimensions, it is often convenient, though not absolutely essential, to use the language of matrix algebra. What we shall need has been summarized in Section 1.5.

2.1 THE GEOMETRY OF REAL-VALUED FUNCTIONS

We launch our investigation of real-valued functions by developing methods of visualizing them. In particular, we shall introduce the notions of graph, level curve, and level surface of such functions.

Let f be a function with domain a subset A of $\mathbb{R}^n$ (i.e., $A \subset \mathbb{R}^n$) and with range contained in $\mathbb{R}^m$. By this we mean that to each $\mathbf{x} = (x_1, \ldots, x_n) \in A$, f assigns a value $f(\mathbf{x})$, an m-tuple in $\mathbb{R}^m$. Such functions f are called vector-valued functions if $m > 1$, and scalar-valued functions if $m = 1$. For example, $f(x, y, z) = (x^2 + y^2 + z^2)^{-3/2}$ maps the set A of $(x, y, z) \neq (0, 0, 0)$ in $\mathbb{R}^3$ ($n = 3$ in this case) to $\mathbb{R}$ ($m = 1$). To denote f we sometimes write

$$f\colon (x, y, z) \mapsto (x^2 + y^2 + z^2)^{-3/2}$$

Note that in $\mathbb{R}^3$ we often use the notation (x, y, z) instead of (x_1, x_2, x_3). In general, the notation $\mathbf{x} \mapsto f(\mathbf{x})$ is useful for indicating the value to which a point $\mathbf{x} \in \mathbb{R}^n$ is sent. We write $f\colon A \subset \mathbb{R}^n \to \mathbb{R}^m$ to signify that A is the domain of f (in $\mathbb{R}^n$) and the range is contained in $\mathbb{R}^m$. We also use the expression f *maps* A into $\mathbb{R}^m$.*

Such functions f are called *functions of several variables* if $A \subset \mathbb{R}^n$, $n > 1$. As another example we can take the function $g\colon \mathbb{R}^6 \to \mathbb{R}^2$ defined by the rule

$$g(\mathbf{x}) = g(x_1, x_2, x_3, x_4, x_5, x_6) = (x_1x_2x_3x_4x_5x_6, \sqrt{x_1^2 + x_6^2})$$

The first coordinate of the value of g at $\mathbf{x}$ is the product of the coordinates of $\mathbf{x}$.

HISTORICAL NOTE

The notion of function was developed over many centuries, with the definition extended to cover more cases as they arose. For example, in 1667 James Gregory defined a function as "a quantity obtained from other quantities by a succession of algebraic operations or by any other operation imaginable". In 1755 Euler gave the following definition: "If some quantities depend on

* Some mathematicians would write such an f in boldface, $\mathbf{f}(\mathbf{x})$, since it is vector-valued. We did not do so, as a matter of personal taste. We use boldface primarily for mappings that are vector fields, introduced later.

others in such a way as to undergo variation when the latter are varied then the former are called functions of the latter."

It is important to be aware at this point that functions from $\mathbb{R}^n$ to $\mathbb{R}^m$ are not just mathematical abstractions. They do arise naturally in problems studied in all the sciences. For example, to specify the temperature T in a region A of space requires a function $T: A \subset \mathbb{R}^3 \to \mathbb{R}$ ($n = 3, m = 1$). Thus $T(x, y, z)$ is the temperature at the point (x, y, z). To specify the velocity of a fluid moving in space requires a map $\mathbf{V}: \mathbb{R}^4 \to \mathbb{R}^3$ where $\mathbf{V}(x, y, z, t)$ is the velocity vector of the fluid at the point (x, y, z) in space at time t (see Figure 2.1.1). To specify the reaction rate of a solution consisting of six reacting chemicals A, B, C, D, E, F in proportions x, y, z, w, u, v requires a map $\sigma: \mathscr{B} \subset \mathbb{R}^6 \to \mathbb{R}$, where $\sigma(x, y, z, w, u, v)$ gives the rate when the chemicals are in the indicated proportions. To specify the cardiac vector (the vector giving the magnitude and direction of electric current flow in the heart) at time t requires a map $\mathbf{c}: \mathbb{R}^3, t \mapsto \mathbf{c}(t)$.

When $f: U \subset \mathbb{R}^n \to \mathbb{R}$, we say that f is a *real-valued function of n variables with domain U*. The reason we say n variables is simply that we regard the

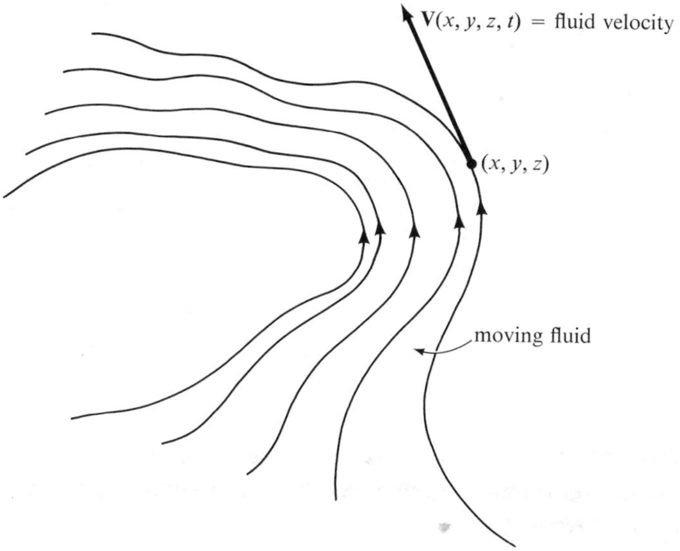

FIGURE 2.1.1
A fluid in motion defines a vector field **V** *by specifying the the velocity of the fluid particles at each point in space and time.*

coordinates of a point $\mathbf{x} = (x_1, \ldots, x_n) \in U$ as n variables, and $f(\mathbf{x}) = f(x_1, \ldots, x_n)$ depends on these variables. We say real-valued because $f(x_1, \ldots, x_n)$ is a real number. A good deal of our work will be with real-valued functions, so we shall give them special attention.

For $f: U \subset \mathbb{R} \to \mathbb{R}$ $(n = 1)$, the graph of f is the subset of $\mathbb{R}^2$ consisting of all points $(x, f(x))$ in the plane for x in U. This subset can be thought of as a curve in $\mathbb{R}^2$. In symbols, we write this as

$$\text{graph } f = \{(x, f(x)) \in \mathbb{R}^2 \mid x \in U\}$$

where the braces { } mean "the set of all" and the vertical bar is read as "such that." Drawing the graph of a function of one variable is a very useful device to help visualize how the function actually behaves. (See Figure 2.1.2.) Thus it would be extremely helpful if we could generalize the idea of a graph to functions of several variables. This leads us to the following definition:

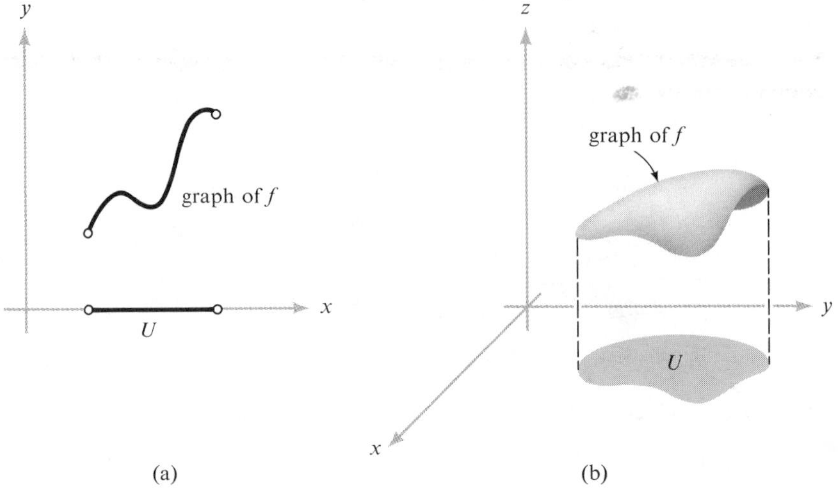

FIGURE 2.1.2
The graphs of (a) *a function of one variable and* (b) *a function of two variables.*

DEFINITION. *Let $f: U \subset \mathbb{R}^n \to \mathbb{R}$. Define the **graph** of f to be the subset of $\mathbb{R}^{n+1}$ consisting of all the points $(x_1, \ldots, x_n, f(x_1, \ldots, x_n))$ in $\mathbb{R}^{n+1}$ for $(x_1, \ldots, x_n)$ in U. In symbols,*

$$\text{graph } f = \{(x_1, \ldots, x_n, f(x_1, \ldots, x_n)) \in \mathbb{R}^{n+1} \mid (x_1, \ldots, x_n) \in U\}$$

For the case $n = 1$ the graph is, intuitively speaking, a curve in $\mathbb{R}^2$, while for $n = 2$ it is a surface in $\mathbb{R}^3$ (see Figure 2.1.2). For $n = 3$ it is difficult to visualize the graph because, being humans living in a three-dimensional world, it is hard for us to envisage sets in $\mathbb{R}^4$. To help overcome this handicap, we introduce the idea of a level set.

Suppose $f(x, y, z) = x^2 + y^2 + z^2$. Then a *level set* is a subset of $\mathbb{R}^3$ on which f is constant; for instance, the set where $x^2 + y^2 + z^2 = 1$ is a level set for f. This we can visualize: It is just a sphere of radius 1 in $\mathbb{R}^3$. The behavior or structure of a function is partially determined by the shape of its level sets; consequently, understanding these sets aids us in understanding the function in question. The concept of level sets is also useful for understanding functions of two variables $f(x, y)$, in which case we speak of level *curves*.

The idea is very similar to that used to prepare contour maps, where one draws lines to represent lines of constant altitude; walking along such a line would mean walking on a level path. In the case of a hill rising from the xy-plane, a graph of all these level curves gives us a good idea of the function $h(x, y)$, which represents the height of the hill at point (x, y) (see Figure 2.1.3).

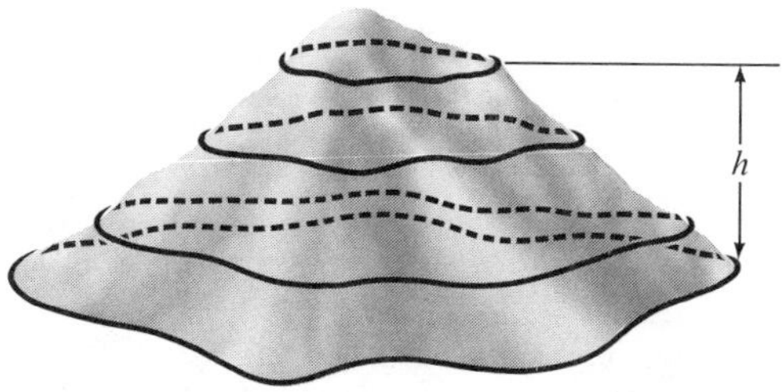

FIGURE 2.1.3
Level contours of a function are defined in the same manner as one obtains contour lines for a topographical map.

DEFINITION. *Let $f: U \subset \mathbb{R}^n \to \mathbb{R}$ and let $c \in \mathbb{R}$. Then the* **level set of value** *c is defined to be those points $\mathbf{x} \in U$ at which $f(\mathbf{x}) = c$. If $n = 2$ we speak of a* **level curve** *(of value c), and if $n = 3$ we speak of a* **level surface**. *In symbols, the level set of value c is written*

$$\{\mathbf{x} \in U \mid f(\mathbf{x}) = c\} \subset \mathbb{R}^n$$

Note that the level set is always in the domain space.

EXAMPLE 1. The constant function $f: \mathbb{R}^2 \to \mathbb{R}$, $(x, y) \mapsto 2$, i.e. $f(x, y) = 2$, has as its graph the horizontal plane $z = 2$ in $\mathbb{R}^3$. The level curve of value c is empty if $c \neq 2$, and is the whole xy-plane if $c = 2$.

EXAMPLE 2. The function $f: \mathbb{R}^2 \to \mathbb{R}$, $(x, y) \mapsto x + y + 2$, has as its graph the inclined plane $z = x + y + 2$. This plane intersects the xy-plane $(z = 0)$ in the line $y = -x - 2$ and the z-axis at the point $(0, 0, 2)$. For any value $c \in \mathbb{R}$, the level curve of value c is the straight line $y = -x + (c - 2)$, or in symbols, the set

$$L_c = \{(x, y) \mid y = -x + (c - 2)\} \subset \mathbb{R}^2.$$

We indicate a few of these level curves in Figure 2.1.4. This is actually a contour map of the function f.

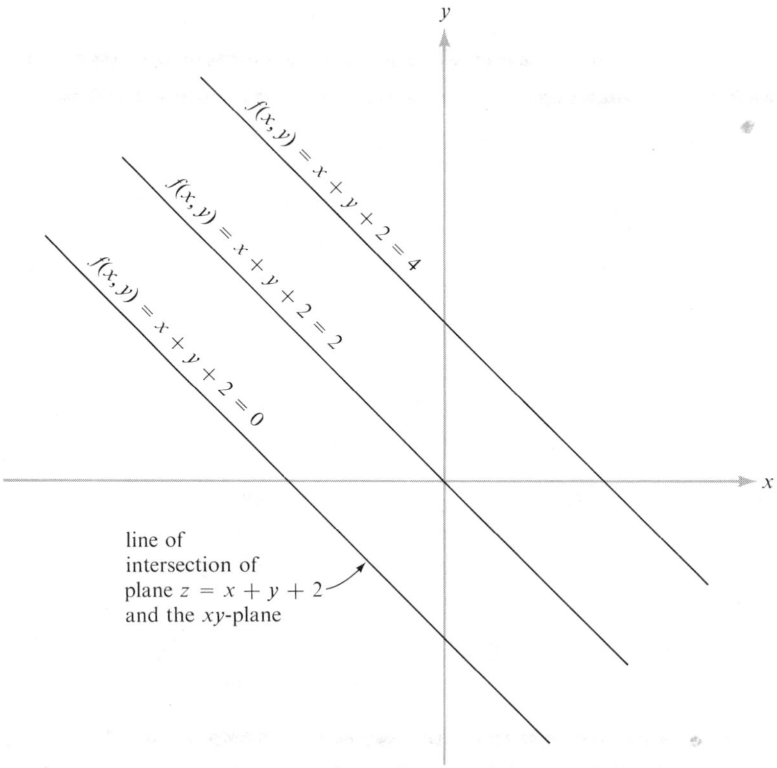

FIGURE 2.1.4
The level curves of $f(x, y) = x + y + 2$ show the behavior of this function.

From the level curves, labelled with the value or "height" of the function, the shape of the graph may be inferred by mentally elevating each level curve to the appropriate height, without stretching, tilting, or sliding it. If this procedure is visualized for all level curves L_c, that is, for all values $c \in \mathbb{R}$, they will assemble to give the entire graph of f, as indicated in Figure 2.1.5 for Example 2. If the graph is visualized for only a finite number of level curves, as is usually the case, a sort of contour model is produced, as in Figure 2.1.4. However, if f is a smooth function its graph will be a smooth surface, so the contour model, mentally smoothed over, gives a good impression of the graph.

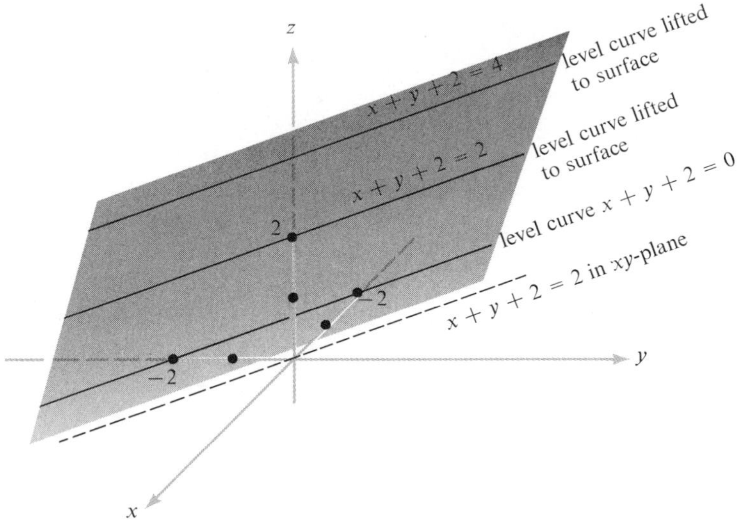

FIGURE 2.1.5

Relationship of level curves of Figure 2.1.4 to the graph of the function $f(x, y) = x + y + 2$, which is the plane $z = x + y + 2$.

EXAMPLE 3. The quadratic function $f\colon \mathbb{R}^2 \to \mathbb{R}, (x, y) \mapsto x^2 + y^2$, has as its graph a paraboloid of revolution, oriented upwards from the origin, around the z-axis. The level curve of value c is empty for $c < 0$; for $c > 0$, if we write $c = a^2$, the level curve of value c is the set $\{(x, y) | x^2 + y^2 = a^2\}$, a circle of radius a centered at the origin. Thus at level c above the xy-plane the level set is a circle of radius $\sqrt{c}$, indicating a parabolic shape (see figures 2.1.6 and 2.1.7).

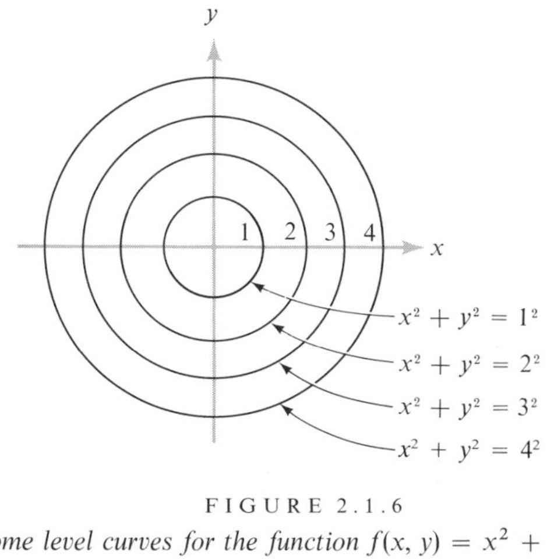

FIGURE 2.1.6

Some level curves for the function $f(x, y) = x^2 + y^2$.

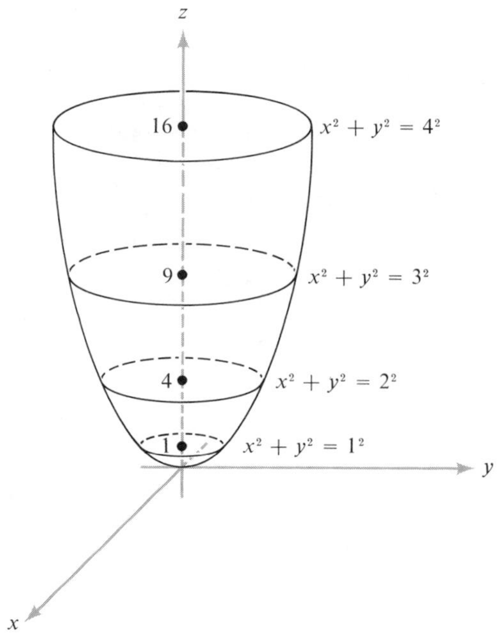

FIGURE 2.1.7
Level curves in Figure 2.1.6 raised to the graph.

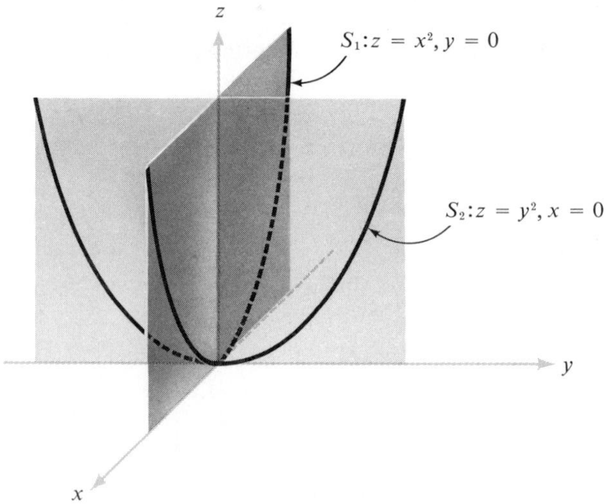

FIGURE 2.1.8
Two sections of the graph of $f(x, y) = x^2 + y^2$.

The shape of a graph may also be determined by the *method of sections*. By a *section* of the graph of f we mean the intersection of the graph and a (vertical) plane. For example, if P_1 is the xz-plane in $\mathbb{R}^3$, defined by $y = 0$, then the section of f in Example 3 is the set

$$P_1 \cap \text{graph } f = \{(x, y, z) | y = 0, z = x^2\}$$

which is a parabola in the xz-plane. Similarly, if P_2 denotes the yz-plane, defined by $x = 0$, then the section

$$P_2 \cap \text{graph } f = \{(x, y, z) | x = 0, z = y^2\}$$

is a parabola in the yz-plane (see Figure 2.1.8). It is usually helpful to compute at least one section to complement the information given by the level sets.

EXAMPLE 4. The quadratic function $f: \mathbb{R}^2 \to \mathbb{R}, (x, y) \mapsto x^2 - y^2$, has as its graph a surface called a *hyperbolic paraboloid*, or *saddle*, centered at the origin. In order to visualize this surface, we first draw the level curves. To determine the level curves, we must solve the equation $x^2 - y^2 = c$. Consider values of c near zero, say $c = 0, \pm 1, \pm 4$. For $c = 0$, we have $y^2 = x^2$, or $y = \pm x$, so this level set consists of two straight lines through the origin. For $c = 1$, the level curve is $x^2 - y^2 = 1$, or $y = \pm\sqrt{x^2 - 1}$, which is a hyperbola that passes vertically through the x-axis at the points $(\pm 1, 0)$ (see Figure 2.1.9). Similarly, for $c = 4$, the level curve is defined by $y = \pm\sqrt{x^2 - 4}$, the hyperbola passing vertically through the x-axis at $(\pm 2, 0)$. For $c = -1$, we obtain the curve $x^2 - y^2 = -1$, that is, $x = \pm\sqrt{y^2 - 1}$, the hyperbola passing horizontally through the y-axis at $(0, \pm 1)$. And for $c = -4$, the hyperbola through $(0, \pm 2)$ is obtained. These level curves are shown in Figure 2.1.9. Since the visualization of the graph of f is not easy from these data alone, we shall compute two sections, as in the previous example. For the section in the xz-plane, we have

$$P_1 \cap \text{graph } f = \{(x, y, z) | y = 0, z = x^2\}$$

which is a parabola opening upward, and for the yz-plane

$$P_2 \cap \text{graph } f = \{(x, y, z) | x = 0, z = -y^2\}$$

which is a parabola opening downward. The graph may now be visualized by lifting the level curves to the appropriate heights, and smoothing out the resulting surface. Their placement is aided by computing the parabolic sections. This procedure generates the hyperbolic saddle indicated in Figure 2.1.10. Compare this with the computer-generated graphs in Figure 2.1.11 (note that the orientation of the axes has been changed).

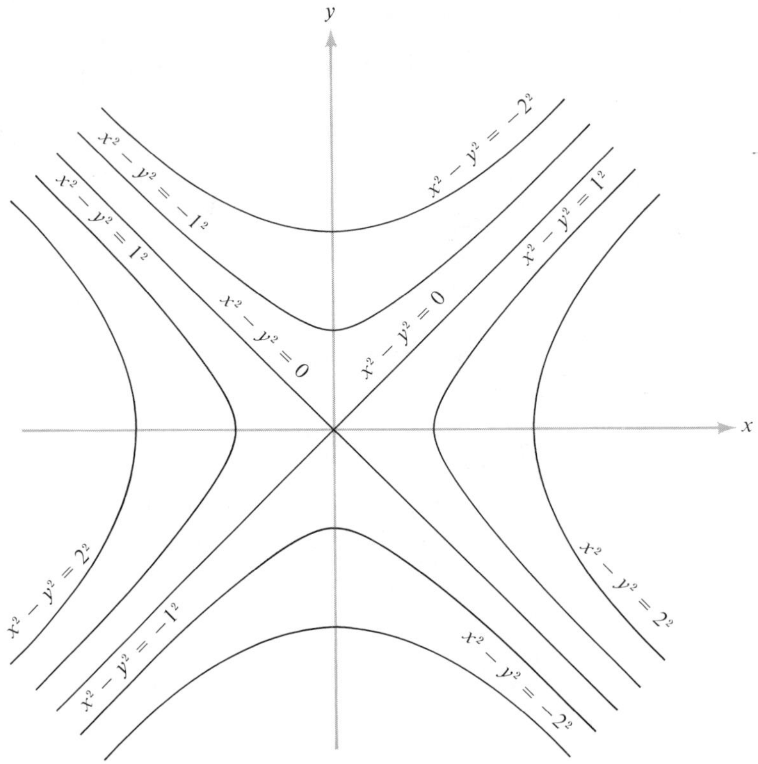

FIGURE 2.1.9
Level curves for the function $f(x, y) = x^2 - y^2$.

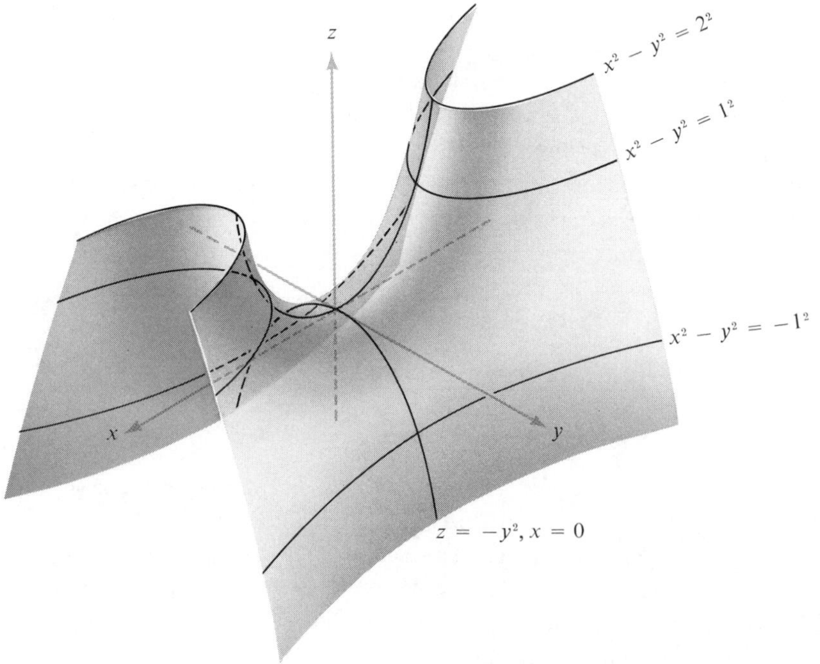

FIGURE 2.1.10
Some level curves on the graph of $f(x, y) = x^2 - y^2$.

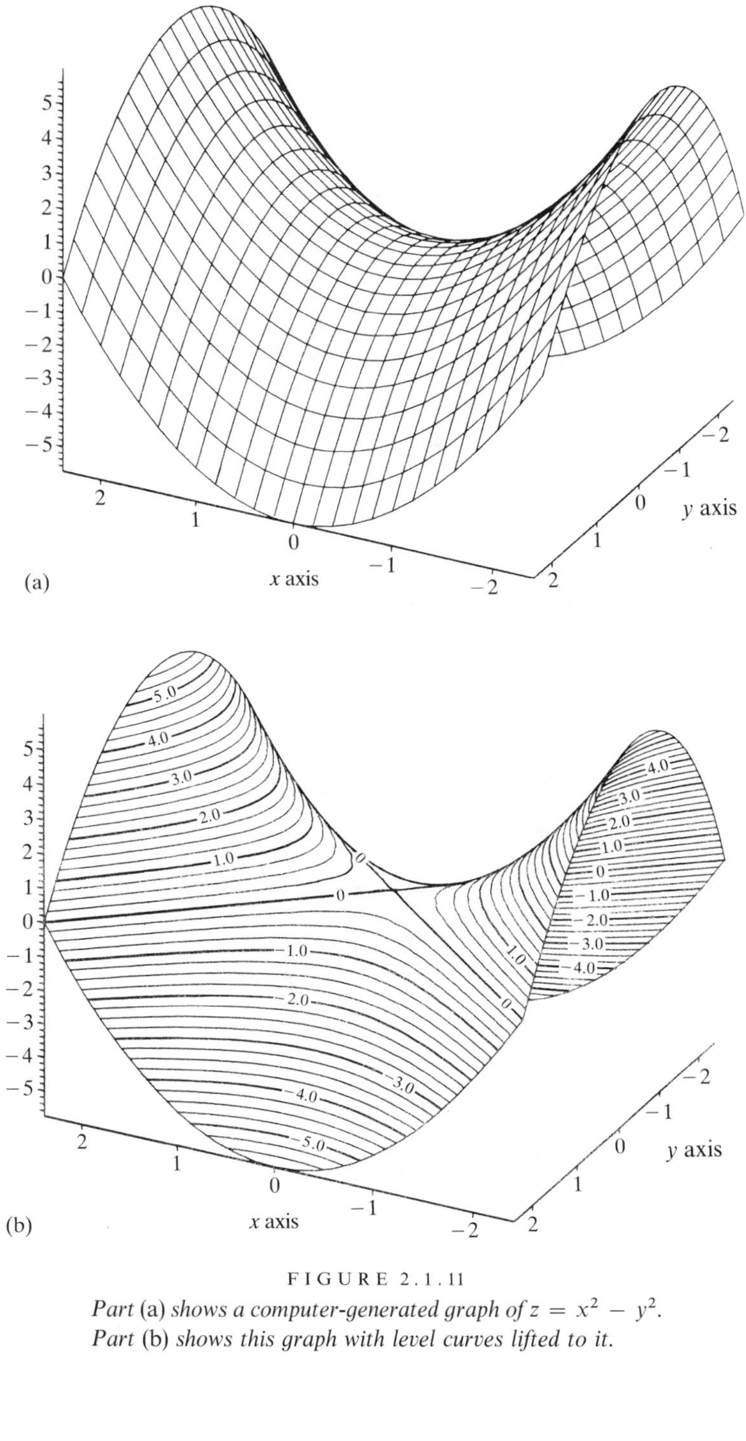

FIGURE 2.1.11

Part (a) *shows a computer-generated graph of* $z = x^2 - y^2$. *Part* (b) *shows this graph with level curves lifted to it.*

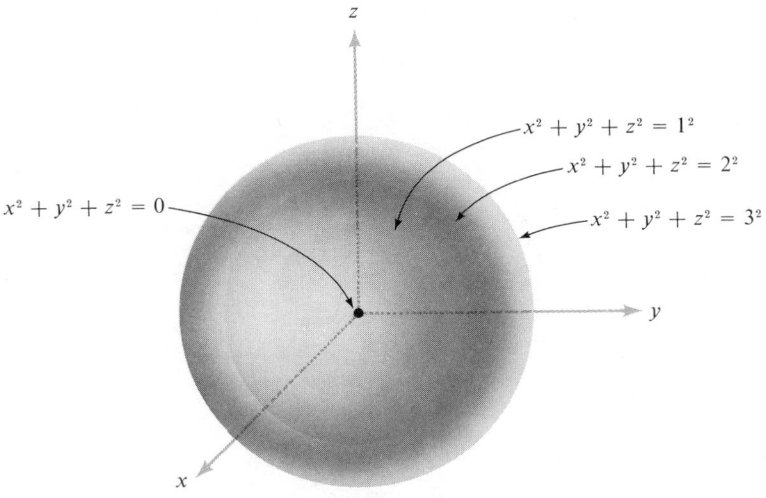

FIGURE 2.1.12
Some level surfaces for $f(x, y, z) = x^2 + y^2 + z^2$.

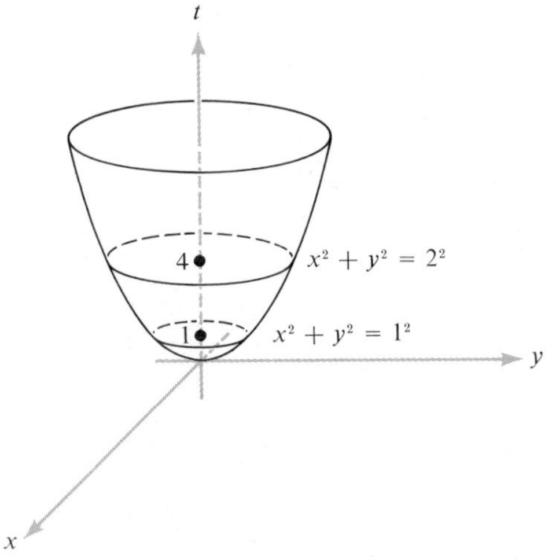

FIGURE 2.1.13
The $z = 0$ section of the graph of $f(x, y, z) = x^2 + y^2 + z^2$.

EXAMPLE 5. Consider the function $f: \mathbb{R}^3 \to \mathbb{R}$, $(x, y, z) \mapsto x^2 + y^2 + z^2$. This is the three-dimensional analogue of Example 3. In this context, level sets are surfaces in the three-dimensional domain $\mathbb{R}^3$. The graph, in $\mathbb{R}^4$, cannot be visualized directly, but sections can nevertheless be computed analytically. The level set with value c is the set

$$L_c = \{(x, y, z) \mid x^2 + y^2 + z^2 = c\}$$

which is clearly the sphere centered at the origin of radius $\sqrt{c}$ for $c > 0$, a single point at the origin for $c = 0$, and is empty for $c < 0$. The level sets for $c = 0, 1, 4$, and 9 are indicated in Figure 2.1.12. Some additional information about the graph is given by computing a section. For example, if we write $S_{z=0} = \{(x, y, z, t) \mid z = 0\}$, then we can look at the section

$$S_{z=0} \cap \text{graph } f = \{(x, y, z, t) \mid t = x^2 + y^2, z = 0\}$$

Since z is held fixed at $z = 0$ here, we can visualize this section of the graph as a surface in $\mathbb{R}^3$ in the variables x, y, t (see Figure 2.1.13). The surface is a *paraboloid of revolution.*

EXAMPLE 6. The function $f: \mathbb{R}^3 \to \mathbb{R}$, $(x, y, z) \mapsto x^2 + y^2 - z^2$, is the three-dimensional analogue of Example 4, and is also called a *saddle*. The level surfaces are defined by

$$L_c = \{(x, y, z) \mid x^2 + y^2 - z^2 = c\}$$

For $c = 0$, this is the cone $z = \pm\sqrt{x^2 + y^2}$ centered around the z-axis. For c negative, say $c = -a^2$, we obtain $z = \pm\sqrt{x^2 + y^2 + a^2}$, which is a hyperboloid of two sheets around the z-axis, passing through the z-axis at the points $(0, 0, \pm a)$. For c positive, say $c = b^2$, the level surface is the *single-sheeted hyperboloid of revolution* around the z-axis defined by $z = \pm\sqrt{x^2 + y^2 - b^2}$, which intersects the xy-plane in the circle of radius $|b|$. These level surfaces are sketched in Figure 2.1.14. Another view of the graph may be obtained from a section. For example, the subspace $S_{y=0} = \{(x, y, z, t) \mid y = 0\}$ intersects the graph in the section

$$S_{y=0} \cap \text{graph } f = \{(x, y, z, t) \mid y = 0, t = x^2 - z^2\}$$

that is, the set of points of the form $(x, 0, z, x^2 - z^2)$, which may be considered, as in the previous example, as a surface in xzt-space (see Figure 2.1.15).

Thus we have seen how methods of sections and level sets can be used to understand the behavior of a function and its graphs; these techniques can be quite useful to people who desire comprehensive visualization of complicated data.

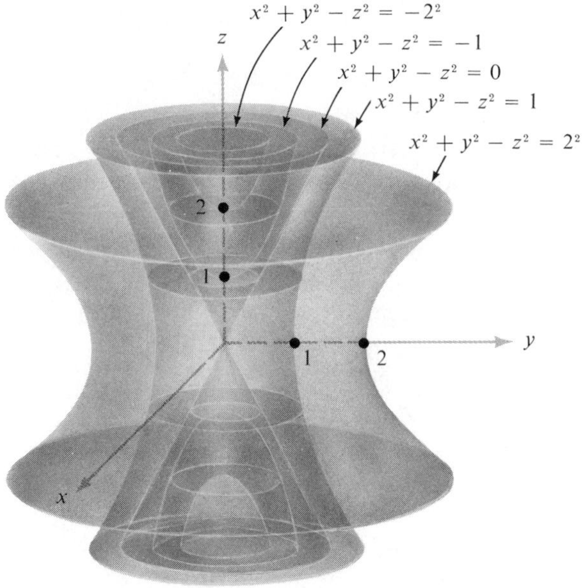

FIGURE 2.1.14
Some level surfaces of the function $f(x, y, z) = x^2 + y^2 - z^2$.

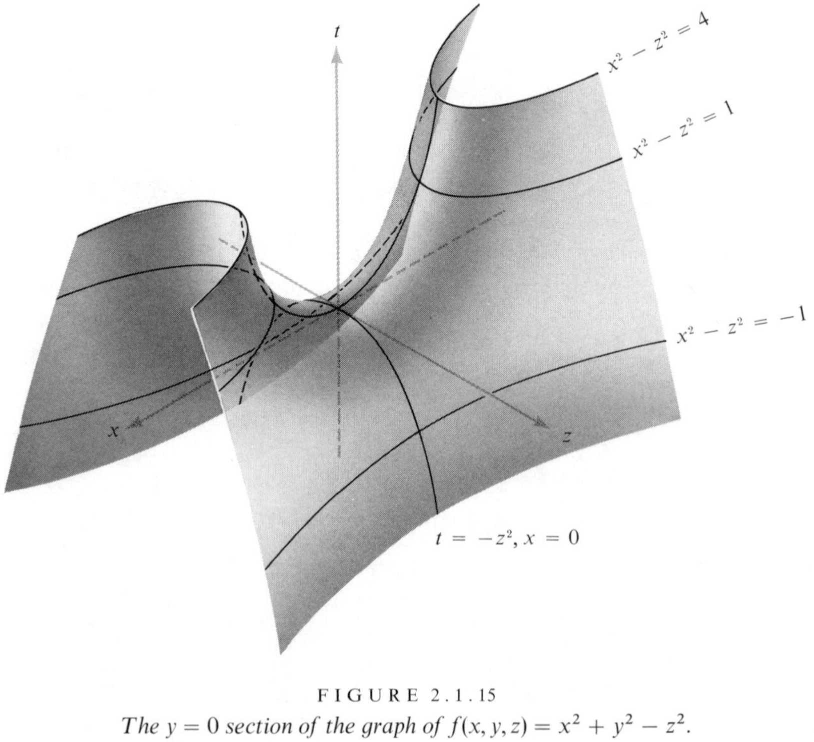

FIGURE 2.1.15
The $y = 0$ section of the graph of $f(x, y, z) = x^2 + y^2 - z^2$.

However, many computer programs are available for plotting a given function. For functions of one variable this is just a matter of calculating selected values of the function and plotting points. For functions of two variables, the method of sections is used. For example, to plot $f(x, y)$, the computer selects sections parallel to the axes by assigning values to, say, y and plotting the corresponding graph, then changing y and repeating. This way a whole piece of the graph can be swept out. Some examples are given in Figures 2.1.16 and 2.1.17. This particular program is also capable of giving the plot perspective so that we can see it from any angle or any distance.

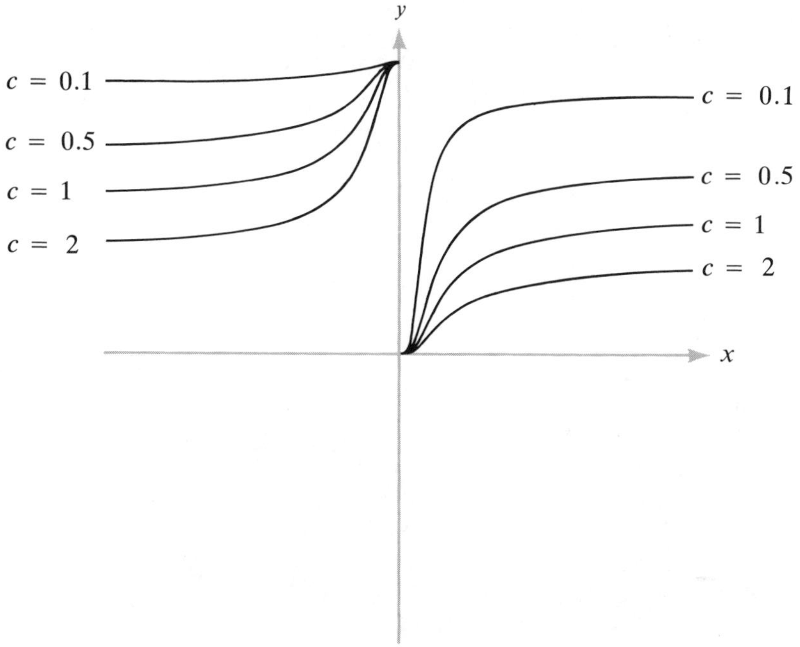

(a) $y = 1/(1 + ce^{1/x})$ *for* $c = 0.1, 0.5, 1,$ *and* 2.

FIGURE 2.1.16
Some computer-generated graphs.

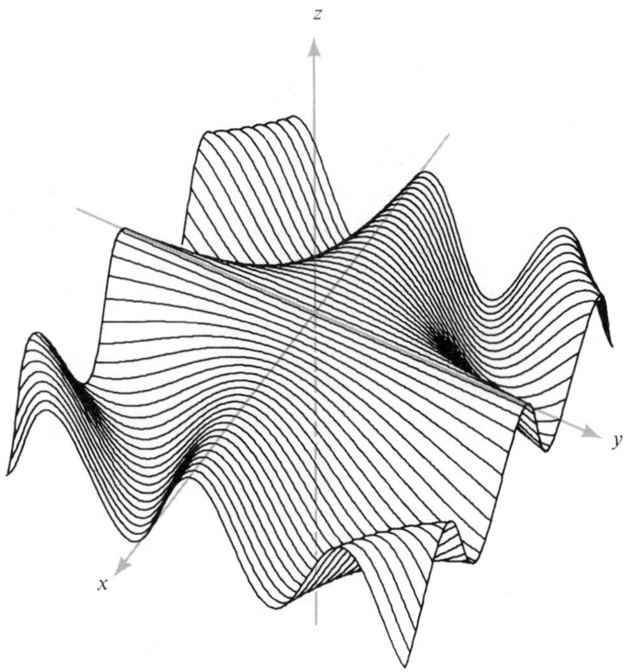

(b) $z = \cos(xy), -3 \leq x \leq 3, -3 \leq y \leq 3.$

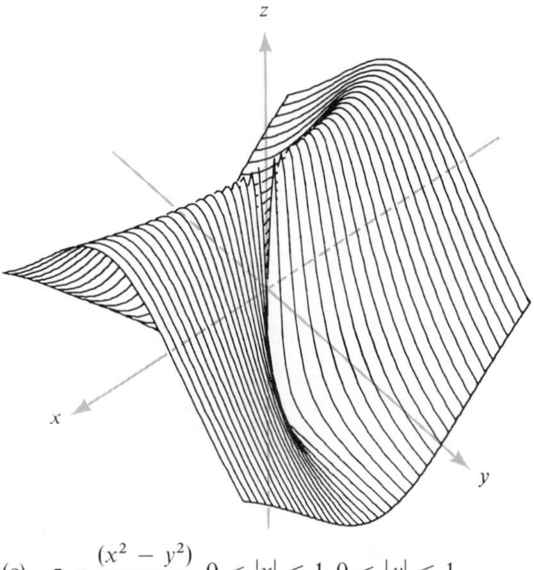

(c) $z = \dfrac{(x^2 - y^2)}{(x^2 + y^2)}, 0 < |x| \leq 1, 0 < |y| \leq 1.$

FIGURE 2.1.16 (*continued*)

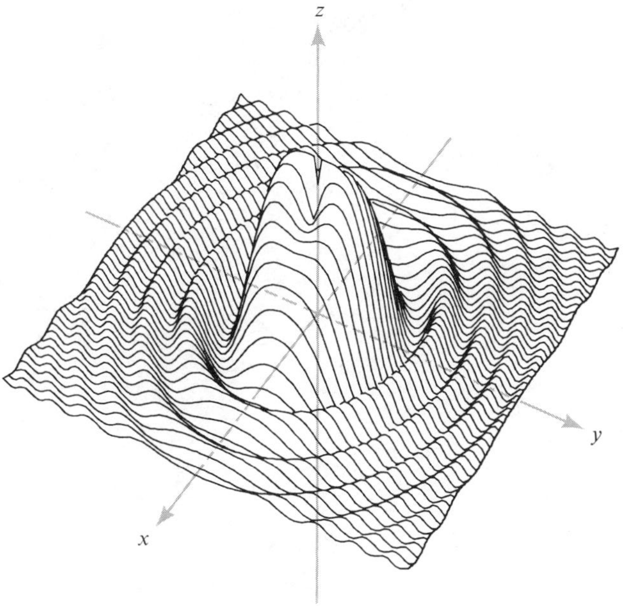

(d) $z = \dfrac{[\sin(2x^2 + 3y^2)]}{[x^2 + y^2]}, 0 < |x| \leq 3, 0 < |y| \leq 3.$

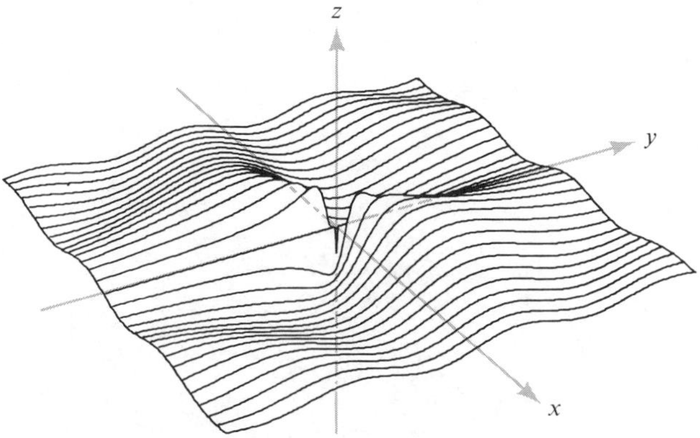

(e) $z = \dfrac{(\sin xy)}{(x^2 + y^2)}, 0 < |x| \leq 3, 0 < |y| \leq 3.$

FIGURE 2.1.16 (*continued*)

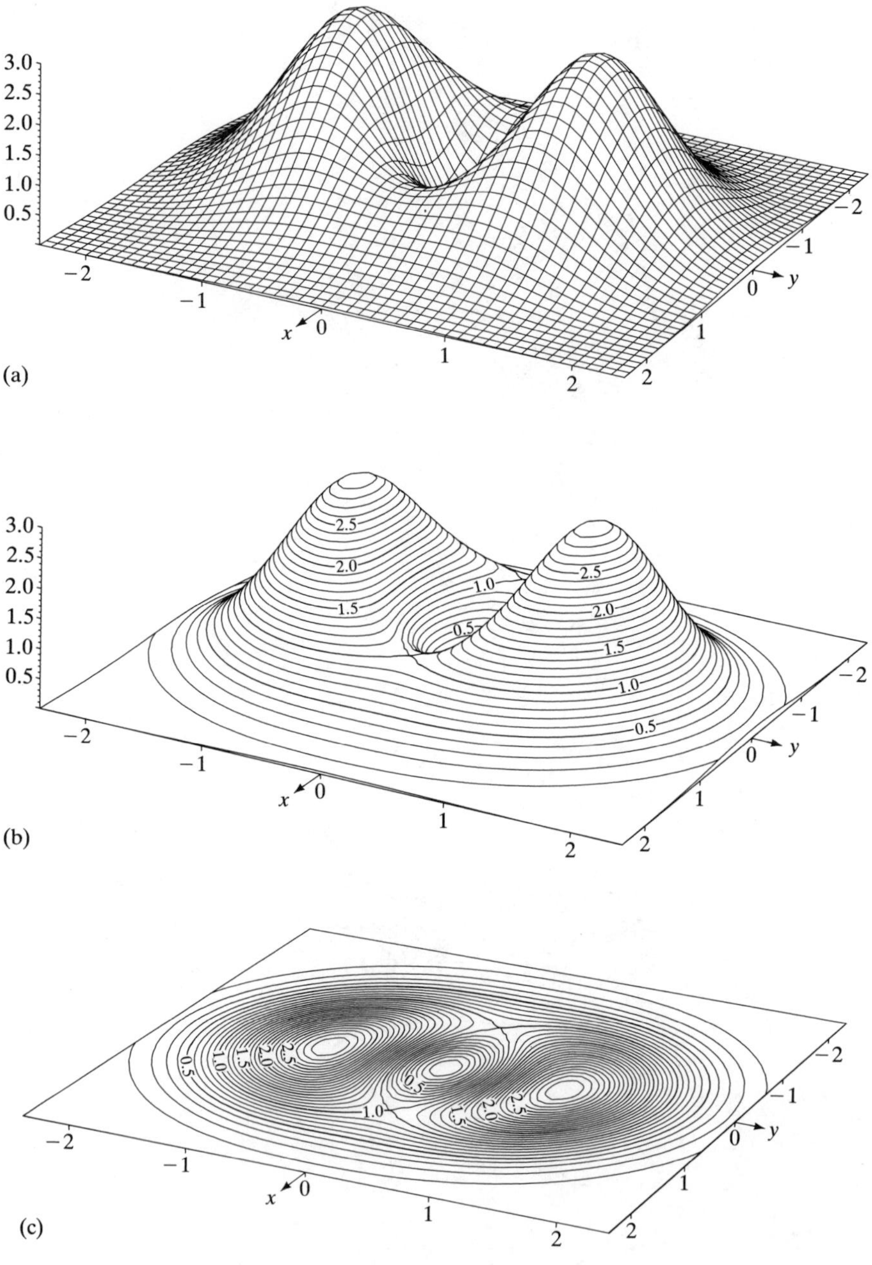

FIGURE 2.1.17

Computer-generated graph of $z = (x^2 + 3y^2)\exp(1 - x^2 - y^2)$ *represented in four ways:* (a) *by sections,* (b) *by level curves on graph,* (c) *by level curves in the xy-plane seen in perspective, and* (d) *by level curves in the xy-plane seen from above.*

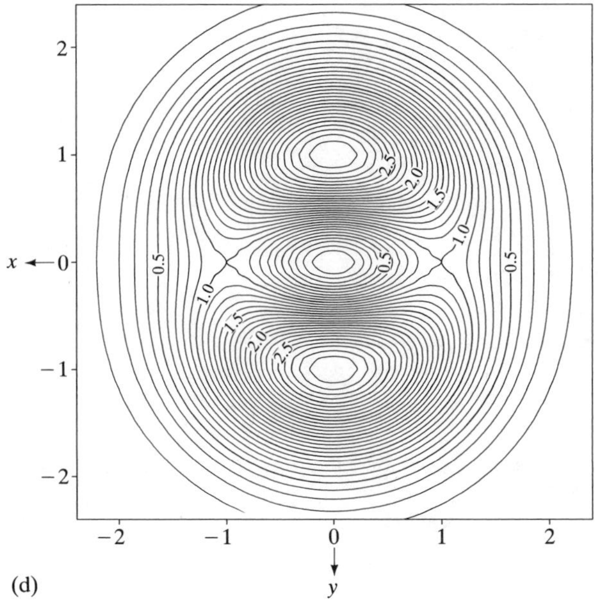

(d)

FIGURE 2.1.17 *(continued)*

EXERCISES

1. Sketch the level curves and graphs of the following functions.
   (a) $f\colon \mathbb{R}^2 \to \mathbb{R}, (x, y) \mapsto x - y + 2$
   (b) $f\colon \mathbb{R}^2 \to \mathbb{R}, (x, y) \mapsto x^2 + 4y^2$
   (c) $f\colon \mathbb{R}^2 \to \mathbb{R}, (x, y) \mapsto xy$

2. For Examples 2, 3, and 4, compute the section of the graph defined by the plane

$$S_\theta = \{(x, y, z) \mid y = x \tan \theta\}$$

for a given constant θ. Determine which of these functions f have the property that the shape of $S_\theta \cap \text{graph } f$ is independent of θ.

In Exercises 3 to 8, draw the level curves (in the xy-plane) for the given function f and specified values of c. Sketch the graphs of $z = f(x, y)$.

3. $f(x, y) = (100 - x^2 - y^2)^{1/2}$, $c = 0, 2, 4, 6, 8, 10$
4. $f(x, y) = (x^2 + y^2)^{1/2}$, $c = 0, 1, 2, 3, 4, 5$
5. $f(x, y) = x^2 + y^2$, $c = 0, 1, 2, 3, 4, 5$
6. $f(x, y) = 3x - 7y$, $c = 0, 1, 2, 3, -1, -2, -3$
7. $f(x, y) = x^2 + xy$, $c = 0, 1, 2, 3, -1, -2, -3$
8. $f(x, y) = x/y$, $c = 0, 1, 2, 3, -1, -2, -3$

In Exercises 9 to 11, determine the level surfaces and a section of the graph of each function.

9. $f: \mathbb{R}^3 \to \mathbb{R}, (x, y, z) \mapsto -x^2 - y^2 - z^2$
10. $f: \mathbb{R}^3 \to \mathbb{R}, (x, y, z) \mapsto 4x^2 + y^2 + 9z^2$
11. $f: \mathbb{R}^3 \to \mathbb{R}, (x, y, z) \mapsto x^2 + y^2$

In Exercises 12 to 16, describe the graph of each function by computing some level sets and sections.

12. $f: \mathbb{R}^3 \to \mathbb{R}, (x, y, z) \mapsto xy$
13. $f: \mathbb{R}^3 \to \mathbb{R}, (x, y, z) \mapsto xy + yz$
14. $f: \mathbb{R}^3 \to \mathbb{R}, (x, y, z) \mapsto xy + z^2$
15. $f: \mathbb{R}^2 \to \mathbb{R}, (x, y) \mapsto |y|$
16. $f: \mathbb{R}^2 \to \mathbb{R}, (x, y) \mapsto \max(|x|, |y|)$

Sketch or describe the surfaces of the equations presented in Exercises 17 to 27.

17. $x^2 + y^2 + z^2 + 4x - by + 9z - b = 0$, where b is a constant
18. $x + 2z = 4$
19. $4x^2 + y^2 = 16$
20. $x^2 + y^2 - 2x = 0$
21. $z^2 = y^2 + 4$
22. $\dfrac{y^2}{9} + \dfrac{z^2}{4} = 1 + \dfrac{x^2}{16}$
23. $z = \dfrac{y^2}{4} - \dfrac{x^2}{9}$
24. $y^2 = x^2 + z^2$
25. $4x^2 - 3y^2 + 2z^2 = 0$
26. $\dfrac{x^2}{9} + \dfrac{y^2}{12} + \dfrac{z^2}{9} = 1$
27. $\dfrac{x}{4} = \dfrac{y^2}{4} + \dfrac{z^2}{9}$
28. Describe the level curves of the function

$$f: \mathbb{R}^2 \to \mathbb{R}, (x, y) \mapsto \begin{cases} \dfrac{2xy}{x^2 + y^2} & \text{if } (x, y) \neq (0, 0) \\ 0 & \text{if } (x, y) = (0, 0) \end{cases}$$

(HINT: Use polar coordinates.)

29. Let $f: \mathbb{R}^2\backslash\{\mathbf{0}\} \to \mathbb{R}$ be given in polar coordinates by $f(r, \theta) = (\cos 2\theta)/r^2$. Sketch a few level curves relative to the xy axes. ($\mathbb{R}^2\backslash\{\mathbf{0}\} = \{\mathbf{x} \in \mathbb{R}^2 | \mathbf{x} \neq \mathbf{0}\}$)
30. In Figure 2.1.17(d) the level "curve" $z = 3$ appears to consist of two points. Prove this algebraically.

2.2 LIMITS AND CONTINUITY

This section develops some necessary terminology that will enable us to study differentiation of functions of several variables in Section 2.3. This material centers around the concepts of open sets, limits, and continuity; open sets are needed to understand limits and limits are in turn needed to understand continuity and differentiability.

As in elementary calculus it is not necessary to completely master the limit concept in order to actually work problems in differential calculus. For this reason, teachers will probably wish to treat the following material with varying degrees of rigor. It is therefore important that the student consults with the instructor about the depth of understanding required.

We begin formulating the concept of open set by defining an open disc. Let $\mathbf{x}_0 \in \mathbb{R}^n$ and let r be a positive real number. The *open disc* of radius r and center $\mathbf{x}_0$ is defined to be the set of all points $\mathbf{x}$ such that $\|\mathbf{x} - \mathbf{x}_0\| < r$. This set is denoted $D_r(\mathbf{x}_0)$, and is the set of points $\mathbf{x}$ in $\mathbb{R}^n$ whose distance (see Section 1.5) from $\mathbf{x}_0$ is less than r. Notice that we include only those $\mathbf{x}$ for which *strict* inequality holds. The disc $D_r(\mathbf{x}_0)$ is illustrated in Figure 2.2.1 for

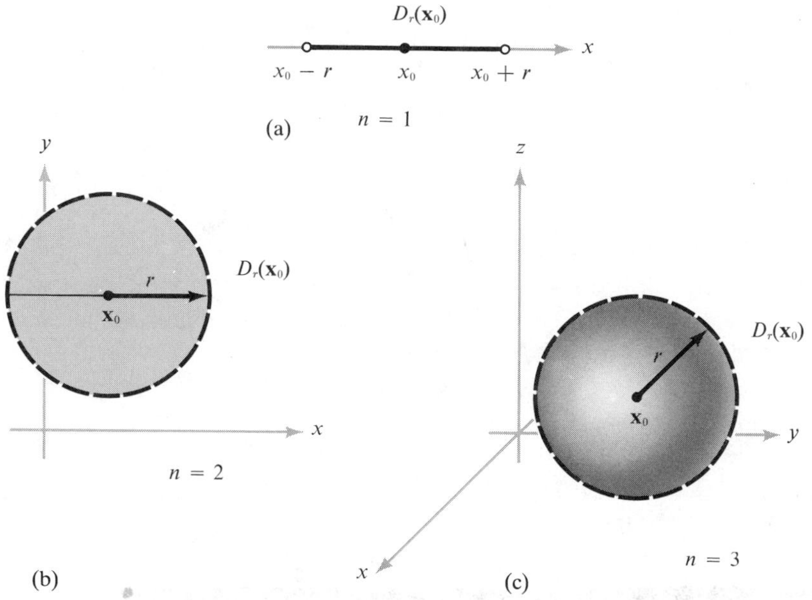

FIGURE 2.2.1
What discs $D_r(\mathbf{x}_0)$ look like in (a) 1, (b) 2, *and* (c) 3 *dimensions.*

$n = 1, 2, 3$. For the case $n = 1$, $x_0 \in \mathbb{R}$, the open disc $D_r(x_0)$ is the open interval $]x_0 - r, x_0 + r[$, which consists of all numbers $x \in \mathbb{R}$ *strictly* between $x_0 - r$ and $x_0 + r$. For the case $n = 2$, $\mathbf{x}_0 \in \mathbb{R}^2$, $D_r(\mathbf{x}_0)$ is the "inside" of the disc of radius r centered at $\mathbf{x}_0$. For the case $n = 3$, $\mathbf{x}_0 \in \mathbb{R}^3$, $D_r(\mathbf{x}_0)$ is the "inside" of the ball of radius r centered at $\mathbf{x}_0$.

We are now ready to define the concept of an open set.

DEFINITION. *Let $U \subset \mathbb{R}^n$ (i.e., U is a subset of $\mathbb{R}^n$). We call U an* **open set** *when for every point $\mathbf{x}_0$ in U there exists some $r > 0$ such that $D_r(\mathbf{x}_0)$ is contained within U; in symbols, $D_r(\mathbf{x}_0) \subset U$ (see Figure 2.2.2).*

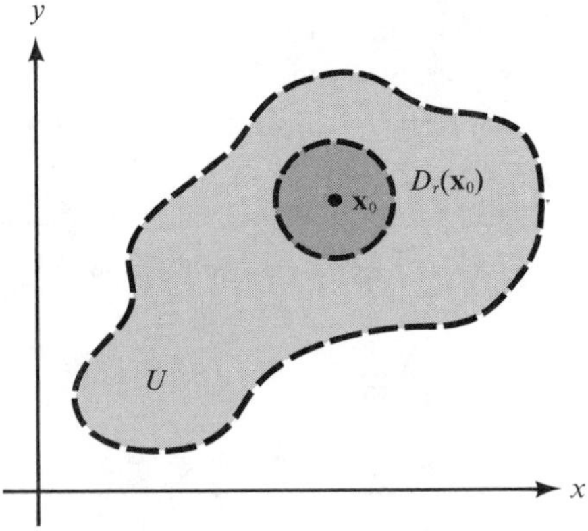

FIGURE 2.2.2

An open set U is one that completely encloses some disc $D_r(\mathbf{x}_0)$ about each of its points $\mathbf{x}_0$.

It is important to realize that the number $r > 0$ depends on the point $\mathbf{x}_0$, and generally r will shrink down as $\mathbf{x}_0$ gets closer to the "edge" of U. Intuitively speaking, a set U is open when the "boundary" points of U do not lie in U. In Figure 2.2.2, the dashed line is *not* included in U.

We shall also establish the convention that the empty set $\varnothing$ (the set consisting of no elements) is open.

We have defined open disc and open set. From our choice of terms it would seem that an open disc should also be an open set. A little thought shows that this fact requires some proof.

THEOREM 1. *For each $\mathbf{x}_0 \in \mathbb{R}^n$ and $r > 0$, $D_r(\mathbf{x}_0)$ is an open set.*

Proof. Let $\mathbf{x} \in D_r(\mathbf{x}_0)$, that is, $\|\mathbf{x} - \mathbf{x}_0\| < r$. According to the definition of open set we must find an $s > 0$ such that $D_s(\mathbf{x}) \subset D_r(\mathbf{x}_0)$. Referring to

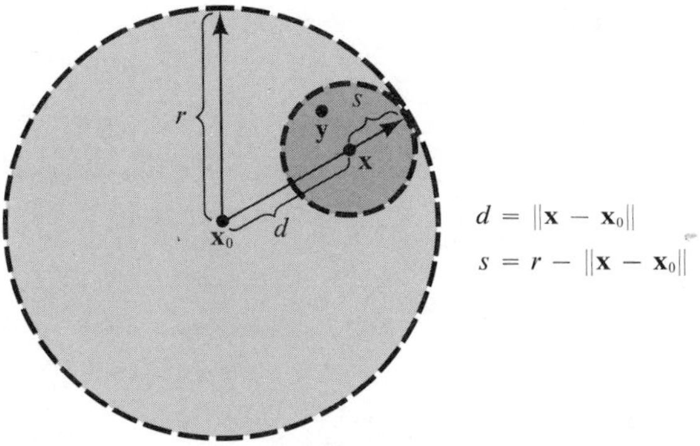

FIGURE 2.2.3
The geometry of the proof that an open disc is an open set.

Figure 2.2.3 we see that $s = r - \|\mathbf{x} - \mathbf{x}_0\|$ is a reasonable choice; note that $s > 0$ but that s becomes smaller if $\mathbf{x}$ is nearer the edge of $D_r(\mathbf{x}_0)$.

To prove $D_s(\mathbf{x}) \subset D_r(\mathbf{x}_0)$, let $\mathbf{y} \in D_s(\mathbf{x})$; that is, $\|\mathbf{y} - \mathbf{x}\| < s$. We want to prove that $\mathbf{y} \in D_r(\mathbf{x}_0)$ as well. Proving this by the definition of an r-disc entails showing that $\|\mathbf{y} - \mathbf{x}_0\| < r$. This is done by using the triangle inequality for vectors in $\mathbb{R}^n$ (see the corollary to Theorem 3, Section 1.5):

$$\begin{aligned}\|\mathbf{y} - \mathbf{x}_0\| &= \|(\mathbf{y} - \mathbf{x}) + (\mathbf{x} - \mathbf{x}_0)\| \\ &\leq \|\mathbf{y} - \mathbf{x}\| + \|\mathbf{x} - \mathbf{x}_0\| < s + \|\mathbf{x} - \mathbf{x}_0\| = r\end{aligned}$$

Hence $\|\mathbf{y} - \mathbf{x}_0\| < r$. ■

The following example illustrates some techniques that are useful in establishing the openness of sets.

EXAMPLE 1. Prove that $A = \{(x, y) \in \mathbb{R}^2 \,|\, x > 0\}$ is an open set.

The set is pictured in Figure 2.2.4. Intuitively, this set is open since no points on the "boundary," $x = 0$, are contained in the set. Such an argument will often suffice after one has gotten used to the ideas. At first, however, we should give all the details. In order to prove A is open, we must show that for every point $(x, y) \in A$ there exists an $r > 0$ such that $D_r(x, y) \subset A$. If $(x, y) \in A$, then $x > 0$. Choose $r = x$. If $(x_1, y_1) \in D_r(x, y)$ we have

$$|x_1 - x| = \sqrt{(x_1 - x)^2} \leq \sqrt{(x_1 - x)^2 + (y_1 - y)^2} < r = x$$

and so $x_1 - x < x$ and $x - x_1 < x$. The latter inequality implies $x_1 > 0$, that is, $(x_1, y_1) \in A$. Hence $D_r(x, y) \subset A$, and therefore A is open (see Figure 2.2.5).

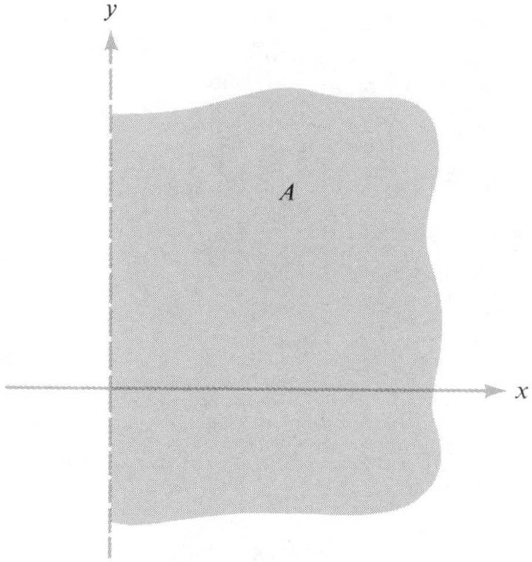

FIGURE 2.2.4
Problem: show that A is an open set.

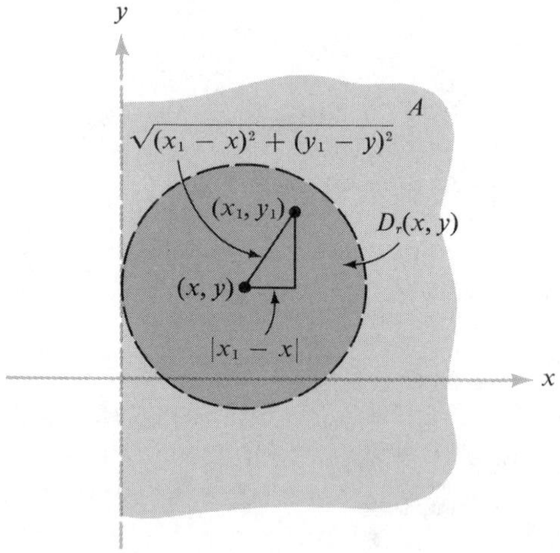

FIGURE 2.2.5
The construction of a disc about a point in A that is completely enclosed in A.

It is useful to have a special name for an open set containing a given point $\mathbf{x}$, since this idea arises very often in the study of limits and continuity. Thus, by a *neighborhood* of $\mathbf{x} \in \mathbb{R}^n$ we merely mean an open set U containing the point $\mathbf{x}$. For example $D_r(\mathbf{x}_0)$ is a neighborhood of $\mathbf{x}_0$ for any $r > 0$. The set A in Example 1 is a neighborhood of $(3, -10)$.

Now let us formally introduce the concept of a boundary point, which we have alluded to in Example 1.

DEFINITION. *Let $A \subset \mathbb{R}^n$. A point $\mathbf{x} \in \mathbb{R}^n$ is called a* ***boundary point*** *of A if every neighborhood of $\mathbf{x}$ contains at least one point in A and at least one point not in A.*

In this definition $\mathbf{x}$ may or may not be in A; if $\mathbf{x} \in A$ then $\mathbf{x}$ is a boundary point if every neighborhood of $\mathbf{x}$ contains at least one point *not* in A (it already contains a point of A, namely $\mathbf{x}$). Similarly, if $\mathbf{x}$ is not in A it is a boundary point if every neighborhood of $\mathbf{x}$ contains at least one point of A.

We shall be particularly interested in boundary points of open sets. By the definition of open set, no point of an open set A can be a boundary point of A. Thus *a point $\mathbf{x}$ is a boundary point of an open set A if and only if $\mathbf{x}$ is not in A and every neighborhood of $\mathbf{x}$ has non-empty intersection with A.*

This expresses in precise terms the intuitive idea that a boundary point of A is a point just on the "edge" of A. In most examples it is perfectly clear what the boundary points are.

EXAMPLE 2. (a) Let $A =]a, b[$ in $\mathbb{R}$. Then the boundary points of A consist of the points a and b. A consideration of Figure 2.2.6 and the definition will make this clear. (The reader will be asked to prove this as Exercise 4.)

(b) Let $A = D_r(x_0, y_0)$ be an r-disc about (x_0, y_0) in the plane. The boundary consists of points (x, y) with $(x - x_0)^2 + (y - y_0)^2 = r^2$ (Figure 2.2.7).

(c) Let $A = \{(x, y) \in \mathbb{R}^2 | x > 0\}$. Then the boundary of A consists of all points on the y-axis (see Figure 2.2.8).

(d) Let A be $D_r(\mathbf{x}_0)$ minus the point $\mathbf{x}_0$ (a "punctured" disc about $\mathbf{x}_0$). Then $\mathbf{x}_0$ is a boundary point of A.

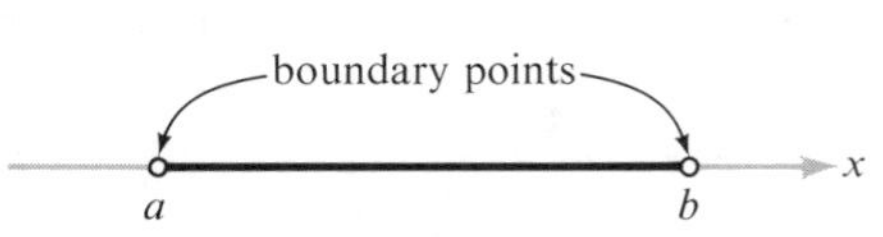

FIGURE 2.2.6
The boundary points of the interval $]a, b[$.

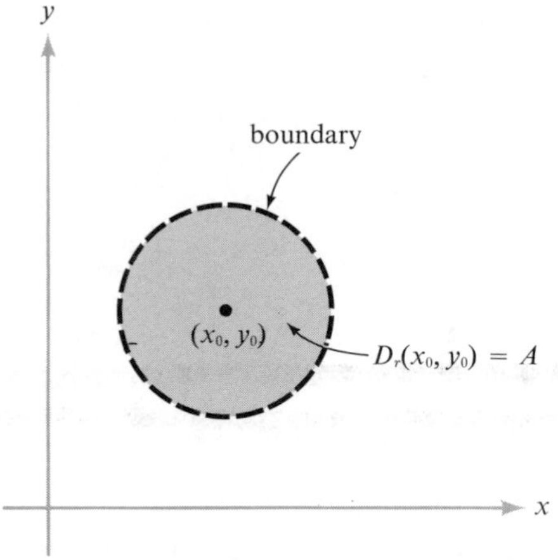

FIGURE 2.2.7
The boundary of A consists of points on the edge of A.

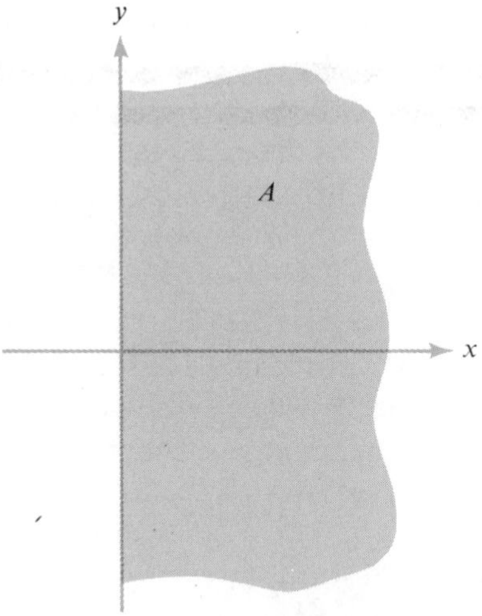

FIGURE 2.2.8
The boundary of A consists of all points on the y-axis.

We shall now turn our attention to the concept of limit. Throughout the following discussions the domain of definition of the function f will be an open set A. We shall be interested in finding the limit of f as $\mathbf{x} \in A$ approaches either a point of A or a boundary point of A.

The reader should appreciate the fact that the limit concept is a basic and useful tool for the analysis of functions; it enables us to study derivatives, and hence maxima and minima, asymptotes, improper integrals, and other important features of functions, as well as being useful for infinite series and sequences. We wish to present, briefly, a theory of limits for functions of several variables that includes the theory for functions of one variable as a special case.

In one-variable calculus the student has probably encountered a definition of $\lim_{x \to x_0} f(x) = l$, for $f\colon A \subset \mathbb{R} \to \mathbb{R}$ a function from a subset A of the real numbers to the real numbers. Intuitively, this means that as x gets closer and closer to x_0, the values $f(x)$ get closer and closer to l. To put this intuitive idea on a firm mathematical foundation, the notions of epsilon (ε) and delta (δ) or the concept of neighborhood are usually introduced—notions that are essential to a precise and workable definition of $\lim_{x \to x_0} f(x)$. The same is true for functions of several variables. In what follows we develop the neighborhood approach to limits. The epsilon-delta approach is left for optional study at the end of this section.

Definition of Limit. *Let $f\colon A \subset \mathbb{R}^n \to \mathbb{R}^m$, where A is an open set. Let $\mathbf{x}_0$ be in A or be a boundary point of A, and let N be a neighborhood of $\mathbf{b} \in \mathbb{R}^m$. We say f is* ***eventually in N as $\mathbf{x}$ approaches $\mathbf{x}_0$*** *if there exists a neighborhood U of $\mathbf{x}_0$ such that $\mathbf{x} \neq \mathbf{x}_0$, $\mathbf{x} \in U$ and $\mathbf{x} \in A$ imply $f(\mathbf{x}) \in N$. (The geometric meaning of this assertion is illustrated in Figure 2.2.9; note that $\mathbf{x}_0$ need not be in the set A, so that $f(\mathbf{x}_0)$ is not necessarily defined.) We say $f(\mathbf{x})$* ***approaches $\mathbf{b}$*** *as $\mathbf{x}$ approaches $\mathbf{x}_0$, or in symbols*

$$\lim_{\mathbf{x} \to \mathbf{x}_0} f(\mathbf{x}) = \mathbf{b} \quad \textit{or} \quad f(\mathbf{x}) \to \mathbf{b} \quad \textit{as} \quad \mathbf{x} \to \mathbf{x}_0$$

when, given ***any*** *neighborhood N of $\mathbf{b}$, f is eventually in N as $\mathbf{x}$ approaches $\mathbf{x}_0$ (that is, "$f(\mathbf{x})$ is close to $\mathbf{b}$ if $\mathbf{x}$ is close to $\mathbf{x}_0$"). It may be that as $\mathbf{x}$ approaches $\mathbf{x}_0$ the values $f(\mathbf{x})$ do not get close to any particular number. In this case we say that* $\lim_{\mathbf{x} \to \mathbf{x}_0} f(\mathbf{x})$ **does not exist**.

Henceforth, whenever we consider the notion $\lim_{\mathbf{x} \to \mathbf{x}_0} f(\mathbf{x})$ we shall always assume that $\mathbf{x}_0$ either belongs to some open set on which f is defined or is on the boundary of such a set.

One reason we insist on $\mathbf{x} \neq \mathbf{x}_0$ in the definition of limit will become clear if we remember from one-variable calculus that we want to be able to define

$$f'(x_0) = \lim_{x \to x_0} \frac{f(x) - f(x_0)}{x - x_0}$$

and this expression is not defined at $x = x_0$.

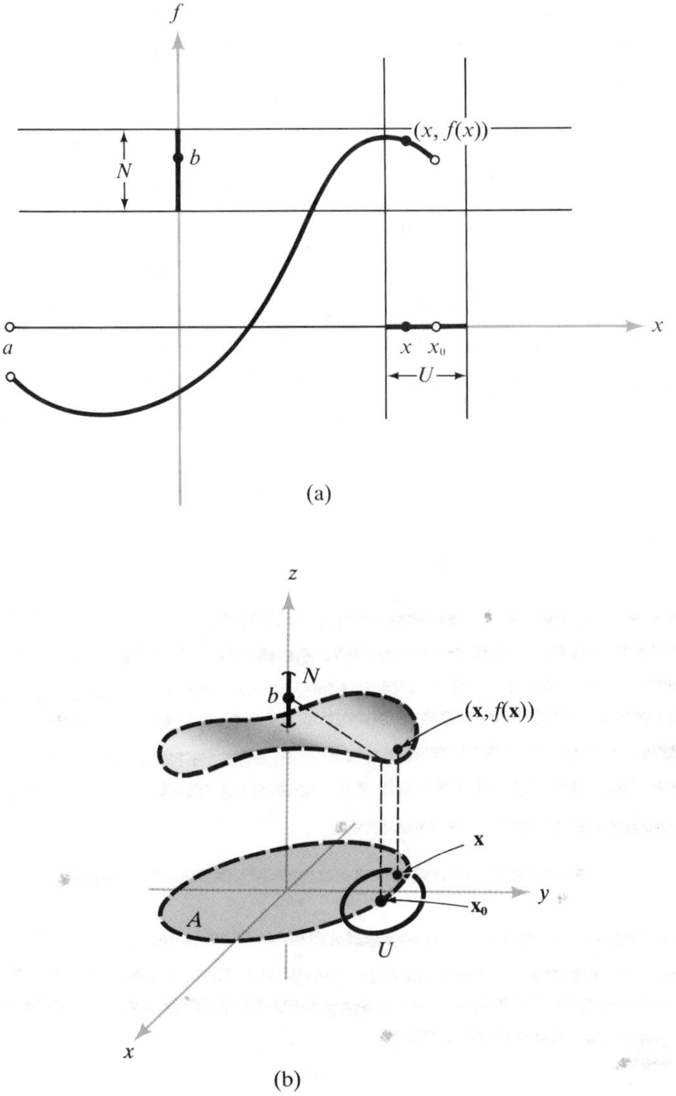

FIGURE 2.2.9

Limits in terms of neighborhoods; if $\mathbf{x}$ *is in* U, *then* $f(\mathbf{x})$ *will be in* N. *(The little open circles denote the fact that* $(a, f(a))$ *and* $(x_0, f(x_0))$ *do not lie on graph* (a).) (a) $f: A =]a, x_0[\to \mathbf{R}$ (b) $f: A = \{(x, y) | x^2 + y^2 < 1\} \to \mathbf{R}$ *(The dashed line is not in the graph of* f.)

EXAMPLE 3. (a) This example illustrates a limit that does not exist. Consider the function $f: \mathbb{R} \to \mathbb{R}$ defined by

$$f(x) = \begin{cases} 1 & \text{if } x > 0 \\ -1 & \text{if } x \leq 0 \end{cases}$$

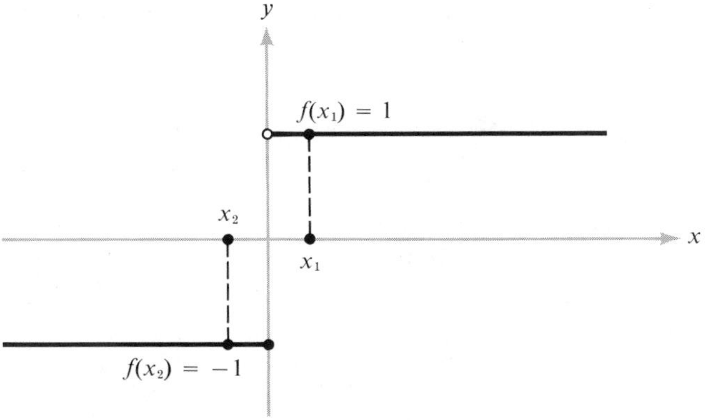

FIGURE 2.2.10
The limit of this function as $x \to 0$ does not exist.

Then $\lim_{x \to 0} f(x)$ does not exist since there are points x_1 arbitrarily close to 0 with $f(x_1) = 1$ and also points x_2 arbitrarily close to 0 with $f(x_2) = -1$; that is, there is no single number that f is close to when x is close to 0 (see Figure 2.2.10). If f is restricted to the domain $]0, 1[$ or $]-1, 0[$ then the limit does exist. Can you see why?

(b) This example illustrates a limit that does exist but whose limiting value does not equal the value of the function at the limiting point. Define $f: \mathbb{R} \to \mathbb{R}$ by

$$f(x) = \begin{cases} 0 & \text{if} \quad x \neq 0 \\ 1 & \text{if} \quad x = 0 \end{cases}$$

It is true that $\lim_{x \to 0} f(x) = 0$, since for any neighborhood U of 0, $x \in U$ and $x \neq 0$ implies that $f(x) = 0$. If we did not insist on $x \neq x_0$, then the limit (assuming we use the above definition of limit without the condition $\mathbf{x} \neq \mathbf{x}_0$) would not exist. Thus we are really interested in what value f approaches as $x \to 0$; as one sees from the graph in Figure 2.2.11, f approaches 0 as

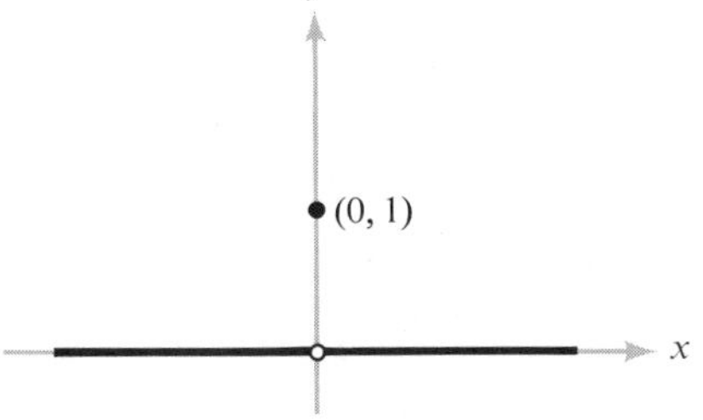

FIGURE 2.2.11
The limit of this function as $x \to 0$ is zero.

$x \to 0$, and we do not care that f happens to take on some other value at 0, or is not defined there.

EXAMPLE 4. Let us use the definition to verify that the "obvious limit" $\lim_{\mathbf{x}\to\mathbf{x}_0} \mathbf{x} = \mathbf{x}_0$ holds, where $\mathbf{x}$ and $\mathbf{x}_0 \in \mathbb{R}^n$. Let f be the function $f: \mathbf{x} \mapsto \mathbf{x}$ and let N be any neighborhood of $\mathbf{x}_0$. We must show that $f(\mathbf{x})$ is eventually in N as $\mathbf{x} \to \mathbf{x}_0$. Thus, by the definition we must find a neighborhood U of $\mathbf{x}_0$ with the property that if $\mathbf{x} \neq \mathbf{x}_0$ and $\mathbf{x} \in U$ then $f(\mathbf{x}) \in N$. Pick $U = N$. If $\mathbf{x} \in U$, then $\mathbf{x} \in N$; and since $\mathbf{x} = f(\mathbf{x})$ it follows immediately that $f(\mathbf{x}) \in N$. Thus we have shown that $\lim_{x\to x_0} \mathbf{x} = \mathbf{x}_0$. In a similar way we have

$$\lim_{(\mathbf{x},\,\mathbf{y})\to(\mathbf{x}_0,\,\mathbf{y}_0)} \mathbf{x} = \mathbf{x}_0, \text{ etc.}$$

In what follows, the student may assume, without proof, the validity of limits from one-variable calculus. For example, $\lim_{x\to 1} \sqrt{x} = \sqrt{1} = 1$ and $\lim_{\theta\to 0} \sin\theta = \sin 0 = 0$.

EXAMPLE 5. This example demonstrates another case in which the limit cannot simply be "read off" from the function. Find $\lim_{x\to 1} g(x)$ where

$$g: x \mapsto \frac{x - 1}{\sqrt{x} - 1}$$

This function is graphed in Figure 2.2.12. We see that $g(1)$ is not defined, since division by zero is not defined. However, if we multiply the numerator and denominator of $g(x)$ by $\sqrt{x} + 1$, we find that for all x in the domain of g

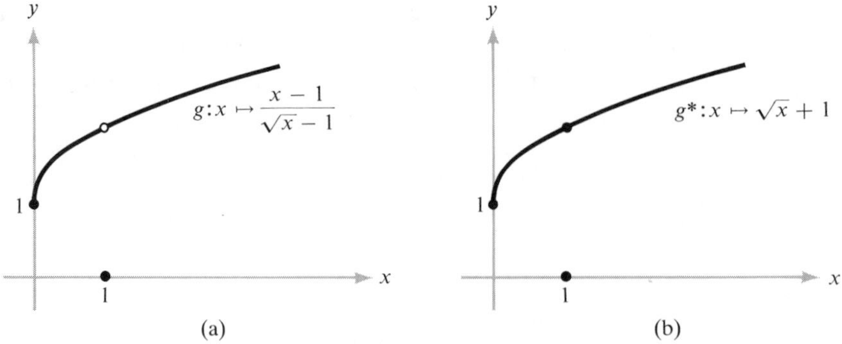

FIGURE 2.2.12

These graphs are the same except that in (a) *g is undefined at* $x = 1$, *while in* (b) g^* *is defined for all* $x \geq 0$.

we have

$$g(x) = \frac{x - 1}{\sqrt{x} - 1} = \sqrt{x} + 1$$

The expression $g^*(x) = \sqrt{x} + 1$ is defined and takes the value 2 at $x = 1$; from one-variable calculus, $g^*(x) \to 2$ as $x \to 1$. But since $g^*(x) = g(x)$ for all $x \geq 0, x \neq 1$, we must have as well that $g(x) \to 2$ as $x \to 1$.

In order that we can properly speak of *the* limit, we should establish that f can have *at most one* limit as $\mathbf{x} \to \mathbf{x}_0$. This is intuitively clear and we now state it formally. (See Appendix A for the proof.)

THEOREM 2 (UNIQUENESS OF LIMITS). *If* $\lim_{\mathbf{x}\to\mathbf{x}_0} f(\mathbf{x}) = \mathbf{b}_1$ *and* $\lim_{\mathbf{x}\to\mathbf{x}_0} f(\mathbf{x}) = \mathbf{b}_2$, *then* $\mathbf{b}_1 = \mathbf{b}_2$.

To be able to carry out practical computations with limits we also need to know some rules that limits obey. For example, we want to prove that the limit of a sum is the sum of the limits. These rules are summarized in the following theorem (see Appendix A for the proof).

THEOREM 3. *Let* $f: A \subset \mathbb{R}^n \to \mathbb{R}^m$, $g: A \subset \mathbb{R}^n \to \mathbb{R}^m$, $\mathbf{x}_0$ *be in* A *or be a boundary point of* A, $\mathbf{b} \in \mathbb{R}^m$, *and* $c \in \mathbb{R}$; *then*

(*i*) *if* $\lim_{\mathbf{x}\to\mathbf{x}_0} f(\mathbf{x}) = \mathbf{b}$, *then* $\lim_{\mathbf{x}\to\mathbf{x}_0} cf(\mathbf{x}) = c\mathbf{b}$, *where* $cf: A \to \mathbb{R}^m$ *is defined by* $\mathbf{x} \mapsto c(f(\mathbf{x}))$.

(*ii*) *if* $\lim_{\mathbf{x}\to\mathbf{x}_0} f(\mathbf{x}) = \mathbf{b}_1$ *and* $\lim_{\mathbf{x}\to\mathbf{x}_0} g(\mathbf{x}) = \mathbf{b}_2$, *then* $\lim_{\mathbf{x}\to\mathbf{x}_0} (f + g)(\mathbf{x}) = \mathbf{b}_1 + \mathbf{b}_2$, *where* $(f + g): A \to \mathbb{R}^m$ *is defined by* $\mathbf{x} \mapsto f(\mathbf{x}) + g(\mathbf{x})$.

(*iii*) *if* $m = 1$, $\lim_{\mathbf{x}\to\mathbf{x}_0} f(\mathbf{x}) = b_1$, *and* $\lim_{\mathbf{x}\to\mathbf{x}_0} g(\mathbf{x}) = b_2$, *then* $\lim_{\mathbf{x}\to\mathbf{x}_0} (fg)(\mathbf{x}) = b_1 b_2$ *where* $(fg): A \to \mathbb{R}$ *is defined by* $\mathbf{x} \mapsto f(\mathbf{x})g(\mathbf{x})$.

(*iv*) *if* $m = 1$, $\lim_{\mathbf{x}\to\mathbf{x}_0} f(\mathbf{x}) = b \neq 0$, *and* $f(\mathbf{x}) \neq 0$ *for all* $\mathbf{x} \in A$, *then* $\lim_{\mathbf{x}\to\mathbf{x}_0} 1/f = 1/b$ *where* $1/f: A \to \mathbb{R}$ *is defined by* $\mathbf{x} \mapsto 1/f(\mathbf{x})$.

(*v*) *if* $f(\mathbf{x}) = (f_1(\mathbf{x}), \ldots, f_m(\mathbf{x}))$ *where* $f_i: A \to \mathbb{R}, i = 1, \ldots, m$, *are the component functions of* f, *then* $\lim_{\mathbf{x}\to\mathbf{x}_0} f(\mathbf{x}) = \mathbf{b} = (b_1, \ldots, b_m)$ *if and only if* $\lim_{\mathbf{x}\to\mathbf{x}_0} f_i(\mathbf{x}) = b_i$ *for each* $i = 1, \ldots, m$.

These results ought to be intuitively clear. For instance (*ii*) says nothing more than if $f(\mathbf{x})$ is close to $\mathbf{b}_1$ and $g(\mathbf{x})$ is close to $\mathbf{b}_2$ when $\mathbf{x}$ is close to $\mathbf{x}_0$, then $f(\mathbf{x}) + g(\mathbf{x})$ is close to $\mathbf{b}_1 + \mathbf{b}_2$ when $\mathbf{x}$ is close to $\mathbf{x}_0$. The following example illustrates how this works.

EXAMPLE 6. Let $f: \mathbb{R}^2 \to \mathbb{R}$, $(x, y) \mapsto x^2 + y^2 + 2$. Compute $\lim_{(x, y)\to(0, 1)} f(x, y)$.

Here f is the sum of the three functions $(x, y) \mapsto x^2$, $(x, y) \mapsto y^2$, and $(x, y) \mapsto 2$. Now the limit of a sum is the sum of the limits, and the limit of a product is the product of the limits (Theorem 3). Hence, using the fact that $\lim_{(x, y)\to(x_0, y_0)} x = x_0$ (Example 4), we obtain: $\lim_{(x, y)\to(x_0, y_0)} x^2 = (\lim_{(x, y)\to(x_0, y_0)} x) \times (\lim_{(x, y)\to(x_0, y_0)} x) = x_0^2$ and, using the same reasoning, $\lim_{(x, y)\to(x_0, y_0)} y^2 = y_0^2$. Consequently $\lim_{(x, y)\to(0, 1)} f(x, y) = 0^2 + 1^2 + 2 = 3$.

In elementary calculus we learned that the idea of a continuous function is based on the intuitive notion of its graph being an unbroken curve, that is, a curve which has no *jumps*, such as would be traced out by a particle in motion or by a moving pencil point that is not lifted from the paper.

To perform a detailed analysis of functions we need concepts more precise than the rather vague notions mentioned above. An example may clarify these ideas. Consider the function $f: \mathbb{R} \to \mathbb{R}$ defined by $f(x) = -1$ if $x \leq 0$ and $f(x) = 1$ if $x > 0$. The graph of f is shown in Figure 2.2.13. (The little open circle denotes the fact that (0, 1) does *not* lie on the graph of f.) Clearly the graph of f is broken at $x = 0$. Consider also the function $g: x \mapsto x^2$. This function is pictured in Figure 2.2.14. The graph of g is not broken at any point. If one examines examples of functions like f above, whose graphs are broken at some point x_0, and functions like g above, whose graphs are not broken, one sees that the principal difference between them is that for a function g whose graph is *unbroken*, the values of $g(x)$ get closer and closer to $g(x_0)$ as x gets closer and closer to x_0. The same idea works for functions of several variables. But the notion of closer and closer does not suffice as a mathematical definition; thus we shall formulate these concepts precisely in terms of limits.

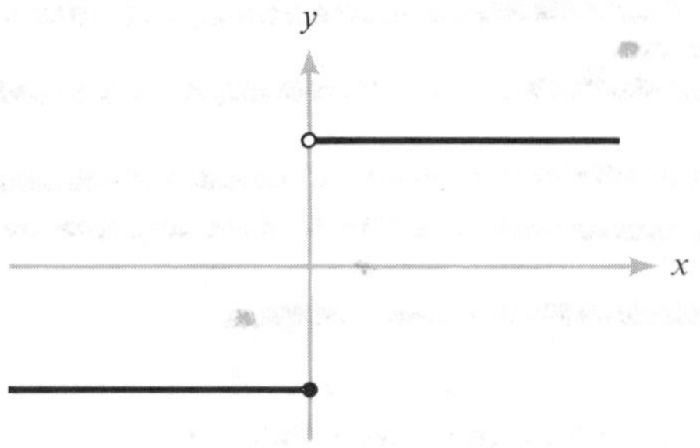

FIGURE 2.2.13
This function is not continuous because its value suddenly jumps as x crosses 0.

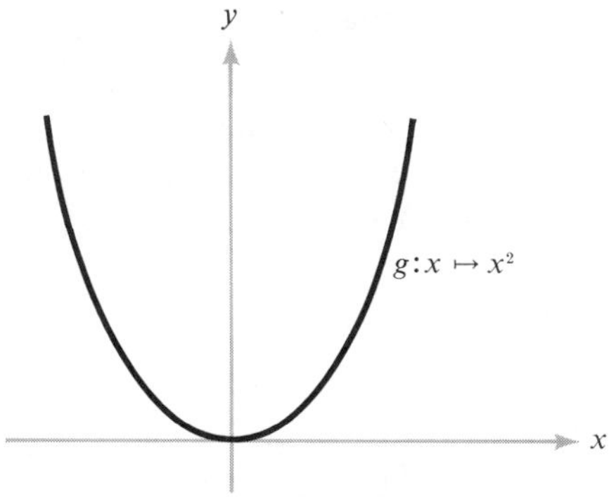

FIGURE 2.2.14
This function is continuous.

DEFINITION. *Let $f: A \subset \mathbb{R}^n \to \mathbb{R}^m$ be a given function with domain A. Let $\mathbf{x}_0 \in A$. We say f is **continuous at** $\mathbf{x}_0$ if and only if*

$$\lim_{\mathbf{x} \to \mathbf{x}_0} f(\mathbf{x}) = f(\mathbf{x}_0)$$

*If we just say that f is **continuous**, we shall mean that f is continuous at each point $\mathbf{x}_0$ of A.*

Since the condition $\lim_{\mathbf{x} \to \mathbf{x}_0} f(\mathbf{x}) = f(\mathbf{x}_0)$ means that $f(\mathbf{x})$ is close to $f(\mathbf{x}_0)$ when $\mathbf{x}$ is close to $\mathbf{x}_0$, we see that our definition does indeed correspond to the requirement that the graph of f is unbroken (see Figure 2.2.15 where we

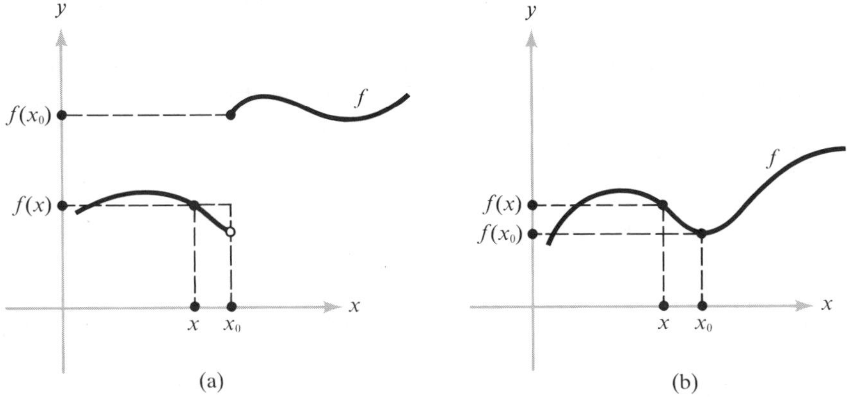

FIGURE 2.2.15
A discontinuous function (a), *for which* $\lim_{x \to x_0} f(x)$ *does not exist, and a continuous function* (b), *for which this limit exists and equals* $f(x_0)$.

illustrate the case $f: \mathbb{R} \to \mathbb{R}$). The case of several variables is easiest to visualize if we deal with real-valued functions, say $f: \mathbb{R}^2 \to \mathbb{R}$. In this case we can visualize f by drawing its graph, which consists of all points (x, y, z) in $\mathbb{R}^3$ with $z = f(x, y)$. The continuity of f thus means that its graph has no "breaks" in it (see Figure 2.2.16).

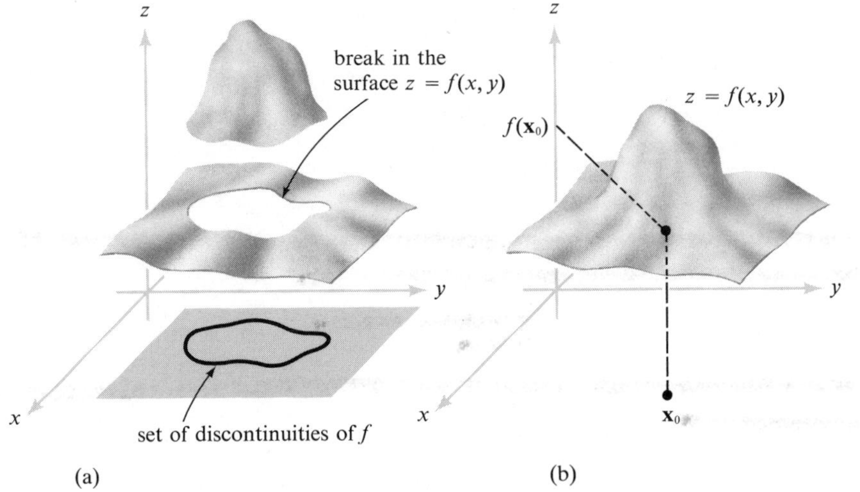

FIGURE 2.2.16
(a) *A discontinuous function of two variables.*
(b) *A continuous function.*

EXAMPLE 7. Any polynomial $p(x) = a_0 + a_1 x + \cdots + a_n x^n$ is continuous from $\mathbb{R}$ to $\mathbb{R}$. Indeed, from Theorem 3 and Example 4

$$\lim_{x \to x_0} (a_0 + a_1 x + \cdots + a_n x^n) = \lim_{x \to x_0} a_0 + \lim_{x \to x_0} a_1 x + \cdots + \lim_{x \to x_0} a_n x^n$$

$$= a_0 + a_1 x_0 + \cdots + a_n x_0^n$$

since

$$\lim_{x \to x_0} x^n = \left(\lim_{x \to x_0} x\right) \cdots \left(\lim_{x \to x_0} x\right) = x_0^n.$$

EXAMPLE 8. Let $f: \mathbb{R}^2 \to \mathbb{R}$, $f(x, y) = xy$. Then f is continuous since, by the limit theorems and Example 4,

$$\lim_{(x, y) \to (x_0, y_0)} xy = \left(\lim_{(x, y) \to (x_0, y_0)} x\right)\left(\lim_{(x, y) \to (x_0, y_0)} y\right) = x_0 y_0$$

One can see by the same method that any polynomial $p(x, y)$ in x and y is continuous.

EXAMPLE 9. The function $f: \mathbb{R}^2 \to \mathbb{R}$ defined by

$$f(x, y) = \begin{cases} 1 & \text{if } x \leq 0 \quad \text{or} \quad y \leq 0 \\ 0 & \text{otherwise} \end{cases}$$

is not continuous at (0, 0) or at any point on the positive x-axis or positive y-axis. Indeed, if $(x_0, y_0) = \mathbf{u}$ is such a point and $\delta > 0$, there are points $(x, y) \in D_\delta(\mathbf{u})$, a neighborhood of $\mathbf{u}$, with $f(x, y) = 1$ and other points $(x, y) \in D_\delta(\mathbf{u})$ with $f(x, y) = 0$. Thus it is *not* true that $f(x, y) \to f(x_0, y_0) = 1$ as $(x, y) \to (x_0, y_0)$.

In order to prove that specific functions are continuous we can avail ourselves of the limit theorems (see Theorem 3 and Example 7). If we transcribe those results in terms of continuity, we are led at once to the following:

THEOREM 4. *Let $f: A \subset \mathbb{R}^n \to \mathbb{R}^m$, $g: A \subset \mathbb{R}^n \to \mathbb{R}^m$, and c a real number.*

(*i*) *If f is continuous at $\mathbf{x}_0$, so is cf, where $(cf)(\mathbf{x}) = c(f(\mathbf{x}))$.*

(*ii*) *If f and g are continuous at $\mathbf{x}_0$, so is $f + g$, where $(f + g)(\mathbf{x}) = f(\mathbf{x}) + g(\mathbf{x})$.*

(*iii*) *If f and g are continuous at $\mathbf{x}_0$ and $m = 1$, then the product function fg defined by $(fg)(\mathbf{x}) = f(\mathbf{x})g(\mathbf{x})$ is continuous at $\mathbf{x}_0$.*

(*iv*) *If $f: A \subset \mathbb{R}^n \to \mathbb{R}$ is continuous at $\mathbf{x}_0$ and nowhere zero on A, then the quotient $1/f$ is continuous at $\mathbf{x}_0$, where $(1/f)(\mathbf{x}) = 1/f(\mathbf{x})$.**

(*v*) *If $f: A \subset \mathbb{R}^n \to \mathbb{R}^m$ and $f(\mathbf{x}) = (f_1(\mathbf{x}), \ldots, f_m(\mathbf{x}))$ then f is continuous at $\mathbf{x}_0$ if and only if each of the real-valued functions $f_1, \ldots, f_m$ are continuous at $\mathbf{x}_0$.*

EXAMPLE 10. Let $f: \mathbb{R}^2 \to \mathbb{R}^2, (x, y) \mapsto (x^2y, (y + x^3)/(1 + x^2))$. Show that f is continuous.

To see this, it is sufficient, by (*v*) above, to show that each component is continuous. So we first show that $(x, y) \mapsto x^2y$ is continuous. Now $(x, y) \mapsto x$ is continuous (see Example 4), so by (*iii*) $(x, y) \mapsto x^2$ is continuous. Since $(x, y) \mapsto y$ is continuous, by (*iii*) again $(x, y) \mapsto x^2y$ is continuous. Since $1 + x^2$ is continuous and non-zero, by (*iv*) $1/(1 + x^2)$ is continuous; hence $(y + x^3)/(1 + x^2)$ is a product of continuous functions, and by (*iii*) is continuous.

Next we discuss another basic operation that can be performed on functions. This operation is called *composition*. If g maps A to B and f maps B to C,

* Another way of stating (*iv*) is often used: If $f(\mathbf{x}_0) \neq 0$ and f is continuous, then $f(\mathbf{x}) \neq 0$ in a neighborhood of $\mathbf{x}_0$ so $1/f$ is defined in that neighborhood, and $1/f$ is continuous at $\mathbf{x}_0$.

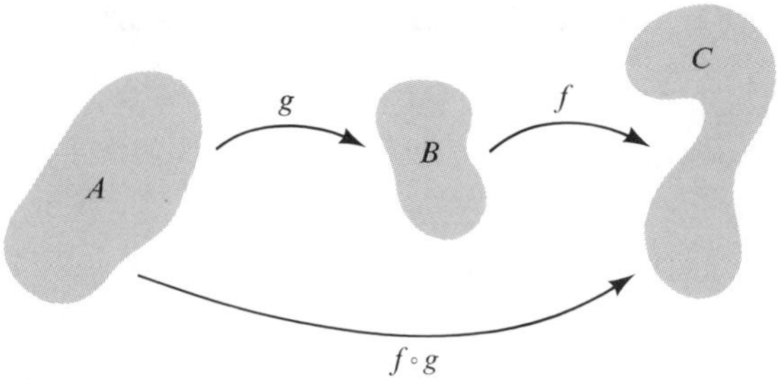

FIGURE 2.2.17
The composition of f on g.

the composition of g with f, or of f on g, denoted by $f \circ g$, maps A to C by sending $\mathbf{x} \mapsto f(g(\mathbf{x}))$ (see Figure 2.2.17). For example, $\sin x^2$ is the composition of $x \mapsto x^2$ with $y \mapsto \sin y$. The reader should be familiar with this idea from one-variable calculus.

THEOREM 5. *Let $g: A \subset \mathbb{R}^n \to \mathbb{R}^m$ and let $f: B \subset \mathbb{R}^m \to \mathbb{R}^p$. Suppose $g(A) \subset B$ so that $f \circ g$ is defined on A. If g is continuous at $\mathbf{x}_0 \in A$ and f is continuous at $\mathbf{y}_0 = g(\mathbf{x}_0)$, then $f \circ g$ is continuous at $\mathbf{x}_0$.*

The intuition behind this is easy; the formal proof in Appendix A follows a similar pattern. Intuitively we must show that as $\mathbf{x}$ gets close to $\mathbf{x}_0$, $f(g(\mathbf{x}))$ gets close to $f(g(\mathbf{x}_0))$. But as $\mathbf{x}$ gets close to $\mathbf{x}_0$, $g(\mathbf{x})$ gets close to $g(\mathbf{x}_0)$ (by continuity of g at $\mathbf{x}_0$) and as $g(\mathbf{x})$ gets close to $g(\mathbf{x}_0)$, $f(g(\mathbf{x}))$ gets close to $f(g(\mathbf{x}_0))$ (by continuity of f at $g(\mathbf{x}_0)$).

EXAMPLE 11. Let $f(x, y, z) = (x^2 + y^2 + z^2)^{30} + \sin z^3$. Show that f is continuous.

Here we can write f as a sum of two functions $(x^2 + y^2 + z^2)^{30}$ and $\sin z^3$, so it suffices to show that each is continuous. The first is the composite of $(x, y, z) \mapsto (x^2 + y^2 + z^2)$ with $u \mapsto u^{30}$ and the second is the composite of $(x, y, z) \mapsto z^3$ with $u \mapsto \sin u$, so we have continuity by Theorem 5.

OPTIONAL

LIMITS IN TERMS OF ε's AND δ's

We shall now state a theorem giving a formulation of the notion of limit in terms of epsilons and deltas. This new formulation is quite useful; it is often taken as the *definition* of limit. It is another way of making precise the

OPTIONAL (*Continued*)

intuitive statement that "$f(\mathbf{x})$ is close to $\mathbf{b}$ when $\mathbf{x}$ is close to $\mathbf{x}_0$." To help understand this formulation, the reader should consider it with regards to each of the examples already presented.

THEOREM 6. *Let $f: A \subset \mathbb{R}^n \to \mathbb{R}^m$ and let $\mathbf{x}_0$ be in A or be a boundary point of A. Then $\lim_{\mathbf{x}\to\mathbf{x}_0} f(\mathbf{x}) = \mathbf{b}$ if and only if for every number $\varepsilon > 0$ there is a $\delta > 0$ such that for any $\mathbf{x} \in A$ satisfying $0 < \|\mathbf{x} - \mathbf{x}_0\| < \delta$, we have $\|f(\mathbf{x}) - \mathbf{b}\| < \varepsilon$ (see Figure 2.2.18).*

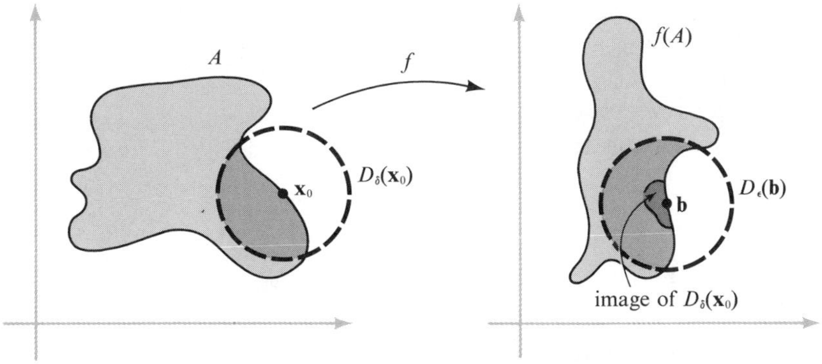

FIGURE 2.2.18
The geometry of the ε-δ definition of limit.

This theorem is proved in Appendix A.

To illustrate the mechanics of the epsilon-delta technique in Theorem 6, we consider the following examples.

EXAMPLE 12. Show that $\lim_{(x,y)\to(0,0)} x = 0$.

Note that if $\delta > 0$, $\|(x, y) - (0, 0)\| = \sqrt{x^2 + y^2} < \delta$ implies $|x - 0| = |x| = \sqrt{x^2} \leq \sqrt{x^2 + y^2} < \delta$. Thus if $\|(x, y) - (0, 0)\| < \delta$, then $|x - 0|$ is also less than δ. Given $\varepsilon > 0$ we must find a $\delta > 0$ (generally depending on ε) with the property that $0 < \|(x, y) - (0, 0)\| < \delta$ implies $|x - 0| < \varepsilon$. What are we to pick as our δ? From the above calculation, we see that if we choose $\delta = \varepsilon$, then $\|(x, y) - (0, 0)\| < \delta$ implies $|x - 0| < \varepsilon$. This shows that $\lim_{(x,y)\to(0,0)} x = 0$. We could have also chosen $\delta = \varepsilon/2$ or $\varepsilon/3$, but it suffices to find just one δ satisfying the requirements of the definition of limit.

OPTIONAL (*Continued*)

More generally, we can see in the same way that

$$\underset{(x, y)\to(x_0, y_0)}{\text{limit}} x = x_0$$

This is just an epsilon-delta reformation of Example 4.

EXAMPLE 13. Consider the function

$$f(x, y) = \frac{\sin(x^2 + y^2)}{x^2 + y^2}$$

Even though f is not defined at (0, 0) we can ask whether $f(x, y)$ approaches some number as (x, y) approaches (0, 0). From elementary calculus we know that

$$\underset{\alpha\to 0}{\text{limit}} \frac{\sin \alpha}{\alpha} = 1$$

Thus it is reasonable to guess that

$$\underset{\mathbf{v}\to(0, 0)}{\text{limit}} f(\mathbf{v}) = \underset{\mathbf{v}\to(0, 0)}{\text{limit}} \frac{\sin\|\mathbf{v}\|^2}{\|\mathbf{v}\|^2} = 1$$

Indeed, since $\underset{\alpha\to 0}{\text{limit}} (\sin \alpha)/\alpha = 1$, given $\varepsilon > 0$ we can find a $\delta > 0$, with

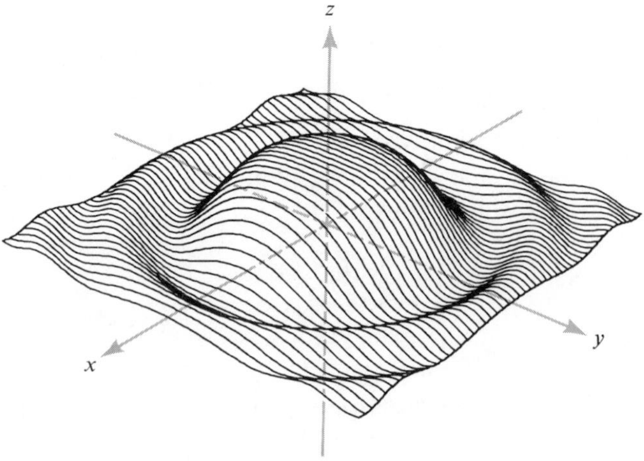

$$z = \frac{\sin(x^2 + y^2)}{(x^2 + y^2)} \qquad -3 \le x \le 3, -3 \le y \le 3$$

FIGURE 2.2.19
Computer-generated graph.

OPTIONAL (*Continued*)

$1 > \delta > 0$, such that $0 < |\alpha| < \delta$ implies that $|(\sin \alpha)/\alpha - 1| < \varepsilon$. If $0 < \|\mathbf{v}\| < \delta$, then $0 < \|\mathbf{v}\|^2 < \delta^2 < \delta$, and therefore

$$|f(\mathbf{v}) - 1| = \left|\frac{\sin\|\mathbf{v}\|^2}{\|\mathbf{v}\|^2} - 1\right| < \varepsilon$$

Thus $\lim_{\mathbf{v}\to(0,0)} f(\mathbf{v}) = 1$. Indeed, if we plot $(\sin(x^2 + y^2))/(x^2 + y^2)$ on a computer, we get a graph that is well-behaved near (0, 0) (Figure 2.2.19).

EXAMPLE 14. Show that

$$\lim_{(x,y)\to(0,0)} \frac{x^2}{\sqrt{x^2 + y^2}} = 0$$

We must show that $x^2/\sqrt{x^2 + y^2}$ is small when (x, y) is close to the origin. To do this we need the following inequality:

$$0 \le \frac{x^2}{\sqrt{x^2 + y^2}} \le \frac{x^2 + y^2}{\sqrt{x^2 + y^2}} \qquad (\text{since } y^2 \ge 0)$$
$$= \sqrt{x^2 + y^2}$$

Given $\varepsilon > 0$, choose $\delta = \varepsilon$. Then $\|(x, y) - (0, 0)\| = \|(x, y)\| = \sqrt{x^2 + y^2}$, so $\|(x, y) - (0, 0)\| < \delta$ implies that

$$\left|\frac{x^2}{\sqrt{x^2 + y^2}} - 0\right| = \frac{x^2}{\sqrt{x^2 + y^2}} \le \sqrt{x^2 + y^2} = \|(x, y) - (0, 0)\| < \delta = \varepsilon$$

Thus the conditions of Theorem 6 have been fulfilled and the limit verified.

EXAMPLE 15. Does $\lim_{(x,y)\to(0,0)} x^2/(x^2 + y^2)$ exist?

If the limit exists, $x^2/(x^2 + y^2)$ should approach a definite value, say a, as (x, y) gets near (0, 0). In particular, if (x, y) approaches zero along any given path then $x^2/(x^2 + y^2)$ should approach the limiting value a. If (x, y) approaches (0, 0) along the line $y = 0$, the limiting value is clearly 1 (just set $y = 0$ in the above expression to get $x^2/x^2 = 1$). If (x, y) approaches (0, 0) along the line $x = 0$, the limiting value is

$$\lim_{(x,y)\to(0,0)} \frac{0^2}{0^2 + y^2} = 0 \ne 1$$

Hence, $\lim_{(x,y)\to(0,0)} x^2/(x^2 + y^2)$ does not exist (see Figure 2.2.20).

OPTIONAL (*Continued*)

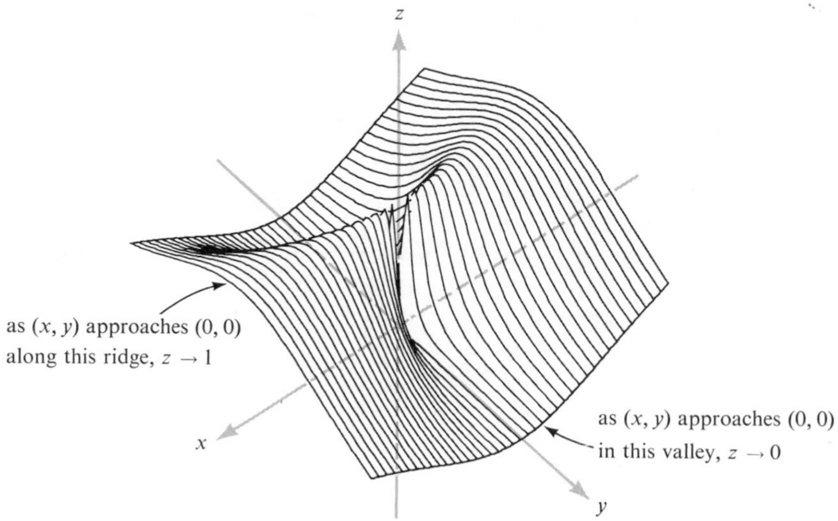

$$z = \frac{x^2}{x^2 + y^2} \qquad -1 \le x \le 1,\ -1 \le y \le 1$$

FIGURE 2.2.20
This function has no limit at (0, 0).

EXAMPLE 16. Prove (see Figure 2.2.21)

$$\operatorname*{limit}_{(x,\,y)\to(0,\,0)} \frac{x^2 y}{x^2 + y^2} = 0$$

Indeed, note that

$$\left|\frac{x^2 y}{x^2 + y^2}\right| \le \left|\frac{x^2 y}{x^2}\right| = |y|$$

Thus, given $\varepsilon > 0$, choose $\delta = \varepsilon$; so $0 < \|(x, y) - (0, 0)\| = \sqrt{x^2 + y^2} < \delta$ implies $|y| < \delta$, and thus

$$\left|\frac{x^2 y}{x^2 + y^2} - 0\right| < \varepsilon$$

For more examples of limits, see Appendix A.

Using the epsilon-delta notation, we are led to the following reformulation of the definition of continuity.

OPTIONAL (*Continued*)

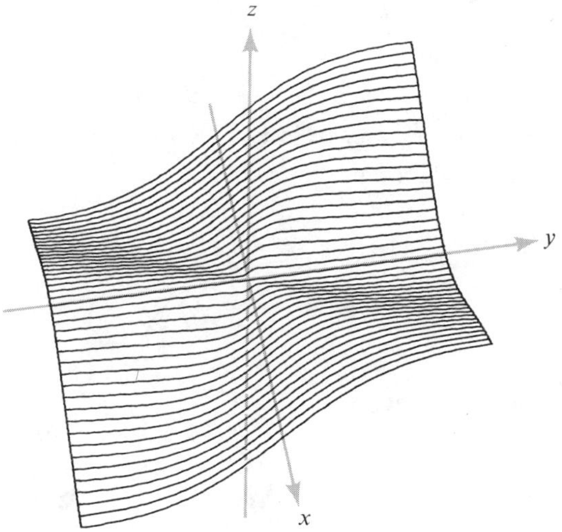

$$z = \frac{x^2y}{x^2 + y^2}$$

FIGURE 2.2.21
This function has limit 0 *at* (0, 0).

THEOREM 7. *Let $f: A \subset \mathbb{R}^n \to \mathbb{R}^m$ be given. Then f is continuous at $\mathbf{x}_0 \in A$ if and only if for every number $\varepsilon > 0$ there is a number $\delta > 0$ such that*

$$\mathbf{x} \in A \textit{ and } \|\mathbf{x} - \mathbf{x}_0\| < \delta \quad \textit{implies} \quad \|f(\mathbf{x}) - f(\mathbf{x}_0)\| < \varepsilon$$

The proof is almost immediate. However, notice that in Theorem 6 we insisted that $0 < \|\mathbf{x} - \mathbf{x}_0\|$, that is, $\mathbf{x} \neq \mathbf{x}_0$. That is *not* imposed here; indeed, the conclusion of Theorem 7 is certainly valid when $\mathbf{x} = \mathbf{x}_0$, so there is no need to exclude this case. Here we do care about the value of f at $\mathbf{x}_0$; we want f at nearby points to be close to this value.

EXERCISES

In the following exercises the reader may assume that the exponential function, sine, and cosine are continuous functions from $\mathbb{R}$ to $\mathbb{R}$.

1. Show that the following subsets of the plane are open
   (a) $A = \{(x, y) | -1 < x < 1, -1 < y < 1\}$
   (b) $B = \{(x, y) | y > 0\}$

 cf Appendix A

(c) $C = \{(x, y) | 2 < x^2 + y^2 < 4\}$
(d) $A \cup B \cup C$
(e) $D = \{(x, y) | x \neq 0 \quad \text{and} \quad y \neq 0\}$

2. Prove that for $\mathbf{x} \in \mathbb{R}^n$ and $s < t$, $D_s(\mathbf{x}) \subset D_t(\mathbf{x})$.
3. Prove that if U and V are neighborhoods of $\mathbf{x} \in \mathbb{R}^n$, then so are $U \cap V$ and $U \cup V$.
4. Prove that the boundary points of $]a, b[\subset \mathbb{R}$ are the points a and b.

*5. Use the ε-δ formulation of limits to prove that $x^2 \to 4$ as $x \to 2$. Give a shorter proof, using Theorem 3.

6. Compute the following limits
(a) $\lim\limits_{(x, y) \to (0, 1)} x^3 y$
(b) $\lim\limits_{(x, y) \to (0, 1)} e^x y$
(c) $\lim\limits_{x \to 0} \dfrac{\sin^2 x}{x}$ $\left(\text{HINT: Recall that } \lim\limits_{x \to 0} \dfrac{\sin x}{x} = 1.\right)$
(d) $\lim\limits_{x \to 0} \dfrac{\sin^2 x}{x^2}$

7. Compute $\lim\limits_{\mathbf{x} \to \mathbf{x}_0} f(\mathbf{x})$, if it exists, for the following cases:
(a) $f: \mathbb{R} \to \mathbb{R}$, $x \mapsto |x|$, $x_0 = 1$
(b) $f: \mathbb{R}^n \to \mathbb{R}$, $\mathbf{x} \mapsto \|\mathbf{x}\|$, arbitrary $\mathbf{x}_0$
(c) $f: \mathbb{R} \to \mathbb{R}^2$, $x \mapsto (x^2, e^x)$, $x_0 = 1$
(d) $f: \mathbb{R}^2 \backslash \{(0, 0)\} \to \mathbb{R}^2$, $(x, y) \mapsto (\sin(x - y),\ e^{x(y+1)} - x - 1)/\|(x, y)\|$, $\mathbf{x}_0 = (0, 0)$

8. Let $A \subset \mathbb{R}^2$ be the unit disc $D_1(0, 0)$ with the point $\mathbf{x}_0 = (1, 0)$ added, and $f: A \to \mathbb{R}$, $\mathbf{x} \mapsto f(\mathbf{x})$ be the constant function $f(\mathbf{x}) = 1$. Show that $\lim\limits_{\mathbf{x} \to \mathbf{x}_0} f(\mathbf{x}) = 1$.

9. Let $f: \mathbb{R}^3 \to \mathbb{R}$, $f(x, y, z) = (x^2 + 3y^2)/(x + 1)$. Compute
$$\lim_{(x, y, z) \to (0, 0, 0)} f(x, y, z).$$

*10. Let $f: A \subset \mathbb{R}^n \to \mathbb{R}$ be given and let $\mathbf{x}_0$ be a boundary point of A. We say that $\lim\limits_{\mathbf{x} \to \mathbf{x}_0} f(\mathbf{x}) = \infty$ if for every $N > 0$ there is a $\delta > 0$ such that $0 < \|\mathbf{x} - \mathbf{x}_0\| < \delta$ implies $f(\mathbf{x}) > N$.
(a) Prove that $\lim\limits_{x \to 1} (x - 1)^{-2} = \infty$.

Prove that $\lim\limits_{x \to 0} 1/|x| = \infty$. Is it true that $\lim\limits_{x \to 0} 1/x = \infty$?

(c) Prove that $\lim\limits_{(x, y) \to (0, 0)} 1/(x^2 + y^2) = \infty$.

*11. Let $f: \mathbb{R} \to \mathbb{R}$ be a function. We write $\lim\limits_{x \to b-} f(x) = L$ and say that L is the *left-hand limit* of f at b, if for every $\varepsilon > 0$, there is a $\delta > 0$ such that $x < b$ and $0 < |x - b| < \delta$ implies $|f(x) - L| < \varepsilon$.
(a) Formulate a definition of *right-hand limit*, or $\lim\limits_{x \to b+} f(x)$.
(b) Find $\lim\limits_{x \to 0-} 1/(1 + e^{1/x})$ and $\lim\limits_{x \to 0+} 1/(1 + e^{1/x})$.
(c) Sketch the graph of $1/(1 + e^{1/x})$.

*12. Prove that there is a number $\delta > 0$ such that if $|a| < \delta$, then $|a^3 + 3a^2 + a| < 1/100$.

(b) Prove that there is a number $\delta > 0$ such that if $x^2 + y^2 < \delta^2$, then

$$|x^2 + y^2 + 3xy + 180xy^5| < 1/10{,}000.$$

13. Compute the following limits (review properties of the relevant functions if necessary).

(a) $\displaystyle\operatorname*{limit}_{x\to 3}(x^2 - 3x + 5)$

(b) $\displaystyle\operatorname*{limit}_{x\to 0}\sin x$

(c) $\displaystyle\operatorname*{limit}_{h\to 0}\frac{(x+h)^2 - x^2}{h}$

(d) $\displaystyle\operatorname*{limit}_{h\to 0}\frac{e^h - 1}{h}$ (HINT: Recall L'Hopital's rule.)

(e) $\displaystyle\operatorname*{limit}_{x\to 0}\frac{\cos x - 1}{x^2}$

14. Compute the following limits.

(a) $\displaystyle\operatorname*{limit}_{(x,y)\to(0,0)}(x^2 + y^2 + 3)$

(b) $\displaystyle\operatorname*{limit}_{(x,y)\to(0,0)}\frac{xy}{x^2 + y^2 + 2}$

(c) $\displaystyle\operatorname*{limit}_{(x,y)\to(0,0)}\frac{e^{xy}}{x + 1}$

15. Prove that $\displaystyle\operatorname*{limit}_{(x,y)\to(0,0)}(\sin xy)/xy = 1$.

16. Show that the map $f\colon \mathbb{R} \to \mathbb{R}$, $x \mapsto x^2e^x/(2 - \sin x)$ is continuous.

*17. Show that f is continuous at $\mathbf{x}_0$ if and only if

$$\operatorname*{limit}_{\mathbf{x}\to\mathbf{x}_0}\|f(\mathbf{x}) - f(\mathbf{x}_0)\| = 0$$

18. Show that $f\colon \mathbb{R} \to \mathbb{R}$, $x \mapsto (1 - x)^8 + \cos(1 + x^3)$ is continuous.

19. If $f\colon \mathbb{R}^n \to \mathbb{R}$ and $g\colon \mathbb{R}^n \to \mathbb{R}$ are continuous, show that the functions

$$f^2g\colon \mathbb{R}^n \to \mathbb{R},\ \mathbf{x} \mapsto (f(\mathbf{x}))^2g(\mathbf{x})$$

and

$$f^2 + g\colon \mathbb{R}^n \to \mathbb{R},\ \mathbf{x} \mapsto (f(\mathbf{x}))^2 + g(\mathbf{x})$$

are continuous.

20. Prove that $f\colon \mathbb{R}^2 \to \mathbb{R}$, $(x, y) \mapsto ye^x + \sin x + (xy)^4$ is continuous.

21. Suppose $\mathbf{x}$ and $\mathbf{y}$ are in $\mathbb{R}^n$ and $\mathbf{x} \neq \mathbf{y}$. Show that there is a continuous function $f\colon \mathbb{R}^n \to \mathbb{R}$ with $f(\mathbf{x}) = 1$, $f(\mathbf{y}) = 0$, and $0 \le f(\mathbf{z}) \le 1$ for every $\mathbf{z}$ in $\mathbb{R}^n$.

*22. Show that $f\colon \mathbb{R}^n \to \mathbb{R}^m$ is continuous at all points if and only if the inverse image of every open set is open.

*23. Let $f\colon A \subset \mathbb{R}^n \to \mathbb{R}^m$ satisfy $\|f(\mathbf{x}) - f(\mathbf{y})\| \le K\|\mathbf{x} - \mathbf{y}\|^\alpha$ for all $\mathbf{x}$ and $\mathbf{y}$ in A for positive constants K and α. Show that f is continuous. (Such functions are called *Hölder-continuous* or, if $\alpha = 1$, *Lipschitz-continuous*.)

2.3 DIFFERENTIATION

In Section 2.1 we considered a few methods for graphing simple functions. However, by these methods alone it is impossible in practice to compute enough information, in a reasonable time, to grasp even the general features of a complicated function. From elementary calculus we know that the idea of the derivative can greatly aid us in this task; for example, it enables

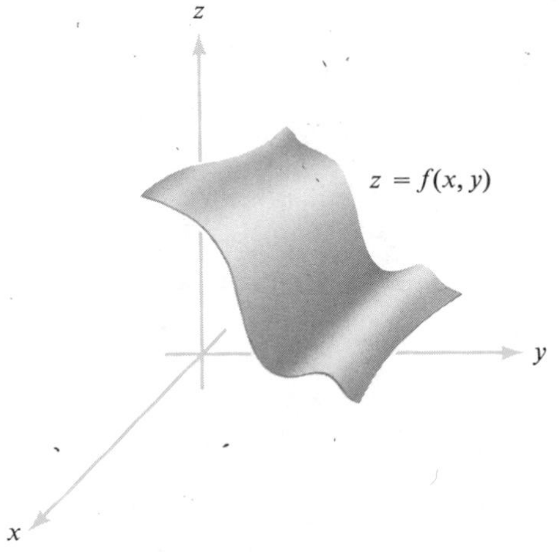

FIGURE 2.3.1
A smooth graph.

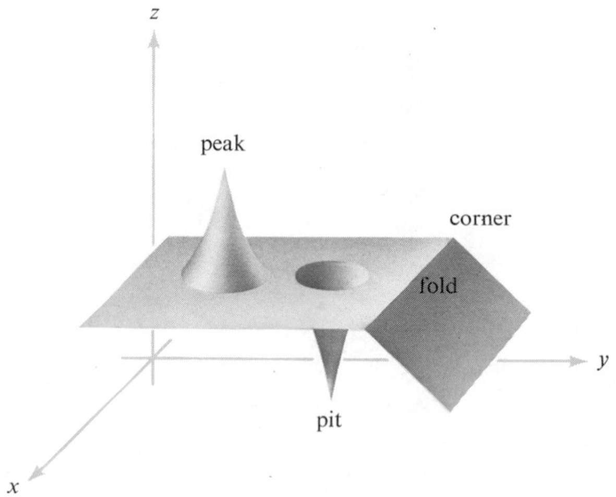

FIGURE 2.3.2
This graph is not smooth.

us to locate maxima and minima and to compute rates of change. The derivative also has many applications beyond this, as the student surely has discovered in elementary calculus.

Intuitively, we know from our work in Section 2.2 that a continuous function is one that has no "breaks" in its graph. A differentiable function from $\mathbb{R}^2$ to $\mathbb{R}$ ought to be such that not only are there no "breaks" in its graph, but there is a well defined tangent plane to the graph at each point. Thus, there must not be any sharp folds, corners, or peaks in the graph (see Figures 2.3.1, 2.3.2). In other words, the graph must be *smooth*.

In order to make these ideas precise, we need a sound definition of what we mean by "$f(x_1, \ldots, x_n)$ is differentiable at $\mathbf{x} = (x_1, \ldots, x_n)$." Actually this is not quite as simple to do as one might think. Towards this end, however, let us introduce the notion of the *partial derivative*. This notion relies only on our knowledge of one-variable calculus. (A quick review of the definition of the derivative in a one-variable calculus text might be advisable at this point.)

DEFINITION. *Let $f: U \subset \mathbb{R}^n \to \mathbb{R}$ be a real-valued function. Then $\partial f/\partial x_1, \ldots, \partial f/\partial x_n$, the* ***partial derivatives*** *of f with respect to the first, second, ..., nth variable are also real-valued functions of n variables, which, at the point $(x_1, \ldots, x_n) = \mathbf{x}$, are defined by*

$$\frac{\partial f}{\partial x_j}(x_1, \ldots, x_n) = \lim_{h \to 0} \frac{f(x_1, x_2, \ldots, x_j + h, \ldots, x_n) - f(x_1, \ldots, x_n)}{h}$$

$$= \lim_{h \to 0} \frac{f(\mathbf{x} + h\mathbf{e}_j) - f(\mathbf{x})}{h}$$

if the limits exist, where $1 \le j \le n$ and $\mathbf{e}_j$ is the jth standard basis vector $\mathbf{e}_j = (0, \ldots, 1, \ldots, 0)$, with 1 in the jth slot (see Section 1.5).

In other words, $\partial f/\partial x_j$ is just the derivative of f with respect to the variable x_j, with the other variables held fixed. If $f: \mathbb{R}^3 \to \mathbb{R}$, we shall often use the notation $\partial f/\partial x$, $\partial f/\partial y$, $\partial f/\partial z$ in place of $\partial f/\partial x_1$, $\partial f/\partial x_2$, $\partial f/\partial x_3$. If $f: U \subset \mathbb{R}^n \to \mathbb{R}^m$, then we can write

$$f(x_1, \ldots, x_n) = (f_1(x_1, \ldots, x_n), \ldots, f_m(x_1, \ldots, x_n))$$

so that we can speak of the partial derivatives of each component; for example, $\partial f_m/\partial x_n$ is the partial derivative of the mth component with respect to x_n, the nth variable.

The existence of the partial derivatives of f at a point $\mathbf{x}$ is the first definition one might think of to give precise meaning to the phrase: "f is differentiable at $\mathbf{x}$." Actually, as we shall see below, to obtain the most satisfactory theory one must strengthen this condition somewhat. Meanwhile, it is essential that the student become thoroughly accustomed to computing partial derivatives. We present some examples to aid in this effort.

EXAMPLE 1. If $f(x, y) = x^2y + y^3$, find $\partial f/\partial x$ and $\partial f/\partial y$.

To find $\partial f/\partial x$ we hold y constant and differentiate only with respect to x; this yields

$$\frac{\partial f}{\partial x} = \frac{d(x^2y + y^3)}{dx} = 2xy$$

Similarly, to find $\partial f/\partial y$ we hold x constant and differentiate only with respect to y.

$$\frac{\partial f}{\partial y} = \frac{d(x^2y + y^3)}{dy} = x^2 + 3y^2$$

If we wish to indicate that a partial derivative is to be evaluated at a particular point, for example at (x_0, y_0), we write

$$\frac{\partial f}{\partial x}(x_0, y_0)$$

When we write $z = f(x, y)$ for the dependent variable, we shall sometimes write $\dfrac{\partial z}{\partial x}$ for $\dfrac{\partial f}{\partial x}$. Strictly speaking, this is an abuse of notation, but it is common practice to use these two notations interchangeably. (See Exercise 22, Section 2.4, for some dangers of sloppy notation.)

EXAMPLE 2. If $z = \cos xy + x \cos y = f(x, y)$, find $(\partial z/\partial x)(x_0, y_0)$ and $(\partial z/\partial y)(x_0, y_0)$.

First we fix y_0 and differentiate with respect to x. So

$$\begin{aligned}\frac{\partial z}{\partial x}(x_0, y_0) &= \left.\frac{d(\cos xy_0 + x \cos y_0)}{dx}\right|_{x=x_0} \\ &= -y_0 \sin xy_0 + \cos y_0\Big|_{x=x_0} \\ &= -y_0 \sin x_0y_0 + \cos y_0\end{aligned}$$

Similarly, we fix x_0 and differentiate with respect to y to obtain

$$\begin{aligned}\frac{\partial z}{\partial y}(x_0, y_0) &= \left.\frac{d(\cos x_0y + x_0 \cos y)}{dy}\right|_{y=y_0} \\ &= -x_0 \sin x_0y - x_0 \sin y\Big|_{y=y_0} \\ &= -x_0 \sin x_0y_0 - x_0 \sin y_0\end{aligned}$$

EXAMPLE 3. Find $\partial f/\partial x$ if $f(x, y) = xy/\sqrt{x^2 + y^2}$.

We have by the quotient rule

$$\frac{\partial f}{\partial x} = \frac{y\sqrt{x^2 + y^2} - xy(x/\sqrt{x^2 + y^2})}{x^2 + y^2}$$

$$= \frac{y(x^2 + y^2) - xy(x)}{(x^2 + y^2)^{3/2}} = \frac{y^3}{(x^2 + y^2)^{3/2}}$$

Let us now consider what problems may arise by defining "differentiability at **x**" to mean just the existence of partial derivatives. Consider the following example.

EXAMPLE 4. Let $f(x, y) = x^{1/3}y^{1/3}$.

By definition

$$\frac{\partial f}{\partial x}(0, 0) = \underset{h \to 0}{\text{limit}} \frac{f(h, 0) - f(0, 0)}{h} = 0$$

and, similarly, $(\partial f/\partial y)(0, 0) = 0$. It is necessary to use the original definition of partial derivative because the functions $x^{1/3}$ and $y^{1/3}$ are not themselves differentiable at 0. But suppose we restrict f to the line $y = x$ to get $f(x, x) = x^{2/3}$ (see Figure 2.3.3). We can view the substitution $y = x$ as the composite of the function $g: \mathbb{R} \to \mathbb{R}^2$, given by $g(x) = (x, x)$, with $f: \mathbb{R}^2 \to \mathbb{R}$, given by $f(x, y) = x^{1/3}y^{1/3}$.

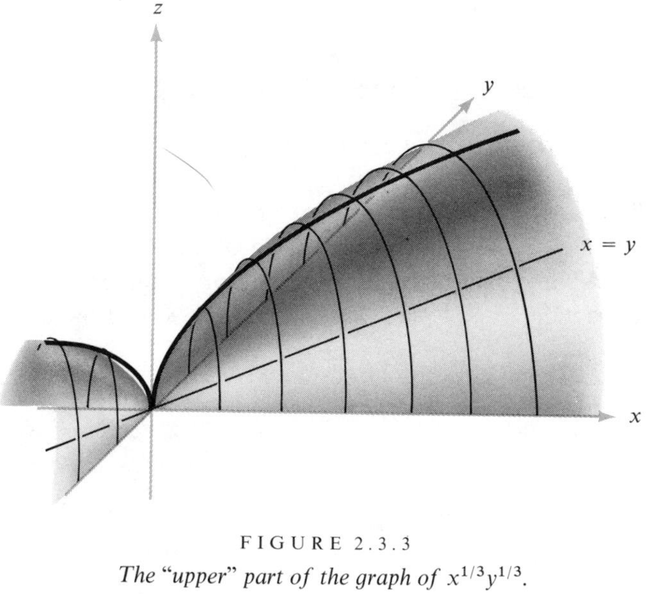

FIGURE 2.3.3

The "upper" part of the graph of $x^{1/3}y^{1/3}$.

Thus $f \circ g: \mathbb{R} \to \mathbb{R}$, with $(f \circ g)(x) = x^{2/3}$. Now g and f are both "differentiable" at 0, but $f \circ g$ is not differentiable at $x = 0$, as the reader may verify. In other words, the composition of differentiable functions would not be differentiable if we used our provisional definition of differentiability, in contrast to what is true of functions of one variable. It would be much more useful to have a definition of differentiability by which the composition of differentiable functions was differentiable.

There is another reason for not wanting to call this function f differentiable at (0, 0). Namely, there is no tangent plane, in any reasonable sense, to the graph at (0, 0). The xy-plane is tangent to the graph along the x- and y-axes because f has slope zero at (0, 0) along these axes; that is, $\partial f/\partial x = 0$ and $\partial f/\partial y = 0$ at (0, 0). Thus, if there is a tangent plane, it must be the xy-plane. However, as is evident from Figure 2.3.3, the xy-plane is not tangent to the graph in other directions, and so cannot be said to be tangent to the graph of f.

To motivate our final definition of differentiability, let us compute what the equation of the plane tangent to the graph of $f: \mathbb{R}^2 \to \mathbb{R}, (x, y) \mapsto f(x, y)$ at (x_0, y_0) ought to be if f is smooth enough. In $\mathbb{R}^3$, a nonvertical plane has an equation of the form

$$z = ax + by + c$$

If it is to be the plane tangent to the graph of f, the slopes along the x- and y-axes must be equal to $\partial f/\partial x$ and $\partial f/\partial y$, the rates of change of f with respect to x and y. Thus, $a = \partial f/\partial x$, $b = \partial f/\partial y$ (evaluated at (x_0, y_0)). Finally, we may determine c from the fact that $z = f(x_0, y_0)$ when $x = x_0$, $y = y_0$. Thus we get

$$z = f(x_0, y_0) + \frac{\partial f}{\partial x}(x_0, y_0)(x - x_0) + \frac{\partial f}{\partial y}(x_0, y_0)(y - y_0) \qquad (1)$$

which should be the equation of the plane tangent to the graph of f at (x_0, y_0), if f is "smooth enough" (see Figure 2.3.4).

Our definition of differentiability will amount to the plane (1) being a "good" approximation of f near (x_0, y_0). To get an idea of what one might mean by a good approximation, let us return for a moment to one-variable calculus. If f is differentiable at a point x_0 then we know that

$$\operatorname*{limit}_{\Delta x \to 0} \frac{f(x_0 + \Delta x) - f(x_0)}{\Delta x} = f'(x_0)$$

Let us rewrite this as

$$\operatorname*{limit}_{x \to x_0} \frac{f(x) - f(x_0)}{x - x_0} = f'(x_0)$$

where $x = x_0 + \Delta x$.

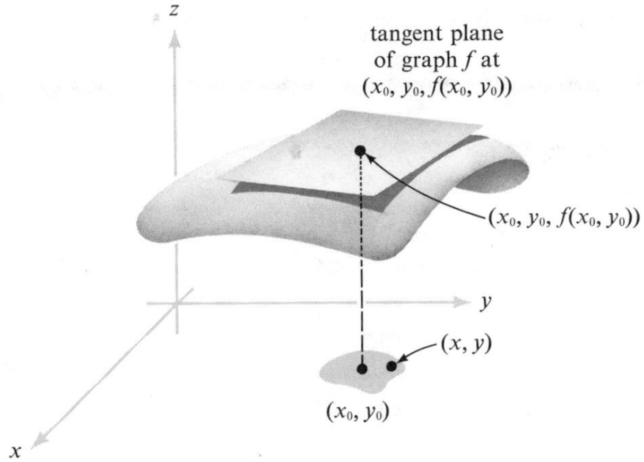

FIGURE 2.3.4
For points (x, y) near (x_0, y_0) the graph of the tangent plane is close to the graph of f.

Since $\lim_{x \to x_0} f'(x_0) = f'(x_0)$ we can rewrite the above equation as

$$\lim_{x \to x_0} \frac{f(x) - f(x_0)}{x - x_0} = \lim_{x \to x_0} f'(x_0)$$

or

$$\lim_{x \to x_0} \left(\frac{f(x) - f(x_0)}{x - x_0} - f'(x_0) \right) = 0$$

or

$$\lim_{x \to x_0} \frac{f(x) - f(x_0) - f'(x_0)(x - x_0)}{x - x_0} = 0$$

Thus the tangent line l through $(x_0, f(x_0))$ with slope $f'(x_0)$ is close to f in the sense that the difference between $f(x)$ and $l(x) = f(x_0) + f'(x_0)(x - x_0)$ goes to zero *even* when divided by $x - x_0$ as x goes to x_0. This is the notion of "good approximation" that we will try to adapt to functions of several variables, with the tangent line replaced by the tangent plane (see equation (1) above).

DEFINITION. *Let $f: \mathbb{R}^2 \to \mathbb{R}$. We say f is **differentiable at** (x_0, y_0), if $\partial f/\partial x$ and $\partial f/\partial y$ exist at (x_0, y_0) **and if***

$$\frac{\left| f(x, y) - f(x_0, y_0) - \dfrac{\partial f}{\partial x}(x_0, y_0)(x - x_0) - \dfrac{\partial f}{\partial y}(x_0, y_0)(y - y_0) \right|}{\|(x, y) - (x_0, y_0)\|} \to 0$$

$$\textit{as } (x, y) \to (x_0, y_0) \qquad (2)$$

This equation expresses what we mean by saying that

$$f(x_0, y_0) + \frac{\partial f}{\partial x}(x_0, y_0)(x - x_0) + \frac{\partial f}{\partial y}(x_0, y_0)(y - y_0)$$

is a ***good approximation*** *to the function f.*

Actually, it is not always easy to use this definition to see if f is differentiable, but we shall easily be able to use another criterion given in Theorem 9 below.

We have used the informal notion of the plane tangent to the graph of a function in order to motivate our definition of differentiability. Now we are ready to adopt a formal definition of the tangent plane.

DEFINITION. *Let $f: \mathbb{R}^2 \to \mathbb{R}$ be differentiable at $\mathbf{x}_0 = (x_0, y_0)$. Then the plane in $\mathbb{R}^3$ defined by equation* (1):

$$z = f(x_0, y_0) + \frac{\partial f}{\partial x}(x_0, y_0)(x - x_0) + \frac{\partial f}{\partial y}(x_0, y_0)(y - y_0)$$

is called the ***tangent plane*** *of the graph of f at the point (x_0, y_0).*

Let us write $Df(x_0, y_0)$ for the row matrix

$$\begin{bmatrix} \dfrac{\partial f}{\partial x}(x_0, y_0) & \dfrac{\partial f}{\partial y}(x_0, y_0) \end{bmatrix}$$

so that the definition of differentiability asserts that

$$f(x_0, y_0) + Df(x_0, y_0)\begin{pmatrix} x - x_0 \\ y - y_0 \end{pmatrix}$$
$$= f(x_0, y_0) + \frac{\partial f}{\partial x}(x_0, y_0)(x - x_0) + \frac{\partial f}{\partial y}(x_0, y_0)(y - y_0) \quad (3)$$

is our good approximation to f near (x_0, y_0). (As above, "good" in the sense that (3) differs from $f(x, y)$ by something small times $\sqrt{(x - x_0)^2 + (y - y_0)^2}$.) We say that (3) is the *best linear approximation* to f near (x_0, y_0).

Now we are ready to give a definition of differentiability for maps f of $\mathbb{R}^n$ to $\mathbb{R}^m$, using the above discussion as motivation. The derivative $Df(\mathbf{x}_0)$ of $f = (f_1, \dots, f_m)$ at a point $\mathbf{x}_0$ is a matrix with elements $t_{ij} = \partial f_i/\partial x_j$ evaluated at $\mathbf{x}_0$.*

* It turns out that we need only postulate the existence of *some* matrix giving the best linear approximation near $\mathbf{x}_0 \in \mathbb{R}^n$, because in fact this matrix is *necessarily* the matrix whose *ij*th entry is $\partial f_i/\partial x_j$ (see Appendix A).

DEFINITION. *Let U be an open set in $\mathbb{R}^n$ and let $f\colon U \subset \mathbb{R}^n \to \mathbb{R}^m$ be a given function. We say that f is* ***differentiable*** *at $\mathbf{x}_0 \in U$ if the partial derivatives of f exist at $\mathbf{x}_0$* ***and*** *if*

$$\lim_{\mathbf{x}\to\mathbf{x}_0} \frac{\|f(\mathbf{x}) - f(\mathbf{x}_0) - T(\mathbf{x} - \mathbf{x}_0)\|}{\|\mathbf{x} - \mathbf{x}_0\|} = 0 \tag{4}$$

where T is the matrix with matrix elements $\partial f_i/\partial x_j$ evaluated at $\mathbf{x}_0$. We call T the ***derivative*** *of f at $\mathbf{x}_0$ and denote it by $Df(\mathbf{x}_0)$.*

We shall always denote the derivative T of f at $\mathbf{x}_0$ by $Df(\mathbf{x}_0)$, although in some books it is denoted $df(\mathbf{x}_0)$ and referred to as the *differential* of f.

In the case $m = 1$, the matrix T is just the row matrix

$$\begin{bmatrix} \dfrac{\partial f}{\partial x_1}(\mathbf{x}_0) & \cdots & \dfrac{\partial f}{\partial x_n}(\mathbf{x}_0) \end{bmatrix}$$

(Sometimes, when there is danger of confusion, we separate the entries by commas.) Setting $n = 2$ and putting this fact back in equation (4) we see that conditions (2) and (4) do agree. Thus, letting $\mathbf{h} = \mathbf{x} - \mathbf{x}_0$, a real-valued function f of n variables is differentiable at a point $\mathbf{x}_0$ if

$$\lim_{\mathbf{h}\to 0} \frac{1}{\|\mathbf{h}\|} \left| f(\mathbf{x}_0 + \mathbf{h}) - f(\mathbf{x}_0) - \sum_{j=1}^{n} \frac{\partial f}{\partial x_j}(\mathbf{x}_0) h_j \right| = 0$$

since

$$T\mathbf{h} = \sum_{j=1}^{n} h_j \frac{\partial f}{\partial x_j}(\mathbf{x}_0)$$

For the general case of f mapping a subset of $\mathbb{R}^n$ to $\mathbb{R}^m$, the derivative is the $m \times n$ matrix given by

$$Df(\mathbf{x}_0) = \begin{bmatrix} \dfrac{\partial f_1}{\partial x_1} & \cdots & \dfrac{\partial f_1}{\partial x_n} \\ \vdots & & \vdots \\ \dfrac{\partial f_m}{\partial x_1} & \cdots & \dfrac{\partial f_m}{\partial x_n} \end{bmatrix}$$

where $\partial f_i/\partial x_j$ is evaluated at $\mathbf{x}_0$. $Df(\mathbf{x}_0)$ is, appropriately, called the matrix of partials of f at $\mathbf{x}_0$.

EXAMPLE 5. Compute the plane tangent to the graph of $z = x^2 + y^4 + e^{xy}$ at the point (1, 0, 2).

Here we use formula (1), with $x_0 = 1$, $y_0 = 0$, and $z_0 = f(x_0, y_0) = 2$. The partial derivatives are

$$\frac{\partial z}{\partial x} = 2x + ye^{xy} \quad \text{and} \quad \frac{\partial z}{\partial y} = 4y^3 + xe^{xy}$$

at (1, 0, 2), they are 2 and 1, respectively. Thus, by (1), the tangent plane is

$$z = 2(x - 1) + 1(y - 0) + 2$$

that is

$$z = 2x + y$$

DEFINITION. *Consider the special case $f: U \subset \mathbb{R}^n \to \mathbb{R}$. Here $Df(\mathbf{x})$ is a $1 \times n$ matrix*

$$Df(\mathbf{x}) = \begin{bmatrix} \dfrac{\partial f}{\partial x_1} & \cdots & \dfrac{\partial f}{\partial x_n} \end{bmatrix}$$

We form the corresponding vector $(\partial f/\partial x_1, \ldots, \partial f/\partial x_n)$, called the **gradient** *of f and denoted* grad f *or* ∇f.

The geometric significance of the gradient will be discussed in Section 2.5. In terms of inner products, we can write the derivative of f as

$$Df(\mathbf{x})(\mathbf{h}) = \nabla f(\mathbf{x}) \cdot \mathbf{h}.$$

EXAMPLE 6. Let $f: \mathbb{R}^3 \to \mathbb{R}$, $f(x, y, z) = xe^y$. Then

$$\text{grad } f = \left(\frac{\partial f}{\partial x}, \frac{\partial f}{\partial y}, \frac{\partial f}{\partial z}\right) = (e^y, xe^y, 0)$$

From the definition we see that for $f: \mathbb{R}^3 \to \mathbb{R}$,

$$\nabla f = \frac{\partial f}{\partial x}\mathbf{i} + \frac{\partial f}{\partial y}\mathbf{j} + \frac{\partial f}{\partial z}\mathbf{k}$$

while for $f: \mathbb{R}^2 \to \mathbb{R}$

$$\nabla f = \frac{\partial f}{\partial x}\mathbf{i} + \frac{\partial f}{\partial y}\mathbf{j}$$

EXAMPLE 7. (a) If $f: \mathbb{R}^2 \to \mathbb{R}$ is given by $(x, y) \mapsto e^{xy} + \sin xy$, then

$$\begin{aligned} \nabla f(x, y) &= (ye^{xy} + y \cos xy)\mathbf{i} + (xe^{xy} + x \cos xy)\mathbf{j} \\ &= (e^{xy} + \cos xy)(y\mathbf{i} + x\mathbf{j}) \end{aligned}$$

(b) If $f: \mathbb{R}^2 \to \mathbb{R}^2$ is defined by $f(x, y) = (e^{x+y} + y, y^2 x)$, then $f_1(x, y) = e^{x+y} + y$ and $f_2(x, y) = y^2 x$. Hence $Df(x, y)$ is the 2×2 matrix

$$Df(x, y) = \begin{bmatrix} e^{x+y} & e^{x+y} + 1 \\ y^2 & 2xy \end{bmatrix}$$

For the rest of this section we shall discuss some general theorems that extend some of the basic results of one-variable calculus to several variables.

In one-variable calculus it is shown that if f is differentiable then f is continuous. We will state in Theorem 8 that this is also true for differentiable functions of several variables. As we know, there are plenty of functions that are continuous but not differentiable, such as $f(x) = |x|$.

Before stating the result, let us give an example of a function whose partial derivatives exist at a point but which is not continuous at that point.

EXAMPLE 8. Let $f: \mathbb{R}^2 \to \mathbb{R}$ be defined by

$$f(x, y) = \begin{cases} 1 & \text{if} \quad x = 0 \quad \text{or if} \quad y = 0 \\ 0 & \text{otherwise} \end{cases}$$

Since f is constant on the x- and y-axes, where it equals 1,

$$\frac{\partial f}{\partial x}(0, 0) = 0 \quad \text{and} \quad \frac{\partial f}{\partial y}(0, 0) = 0$$

But f is not continuous at $(0, 0)$, because $\displaystyle\lim_{(x, y)\to(0, 0)} f(x, y)$ does not exist.

However, it is possible to conclude somewhat more about differentiable functions, as the following theorem states.

THEOREM 8. *Let $f: U \subset \mathbb{R}^n \to \mathbb{R}^m$ be differentiable at $\mathbf{x}_0 \in U$. Then f is continuous at $\mathbf{x}_0$.*

Consult Appendix A for the proof.

As we have seen, it is usually easy to tell when the partial derivatives of a function exist—we just use what we know from one-variable calculus. However the definition of differentiability looks somewhat complicated and the required approximation condition (4) may seem difficult to verify. Fortunately there is a simple criterion, given in the following theorem, that tells us when a function is differentiable.

THEOREM 9. *Let $f: U \subset \mathbb{R}^n \to \mathbb{R}^m$. Suppose the partial derivatives $\partial f_i/\partial x_j$ of f all exist and are continuous in a neighborhood of a point $\mathbf{x} \in U$. Then f is differentiable at $\mathbf{x}$ (continuity of the partials means that for each i, j the mapping $U \to \mathbb{R}$ given by*

$$\mathbf{x} \mapsto \frac{\partial f_i}{\partial x_j}(\mathbf{x})$$

is continuous).

We shall give the proof in Appendix A. Notice the following hierarchy:

$$\begin{array}{ccccc} & & \text{Theorem 9} & & \text{Definition of derivative} \\ & & \downarrow & & \downarrow \\ \text{Continuous partials} & \Rightarrow & \text{Differentiable} & \Rightarrow & \text{Partials exist} \end{array}$$

Each converse statement, obtained by reversing an implication, is invalid.*

A function whose partial derivatives exist and are continuous is said to be of *class* C^1. Thus Theorem 9 says that *any* C^1 *function is differentiable.*

EXAMPLE 9. Let

$$f(x, y) = \frac{\cos x + e^{xy}}{x^2 + y^2}$$

Show that f is differentiable at all points $(x, y) \neq (0, 0)$.

This is so because, by what we know from Section 2.2, the partials

$$\frac{\partial f}{\partial x} = \frac{(x^2 + y^2)(ye^{xy} - \sin x) - 2x(\cos x + e^{xy})}{(x^2 + y^2)^2}$$

$$\frac{\partial f}{\partial y} = \frac{(x^2 + y^2)xe^{xy} - 2y(\cos x + e^{xy})}{(x^2 + y^2)^2}$$

are continuous except when $x = 0$ and $y = 0$.

EXERCISES

1. Find $\partial f/\partial x$, $\partial f/\partial y$ if
   (a) $f(x, y) = xy$
   (b) $f(x, y) = e^{xy}$
   (c) $f(x, y) = x \cos x \cos y$
   (d) $f(x, y) = (x^2 + y^2)\log(x^2 + y^2)$
2. Evaluate the partial derivatives $\partial z/\partial x$, $\partial z/\partial y$ for the given function at the indicated points.
   (a) $z = \sqrt{a^2 - x^2 - y^2}$; $(0, 0)$, $(a/2, a/2)$
   (b) $z = \log \sqrt{1 + xy}$; $(1, 2)$, $(0, 0)$
   (c) $z = e^{ax} \cos(bx + y)$; $(2\pi/b, 0)$
3. In each case following, find the partial derivatives $\partial w/\partial x$, $\partial w/\partial y$.
   (a) $w = xe^{x^2+y^2}$
   (b) $w = \dfrac{x^2 + y^2}{x^2 - y^2}$

* For a counter-example to the converse of the first implication, use $f(x) = x^2 \sin(1/x)$, $f(0) = 0$; for the second, see Example 2 in Appendix A.

(c) $w = e^{xy} \log(x^2 + y^2)$ (d) $w = x/y$
(e) $w = \cos(ye^{xy})\sin x$

4. Show that each of the following functions is differentiable at each point in its domain. Decide which of the functions are C^1.

(a) $f(x, y) = \dfrac{2xy}{(x^2 + y^2)^2}$ (b) $f(x, y) = \dfrac{x}{y} + \dfrac{y}{x}$

(c) $f(r, \theta) = \frac{1}{2} r \sin 2\theta, r > 0$ (d) $f(x, y) = \dfrac{xy}{\sqrt{x^2 + y^2}}$

(e) $f(x, y) = \dfrac{x^2 y}{x^4 + y^2}$

5. Show that the equation of the plane tangent to the surface $z = x^2 + y^3$ at $(3, 1, 10)$ is $z = 6x + 3y - 11$.

6. Compute the plane tangent to the graphs of the functions in Exercise 1 at the indicated points.
(a) $(0, 0)$ (b) $(0, 1)$ (c) $(0, \pi)$ (d) $(0, 1)$

7. Compute the derivatives of the following functions.
(a) $f\colon \mathbb{R}^2 \to \mathbb{R}^2, f(x, y) = (x, y)$
(b) $f\colon \mathbb{R}^2 \to \mathbb{R}^3, f(x, y) = (xe^y + \cos y, x, x + e^y)$
(c) $f\colon \mathbb{R}^3 \to \mathbb{R}^2, f(x, y, z) = (x + e^z + y, yx^2)$

8. Why should the graphs of $f(x, y) = x^2 + y^2$ and $g(x, y) = -x^2 - y^2 + xy^3$ be tangent at $(0, 0)$?

9. Let $f(x, y) = e^{xy}$. Show that $x(\partial f/\partial x) = y(\partial f/\partial y)$.

10. Approximate $(.99\, e^{.02})^8$ using the expression (1), p. 108, to approximate a suitable function $f(x, y)$.

11. Compute the gradients of the following functions.
(a) $f(x, y, z) = x \exp(-x^2 - y^2 - z^2)$ (Note that $\exp u = e^u$.)

(b) $f(x, y, z) = \dfrac{xyz}{x^2 + y^2 + z^2}$

12. Suppose $f\colon \mathbb{R}^n \to \mathbb{R}^m$ is a linear map. What is the derivative of f?

*13. Describe all Hölder-continuous functions with $\alpha > 1$ (see Exercise 23, Section 2.2). *What is the derivative* of such a function?

2.4 PROPERTIES OF THE DERIVATIVE

In elementary calculus we learned how to differentiate sums, products, quotients, and composite functions. We now want to generalize these ideas to functions of several variables, and we want to pay particular attention to the differentiation of composite functions. The rule for differentiating composites, called the Chain Rule, takes on a more profound form for functions of several variables than for those of one variable. Thus, for example, if f is a real-valued function of one variable, written as $z = f(y)$, and y is a function of x, $y = g(x)$, then z becomes a function of x through substitution, namely, $z = f(g(x))$, and we have the familiar formula

$$\frac{dz}{dx} = \frac{dz}{dy}\frac{dy}{dx}$$

or equivalently

$$z'(x) = f'(g(x))g'(x)$$

If f is a real-valued function of three variables u, v, and w, written in the form $z = f(u, v, w)$, and the variables u, v, w are each functions of x, $u = g(x)$, $v = h(x)$, $w = k(x)$, then by substituting $g(x)$, $h(x)$, $k(x)$ for u, v, and w, z becomes a function of x, $z = f(g(x), h(x), k(x))$, and we get

$$\frac{dz}{dx} = \frac{\partial z}{\partial u}\frac{du}{dx} + \frac{\partial z}{\partial v}\frac{dv}{dx} + \frac{\partial z}{\partial w}\frac{dw}{dx}$$

One of the objects of this section is to explain such formulas in detail.

We begin with the differentiation rules for sums, products, and quotients.

THEOREM 10.

(i) ***Constant Multiple Rule.*** *Let $f\colon U \subset \mathbb{R}^n \to \mathbb{R}^m$ be differentiable at $\mathbf{x}_0$ and let c be a real number. Then $h(\mathbf{x}) = cf(\mathbf{x})$ is differentiable at $\mathbf{x}_0$ and*

$$Dh(\mathbf{x}_0) = cDf(\mathbf{x}_0) \quad \text{(equality of matrices)}$$

(ii) ***Sum Rule.*** *Let $f\colon U \subset \mathbb{R}^n \to \mathbb{R}^m$ and $g\colon U \subset \mathbb{R}^n \to \mathbb{R}^m$ be differentiable at $\mathbf{x}_0$. Then $h(\mathbf{x}) = f(\mathbf{x}) + g(\mathbf{x})$ is differentiable at $\mathbf{x}_0$ and*

$$Dh(\mathbf{x}_0) = Df(\mathbf{x}_0) + Dg(\mathbf{x}_0) \quad \text{(sum of matrices)}$$

(iii) ***Product Rule.*** *Let $f\colon U \subset \mathbb{R}^n \to \mathbb{R}$ and $g\colon U \subset \mathbb{R}^n \to \mathbb{R}$ be differentiable at $\mathbf{x}_0$ and let $h(\mathbf{x}) = g(\mathbf{x})f(\mathbf{x})$. Then $h\colon U \subset \mathbb{R}^n \to \mathbb{R}$ is differentiable at $\mathbf{x}_0$ and*

$$Dh(\mathbf{x}_0) = g(\mathbf{x}_0)Df(\mathbf{x}_0) + f(\mathbf{x}_0)Dg(\mathbf{x}_0)$$

(Note that each side of this equation is an $1 \times n$ matrix; A more general product rule is presented in Exercise 23 at the end of this section.)

(iv) ***Quotient Rule.*** *With the same hypotheses as (iii), let $h(\mathbf{x}) = f(\mathbf{x})/g(\mathbf{x})$ and suppose g is never zero on U. Then h is differentiable at $\mathbf{x}_0$ and*

$$Dh(\mathbf{x}_0) = \frac{g(\mathbf{x}_0)Df(\mathbf{x}_0) - f(\mathbf{x}_0)Dg(\mathbf{x}_0)}{(g(\mathbf{x}_0))^2}$$

Proof. The proofs of (*i*) through (*iv*) proceed almost exactly as in the one-variable case (although with a slight difference in notation). To show this, we shall prove (*i*) and (*ii*), leaving the proofs of (*iii*) and (*iv*) as an exercise (Exercise 15).

(*i*) To show that $Dh(\mathbf{x}_0) = cDf(\mathbf{x}_0)$, we must show

$$\lim_{\mathbf{x}\to\mathbf{x}_0} \frac{\|h(\mathbf{x}) - h(\mathbf{x}_0) - cDf(\mathbf{x}_0)(\mathbf{x} - \mathbf{x}_0)\|}{\|\mathbf{x} - \mathbf{x}_0\|} = 0$$

that is

$$\lim_{\mathbf{x}\to\mathbf{x}_0} \frac{\|cf(\mathbf{x}) - cf(\mathbf{x}_0) - cDf(\mathbf{x}_0)(\mathbf{x} - \mathbf{x}_0)\|}{\|\mathbf{x} - \mathbf{x}_0\|} = 0$$

(see Equation (4) of Section 2.3). This is certainly true since f is differentiable and the constant c can be factored out (see Theorem 3(*i*), Section 2.2).

(*ii*) By the triangle inequality, we may write

$$\frac{\|h(\mathbf{x}) - h(\mathbf{x}_0) - (Df(\mathbf{x}_0) + Dg(\mathbf{x}_0))(\mathbf{x} - \mathbf{x}_0)\|}{\|\mathbf{x} - \mathbf{x}_0\|}$$

$$= \frac{\|f(\mathbf{x}) - f(\mathbf{x}_0) - Df(\mathbf{x}_0)(\mathbf{x} - \mathbf{x}_0) + g(\mathbf{x}) - g(\mathbf{x}_0) - Dg(\mathbf{x}_0)(\mathbf{x} - \mathbf{x}_0)\|}{\|\mathbf{x} - \mathbf{x}_0\|}$$

$$\leq \frac{\|f(\mathbf{x}) - f(\mathbf{x}_0) - Df(\mathbf{x}_0)(\mathbf{x} - \mathbf{x}_0\|}{\|\mathbf{x} - \mathbf{x}_0\|}$$

$$+ \frac{\|g(\mathbf{x}) - g(\mathbf{x}_0) - Dg(\mathbf{x}_0)(\mathbf{x} - \mathbf{x}_0)\|}{\|\mathbf{x} - \mathbf{x}_0\|}$$

and each term approaches 0 as $\mathbf{x} \to \mathbf{x}_0$. Hence (*ii*) holds. ■

EXAMPLE 1. Verify the formula for Dh in (*iv*) above with $f(x, y, z) = x^2 + y^2 + z^2$ and $g(x, y, z) = x^2 + 1$.

Here
$$h(x, y, z) = \frac{x^2 + y^2 + z^2}{x^2 + 1}$$

so that directly

$$\begin{aligned} Dh(x, y, z) &= \left[\frac{\partial h}{\partial x}, \frac{\partial h}{\partial y}, \frac{\partial h}{\partial z}\right] \\ &= \left[\frac{(x^2 + 1)2x - (x^2 + y^2 + z^2)2x}{(x^2 + 1)^2}, \frac{2y}{x^2 + 1}, \frac{2z}{x^2 + 1}\right] \\ &= \left[\frac{2x(1 - y^2 - z^2)}{(x^2 + 1)^2}, \frac{2y}{x^2 + 1}, \frac{2z}{x^2 + 1}\right] \end{aligned}$$

By the formula in (*iv*) we get

$$Dh = \frac{gDf - fDg}{g^2} = \frac{(x^2 + 1)[2x, 2y, 2z] - (x^2 + y^2 + z^2)[2x, 0, 0]}{(x^2 + 1)^2}$$

which is the same as what we obtained directly.

As we mentioned above, it is in the differentiation of composite functions that we meet apparently substantial alterations of the one-variable formula. However if we use the D notation, that is, matrix notation, the several-variable Chain Rule looks similar to the one-variable rule. Recall (see Example 4, Section 2.3) that it was the nonvalidity of the Chain Rule that forced us to change our provisional definition of differentiability.

THEOREM 11 (CHAIN RULE). *Let $U \subset \mathbb{R}^n$ and $V \subset \mathbb{R}^m$ be open. Let $g: U \subset \mathbb{R}^n \to \mathbb{R}^m$ and $f: V \subset \mathbb{R}^m \to \mathbb{R}^p$ be given functions such that g maps U into V, so that $f \circ g$ is defined. Supposed g is differentiable at $\mathbf{x}_0$ and f is differentiable at $\mathbf{y}_0 = g(\mathbf{x}_0)$. Then $f \circ g$ is differentiable at $\mathbf{x}_0$ and*

$$D(f \circ g)(\mathbf{x}_0) = Df(\mathbf{y}_0)Dg(\mathbf{x}_0) \tag{1}$$

The right-hand side is a matrix product.

We shall now give a proof of the chain rule *under the additional assumption that the partial derivatives of f are continuous*, building up to the general case by developing two special cases that are themselves important. (The complete proof of Theorem 11 without the additional assumption of continuity is given in Appendix A.)

FIRST SPECIAL CASE OF THE CHAIN RULE. *Let $\mathbf{c}: \mathbb{R} \to \mathbb{R}^3$ and $f: \mathbb{R}^3 \to \mathbb{R}$. Let $h(t) = f(\mathbf{c}(t)) = f(x(t), y(t), z(t))$ where $\mathbf{c}(t) = (x(t), y(t), z(t))$. Then*

$$\frac{dh}{dt} = \frac{\partial f}{\partial x}\frac{dx}{dt} + \frac{\partial f}{\partial y}\frac{dy}{dt} + \frac{\partial f}{\partial z}\frac{dz}{dt} \tag{2}$$

That is,

$$\frac{dh}{dt} = \nabla f(\mathbf{c}(t)) \cdot \mathbf{c}'(t)$$

where $\mathbf{c}'(t) = (x'(t), y'(t), z'(t))$.

This is indeed the special case of Theorem 11 in which we take $\mathbf{c} = g$, f to be real-valued and $m = 3$. Notice that

$$\nabla f(\mathbf{c}(t)) \cdot \mathbf{c}'(t) = Df(\mathbf{c}(t))D\mathbf{c}(t)$$

where we regard $Df(\mathbf{c}(t))$ as a row matrix and $D\mathbf{c}(t)$ as a column matrix. The vectors $\nabla f(\mathbf{c}(t))$ and $\mathbf{c}'(t)$ have the same components as their matrix equivalents; the notational change indicates the switch from matrices to vectors. The following diagrams may help the reader understand this relationship:

Proof of equation (2). By definition,

$$\frac{dh}{dt}(t_0) = \lim_{t \to t_0} \frac{h(t) - h(t_0)}{t - t_0}$$

We write

$$\begin{aligned}\frac{h(t) - h(t_0)}{t - t_0} &= \frac{f(x(t), y(t), z(t)) - f(x(t_0), y(t_0), z(t_0))}{t - t_0} \\ &= \frac{f(x(t), y(t), z(t)) - f(x(t_0), y(t), z(t))}{t - t_0} \\ &\quad + \frac{f(x(t_0), y(t), z(t)) - f(x(t_0), y(t_0), z(t))}{t - t_0} \\ &\quad + \frac{f(x(t_0), y(t_0), z(t)) - f(x(t_0), y(t_0), z(t_0))}{t - t_0}\end{aligned}$$

Now we invoke the *Mean Value Theorem* from one-variable calculus, which states: *if $g: [a, b] \to \mathbb{R}$ is continuous and is differentiable on the open interval $]a, b[$, then there is a point c in $]a, b[$ such that $g(b) - g(a) = g'(c)(b - a)$.*

Thus, by applying the mean value theorem to f as a function of x, we can assert that for some c between x and x_0,

$$f(x, y, z) - f(x_0, y, z) = \frac{\partial f}{\partial x}(c, y, z)(x - x_0)$$

In this way, we find that

$$\begin{aligned}\frac{h(t) - h(t_0)}{t - t_0} &= \frac{\partial f}{\partial x}(c, y(t), z(t)) \frac{x(t) - x(t_0)}{t - t_0} \\ &\quad + \frac{\partial f}{\partial y}(x(t), d, z(t)) \frac{y(t) - y(t_0)}{t - t_0} \\ &\quad + \frac{\partial f}{\partial z}(x(t), y(t), e) \frac{z(t) - z(t_0)}{t - t_0}\end{aligned}$$

where c; d; and e lie between $x(t)$, $x(t_0)$; $y(t)$, $y(t_0)$; and $z(t)$, $z(t_0)$ respectively. By taking the limit $t \to t_0$ and using the continuity of the partials $\frac{\partial f}{\partial x}, \frac{\partial f}{\partial y}, \frac{\partial f}{\partial z}$ and the fact that c, d, and e converge to $x(t_0)$, $y(t_0)$, and $z(t_0)$ respectively, we obtain formula (2). ■

SECOND SPECIAL CASE OF THE CHAIN RULE. *Let $f: \mathbb{R}^3 \to \mathbb{R}$ and let $g: \mathbb{R}^3 \to \mathbb{R}^3$. Write $g(x, y, z) = (u(x, y, z), v(x, y, z), w(x, y, z))$ and let $h(x, y, z) = f(u(x, y, z), v(x, y, z), w(x, y, z))$. Then*

$$\begin{bmatrix} \frac{\partial h}{\partial x} & \frac{\partial h}{\partial y} & \frac{\partial h}{\partial z} \end{bmatrix} = \begin{bmatrix} \frac{\partial f}{\partial u} & \frac{\partial f}{\partial v} & \frac{\partial f}{\partial w} \end{bmatrix} \begin{bmatrix} \frac{\partial u}{\partial x} & \frac{\partial u}{\partial y} & \frac{\partial u}{\partial z} \\ \frac{\partial v}{\partial x} & \frac{\partial v}{\partial y} & \frac{\partial v}{\partial z} \\ \frac{\partial w}{\partial x} & \frac{\partial w}{\partial y} & \frac{\partial w}{\partial z} \end{bmatrix} \tag{3}$$

In this special case we have taken $n = m = 3$ and $p = 1$ for concreteness, $U = \mathbb{R}^3$ and $V = \mathbb{R}^3$ for simplicity, and have written out the matrix product $Df(\mathbf{y}_0) \cdot Dg(\mathbf{x}_0)$ explicitly (with the arguments $\mathbf{x}_0$ and $\mathbf{y}_0$ suppressed in the matrices).

Proof of the Second Special Case of the Chain Rule. By definition $\frac{\partial h}{\partial x}$ is obtained by differentiating h with respect to x, holding y and z fixed. But then $(u(x, y, z), v(x, y, z), w(x, y, z))$ may be regarded as a vector function of the single variable x. The first special case applies to this situation and, after renaming the variables, gives

$$\frac{\partial h}{\partial x} = \frac{\partial f}{\partial u}\frac{\partial u}{\partial x} + \frac{\partial f}{\partial v}\frac{\partial v}{\partial x} + \frac{\partial f}{\partial w}\frac{\partial w}{\partial x} \tag{3$'$}$$

Similarly

$$\frac{\partial h}{\partial y} = \frac{\partial f}{\partial u}\frac{\partial u}{\partial y} + \frac{\partial f}{\partial v}\frac{\partial v}{\partial y} + \frac{\partial f}{\partial w}\frac{\partial w}{\partial y} \tag{3$''$}$$

and

$$\frac{\partial h}{\partial z} = \frac{\partial f}{\partial u}\frac{\partial u}{\partial z} + \frac{\partial f}{\partial v}\frac{\partial v}{\partial z} + \frac{\partial f}{\partial w}\frac{\partial w}{\partial z}. \tag{3$'''$}$$

These equations are exactly what would be obtained by multiplying out the matrices in equation (3). ■

Proof of Theorem 11. The general case in equation (1) may be proved by (a) generalizing equation (2) to m variables, that is, for $f(x_1, \ldots, x_m)$ and $\mathbf{c}(t) = (x_1(t), \ldots, x_m(t))$, one has

$$\frac{dh}{dt} = \sum_{i=1}^{m} \frac{\partial f}{\partial x_i} \frac{dx_i}{dt}$$

where $h(t) = f(x_1(t), \ldots, x_m(t))$; and (b) applying (a) to obtain the formula

$$\frac{\partial h_j}{\partial x_i} = \sum_{k=1}^{m} \frac{\partial f_j}{\partial y_k} \frac{\partial y_k}{\partial x_i}$$

where $f = (f_1, \ldots, f_p)$ is a vector function of arguments $y_1, \ldots, y_m$, $g(x_1, \ldots, x_n) = (y_1(x_1, \ldots, x_n), \ldots, y_m(x_1, \ldots, x_n))$, and $h_j(x_1, \ldots, x_n) = f_j(y_1(x_1, \ldots, x_n), \ldots, y_m(x_1, \ldots, x_n))$. (Using the letter y for both functions and arguments is an abuse of notation, but it can help one remember the formula.)

This formula is equivalent to formula (1) after the matrices are multiplied out.

Since the proof of steps (a) and (b) closely resembles those just presented, we can omit the details. This completes the proof of formula (1) and hence of Theorem 11. ■

The pattern of the Chain Rule will become clear once the student has worked some additional examples. For instance,

$$\frac{\partial}{\partial x} f(u(x, y), v(x, y), w(x, y), z(x, y)) = \frac{\partial f}{\partial u}\frac{\partial u}{\partial x} + \frac{\partial f}{\partial v}\frac{\partial v}{\partial x} + \frac{\partial f}{\partial w}\frac{\partial w}{\partial x} + \frac{\partial f}{\partial z}\frac{\partial z}{\partial x}$$

with a similar formula for $\partial f/\partial y$.

The map $\mathbf{c}$ in the first special case of the Chain Rule represents a curve (Figure 2.4.1) and $\mathbf{c}'(t)$ can be thought of as a tangent vector (or velocity vector) of the curve. Although this idea is studied in greater detail in Chapter 3, we can indicate here the reason for this interpretation. Using the definition of the derivative of a function of a single variable, we see that

$$\mathbf{c}'(t) = \operatorname*{limit}_{h \to 0} \frac{\mathbf{c}(t + h) - \mathbf{c}(t)}{h}$$

The quotient represents a secant that approximates a tangent vector as $h \to 0$ (see Figure 2.4.2).

The Chain Rule can help us understand the relationship between the geometry of a mapping $f\colon \mathbb{R}^2 \to \mathbb{R}^2$ and the geometry of curves in $\mathbb{R}^2$. (Similar statements may be made about $\mathbb{R}^3$ or, generally, $\mathbb{R}^n$.) If $\mathbf{c}(t)$ is a curve in the plane, then $\mathbf{c}'(t)$ represents the tangent (or velocity) vector of the curve $\mathbf{c}(t)$, and, as demonstrated in Figure 2.4.1, this tangent (or velocity) vector

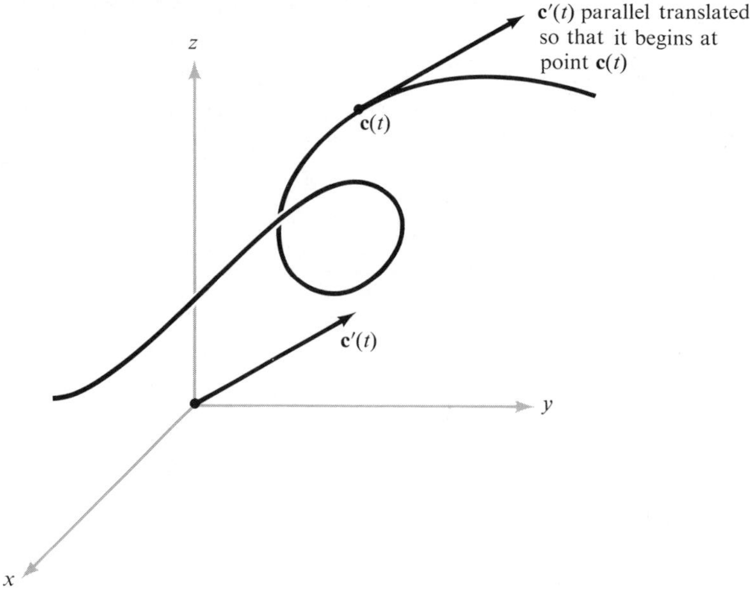

FIGURE 2.4.1

The vector $\mathbf{c}'(t)$ *represents the tangent vector (or velocity vector) of the curve* $\mathbf{c}(t)$.

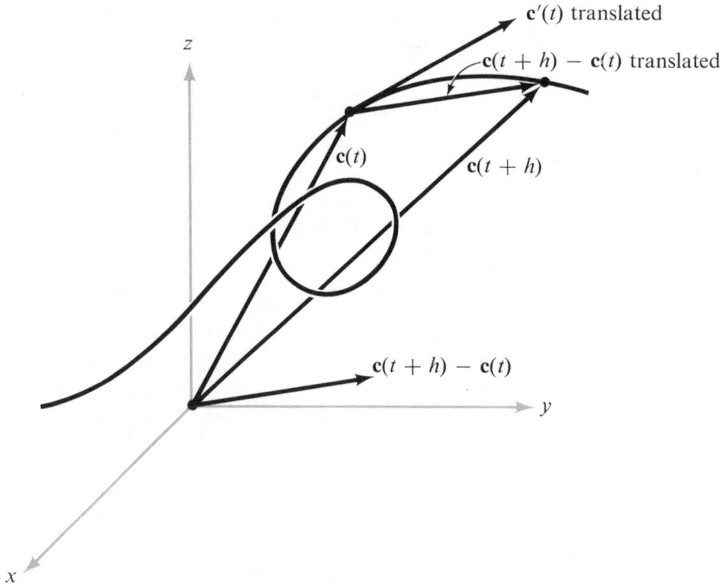

FIGURE 2.4.2

Geometry associated with the formula

$$\lim_{h\to 0} \frac{(\mathbf{c}(t+h) - \mathbf{c}(t)}{h} = \mathbf{c}'(t).$$

can be thought of as beginning at $\mathbf{c}(t)$. Now let $\boldsymbol{\sigma}(t) = f(\mathbf{c}(t))$ where $f\colon \mathbb{R}^2 \to \mathbb{R}^2$. The curve $\boldsymbol{\sigma}$ represents the image of the curve $\mathbf{c}(t)$ under the mapping f. The tangent vector to $\boldsymbol{\sigma}$ is given by the Chain Rule:

$$\boldsymbol{\sigma}'(t) = \underbrace{\nabla f(\mathbf{c}(t))}_{\text{row vector}} \overset{\text{dot product}}{\cdot} \underbrace{\mathbf{c}'(t)}_{\text{row matrix}} = \underbrace{Df(\mathbf{c}(t))}_{\text{row vector}} \overset{\text{matrix multiplication}}{\,} \underbrace{D\mathbf{c}(t)}_{\text{column matrix}}$$

In other words, *the derivative matrix of f maps the tangent (or velocity) vector of a curve to the tangent (or velocity) vector of the corresponding image curve* (see Figure 2.4.3). Thus, points are mapped by f, while tangent vectors to curves are mapped by the derivative of f, evaluated at the base point of the tangent vector in the domain.

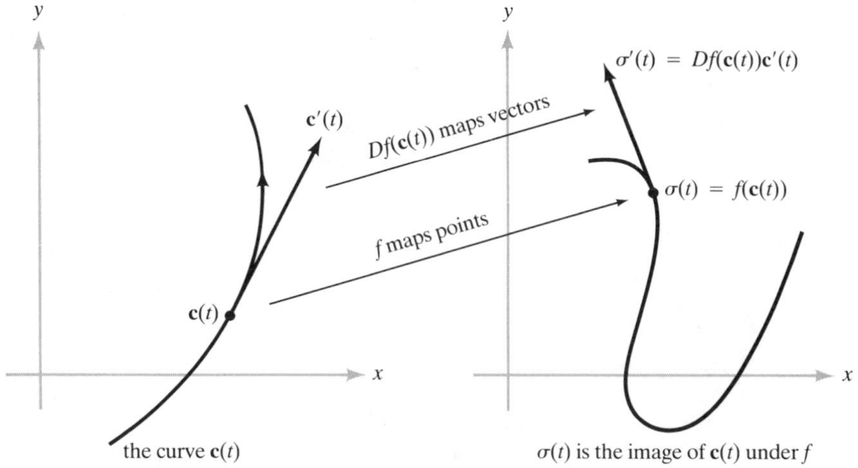

FIGURE 2.4.3
Tangent vectors are mapped by the derivative matrix.

EXAMPLE 2. Compute a tangent vector to the curve $\mathbf{c}(t) = (t, t^2, e^t)$ at $t = 0$.
Here $\mathbf{c}'(t) = (1, 2t, e^t)$, so at $t = 0$ a tangent vector is $(1, 0, 1)$.

EXAMPLE 3. Verify the Chain Rule in the form of formula (3′) for

$$f(u, v, w) = u^2 + v^2 - w$$

$$u(x, y, z) = x^2 y$$

$$v(x, y, z) = y^2$$

$$w(x, y, z) = e^{-xz}$$

Here

$$\begin{aligned} h(x, y, z) &= f(u(x, y, z), v(x, y, z), w(x, y, z)) \\ &= (x^2y)^2 + y^4 - e^{-xz} \\ &= x^4y^2 + y^4 - e^{-xz} \end{aligned}$$

Thus, differentiating directly,

$$\frac{\partial h}{\partial x} = 4x^3y^2 + ze^{-xz}$$

On the other hand, using the Chain Rule,

$$\begin{aligned} \frac{\partial h}{\partial x} = \frac{\partial f}{\partial u}\frac{\partial u}{\partial x} + \frac{\partial f}{\partial v}\frac{\partial v}{\partial x} + \frac{\partial f}{\partial w}\frac{\partial w}{\partial x} &= 2u(2xy) + 2v \cdot 0 + ze^{-xz} \\ &= (2x^2y)(2xy) + ze^{-xz} \end{aligned}$$

which is the same as the above.

Example 4. Given $g(x, y) = (x^2 + 1, y^2)$ and $f(u, v) = (u + v, u, v^2)$, compute the derivative of $f \circ g$ at (1, 1) using the Chain Rule.

Indeed,

$$Df(u, v) = \begin{bmatrix} \dfrac{\partial f_1}{\partial u} & \dfrac{\partial f_1}{\partial v} \\ \dfrac{\partial f_2}{\partial u} & \dfrac{\partial f_2}{\partial v} \\ \dfrac{\partial f_3}{\partial u} & \dfrac{\partial f_3}{\partial v} \end{bmatrix} = \begin{bmatrix} 1 & 1 \\ 1 & 0 \\ 0 & 2v \end{bmatrix} \quad \text{and} \quad Dg(x, y) = \begin{bmatrix} 2x & 0 \\ 0 & 2y \end{bmatrix}$$

When $(x, y) = (1, 1)$, $g(x, y) = (u, v) = (2, 1)$. Hence

$$D(f \circ g)(1, 1) = Df(2, 1)Dg(1, 1) = \begin{bmatrix} 1 & 1 \\ 1 & 0 \\ 0 & 2 \end{bmatrix}\begin{bmatrix} 2 & 0 \\ 0 & 2 \end{bmatrix} = \begin{bmatrix} 2 & 2 \\ 2 & 0 \\ 0 & 4 \end{bmatrix}$$

is the required derivative.

Example 5. Let $f(x, y)$ be given and make the substitution $x = r\cos\theta$, $y = r\sin\theta$ (polar coordinates). Write a formula for $\partial f/\partial\theta$.

By the Chain Rule

$$\frac{\partial f}{\partial \theta} = \frac{\partial f}{\partial x}\frac{\partial x}{\partial \theta} + \frac{\partial f}{\partial y}\frac{\partial y}{\partial \theta}$$

that is

$$\frac{\partial f}{\partial \theta} = -r\sin\theta\,\frac{\partial f}{\partial x} + r\cos\theta\,\frac{\partial f}{\partial y}$$

which is the formula sought.

OPTIONAL

EXAMPLE 6. Let $f\colon U \subset \mathbb{R}^n \to \mathbb{R}^m$ be differentiable, with $f = (f_1, \ldots, f_m)$, and let $g(\mathbf{x}) = \sin(f(\mathbf{x}) \cdot f(\mathbf{x}))$. Compute $Dg(\mathbf{x})$.

By the Chain Rule, $Dg(\mathbf{x}) = \cos(f(\mathbf{x}) \cdot f(\mathbf{x}))Dh(\mathbf{x})$ where $h(\mathbf{x}) = (f(\mathbf{x}) \cdot f(\mathbf{x})) = f_1^2(\mathbf{x}) + \cdots + f_m^2(\mathbf{x})$. Then

$$\begin{aligned} Dh(\mathbf{x}) &= \left[\frac{\partial h}{\partial x_1} \cdots \frac{\partial h}{\partial x_n}\right] \\ &= \left[2f_1 \frac{\partial f_1}{\partial x_1} + \cdots + 2f_m \frac{\partial f_m}{\partial x_1} \quad \cdots \quad 2f_1 \frac{\partial f_1}{\partial x_n} + \cdots + 2f_m \frac{\partial f_m}{\partial x_n}\right] \end{aligned}$$

which can be written as $2f(\mathbf{x})Df(\mathbf{x})$, where we regard f as a row matrix,

$$f = [f_1 \cdots f_m] \quad \text{and} \quad Df = \begin{bmatrix} \dfrac{\partial f_1}{\partial x_1} & \cdots & \dfrac{\partial f_1}{\partial x_n} \\ \vdots & & \vdots \\ \dfrac{\partial f_m}{\partial x_1} & \cdots & \dfrac{\partial f_m}{\partial x_n} \end{bmatrix}$$

Thus $Dg(\mathbf{x}) = 2(\cos(f(\mathbf{x}) \cdot f(\mathbf{x}))f(\mathbf{x})Df(\mathbf{x})$.

EXERCISES

1. If $f\colon U \subset \mathbb{R}^n \to \mathbb{R}$ is differentiable, prove that $\mathbf{x} \mapsto f^2(\mathbf{x}) + 2f(\mathbf{x})$ is differentiable as well, and compute its derivative in terms of $Df(\mathbf{x})$.
2. Prove that the following functions are differentiable, and find their derivatives at an arbitrary point.
   (a) $f\colon \mathbb{R}^2 \to \mathbb{R}, (x, y) \mapsto 2$
   (b) $f\colon \mathbb{R}^2 \to \mathbb{R}, (x, y) \mapsto x + y$
   (c) $f\colon \mathbb{R}^2 \to \mathbb{R}, (x, y) \mapsto 2 + x + y$
   (d) $f\colon \mathbb{R}^2 \to \mathbb{R}, (x, y) \mapsto x^2 + y^2$
   (e) $f\colon \mathbb{R}^2 \to \mathbb{R}, (x, y) \mapsto e^{xy}$
   (f) $f\colon U \to \mathbb{R}, (x, y) \mapsto \sqrt{1 - x^2 - y^2}, U = \{(x, y) | x^2 + y^2 < 1\}$
   (g) $f\colon \mathbb{R}^2 \to \mathbb{R}, (x, y) \mapsto x^4 - y^4$
3. Write out the Chain Rule for each of the following functions.
   (a) $\dfrac{\partial h}{\partial x}$ where $h(x, y) = f(x, u(x, y))$
   (b) $\dfrac{dh}{dx}$ where $h(x) = f(x, u(x), v(x))$
   (c) $\dfrac{\partial h}{\partial x}$ where $h(x, y, z) = f(u(x, y, z), v(x, y), w(x))$

   Now justify your answer in each case by using Theorem 11.

4. Verify the Chain Rule for $\partial h/\partial x$ where $h(x, y) = f(u(x, y), v(x, y))$ and

$$f(u, v) = \frac{u^2 + v^2}{u^2 - v^2}, \qquad u(x, y) = e^{-x-y}, \qquad v(x, y) = e^{xy}$$

5. Verify the first special case of the Chain Rule for the composition $f \circ \mathbf{c}$ in each of the cases below.
   (a) $f(x, y) = xy$, $\mathbf{c}(t) = (e^t, \cos t)$
   (b) $f(x, y) = e^{xy}$, $\mathbf{c}(t) = (3t^2, t^3)$
   (c) $f(x, y) = (x^2 + y^2)\log\sqrt{x^2 + y^2}$, $\mathbf{c}(t) = (e^t, e^{-t})$
   (d) $f(x, y) = x\exp(x^2 + y^2)$, $\mathbf{c}(t) = (t, -t)$

6. What is the velocity vector for each curve $\mathbf{c}(t)$ in Exercise 5?

7. Let $f\colon \mathbb{R}^3 \to \mathbb{R}$, $g\colon \mathbb{R}^3 \to \mathbb{R}$ be differentiable. Prove that

$$\nabla(fg) = f\nabla g + g\nabla f$$

8. Let $f\colon \mathbb{R}^3 \to \mathbb{R}$ be differentiable. Making the substitution

$$x = r\cos\theta\sin\phi, \qquad y = r\sin\theta\sin\phi, \qquad z = r\cos\phi$$

(spherical coordinates) into $f(x, y, z)$, compute $\partial f/\partial r$, $\partial f/\partial \theta$, and $\partial f/\partial \phi$.

*9. This exercise gives another example of the fact that the Chain Rule is not applicable if f is not differentiable. Consider the function

$$f(x, y) = \begin{cases} \dfrac{xy^2}{x^2 + y^2} & (x, y) \neq (0, 0) \\ 0 & (x, y) = (0, 0) \end{cases}$$

Show that

(a) $\dfrac{\partial f}{\partial x}(0, 0)$ and $\dfrac{\partial f}{\partial y}(0, 0)$ exist;

(b) if $\mathbf{g}(t) = (at, bt)$ then $f \circ \mathbf{g}$ is differentiable and $(f \circ \mathbf{g})'(0) = ab^2/(a^2 + b^2)$, but $\nabla f(0, 0) \cdot \mathbf{g}'(0) = 0$.

10. Find $(\partial/\partial s)(f \circ T)(0, 0)$, where $f(u, v) = \cos u \sin v$ and $T(s, t) = (\cos t^2 s, \log\sqrt{1 + s^2})$.

11. (a) Let $y(x)$ be defined implicitly by $G(x, y(x)) = 0$, where G is a given function of two variables. Prove that if $y(x)$ and G are differentiable, then

$$\frac{dy}{dx} = -\frac{\partial G/\partial x}{\partial G/\partial y} \quad \text{if} \quad \frac{\partial G}{\partial y} \neq 0$$

(b) Obtain a formula analogous to that in (a) if y_1, y_2 are defined implicitly by

$$G_1(x, y_1(x), y_2(x)) = 0$$

$$G_2(x, y_1(x), y_2(x)) = 0$$

(c) Let y be defined implicitly by

$$x^2 + y^3 + e^y = 0$$

Compute dy/dx in terms of x and y.

12. Let $f: \mathbb{R}^n \to \mathbb{R}^m$ be a linear mapping so that by Exercise 12, Section 2.3 $Df(\mathbf{x})$ is the matrix of f. Check the validity of the Chain Rule directly for linear mappings.

*13. Prove that if $f: U \subset \mathbb{R}^n \to \mathbb{R}$ is differentiable at $\mathbf{x}_0 \in U$, there is a neighborhood V of $\mathbf{0} \in \mathbb{R}^n$ and a function $R_1: V \to \mathbb{R}$ such that for all $\mathbf{h} \in V$, $\mathbf{x}_0 + \mathbf{h} \in U$,

$$f(\mathbf{x}_0 + \mathbf{h}) = f(\mathbf{x}_0) + Df(\mathbf{x}_0) \cdot \mathbf{h} + R_1(\mathbf{h})$$

and

$$R_1(\mathbf{h})/\|\mathbf{h}\| \to 0 \quad \text{as} \quad \mathbf{h} \to \mathbf{0}$$

*14. For what integers $p > 0$ is

$$f(x) = \begin{cases} x^p \sin(1/x) & x \neq 0 \\ 0 & x = 0 \end{cases}$$

differentiable? For what p is the derivative continuous?

*15. Prove (*iii*) and (*iv*) of Theorem 10. (HINT: Use the same addition and subtraction tricks as in the one-variable case and Theorem 8.)

16. Let $f(x, y) = 1/\sqrt{x^2 + y^2}$. Compute $\nabla f(x, y)$. In what direction is this vector pointing?

17. Dieteriei's equation of state for a gas is

$$P(V - b)e^{a/RVT} = RT$$

where a, b, and R are constants. Regard volume V as a function of temperature T and pressure P and prove that

$$\frac{\partial V}{\partial T} = \left(R + \frac{a}{TV}\right) \bigg/ \left(\frac{RT}{V - b} - \frac{a}{V^2}\right)$$

18. Thermodynamics texts* use the relationship

$$\left(\frac{\partial y}{\partial x}\right)\left(\frac{\partial z}{\partial y}\right)\left(\frac{\partial x}{\partial z}\right) = -1$$

Explain the meaning of this equation and prove that it is true. (HINT: Start with a relationship $F(x, y, z) = 0$ that defines $x = f(y, z)$, $y = g(x, z)$, and $z = h(x, y)$ and differentiate implicitly.)

19. Let $f: \mathbb{R}^2 \to \mathbb{R}^2$; $(x, y) \mapsto (e^{x+y}, e^{x-y})$. Let $\mathbf{c}(t)$ be a curve with $\mathbf{c}(0) = (0, 0)$ and $\mathbf{c}'(0) = (1, 1)$. What is the tangent vector to the image of $\mathbf{c}(t)$ under f at $t = 0$?

*20. Suppose $\mathbf{x}_0 \in \mathbb{R}^n$ and $0 \leq r_1 < r_2$. Show there is a C^1 function $f: \mathbb{R}^n \to \mathbb{R}$ such that $f(\mathbf{x}) = 0$ for $\|\mathbf{x} - \mathbf{x}_0\| \geq r_2$; $0 < f(\mathbf{x}) < 1$ for $r_1 < \|\mathbf{x} - \mathbf{x}_0\| < r_2$; and $f(\mathbf{x}) = 1$ for $\|\mathbf{x} - \mathbf{x}_0\| \leq r_1$.
(HINT: Apply a cubic polynomial with $g(r_1^2) = 1$ and $g(r_2^2) = g'(r_2^2) = g'(r_1^2) = 0$ to $\|\mathbf{x} - \mathbf{x}_0\|^2$ when $r_1 < \|\mathbf{x} - \mathbf{x}_0\| < r_2$.)

*21. Find a C^1 mapping $f: \mathbb{R}^3 \to \mathbb{R}^3$ that takes the vector $\mathbf{i} + \mathbf{j} + \mathbf{k}$ emanating from the origin to $\mathbf{i} - \mathbf{j}$ emanating from $(1, 1, 0)$ and takes $\mathbf{k}$ emanating from $(1, 1, 0)$ to $\mathbf{k} - \mathbf{i}$ emanating from the origin.

* See S. M. Binder, "Mathematical Methods in Elementary Thermodynamics," *J. Chem. Educ.* 43 (1966):85–92. A proper understanding of partial differentiation can be of significant use in applications; for example, see M. Feinberg, "Constitutive equation for ideal gas mixtures and ideal solutions as consequences of simple postulates" *Chem. Eng. Sci.* 32 (1977): 75–78.

22. What is wrong with the following argument? Suppose $w = f(x, y, z)$ and $z = g(x, y)$. Then by the Chain Rule,

$$\frac{\partial w}{\partial x} = \frac{\partial w}{\partial x} + \frac{\partial w}{\partial z}\frac{\partial z}{\partial x}$$

Hence

$$0 = \frac{\partial w}{\partial z}\frac{\partial z}{\partial x}$$

so $\dfrac{\partial w}{\partial z} = 0$ or $\dfrac{\partial z}{\partial x} = 0$, which is, in general, absurd.

23. Suppose $f: \mathbb{R}^n \to \mathbb{R}$ and $g: \mathbb{R}^n \to \mathbb{R}^m$ are differentiable. Show that the product function $h(\mathbf{x}) = f(\mathbf{x})g(\mathbf{x})$ from $\mathbb{R}^n$ to $\mathbb{R}^m$ is differentiable and that if $\mathbf{x}_0$ and $\mathbf{y}$ are in $\mathbb{R}^n$, then $(Dh(\mathbf{x}_0))(\mathbf{y}) = f(\mathbf{x}_0)Dg(\mathbf{x}_0)(\mathbf{y}) + (Df(\mathbf{x}_0))(\mathbf{y}) \cdot g(\mathbf{x}_0)$.

*24. Show that $h: \mathbb{R}^n \to \mathbb{R}^m$ is differentiable if and only if each of the m component functions $h_i: \mathbb{R}^n \to \mathbb{R}$ is differentiable. (HINT: Use the coordinate projection function and the Chain Rule for one implication and consider $(\|h(\mathbf{x}) - h(\mathbf{x}_0) - Dh(\mathbf{x}_0)(\mathbf{x} - \mathbf{x}_0)\|/\|\mathbf{x} - \mathbf{x}_0\|)^2 = \sum_{i=1}^m (h_i(\mathbf{x}) - h_i(\mathbf{x}_0)Dh_i(\mathbf{x}_0)(\mathbf{x} - \mathbf{x}_0))^2/\|\mathbf{x} - \mathbf{x}_0\|^2$ to obtain the other.)

2.5 GRADIENTS AND DIRECTIONAL DERIVATIVES

In Section 2.1 we studied the graphs of real-valued functions. Now we take up this study again, using the methods of calculus. We shall use gradients to obtain a formula for the plane tangent to a level surface. Let us begin by recalling how the gradient is defined.

DEFINITION. *If $f: U \subset \mathbb{R}^3 \to \mathbb{R}$ is differentiable, the* ***gradient*** *of f at (x, y, z) is the vector in space $\mathbb{R}^3$ given by*

$$\text{grad } f = \left(\frac{\partial f}{\partial x}, \frac{\partial f}{\partial y}, \frac{\partial f}{\partial z}\right)$$

This vector is also denoted ∇f or $\nabla f(x, y, z)$. Thus ∇f is just the matrix of the derivative Df, written as a vector.

EXAMPLE 1. Let $f(x, y, z) = \sqrt{x^2 + y^2 + z^2} = r$, the distance from $\mathbf{0}$ to (x, y, z). Then

$$\nabla f(x, y, z) = \left(\frac{\partial f}{\partial x}, \frac{\partial f}{\partial y}, \frac{\partial f}{\partial z}\right)$$

$$= \left(\frac{x}{\sqrt{x^2 + y^2 + z^2}}, \frac{y}{\sqrt{x^2 + y^2 + z^2}}, \frac{z}{\sqrt{x^2 + y^2 + z^2}}\right) = \frac{\mathbf{r}}{r}$$

where $\mathbf{r}$ is the point (x, y, z). Thus ∇f is the unit vector in the direction of (x, y, z).

EXAMPLE 2. If $f(x, y, z) = xy + z$, then

$$\nabla f(x, y, z) = \left(\frac{\partial f}{\partial x}, \frac{\partial f}{\partial y}, \frac{\partial f}{\partial z}\right) = (y, x, 1)$$

Suppose $f\colon \mathbb{R}^3 \to \mathbb{R}$ is a real-valued function. Let $\mathbf{v}$ and $\mathbf{x} \in \mathbb{R}^3$ be fixed vectors and consider the function from $\mathbb{R}$ to $\mathbb{R}$ defined by $t \mapsto f(\mathbf{x} + t\mathbf{v})$. The set of points of the form $\mathbf{x} + t\mathbf{v}$, $t \in \mathbb{R}$, is the line L through the point $\mathbf{x}$ parallel to the vector $\mathbf{v}$ (see Figure 2.5.1). Therefore, the function $t \mapsto f(\mathbf{x} + t\mathbf{v})$ represents the function f restricted to the line L. We may ask: How fast are the values of f changing along the line L at the point $\mathbf{x}$? Since the rate of change of a function is given by a derivative we could say that the answer to this question is the value of the derivative of this function of t at $t = 0$ (when $t = 0$, $\mathbf{x} + t\mathbf{v}$ reduces to $\mathbf{x}$). This would be the derivative of f at the point $\mathbf{x}$ in the direction of L, that is, of $\mathbf{v}$. We can formalize this concept as follows.

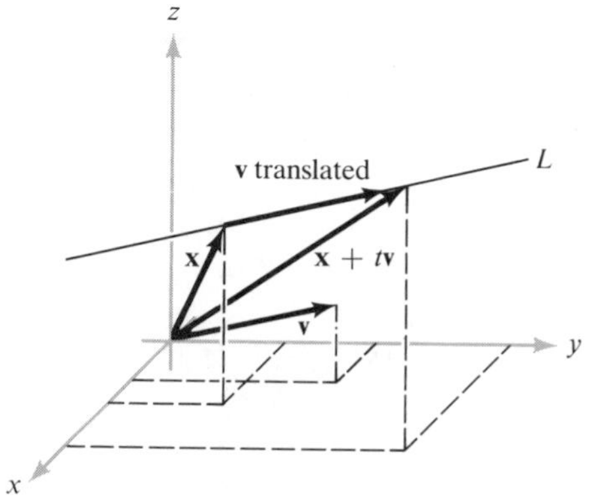

FIGURE 2.5.1
The equation of L is $\mathbf{l}(t) = \mathbf{x} + t\mathbf{v}$.

DEFINITION. *If* $f\colon \mathbb{R}^3 \to \mathbb{R}$, *the* ***directional derivative*** *of* f *at* $\mathbf{x}$ *in the direction of a unit vector* $\mathbf{v}$ *is given by*

$$\left.\frac{d}{dt} f(\mathbf{x} + t\mathbf{v})\right|_{t=0}$$

if this exists.

From the definition we can see that the directional derivative can also be defined by the formula

$$\operatorname*{limit}_{h \to 0} \frac{f(\mathbf{x} + h\mathbf{v}) - f(\mathbf{x})}{h}$$

THEOREM 12. *If $f: \mathbb{R}^3 \to \mathbb{R}$ is differentiable, then all directional derivatives exist. The directional derivative at* **x** *in direction* **v** *is given by*

$$Df(\mathbf{x})\mathbf{v} = \text{grad } f(\mathbf{x}) \cdot \mathbf{v} = \nabla f(\mathbf{x}) \cdot \mathbf{v}$$

$$= \frac{\partial f}{\partial x}(\mathbf{x})v_1 + \frac{\partial f}{\partial y}(\mathbf{x})v_2 + \frac{\partial f}{\partial z}(\mathbf{x})v_3$$

where $\mathbf{v} = (v_1, v_2, v_3)$ *and* $\|\mathbf{v}\| = 1$.

Proof. Let $\mathbf{c}(t) = \mathbf{x} + t\mathbf{v}$, so that $f(\mathbf{x} + t\mathbf{v}) = f(\mathbf{c}(t))$. By the first special case of the Chain Rule, $(d/dt)f(\mathbf{c}(t)) = \nabla f(\mathbf{c}(t)) \cdot \mathbf{c}'(t)$. However, $\mathbf{c}(0) = \mathbf{x}$ and $\mathbf{c}'(0) = \mathbf{v}$, so

$$\left.\frac{d}{dt} f(\mathbf{x} + t\mathbf{v})\right|_{t=0} = \nabla f(\mathbf{x}) \cdot \mathbf{v}$$

as we were required to prove. ■

We should explain why in the definition of directional derivative we have chosen **v** to be a unit vector. There are two reasons. First of all, if α is any positive real number, $\alpha\mathbf{v}$ is a vector that points in the same direction as **v**, but may be either longer (if $\alpha > 1$) or shorter than **v** (if $\alpha < 1$) (see Figure 2.5.2). By Theorem 12 the directional derivative of f in the direction **v** is

$$\nabla f(\mathbf{x}) \cdot \mathbf{v} = \frac{\partial f}{\partial x}(\mathbf{x})v_1 + \frac{\partial f}{\partial y}(\mathbf{x})v_2 + \frac{\partial f}{\partial z}(\mathbf{x})v_3$$

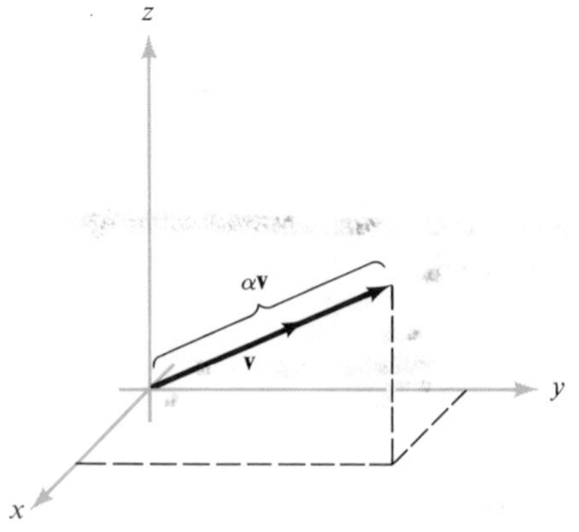

FIGURE 2.5.2
Multiplication of a vector **v** *by a scalar* α *rescales the length of* **v**.

The derivative of f "in the direction" $\alpha\mathbf{v}$ is $\langle\nabla f(\mathbf{x}), \alpha\mathbf{v}\rangle = \alpha\langle\nabla f(\mathbf{x}), \mathbf{v}\rangle$, which is α times the directional derivative in direction $\mathbf{v}$, and therefore not equal to it. So the directional derivative, if defined for *all* $\alpha\mathbf{v}$, would not depend only on a point $\mathbf{x}$ and a direction. To resolve this problem *we have required that the vector* $\mathbf{v}$ *be of standard length*, that is, of length 1. Then the vector $\mathbf{v}$ determines a direction, the same direction determined by $\alpha\mathbf{v}$ if $\alpha > 0$, but now the directional derivative is *uniquely* defined by $\nabla f(\mathbf{x}) \cdot \mathbf{v}$.

Secondly, we may interpret $\nabla f(\mathbf{x}) \cdot \mathbf{v}$ as the rate of change of f in the direction $\mathbf{v}$, for when $\|\mathbf{v}\| = 1$, the point $\mathbf{x} + t\mathbf{v}$ moves a distance s when t is increased by s; so we have, in effect, chosen a scale on L in Figure 2.5.1.

Using Theorem 12, it is easy to compute the directional derivative in terms of the partial derivatives.

EXAMPLE 3. Let $f(x, y, z) = x^2e^{-yz}$. Compute the rate of change of f in the direction

$$\mathbf{v} = \left(\frac{1}{\sqrt{3}}, \frac{1}{\sqrt{3}}, \frac{1}{\sqrt{3}}\right) \qquad \text{at } (1, 0, 0)$$

(Note that $\|\mathbf{v}\| = 1$.)

The required rate of change is

$$\operatorname{grad} f \cdot \mathbf{v} = (2xe^{-yz}, -x^2ze^{-yz}, -x^2ye^{-yz}) \cdot (1/\sqrt{3}, 1/\sqrt{3}, 1/\sqrt{3})$$

which at (1, 0, 0) becomes

$$(2, 0, 0) \cdot (1/\sqrt{3}, 1/\sqrt{3}, 1/\sqrt{3}) = 2/\sqrt{3}$$

From Theorem 12 we can also obtain the geometrical significance of the gradient:

THEOREM 13. *Assume* $\operatorname{grad} f(\mathbf{x}) \neq \mathbf{0}$. *Then* $\operatorname{grad} f(\mathbf{x})$ *points in the direction along which f is increasing the fastest.*

Proof. If $\mathbf{n}$ is a unit vector, the rate of change of f in direction $\mathbf{n}$ is $\operatorname{grad} f(\mathbf{x}) \cdot \mathbf{n} = \|\operatorname{grad} f(\mathbf{x})\| \cos\theta$ where θ is the angle between $\mathbf{n}$ and $\operatorname{grad} f(\mathbf{x})$. This is maximum when $\theta = 0$; that is, when $\mathbf{n}$ and $\operatorname{grad} f$ are parallel. (If $\operatorname{grad} f(\mathbf{x}) = \mathbf{0}$ this rate of change is 0 for any $\mathbf{n}$.) ∎

In other words, if one wishes to move in a direction which f will increase most quickly, one should proceed in the direction $\nabla f(\mathbf{x})$. This idea will be discussed more fully following Example 4.

Now we can find the relationship of the gradient of a function f to the level surfaces of f. The gradient points in the direction in which the values of f change most rapidly whereas a level surface lies in the directions in which they do not change at all. If f is reasonably well behaved, the gradient and the level surface will be perpendicular.

THEOREM 14. *Let $f: \mathbb{R}^3 \to \mathbb{R}$ be a C^1 map and let (x_0, y_0, z_0) lie on the level surface S defined by $f(x, y, z) = k$, for k a constant. Then* grad $f(x_0, y_0, z_0)$ *is normal to the level surface in the following sense: If $\mathbf{v}$ is the tangent vector at $t = 0$ of a path $\mathbf{c}(t)$ in S with $\mathbf{c}(0) = (x_0, y_0, z_0)$, then* grad $f \cdot \mathbf{v} = 0$ *(see Figure 2.5.3)*

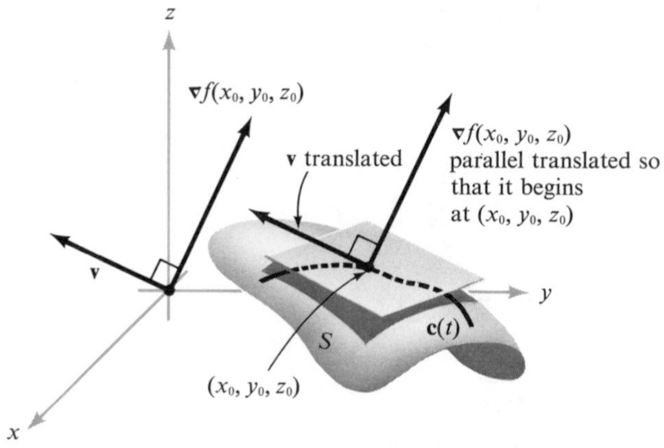

FIGURE 2.5.3

Geometric significance of the gradient: ∇f is orthogonal to the surface S on which f is constant.

Proof. Let $\mathbf{c}(t)$ lie in S; then $f(\mathbf{c}(t)) = k$. Let $\mathbf{v}$ be as in the hypothesis; then $\mathbf{v} = \mathbf{c}'(0)$. Hence

$$\left.\frac{d}{dt} f(\mathbf{c}(t))\right|_{t=0} = 0$$

By the Chain Rule, we get $\frac{d}{dt} f(\mathbf{c}(t)) = Df(x_0, y_0, z_0)\mathbf{v} = 0$, that is, $\nabla f \cdot \mathbf{v} = 0$. ■

If we study the conclusion of Theorem 14, we see that it is not unreasonable to *define* the plane tangent to S as follows:

DEFINITION. *Let S be the surface consisting of those (x, y, z) such that $f(x y, z) = k$, for k a constant. The* **tangent plane** *of S at a point (x_0, y_0, z_0)*

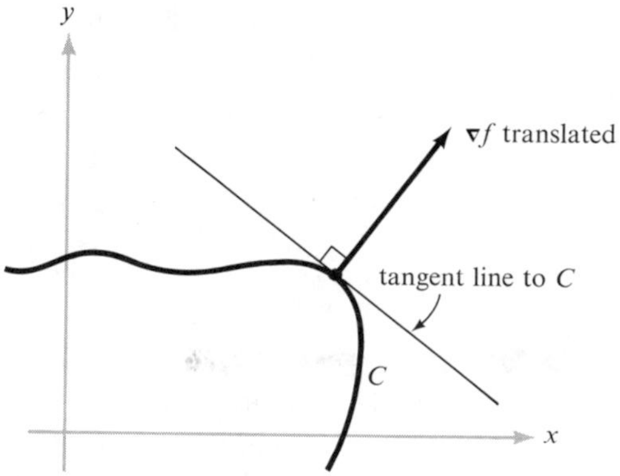

FIGURE 2.5.4
In the plane, the gradient ∇f *is orthogonal to the curve* f = *constant.*

of S is defined by the equation

$$\nabla f(x_0, y_0, z_0) \cdot (x - x_0, y - y_0, z - z_0) = 0 \tag{1}$$

if $\nabla f(x_0, y_0, z_0) \neq \mathbf{0}$. *That is, the tangent plane is the set of points* (x, y, z) *that satisfy (1).*

This extends the definition we gave earlier for the tangent plane of the graph of a function (see Exercise 7 on p. 137).*

In Theorem 14 and the above definition we could have just as well worked in two dimensions as in three. Thus if we have $f: \mathbb{R}^2 \to \mathbb{R}$ and consider a level curve

$$C = \{(x, y) | f(x, y) = k\}$$

then $\nabla f(x_0, y_0)$ is perpendicular to C for any point (x_0, y_0) on C. Likewise, the tangent line to C at (x_0, y_0) has the equation

$$\nabla f(x_0, y_0) \cdot (x - x_0, y - y_0) = 0 \tag{2}$$

if $\nabla f(x_0, y_0) \neq 0$; that is, the tangent line is the set of points (x, y) that satisfy equation (2) (see Figure 2.5.4).

* At this point the reader may want a specific definition of a surface. So far we have used the term "surface" in two ways: first, as the graph of a function of two variables, and second, as the level set of a real valued function of three variables. We also defined the tangent planes to such surfaces. In Section 6.3 we shall discuss parameterized surfaces. At first glance, these various definitions seem to be different. However, in Section 4.4 we show that there is indeed a central concept that unifies these notions, namely, that surfaces are "locally" graphs of functions. Thus the level surface S in Theorem 14 will be shown to be the graph of a function of two variables in the neighborhood of any point (x_0, y_0, z_0) at which $\nabla f(x_0, y_0, z_0) \neq 0$; furthermore, the two definitions of tangent planes will be shown to coincide.

EXAMPLE 4. Compute the equation of the plane tangent to the surface defined by $3xy + z^2 = 4$ at (1, 1, 1).

Here $f(x, y, z) = 3xy + z^2$ and $\nabla f = (3y, 3x, 2z)$ which at (1, 1, 1) is the vector (3, 3, 2). Thus the tangent plane is

$$(x - 1, y - 1, z - 1) \cdot (3, 3, 2) = 0$$

$$3x + 3y + 2z = 8$$

We often speak of ∇f as a *gradient vector field*. Note that ∇f assigns a vector to each point in the domain of f (see Figure 2.5.5). In this figure we describe the function ∇f not by drawing its graph, which would be a subset of $\mathbb{R}^6$, but by representing $\nabla f(\mathrm{P})$, for each point P, as a vector emanating from the point P rather than from the origin. Like a graph, this pictorial method of depicting ∇f contains the point P and the value $\nabla f(\mathrm{P})$ in the same picture.

The gradient vector field has important geometric significance. It shows the direction in which f is increasing the fastest and the direction that is orthogonal to the level surfaces of f. That it does both of these at once is intuitively quite plausible. To see this, imagine a hill as shown in Figure 2.5.6(a). Let h be the height function, a function of two variables. If we draw level curves of h, these are just level contours of the hill. We could imagine them as level paths on the hill (see Figure 2.5.6(b)). One thing should be obvious to anyone who

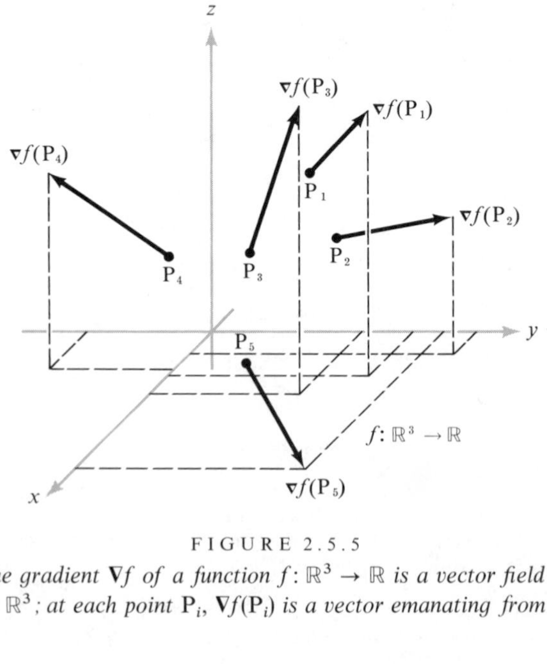

FIGURE 2.5.5

The gradient ∇f of a function $f: \mathbb{R}^3 \to \mathbb{R}$ is a vector field on $\mathbb{R}^3$; at each point P_i, $\nabla f(\mathrm{P}_i)$ is a vector emanating from P_i.

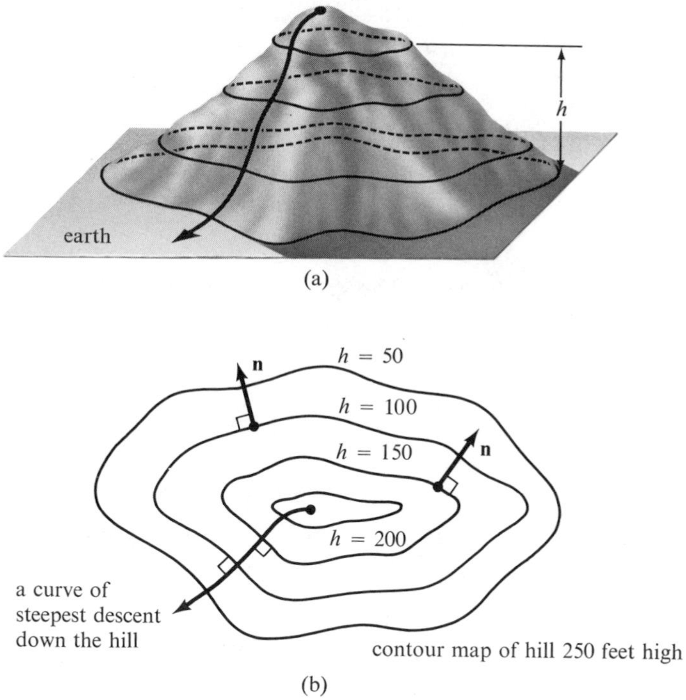

FIGURE 2.5.6
A physical illustration of the two facts (a) ∇f *is the direction of fastest increase of* f *and* (b) ∇f *is orthogonal to the level curves.*

has gone for a hike: To get to the top of the hill the fastest, we should walk perpendicular to level contours. This is consistent with the above theorems, which state that the direction of fastest increase (the gradient) is orthogonal to the level curves.

EXAMPLE 5. The gravitational force on a unit mass at (x, y, z) produced by a mass M at the origin in $\mathbb{R}^3$ is, according to Newton's law of gravitation, given by

$$\mathbf{F} = -\frac{GM}{r^2}\mathbf{n}$$

where G is a constant, $r = \|\mathbf{r}\| = \sqrt{x^2 + y^2 + z^2}$ is the distance of (x, y, z) from the origin, and $\mathbf{n} = \mathbf{r}/r$, the unit vector in the direction of $\mathbf{r} = x\mathbf{i} + y\mathbf{j} + z\mathbf{k}$, the position vector from the origin to (x, y, z).

Let us note that $\mathbf{F} = \nabla(GM/r) = -\nabla V$, that is, $\mathbf{F}$ is minus the gradient of the gravitational potential $V = -GM/r$. This can be verified as in Example 1.

Notice that **F** is directed inwards towards the origin. Also, the level surfaces of V are spheres. **F** is normal to these spheres, which confirms the result of Theorem 14.

EXAMPLE 6. Find a unit vector normal to the surface S given by $z = x^2y^2 + y + 1$ at the point (0, 0, 1).

Let $f(x, y, z) = x^2y^2 + y + 1 - z$, and consider the surface defined by $f(x, y, z) = 0$. Since this is the set of points (x, y, z) with $z = x^2y^2 + y + 1$ we see that this is the surface S. The gradient is given by

$$\begin{aligned}\nabla f(x, y, z) &= \frac{\partial f}{\partial x}\mathbf{i} + \frac{\partial f}{\partial y}\mathbf{j} + \frac{\partial f}{\partial z}\mathbf{k} \\ &= 2xy^2\mathbf{i} + (2x^2y + 1)\mathbf{j} - \mathbf{k}\end{aligned}$$

and so

$$\nabla f(0, 0, 1) = \mathbf{j} - \mathbf{k}$$

This vector is perpendicular to S at (0, 0, 1), so to find a unit normal **n** we divide this vector by its length to obtain

$$\mathbf{n} = \frac{\nabla f(0, 0, 1)}{\|\nabla f(0, 0, 1)\|} = \frac{1}{\sqrt{2}}(\mathbf{j} - \mathbf{k})$$

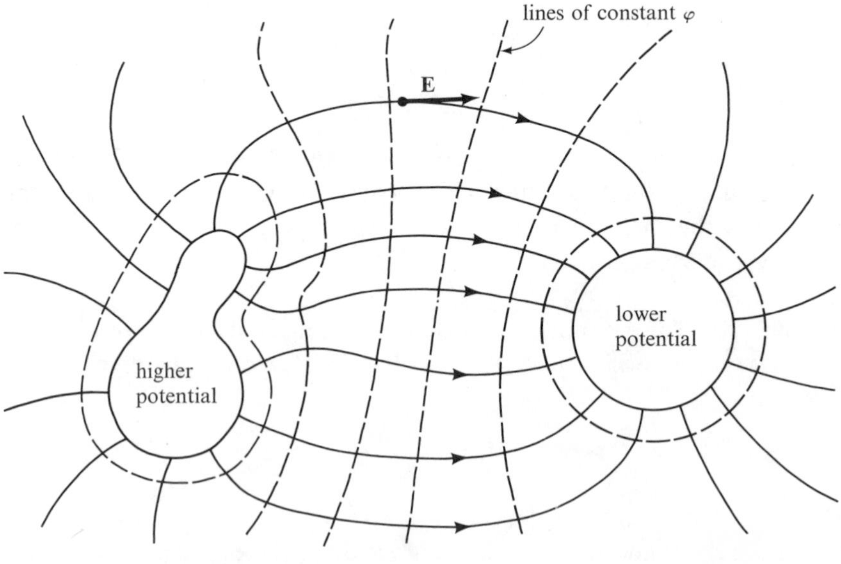

FIGURE 2.5.7
Equipotential surfaces are orthogonal to the electric force field **E**.

EXAMPLE 7. Consider two conductors, one charged positively and the other negatively. Between them, there is an electric potential set up. This potential is a function $\phi: \mathbb{R}^3 \to \mathbb{R}$. The electric field is given by $\mathbf{E} = -\nabla\phi$. From Theorem 14 we know that $\mathbf{E}$ is perpendicular to level surfaces of ϕ. These level surfaces are called equipotential surfaces since the potential is constant on them (see Figure 2.5.7).

EXERCISES

1. Prove that the directional derivative of $f(x, y, z) = z^2x + y^3$ at $(1, 1, 2)$ in the direction $(1/\sqrt{5})\mathbf{i} + (2/\sqrt{5})\mathbf{j}$ is $2\sqrt{5}$.

2. Compute the directional derivatives of the following functions at the indicated points in the given directions.
   (a) $f(x, y) = x + 2xy - 3y^2, (x_0, y_0) = (1, 2), \mathbf{v} = \frac{3}{5}\mathbf{i} + \frac{4}{5}\mathbf{j}$ $\nabla f(x_0 y_0) \times V.$
   (b) $f(x, y) = \log(\sqrt{x^2 + y^2}), (x_0, y_0) = (1, 0), \mathbf{v} = (1/\sqrt{5})(2\mathbf{i} + \mathbf{j})$
   (c) $f(x, y) = e^x \cos(\pi y), (x_0, y_0) = (0, -1), \mathbf{v} = -(1/\sqrt{5})\mathbf{i} + (2/\sqrt{5})\mathbf{j}$
   (d) $f(x, y) = xy^2 + x^3y, (x_0, y_0) = (4, -2), \mathbf{v} = (1/\sqrt{10})\mathbf{i} + (3/\sqrt{10})\mathbf{j}$
   (e) $f(x, y) = x^y, (x_0, y_0) = (e, e), \mathbf{v} = \frac{5}{13}\mathbf{i} + \frac{12}{13}\mathbf{j}$
   (f) $f(x, y, z) = e^x + yz, (x_0, y_0, z_0) = (1, 1, 1), \mathbf{v} = (1/\sqrt{3}, -1/\sqrt{3}, 1/\sqrt{3})$
   (g) $f(x, y, z) = xyz, (x_0, y_0, z_0) = (1, 0, 1), \mathbf{v} = (1/\sqrt{2}, 0, -1/\sqrt{2})$

3. Find the planes tangent to the following surfaces at the indicated points.
   (a) $x^2 + 2y^2 + 3zx = 10, (1, 2, \frac{1}{3})$
   (b) $y^2 - x^2 = 3, (1, 2, 8)$
   (c) $xyz = 1, (1, 1, 1)$
   (d) $z = x^3 + y^3 - 6xy, (1, 2, -3)$
   (e) $z = (\cos x)(\cos y), (0, \pi/2, 0)$
   (f) $z = (\cos x)(\sin y), (0, \pi/2, 1)$

*4. Let $f: \mathbb{R}^3 \to \mathbb{R}$ and regard $Df(x, y, z)$ as a linear map of $\mathbb{R}^3$ to $\mathbb{R}$. Show that its kernel (null space) is the linear subspace of $\mathbb{R}^3$ orthogonal to ∇f.

5. Verify Theorems 13 and 14 for $f(x, y, z) = x^2 + y^2 + z^2$.

6. Show that a unit normal to the surface $x^3y^3 + y - z + 2 = 0$ at $(0, 0, 2)$ is $\mathbf{n} = (1/\sqrt{2})(\mathbf{j} - \mathbf{k})$.

7. Show that the definition following Theorem 14 yields, as a special case, the formula for the plane tangent to the graph of $f(x, y)$ by regarding the graph as a level surface of $F(x, y, z) = f(x, y) - z$ (see Section 2.3).

8. Compute the gradient ∇f for each of the following functions.
   (a) $f(x, y, z) = \sqrt{x^2 + y^2 + z^2}$
   (b) $f(x, y, z) = xy + yz + xz$
   (c) $f(x, y, z) = \dfrac{1}{x^2 + y^2 + z^2}$

9. For the following functions $f: \mathbb{R}^3 \to \mathbb{R}$ and $\mathbf{g}: \mathbb{R} \to \mathbb{R}^3$ find ∇f and $\mathbf{g}'$ and evaluate $(f \circ \mathbf{g})'(1)$.
   (a) $f(x, y, z) = xz + yz + xy, \mathbf{g}(t) = (e^t, \cos t, \sin t)$

(b) $f(x, y, z) = e^{xyz}$, $\mathbf{g}(t) = (6t, 3t^2, t^3)$
(c) $f(x, y, z) = (x^2 + y^2 + z^2)\log\sqrt{x^2 + y^2 + z^2}$, $\mathbf{g}(t) = (e^t, e^{-t}, t)$

10. Compute the directional derivative of f in the given directions at the given points.
(a) $f(x, y, z) = xy^2 + y^2z^3 + z^3x$, $\mathrm{P} = (4, -2, -1)$, $\mathbf{v} = 1/\sqrt{14}(\mathbf{i} + 3\mathbf{j} + 2\mathbf{k})$
(b) $f(x, y, z) = x^{yz}$, $\mathrm{P} = (e, e, 0)$, $\mathbf{v} = \frac{12}{13}\mathbf{i} + \frac{3}{13}\mathbf{j} + \frac{4}{13}\mathbf{k}$

11. Let $\mathbf{r} = x\mathbf{i} + y\mathbf{j} + z\mathbf{k}$ and $r = \|\mathbf{r}\|$. Prove that

$$\nabla\left(\frac{1}{r}\right) = -\frac{\mathbf{r}}{r^3}$$

12. Captain Ralph finds himself on the sunny side of Mercury and notices his space suit is melting. The temperature in a rectangular coordinate system in his vicinity is

$$T(x, y, z) = e^{-x} + e^{-2y} + e^{-3z}$$

If he is at (1, 1, 1), in what direction should he start to move in order to cool down the fastest?

13. A function $f\colon \mathbb{R}^2 \to \mathbb{R}$ is independent of the second variable if and only if there is a function $g\colon \mathbb{R} \to \mathbb{R}$ such that $f(x, y) = g(x)$ for all x in $\mathbb{R}$. In this case, calculate ∇f in terms of g'.

14. Let f and g be functions from $\mathbb{R}^3$ to $\mathbb{R}$. Suppose f is differentiable and $\nabla f(\mathbf{x}) = g(\mathbf{x})\mathbf{x}$. Show that spheres centered at the origin are contained in the level sets for f; that is, f is constant on such spheres.

15. A function $f\colon \mathbb{R}^n \to \mathbb{R}$ is called *even* if $f(\mathbf{x}) = f(-\mathbf{x})$ for every $\mathbf{x}$ in $\mathbb{R}^n$. If f is differentiable and even, find Df at the origin.

2.6 ITERATED PARTIAL DERIVATIVES

In the preceding sections we have developed considerable information concerning the derivative of a map, and we have investigated the geometry associated with the derivative of real-valued functions by making use of the gradient. We shall now proceed to study higher-order derivatives. Although this topic will be taken up in detail in Chapter 4, we shall consider it briefly here. Our main goal is to prove a theorem that asserts the equality of the "mixed partial derivatives." This result holds for functions of any number of variables, but we shall confine our attention here to real-valued functions of two or three variables.

Let $f\colon \mathbb{R}^3 \to \mathbb{R}$ be of class C^1. Remember this means that $\partial f/\partial x$, $\partial f/\partial y$, and $\partial f/\partial z$ exist and are continuous; and this implies f is differentiable (Theorem 9). If these derivatives, in turn, have continuous partials, we say f is of *class* C^2, or is *twice continuously differentiable.* A few examples of how these higher derivatives are written follow:

$$\frac{\partial^2 f}{\partial x^2} = \frac{\partial}{\partial x}\left(\frac{\partial f}{\partial x}\right)$$

$$\frac{\partial^2 f}{\partial x\,\partial y} = \frac{\partial}{\partial x}\left(\frac{\partial f}{\partial y}\right)$$

$$\frac{\partial^2 f}{\partial z\,\partial y} = \frac{\partial}{\partial z}\left(\frac{\partial f}{\partial y}\right)$$

The process can of course be repeated for third-order derivatives and so on. If f is a function of only x and y and $\partial f/\partial x$, $\partial f/\partial y$ are continuously differentiable, then by taking second partials we get the four functions

$$\frac{\partial^2 f}{\partial x^2}, \frac{\partial^2 f}{\partial y^2}, \frac{\partial^2 f}{\partial x\,\partial y}, \quad \text{and} \quad \frac{\partial^2 f}{\partial y\,\partial x}$$

EXAMPLE 1. Let $f(x, y) = xy + (x + 2y)^2$. Then

$$\frac{\partial f}{\partial x} = y + 2(x + 2y), \qquad \frac{\partial f}{\partial y} = x + 4(x + 2y)$$

$$\frac{\partial^2 f}{\partial x^2} = 2, \qquad \frac{\partial^2 f}{\partial y^2} = 8$$

$$\frac{\partial^2 f}{\partial x\,\partial y} = 5, \qquad \frac{\partial^2 f}{\partial y\,\partial x} = 5$$

EXAMPLE 2. If $f(x, y) = \sin x \sin^2 y$, then

$$\frac{\partial f}{\partial x} = \cos x \sin^2 y, \qquad \frac{\partial f}{\partial y} = 2 \sin x \sin y \cos y = \sin x \sin 2y$$

$$\frac{\partial^2 f}{\partial x^2} = -\sin x \sin^2 y, \qquad \frac{\partial^2 f}{\partial y^2} = 2 \sin x \cos 2y$$

$$\frac{\partial^2 f}{\partial x\,\partial y} = \cos x \sin 2y, \qquad \frac{\partial^2 f}{\partial y\,\partial x} = 2 \cos x \sin y \cos y = \cos x \sin 2y$$

EXAMPLE 3. Let $f(x, y, z) = e^{xy} + z \cos x$. Then, for example,

$$\frac{\partial f}{\partial x} = ye^{xy} - z \sin x, \qquad \frac{\partial f}{\partial y} = xe^{xy}, \qquad \frac{\partial f}{\partial z} = \cos x$$

$$\frac{\partial^2 f}{\partial z\,\partial x} = -\sin x, \qquad \frac{\partial^2 f}{\partial x\,\partial z} = -\sin x$$

In all these examples note that the mixed partials, like $\partial^2 f/(\partial x\,\partial y)$ and $\partial^2 f/(\partial y\,\partial x)$, or $\partial^2 f/(\partial z\,\partial x)$ and $\partial^2 f/(\partial x\,\partial z)$, are equal. It is a very basic and perhaps surprising fact that this is usually the case. We shall prove this in the next theorem for functions $f(x, y)$ of two variables, but the proof can easily be extended to functions of n variables.

THEOREM 15. *If f is C^2 (twice continuously differentiable) then the mixed partials are equal; that is,*

$$\frac{\partial^2 f}{\partial x\, \partial y} = \frac{\partial^2 f}{\partial y\, \partial x}$$

Proof. Consider the expression

$$f(x_0 + \Delta x, y_0 + \Delta y) - f(x_0 + \Delta x, y_0) - f(x_0, y_0 + \Delta y) + f(x_0, y_0) \quad (1)$$

We fix y_0 and Δy, and introduce the function

$$g(x) = f(x, y_0 + \Delta y) - f(x, y_0)$$

so that (1) equals $g(x_0 + \Delta x) - g(x_0)$. By the Mean Value Theorem for functions of one variable (p. 119) this equals $g'(\bar{x})\, \Delta x$ for some $\bar{x}$ between x_0 and $x_0 + \Delta x$. Hence (1) equals

$$\left[\frac{\partial f}{\partial x}(\bar{x}, y_0 + \Delta y) - \frac{\partial f}{\partial x}(\bar{x}, y_0)\right] \Delta x$$

Applying the Mean Value Theorem again we get for (1)

$$\frac{\partial^2 f}{\partial y\, \partial x}(\bar{x}, \bar{y})\, \Delta x\, \Delta y$$

Since $\partial^2 f/\partial y\, \partial x$ is continuous, it follows that

$$\frac{\partial^2 f}{\partial y\, \partial x}(x_0, y_0) = \lim_{(\Delta x, \Delta y)\to(0,0)} \frac{1}{\Delta x\, \Delta y}[f(x_0 + \Delta x, y_0 + \Delta y) - f(x_0 + \Delta x, y_0) - f(x_0, y_0 + \Delta y) + f(x_0, y_0)] \quad (2)$$

In a similar manner, one shows that $\partial^2 f/\partial x\, \partial y$ is given by the same limit formula (2), which proves the result. ■

This theorem was first proved by Leonhard Euler in 1734 in connection with his studies of hydrodynamics. It follows from Theorem 15 that, for example, for a C^3 function

$$\frac{\partial^3 f}{\partial x\, \partial y\, \partial z} = \frac{\partial^3 f}{\partial z\, \partial y\, \partial x} = \frac{\partial^3 f}{\partial y\, \partial z\, \partial x} \text{ etc.}$$

In other words, we may compute iterated partial derivatives in any order we please.

EXAMPLE 4. Verify the equality of the mixed second partial derivatives for the function

$$f(x, y) = xe^y + yx^2$$

Here

$$\frac{\partial f}{\partial x} = e^y + 2xy, \qquad \frac{\partial f}{\partial y} = xe^y + x^2$$

$$\frac{\partial^2 f}{\partial y\, \partial x} = e^y + 2x, \qquad \frac{\partial^2 f}{\partial x\, \partial y} = e^y + 2x$$

and so we have

$$\frac{\partial^2 f}{\partial y\, \partial x} = \frac{\partial^2 f}{\partial x\, \partial y}$$

HISTORICAL NOTE: SOME PARTIAL DIFFERENTIAL EQUATIONS

"Philosophy [nature] is written in that great book which ever is before our eyes—I mean the universe—but we cannot understand it if we do not first learn the language and grasp the symbols in which it is written. The book is written in mathematical language, and the symbols are triangles, circles and other geometrical figures, without whose help it is impossible to comprehend a single word of it; without which one wanders in vain through a dark labyrinth."

RENÉ DESCARTES

This quote illustrates the belief, popular in Descartes' age, that much of nature could be reduced to mathematics. In the latter part of the seventeenth century this thinking was dramatically reinforced when Newton used his law of gravitation and the new calculus to derive Kepler's three laws of celestial motion (see Section 3.1). The impact of this philosophy on mathematics was substantial, and many mathematicians sought to "mathematize" nature. The extent to which mathematics pervades the physical sciences today (and, to an increasing amount, economics and the social and life sciences) is a testament to the success of these endeavors. Correspondingly, the attempts to mathematize nature have often led to new mathematical discoveries.

HISTORICAL NOTE (*Continued*)

Many of the laws of nature were described in terms of either ordinary differential equations (equations involving the derivatives of functions of one variable alone, such as Newton's law of gravitation) or partial differential equations, that is, equations involving partial derivatives of functions. In order to give the reader some historical perspective and offer motivation for studying partial derivatives, we now present a brief description of three of the most famous partial differential equations: the heat equation, the wave equation, and the potential (or Laplace's) equation. All three of these will be discussed at greater depth in Section 7.5.

THE HEAT EQUATION. In the early part of the nineteenth century the French mathematician Joseph Fourier (1768–1830) took up the study of heat. Heat-flow had obvious applications to both industrial and scientific problems: A better understanding of it would, for example, make possible more efficient smelting of metals and enable scientists to determine the temperature of a body given the temperature at its boundary and thereby approximate the temperature of the earth's interior.

Let a homogeneous body $B \subset \mathbb{R}^3$ (Figure 2.6.1) be represented by some region in 3-space. Let $T(x, y, z, t)$ denote the temperature of the body at the point (x, y, z) at time t. Fourier proved, on the basis of the physical principles described in Section 7.5, that T must satisfy the partial differential equation

$$k\left(\frac{\partial^2 T}{\partial x^2} + \frac{\partial^2 T}{\partial y^2} + \frac{\partial^2 T}{\partial z^2}\right) = \frac{\partial T}{\partial t} \tag{1}$$

called the "heat equation", where k is a constant whose value depends on the conductivity of the material comprising the body.

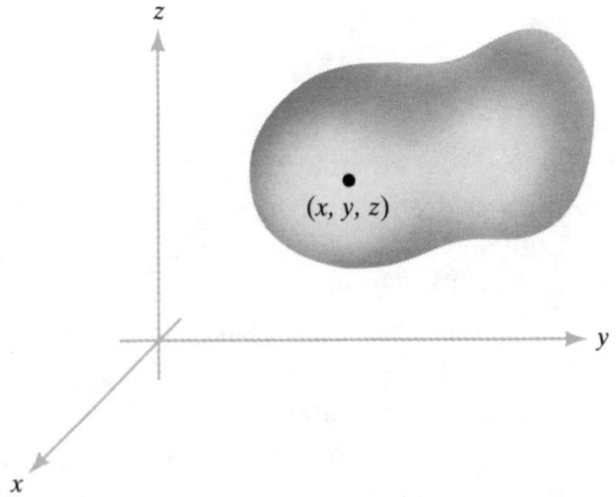

FIGURE 2.6.1
A homogeneous body in space.

HISTORICAL NOTE (*Continued*)

Fourier used this equation to solve problems in heat conduction. In fact, his investigations into the solutions of (1) led him to a new mathematical discovery now called Fourier series.

THE POTENTIAL EQUATION. In Example 5 of Section 2.5 we introduced the gravitational potential V (often called Newton's potential) caused by a point mass situated at the origin. This potential is given by $V = -GM/r$, where $r = \sqrt{x^2 + y^2 + z^2}$. The potential V satisfies the equation

$$\frac{\partial^2 V}{\partial x^2} + \frac{\partial^2 V}{\partial y^2} + \frac{\partial^2 V}{\partial z^2} = 0 \tag{2}$$

everywhere except at the origin, as the reader can check by doing Exercise 11. This equation is known as *Laplace's equation.* Pierre-Simon de Laplace (1749–1827) had worked on the gravitational attraction of non-point masses and was the first to consider equation (2) with regard to gravitational attraction. He gave arguments (later shown to be incorrect) that equation (2) held for any body and any point whether inside or outside of that body. However, Laplace was not the first person to write down equation (2). The potential equation appeared for the first time in one of Euler's major papers in 1752, "Principles of the Motions of Fluids," in which he derived the potential equation with regard to the motion of (incompressible) fluids. Euler remarked that he had no idea how to solve (2). We shall study the equations of fluid mechanics in Section 7.5.

Poisson later showed that if (x, y, z) lies inside an attracting body then V satisfies

$$\frac{\partial^2 V}{\partial x^2} + \frac{\partial^2 V}{\partial y^2} + \frac{\partial^2 V}{\partial z^2} = -4\pi\rho \tag{3}$$

where ρ is the density of the attracting body. Equation (3) is now called **Poisson's equation**. Poisson was also the first to point out the importance of this equation for problems involving electric fields.

Notice that if the temperature T is constant in time then the heat equation reduces to Laplace's equation (why?).

THE WAVE EQUATION. The wave equation in space has the form

$$\frac{\partial^2 f}{\partial x^2} + \frac{\partial^2 f}{\partial y^2} + \frac{\partial^2 f}{\partial z^2} = c^2 \frac{\partial^2 f}{\partial t^2}. \tag{4}$$

The one-dimensional wave equation

$$\frac{\partial^2 f}{\partial x^2} = c^2 \frac{\partial^2 f}{\partial t^2} \tag{4'}$$

HISTORICAL NOTE (*Continued*)

had been derived in about 1727 by John Bernoulli and several years later by Jean Le Rond d'Alembert in the study of how to determine the motion of a vibrating string (such as a violin string). Equation (4) became useful in the study of both vibrating bodies and elasticity. As we shall see when we consider Maxwell's equations of electromagnetism in Section 7.5, this equation also arises in the study of the propagation of electromagnetic radiation and sound waves.

EXERCISES

1. Compute the second partial derivatives $\partial^2 f/\partial x^2$, $\partial^2 f/\partial x\,\partial y$, $\partial^2 f/\partial y\,\partial x$, $\partial^2 f/\partial y^2$ for each of the following functions. Verify Theorem 15 in each case:
   (a) $f(x, y) = 2xy/((x^2 + y^2)^2)$
   (b) $f(x, y, z) = e^z + 1/x + xe^{-y}$
   (c) $f(x, y) = \cos(xy^2)$
   (d) $f(x, y) = e^{-xy^2} + y^3x^4$
   (e) $f(x, y) = 1/(\cos^2 x + e^{-y})$

2. Let
$$f(x, y) = \begin{cases} xy(x^2 - y^2)/(x^2 + y^2), & (x, y) \neq (0, 0) \\ 0, & (x, y) = (0, 0) \end{cases}$$
(see Figure 2.6.2).

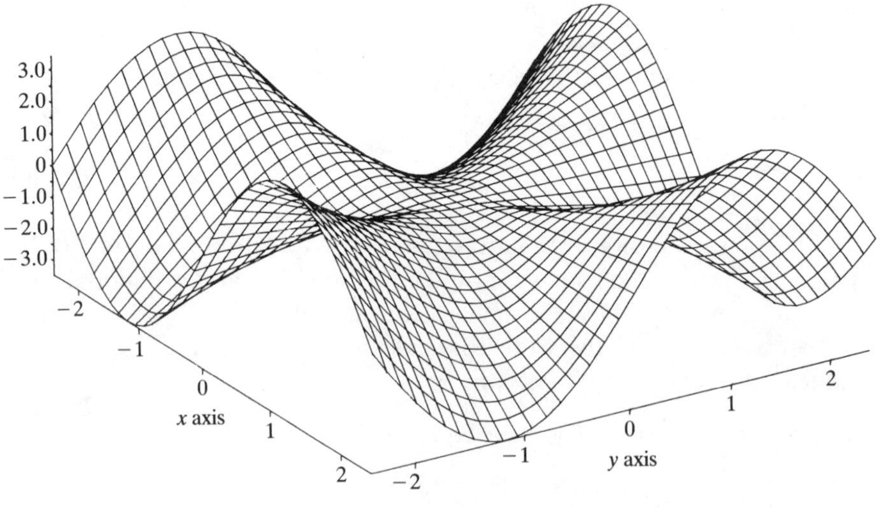

FIGURE 2.6.2
Computer-generated graph of the function in Exercise 2.

(a) If $(x, y) \neq (0, 0)$ calculate $\partial f/\partial x$ and $\partial f/\partial y$.
(b) Show that $(\partial f/\partial x)(0, 0) = 0 = (\partial f/\partial y)(0, 0)$
(c) Show that $(\partial^2 f/\partial x\,\partial y)(0, 0) = 1$, $(\partial^2 f/\partial y\,\partial x)(0, 0) = -1$
(d) What went wrong? Why are the mixed partials not equal?

3. Let $f: \mathbb{R}^2 \to \mathbb{R}$ be a C^2 function and let $\mathbf{c}(t)$ be a C^2 curve in $\mathbb{R}^2$. Write a formula for $(d^2/dt^2)((f \circ \mathbf{c})(t))$ using the Chain Rule twice.

4. Show that $u(x, y) = e^x \sin y$ satisfies Laplace's equation

$$\frac{\partial^2 u}{\partial x^2} + \frac{\partial^2 u}{\partial y^2} = 0$$

5. Find the equation of the plane tangent to $z = x^2 + 2y^3$ at $(1, 1, 3)$.

6. Verify that

$$\frac{\partial^3 f}{\partial x\,\partial y\,\partial z} = \frac{\partial^3 f}{\partial z\,\partial y\,\partial x}$$

for $f(x, y, z) = ze^{xy} + yz^3x^2$.

7. Evaluate all first and second partial derivatives of the following functions.
(a) $f(x, y) = x \arctan(x/y)$
(b) $f(x, y) = \cos\sqrt{x^2 + y^2}$
(c) $f(x, y) = \exp(-x^2 - y^2)$

8. Find the directional derivatives of the functions in Exercise 7 at the point $(1, 1)$ in the direction $(1/\sqrt{2}, 1/\sqrt{2})$.

9. Use Theorem 15 to show that if $f(x, y, z)$ is of class C^3, then

$$\frac{\partial^3 f}{\partial x\,\partial y\,\partial z} = \frac{\partial^3 f}{\partial y\,\partial z\,\partial x}$$

10. Let g and f be twice-differentiable functions from $\mathbb{R}$ to $\mathbb{R}$. Set $\phi(x, t) = f(x - t) + g(x + t)$. Prove that ϕ satisfies the wave equation

$$\frac{\partial^2 \phi}{\partial t^2} = \frac{\partial^2 \phi}{\partial x^2}$$

11. Show that Newton's potential V (see Example 5, Section 2.5) satisfies Laplace's equation

$$\frac{\partial^2 V}{\partial x^2} + \frac{\partial^2 V}{\partial y^2} + \frac{\partial^2 V}{\partial z^2} = 0 \text{ for } (x, y, z) \neq (0, 0, 0).$$

REVIEW EXERCISES FOR CHAPTER 2

1. Describe the graphs of:
(a) $f(x, y) = 3x^2 + y^2$
(b) $f(x, y) = xy + 3x$

2. Describe some appropriate level surfaces and sections of the graphs of:
   (a) $f(x, y, z) = 2x^2 + y^2 + z^2$
   (b) $f(x, y, z) = x^2$
   (c) $f(x, y, z) = xyz$

3. Compute the derivative $Df(\mathbf{x})$ of each of the following functions.
   (a) $f(x, y) = (x^2y, e^{-xy})$
   (b) $f(x) = (x, x)$
   (c) $f(x, y, z) = e^x + e^y + e^z$
   (d) $f(x, y, z) = (x, y, z)$

4. Suppose $f(x, y) = f(y, x)$ for all (x, y). Prove that

$$(\partial f/\partial x)(a, b) = (\partial f/\partial y)(b, a).$$

5. Let $f(x, y) = (1 - x^2 - y^2)^{1/2}$. Show that the plane tangent to the graph of f at $(x_0, y_0, f(x_0, y_0))$ is orthogonal to the vector $(x_0, y_0, f(x_0, y_0))$. Interpret geometrically.

6. Suppose $F(x, y) = (\partial f/\partial x) - (\partial f/\partial y)$ and f is C^2. Prove that

$$\frac{\partial F}{\partial x} + \frac{\partial F}{\partial y} = \frac{\partial^2 f}{\partial x^2} - \frac{\partial^2 f}{\partial y^2}$$

7. Find an equation for the tangent plane of the graph of f at the point $(x_0, y_0, f(x_0, y_0))$ for:
   (a) $f: \mathbb{R}^2 \to \mathbb{R}, (x, y) \mapsto x - y + 2,\quad (x_0, y_0) = (1, 1)$
   (b) $f: \mathbb{R}^2 \to \mathbb{R}, (x, y) \mapsto x^2 + 4y^2,\quad (x_0, y_0) = (2, -1)$
   (c) $f: \mathbb{R}^2 \to \mathbb{R}, (x, y) \mapsto xy,\quad (x_0, y_0) = (-1, -1)$
   (d) $f(x, y) = \log(x + y) + x\cos y + \arctan(x + y),\quad (x_0, y_0) = (1, 0)$
   (e) $f(x, y) = \sqrt{x^2 + y^2},\quad (x_0, y_0) = (1, 1)$
   (f) $f(x, y) = xy,\quad (x_0, y_0) = (2, 1)$

8. Compute an equation for the tangent planes of the following surfaces at the indicated points.
   (a) $x^2 + y^2 + z^2 = 3,\quad (1, 1, 1)$
   (b) $x^3 - 2y^3 + z^3 = 0,\quad (1, 1, 1)$
   (c) $(\cos x)(\cos y)e^z = 0,\quad (\pi/2, 1, 0)$
   (d) $e^{xyz} = 1,\quad (1, 1, 0)$

9. Draw some level curves for the following functions.
   (a) $f(x, y) = 1/xy$
   (b) $f(x, y) = x^2 - xy - y^2$

10. Consider a temperature function $T(x, y) = x \sin y$. Plot a few level curves. Compute ∇T and explain its meaning.

11. Find the following limits if they exist.
   (a) $\lim_{(x, y)\to(0, 0)} \dfrac{\cos xy - 1}{x}$
   (b) $\lim_{(x, y)\to(0, 0)} \sqrt{|(x + y)/(x - y)|},\ x \neq y$

12. Compute the partial derivatives and gradients of the following functions.
    (a) $f(x, y, z) = xe^z + y \cos x$
    (b) $f(x, y, z) = (x + y + z)^{10}$
    (c) $f(x, y, z) = (x^2 + y)/z$

13. Let $\mathbf{F} = F_1(x, y)\mathbf{i} + F_2(x, y)\mathbf{j}$ be a C^1 vector field. Show that if $\mathbf{F} = \nabla f$ for some f then $\partial F_1/\partial y = \partial F_2/\partial x$. Show that $\mathbf{F} = y(\cos x)\mathbf{i} + x(\sin y)\mathbf{j}$ is not a gradient vector field (see p. 134).

14. Let $y(x)$ be a differentiable function defined implicitly by $F(x, y(x)) = 0$. From problem 11(a), Section 2.4, we know that

$$\frac{dy}{dx} = -\frac{\partial F/\partial x}{\partial F/\partial y}$$

Consider the surface $z = F(x, y)$, and suppose F is increasing as a function of x and as a function of y; i.e., $\partial F/\partial x > 0$ and $\partial F/\partial y > 0$. By considering the graph and the plane $z = 0$, show that for z fixed at $z = 0$ y should *decrease* as x increases and x should *decrease* as y increases. Does this agree with the minus sign in the formula for dy/dx?

15. (a) Consider the graph of a function $f(x, y)$ (Figure 2.R.1). Let (x_0, y_0) lie on a level curve C, so $\nabla f(x_0, y_0)$ is perpendicular to this curve. Show that the tangent plane of the graph is the plane that (*i*) contains the line perpendicular to $\nabla f(x_0, y_0)$ and lying in the horizontal plane $z = f(x_0, y_0)$, and (*ii*) has slope $\|\nabla f(x_0, y_0)\|$ relative to the xy-plane. By the slope of a plane P relative to the xy-plane we mean the tangent of the angle θ, $0 \leq \theta \leq \pi$, between the upward pointing normal $\mathbf{p}$ to P and the unit vector $\mathbf{k}$.

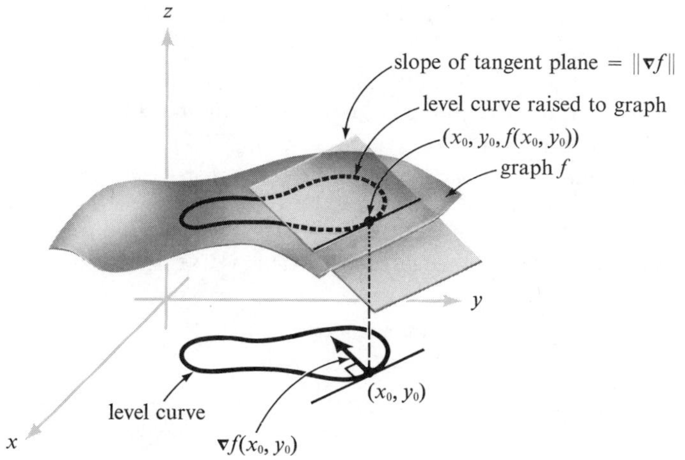

FIGURE 2.R.1

Relationship between the gradient of a function and the plane tangent to the function's graph (Exercise 15(a)).

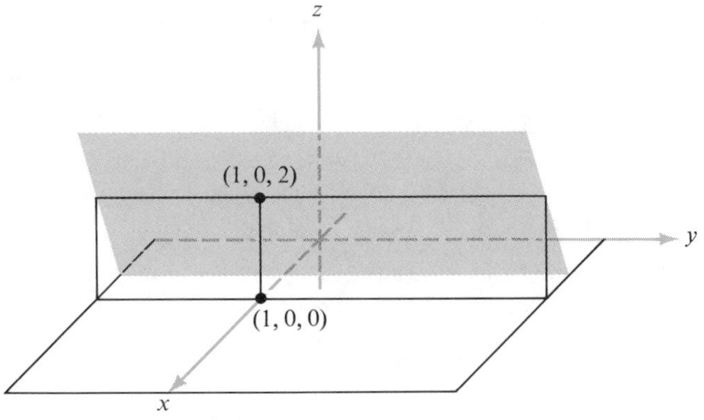

FIGURE 2.R.2
The plane referred to in Exercise 15(b).

(b) Use this method to show that the tangent plane of the graph of $f(x, y) = (x + \cos y)x^2$ at $(1, 0, 2)$ is as sketched in Figure 2.R.2.

16. Find the plane tangent to the surface $z = x^2 + y^2$ at the point $(1, -2, 5)$. Explain the geometrical significance, for this surface, of the gradient of $f(x, y) = x^2 + y^2$ (see Exercise 15).

17. In which direction is the directional derivative of $f(x, y) = (x^2 - y^2)/(x^2 + y^2)$ at $(1, 1)$ equal to zero?

18. Find the directional derivative of the given function at the given point and in the direction of the given vector.
(a) $f(x, y, z) = e^x \cos(yz)$, $p_0 = (0, 0, 0)$, $\mathbf{v} = (2, 1, -2)$
(b) $f(x, y, z) = xy + yz + zx$, $p_0 = (1, 1, 2)$, $\mathbf{v} = (10, -1, 2)$

19. Find the tangent plane and normal to the hyperboloid $x^2 + y^2 - z^2 = 18$ at $(3, 5, -4)$.

20. Find the direction in which the function $w = x^2 + xy$ increases most rapidly at the point $(-1, 1)$. What is the magnitude of ∇w in this direction? Interpret this magnitude geometrically.

21. Let f be defined on an open set S in $\mathbb{R}^n$. We say that f is *homogeneous of degree* p over S if $f(\lambda\mathbf{x}) = \lambda^p f(\mathbf{x})$ for every real λ and for every $\mathbf{x}$ in S for which $\lambda\mathbf{x} \in S$. If such a function is differentiable at $\mathbf{x}$, show that $\mathbf{x} \cdot \nabla f(\mathbf{x}) = pf(\mathbf{x})$. This is known as *Euler's Theorem* for homogeneous functions. (HINT: For fixed $\mathbf{x}$, define $g(\lambda) = f(\lambda\mathbf{x})$ and compute $g'(1)$).

22. Prove that if $f(x, y)$ satisfies the Laplace equation

$$\frac{\partial^2 f}{\partial x^2} + \frac{\partial^2 f}{\partial y^2} = 0$$

so does

$$\phi(x, y) = f\left(\frac{x}{x^2 + y^2}, \frac{y}{x^2 + y^2}\right) \quad \text{for} \quad (x, y) \neq (0, 0)$$

(A function that satisfies Laplace's equation is said to be *harmonic*.)

23. Prove that the functions

(a) $f(x, y) = \log(x^2 + y^2)$

(b) $g(x, y, z) = \dfrac{1}{(x^2 + y^2 + z^2)^{1/2}}$

(c) $h(x, y, z, w) = \dfrac{1}{x^2 + y^2 + z^2 + w^2}$

satisfy the respective Laplace equations:

(a) $f_{xx} + f_{yy} = 0$

(b) $g_{xx} + g_{yy} + g_{zz} = 0$

(c) $h_{xx} + h_{yy} + h_{zz} + h_{ww} = 0$

where $f_{xx} = \partial^2 f/\partial x^2$ etc.

24. If $z = f(x - y)/y$ show that $z + y(\partial z/\partial x) + y(\partial z/\partial y) = 0$

25. Given $w = f(x, y)$ with $x = u + v$, $y = u - v$, and f a C^2 function, show that

$$\frac{\partial^2 w}{\partial u\, \partial v} = \frac{\partial^2 w}{\partial x^2} - \frac{\partial^2 w}{\partial y^2}$$

26. Let f have partial derivatives $\partial f(\mathbf{x})/\partial x_i$, where $i = 1, 2, \ldots, n$, at each point $\mathbf{x}$ of an open set U in $\mathbb{R}^n$. If f has a local maximum or a local minimum at the point $\mathbf{x}_0$ in U, show that $\partial f(\mathbf{x}_0)/\partial x_i = 0$ for each i.

27. Consider the functions defined in $\mathbb{R}^2$ by the following formulas:

(*i*) $f(x, y) = xy/(x^2 + y^2)$ if $(x, y) \neq (0, 0)$, $f(0, 0) = 0$

(*ii*) $f(x, y) = x^2y^2/(x^2 + y^4)$ if $(x, y) \neq (0, 0)$, $f(0, 0) = 0$

(a) In each case show that the partial derivatives $\partial f(x, y)/\partial x$ and $\partial f(x, y)/\partial y$ exist for every (x, y) in $\mathbb{R}^2$ and evaluate these derivatives explicitly in terms of x and y.

(b) Explain why the functions described in (*i*) and (*ii*) are or are not differentiable at (0, 0).

28. Compute the gradient vector $\nabla f(x, y)$ at all points (x, y) in $\mathbb{R}^2$ for each of the following functions.

(a) $f(x, y) = x^2y^2 \log(x^2 + y^2)$ if $(x, y) \neq (0, 0)$, $f(0, 0) = 0$

(b) $f(x, y) = xy \sin(1/(x^2 + y^2))$ if $(x, y) \neq (0, 0)$, $f(0, 0) = 0$

29. Given a function f defined in $\mathbb{R}^2$, let $F(r, \theta) = f(r \cos\theta, r \sin\theta)$.

(a) Assume appropriate differentiability properties of f and show that

$$\frac{\partial F}{\partial r}(r, \theta) = \cos\theta \frac{\partial}{\partial x} f(x, y) + \sin\theta \frac{\partial}{\partial y} f(x, y)$$

$$\frac{\partial^2}{\partial r^2} F(r, \theta) = \cos^2\theta \frac{\partial^2}{\partial x^2} f(x, y) + 2 \sin\theta \cos\theta \frac{\partial^2}{\partial x\, \partial y} f(x, y) + \sin^2\theta \frac{\partial^2}{\partial y^2} f(x, y)$$

where $x = r\cos\theta$, $y = r \sin\theta$.

(b) Verify the formula

$$\|\nabla f(r\cos\theta, r\sin\theta)\|^2 = \left(\frac{\partial}{\partial r} F(r, \theta)\right)^2 + \frac{1}{r^2}\left(\frac{\partial}{\partial \theta} F(r, \theta)\right)^2$$

30. Let $\mathbf{u} = \mathbf{i} - 2\mathbf{j} + 2\mathbf{k}$ and $\mathbf{v} = 2\mathbf{i} + \mathbf{j} - 3\mathbf{k}$. Find: $\|\mathbf{u}\|$, $\mathbf{u} \cdot \mathbf{v}$, $\mathbf{u} \times \mathbf{v}$, and a vector in the same direction as $\mathbf{u}$, but of unit length.

31. Let $h(x, y) = 2e^{-x^2} + e^{-3y^2}$ denote the height on a mountain at position (x, y). In what direction from $(1, 0)$ should one begin walking in order to climb the fastest?

32. Compute an equation for the plane tangent to the graph of

$$f(x, y) = \frac{e^x}{x^2 + y^2}$$

at $x = 1, y = 2$.

33. (a) Give a careful statement of the general form of the Chain Rule.
(b) Let $f(x, y) = x^2 + y$ and let $\mathbf{h}(u) = (\sin 3u, \cos 8u)$. Let $g(u) = f(\mathbf{h}(u))$. Compute dg/du at $u = 0$ both directly *and* by using the Chain Rule.

34. (a) Sketch the level curves of $f(x, y) = -x^2 - 9y^2$ for $c = 0, -1, -10$.
(b) On your sketch, draw in ∇f at $(1, 1)$. Discuss.

CHAPTER 3

VECTOR-VALUED FUNCTIONS

Who by vigor of mind almost divine, the motions and figures of the planets, the paths of comets, and the tides of the seas first demonstrated.

NEWTON'S EPITAPH

One of our main concerns in Chapter 2 was the study of real-valued functions. This chapter deals with functions whose values are vectors. We shall begin in Section 3.1 with paths, which are maps from $\mathbb{R}$ to $\mathbb{R}^n$. Then we shall go on to study vector fields, and to introduce the main operations of vector differential calculus other than the gradient, namely, the divergence and the curl. We shall consider here some of the geometry associated with these operations, just as we did for the gradient, but the results of the most physical significance will have to wait until we have studied integration theory.

3.1 PATHS AND VELOCITY

One often thinks of a curve as a line drawn on paper, such as a straight line, a circle, or a sine curve. To deal with such objects effectively, it is convenient to think of a curve in $\mathbb{R}^n$ as the image, or set of values, of a function (mapping) on an interval $[a, b]$ (or occasionally on all of $\mathbb{R}$) into $\mathbb{R}^n$. We shall call such a map a *path*. A path is in the plane if $n = 2$, and in space if $n = 3$. The image of the path may then correspond to a line we see on paper (see Figure 3.1.1).

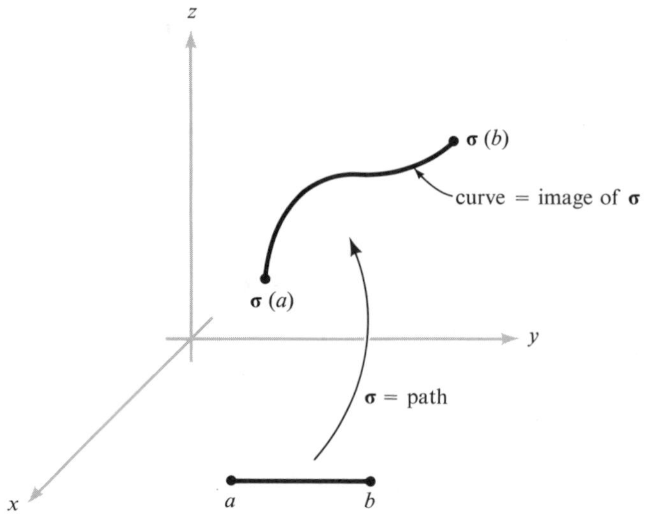

FIGURE 3.1.1
The map $\boldsymbol{\sigma}$ *is the path; its image is the curve we "see."*

In this section we define precisely the notion of path, give some examples, and show how paths can model trajectories of moving objects. We then define the velocity and acceleration of paths and apply these concepts to Newton's second law of motion and to the motion of planets in circular orbits.

DEFINITION. *A **path** in* $\mathbb{R}^n$ *is a map* $\boldsymbol{\sigma}: [a, b] \to \mathbb{R}^n$. *If* $\boldsymbol{\sigma}$ *is differentiable (respectively, of class* C^1*) we say* $\boldsymbol{\sigma}$ *is a **differentiable** (respectively,* C^1*) path. The points* $\boldsymbol{\sigma}(a)$ *and* $\boldsymbol{\sigma}(b)$ *are called the **endpoints** of the path.*

It is useful to write t for the variable and to think of $\boldsymbol{\sigma}(t)$ as tracing out a curve in $\mathbb{R}^n$ as t varies. Often we imagine t to be the time and $\boldsymbol{\sigma}(t)$ as being the position of a moving particle at time t.

If $\boldsymbol{\sigma}$ is a path in $\mathbb{R}^3$, we can write $\boldsymbol{\sigma}(t) = (x(t), y(t), z(t))$ and we call $x(t)$, $y(t)$, and $z(t)$ the *component functions* of $\boldsymbol{\sigma}$. Of course we can similarly form component functions in $\mathbb{R}^2$ or, generally, in $\mathbb{R}^n$.

EXAMPLE 1. The straight line L in $\mathbb{R}^3$ through the point (x_0, y_0, z_0) in the direction of vector $\mathbf{v}$ is the image of the path

$$\boldsymbol{\sigma}(t) = (x_0, y_0, z_0) + t\mathbf{v}$$

for $t \in \mathbb{R}$ (see Figure 3.1.2).

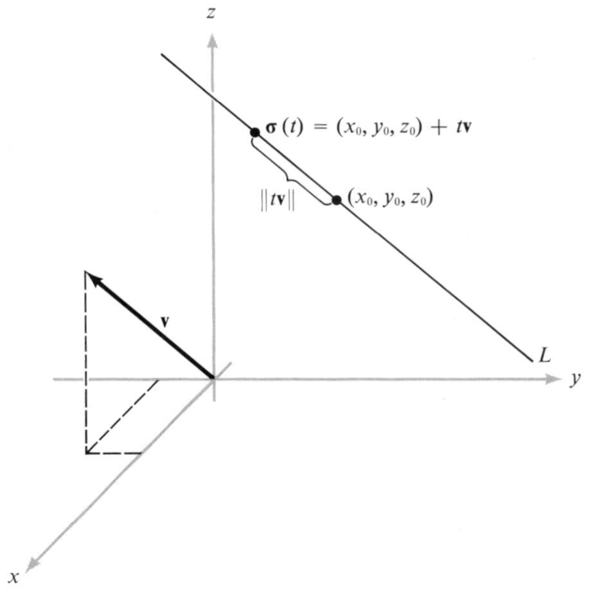

FIGURE 3.1.2
L is the straight line in space through (x_0, y_0, z_0) and in direction $\mathbf{v}$; its equation is $\boldsymbol{\sigma}(t) = (x_0, y_0, z_0) + t\mathbf{v}$.

EXAMPLE 2. The unit circle in the plane is represented by the path

$$\boldsymbol{\sigma}: \mathbb{R} \to \mathbb{R}^2, \qquad \boldsymbol{\sigma}(t) = (\cos t, \sin t)$$

(see Figure 3.1.3). The image of $\boldsymbol{\sigma}$, that is, the set of points $\boldsymbol{\sigma}(t) \in \mathbb{R}^2$, for $t \in \mathbb{R}$, is the unit circle.

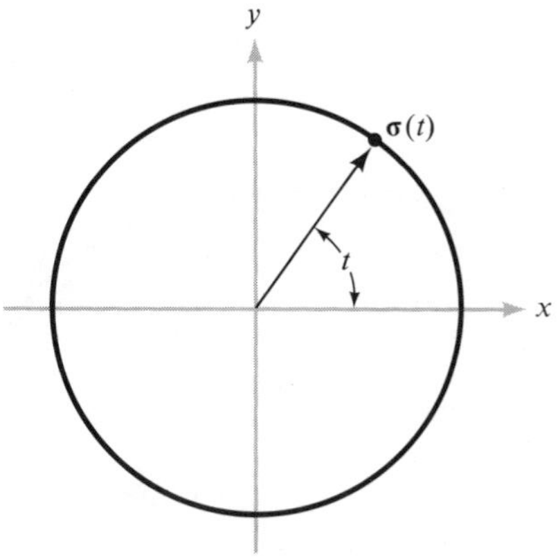

FIGURE 3.1.3
$\boldsymbol{\sigma}(t) = (\cos t, \sin t)$ is a path whose image is the unit circle.

EXAMPLE 3. The path $\boldsymbol{\sigma}(t) = (t, t^2)$ has an image that is a parabolic arc. The curve coincides with the graph of $f(x) = x^2$ (see Figure 3.1.4).

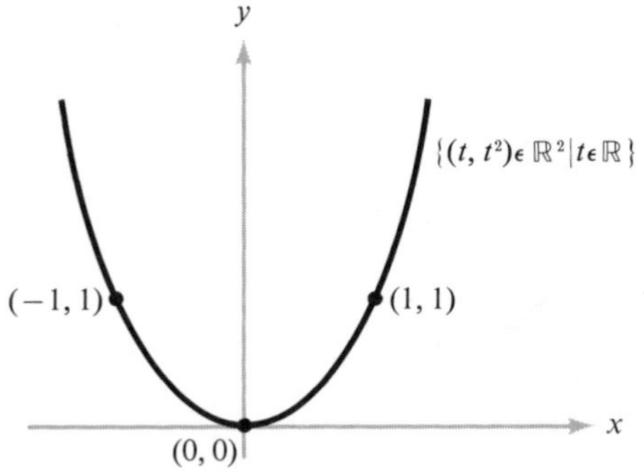

FIGURE 3.1.4
The image of $\boldsymbol{\sigma}(t) = (t, t^2)$ is the parabola $y = x^2$.

EXAMPLE 4. The function $\boldsymbol{\sigma}: t \mapsto (t - \sin t, 1 - \cos t)$ is the position function of a point on a rolling circle of radius 1. The circle lies in the xy-plane and rolls along the x-axis at constant speed; that is, the center of the circle is moving to the right along the line $y = 1$ at a constant speed of 1 radian per unit of time. The motion of the point $\boldsymbol{\sigma}(t)$ is more complicated; its locus is known as the *cycloid* (see Figure 3.1.5).

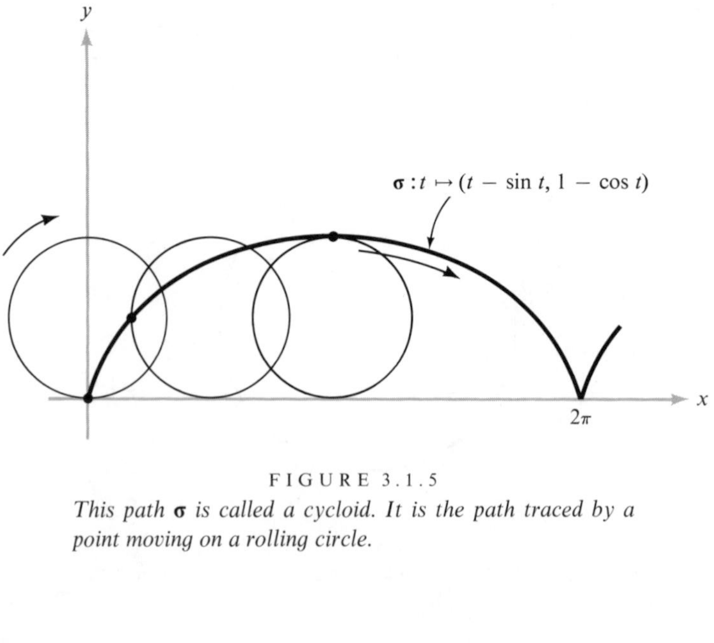

FIGURE 3.1.5
This path $\boldsymbol{\sigma}$ is called a cycloid. It is the path traced by a point moving on a rolling circle.

Usually particles that move in space do so on smooth curves. For example, particles do not usually disappear and spontaneously reappear at another point or change velocity suddenly. Thus we shall restrict our attention to sufficiently smooth paths, say C^1, for the rest of this section.

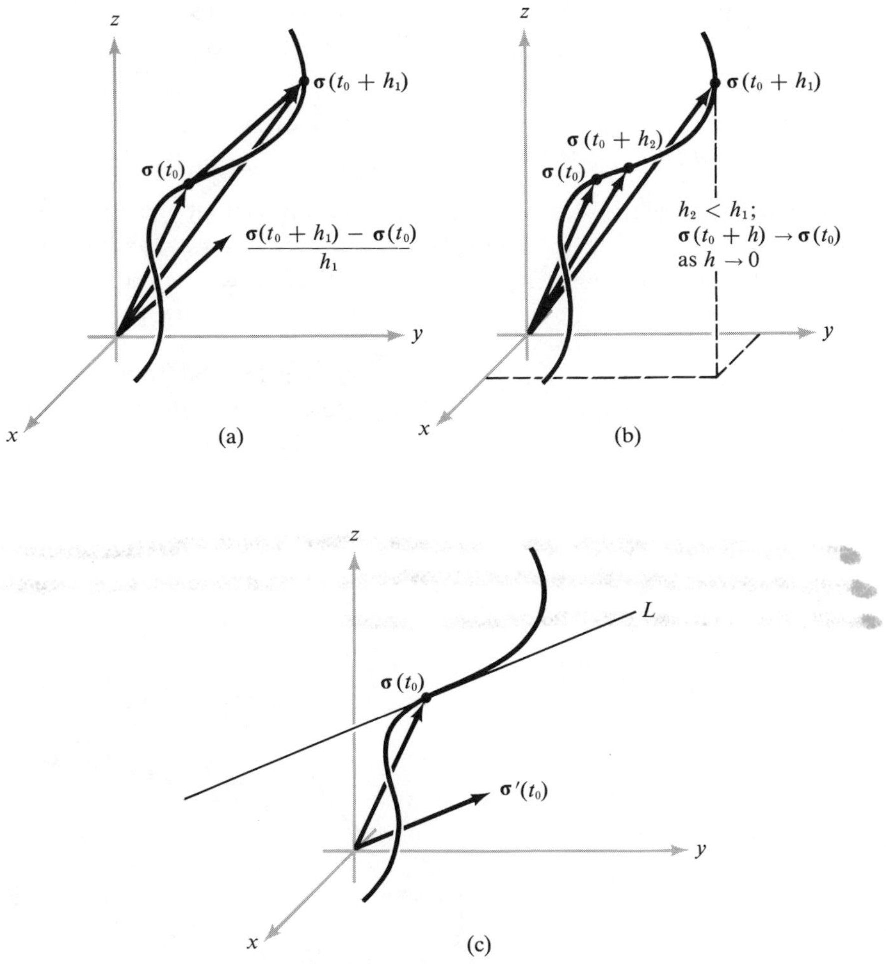

FIGURE 3.1.6

Illustrating the geometry of the definition of the derivative of a path: $\sigma'(t_0) = \lim_{h\to 0} (\sigma(t_0 + h) - \sigma(t_0))/h$.

(a) $[\sigma(t_0 + h) - \sigma(t_0)]/h$ *is a vector from* $\sigma(t_0)$ *to* $\sigma(t_0 + h)$ *for* $h = h_1$.

(b) *The same vector for a smaller increment* $h = h_2$.

(c) *The limiting case* $h \to 0$.

From our work in Chapter 2 we know that if $\boldsymbol{\sigma}$ is a path in $\mathbb{R}^3$, the derivative of $\boldsymbol{\sigma}$, $\boldsymbol{D\sigma}(t)$, is a 3×1 matrix.

$$\boldsymbol{D\sigma}(t) = \begin{bmatrix} x'(t) \\ y'(t) \\ z'(t) \end{bmatrix}$$

Let us write $\boldsymbol{\sigma}'(t)$ for the corresponding (row) vector. Thus

$$\boldsymbol{\sigma}'(t) = (x'(t), y'(t), z'(t))$$

and so

$$\boldsymbol{\sigma}'(t_0) = \operatorname*{limit}_{h \to 0} \frac{\boldsymbol{\sigma}(t_0 + h) - \boldsymbol{\sigma}(t_0)}{h}$$

Referring to Figure 3.1.6, we can argue intuitively that the vector $\boldsymbol{\sigma}'(t_0)$ ought to be parallel to the line L tangent to the path $\boldsymbol{\sigma}$ at the point $\boldsymbol{\sigma}(t_0)$, and that it ought to represent the velocity of the particle. Indeed,

$$(\boldsymbol{\sigma}(t_0 + h) - \boldsymbol{\sigma}(t_0))/h$$

represents the average directed velocity in the time interval from t_0 to $t_0 + h$ (that is, total displacement/elapsed time). Hence as $h \to 0$, this expression approaches the instantaneous velocity vector. This leads us to the following definition.

DEFINITION. *Let $\boldsymbol{\sigma}: \mathbb{R} \to \mathbb{R}^3$ be a C^1 path. The* ***velocity vector*** *at $\boldsymbol{\sigma}(t)$ is given by $\mathbf{v}(t) = \boldsymbol{\sigma}'(t) = (x'(t), y'(t), z'(t))$, and the* ***speed*** *of the particle is given by $S(t) = \|\boldsymbol{\sigma}'(t)\|$, the length of the vector $\boldsymbol{\sigma}'(t)$.*

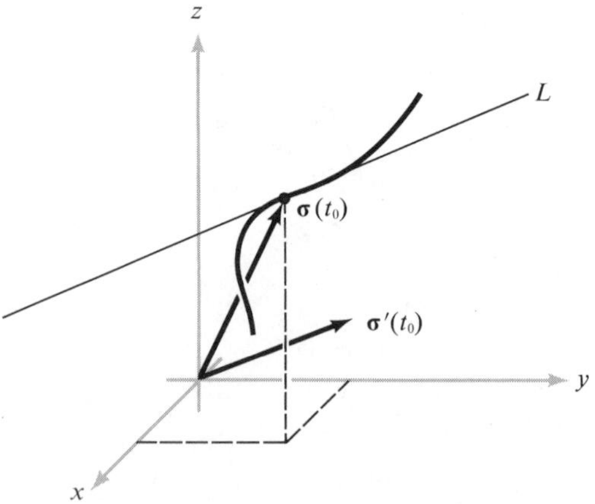

FIGURE 3.1.7

The line L tangent to a path $\boldsymbol{\sigma}$ at $\boldsymbol{\sigma}(t_0)$ has the equation $\mathbf{l}(\lambda) = \boldsymbol{\sigma}(t_0) + \lambda\boldsymbol{\sigma}'(t_0)$.

We know that the velocity vector $\boldsymbol{\sigma}'(t_0)$ is parallel to the line L tangent to the path $t \mapsto \boldsymbol{\sigma}(t)$ at the point $\boldsymbol{\sigma}(t_0)$ (Figure 3.1.7). Thus an equation of the line L tangent to $t \mapsto \boldsymbol{\sigma}(t)$ at $\boldsymbol{\sigma}(t_0)$ would be given by the formula $\lambda \mapsto \boldsymbol{\sigma}(t_0) + \lambda\boldsymbol{\sigma}'(t_0)$, where the parameter λ ranges over the real numbers (see p. 12).

DEFINITION. *Let $\boldsymbol{\sigma}$ be a C^1 curve in $\mathbb{R}^3$. The* ***tangent line*** *of $\boldsymbol{\sigma}$ at $\boldsymbol{\sigma}(t_0)$ is given in parametric form by**

$$\mathbf{l}(\lambda) = \boldsymbol{\sigma}(t_0) + \lambda\boldsymbol{\sigma}'(t_0)$$

EXAMPLE 5. If $\boldsymbol{\sigma}\colon t \mapsto (\cos t, \sin t, t)$, then the velocity vector is $\mathbf{v}(t) = \boldsymbol{\sigma}'(t) = (-\sin t, \cos t, 1)$. The speed of a point is the magnitude of the velocity:

$$S(t) = \|\mathbf{v}(t)\| = (\sin^2 t + \cos^2 t + 1)^{1/2} = \sqrt{2}$$

Thus the point moves at constant speed, although its velocity is not constant since it continually changes direction. The trajectory of the point whose motion is given by $\boldsymbol{\sigma}$ is called a (right-circular) helix (see Figure 3.1.8). The helix lies on a right-circular cylinder.

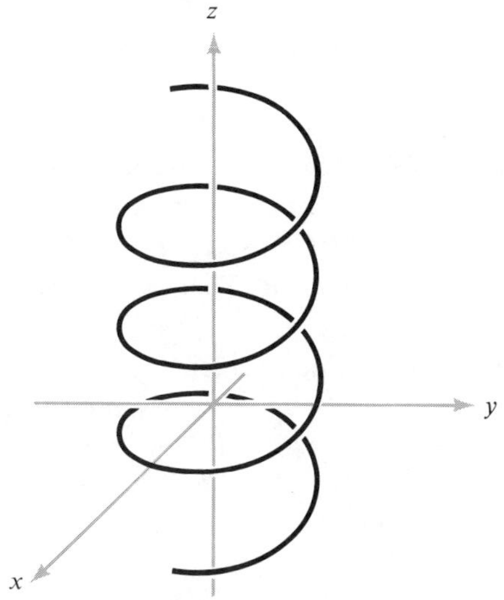

FIGURE 3.1.8
$\boldsymbol{\sigma}(t) = (\cos t, \sin t, t)$ *is a right-circular helix.*

EXAMPLE 6. Consider a particle moving on the path described in Example 5, where t is the time. At time $t = \pi$ the particle leaves the path and flies off on a tangent (as mud leaves a bicycle wheel). Find the location of the particle

* Strictly speaking, this definition should only be used if $\boldsymbol{\sigma}'(t_0) \neq \mathbf{0}$. See p. 165 for the reasons.

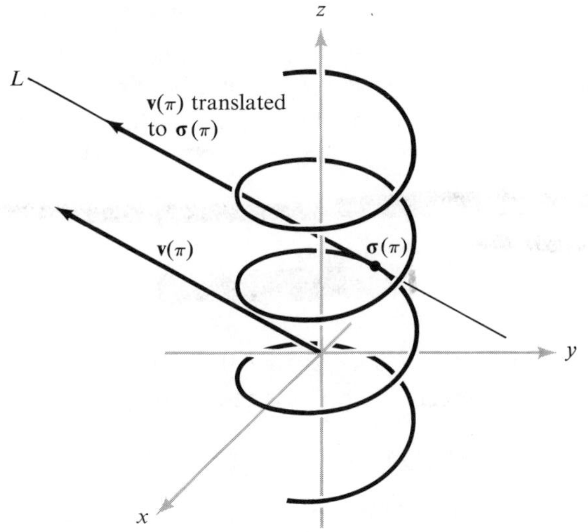

FIGURE 3.1.9
Finding the velocity vector of the right-circular helix. (The drawing is not to scale.)

at time $t = 2\pi$. Assume no forces act on it after it leaves the helix (see Figure 3.1.9).

To do this, we note that $\mathbf{v}(\pi) = (0, -1, 1)$ so that the particle, after leaving the first curve, travels in a straight path along a line L that is parallel to the velocity vector $\mathbf{v}(\pi) = \boldsymbol{\sigma}'(\pi)$. If $t \mapsto \mathbf{c}(t)$ represents the path of the particle for $t \geq \pi$, the velocity vector, $\mathbf{c}'(t)$, must be constant, since after the particle leaves the curve no forces act on it. Then $\mathbf{c}'(t) = \boldsymbol{\sigma}'(\pi) = \mathbf{v}(\pi) = (0, -1, 1)$ and $\mathbf{c}(\pi) = \boldsymbol{\sigma}(\pi) = (-1, 0, \pi)$.

Since $t \mapsto \mathbf{c}(t)$ is a straight path parallel to $\mathbf{v}(\pi)$, its equation is given by $t \mapsto \mathbf{w} + t\mathbf{v}(\pi) = \mathbf{w} + t(0, -1, 1)$, where $\mathbf{w}$ is some constant vector. To find $\mathbf{w}$ we note that $\mathbf{c}(\pi) = \mathbf{w} + \pi(0, -1, 1) = \boldsymbol{\sigma}(\pi) = (-1, 0, \pi)$, so $\mathbf{w} = (-1, 0, \pi) - (0, -\pi, \pi) = (-1, \pi, 0)$. Thus $\mathbf{c}(t) = (-1, \pi, 0) + t(0, -1, 1)$. Consequently, $\mathbf{c}(2\pi) = (-1, \pi, 0) + 2\pi(0, -1, 1) = (-1, -\pi, 2\pi)$.

Given a particle moving on a path $\boldsymbol{\sigma}(t)$, it is natural to define the rate of change of the velocity vector to be the *acceleration*. Thus

$$\mathbf{a}(t) = \boldsymbol{\sigma}''(t) = (x''(t), y''(t), z''(t))$$

If a particle of mass m moves in $\mathbb{R}^3$, the force $\mathbf{F}$ acting on it at the point $\boldsymbol{\sigma}(t)$ is related to the acceleration by *Newton's second law*:

$$\mathbf{F}(\boldsymbol{\sigma}(t)) = m\mathbf{a}(t)$$

In the very interesting problem of determining the path $\boldsymbol{\sigma}(t)$ of a particle, given its mass, initial position, and velocity and given a force, Newton's law becomes a differential equation for $\boldsymbol{\sigma}(t)$, and techniques of differential equations can be used to solve it. For example, a planet moving round the sun (considered to be located at the origin in $\mathbb{R}^3$) in a path $\mathbf{r}(t)$ obeys the law

$$m\mathbf{r}''(t) = -\frac{GmM}{\|\mathbf{r}(t)\|^3}\mathbf{r}(t) = -\frac{GmM}{r^3(t)}\mathbf{r}(t)$$

where M is the mass of the sun, m that of the planet, $r = \|\mathbf{r}\|$, and G is the gravitational constant. The fact that the force is given by $\mathbf{F} = -GmM\mathbf{r}/r^3$ is called *Newton's law of gravitation* (see Figure 3.1.10). We shall not investigate the solution of such equations in this book, but content ourselves with the following special case.

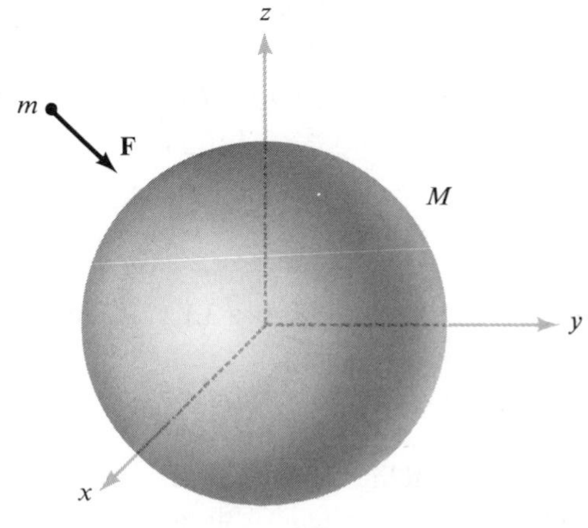

FIGURE 3.1.10

A mass M attracts a mass m with a force **F** *given by Newton's law of gravitation:* $\mathbf{F} = -GmM\mathbf{r}/r^3$.

EXAMPLE 7. Consider a particle of mass m moving at constant speed S in a circular path of radius r_0. Then, supposing it moves in the xy-plane, we can supress the third component and write

$$\mathbf{r}(t) = \left(r_0 \cos\left(\frac{tS}{r_0}\right), r_0 \sin\left(\frac{tS}{r_0}\right)\right)$$

since this is a circle of radius r_0 and $\|\mathbf{r}'(t)\| = S$. Then we can see that

$$\mathbf{a}(t) = \mathbf{r}''(t) = \left(-\frac{S^2}{r_0}\cos\left(\frac{tS}{r_0}\right), -\frac{S^2}{r_0}\sin\left(\frac{tS}{r_0}\right)\right) = -\frac{S^2}{r_0^2}\mathbf{r}(t)$$

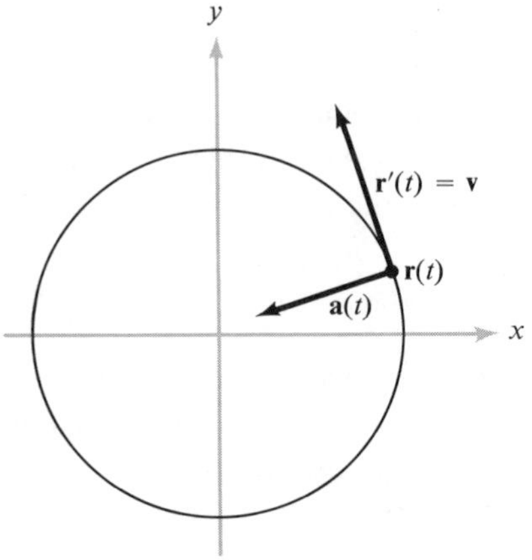

FIGURE 3.1.11
The position of a particle moving with speed S on a circle of radius r_0 is given by the equation

$$\mathbf{r}(t) = (r_0 \cos(tS/r_0), r_0 \sin(tS/r_0)),$$

and its acceleration by $\mathbf{a}(t) = -S^2\mathbf{r}(t)/r_0^2$.

Thus, the acceleration is in a direction opposite to $\mathbf{r}(t)$; that is, it is directed towards the center of the circle (see Figure 3.1.11). This acceleration multiplied by the mass of the particle is called the *centripetal force*. Note that even though the speed is constant, the direction of the velocity is continually changing, which is why an acceleration results.

Now suppose we have a satellite of mass m moving with a speed S around a central body with mass M in a *circular* orbit of radius r_0. Then by Newton's laws,

$$-\frac{S^2 m}{r_0^2}\mathbf{r}(t) = -\frac{GmM}{r_0^3}\mathbf{r}(t)$$

The lengths of the vectors on both sides of this equation must be equal. Hence

$$S^2 = \frac{GM}{r_0}$$

If T is the period of one revolution, then $2\pi r_0/T = S$; substituting this value for S in the above equation and solving for T, we obtain the rule:

$$T^2 = r_0^3 \frac{(2\pi)^2}{GM}$$

Thus the square of the period is proportional to the cube of the radius.

HISTORICAL NOTE

Example 7 illustrates one of the three famous laws of Kepler, which he observed before Newton's laws were formulated; it enables one to compute the period of a satellite with a given radius, and vice versa. Newton was able to derive from his law of gravitation Kepler's three celestial laws. The neat mathematical order of the universe that these people developed had great impact on eighteenth-century thought.

Newton never wrote down his laws as analytical equations. This was first done by Euler around 1750. (See C. Truesdell, Essays in the History of Mechanics, Springer, New York, 1968.) Newton made his deductions (at least in published form) by geometric methods alone.

We have defined two basic concepts associated with a path: its velocity and acceleration. Both involve *differential* calculus. The basic concept of the length of a path, which involves *integral* calculus, will be taken up in the next section.

EXERCISES

1. Find $\boldsymbol{\sigma}'(t)$ and $\boldsymbol{\sigma}'(0)$ in each of the following cases.
   (a) $\boldsymbol{\sigma}(t) = (\sin 2\pi t, \cos 2\pi t, 2t - t^2)$
   (b) $\boldsymbol{\sigma}(t) = (e^t, \cos t, \sin t)$
   (c) $\boldsymbol{\sigma}(t) = (t^2, t^3 - 4t, 0)$
   (d) $\boldsymbol{\sigma}(t) = (\sin 2t, \log(1 + t), t)$
2. For each of the following curves, determine the velocity and acceleration vectors, and the equation of the tangent line at the specified value of t.
   (a) $\mathbf{r}(t) = 6t\mathbf{i} + 3t^2\mathbf{j} + t^3\mathbf{k}$, $t = 0$
   (b) $\boldsymbol{\sigma}(t) = (\sin 3t, \cos 3t, 2t^{3/2})$, $t = 1$
   (c) $\boldsymbol{\sigma}(t) = (\cos^2 t, 3t - t^3, t)$, $t = 0$
   (d) $\boldsymbol{\sigma}(t) = (0, 0, t)$, $t = 1$
   (e) $\boldsymbol{\sigma}(t) = (t \sin t, t \cos t, \sqrt{3}t)$, $t = 0$
   (f) $\mathbf{r}(t) = \sqrt{2}t\mathbf{i} + e^t\mathbf{j} + e^{-t}\mathbf{k}$, $t = 0$
   (g) $\boldsymbol{\sigma}(t) = t\mathbf{i} + t\mathbf{j} + \frac{2}{3}t^{3/2}\mathbf{k}$, $t = 9$
3. In Exercise 2(a), what force acts on a particle of mass m at $t = 0$?
4. Let $\boldsymbol{\sigma}$ be a path in $\mathbb{R}^3$ with zero acceleration. Prove that $\boldsymbol{\sigma}$ is a straight line or a point.
5. Find the path $\boldsymbol{\sigma}$ such that $\boldsymbol{\sigma}(0) = (0, -5, 1)$ and $\boldsymbol{\sigma}'(t) = (t, e^t, t^2)$.
6. Find paths that represent the following curves or trajectories, and sketch.
   (a) $\{(x, y) | y = e^x\}$
   (b) $\{(x, y) | 4x^2 + y^2 = 1\}$
   (c) A straight line in $\mathbb{R}^3$ passing through the origin and the point (a, b, c).

7. A satellite orbits 500 kilometers above the Earth in a circular orbit. Compute its period. ($G = 6.67 \times 10^{-11}$, $M = 5.98 \times 10^{24}$ kilograms = mass of Earth, radius of Earth = 6.37×10^3 kilometers. Here G is given in KMS units—kilograms, meters, seconds.)
8. Suppose a particle follows the path $\boldsymbol{\sigma}(t) = (e^t, e^{-t}, \cos t)$ until it flies off on a tangent at $t = 1$. Where is it at $t = 2$?
9. Prove the following rules for differentiable paths (in $\mathbb{R}^n$ or $\mathbb{R}^3$ as appropriate).

(a) $$\frac{d}{dt}(\boldsymbol{\sigma}(t) \cdot \boldsymbol{\rho}(t)) = \frac{d\boldsymbol{\sigma}}{dt} \cdot \boldsymbol{\rho}(t) + \boldsymbol{\sigma}(t) \cdot \frac{d\boldsymbol{\rho}}{dt}$$

(b) $$\frac{d}{dt}(\boldsymbol{\sigma}(t) \times \boldsymbol{\rho}(t)) = \frac{d\boldsymbol{\sigma}}{dt} \times \boldsymbol{\rho}(t) + \boldsymbol{\sigma}(t) \times \frac{d\boldsymbol{\rho}}{dt}$$

(c) $$\frac{d}{dt}\{\boldsymbol{\sigma}(t) \cdot [\boldsymbol{\rho}(t) \times \boldsymbol{\tau}(t)]\} = \frac{d\boldsymbol{\sigma}}{dt} \cdot [\boldsymbol{\rho}(t) \times \boldsymbol{\tau}(t)] + \boldsymbol{\sigma}(t) \cdot \left[\frac{d\boldsymbol{\rho}}{dt} \times \boldsymbol{\tau}(t)\right] + \boldsymbol{\sigma}(t) \cdot \left[\boldsymbol{\rho}(t) \times \frac{d\boldsymbol{\tau}}{dt}\right]$$

10. Let $\boldsymbol{\sigma}(t)$ be a path, $\mathbf{v}(t)$ the velocity, and $\mathbf{a}(t)$ the acceleration. Suppose $\mathbf{F}$ is a vector field, $m > 0$, and $\mathbf{F}(\boldsymbol{\sigma}(t)) = m\mathbf{a}(t)$ (Newton's second law). Prove that

$$\frac{d}{dt}(m\boldsymbol{\sigma}(t) \times \mathbf{v}(t)) = \boldsymbol{\sigma}(t) \times \mathbf{F}(\boldsymbol{\sigma}(t))$$

(i.e., "rate of change of angular momentum = torque"). What can you conclude if $\mathbf{F}(\boldsymbol{\sigma}(t))$ is parallel to $\boldsymbol{\sigma}(t)$? Is this the case in planetary motion?

11. Continue the investigations in Exercise 10 to prove Kepler's law that a planet moving about the sun does so in a fixed plane.

3.2 ARC LENGTH

Consider a given path $\boldsymbol{\sigma}(t)$. We can think of $\boldsymbol{\sigma}(t)$ as the path of a particle with speed $S(t) = \|\boldsymbol{\sigma}'(t)\|$; this path traces out a curve in space. What is the length of this curve as t ranges from, say, a to b? Intuitively, this ought to be nothing more than the total distance travelled, that is, $\int_a^b S(t)\, dt$. This leads us to the following.

DEFINITION. *Let $\boldsymbol{\sigma}: [a, b] \to \mathbb{R}^n$ be a C^1 path. The **length** of $\boldsymbol{\sigma}$ is defined to be*

$$l(\boldsymbol{\sigma}) = \int_a^b \|\boldsymbol{\sigma}'(t)\|\, dt$$

In $\mathbb{R}^3$ our formula reads

$$l(\boldsymbol{\sigma}) = \int_a^b \sqrt{(x'(t))^2 + (y'(t))^2 + (z'(t))^2}\, dt$$

and for curves in $\mathbb{R}^2$, the formula reads

$$l(\boldsymbol{\sigma}) = \int_a^b \sqrt{(x'(t))^2 + y'(t)^2}\, dt$$

EXAMPLE 1. The arc length of the curve $\boldsymbol{\sigma}(t) = (r \cos t, r \sin t)$, for $0 \le t \le 2\pi$, is

$$l = \int_0^{2\pi} \sqrt{(-r \sin t)^2 + (r \cos t)^2}\, dt = 2\pi r$$

which is nothing more than the circumference of a circle of radius r. If we allowed $0 \le t \le 4\pi$, we would have gotten $4\pi r$ because the path traverses the same circle twice (Figure 3.2.1).

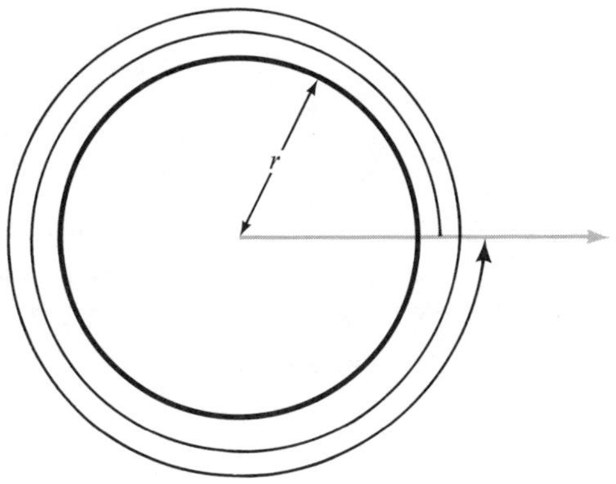

FIGURE 3.2.1
The arc length of a circle traversed twice is $4\pi r$.

EXAMPLE 2. Find the arc length of the helix $= \boldsymbol{\rho}: [0, 4\pi] \to \mathbb{R}^3, t \mapsto (\cos 2t, \sin 2t, \sqrt{5}t)$. The velocity vector is $\boldsymbol{\rho}'(t) = (-2 \sin 2t, 2 \cos 2t, \sqrt{5})$ which has magnitude

$$\|\boldsymbol{\rho}'(t)\| = \sqrt{4(\sin 2t)^2 + 4(\cos 2t)^2 + 5} = \sqrt{9} = 3$$

The arc length of $\boldsymbol{\rho}$ is therefore

$$l(\boldsymbol{\rho}) = \int_0^{4\pi} \|\boldsymbol{\rho}'(t)\|\, dt = \int_0^{4\pi} 3\, dt = 12\pi$$

EXAMPLE 3. Consider the point with position function $\boldsymbol{\sigma}: t \mapsto (t - \sin t, 1 - \cos t)$ tracing out the cycloid discussed in Example 4, Section 3.1. The velocity vector is $\boldsymbol{\sigma}'(t) = (1 - \cos t, \sin t)$ and the speed of the point $\boldsymbol{\sigma}(t)$ is

$$\|\boldsymbol{\sigma}'(t)\| = \sqrt{(1 - \cos t)^2 + \sin^2 t} = \sqrt{2 - 2 \cos t}$$

Hence, $\boldsymbol{\sigma}(t)$ moves at variable speed although, as we discovered earlier, the circle rolls at constant speed. Furthermore, the speed of $\boldsymbol{\sigma}(t)$ is zero when t is an integral multiple of 2π. At these values of t the y-coordinate of the point $\boldsymbol{\sigma}(t)$ is zero and so the point lies on the x-axis. The arc length of one cycle is $l(\boldsymbol{\sigma}) = \int_0^{2\pi} \sqrt{2 - 2 \cos t}\, dt$. This is an example of a formula for arc length that cannot be integrated by the methods of elementary calculus.

OPTIONAL

(This paragraph assumes an acquaintance with the definite integral defined in terms of Riemann sums. If your background in this is weak, the discussion may be postponed until after Chapter 5.)

In $\mathbb{R}^3$ there is another way to justify the formula for $l(\boldsymbol{\sigma})$ given in the above definition. This method is based on polygonal approximations and proceeds as follows. We partition the interval $[a, b]$ into N subintervals of equal length:

$$a = t_0 < t_1 < \cdots < t_N = b$$

$$t_{i+1} - t_i = \frac{b-a}{N} \quad \text{for} \quad 0 \le i \le N-1$$

We then consider the polygonal line obtained by joining the successive pairs of points $\boldsymbol{\sigma}(t_i), \boldsymbol{\sigma}(t_{i+1})$ for $0 \le i \le N-1$. This yields a polygonal approximation to $\boldsymbol{\sigma}$ as in Figure 3.2.2. By the formula for distance in $\mathbb{R}^3$, it follows that the line segment from $\boldsymbol{\sigma}(t_i)$ to $\boldsymbol{\sigma}(t_{i+1})$ has length

$$\|\boldsymbol{\sigma}(t_{i+1}) - \boldsymbol{\sigma}(t_i)\| = \sqrt{(x(t_{i+1}) - x(t_i))^2 + (y(t_{i+1}) - y(t_i))^2 + (z(t_{i+1}) - z(t_i))^2}$$

where $\boldsymbol{\sigma}(t) = (x(t), y(t), z(t))$. Applying the Mean Value Theorem to $x(t)$, $y(t)$, and $z(t)$ on $[t_i, t_{i+1}]$ we obtain three points t_i^*, t_i^{**}, and t_i^{***} such that

$$x(t_{i+1}) - x(t_i) = x'(t_i^*)(t_{i+1} - t_i)$$

$$y(t_{i+1}) - y(t_i) = y'(t_i^{**})(t_{i+1} - t_i)$$

and

$$z(t_{i+1}) - z(t_i) = z'(t_i^{***})(t_{i+1} - t_i)$$

Thus the line segment from $\boldsymbol{\sigma}(t_i)$ to $\boldsymbol{\sigma}(t_{i+1})$ has length

$$\sqrt{(x'(t_i^*))^2 + (y'(t_i^{**}))^2 + (z'(t_i^{***}))^2}(t_{i+1} - t_i)$$

Therefore the length of our approximating polygonal line is

$$S_N = \sum_{i=0}^{N-1} \sqrt{(x'(t_i^*))^2 + (y'(t_i^{**}))^2 + (z'(t_i^{***}))^2}(t_{i+1} - t_i)$$

As $N \to \infty$, this polygonal line approximates the image of $\boldsymbol{\sigma}$ more closely. Therefore, we define the arc length of $\boldsymbol{\sigma}$ as the limit, if it exists, of the sequence S_N as $N \to \infty$. Since the derivatives x', y', and z' are all assumed to be continuous on $[a, b]$, we can conclude that, in fact, the limit does exist and is given by

$$\lim_{N \to \infty} S_N = \int_a^b \sqrt{(x'(t))^2 + (y'(t))^2 + (z'(t))^2}\, dt$$

OPTIONAL (*Continued*)

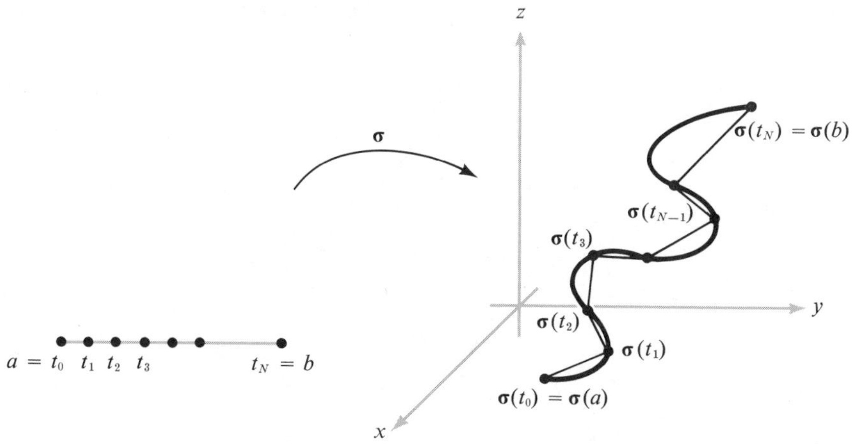

FIGURE 3.2.2
A path $\boldsymbol{\sigma}$ *may be approximated by a polygonal path obtained by joining each* $\boldsymbol{\sigma}(t_i)$ *to* $\boldsymbol{\sigma}(t_{i+1})$ *by a straight line.*

(The theory of integration relates the integral to sums by the formula

$$\int_a^b f(t)\,dt = \lim_{N\to\infty} \sum_{i=0}^{N-1} f(t_i^*)(t_{i+1} - t_i),$$

where $t_0, \ldots, t_N$ is a partition of $[a, b]$, $t_i^* \in [t_i, t_{i+1}]$ is arbitrary, and f is a continuous function. Here we have *possibly different points* t_i^*, t_i^{**}, and t_i^{***}, so this formula must be extended slightly.)

The image of a C^1 path is not necessarily "very smooth"; indeed it may have sharp bends or changes of direction. For instance, the cycloid in the above example has cusps at all points where $\boldsymbol{\sigma}(t)$ touches the x-axis (that is, where $t = 2\pi n$, $n = 0, \pm 1, \ldots$). Another example is the *hypocycloid of four cusps*, $\boldsymbol{\sigma}\colon [0, 2\pi] \to \mathbb{R}^2$, $t \mapsto (\cos^3 t, \sin^3 t)$, which has cusps at four points (Figure 3.2.3). However, at all such points $\boldsymbol{\sigma}'(t) = \mathbf{0}$, the tangent line is not well defined, and the speed of the point $\boldsymbol{\sigma}(t)$ is zero. Evidently the direction of $\boldsymbol{\sigma}(t)$ may change abruptly at points where it slows to rest.

If we have a path $\boldsymbol{\sigma}(t) = (x(t), y(t), z(t))$ in $\mathbb{R}^3$, it is sometimes conventional to denote it by $\mathbf{s}(t) = \boldsymbol{\sigma}(t)$, so

$$\boldsymbol{\sigma}'(t) = \frac{d\mathbf{s}}{dt} = \frac{dx}{dt}\mathbf{i} + \frac{dy}{dt}\mathbf{j} + \frac{dz}{dt}\mathbf{k}$$

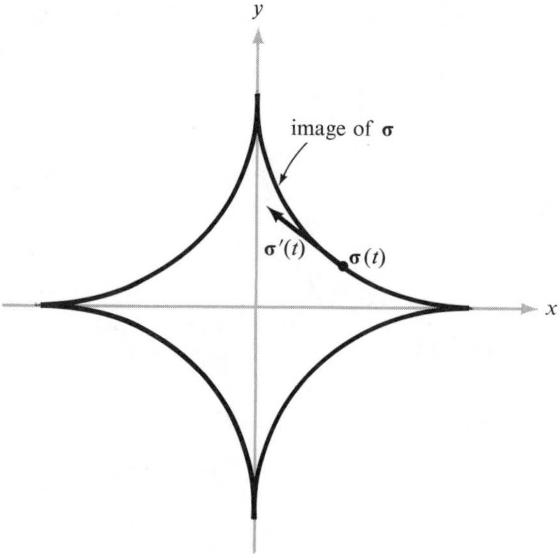

FIGURE 3.2.3
The image of the smooth path $\boldsymbol{\sigma}(t) = (\cos^3 t, \sin^3 t)$, *a hypocycloid, does not "look smooth."*

Thus

$$\|\boldsymbol{\sigma}'(t)\| = \left\|\frac{d\mathbf{s}}{dt}\right\| = \sqrt{\left(\frac{dx}{dt}\right)^2 + \left(\frac{dy}{dt}\right)^2 + \left(\frac{dz}{dt}\right)^2}$$

The arc length of $\boldsymbol{\sigma}$ may be written in this notation as

$$l(\boldsymbol{\sigma}) = \int_a^b \|d\mathbf{s}/dt\|\, dt$$

It is also common to introduce the *arc-length function* $s(t)$ by

$$s(t) = \int_a^t \|\boldsymbol{\sigma}'(\tau)\|\, d\tau,$$

so

$$s'(t) = \|\boldsymbol{\sigma}'(t)\| = \|d\mathbf{s}/dt\|$$

and

$$l(\boldsymbol{\sigma}) = \int_a^b s'(t)\, dt.$$

EXAMPLE 4. A particle moves along the hypocycloid according to the equations

$$x = \cos^3 t \qquad y = \sin^3 t \qquad a \le t \le b$$

The velocity vector of the particle is

$$\frac{d\mathbf{s}}{dt} = \frac{dx}{dt}\mathbf{i} + \frac{dy}{dt}\mathbf{j} = -(3 \sin t \cos^2 t)\mathbf{i} + (3 \cos t \sin^2 t)\mathbf{j}$$

and its speed is

$$S(t) = s'(t) = \left\|\frac{d\mathbf{s}}{dt}\right\| = (9\sin^2 t\cos^4 t + 9\cos^2 t\sin^4 t)^{1/2}$$
$$= 3|\sin t|\,|\cos t|$$

If we differentiate the velocity vector

$$\mathbf{v}(t) = \frac{d\mathbf{s}}{dt} = \frac{dx}{dt}\mathbf{i} + \frac{dy}{dt}\mathbf{j} + \frac{dz}{dt}\mathbf{k}$$

of a moving particle, we obtain its acceleration vector, as we have seen. In this notation the acceleration vector is given by the formula

$$\mathbf{a}(t) = \frac{d\mathbf{v}}{dt} = \frac{d^2x}{dt^2}\mathbf{i} + \frac{d^2y}{dt^2}\mathbf{j} + \frac{d^2z}{dt^2}\mathbf{k}$$

EXAMPLE 5. Let $\boldsymbol{\sigma}(t) = (e^t, t^2, \cos t)$. Then the acceleration vector is $\mathbf{a}(t) = e^t\mathbf{i} + 2\mathbf{j} - (\cos t)\mathbf{k}$.

The definition of arc length can be extended to include paths that are not C^1 but that are formed by piecing together a finite number of C^1 paths. A path $\boldsymbol{\sigma}\colon [a, b] \to \mathbb{R}^3$, $t \mapsto (x(t), y(t), z(t))$ is called *piecewise* C^1 if there is a partition of $[a, b]$

$$a = t_0 < t_1 < \cdots < t_N = b$$

such that the function $\boldsymbol{\sigma}$ restricted to each interval $[t_i, t_{i+1}]$, $0 \le i \le N - 1$, is continuously differentiable. By this we mean that the derivative exists on (t_i, t_{i+1}) and is continuous on $[t_i, t_{i+1}]$. The derivatives at the endpoints of each interval are computed by using limits from within the interval (i.e., one-sided limits, as on p. 102). In the case of a path that is piecewise C^1, we define the arc length of the path to be the sum of the arc lengths of the C^1 paths that make it up. That is, if the partition

$$a = t_0 < t_1 < \cdots < t_N = b$$

satisfies the above conditions, we define

$$\text{arc length of } \boldsymbol{\sigma} = \sum_{i=0}^{N-1} (\text{arc length of } \boldsymbol{\sigma} \text{ from } t_i \text{ to } t_{i+1})$$

EXAMPLE 6. The path $\boldsymbol{\sigma}\colon [-1, 1] \to \mathbb{R}^3$, $t \mapsto (|t|, |t - \frac{1}{2}|, 0)$ is not C^1 because $\sigma_1\colon t \mapsto |t|$ is not differentiable at 0, nor is $\sigma_2\colon t \mapsto |t - \frac{1}{2}|$ differentiable at $\frac{1}{2}$. However, if we take the partition

$$-1 = t_0 < 0 = t_1 < \tfrac{1}{2} = t_2 < 1 = t_3$$

we see that each of the σ_i is continuously differentiable on each of the intervals $[-1, 0]$, $[0, \frac{1}{2}]$, and $[\frac{1}{2}, 1]$, and therefore $\boldsymbol{\sigma}$ is continuously differentiable on

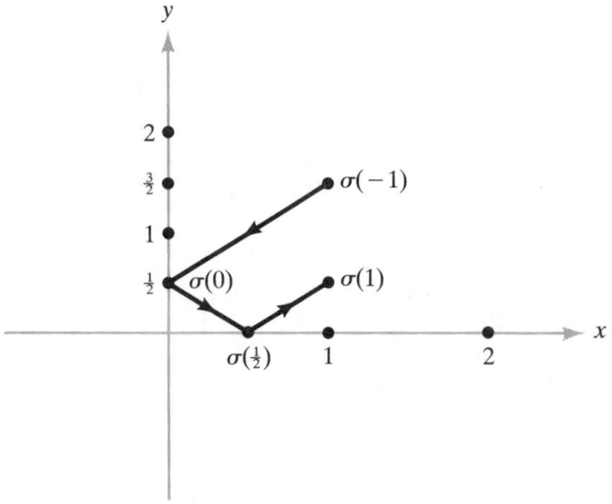

FIGURE 3.2.4
A specific C^1 path: $\boldsymbol{\sigma}(t) = |t|\mathbf{i} + |t - \frac{1}{2}|\mathbf{j}$, $-1 \leq t \leq 1$.

each interval. Thus, $\boldsymbol{\sigma}$ is piecewise C^1. (See Figure 3.2.4.)

On $[-1, 0]$, we have $x(t) = -t$, $y(t) = -t + \frac{1}{2}$, $z(t) = 0$, $\|d\mathbf{s}/dt\| = \sqrt{2}$; hence, the arc length of $\boldsymbol{\sigma}$ between -1 and 0 is $\int_{-1}^{0} \sqrt{2}\, dt = \sqrt{2}$. Similarly, on $[0, \frac{1}{2}]$, $x(t) = t$, $y(t) = -t + \frac{1}{2}$, $z(t) = 0$, and again $\|d\mathbf{s}/dt\| = \sqrt{2}$, so that the arc length of $\boldsymbol{\sigma}$ between 0 and $\frac{1}{2}$ is $\frac{1}{2}\sqrt{2}$. Finally, on $[\frac{1}{2}, 1]$ we have $x(t) = t$, $y(t) = t - \frac{1}{2}$, $z(t) = 0$, and the arc length of $\boldsymbol{\sigma}$ between $\frac{1}{2}$ and 1 is $\frac{1}{2}\sqrt{2}$. Thus the total arc length of $\boldsymbol{\sigma}$ is $2\sqrt{2}$.

EXERCISES

1. Calculate the arc length of the given curve on the specified interval.*
   (a) The path in Exercise 2(a), Section 3.1, $[0, 1]$
   (b) The path in Exercise 2(b), Section 3.1, $[0, 1]$
   (c) $\mathbf{s}(t) = t\mathbf{i} + t(\sin t)\mathbf{j} + t(\cos t)\mathbf{k}$, $[0, \pi]$
   (d) $\mathbf{s}(t) = 2t\mathbf{i} + t\mathbf{j} + t^2\mathbf{k}$, $[0, 2]$
   (e) The path in Exercise 2(e), Section 3.1, $[0, 1]$
   (f) The path in Exercise 2(f), Section 3.1, $[-1, 1]$
   (g) The path in Exercise 2(g), Section 3.1, $[t_0, t_1]$

* Several of these problems make use of the formula

$$\int \sqrt{x^2 + a^2}\, dx = \tfrac{1}{2}\left(x\sqrt{x^2 + a^2} + a^2 \log(x + \sqrt{x^2 + a^2})\right) + C$$

from the table of integrals in the back of the book.

2. The arc-length function $s(t)$ for a given path $\boldsymbol{\sigma}(t)$, defined by $s(t) = \int_a^t \|\boldsymbol{\sigma}'(\tau)\|\, d\tau$, represents the distance a particle traversing the trajectory of $\boldsymbol{\sigma}$ will have travelled by time t if it starts out at time a; that is, the length of $\boldsymbol{\sigma}$ between $\boldsymbol{\sigma}(a)$ and $\boldsymbol{\sigma}(t)$. Find the arc-length functions for the curves $\boldsymbol{\alpha}(t) = (\cosh t, \sinh t, t)$ and $\boldsymbol{\beta}(t) = (\cos t, \sin t, t)$, with $a = 0$.

3. Let $\boldsymbol{\sigma}$ be the path $\boldsymbol{\sigma}(t) = (2t, t^2, \log t)$, defined for $t > 0$. Find the arc length of $\boldsymbol{\sigma}$ between the points $(2, 1, 0)$ and $(4, 4, \log 2)$.

4. Suppose a particle following the path $\boldsymbol{\sigma}(t) = (t^2, t^3 - 4t, 0)$ flies off on a tangent at $t = 2$. Compute the position of the particle at $t = 3$.

5. Let $\mathbf{c}(t)$ be a given path, $a \le t \le b$. Let $s = \alpha(t)$ be a new variable, where α is a strictly increasing C^1 function given on $[a, b]$. For each s in $[\alpha(a), \alpha(b)]$ there is a unique t with $\alpha(t) = s$. Define the function $\mathbf{d}\colon [\alpha(a), \alpha(b)] \to \mathbb{R}^3$ by $\mathbf{d}(s) = \mathbf{c}(t)$.
   (a) Argue that the image curves of $\mathbf{c}$ and $\mathbf{d}$ are the same.
   (b) Show that $\mathbf{c}$ and $\mathbf{d}$ have the same arc length.
   (c) Let $s = \alpha(t) = \int_a^t \|\mathbf{c}'(\tau)\|\, d\tau$. Define $\mathbf{d}$ as above by $\mathbf{d}(s) = \mathbf{c}(t)$. Show that

$$\left\| \frac{d}{ds}\mathbf{d}(s) \right\| = 1.$$

   The path $s \mapsto \mathbf{d}(s)$ is said to be a *reparametrization* of $\mathbf{c}$.

Exercises 6 to 11 are intended to develop some of the classical differential geometry of curves.

6. Let $\boldsymbol{\sigma}\colon [a, b] \to \mathbb{R}^3$ be an infinitely differentiable path (derivatives of all orders exist). Assume $\boldsymbol{\sigma}'(t) \neq \mathbf{0}$ for any t. The vector $\boldsymbol{\sigma}'(t)/\|\boldsymbol{\sigma}'(t)\| = \mathbf{T}(t)$ is tangent to $\boldsymbol{\sigma}$ at $\boldsymbol{\sigma}(t)$, and, since $\|\mathbf{T}(t)\| = 1$, $\mathbf{T}$ is called the *unit tangent* to $\boldsymbol{\sigma}$.
   (a) Show that $\mathbf{T}'(t) \cdot \mathbf{T}(t) = 0$. (HINT: Differentiate $\mathbf{T}(t) \cdot \mathbf{T}(t) = 1$.)
   (b) Write down a formula for $\mathbf{T}'(t)$.

7. (a) A path $\boldsymbol{\sigma}(s)$ is said to be *parametrized by arc length* or, what is the same thing, to have *unit speed* if $\|\boldsymbol{\sigma}'(s)\| = 1$. Show that for a path parametrized by arc length on $[a, b]$, $l(\boldsymbol{\sigma}) = b - a$.
   (b) The *curvature* at a point $\boldsymbol{\sigma}(s)$ on a path is defined by $k = \|\mathbf{T}'(s)\|$ when the path is parametrized by arc length (see Exercises 6 and 7(a)). Show that $k = \|\boldsymbol{\sigma}''(s)\|$.
   (c) If $\boldsymbol{\sigma}$ is given in terms of some other parameter t and $\boldsymbol{\sigma}'(t)$ is never $\mathbf{0}$, show that $k = \|\boldsymbol{\sigma}'(t) \times \boldsymbol{\sigma}''(t)\|/\|\boldsymbol{\sigma}'(t)\|^3$, using Exercise 5.
   (d) Calculate the curvature of the helix $\boldsymbol{\sigma}(t) = (1/\sqrt{2})(\cos t, \sin t, t)$. (This $\boldsymbol{\sigma}$ is just a scalar multiple of the right-circular helix depicted in Figure 3.1.8.)

8. If $\mathbf{T}(t) \neq \mathbf{0}$ it follows from Exercise 6 that $\mathbf{N}(t) = \mathbf{T}'(t)/\|\mathbf{T}'(t)\|$ is normal (i.e., perpendicular) to $\boldsymbol{\sigma}$; $\mathbf{N}$ is called the *principal normal vector*. Let a third unit vector that is perpendicular to both $\mathbf{T}$ and $\mathbf{N}$ be defined by $\mathbf{B} = \mathbf{T} \times \mathbf{N}$; $\mathbf{B}$ is called the *binormal vector*. Together, $\mathbf{T}$, $\mathbf{N}$, and $\mathbf{B}$ form a right-handed system of mutually orthogonal vectors that may be thought of as moving along the path.
   (a) Show that $\dfrac{d\mathbf{B}}{dt} \cdot \mathbf{B} = 0$.
   (b) Show that $\dfrac{d\mathbf{B}}{dt} \cdot \mathbf{T} = 0$.

(c) Show that $\frac{d\mathbf{B}}{dt}$ is a scalar multiple of $\mathbf{N}$.

9. If $\boldsymbol{\sigma}(s)$ is parametrized by arc length, we use the result of Exercise 8(c) to define a scalar-valued function τ, *torsion*, by

$$\frac{d\mathbf{B}}{ds} = -\tau\mathbf{N}$$

(a) Show that $\tau = [\boldsymbol{\sigma}'(s) \times \boldsymbol{\sigma}''(s)] \cdot \boldsymbol{\sigma}'''(s)/\|\boldsymbol{\sigma}''(s)\|^2$.
(b) Show that if $\boldsymbol{\sigma}$ is given in terms of some other parameter t,

$$\tau = \frac{[\boldsymbol{\sigma}'(t) \times \boldsymbol{\sigma}''(t)] \cdot \boldsymbol{\sigma}'''(t)}{\|\boldsymbol{\sigma}'(t) \times \boldsymbol{\sigma}''(t)\|^2}$$

Compare with 7(c).
(c) Compute the torsion of the helix $\boldsymbol{\sigma}(t) = (1/\sqrt{2})(\cos t, \sin t, t)$.

*10. Show that if a path lies in a plane then the torsion is zero. Conclude by demonstrating that $\mathbf{B}$ is constant and is a normal vector to the plane in which $\boldsymbol{\sigma}$ lies. (If the torsion is not zero, it gives a measure of how fast the curve is twisting out of the plane of $\mathbf{T}$ and $\mathbf{N}$.)

*11. Use the results of Exercises 8 and 9 to prove the following Frenet formulas for a unit-speed curve:

$$\frac{d\mathbf{T}}{ds} = k\mathbf{N}; \qquad \frac{d\mathbf{N}}{ds} = -k\mathbf{T} + \tau\mathbf{B}; \qquad \frac{d\mathbf{B}}{ds} = -\tau\mathbf{N}$$

3.3 VECTOR FIELDS

In Chapter 2 we introduced vector fields through the idea of the gradient vector field. In this section we would like to study some general properties of vector fields, including their geometrical and physical significance. A clear understanding of this is important for our work in Sections 3.4 and 3.5 and for our studies in Chapter 7.

DEFINITION. *A **vector field** on $\mathbb{R}^n$ is a map $\mathbf{F}: A \subset \mathbb{R}^n \to \mathbb{R}^n$ that assigns to each point $\mathbf{x}$ in its domain A a vector $\mathbf{F}(\mathbf{x})$.*

Graphically we may picture $\mathbf{F}$ as attaching an arrow to each point (Figure 3.3.1). Similarly a map $f: A \subset \mathbb{R}^n \to \mathbb{R}$ that assigns a number to each point is called a *scalar field.* For example, a vector field $\mathbf{F}(x, y, z)$ on $\mathbb{R}^3$ has three component scalar fields F_1, F_2, and F_3, so that $\mathbf{F}(x, y, z) = (F_1(x, y, z), F_2(x, y, z), F_3(x, y, z))$.

It is convenient and quite natural to draw the arrow representing $\mathbf{F}(\mathbf{x})$ with its foot at $\mathbf{x}$ rather than at the origin (which is the normal custom for drawing vectors). We regard this displaced vector with foot at $\mathbf{x}$ and the

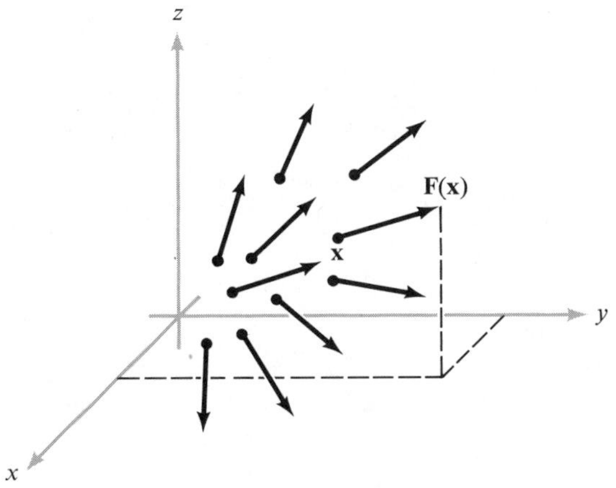

FIGURE 3.3.1
A vector field **F** *assigns a vector (arrow)* **F(x)** *to each point* **x** *of its domain.*

corresponding vector with foot at **0** as equivalent. In the remainder of this book we shall be concerned primarily with vector fields on $\mathbb{R}^2$ and $\mathbb{R}^3$, so that the term *vector field* should be taken to mean a vector field on $\mathbb{R}^2$ or $\mathbb{R}^3$ unless otherwise stated.

EXAMPLE 1. Imagine fluid moving down a pipe in a steady flow. If we attach to each point the fluid velocity at that point we obtain the velocity field **V** of the fluid (see Figure 3.3.2). Notice that the length of the arrows (the speed), as well as the direction of flow, can change from point to point.

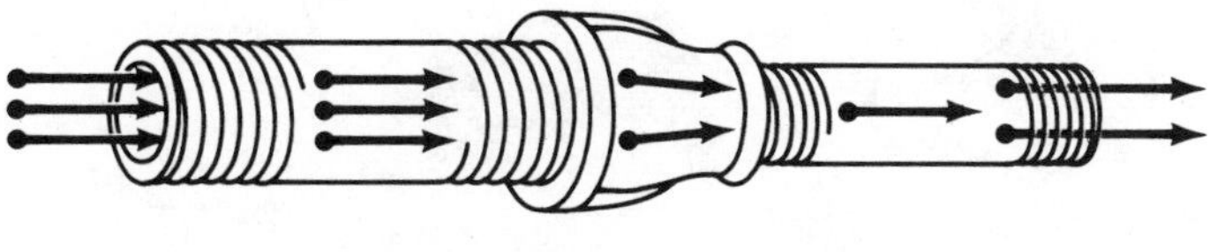

FIGURE 3.3.2
A vector field describing the velocity of flow in a pipe.

EXAMPLE 2. Consider a piece of material that is heated on one side and cooled on another. The temperature at each point within the body yields a scalar field $T(x, y, z)$.

The actual flow of heat may be marked by a field of arrows indicating the direction and magnitude of the flow (Figure 3.3.3). This *energy flux vector field* is given by $\mathbf{J} = -k\,\nabla T$, where $k > 0$ is a constant called the *conductivity*, and ∇T is the gradient of the real-valued function T. Note that the

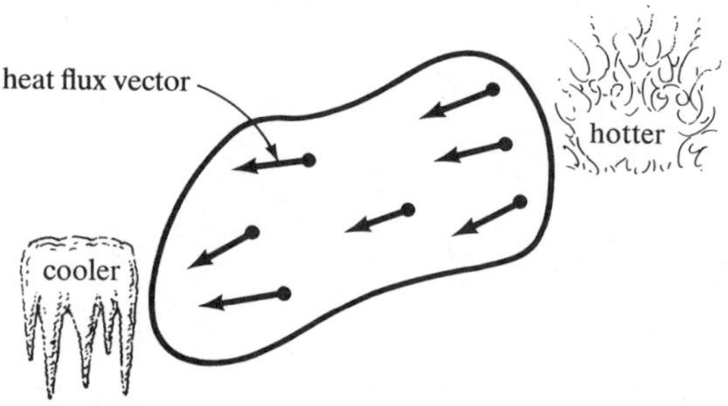

FIGURE 3.3.3
A vector field describing the direction and magnitude of heat flow.

heat flows, as it should, from hot regions toward cold ones, since $-\nabla T$ points in a direction of decreasing T (see Section 2.5).

EXAMPLE 3. The force of attraction of the Earth on a mass m can be described by a vector field on $\mathbb{R}^3$, the gravitational force field. According to Newton's law, this field is given by (see Sections 2.5 and 3.1)

$$\mathbf{F} = -\frac{mMG}{r^3}\mathbf{r}$$

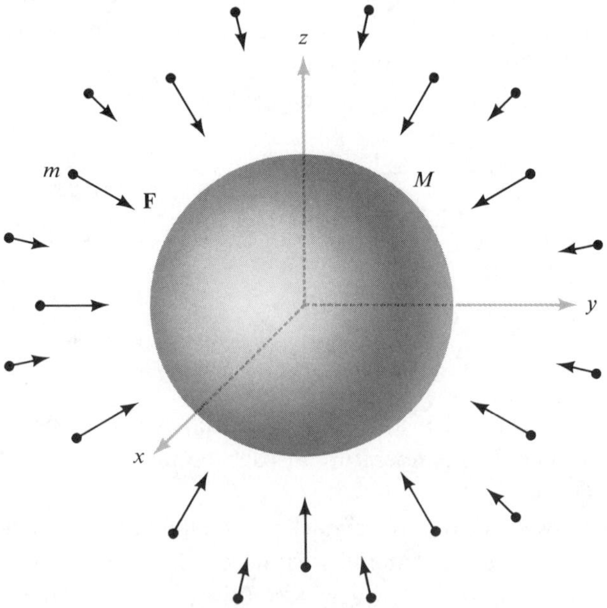

FIGURE 3.3.4
The vector field **F** *given by Newton's law of gravitation.*

where $\mathbf{r}(x, y, z) = (x, y, z)$, and $r = \|\mathbf{r}\|$ (see Figure 3.3.4). As we saw earlier (Example 5, Section 2.5), $\mathbf{F}$ is actually a gradient field, $\mathbf{F} = -\nabla V$ where $V = -(mMG)/r$. Note again that $\mathbf{F}$ points in the direction of decreasing V. Writing $\mathbf{F}$ out in terms of components we see that

$$\mathbf{F}(x, y, z) = \left(\frac{-mMG}{r^3}x, \frac{-mMG}{r^3}y, \frac{-mMG}{r^3}z\right)$$

EXAMPLE 4. In the plane, $\mathbb{R}^2$, the function $\mathbf{V}\colon \mathbb{R}^2\backslash\{\mathbf{0}\} \to \mathbb{R}^2$ defined by

$$\mathbf{V}(x, y) = \frac{y\mathbf{i}}{x^2 + y^2} - \frac{x\mathbf{j}}{x^2 + y^2} = \left(\frac{y}{x^2 + y^2}, -\frac{x}{x^2 + y^2}\right)$$

is a vector field on $\mathbb{R}^2$. This is the velocity field approximating the velocity field of water in "circular" motion such as occurs, for example, when you pull the plug in a tub of water (Figure 3.3.5).

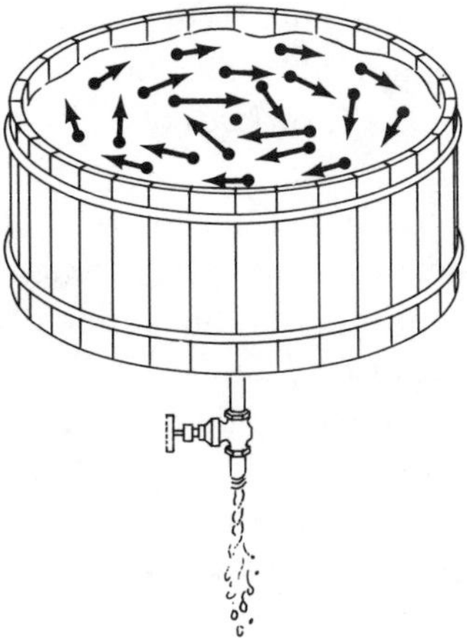

FIGURE 3.3.5
The vector field describing circular flow in a tub.

EXAMPLE 5. According to Coulomb's law, the force acting on a charge e at $\mathbf{r}$ due to a charge Q at the origin is

$$\mathbf{F} = \frac{\varepsilon Qe}{r^3}\mathbf{r} = -\nabla V$$

where $V = \varepsilon Qe/r$ and ε is a constant that depends on the units used. For $Qe > 0$ (like charges) the force is repulsive (Figure 3.3.6(a)) and for $Qe < 0$ (unlike charges) the force is attractive. In this example, as in Example 3, the level surfaces of V are called *equipotential surfaces* since, on them, the potential V is constant. Note that the force field is orthogonal to the equipotential

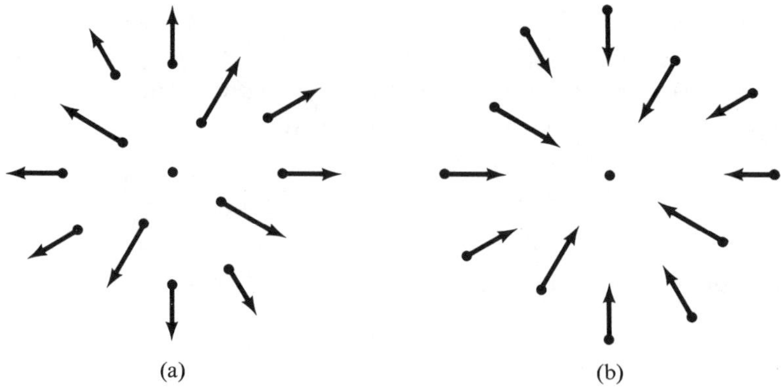

FIGURE 3.3.6
The vector fields associated with (a) like charges ($Qe > 0$) and (b) unlike charges ($Qe < 0$).

surfaces (the force field is radial and the equipotential surfaces are spheres). This agrees with our general result in Section 2.5. In the case of temperature gradients in which $\mathbf{F} = -k\,\nabla T$, surfaces of constant T are called *isotherms*.

We remark that, in general, a vector field need not be a gradient field; that is, a vector field need not be the gradient of a real-valued function. This will become clearer in later chapters. Thus the notion of an equipotential surface only makes sense if the vector field happens to be a gradient field.

Another concept of importance is the notion of a *flow line*. This idea is easiest to visualize in the context of Example 1. In that case, a flow line is just a path followed by a small particle suspended in the fluid (Figure 3.3.7). These lines are also appropriately called *streamlines* or *integral curves*.

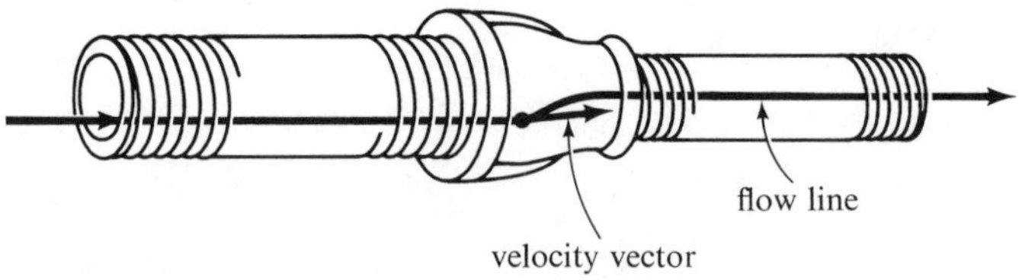

FIGURE 3.3.7
The velocity vector of a fluid is tangent to a flow line.

DEFINITION. *If* $\mathbf{F}$ *is a vector field, a* ***flow line*** *for* $\mathbf{F}$ *is a path* $\boldsymbol{\sigma}(t)$ *such that*

$$\boldsymbol{\sigma}'(t) = \mathbf{F}(\boldsymbol{\sigma}(t)) \tag{1}$$

That is, $\mathbf{F}$ *yields the velocity field of the path* $\boldsymbol{\sigma}(t)$.

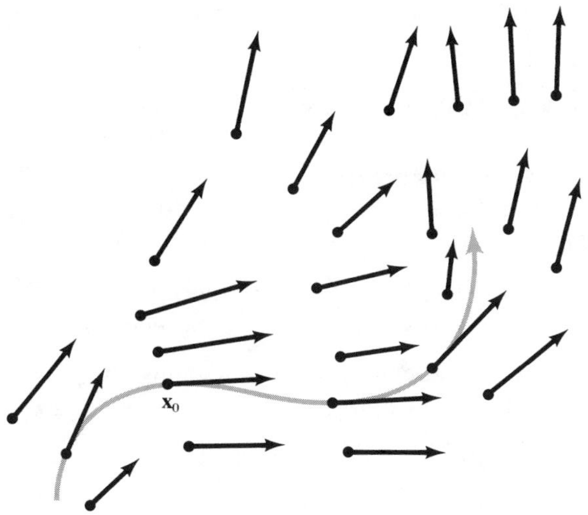

FIGURE 3.3.8
A flow line threading its way through a vector field in the plane.

Geometrically, the problem of finding a flow line through a given point $\mathbf{x}_0$ for a given vector field $\mathbf{F}$ is that of threading a curve through the vector field in such a way that the tangent vector of the curve coincides with vector field, as in Figure 3.3.8.

Analytically, the problem of finding a flow line that passes through $\mathbf{x}_0$ at time $t = 0$ involves solving the differential equation (1) with initial condition $\mathbf{x}_0$; that is,

$$\boldsymbol{\sigma}'(t) = \mathbf{F}(\boldsymbol{\sigma}(t)); \qquad \boldsymbol{\sigma}(0) = \mathbf{x}_0$$

Using coordinates (x, y, z), the preceding equation can be written as the simultaneous equations

$$x'(t) = F_1(x(t), y(t), z(t))$$
$$y'(t) = F_2(x(t), y(t), z(t))$$
$$z'(t) = F_3(x(t), y(t), z(t))$$

with initial conditions

$$(x(0), y(0), z(0)) = (x_0, y_0, z_0)$$

where $\mathbf{F} = (F_1, F_2, F_3)$. In courses on differential equations it is proved that if F is C^1 there is a unique solution for each $\mathbf{x}_0$ (but not necessarily defined for all t).

OPTIONAL

FLOWS OF VECTOR FIELDS

It is convenient for the unique solution through a given point at time 0 to have a special notation, which will be used in optional material in the next section. We let

$$\phi(\mathbf{x}, t) = \begin{Bmatrix} \text{the position of the point on the flow line} \\ \text{through } \mathbf{x} \text{ after time } t \text{ has elapsed.} \end{Bmatrix}$$

Thus, with $\mathbf{x}$ as the initial condition, follow along the flow line for a time period t until the new position $\phi(\mathbf{x}, t)$ is reached (see Figure 3.3.9). Analytically, $\phi(\mathbf{x}, t)$ is defined by:

$$\left.\begin{aligned} \frac{\partial}{\partial t}\phi(\mathbf{x}, t) &= \mathbf{F}(\phi(\mathbf{x}, t)) \\ \phi(\mathbf{x}, 0) &= \mathbf{x} \end{aligned}\right\} \tag{2}$$

We call the mapping ϕ, which is regarded as the function of the variables $\mathbf{x}$ and t, the *flow* of $\mathbf{F}$.

Let $D_{\mathbf{x}}$ denote differentiation with respect to $\mathbf{x}$, holding t fixed. It is proved in courses on differential equations that ϕ is a differentiable function of $\mathbf{x}$. Thus by differentiating (2) with respect to $\mathbf{x}$,

$$D_{\mathbf{x}}\frac{\partial}{\partial t}\phi(\mathbf{x}, t) = D_{\mathbf{x}}[\mathbf{F}(\phi(\mathbf{x}, t))]$$

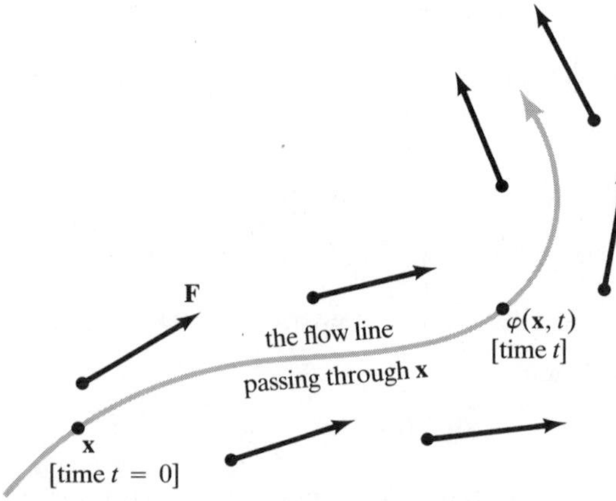

FIGURE 3.3.9
The definition of the flow $\phi(\mathbf{x}, t)$ of $\mathbf{F}$.

OPTIONAL (*Continued*)

The equality of mixed partials may be used on the left-hand side of this equation and the Chain Rule applied to the right-hand side, yielding

$$\frac{\partial}{\partial t} D_{\mathbf{x}}\phi(\mathbf{x}, t) = D_{\mathbf{x}}\mathbf{F}(\phi(\mathbf{x}, t)) \cdot D_{\mathbf{x}}\phi(\mathbf{x}, t)) \tag{3}$$

This, a linear differential equation for $D_{\mathbf{x}}\phi(\mathbf{x}, t)$, is called the *equation of first variation.* It will be useful in our discussion of divergence and curl in the next section. Notice that in space both $D_{\mathbf{x}}\mathbf{F}(\phi)$ and $D_{\mathbf{x}}\phi$ are 3×3 matrices since $\mathbf{F}$ and ϕ take values in $\mathbb{R}^3$ and are differentiated with respect to $\mathbf{x} \in \mathbb{R}^3$. Similarly, for vector fields in the plane, they would be 2×2 matrices.

EXERCISES

1. Let a particle of mass m move on a path $\mathbf{r}(t)$ in a force field $\mathbf{F} = -\nabla V$ on $\mathbb{R}^3$ according to Newton's law.
   (a) Prove that the energy $E = \frac{1}{2}m\|\mathbf{r}'(t)\|^2 + V(\mathbf{r}(t))$ is constant in time. (HINT: Carry out the differentiation dE/dt.)
   (b) If the particle moves on an equipotential surface show that its speed is constant.
2. Sketch a few flow lines of the vector fields
   (a) $\mathbf{F}(x, y) = (y, -x)$
   (b) $\mathbf{F}(x, y) = (x, -y)$
   (c) $\mathbf{F}(x, y) = (x, x^2)$
3. Let $\mathbf{c}(t)$ be a flow line of a gradient field $\mathbf{F} = -\nabla V$. Prove $V(\mathbf{c}(t))$ is a decreasing function of t. Explain.
4. Sketch the gradient field $-\nabla V$ for $V(x, y) = (x + y)/(x^2 + y^2)$. Sketch the equipotential surface $V = 1$.
5. Suppose that the isotherms in a region are all concentric spheres centered at the origin. Prove that the energy flux vector field points either toward or away from the origin.

*6. (a) Assuming uniqueness of flow lines through a given point at a given time, prove the following property of the flow $\phi(\mathbf{x}, t)$ of a vector field $\mathbf{F}$:

$$\phi(\mathbf{x}, t + s) = \phi(\phi(\mathbf{x}, s), t)$$

   (b) What is the corresponding property for $D_{\mathbf{x}}\phi$?

*7. If $f(\mathbf{x}, t)$ is a real-valued function of $\mathbf{x}$ and t, let us define the *material derivative* of f relative to a vector field $\mathbf{F}$ as

$$\frac{Df}{Dt} = \frac{\partial f}{\partial t} + \nabla f(x) \cdot \mathbf{F}$$

Show that Df/Dt is the t-derivative of $f(\phi(\mathbf{x}, t), t)$ (i.e., the t-derivative of f transported by the flow of $\mathbf{F}$).

3.4 DIVERGENCE AND CURL OF A VECTOR FIELD

This section deals with two basic operations that can be performed on C^1 vector fields.

First let us consider the curl of a vector field. This associates with any vector field on $\mathbb{R}^3$, $\mathbf{F} = F_1\mathbf{i} + F_2\mathbf{j} + F_3\mathbf{k} = (F_1, F_2, F_3)$, the vector field curl $\mathbf{F}$ defined by

$$\text{curl } \mathbf{F} = \left(\frac{\partial F_3}{\partial y} - \frac{\partial F_2}{\partial z}\right)\mathbf{i} + \left(\frac{\partial F_1}{\partial z} - \frac{\partial F_3}{\partial x}\right)\mathbf{j} + \left(\frac{\partial F_2}{\partial x} - \frac{\partial F_1}{\partial y}\right)\mathbf{k} \tag{1}$$

The lengthy formula is much easier to remember if we rewrite it using "operator" notation. Let us formally introduce the symbol "del"

$$\nabla = \mathbf{i}\frac{\partial}{\partial x} + \mathbf{j}\frac{\partial}{\partial y} + \mathbf{k}\frac{\partial}{\partial z},$$

∇ is an operator; that is, it makes sense when it acts or operates on real-valued functions. Specifically, ∇f, ∇ operating on f, is given by

$$\nabla f = \mathbf{i}\frac{\partial f}{\partial x} + \mathbf{j}\frac{\partial f}{\partial y} + \mathbf{k}\frac{\partial f}{\partial z}, \tag{2}$$

the *gradient* of f. This formal notation is quite useful; if we view ∇ as a vector with components $\partial/\partial x$, $\partial/\partial y$, $\partial/\partial z$, then we can also take the formal cross product

$$\nabla \times \mathbf{F} = \begin{vmatrix} \mathbf{i} & \mathbf{j} & \mathbf{k} \\ \dfrac{\partial}{\partial x} & \dfrac{\partial}{\partial y} & \dfrac{\partial}{\partial z} \\ F_1 & F_2 & F_3 \end{vmatrix}$$

$$= \left(\frac{\partial F_3}{\partial y} - \frac{\partial F_2}{\partial z}\right)\mathbf{i} + \left(\frac{\partial F_1}{\partial z} - \frac{\partial F_3}{\partial x}\right)\mathbf{j} + \left(\frac{\partial F_2}{\partial x} - \frac{\partial F_1}{\partial y}\right)\mathbf{k}$$

$$= \text{curl } \mathbf{F}$$

Thus curl $\mathbf{F} = \nabla \times \mathbf{F}$, and we shall often use the latter expression.

EXAMPLE 1. Let $\mathbf{F}(x, y, z) = x\mathbf{i} + xy\mathbf{j} + \mathbf{k}$. Find $\nabla \times \mathbf{F}$.

We have

$$\nabla \times \mathbf{F} = \begin{vmatrix} \mathbf{i} & \mathbf{j} & \mathbf{k} \\ \dfrac{\partial}{\partial x} & \dfrac{\partial}{\partial y} & \dfrac{\partial}{\partial z} \\ x & xy & 1 \end{vmatrix} = (0 - 0)\mathbf{i} - (0 - 0)\mathbf{j} + (y - 0)\mathbf{k}$$

Thus $\nabla \times \mathbf{F} = y\mathbf{k}$.

The following theorem states a basic property of the curl. The result should be compared with the fact that for any vector **v**, we have $\mathbf{v} \times \mathbf{v} = \mathbf{0}$.

THEOREM 1. *For any C^2 function f we have*

$$\nabla \times (\nabla f) = \mathbf{0}$$

that is, the curl of any gradient is the zero vector.

Proof. Let us write out the components of $\nabla \times (\nabla f)$. Since $\nabla f = (\partial f/\partial x, \partial f/\partial y, \partial f/\partial z)$ we have, by definition,

$$\nabla \times \nabla f = \begin{vmatrix} \mathbf{i} & \mathbf{j} & \mathbf{k} \\ \dfrac{\partial}{\partial x} & \dfrac{\partial}{\partial y} & \dfrac{\partial}{\partial z} \\ \dfrac{\partial f}{\partial x} & \dfrac{\partial f}{\partial y} & \dfrac{\partial f}{\partial z} \end{vmatrix} = \left(\frac{\partial^2 f}{\partial y\, \partial z} - \frac{\partial^2 f}{\partial z\, \partial y}\right)\mathbf{i} + \left(\frac{\partial^2 f}{\partial z\, \partial x} - \frac{\partial^2 f}{\partial x\, \partial z}\right)\mathbf{j} + \left(\frac{\partial^2 f}{\partial x\, \partial y} - \frac{\partial^2 f}{\partial y\, \partial x}\right)\mathbf{k}$$

Each component is zero because of the symmetry property of mixed partial derivatives; hence, the desired result follows. ■

The full physical significance of the curl will be discussed below and also in Chapter 7, when we study Stokes' Theorem. However, we can now consider a simple situation that shows why the curl is associated with rotations.

EXAMPLE 2. Suppose we have a rigid body B rotating about an axis L. The rotational motion of the body can be completely described by a vector **w** along the axis of rotation, the direction being chosen so that the body rotates about **w** as in Figure 3.4.1. Moreover, we take the length $\omega = \|\mathbf{w}\|$ to be the angular speed of the body B—that is, the tangential speed of any point in B divided by its distance to the axis L of rotation.

We also assume that we have selected a coordinate system so that L is the z-axis. Let Q be any point in B and let α be the distance from Q to L. Clearly,

$$\alpha = \|\mathbf{r}\|\sin\theta$$

where **r** is the vector whose initial point is the origin and whose terminal point is Q. Then the tangential velocity **v** of Q is directed counterclockwise along the tangent to a circle parallel to the xy-plane with radius α. The magnitude of this velocity is

$$\|\mathbf{v}\| = \omega\alpha = \omega\|\mathbf{r}\|\sin\theta = \|\mathbf{w}\|\,\|\mathbf{r}\|\sin\theta$$

We have seen (p. 32) that the direction and magnitude of **v** imply that

$$\mathbf{v} = \mathbf{w} \times \mathbf{r}$$

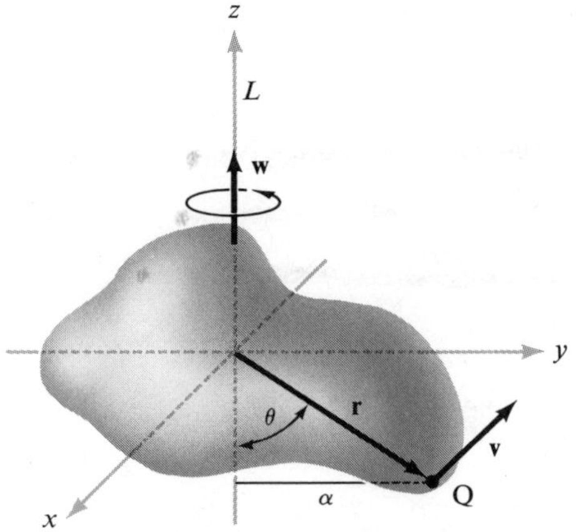

FIGURE 3.4.1
The velocity **v** *and angular velocity* **w** *of a rotating body are related by* $\mathbf{v} = \mathbf{w} \times \mathbf{r}$.

Because of our choice of axes, we can write $\mathbf{w} = \omega\mathbf{k}$, $\mathbf{r} = x\mathbf{i} + y\mathbf{j} + z\mathbf{k}$, so that

$$\mathbf{v} = \mathbf{w} \times \mathbf{r} = -\omega y\mathbf{i} + \omega x\mathbf{j}$$

and moreover

$$\text{curl } \mathbf{v} = \begin{vmatrix} \mathbf{i} & \mathbf{j} & \mathbf{k} \\ \dfrac{\partial}{\partial x} & \dfrac{\partial}{\partial y} & \dfrac{\partial}{\partial z} \\ -\omega y & \omega x & 0 \end{vmatrix} = 2\omega\mathbf{k} = 2\mathbf{w}$$

Hence, for the rotation of a rigid body, the curl of the velocity vector field is a vector field directed along the axis of rotation with magnitude twice the angular speed.

If a vector field **F** represents the flow of a fluid (see Example 1, Section 3.3), then curl $\mathbf{F} = \mathbf{0}$ at P means physically that the fluid is free from rotations or is irrotational at P; that is, it has no whirlpools. Justification of this idea and, therefore, of the use of the word irrotational, depends on Stokes' Theorem, or the (optional) discussion below. However, we can say informally that curl $\mathbf{F} = \mathbf{0}$ means that if a *small* paddle wheel is placed in the fluid, it will move with the fluid, but will not rotate around its axis. For example, it has been determined from experiments that fluid draining from a tub is usually irrotational except right at the center, even though the fluid is "rotating" around the drain (see Figure 3.4.2). Thus the reader should be warned of the possible confusion the word "irrotational" can cause. Let us consider some examples of rotational and irrotational fields.

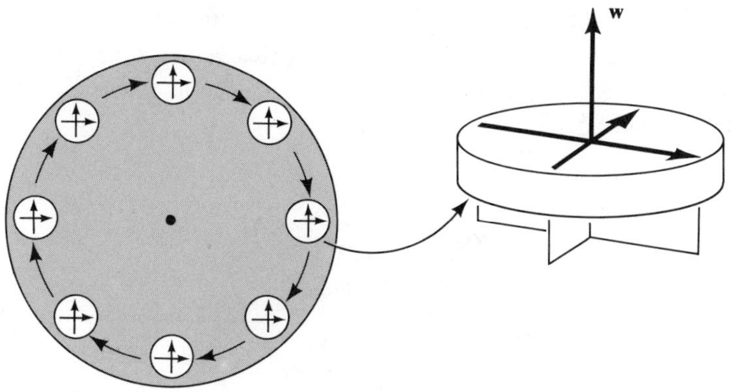

FIGURE 3.4.2
The velocity field $\mathbf{V}(x, y, z) = (y\mathbf{i} - x\mathbf{j})/(x^2 + y^2)$ *is irrotational; a small "paddle wheel" moving in the fluid will not rotate around its axis* $\mathbf{w}$.

EXAMPLE 3. Let us verify that the vector field of Example 4, Section 3.3, is irrotational at each point $(x, y) \neq (0, 0)$. Indeed,

$$
\begin{aligned}
\nabla \times \mathbf{V} &= \begin{vmatrix} \mathbf{i} & \mathbf{j} & \mathbf{k} \\ \dfrac{\partial}{\partial x} & \dfrac{\partial}{\partial y} & \dfrac{\partial}{\partial z} \\ \dfrac{y}{x^2 + y^2} & \dfrac{-x}{x^2 + y^2} & 0 \end{vmatrix} \\
&= 0\mathbf{i} + 0\mathbf{j} + \left(\frac{\partial}{\partial x}\left(\frac{-x}{x^2 + y^2}\right) - \frac{\partial}{\partial y}\left(\frac{y}{x^2 + y^2}\right)\right)\mathbf{k} \\
&= \left(\frac{-(x^2 + y^2) + 2x^2}{(x^2 + y^2)^2} + \frac{-(x^2 + y^2) + 2y^2}{(x^2 + y^2)^2}\right)\mathbf{k} \\
&= \mathbf{0}
\end{aligned}
$$

EXAMPLE 4. Let $\mathbf{V}(x, y, z) = y\mathbf{i} - x\mathbf{j}$. Show that $\mathbf{V}$ is not a gradient field.

Indeed, if $\mathbf{V}$ were a gradient field, then by Theorem 1 we would have curl $\mathbf{V} = \mathbf{0}$. But

$$
\operatorname{curl} \mathbf{V} = \begin{vmatrix} \mathbf{i} & \mathbf{j} & \mathbf{k} \\ \dfrac{\partial}{\partial x} & \dfrac{\partial}{\partial y} & \dfrac{\partial}{\partial z} \\ y & -x & 0 \end{vmatrix} = -2\mathbf{k} \neq \mathbf{0}
$$

The flow lines for this vector field, just as for the one in Example 3, are circles about the origin in the xy-plane, but this velocity field has rotation. In such a flow, a small paddle wheel rotates once as it circulates around the origin (Figure 3.4.3).

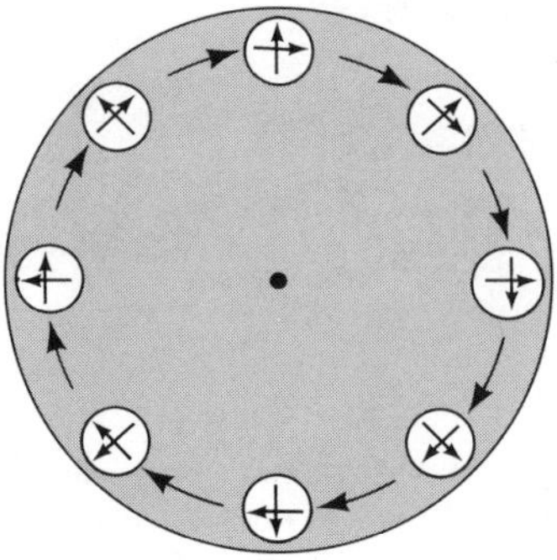

FIGURE 3.4.3
The velocity field $\mathbf{V}(x, y, z) = y\mathbf{i} - x\mathbf{j}$ *is rotational; a small "paddle wheel" moving in the fluid rotates around its axis* $\mathbf{w}$ *(see Figure 3.4.2).*

Another basic operation is the *divergence*, defined by

$$\operatorname{div} \mathbf{F} = \nabla \cdot \mathbf{F} = \frac{\partial F_1}{\partial x} + \frac{\partial F_2}{\partial y} + \frac{\partial F_3}{\partial z} \tag{3}$$

In operator notation, div $\mathbf{F}$ is the dot product of ∇ and $\mathbf{F}$. Note that $\nabla \times \mathbf{F}$ is a vector field whereas $\nabla \cdot \mathbf{F}\colon \mathbb{R}^3 \to \mathbb{R}$, so that $\nabla \cdot \mathbf{F}$ is a scalar field. We read $\nabla \cdot \mathbf{F}$ as "the divergence of $\mathbf{F}$."

The full significance of the divergence is given below in an (optional) discussion and is also presented in connection with Gauss' Theorem in Chapter 7, but we can give some of its physical meaning here. If we imagine $\mathbf{F}$ as the velocity field of a gas (or a fluid), then div $\mathbf{F}$ represents the rate of expansion per unit volume of the gas (or fluid). For example, if $\mathbf{F}(x, y, z) = x\mathbf{i} + y\mathbf{j} + z\mathbf{k}$, then div $\mathbf{F} = 3$; this means the gas is expanding at the rate of 3 cubic units per unit of volume per unit of time. This is reasonable, because in this case $\mathbf{F}$ is an outward radial vector, and as the gas moves outward along the flow lines it expands. (See Section 3.3 for a discussion of flow lines).

The next theorem is analogous to Theorem 1.

THEOREM 2. *For any C^2 vector field* $\mathbf{F}$,

$$\operatorname{div} \operatorname{curl} \mathbf{F} = \nabla \cdot (\nabla \times \mathbf{F}) = 0$$

that is, the divergence of any curl is zero.

The proof is similar to that of Theorem 1, in that it rests on the equality of the mixed partial derivatives. The student should write out the details.

We have seen that $\nabla \times \mathbf{F}$ is related to rotations and $\nabla \cdot \mathbf{F}$ is related to compressions and expansions. This leads to the following common terminology. If $\nabla \cdot \mathbf{F} = 0$ we say $\mathbf{F}$ is *incompressible*, just as we call $\mathbf{F}$ irrotational if

$$\nabla \times \mathbf{F} = \mathbf{0}.$$

EXAMPLE 5. Compute the divergence of

$$\mathbf{F} = x^2y\mathbf{i} + z\mathbf{j} + xyz\mathbf{k}$$

Indeed

$$\operatorname{div} \mathbf{F} = \frac{\partial}{\partial x}(x^2y) + \frac{\partial}{\partial y}(z) + \frac{\partial}{\partial z}(xyz) = 2xy + 0 + xy = 3xy$$

EXAMPLE 6. From Theorem 2 we can conclude that $\mathbf{F}$ in Example 5 cannot be the curl of another vector field, or else it would have zero divergence.

Finally, let us mention the *Laplace operator* ∇^2, which operates on functions f as follows:

$$\nabla^2 f = \nabla \cdot (\nabla f) = \frac{\partial^2 f}{\partial x^2} + \frac{\partial^2 f}{\partial y^2} + \frac{\partial^2 f}{\partial z^2}$$

If $\mathbf{F} = F_1\mathbf{i} + F_2\mathbf{j} + F_3\mathbf{k}$ is a C^2 vector field we can also define $\nabla^2\mathbf{F}$ in terms of components

$$\nabla^2\mathbf{F} = \nabla^2 F_1\mathbf{i} + \nabla^2 F_2\mathbf{j} + \nabla^2 F_3\mathbf{k}$$

As was discussed in Section 2.6, this operator plays an important role in many physical laws. We shall discuss these further in Chapter 7.

OPTIONAL

THE GEOMETRY OF THE DIVERGENCE AND CURL

We shall now study the geometric meaning of the divergence and curl in more detail. This discussion depends on the concept of the flow of a vector field given at the end of Section 3.3.

Fix a point $\mathbf{x}$ and consider the three standard basis vectors $\mathbf{i}, \mathbf{j}, \mathbf{k}$ emanating from $\mathbf{x}$. Let $\varepsilon > 0$ be small and consider the basis vectors $\mathbf{v}_1 = \varepsilon\mathbf{i}$, $\mathbf{v}_2 = \varepsilon\mathbf{j}$, $\mathbf{v}_3 = \varepsilon\mathbf{k}$. These vectors span a parallelepiped $P(0)$. As time increases or decreases, the flow $\phi(\mathbf{x}, t)$ carries $P(0)$ into some object. For *fixed* time, ϕ is a differentiable function of $\mathbf{x}$ (i.e., ϕ is a differentiable map of $\mathbb{R}^3$ to $\mathbb{R}^3$).

OPTIONAL (*Continued*)

When ε is small the image of $P(0)$ under ϕ can be approximated by its image under the derivative of ϕ with respect to $\mathbf{x}$. (See the discussion of the linear approximation to a mapping in Section 2.3.) Thus for fixed time and small positive ε, $P(0)$ is approximately carried into a parallelepiped spanned by the vectors $\mathbf{v}_1(t)$, $\mathbf{v}_2(t)$, $\mathbf{v}_3(t)$ given by

$$\mathbf{v}_1(t) = D_{\mathbf{x}}\phi(\mathbf{x}, t) \cdot \mathbf{v}_1$$

$$\mathbf{v}_2(t) = D_{\mathbf{x}}\phi(\mathbf{x}, t) \cdot \mathbf{v}_2$$

$$\mathbf{v}_3(t) = D_{\mathbf{x}}\phi(\mathbf{x}, t) \cdot \mathbf{v}_3$$

Since $\phi(\mathbf{x}, 0) = \mathbf{x}$ for all $\mathbf{x}$, it follows that $\mathbf{v}_1(0) = \mathbf{v}_1$, $\mathbf{v}_2(0) = \mathbf{v}_2$, and $\mathbf{v}_3(0) = \mathbf{v}_3$. (This formula for mapping vectors is discussed on p. 123.) The vectors $\mathbf{v}_1(t), \mathbf{v}_2(t), \mathbf{v}_3(t)$ span a parallelepiped $P(t)$ that moves in time (see Figure 3.4.4).

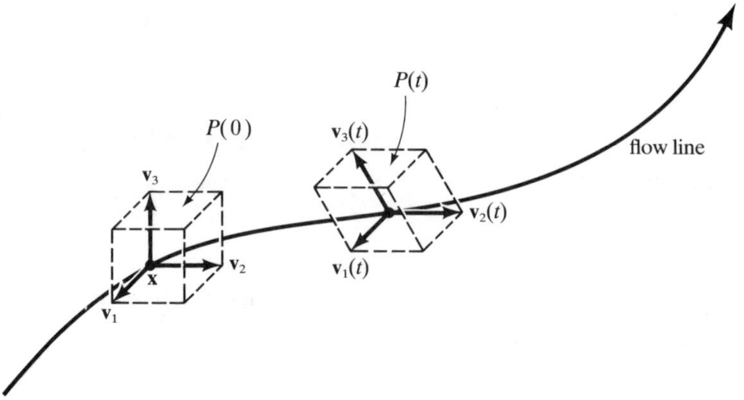

FIGURE 3.4.4
The moving basis $\mathbf{v}_1(t)$, $\mathbf{v}_2(t)$, $\mathbf{v}_3(t)$ *and the associated parallelepiped.*

Let the volume of $P(t)$ be denoted by $\mathscr{V}(t)$. The main geometric meaning of divergence is given by the following theorem.

THEOREM 3. $\operatorname{div} \mathbf{F}(\mathbf{x}) = \dfrac{1}{\mathscr{V}(0)} \dfrac{d}{dt} \mathscr{V}(t)\Big|_{t=0}$

OPTIONAL (*Continued*)

Proof. By equation (3) of the previous section

$$\frac{d}{dt}\mathbf{v}_i(t) = D_{\mathbf{x}}\mathbf{F}(\phi(\mathbf{x}, t)) \cdot D_x\phi(\mathbf{x}, t) \cdot \mathbf{v}_i \tag{4}$$

for $i = 1, 2, 3$.

Since $\phi(\mathbf{x}, 0) = \mathbf{x}$, $D_{\mathbf{x}}\phi(\mathbf{x}, 0)$ is the identity matrix so evaluation at $t = 0$ gives

$$\left.\frac{d}{dt}\mathbf{v}_i(t)\right|_{t=0} = D_{\mathbf{x}}\mathbf{F}(\mathbf{x}) \cdot \mathbf{v}_i$$

The volume $\mathscr{V}(t)$ is given by the triple product (see p. 34)

$$\mathscr{V}(t) = \mathbf{v}_1(t) \cdot [\mathbf{v}_2(t) \times \mathbf{v}_3(t)]$$

Using Exercise 9, Section 3.1 and the identities $\mathbf{v}_1 \cdot [\mathbf{v}_2 \times \mathbf{v}_3] = \mathbf{v}_2 \cdot [\mathbf{v}_3 \times \mathbf{v}_1] = \mathbf{v}_3 \cdot [\mathbf{v}_1 \times \mathbf{v}_2]$, we get

$$\begin{aligned}\frac{d\mathscr{V}}{dt} &= \frac{d\mathbf{v}_1}{dt} \cdot [\mathbf{v}_2(t) \times \mathbf{v}_3(t)] + \mathbf{v}_1(t) \cdot \left[\frac{d\mathbf{v}_2}{dt} \times \mathbf{v}_3(t)\right] + \mathbf{v}_1(t) \cdot \left[\mathbf{v}_2(t) \times \frac{d\mathbf{v}_3}{dt}\right] \\ &= \frac{d\mathbf{v}_1}{dt} \cdot [\mathbf{v}_2(t) \times \mathbf{v}_3(t)] + \frac{d\mathbf{v}_2}{dt} \cdot [\mathbf{v}_1(t) \times \mathbf{v}_3(t)] + \frac{d\mathbf{v}_3}{dt} \cdot [\mathbf{v}_1(t) \times \mathbf{v}_2(t)]\end{aligned}$$

At $t = 0$, substitution from formula (4) and the fact that $\mathbf{v}_1 \times \mathbf{v}_2 = \mathbf{v}_3$, $\mathbf{v}_3 \times \mathbf{v}_1 = \mathbf{v}_2$, and $\mathbf{v}_2 \times \mathbf{v}_3 = \mathbf{v}_1$ gives

$$\left.\frac{d\mathscr{V}}{dt}\right|_{t=0} = \varepsilon^3[D_{\mathbf{x}}\mathbf{F}(\mathbf{x})\mathbf{i}] \cdot \mathbf{i} + \varepsilon^3[D_{\mathbf{x}}\mathbf{F}(\mathbf{x})\mathbf{j}] \cdot \mathbf{j} + \varepsilon^3[D_{\mathbf{x}}\mathbf{F}(\mathbf{x})\mathbf{k}] \cdot \mathbf{k} \tag{5}$$

But $\mathscr{V}(0) = \varepsilon^3$, $\mathbf{F} = F_1\mathbf{i} + F_2\mathbf{j} + F_3\mathbf{k}$, $[D_{\mathbf{x}}\mathbf{F}(\mathbf{x})\mathbf{i}] \cdot \mathbf{i} = \partial F_1/\partial x$, and, similarly, the second and third terms of (5) are $\varepsilon^3(\partial F_2/\partial y)$ and $\varepsilon^3(\partial F_3/\partial z)$. Substituting these into (5) and dividing by ε^3 proves the theorem. ■

The reader who is familiar with a little more linear algebra can prove this generalization of Theorem 3:* Let $\mathbf{v}_1$, $\mathbf{v}_2$, and $\mathbf{v}_3$ be any three noncoplanar vectors emanating from $\mathbf{x}$ that flow according to the formula

$$\mathbf{v}_i(t) = D_{\mathbf{x}}\phi(\mathbf{x}, t) \cdot \mathbf{v}_i, \qquad i = 1, 2, 3$$

* The reader will need to know how to write the matrix of a linear transformation with respect to a given basis and be familiar with the fact that the trace of a matrix is independent of the basis.

OPTIONAL (*Continued*)

The vectors $\mathbf{v}_1(t)$, $\mathbf{v}_2(t)$, $\mathbf{v}_3(t)$ span a parallelepiped $P(t)$ with volume $\mathscr{V}(t)$. We have the result

$$\frac{1}{\mathscr{V}(0)} \frac{d\mathscr{V}}{dt}\bigg|_{t=0} = \operatorname{div} \mathbf{F}(\mathbf{x}) \tag{6}$$

In other words, the divergence of $\mathbf{F}$ at $\mathbf{x}$ is the rate at which the volumes change, per unit volume. "Rate" refers to the rate of change with respect to time as the volumes are transported by the flow.

Next we examine the curl of $\mathbf{F}$ at a point $\mathbf{x}$, which, without loss of generality, we can take to be the origin. The curl measures how the flow rotates vectors. The following requires an understanding of Example 2, which appears early in the nonoptional part of this section.

Rotations preserve the inner product in $\mathbb{R}^3$. To study "how much" rotation is in the flow, let us begin by studying how the flow affects the inner product. Let $\mathbf{v}$ and $\mathbf{w}$ be two vectors emanating from the origin and let them be moved by the derivative of the flow as above:

$$\mathbf{v}(t) = D_\mathbf{x}\phi(\mathbf{0}, t)\mathbf{v}, \qquad \mathbf{w}(t) = D_\mathbf{x}\phi(\mathbf{0}, t)\mathbf{w}$$

so that at time $t = 0$ and at the origin $\mathbf{0}$ in $\mathbb{R}^3$

$$\frac{d\mathbf{v}}{dt}\bigg|_{t=0} = D_\mathbf{x}\mathbf{F}(\mathbf{0}) \cdot \mathbf{v} \quad \text{and} \quad \frac{d\mathbf{w}}{dt}\bigg|_{t=0} = D_\mathbf{x}\mathbf{F}(\mathbf{0}) \cdot \mathbf{w} \tag{7}$$

Then

$$\frac{d}{dt}\mathbf{v} \cdot \mathbf{w} = \frac{d\mathbf{v}}{dt} \cdot \mathbf{w} + \mathbf{v} \cdot \frac{d\mathbf{w}}{dt}$$

and therefore

$$\frac{d}{dt}\mathbf{v} \cdot \mathbf{w}\bigg|_{t=0} = [D_\mathbf{x}\mathbf{F}(\mathbf{0}) \cdot \mathbf{v}] \cdot \mathbf{w} + \mathbf{v} \cdot [D_\mathbf{x}\mathbf{F}(\mathbf{0}) \cdot \mathbf{w}]$$

The transpose A^T of a matrix satisfies $A^T\mathbf{v} \cdot \mathbf{w} = \mathbf{v} \cdot A\mathbf{w}$ (see Exercise 16, Section 1.5), so

$$\frac{d}{dt}\mathbf{v} \cdot \mathbf{w}\bigg|_{t=0} = [(D_\mathbf{x}\mathbf{F}(\mathbf{0}) + [D_\mathbf{x}\mathbf{F}(\mathbf{0})]^T)\mathbf{v}] \cdot \mathbf{w} \tag{8}$$

Thus the rate of change of the inner product at the origin at $t = 0$ is determined by the matrix $(D_\mathbf{x}\mathbf{F}(\mathbf{0}) + [D_\mathbf{x}\mathbf{F}(\mathbf{0})]^T)$.

Any matrix A can be written (uniquely) as the sum of a symmetric ($S^T = S$) and an antisymmetric ($W^T = -W$) matrix as follows:

$$A = \tfrac{1}{2}(A + A^T) + \tfrac{1}{2}(A - A^T) = S + W$$

OPTIONAL (*Continued*)

In particular, for $A = D_{\mathbf{x}}\mathbf{F}(\mathbf{0})$,

$$S = \tfrac{1}{2}(D_{\mathbf{x}}\mathbf{F}(\mathbf{0}) + [D_{\mathbf{x}}\mathbf{F}(\mathbf{0})]^T)$$

and

$$W = \tfrac{1}{2}(D_{\mathbf{x}}\mathbf{F}(\mathbf{0}) - [D_{\mathbf{x}}\mathbf{F}(\mathbf{0})]^T)$$

We call S the *deformation matrix* and W the *rotation matrix*. From (8) we see that it is the deformation matrix that contributes to changing the inner product, and it is therefore a reasonable guess to expect W to correspond to a rotation. For a general matrix A, the elements of W are

$$w_{ij} = \tfrac{1}{2}(a_{ij} - a_{ji})$$

and for $A = D_{\mathbf{x}}\mathbf{F}(\mathbf{0})$ this becomes

$$w_{ij} = \frac{1}{2}\left(\frac{\partial F_i}{\partial x_j} - \frac{\partial F_j}{\partial x_i}\right)$$

using the notation (x_1, x_2, x_3) in place of (x, y, z). Explicitly, $w_{12} = -\tfrac{1}{2}(\text{curl } \mathbf{F})_3$, $w_{23} = -\tfrac{1}{2}(\text{curl } \mathbf{F})_1$, and $w_{31} = -\tfrac{1}{2}(\text{curl } \mathbf{F})_2$.

THEOREM 4. *Let* $\mathbf{w} = \tfrac{1}{2}(\nabla \times \mathbf{F})(\mathbf{0})$. *Let us assume that axes have been chosen so that* $\mathbf{w}$ *is parallel to the z-axis and points in the direction of* $\mathbf{k}$. *Furthermore, let the vector field* $\mathbf{v}$ *be given by* $\mathbf{v} = \mathbf{w} \times \mathbf{r}$, *where* $\mathbf{r} = x\mathbf{i} + y\mathbf{j} + z\mathbf{k}$ *as in Example 2; thus* $\mathbf{v}$ *is the velocity field of a rotation about the axis* $\mathbf{w}$ *with angular velocity* $\omega = \|\mathbf{w}\|$ *and with curl* $\mathbf{v} = 2\mathbf{w}$. *Then since* $\mathbf{r}$ *is a function of* (x, y, z), $\mathbf{v}$ *is also a function of* (x, y, z). *The derivative of* $\mathbf{v}$ *at the origin is given by*

$$D\mathbf{v}(0) = W = \begin{pmatrix} 0 & -\omega & 0 \\ \omega & 0 & 0 \\ 0 & 0 & 0 \end{pmatrix} \tag{9}$$

This theorem warrants some explanation before it is proved. The vector field $\mathbf{v}$ represents rotation around a fixed axis $\mathbf{w}$. The flow $\psi(\mathbf{x}, t)$ of $\mathbf{v}$ rotates points in this field, and, for fixed t, its derivative $D_{\mathbf{x}}\psi(\mathbf{x}, t)$ rotates vectors as well (again see the discussion on derivatives of mappings on p. 123). Let $\mathbf{Y}$ be an arbitrary vector and set $\mathbf{Y}(t) = D_{\mathbf{x}}\psi(\mathbf{x}, t)\mathbf{Y}$. As t increases or decreases, $\mathbf{Y}(t)$ rotates around $\mathbf{w}$, and we have

$$\left.\frac{d\mathbf{Y}}{dt}\right|_{t=0} = D_{\mathbf{x}}\mathbf{v}(\mathbf{0})\mathbf{Y} \tag{10}$$

(see formula 7). Formula (10) gives the rate of change of $\mathbf{Y}$ as it is transported

OPTIONAL (*Continued*)

(rotated) by $D_{\mathbf{x}}\psi$. However, again by formula (7), the rate of change of any vector $\mathbf{X}$ at the origin under transport by the derivative of the flow $\phi(x, t)$ of $\mathbf{F}$ is given by

$$\left.\frac{d\mathbf{X}}{dt}\right|_{t=0} = D_{\mathbf{x}}\mathbf{F}(\mathbf{0})\mathbf{X} = (S + W)\mathbf{X}$$

Thus this rate of change of $\mathbf{X}$ has two components: the deformation matrix, which affects inner products, and the W matrix. By formulas (9) and (10) *the W matrix is precisely the rate of change of vectors as they undergo an infinitesimal rotation around the axis* (curl $\mathbf{F})(\mathbf{0}) = (\nabla \times \mathbf{F})(\mathbf{0})$ by the mapping $D_{\mathbf{x}}\psi(\mathbf{x}, t)$.

Proof of Theorem 4. By our choice of axes,

$$\mathbf{w} = \tfrac{1}{2}(\nabla \times \mathbf{F})(\mathbf{0}) = \omega\mathbf{k}$$

By Example 2,

$$\mathbf{v} = -\omega y\mathbf{i} + \omega x\mathbf{j}$$

and therefore

$$D\mathbf{v}(\mathbf{0}) = \begin{bmatrix} 0 & -\omega & 0 \\ \omega & 0 & 0 \\ 0 & 0 & 0 \end{bmatrix}$$

On the other hand, by the definitions of W and $\nabla \times \mathbf{F}$,

$$W = \begin{bmatrix} w_{11} & w_{12} & w_{13} \\ w_{21} & w_{22} & w_{23} \\ w_{31} & w_{32} & w_{33} \end{bmatrix}$$

$$= \begin{bmatrix} 0 & \frac{1}{2}\left(\frac{\partial F_1}{\partial y} - \frac{\partial F_2}{\partial x}\right) & \frac{1}{2}\left(\frac{\partial F_1}{\partial z} - \frac{\partial F_3}{\partial x}\right) \\ \frac{1}{2}\left(\frac{\partial F_2}{\partial x} - \frac{\partial F_1}{\partial y}\right) & 0 & \frac{1}{2}\left(\frac{\partial F_2}{\partial z} - \frac{\partial F_3}{\partial y}\right) \\ \frac{1}{2}\left(\frac{\partial F_3}{\partial x} - \frac{\partial F_1}{\partial z}\right) & \frac{1}{2}\left(\frac{\partial F_3}{\partial y} - \frac{\partial F_2}{\partial z}\right) & 0 \end{bmatrix}$$

$$= \frac{1}{2}\begin{bmatrix} 0 & -(\nabla \times \mathbf{F})_z & (\nabla \times \mathbf{F})_y \\ (\nabla \times \mathbf{F})_z & 0 & -(\nabla \times \mathbf{F})_x \\ -(\nabla \times \mathbf{F})_y & (\nabla \times \mathbf{F})_x & 0 \end{bmatrix}$$

OPTIONAL (*Continued*)

Our choice of coordinate axes gives

$$W = \begin{bmatrix} 0 & -\omega & 0 \\ \omega & 0 & 0 \\ 0 & 0 & 0 \end{bmatrix}$$

which proves the theorem. ■

The deformation matrix S incorporates all the length and angle changes caused by the flow, as we have remarked. In particular, volume changes are contained in S. In fact, the trace of S is the divergence: $\operatorname{tr} S = \operatorname{div} \mathbf{F}(\mathbf{x})$. (The trace of a matrix is the sum of its diagonal entries.) The trace-free part of S,

$$S' = (S - \tfrac{1}{3}(\operatorname{tr} S)I)$$

where I is the 3×3 identity, is called the *shear*.

EXERCISES

1. Compute the gradient ∇f for the following functions.
   (a) $f(x, y, z) = \sqrt{x^2 + y^2 + z^2}$
   (b) $f(x, y, z) = xy + yz + xz$
   (c) $f(x, y, z) = 1/(x^2 + y^2 + z^2)$
2. Compute the divergence $\nabla \cdot \mathbf{F}$ of the following vector fields.
   (a) $\mathbf{F}(x, y, z) = x\mathbf{i} + y\mathbf{j} + z\mathbf{k}$
   (b) $\mathbf{F}(x, y, z) = yz\mathbf{i} + xz\mathbf{j} + xy\mathbf{k}$
   (c) $\mathbf{F}(x, y, z) = (x^2 + y^2 + z^2)(3\mathbf{i} + 4\mathbf{j} + 5\mathbf{k})$
3. Compute the curl, $\nabla \times \mathbf{F}$, of each vector field in Exercise 2.
4. Verify that the vector field of Example 4, Section 3.3, is incompressible. Can you interpret this result physically?
5. Verify that $\mathbf{F} = y\mathbf{i} + x\mathbf{j}$ is incompressible.
6. Let $\mathbf{F}(x, y, z) = 3x^2y\mathbf{i} + (x^3 + y^3)\mathbf{j}$.
   (a) Verify that $\operatorname{curl} \mathbf{F} = \mathbf{0}$.
   (b) Find a function f such that $\mathbf{F} = \nabla f$. (Specific techniques for constructing f in general are given in Chapter 7. The one in this problem should be sought directly.)
   (c) Is it true that for a vector field $\mathbf{F}$ such a function f can exist only if $\operatorname{curl} \mathbf{F} = \mathbf{0}$?
7. Let $f(x, y, z) = x^2y^2 + y^2z^2$. Verify directly that $\nabla \times \nabla f = \mathbf{0}$.
8. Show that the real and imaginary parts of each of the following complex functions form the components of an irrotational and incompressible vector field in the plane.
   (a) $(x - iy)^2$
   (b) $(x - iy)^3$
   (c) $e^{x-iy} = e^x(\cos y - i \sin y)$

9. Show that $\mathbf{F} = y(\cos x)\mathbf{i} + x(\sin y)\mathbf{j}$ is *not* a gradient vector field.
10. Let $\mathbf{r}(x, y, z) = x\mathbf{i} + y\mathbf{j} + z\mathbf{k}$. From Exercise 3(a), we know that $\nabla \times \mathbf{r} = \mathbf{0}$. Can you see why this is so physically, by visualizing $\mathbf{r}$ as the velocity field of a fluid?

3.5 VECTOR DIFFERENTIAL CALCULUS

We now have these basic operations on hand: gradient, divergence, curl, and the Laplace operator. This section develops their properties and the relationships between them a little further.

In Table 3.1 are summarized some basic general formulas that are useful when computing with vector fields. Some of these, such as formulas 10

Table 3.1. Some Common Formulas of Vector Analysis

1. $\nabla(f + g) = \nabla f + \nabla g$
2. $\nabla(cf) = c\nabla f$, for a constant c
3. $\nabla(fg) = f\nabla g + g\nabla f$
4. $\nabla(f/g) = (g\nabla f - f\nabla g)/g^2$, at points where $g(\mathbf{x}) \neq 0$
5. $\operatorname{div}(\mathbf{F} + \mathbf{G}) = \operatorname{div} \mathbf{F} + \operatorname{div} \mathbf{G}$
6. $\operatorname{curl}(\mathbf{F} + \mathbf{G}) = \operatorname{curl} \mathbf{F} + \operatorname{curl} \mathbf{G}$
7. $\nabla(\mathbf{F} \cdot \mathbf{G}) = (\mathbf{F} \cdot \nabla)\mathbf{G} + (\mathbf{G} \cdot \nabla)\mathbf{F} + \mathbf{F} \times \operatorname{curl} \mathbf{G} + \mathbf{G} \times \operatorname{curl} \mathbf{F}$
8. $\operatorname{div}(f\mathbf{F}) = f \operatorname{div} \mathbf{F} + \mathbf{F} \cdot \nabla f$
9. $\operatorname{div}(\mathbf{F} \times \mathbf{G}) = \mathbf{G} \cdot \operatorname{curl} \mathbf{F} - \mathbf{F} \cdot \operatorname{curl} \mathbf{G}$
10. $\operatorname{div} \operatorname{curl} \mathbf{F} = 0$
11. $\operatorname{curl}(f\mathbf{F}) = f \operatorname{curl} \mathbf{F} + \nabla f \times \mathbf{F}$
12. $\operatorname{curl}(\mathbf{F} \times \mathbf{G}) = \mathbf{F} \operatorname{div} \mathbf{G} - \mathbf{G} \operatorname{div} \mathbf{F} + (\mathbf{G} \cdot \nabla)\mathbf{F} - (\mathbf{F} \cdot \nabla)\mathbf{G}$
13. $\operatorname{curl} \operatorname{curl} \mathbf{F} = \operatorname{grad} \operatorname{div} \mathbf{F} - \nabla^2 \mathbf{F}$
14. $\operatorname{curl} \nabla f = \mathbf{0}$
15. $\nabla(\mathbf{F} \cdot \mathbf{F}) = 2(\mathbf{F} \cdot \nabla)\mathbf{F} + 2\mathbf{F} \times (\operatorname{curl} \mathbf{F})$
16. $\nabla^2(fg) = f\nabla^2 g + g\nabla^2 f + 2(\nabla f \cdot \nabla g)$
17. $\operatorname{div}(\nabla f \times \nabla g) = 0$
18. $\nabla \cdot (f\nabla g - g\nabla f) = f\nabla^2 g - g\nabla^2 f$
19. $\mathbf{H} \cdot (\mathbf{F} \times \mathbf{G}) = \mathbf{G} \cdot (\mathbf{H} \times \mathbf{F}) = \mathbf{F} \cdot (\mathbf{G} \times \mathbf{H})$
20. $\mathbf{H} \cdot ((\mathbf{F} \times \nabla) \times \mathbf{G}) = ((\mathbf{H} \cdot \nabla)\mathbf{G}) \cdot \mathbf{F} - (\mathbf{H} \cdot \mathbf{F})(\nabla \cdot \mathbf{G})$
21. $\mathbf{F} \times (\mathbf{G} \times \mathbf{H}) = (\mathbf{F} \cdot \mathbf{H})\mathbf{G} - \mathbf{H}(\mathbf{F} \cdot \mathbf{G})$

NOTE: f and g denote scalar fields; **F**, **G**, and **H** denote vector fields.

and 14, were given in Section 3.4. Others are proven in the examples and exercises.

Some expressions in this table require explanation. First,

$$\mathbf{V} = (\mathbf{F} \cdot \nabla)\mathbf{G}$$

has, by definition, components $V_i = \mathbf{F} \cdot (\nabla G_i)$, for $i = 1, 2, 3$, where $\mathbf{G} = (G_1, G_2, G_3)$. Second, $\nabla^2 \mathbf{F}$ has components $\nabla^2 F_i$, where $\mathbf{F} = (F_1, F_2, F_3)$. In the expression $(\mathbf{F} \times \nabla) \times \mathbf{G}$, the ∇ is to operate only on $\mathbf{G}$, namely, $\mathbf{U} = (\mathbf{F} \times \nabla) \times \mathbf{G}$ has components $U_i = \mathbf{F} \times (\nabla G_i)$, $i = 1, 2, 3$.

EXAMPLE 1. Prove Formula 8 in Table 3.1.

$f\mathbf{F}$ has components fF_i, for $i = 1, 2, 3$, so

$$\operatorname{div}(f\mathbf{F}) = \frac{\partial}{\partial x}(fF_1) + \frac{\partial}{\partial y}(fF_2) + \frac{\partial}{\partial z}(fF_3)$$

However, $\partial/\partial x(fF_1) = f\, \partial F_1/\partial x + F_1\, \partial f/\partial x$, with similar expressions for the other terms. Therefore

$$\begin{aligned}\operatorname{div}(f\mathbf{F}) &= f\left(\frac{\partial F_1}{\partial x} + \frac{\partial F_2}{\partial y} + \frac{\partial F_3}{\partial z}\right) + F_1\frac{\partial f}{\partial x} + F_2\frac{\partial f}{\partial y} + F_3\frac{\partial f}{\partial z} \\ &= f(\nabla \cdot \mathbf{F}) + \mathbf{F} \cdot \nabla f\end{aligned}$$

EXAMPLE 2. Let $\mathbf{r}$ be the vector field $\mathbf{r}(x, y, z) = (x, y, z)$ (the position vector), and let $r = \|\mathbf{r}\|$. Compute ∇r and $\nabla \cdot (r\mathbf{r})$.

We have

$$\nabla r = \left(\frac{\partial r}{\partial x}, \frac{\partial r}{\partial y}, \frac{\partial r}{\partial z}\right)$$

Now $r(x, y, z) = \sqrt{x^2 + y^2 + z^2}$ so for $r \neq 0$ we get

$$\frac{\partial r}{\partial x} = \frac{x}{\sqrt{x^2 + y^2 + z^2}} = \frac{x}{r}$$

Thus

$$\nabla r = \left(\frac{x}{r}, \frac{y}{r}, \frac{z}{r}\right) = \frac{\mathbf{r}}{r}$$

For the second part, use Formula 8 to write

$$\nabla \cdot (r\mathbf{r}) = r(\nabla \cdot \mathbf{r}) + \mathbf{r} \cdot \nabla r$$

Now $\nabla \cdot \mathbf{r} = \partial x/\partial x + \partial y/\partial y + \partial z/\partial z = 3$, and we have computed $\nabla r = \mathbf{r}/r$ above. Since $\mathbf{r} \cdot \mathbf{r}/r = r^2/r$,

$$\nabla \cdot (r\mathbf{r}) = 3r + \frac{r^2}{r} = 4r$$

EXAMPLE 3. Show that $\nabla f \times \nabla g$ is always incompressible. In fact, deduce Formula 17 of Table 3.1 from Formula 9.

By Formula 9,

$$\operatorname{div}(\nabla f \times \nabla g) = \nabla g \cdot (\nabla \times \nabla f) - \nabla f \cdot (\nabla \times \nabla g)$$

which is zero, since $\nabla \times \nabla f = \mathbf{0}$ and $\nabla \times \nabla g = \mathbf{0}$.

In Exercises 2 to 6 at the end of this section, the reader will get practice with this kind of manipulation. Later in the text we shall have use for the identities in Exercise 8.

We shall now discuss the expressions for gradient, divergence, and curl in cylindrical and spherical coordinates, first stating the results and then verifying some of them in examples (the rest will be left for exercises). The verifications below are, however, somewhat messy and may be omitted if desired.

THEOREM 5′. *The following formulas hold in cylindrical coordinates:*

(*i*)
$$\nabla f = \frac{\partial f}{\partial r}\mathbf{e}_r + \frac{1}{r}\frac{\partial f}{\partial \theta}\mathbf{e}_\theta + \frac{\partial f}{\partial z}\mathbf{e}_z$$

(*ii*)
$$\nabla \cdot \mathbf{F} = \frac{1}{r}\left[\frac{\partial}{\partial r}(rF_r) + \frac{\partial F_\theta}{\partial \theta} + \frac{\partial}{\partial z}(rF_z)\right]$$

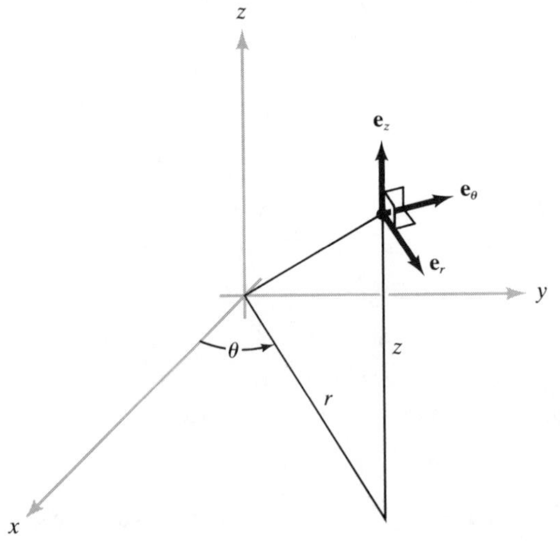

FIGURE 3.5.1

Orthonormal vectors $\mathbf{e}_r$, $\mathbf{e}_\theta$, *and* $\mathbf{e}_z$ *associated with cylindrical coordinates.*

and

$$(iii)\qquad \nabla \times \mathbf{F} = \frac{1}{r}\begin{vmatrix} \mathbf{e}_r & r\mathbf{e}_\theta & \mathbf{e}_z \\ \dfrac{\partial}{\partial r} & \dfrac{\partial}{\partial \theta} & \dfrac{\partial}{\partial z} \\ F_r & rF_\theta & F_z \end{vmatrix}$$

where $\mathbf{e}_r$, $\mathbf{e}_\theta$, $\mathbf{e}_z$ *are the unit orthonormal vectors shown in Figure 3.5.1 and* $\mathbf{F} = F_r\mathbf{e}_r + F_\theta\mathbf{e}_\theta + F_z\mathbf{e}_z$.

THEOREM 5″. *In spherical coordinates (see Section 1.4)*

$$(i)\qquad \nabla f = \frac{\partial f}{\partial \rho}\hat{\mathbf{e}}_\rho + \frac{1}{\rho}\frac{\partial f}{\partial \phi}\hat{\mathbf{e}}_\phi + \frac{1}{\rho \sin \phi}\frac{\partial f}{\partial \theta}\hat{\mathbf{e}}_\theta$$

$$(ii)\qquad \nabla \cdot \mathbf{F} = \frac{1}{\rho^2}\frac{\partial}{\partial \rho}(\rho^2 F_\rho) + \frac{1}{\rho \sin \phi}\frac{\partial}{\partial \phi}(\sin \phi\, F_\phi) + \frac{1}{\rho \sin \phi}\frac{\partial F_\theta}{\partial \theta}$$

$$(iii)\qquad \nabla \times \mathbf{F} = \left(\frac{1}{\rho \sin \phi}\frac{\partial}{\partial \phi}(\sin \phi\, F_\theta) - \frac{1}{\rho \sin \phi}\frac{\partial F_\phi}{\partial \theta}\right)\hat{\mathbf{e}}_\rho + \left(\frac{1}{\rho \sin \phi}\frac{\partial F_\rho}{\partial \theta} - \frac{1}{\rho}\frac{\partial}{\partial \rho}(\rho F_\theta)\right)\hat{\mathbf{e}}_\phi + \left(\frac{1}{\rho}\frac{\partial}{\partial \rho}(\rho F_\phi) - \frac{1}{\rho}\frac{\partial F_\rho}{\partial \phi}\right)\hat{\mathbf{e}}_\theta$$

where $\hat{\mathbf{e}}_\rho$, $\hat{\mathbf{e}}_\phi$, $\hat{\mathbf{e}}_\theta$ *are as shown in Figure 3.5.2.*

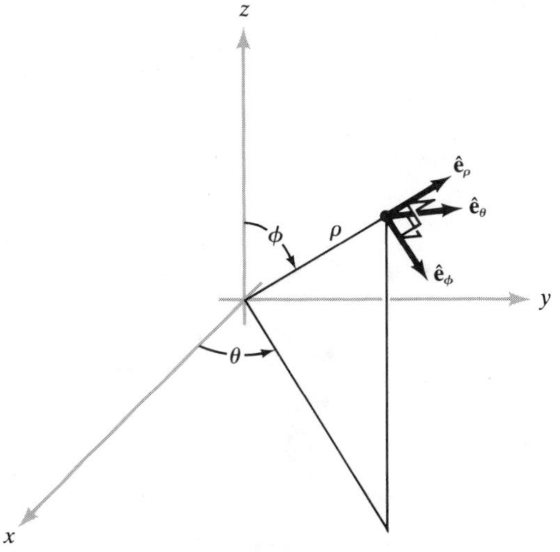

FIGURE 3.5.2

Orthonormal vectors $\hat{\mathbf{e}}_\rho$, $\hat{\mathbf{e}}_\phi$, *and* $\hat{\mathbf{e}}_\theta$ *associated with spherical coordinates.*

EXAMPLE 4. Prove (*i*) of Theorem 5′.

We have

$$\nabla f = \frac{\partial f}{\partial x}\mathbf{i} + \frac{\partial f}{\partial y}\mathbf{j} + \frac{\partial f}{\partial z}\mathbf{k} \tag{1}$$

From Figure 3.5.1, we have

$$\mathbf{e}_r = \frac{x\mathbf{i} + y\mathbf{j}}{\sqrt{x^2 + y^2}} = \cos\theta\mathbf{i} + \sin\theta\mathbf{j}$$

$$\mathbf{e}_\theta = \frac{-y\mathbf{i} + x\mathbf{j}}{\sqrt{x^2 + y^2}} = -\sin\theta\mathbf{i} + \cos\theta\mathbf{j}$$

($\mathbf{e}_\theta$ and $\mathbf{e}_r$ are orthogonal in the plane)

and

$$\mathbf{e}_z = \mathbf{k}$$

(Note the differences between the above and Exercise 6, Section 1.4.) Solving for **i**, **j**, and **k**, we get

$$\begin{aligned} \mathbf{i} &= \mathbf{e}_r \cos\theta - \mathbf{e}_\theta \sin\theta \\ \mathbf{j} &= \mathbf{e}_r \sin\theta + \mathbf{e}_\theta \cos\theta \\ \mathbf{k} &= \mathbf{e}_z \end{aligned} \tag{2}$$

By the Chain Rule,

$$\frac{\partial f}{\partial r} = \frac{\partial f}{\partial x}\frac{\partial x}{\partial r} + \frac{\partial f}{\partial y}\frac{\partial y}{\partial r} + \frac{\partial f}{\partial z}\frac{\partial z}{\partial r}$$

that is,

$$\frac{\partial f}{\partial r} = \cos\theta\frac{\partial f}{\partial x} + \sin\theta\frac{\partial f}{\partial y}$$

Similarly

$$\frac{\partial f}{\partial \theta} = -r\sin\theta\frac{\partial f}{\partial x} + r\cos\theta\frac{\partial f}{\partial y} \quad \text{and} \quad \frac{\partial f}{\partial z} = \frac{\partial f}{\partial z}$$

Solving,

$$\left.\begin{aligned} \frac{\partial f}{\partial x} &= \cos\theta\frac{\partial f}{\partial r} - \frac{\sin\theta}{r}\frac{\partial f}{\partial \theta} \\ \frac{\partial f}{\partial y} &= \sin\theta\frac{\partial f}{\partial r} + \frac{\cos\theta}{r}\frac{\partial f}{\partial \theta} \\ \frac{\partial f}{\partial z} &= \frac{\partial f}{\partial z} \end{aligned}\right\} \tag{3}$$

Substituting (2) and (3) into (1) and simplifying yield the desired result.

EXAMPLE 5. Prove (*ii*) of Theorem 5″.

We have

$$\nabla \cdot \mathbf{F} = \frac{\partial F_1}{\partial x} + \frac{\partial F_2}{\partial y} + \frac{\partial F_3}{\partial z} \tag{4}$$

By the Chain Rule

$$\frac{\partial F_1}{\partial x} = \frac{\partial F_1}{\partial \rho}\frac{\partial \rho}{\partial x} + \frac{\partial F_1}{\partial \theta}\frac{\partial \theta}{\partial x} + \frac{\partial F_1}{\partial \phi}\frac{\partial \phi}{\partial x} \tag{5}$$

Differentiating the relations

$$x = \rho \sin \phi \cos \theta$$

$$y = \rho \sin \phi \sin \theta$$

$$z = \rho \cos \phi$$

implicitly with respect to x gives

$$1 = \frac{\partial \rho}{\partial x} \sin \phi \cos \theta + \rho \cos \phi \cos \theta \frac{\partial \phi}{\partial x} - \rho \sin \phi \sin \theta \frac{\partial \theta}{\partial x}$$

$$0 = \frac{\partial \rho}{\partial x} \sin \phi \sin \theta + \rho \cos \phi \sin \theta \frac{\partial \phi}{\partial x} + \rho \sin \phi \cos \theta \frac{\partial \theta}{\partial x}$$

and

$$0 = \frac{\partial \rho}{\partial x} \cos \phi - \rho \sin \phi \frac{\partial \phi}{\partial x}$$

Solving these yields,

$$\left.\begin{aligned} \frac{\partial \phi}{\partial x} &= \cos \theta \sin \phi \\ \frac{\partial \rho}{\partial x} &= \frac{\cos \theta \cos \phi}{\rho} \\ \frac{\partial \theta}{\partial x} &= -\frac{\sin \theta}{\rho \sin \phi} \end{aligned}\right\} \tag{6}$$

Substituting (6) into (5) gives

$$\frac{\partial F_1}{\partial x} = \frac{\partial F_1}{\partial \rho} \cos \theta \sin \phi + \frac{\partial F_1}{\partial \phi} \frac{\cos \theta \cos \phi}{\rho} - \frac{\partial F_1}{\partial \theta} \frac{\sin \theta}{\rho \sin \phi} \tag{7}$$

with similar expressions for $\partial F_2/\partial y$ and $\partial F_3/\partial z$ (which the reader should work out). Substitution of these into (4) and simplification gives the desired answer.

EXERCISES

1. Suppose $\nabla \cdot \mathbf{F} = 0$ and $\nabla \cdot \mathbf{G} = 0$. Which of the following necessarily have zero divergence?
   (a) $\mathbf{F} + \mathbf{G}$
   (b) $\mathbf{F} \times \mathbf{G}$
   (c) $\mathbf{F}(\mathbf{F} \cdot \mathbf{G})$

2. Prove formulas 1 to 6 of Table 3.1.

3. Prove formulas 7, 9, and 11 of Table 3.1.

4. Prove formulas 12, 13, 15, and 16 of Table 3.1.

5. Let $\mathbf{F} = 2xz^2\mathbf{i} + \mathbf{j} + y^3zx\mathbf{k}$, $\mathbf{G} = x^2\mathbf{i} + y^2\mathbf{j} + z^2\mathbf{k}$ and $f = x^2y$. Compute the following quantities.
   (a) ∇f (b) $\nabla \times \mathbf{F}$
   (c) $(\mathbf{F} \cdot \nabla)\mathbf{G}$ (d) $\mathbf{F} \cdot (\nabla f)$
   (e) $\mathbf{F} \times \nabla f$

6. Prove formulas 18 to 21 of Table 3.1.

7. Let $\mathbf{F}$ be a general vector field. Does $\nabla \times \mathbf{F}$ have to be perpendicular to $\mathbf{F}$?

8. Let $\mathbf{r}(x, y, z) = (x, y, z)$ and $r = \sqrt{x^2 + y^2 + z^2} = \|\mathbf{r}\|$. Prove the following identities (make use of Table 3.1 as much as possible).
   (a) $\nabla(1/r) = -\mathbf{r}/r^3$, $r \neq 0$; and, in general, $\nabla(r^n) = nr^{n-2}\mathbf{r}$ and $\nabla(\log r) = \mathbf{r}/r^2$
   (b) $\nabla^2(1/r) = 0$, $r \neq 0$; and, in general, $\nabla^2 r^n = n(n+1)r^{n-2}$
   (c) $\nabla \cdot (\mathbf{r}/r^3) = 0$; and, in general, $\nabla \cdot (r^n\mathbf{r}) = (n+3)r^n$
   (d) $\nabla \times \mathbf{r} = \mathbf{0}$; and, in general, $\nabla \times (r^n\mathbf{r}) = \mathbf{0}$.

9. (a) Prove (*ii*) of Theorem 5′.
   (b) Prove (*iii*) of Theorem 5′.

*10. (a) Prove (*i*) of Theorem 5″.
   (b) Prove (*iii*) of Theorem 5″.

11. Show that in polar coordinates (r, θ) in $\mathbb{R}^2$, Laplace's equation

$$\nabla^2 u = \frac{\partial^2 u}{\partial x^2} + \frac{\partial^2 u}{\partial y^2} = 0$$

takes the form

$$\frac{\partial^2 u}{\partial r^2} + \frac{1}{r^2}\frac{\partial^2 u}{\partial \theta^2} + \frac{1}{r}\frac{\partial u}{\partial r} = 0$$

*12. Show that in spherical coordinates (ρ, θ, ϕ) Laplace's equation $\nabla^2 V = 0$ takes the form

$$\frac{\partial}{\partial \mu}\left((1 - \mu^2)\frac{\partial V}{\partial \mu}\right) + \frac{1}{1 - \mu^2}\frac{\partial^2 V}{\partial \theta^2} + \rho\frac{\partial^2(rV)}{\partial \rho^2} = 0$$

where $\mu = \cos \phi$. (This was Laplace's original form of the Laplace equation.) Compare this with the expression in Problem 11 when V is constant in ϕ.

REVIEW EXERCISES FOR CHAPTER 3

1. Compute the divergence of the following vector fields at the points indicated.
   (a) $\mathbf{F}(x, y, z) = x\mathbf{i} + 3xy\mathbf{j} + z\mathbf{k}$, $(0, 1, 0)$
   (b) $\mathbf{F}(x, y, z) = y\mathbf{i} + z\mathbf{j} + x\mathbf{k}$, $(1, 1, 1)$
   (c) $\mathbf{F}(x, y, z) = (x + y)^3\mathbf{i} + (\sin xy)\mathbf{j} + (\cos xyz)\mathbf{k}$, $(2, 0, 1)$
2. Compute the curl of each vector field in Exercise 1 at the given point.
3. (a) Let $f(x, y, z) = xyz^2$; compute ∇f.
   (b) Let $\mathbf{F}(x, y, z) = xy\mathbf{i} + yz\mathbf{j} + zy\mathbf{k}$; compute $\nabla \times \mathbf{F}$.
   (c) Compute $\nabla \times (f\mathbf{F})$ using Formula 11 of Table 3.1. Compare with a direct computation.
4. Compute $\nabla \cdot \mathbf{F}$ and $\nabla \times \mathbf{F}$ for the following vector fields.
   (a) $\mathbf{F} = 2x\mathbf{i} + 3y\mathbf{j} + 4z\mathbf{k}$
   (b) $\mathbf{F} = x^2\mathbf{i} + y^2\mathbf{j} + z^2\mathbf{k}$
   (c) $\mathbf{F} = (x + y)\mathbf{i} + (y + z)\mathbf{j} + (z + x)\mathbf{k}$
5. Find the equation of the plane tangent to each surface at the indicated point.
   (a) $z = x^2 + y^2$, $(0, 0, 0)$
   (b) $z = x^2 - y^2 + x$, $(1, 0, 2)$
   (c) $z = (x + y)^2$, $(3, 2, 25)$
6. Let $\boldsymbol{\sigma}: \mathbb{R} \to \mathbb{R}^3$ be a path and $h: \mathbb{R} \to \mathbb{R}$ a strictly increasing differentiable function. The composition $\boldsymbol{\sigma} \circ h: \mathbb{R} \to \mathbb{R}^3$ is called the *reparametrization* of $\boldsymbol{\sigma}$ by h; argue that $\boldsymbol{\sigma} \circ h$ has the same trajectory as $\boldsymbol{\sigma}$, and prove that if $\boldsymbol{\alpha} = \boldsymbol{\sigma} \circ h$ then $\boldsymbol{\alpha}'(t) = h'(t)\boldsymbol{\sigma}'(h(t))$. (See Exercise 5(c), §3.2.)
7. What altitude must a satellite have in order that it appear stationary in the sky when viewed from Earth? (See Exercise 7, Section 3.1, for units.)
8. (a) Let $\boldsymbol{\alpha}$ be any differentiable path whose speed is never zero. Let $s(t)$ be the arc-length function for $\boldsymbol{\alpha}$, $s(t) = \int_a^t \|\boldsymbol{\alpha}'(\tau)\|\, d\tau$. Let $t(s)$ be the inverse function of s. Prove that the curve $\boldsymbol{\beta} = \boldsymbol{\alpha} \circ t$ has unit speed; i.e., $\|\boldsymbol{\beta}'(s)\| = 1$.
   (b) Let $\boldsymbol{\sigma}$ be the path $\boldsymbol{\sigma}(t) = (a \cos t, a \sin t, bt)$. Find a path $\boldsymbol{\alpha}$ that has the same trajectory as $\boldsymbol{\sigma}$, but has unit speed, $\|\boldsymbol{\alpha}'(t)\| = 1$; i.e., find a unit-speed reparametrization of $\boldsymbol{\sigma}$. (See Exercises 5 and 7 in §3.2.)
9. Let a particle of mass m move on the path $\boldsymbol{\sigma}(t) = (t^2, \sin t, \cos t)$. Compute the force acting on the particle at $t = 0$.
10. (a) Let $\mathbf{c}(t)$ be a path with $\|\mathbf{c}(t)\| = $ constant; i.e., the curve lies on a sphere. Prove that $\mathbf{c}'(t)$ is orthogonal to $\mathbf{c}(t)$.
    (b) Let $\mathbf{c}$ be a path whose speed is never zero. Show that $\mathbf{c}$ has constant speed iff the acceleration vector $\mathbf{c}''$ is always perpendicular to the velocity vector $\mathbf{c}'$.
11. Let a particle travel on the path $\mathbf{c}(t) = (t, t^2, t \cos t)$ and, at $t = \pi$, leave this curve on a tangent. Where is the particle at time $t = 2\pi$?
12. (a) For the following functions $f: \mathbb{R}^3 \to \mathbb{R}$ and $\mathbf{g}: \mathbb{R} \to \mathbb{R}^3$ find ∇f and $\mathbf{g}'$ and evaluate $(f \circ \mathbf{g})'(1)$.
    (*i*) $f(x, y, z) = xyz$, $\mathbf{g}(t) = (t, \cos t, \sin t)$
    (*ii*) $f(x, y, z) = xyz$, $\mathbf{g}(t) = (6t, 3t^2, t^3)$
    (b) Let $f(x, y) = (e^x, x + y)$ and $g(u, v) = (u, \cos v, v + u)$. Compute the derivative of $g \circ f$ at $(0, 0)$ in two ways.

13. Compute the directional derivative of the following functions in the given direction at the given point.
    (a) $f(x, y, z) = xyz$, $\mathbf{v} = (1/\sqrt{14})(\mathbf{i} + 3\mathbf{j} + 2\mathbf{k})$, $(1, 1, 1)$
    (b) $f(x, y, z) = x^2 + y$, $\mathbf{v} = \frac{12}{13}\mathbf{i} + \frac{3}{13}\mathbf{j} + \frac{4}{13}\mathbf{k}$, $(1, 0, 0)$

14. (a) Review the proof that $\nabla f(x, y, z)$ is perpendicular to the surface $f(x, y, z) =$ constant.
    (b) Find a unit normal to the surface $x^3y + xz = 1$ at the point $(1, 2, -1)$.
    (c) Find an equation for the plane tangent to the surface in (b) at the indicated point.
    (d) Find the angle between the surfaces $x^2 + y^2 + z^2 = 3$ and $x = z^2 + y^2 - 3$ at the point $(-1, 1, -1)$.

15. Let $\mathbf{F}(x, y, z) = (x^2, 0, z(1 + x))$. Show that $\boldsymbol{\sigma}(t) = (1/(1 - t), 0, e^t/(1 - t))$ is a flow line of $\mathbf{F}$.

16. Let $\mathbf{v} = 3\mathbf{i} + 2\mathbf{j} + \mathbf{k}$ and $\mathbf{w} = \mathbf{i} - \mathbf{j}$.
    (a) Compute $\mathbf{v} + 3\mathbf{w}$, $\mathbf{v} \cdot \mathbf{w}$, and $\mathbf{v} \times \mathbf{w}$.
    (b) Compute the area of the parallelogram spanned by $\mathbf{v}$ and $\mathbf{w}$.

17. (a) A bug, finding itself in a toxic environment caused by chemical AEF, decides to move in a direction that will decrease the concentration of AEF the fastest. If the concentration of AEF is given by $\sigma(x, y, z) = e^{-3x} + \sin(yz) + e^{-z^2}$, with the bug at $(0, 0, 0)$, in which direction should it swim?
    (b) Discuss briefly the theory behind your answer to (a).

18. (a) Does $\lim_{(x, y)\to(0, 0)} (x^2 - y^2)/(x^2 + y^2)$ exist?
    (b) Explain what it means for $f: \mathbb{R}^n \to \mathbb{R}^m$ to be continuous at $\mathbf{x}_0 \in \mathbb{R}^n$. (State the precise definition.)
    (c) Explain what it means for $f: \mathbb{R}^n \to \mathbb{R}^m$ to be differentiable at $\mathbf{x}_0 \in \mathbb{R}^n$. (State the precise definition.)

19. Compute the tangent planes of the following surfaces at the indicated points.
    (a) $z = x^2 + 3y^3 + \sin(xy)$; $x = 1$, $y = 0$
    (b) $x^2 + 3y^2 + 4z^2 = 10$; $(0, \sqrt{2}, 1)$

20. Verify the identities in Exercise 8, Section 3.5 using the expressions for div, grad, and curl in spherical coordinates.

21. (a) Let $\mathbf{F} = 2xye^z\mathbf{i} + e^zx^2\mathbf{j} + (x^2ye^z + z^2)\mathbf{k}$. Compute $\nabla \cdot \mathbf{F}$ and $\nabla \times \mathbf{F}$.
    (b) Find a function $f(x, y, z)$ such that $\mathbf{F} = \nabla f$. Discuss briefly.

CHAPTER 4

HIGHER-ORDER DERIVATIVES; MAXIMA AND MINIMA

For since the fabric of the Universe is most perfect and the works of a most wise creator, nothing at all takes place in the Universe in which some rule of maximum or minimum does not appear.

LEONHARD EULER

In one-variable calculus, to test a function $f(x)$ for a local maximum or minimum we look for critical points x_0, that is, points x_0 for which $f'(x_0) = 0$, and at each such point we check the sign of the second derivative $f''(x_0)$. If $f''(x_0) < 0$, $f(x_0)$ is a local maximum of f; if $f''(x_0) > 0$, $f(x_0)$ is a local minimum of f; if $f''(x_0) = 0$, the test fails.

One of the goals of this chapter is to extend these methods to real-valued functions of several variables. We shall begin in Section 4.1 with a discussion of Taylor's Theorem; this will then be used in Section 4.2 to derive tests for maxima, minima, and saddle points. As with functions of one variable, such methods help one to visualize the shape of a graph.

In Section 4.3 we shall study the problem of maximizing a real-valued function subject to supplementary conditions, also referred to as constraints. For example, we might wish to maximize $f(x, y, z)$ among those (x, y, z) constrained to lie on the unit sphere, $x^2 + y^2 + z^2 = 1$. Section 4.4 gives a technical theorem (the implicit function theorem) useful for studying constraints. It will also be useful later in our study of surfaces.

In Section 4.5 we shall describe a few applications of the preceding material, relating to geometry, to economics, and to equilibrium points of physical systems and their stability.

4.1 TAYLOR'S THEOREM

We shall use Taylor's Theorem in several variables to derive a test for different types of extrema, finally obtaining a test much like the second-derivative test that is learned in one-variable calculus. There are other important applications of this theorem as well. Basically, Taylor's Theorem gives us "higher-order" approximations to a function by using more than just the first derivative of the function.

For smooth functions of one variable $f\colon \mathbb{R} \to \mathbb{R}$, Taylor's Theorem asserts that

$$f(x) = f(a) + f'(a)(x - a) + \frac{f''(a)}{2!}(x - a)^2 + \cdots + \frac{f^{(k)}(a)}{k!}(x - a)^k + R_k(x, a) \tag{1}$$

where

$$R_k(x, a) = \int_a^x \frac{(x - t)^k}{k!} f^{(k+1)}(t)\, dt \tag{1'}$$

is the remainder. For x near a this error $R_k(x, a)$ is small to "order k." This means that

$$\frac{R_k(x, a)}{(x - a)^k} \to 0 \quad \text{as} \quad x \to a \tag{2}$$

In other words, $R_k(x, a)$ is small compared to the (already small) quantity $(x - a)^k$.

Our goal in this section is to prove an analogous theorem valid for functions of several variables. The theorem for functions of one variable will then follow as a corollary.

We already know a first-order version, that is, when $k = 1$. Indeed, if $f\colon \mathbb{R}^n \to \mathbb{R}$ is differentiable at $\mathbf{x}_0$, and we define

$$R_1(\mathbf{x}, \mathbf{x}_0) = f(\mathbf{x}) - f(\mathbf{x}_0) - Df(\mathbf{x}_0) \cdot (\mathbf{x} - \mathbf{x}_0)$$

so

$$f(\mathbf{x}) = f(\mathbf{x}_0) + Df(\mathbf{x}_0)(\mathbf{x} - \mathbf{x}_0) + R_1(\mathbf{x}, \mathbf{x}_0)$$

then by the definition of differentiability,

$$\frac{|R_1(\mathbf{x}, \mathbf{x}_0)|}{\|\mathbf{x} - \mathbf{x}_0\|} \to 0 \quad \text{as} \quad \mathbf{x} \to \mathbf{x}_0$$

that is, $R_1(\mathbf{x}, \mathbf{x}_0)$ vanishes to first order at $\mathbf{x}_0$. Let us summarize, writing $\mathbf{h} = \mathbf{x} - \mathbf{x}_0$, and $R_1(\mathbf{x}, \mathbf{x}_0) = R_1(\mathbf{h}, \mathbf{x}_0)$ (an admitted abuse of notation!).

THEOREM 1. *(First-order Taylor's Theorem). Let $f: U \subset \mathbb{R}^n \to \mathbb{R}$ be differentiable at $\mathbf{x}_0 \in U$. Then we may write*

$$f(\mathbf{x}_0 + \mathbf{h}) = f(\mathbf{x}_0) + \sum_{i=1}^{n} h_i \frac{\partial f}{\partial x_i}(\mathbf{x}_0) + R_1(\mathbf{h}, \mathbf{x}_0)$$

where $R_1(\mathbf{h}, \mathbf{x}_0)/\|\mathbf{h}\| \to 0$ as $\mathbf{h} \to \mathbf{0}$ in $\mathbb{R}^n$.

The *second-order Taylor formula* is as follows:

THEOREM 2. *Let $f: U \subset \mathbb{R}^n \to \mathbb{R}$ have continuous partials of third order.* Then we may write*

$$f(\mathbf{x}_0 + \mathbf{h}) = f(\mathbf{x}_0) + \sum_{i=1}^{n} h_i \frac{\partial f}{\partial x_i}(\mathbf{x}_0) + \frac{1}{2} \sum_{i,j=1}^{n} h_i h_j \frac{\partial^2 f}{\partial x_i \, \partial x_j}(\mathbf{x}_0) + R_2(\mathbf{h}, \mathbf{x}_0)$$

where $R_2(\mathbf{h}, \mathbf{x}_0)/\|\mathbf{h}\|^2 \to 0$ as $\mathbf{h} \to \mathbf{0}$.

In the course of the proof, we shall obtain a useful explicit formula (see (5′) below) for the remainder R_2. This formula is a generalization of formula (1′).

Proof of Theorem 2. If we notice that, by the Chain Rule,

$$\frac{d}{dt} f(\mathbf{x}_0 + t\mathbf{h}) = Df(\mathbf{x}_0 + t\mathbf{h}) \cdot \mathbf{h} = \sum_{i=1}^{n} \frac{\partial f}{\partial x_i}(\mathbf{x}_0 + t\mathbf{h}) h_i$$

then we can integrate both sides from $t = 0$ to $t = 1$ to obtain

$$f(\mathbf{x}_0 + \mathbf{h}) - f(\mathbf{x}_0) = \int_0^1 \sum_{i=1}^{n} \frac{\partial f}{\partial x_i}(\mathbf{x}_0 + t\mathbf{h}) h_i \, dt$$

We shall now integrate the expression on the right-hand side by parts. Remember the general formula

$$\int_0^1 u \frac{dv}{dt} \, dt = -\int_0^1 v \frac{du}{dt} \, dt + uv \Big|_0^1$$

In this case, let $u = \partial f/\partial x_i(\mathbf{x}_0 + t\mathbf{h}) h_i$ and let $v = t - 1$. Therefore

$$\sum_{i=1}^{n} \int_0^1 \frac{\partial f}{\partial x_i}(\mathbf{x}_0 + t\mathbf{h}) h_i \, dt = \sum_{i,j=1}^{n} \int_0^1 (1 - t) \frac{\partial^2 f}{\partial x_i \, \partial x_j}(\mathbf{x}_0 + t\mathbf{h}) h_i h_j \, dt + \sum_{i=1}^{n} h_i \frac{\partial f}{\partial x_i}(\mathbf{x}_0)$$

* For the statement of the theorem as given here, f actually needs only to be of class C^2, but for a convenient form of the remainder we assume f is C^3. If one *assumes* the one-variable version, then application of it to $g(t) = f(\mathbf{x}_0 + t\mathbf{h})$ yields the version given here for several variables.

since

$$\frac{du}{dt} = \sum_{j=1}^{n} \frac{\partial^2 f}{\partial x_i\, \partial x_j}(\mathbf{x}_0 + t\mathbf{h})h_i h_j$$

by the Chain Rule, and

$$uv\Big|_0^1 = (t-1)\frac{\partial f}{\partial x_i}(\mathbf{x}_0 + t\mathbf{h})h_i\Big|_{t=0}^1 = \frac{\partial f}{\partial x_i}(\mathbf{x}_0)h_i$$

Thus we have proved the identity

$$\left.\begin{aligned} f(\mathbf{x}_0 + \mathbf{h}) - f(\mathbf{x}_0) &= \sum_{i=1}^{n} \frac{\partial f}{\partial x_i}(\mathbf{x}_0)h_i + R_1(\mathbf{h}, \mathbf{x}_0) \\ \text{where} \qquad & \\ R_1(\mathbf{h}, \mathbf{x}_0) &= \sum_{i,\,j=1}^{n} \int_0^1 (1-t)\frac{\partial^2 f}{\partial x_i\, \partial x_j}(\mathbf{x}_0 + t\mathbf{h})h_i h_j\, dt \end{aligned}\right\} \qquad (3)$$

(Formula (3) gives an explicit formula for the remainder in Theorem 1.)

If we integrate the expression for $R_1(\mathbf{h}, \mathbf{x}_0)$ by parts, with

$$u = \frac{\partial^2 f}{\partial x_i\, \partial x_j}(\mathbf{x}_0 + t\mathbf{h})h_i h_j \quad \text{and} \quad v = -(t-1)^2/2$$

we get

$$\begin{aligned} R_1(\mathbf{h}, \mathbf{x}_0) &= \sum_{i,\,j,\,k=1}^{n} \int_0^1 \frac{(t-1)^2}{2} \frac{\partial^3 f}{\partial x_i\, \partial x_j\, \partial x_k}(\mathbf{x}_0 + t\mathbf{h})h_i h_j h_k\, dt \\ &\quad + \sum_{i,\,j=1}^{n} \frac{1}{2}\frac{\partial^2 f}{\partial x_i\, \partial x_j}(\mathbf{x}_0)h_i h_j \end{aligned}$$

Thus we have proved that

$$\left.\begin{aligned} f(\mathbf{x}_0 + \mathbf{h}) &= f(\mathbf{x}_0) + \sum_{i=1}^{n} h_i \frac{\partial f}{\partial x_i}(\mathbf{x}_0) \\ &\quad + \frac{1}{2}\sum_{i,\,j=1}^{n} h_i h_j \frac{\partial^2 f}{\partial x_i\, \partial x_j}(\mathbf{x}_0) + R_2(\mathbf{h}, \mathbf{x}_0) \\ \text{where} \qquad & \\ R_2(\mathbf{h}, \mathbf{x}_0) &= \sum_{i,\,j,\,k=1}^{n} \int_0^1 \frac{(t-1)^2}{2} \frac{\partial^3 f}{\partial x_i\, \partial x_j\, \partial x_k}(\mathbf{x}_0 + t\mathbf{h})h_i h_j h_k\, dt \end{aligned}\right\} \qquad (4)$$

The integrand is a continuous function of t and is therefore bounded on a small neighborhood of $\mathbf{x}_0$ (since it has to be close to its value at $\mathbf{x}_0$). Thus for a constant $M \geq 0$ we get, for $\|\mathbf{h}\|$ small,

$$|R_2(\mathbf{h}, \mathbf{x}_0)| \leq \|\mathbf{h}\|^3 M$$

In particular,

$$\frac{|R_2(\mathbf{h}, \mathbf{x}_0)|}{\|\mathbf{h}\|^2} \le \|\mathbf{h}\| M \to 0 \quad \text{as} \quad \mathbf{h} \to \mathbf{0},$$

as required by the theorem. A similar argument for R_1 shows that $|R_1(\mathbf{h}, \mathbf{x}_0)|/\|\mathbf{h}\| \to 0$ as $\mathbf{h} \to \mathbf{0}$, although this also follows from the definition of differentiability, as noted on p. 200. ∎

COROLLARY (EXPLICIT FORM OF THE REMAINDER).

(*i*) *In Theorem 1*

$$\begin{aligned} R_1(\mathbf{h}, \mathbf{x}_0) &= \sum_{i,j=1}^{n} \int_0^1 (1-t) \frac{\partial^2 f}{\partial x_i\, \partial x_j}(\mathbf{x}_0 + t\mathbf{h}) h_i h_j \, dt \\ &= \sum_{i,j=1}^{n} \frac{1}{2} \frac{\partial^2 f}{\partial x_i\, \partial x_j}(\mathbf{c}_{ij}) h_i h_j \end{aligned} \tag{5}$$

where $\mathbf{c}_{ij}$ *lies somewhere on the line joining* $\mathbf{x}_0$ *to* $\mathbf{x}_0 + \mathbf{h}$

(*ii*) *In Theorem 2*

$$\begin{aligned} R_2(\mathbf{h}, \mathbf{x}_0) &= \sum_{i,j,k=1}^{n} \int_0^1 \frac{(t-1)^2}{2} \frac{\partial^3 f}{\partial x_i\, \partial x_j\, \partial x_k}(\mathbf{x}_0 + t\mathbf{h}) h_i h_j h_k \, dt \\ &= \sum_{i,j,k=1}^{n} \frac{1}{3!} \frac{\partial^3 f}{\partial x_i\, \partial x_j\, \partial x_k}(\mathbf{c}_{ijk}) h_i h_j h_k \end{aligned} \tag{5'}$$

where $\mathbf{c}_{ijk}$ *lies somewhere on the line joining* $\mathbf{x}_0$ *to* $\mathbf{x}_0 + \mathbf{h}$.

These integral formulas were obtained in the course of the proof of Theorem 2 (see formulas (3) and (4)). The formulas involving $\mathbf{c}_{ij}$ and $\mathbf{c}_{ijk}$ (Lagrange's form of the remainder) are obtained from the Second Mean Value Theorem for Integrals. This states that

$$\int_a^b h(t)g(t)\, dt = h(c) \int_a^b g(t)\, dt$$

provided h *and* g *are continuous and* $g \ge 0$ *on* $[a, b]$; *here* c *is some number between* a *and* b.* This is applied in (*i*) with $h(t) = ([\partial^2 f/\partial x_i\, \partial x_j])(x_0 + t\mathbf{h})$ and $g(t) = 1 - t$.

* *Proof:* *If* $g = 0$ *the result is trivial, so we can suppose* $g \ne 0$; *thus we can assume* $\int_a^b g(t)\, dt > 0$. *Let* M *and* m *be the maximum and minimum values of* h, *achieved at* t_M *and* t_m *respectively. Since* $g(t) \ge 0$,

$$m \int_a^b g(t)\, dt \le \int_a^b h(t)g(t)\, dt \le M \int_a^b g(t)\, dt$$

Thus $(\int_a^b h(t)g(t)\, dt)/(\int_a^b g(t)\, dt)$ *lies between* $m = h(t_m)$ *and* $M = h(t_M)$ *and therefore, by the Intermediate Value Theorem, equals* $h(c)$ *for some intermediate* c.

It is not hard to guess the general form of Taylor's Theorem. For example, the third-order Taylor formula is:

$$f(\mathbf{x}_0 + \mathbf{h}) = f(\mathbf{x}_0) + \sum_{i=1}^{n} h_i \frac{\partial f}{\partial x_i}(\mathbf{x}_0) + \frac{1}{2}\sum_{i,j=1}^{n} h_i h_j \frac{\partial^2 f}{\partial x_i\,\partial x_j}(\mathbf{x}_0)$$

$$+ \frac{1}{3!}\sum_{i,j,k=1}^{n} h_i h_j h_k \frac{\partial^3 f}{\partial x_i\,\partial x_j\,\partial x_k}(\mathbf{x}_0) + R_3(\mathbf{h}, \mathbf{x}_0)$$

where $R_3(\mathbf{h}, \mathbf{x}_0)/\|\mathbf{h}\|^3 \to 0$ as $\mathbf{h} \to \mathbf{0}$, and so on. The general formula can be proved by induction, using the method of proof given on pages 201–202.

EXAMPLE 1. Compute the second-order Taylor formula for $f(x, y) = \sin(x + 2y)$, $\mathbf{x}_0 = (0, 0)$.

Notice that

$$f(0, 0) = 0$$

$$\frac{\partial f}{\partial x}(0, 0) = \cos(0 + 2 \cdot 0) = 1, \qquad \frac{\partial f}{\partial y}(0, 0) = 2\cos(0 + 2 \cdot 0) = 2$$

$$\frac{\partial^2 f}{\partial x^2}(0, 0) = 0, \qquad \frac{\partial^2 f}{\partial y^2}(0, 0) = 0, \qquad \frac{\partial^2 f}{\partial x\,\partial y}(0, 0) = 0$$

Thus

$$f(\mathbf{h}) = f(h_1, h_2) = h_1 + 2h_2 + R_2(\mathbf{h}, \mathbf{0})$$

where

$$R_2(\mathbf{h}, \mathbf{0})/\|\mathbf{h}\|^2 \to 0 \quad \text{as} \quad \mathbf{h} \to \mathbf{0}$$

EXAMPLE 2. Compute the second-order Taylor formula for $f(x, y) = e^x \cos y$, $x_0 = 0$, $y_0 = 0$.

Here

$$f(0, 0) = 1, \qquad \frac{\partial f}{\partial x}(0, 0) = 1, \qquad \frac{\partial f}{\partial y}(0, 0) = 0$$

$$\frac{\partial^2 f}{\partial x^2}(0, 0) = 1, \qquad \frac{\partial^2 f}{\partial y^2}(0, 0) = -1, \qquad \frac{\partial^2 f}{\partial x\,\partial y}(0, 0) = 0$$

so

$$f(\mathbf{h}) = f(h_1, h_2) = 1 + h_1 + \tfrac{1}{2}h_1^2 - \tfrac{1}{2}h_2^2 + R_2(\mathbf{h}, \mathbf{0})$$

where

$$R_2(\mathbf{h}, \mathbf{0})/\|\mathbf{h}\|^2 \to 0 \quad \text{as} \quad \mathbf{h} \to \mathbf{0}$$

In the case of functions of one variable, one can develop $f(x)$ in an infinite power series, called the *Taylor series*:

$$f(x_0 + h) = f(x_0) + f'(x_0)h + \frac{f''(x_0)h^2}{2} + \cdots + \frac{f^{(k)}(x_0)h^k}{k!} + \cdots$$

provided one can show that $R_k(h, x_0) \to 0$ as $k \to \infty$. Similarly, for functions of several variables the above terms are replaced by the corresponding ones involving partial derivatives, as we have seen in Theorem 2. Again, one can represent such a function by its Taylor series provided one can show that $R_k \to 0$ as $k \to \infty$. This point is examined further in Exercise 7.

EXERCISES

In each of Exercises 1 to 6 determine the second-order Taylor formula for the given function about the given point (x_0, y_0).

1. $f(x, y) = (x + y)^2$, $x_0 = 0$, $y_0 = 0$
2. $f(x, y) = 1/(x^2 + y^2 + 1)$, $x_0 = 0$, $y_0 = 0$
3. $f(x, y) = e^{x+y}$, $x_0 = 0$, $y_0 = 0$
4. $f(x, y) = e^{-x^2-y^2} \cos(xy)$, $x_0 = 0$, $y_0 = 0$
5. $f(x, y) = \sin(xy) + \cos(xy)$, $x_0 = 0$, $y_0 = 0$
6. $f(x, y) = e^{(x-1)^2} \cos y$, $x_0 = 1$, $y_0 = 0$

*7. A function $f: \mathbb{R} \to \mathbb{R}$ is called *analytic* provided

$$f(x + h) = f(x) + f'(x)h + \cdots + \frac{f^{(k)}(x)}{k!} h^k + \cdots$$

(i.e., the series on the right-hand side converges and equals $f(x + h)$).

(a) Suppose f satisfies the following condition: on any closed interval $[a, b]$ there is a constant M such that for all $k = 1, 2, 3, \ldots$, $|f^{(k)}(x)| \le M^k$ for all $x \in [a, b]$. Prove that f is analytic.

(b) Let $f(x) = \begin{cases} e^{-1/x} & x > 0 \\ 0 & x \le 0 \end{cases}$

Show that f is a C^∞ function, but f is not analytic.

(c) Give a definition of analytic functions from $\mathbb{R}^n$ to $\mathbb{R}$. Generalize (a) to this class of functions.

(d) Develop $f(x, y) = e^{x+y}$ in a power series about $x_0 = 0$, $y_0 = 0$.

4.2 EXTREMA OF REAL-VALUED FUNCTIONS

Among the most basic geometric features of the graph of a function are its extreme points, at which the function attains its greatest and least values. In this section, we shall derive a method for determining these points. In fact, the method reveals local extrema as well. These are points at which the function attains a maximum or minimum value relative only to nearby points. Let us begin by defining our terms.

DEFINITION. *If $f: U \subset \mathbb{R}^n \to \mathbb{R}$ is a given scalar function, a point $\mathbf{x}_0 \in U$ is called a **local minimum** of f if there is a neighborhood V of $\mathbf{x}_0$ such that for*

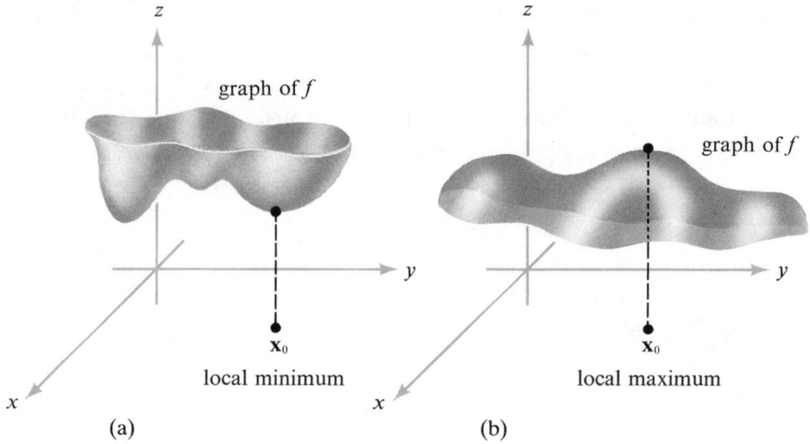

FIGURE 4.2.1
Local minimum (a) and local maximum (b) points for a function of two variables.

all points $\mathbf{x} \in V$, $f(\mathbf{x}) \geq f(\mathbf{x}_0)$. *(See Figure 4.2.1.) Similarly,* $\mathbf{x}_0 \in U$ *is a* ***local maximum*** *if there is a neighborhood* V *of* $\mathbf{x}_0$ *such that* $f(\mathbf{x}) \leq f(\mathbf{x}_0)$ *for all* $\mathbf{x} \in V$. *And* $\mathbf{x}_0 \in U$ *is said to be a* ***local**, **or relative**, **extremum*** *if it is either a local minimum or a local maximum. A point* $\mathbf{x}_0$ *is a* ***critical point*** *of* f *if* $Df(\mathbf{x}_0) = 0$. *A critical point that is not a local extremum is called a* ***saddle point***.*

The location of extrema is based on the following fact, which should be familiar from one-variable calculus (the case $n = 1$): Every extremum is a critical point.

THEOREM 3. *If* $U \subset \mathbb{R}^n$ *is open and* $f: U \subset \mathbb{R}^n \to \mathbb{R}$ *is differentiable and* $\mathbf{x}_0 \in U$ *is a local extremum, then* $Df(\mathbf{x}_0) = 0$; *that is,* $\mathbf{x}_0$ *is a critical point of* f.

Proof. Suppose that f achieves a local maximum at $\mathbf{x}_0$. Then for any $\mathbf{h} \in \mathbb{R}^n$, the function $g(t) = f(\mathbf{x}_0 + t\mathbf{h})$ has a local maximum at $t = 0$. Thus from one-variable calculus $g'(0) = 0^{\dagger}$. On the other hand, by the Chain Rule,

$$g'(0) = Df(\mathbf{x}_0) \cdot \mathbf{h}$$

* The term *saddle point* is sometimes not used this generally.

† *Proof. Since* $g(0)$ *is a local maximum,* $g(t) \leq g(0)$ *for small* $t > 0$, *so* $g(t) - g(0) \leq 0$, *and hence* $g'(0) = \lim\limits_{t \to 0^+} (g(t) - g(0))/(t) \leq 0$, *where* $\lim\limits_{t \to 0^+}$ *means the limit as* $t \to 0$, $t > 0$. *For small* $t < 0$, *we similarly have* $g'(0) = \lim\limits_{t \to 0^-} (g(t) - g(0))/(t) \geq 0$, *so* $g'(0) = 0$.

Thus $Df(\mathbf{x}_0) \cdot \mathbf{h} = 0$ for every $\mathbf{h}$, and so $Df(\mathbf{x}_0) = 0$. The case in which f achieves a local minimum at $\mathbf{x}_0$ is entirely analogous. ■

If we remember that $Df(\mathbf{x}_0) = 0$ means that all the components of $Df(\mathbf{x}_0)$ are zero, we can rephrase the result of Theorem 3: If $\mathbf{x}_0$ is a local extremum, then

$$\frac{\partial f}{\partial x_i}(\mathbf{x}_0) = 0, \qquad i = 1, \ldots, n$$

that is, each partial derivative is zero at $\mathbf{x}_0$. In other words, $\nabla f(\mathbf{x}_0) = \mathbf{0}$, where ∇f is the gradient of f.

If we seek to find the extrema or local extrema of a function, then Theorem 3 states that we should look among the critical points. Sometimes these can be tested by inspection, but usually we use tests (to be developed below) analogous to the second-derivative test in one-variable calculus.

EXAMPLE 1. Consider the function $f: \mathbb{R}^2 \to \mathbb{R}, (x, y) \mapsto x^2 + y^2$. Of course, we already know that this function has a single minimum at the origin, but let us ignore this and apply the method described above.

We must identify the critical points of f by solving the equations $\partial f(x, y)/\partial x = 0, \partial f(x, y)/\partial y = 0$, for x and y. But

$$\frac{\partial}{\partial x} f(x, y) = 2x \quad \text{and} \quad \frac{\partial}{\partial y} f(x, y) = 2y$$

so the only critical point is the origin $(0, 0)$, where the value of the function is zero. As $f(x, y) \geq 0$, this point is a relative minimum—in fact, an absolute minimum—of f.

EXAMPLE 2. Consider the function of Example 4 of Section 2.1, $f: \mathbb{R}^2 \to \mathbb{R}$, $(x, y) \mapsto x^2 - y^2$. Ignoring for the moment that this function has a saddle and no extrema, let us apply the method of Theorem 3 for the location of extrema.

As in Example 1, we find that f has only one critical point, at the origin, and the value of f there is zero. Examining values of f directly for points near the origin, we see that $f(x, 0) \geq f(0, 0)$ and $f(0, y) \leq f(0, 0)$. As x or y can be taken arbitrarily small, the origin cannot be either a relative minimum or a relative maximum. Therefore this function can have no relative extrema (see Figure 4.2.2).

The phenomenon in this example, called a critical point of saddle type, or a saddle point, may also occur in case $n = 1$. In that case, when a point is a critical point without being a local extremum, it is called an *inflection point*. For example, $f(x) = x^3$ has a critical point at $x = 0$, but $x = 0$ is not a relative extremum.

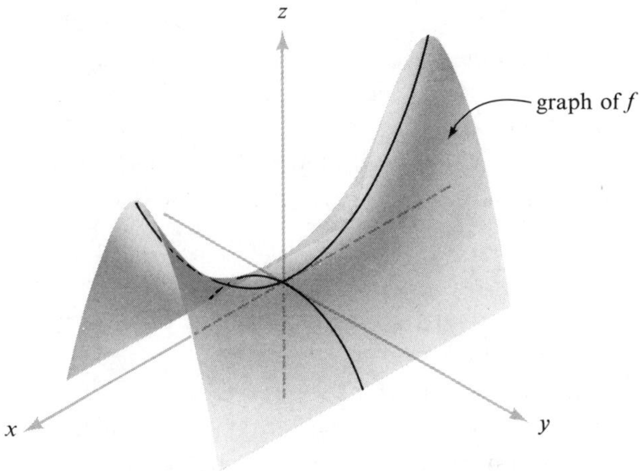

FIGURE 4.2.2
A function of two variables with a saddle point.

EXAMPLE 3. Find all the critical points of $z = x^2y + y^2x$.

Differentiating, we obtain

$$\frac{\partial z}{\partial x} = 2xy + y^2, \qquad \frac{\partial z}{\partial y} = 2xy + x^2$$

Equating the partials to zero yields

$$2xy + y^2 = 0, \qquad 2xy + x^2 = 0$$

Subtracting, we obtain $x^2 = y^2$. Thus $x = \pm y$. Substituting $x = +y$ in the first equation above, we find that

$$2y^2 + y^2 = 3y^2 = 0$$

so $y = 0$ and thus $x = 0$. If $x = -y$ then

$$-2y^2 + y^2 = -y^2 = 0$$

so $y = 0$ and therefore $x = 0$.

Hence the only critical point is (0, 0). For $x = y$, $z = 2x^3$, which is both positive and negative for x near zero. Thus (0, 0) is not a relative extremum.

The remainder of this section is devoted to deriving a criterion, depending on the second derivative, for a critical point to be a relative extremum. In the special case $n = 1$, our criterion will reduce to the familiar condition $f''(x) > 0$ for a minimum and $f''(x) < 0$ for a maximum. But in the general context, the second derivative is a fairly complicated mathematical object. To state our criterion, we will introduce a version of the second derivative called the Hessian.

The concept we wish to introduce involves the notion of a *quadratic function.* Quadratic functions are functions $g\colon \mathbb{R}^n \to \mathbb{R}$ that have the form

$$g(h_1, \ldots, h_n) = \sum_{i,j=1}^{n} a_{ij}h_ih_j$$

for a matrix a_{ij}. In terms of matrix multiplication we can write

$$g(h_1, \ldots, h_n) = [h_1 \cdots h_n]\begin{bmatrix} a_{11} & a_{12} & \cdots & a_{1n} \\ \vdots & \vdots & & \vdots \\ a_{n1} & a_{n2} & \cdots & a_{nn} \end{bmatrix}\begin{bmatrix} h_1 \\ \vdots \\ h_n \end{bmatrix}$$

We can, if we wish, assume a_{ij} is symmetric; in fact, g is unchanged if we replace a_{ij} by $b_{ij} = \frac{1}{2}(a_{ij} + a_{ji})$, since $h_ih_j = h_jh_i$ and the sum is over all i and j. The *quadratic* nature of g is reflected in the identity

$$g(\lambda h_1, \ldots, \lambda h_n) = \lambda^2 g(h_1, \ldots, h_n)$$

which follows from the definition.

Now we are ready to define Hessian functions (named after Ludwig Otto Hesse who introduced them in 1844).

DEFINITION. *Suppose $f\colon U \subset \mathbb{R}^n \to \mathbb{R}$ has second-order partial derivatives $(\partial^2 f/\partial x_i\, \partial x_j)(\mathbf{x}_0)$, for $i, j = 1, \ldots, n$, at a point $\mathbf{x}_0 \in U$. Then the* ***Hessian of f at $\mathbf{x}_0$*** *is the quadratic function defined by*

$$Hf(\mathbf{x}_0)(\mathbf{h}) = \frac{1}{2}\sum_{i,j=1}^{n} \frac{\partial^2 f}{\partial x_i\, \partial x_j}(\mathbf{x}_0)h_ih_j$$

This function is usually used at critical points $\mathbf{x}_0 \in U$. In this case, $Df(\mathbf{x}_0) = 0$, and the Taylor formula (see Theorem 2, Section 4.1) may be written in the form

$$f(\mathbf{x}_0 + \mathbf{h}) = f(\mathbf{x}_0) + Hf(\mathbf{x}_0)(\mathbf{h}) + R_2(\mathbf{h}, \mathbf{x}_0)$$

Thus at a critical point the Hessian equals the first nonconstant term in the Taylor series of f.

A quadratic function $g\colon \mathbb{R}^n \to \mathbb{R}$ is called *positive definite* if $g(\mathbf{h}) \geq 0$ for all $\mathbf{h} \in \mathbb{R}^n$, and $g(\mathbf{h}) = 0$ only for $\mathbf{h} = \mathbf{0}$. Similarly, g is *negative definite* if $g(\mathbf{h}) \leq 0$, and $g(\mathbf{h}) = 0$ for $\mathbf{h} = \mathbf{0}$ only.

Note that if $n = 1$, $Hf(x_0)(h) = \frac{1}{2}f''(x_0)h^2$, which is positive definite if $f''(x_0) > 0$. We are now ready to state the criterion for relative extrema.

THEOREM 4. *If $f\colon U \subset \mathbb{R}^n \to \mathbb{R}$ is of class C^3, $\mathbf{x}_0 \in U$ is a critical point of f, and the Hessian $Hf(\mathbf{x}_0)$ is positive definite, then $\mathbf{x}_0$ is a relative minimum of f. Similarly, if $Hf(\mathbf{x}_0)$ is negative definite, then $\mathbf{x}_0$ is a relative maximum.*

Actually, we shall prove that the extrema are *strict*. A relative maximum $\mathbf{x}_0$ is called *strict* if $f(\mathbf{x}) < f(\mathbf{x}_0)$ for nearby $\mathbf{x} \neq \mathbf{x}_0$. A strict relative minimum is defined similarly.

OPTIONAL

The proof of Theorem 4 requires Taylor's Theorem and the following result from linear algebra.

LEMMA 1. *If $B = (b_{ij})$ is an $n \times n$ real matrix, and if the associated quadratic function*

$$H: \mathbb{R}^n \to \mathbb{R}, (h_1, \ldots, h_n) \mapsto \frac{1}{2} \sum_{i,j=1}^{n} b_{ij} h_i h_j$$

is positive definite, then there exists a constant $M > 0$ such that for all $\mathbf{h} \in \mathbb{R}^n$,

$$H(\mathbf{h}) \geq M\|\mathbf{h}\|^2$$

Proof. For $\|\mathbf{h}\| = 1$ set $g(\mathbf{h}) = H(\mathbf{h})$. Then g is a continuous function of $\mathbf{h}$ for $\|\mathbf{h}\| = 1$ and so achieves a minimum value, say M.* Since H is quadratic, we have

$$H(\mathbf{h}) = H\left(\frac{\mathbf{h}}{\|\mathbf{h}\|}\|\mathbf{h}\|\right) = H\left(\frac{\mathbf{h}}{\|\mathbf{h}\|}\right)\|\mathbf{h}\|^2 = g\left(\frac{\mathbf{h}}{\|\mathbf{h}\|}\right)\|\mathbf{h}\|^2 \geq M\|\mathbf{h}\|^2$$

for any $\mathbf{h} \neq \mathbf{0}$. ■

Note that the quadratic function associated with the matrix $\frac{1}{2}(\partial^2 f/\partial x_i\, \partial x_j)$ is exactly the Hessian.

Proof of Theorem 4. Recall that if $f: U \subset \mathbb{R}^n \to \mathbb{R}$ is of class C^3 and $\mathbf{x}_0 \in U$ is a critical point, Taylor's Theorem may be expressed in the form

$$f(\mathbf{x}_0 + \mathbf{h}) - f(\mathbf{x}_0) = Hf(\mathbf{x}_0)(\mathbf{h}) + R_2(\mathbf{h}, \mathbf{x}_0)$$

where

$$R_2(\mathbf{h}, \mathbf{x}_0)/\|\mathbf{h}\|^2 \to 0$$

as $\mathbf{h} \to \mathbf{0}$.

Since $Hf(\mathbf{x}_0)$ is positive definite by hypothesis, there is a constant $M > 0$ such that for all $\mathbf{h} \in \mathbb{R}^n$

$$Hf(\mathbf{x}_0)(\mathbf{h}) \geq M\|\mathbf{h}\|^2$$

Since $R_2(\mathbf{h}, \mathbf{x}_0)/\|\mathbf{h}\|^2 \to 0$ as $\mathbf{h} \to \mathbf{0}$, there is a $\delta > 0$ such that for $0 < \|\mathbf{h}\| < \delta$

$$|R_2(\mathbf{h}, \mathbf{x}_0)| < M\|\mathbf{h}\|^2$$

Thus $0 < Hf(\mathbf{x}_0)(\mathbf{h}) + R_2(\mathbf{h}, \mathbf{x}_0) = f(\mathbf{x}_0 + \mathbf{h}) - f(\mathbf{x}_0)$ for $0 < \|\mathbf{h}\| < \delta$, so $\mathbf{x}_0$ is a relative minimum, in fact, a strict relative minimum.

The proof in the negative-definite case is similar, or else follows by applying the above to $-f$, and is left as an exercise. ■

* Here we are using, without proof, a theorem analogous to a theorem in calculus that states that every continuous function on an interval $[a, b]$ achieves a maximum and minimum.

EXAMPLE 4. Consider again the function $f\colon \mathbb{R}^2 \to \mathbb{R}$, $(x, y) \mapsto x^2 + y^2$. Then $(0, 0)$ is a critical point, and f is already in the form of Taylor's Theorem

$$f((0, 0) + (h_1, h_2)) = f(0, 0) + (h_1^2 + h_2^2) + 0$$

We can see directly that the Hessian at $(0, 0)$ is

$$Hf(\mathbf{0})(\mathbf{h}) = h_1^2 + h_2^2$$

which is clearly positive definite. Thus $(0, 0)$ is a relative minimum. This simple case can, of course, be done without Calculus. Indeed, it is clear that $f(x, y) > 0$ for all $(x, y) \neq (0, 0)$.

For functions of two variables $f(x, y)$, the Hessian may be written as follows:

$$Hf(x, y)(\mathbf{h}) = \tfrac{1}{2}[h_1, h_2]\begin{bmatrix} \dfrac{\partial^2 f}{\partial x^2} & \dfrac{\partial^2 f}{\partial y\, \partial x} \\[2ex] \dfrac{\partial^2 f}{\partial x\, \partial y} & \dfrac{\partial^2 f}{\partial y^2} \end{bmatrix}\begin{bmatrix} h_1 \\ h_2 \end{bmatrix}$$

Now we shall give a useful criterion for when a quadratic function defined by such a 2×2 matrix is positive definite. This will then be applied to Theorem 4.

LEMMA 2. *Let*

$$B = \begin{bmatrix} a & b \\ b & c \end{bmatrix} \quad \textit{and} \quad H(\mathbf{h}) = \tfrac{1}{2}[h_1, h_2]B\begin{bmatrix} h_1 \\ h_2 \end{bmatrix}$$

Then $H(\mathbf{h})$ is positive definite if and only if $a > 0$ and $\det B = ac - b^2 > 0$.

Proof. We have

$$H(\mathbf{h}) = \tfrac{1}{2}[h_1, h_2]\begin{bmatrix} ah_1 + bh_2 \\ bh_1 + ch_2 \end{bmatrix} = \tfrac{1}{2}(ah_1^2 + 2bh_1h_2 + ch_2^2)$$

Let us complete the square, writing

$$H(\mathbf{h}) = \frac{1}{2}\left(a\left(h_1 + \frac{b}{a}h_2\right)^2 + \left(c - \frac{b^2}{a}\right)h_2^2\right)$$

Suppose H is positive definite. Setting $h_2 = 0$ we see that $a > 0$. Setting $h_1 = -(b/a)h_2$ we get $c - b^2/a > 0$ or $ac - b^2 > 0$. Conversely, if $a > 0$ and $c - b^2/a > 0$, $H(\mathbf{h})$ is a sum of squares, so $H(\mathbf{h}) \geq 0$. If $H(\mathbf{h}) = 0$ then each square must be zero. This implies that both h_1 and h_2 must be zero, so $H(\mathbf{h})$ is positive definite. ■

Similarly, $H(\mathbf{h})$ is negative definite if and only if $a < 0$, $ac - b^2 > 0$. There are similar criteria for an $n \times n$ symmetric matrix B. Consider the n square

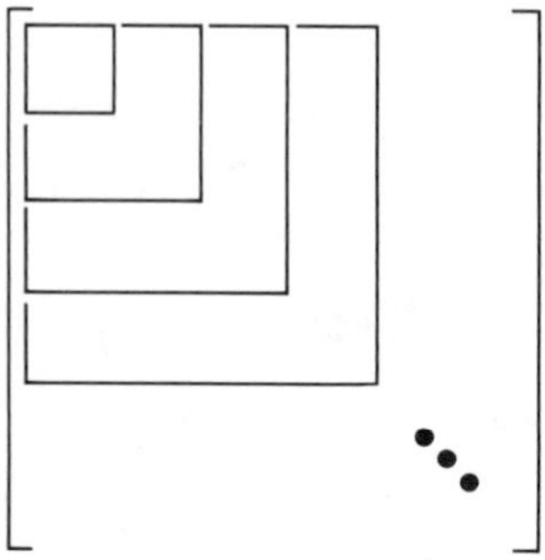

FIGURE 4.2.3
"Diagonal" submatrices are used in the criterion for positive definiteness; they must all have determinant > 0.

submatrices along the diagonal (see Figure 4.2.3). Then B is positive definite (that is, the quadratic function associated with B is positive definite) if and only if the determinants of these diagonal submatrices are all greater than zero. For negative definite the signs should be alternately <0 and >0. We shall not prove this general case here.*

We can use Lemma 2 and Theorem 4 to immediately obtain the following result.

THEOREM 5. *Let $f(x, y)$ be of class C^3 on an open set U in $\mathbb{R}^2$. A point (x_0, y_0) is a (strict) local minimum of f provided the following three conditions hold:*

(*i*) $\dfrac{\partial f}{\partial x}(x_0, y_0) = \dfrac{\partial f}{\partial y}(x_0, y_0) = 0$

(*ii*) $\dfrac{\partial^2 f}{\partial x^2}(x_0, y_0) > 0$

(*iii*) $D = \left(\dfrac{\partial^2 f}{\partial x^2}\right)\left(\dfrac{\partial^2 f}{\partial y^2}\right) - \left(\dfrac{\partial^2 f}{\partial x\, \partial y}\right)^2 > 0$ *at* (x_0, y_0)

(D is called the discriminant.) If in (ii) we have <0 instead of >0 and condition (iii) is unchanged, then we have a (strict) local maximum.

EXAMPLE 5. Consider the function $f: \mathbb{R}^2 \to \mathbb{R}$, $(x, y) \mapsto x^2 - 2xy + 2y^2$. As in Example 4, we easily compute that the origin is the only critical point,

* This is proved in, for example, K. Hoffman and R. Kunze, *Linear Algebra*, Prentice-Hall, Englewood Cliffs, N.J., 1961, pp. 249–251.

$f(0, 0) = 0$, and the Hessian is

$$Hf(\mathbf{0})(\mathbf{h}) = h_1^2 - 2h_1h_2 + 2h_2^2 = (h_1 - h_2)^2 + h_2^2$$

which is clearly positive definite. Thus f has a relative minimum at (0, 0). Alternatively, we can apply Theorem 5. At (0, 0), $\partial^2 f/\partial x^2 = 2$, $\partial^2 f/\partial y^2 = 4$, and $\partial^2 f/\partial x\, \partial y = -2$. Conditions (*i*), (*ii*), and (*iii*) hold, so f has a relative minimum at (0, 0).

If, in Theorem 5, $D < 0$ then we have a saddle point. In fact one can prove that $f(x, y)$ is larger than $f(x_0, y_0)$ as we move away from (x_0, y_0) in some direction and smaller in the orthogonal direction (see Exercise 14). The general appearance is thus similar to that shown in Figure 4.2.2. The appearance of the case $D = 0$ must be examined by further analysis.

In summary: When, in an example dealing with functions of two variables, all critical points have been found and their associated Hessians computed, some of these Hessians may be positive definite, indicating relative minima; some may be negative definite, indicating relative maxima; and some may take both positive and negative values, indicating saddle points. The shape of the graph at a saddle point *where* $D < 0$ is like that in Figure 4.2.2. Critical points for which $D \neq 0$ (or the Hessian is positive or negative definite) are called non-degenerate. Thus there are nondegenerate maxima, minima, and saddle points. The remaining critical points, where $D = 0$, may be tested directly, with level sets and sections (Section 2.1) or by some other method. Such critical points are called *degenerate*, and the methods developed in this chapter fail to provide a picture of the behavior of a function near such points.

EXAMPLE 6. Locate the maxima, minima, and saddle points of the function

$$f(x, y) = \log(x^2 + y^2 + 1)$$

We must first locate the critical points of this function; so, according to Theorem 3, we calculate

$$\nabla f(x, y) = \frac{2x}{x^2 + y^2 + 1}\mathbf{i} + \frac{2y}{x^2 + y^2 + 1}\mathbf{j}$$

Thus $\nabla f(x, y) = \mathbf{0}$ if and only if $(x, y) = (0, 0)$, and so the only critical point of f is (0, 0). Now we must determine whether this is a maximum, a minimum, or a saddle point. The second partials are

$$\frac{\partial^2 f}{\partial x} = \frac{2(x^2 + y^2 + 1) - (2x)(2x)}{(x^2 + y^2 + 1)^2}$$

$$\frac{\partial^2 f}{\partial y^2} = \frac{2(x^2 + y^2 + 1) - (2y)(2y)}{(x^2 + y^2 + 1)^2}$$

and

$$\frac{\partial^2 f}{\partial x\,\partial y} = \frac{-2x(2y)}{(x^2 + y^2 + 1)^2}$$

So

$$\frac{\partial^2 f}{\partial x^2}(0, 0) = 2 = \frac{\partial^2 f}{\partial y^2}(0, 0)$$

and

$$\frac{\partial^2 f}{\partial x\,\partial y}(0, 0) = 0$$

which yields

$$D = 2 \cdot 2 = 4 > 0$$

Since $(\partial^2 f/\partial x^2)(0, 0) > 0$ we conclude by Theorem 5 that $(0, 0)$ is a local minimum. (Can you show this just from the fact that $\log t$ is an increasing function of $t > 0$?)

EXAMPLE 7. The graph of the function $g(x, y) = 1/xy$ is a surface S in $\mathbb{R}^3$. Find the points on S that are closest to the origin $(0, 0, 0)$.

The distance from (x, y, z) to $(0, 0, 0)$ is given by the formula

$$d(x, y, z) = \sqrt{x^2 + y^2 + z^2}$$

If $(x, y, z) \in S$ then d can be expressed as a function $d_*(x, y) = d(x, y, 1/xy)$ of two variables:

$$d_*(x, y) = \sqrt{x^2 + y^2 + \frac{1}{x^2 y^2}}$$

Note that the minimum (if it exists) cannot occur "too near" the x-axis or y-axis because d_* is not defined if either $x = 0$ or $y = 0$, and d_* gets very large as x or y approaches the x-axis or the y-axis.

Since $d_* > 0$ it will be minimized when $d_*^2(x, y) = x^2 + y^2 + (1/x^2y^2) = f(x, y)$ is minimized. (This function f is much easier to deal with.) We calculate the gradient

$$\nabla f(x, y) = \nabla d_*^2(x, y) = \left(2x - \frac{2}{x^3 y^2}\right)\mathbf{i} + \left(2y - \frac{2}{y^3 x^2}\right)\mathbf{j}$$

This is $\mathbf{0}$ if and only if

$$\left(2x - \frac{2}{x^3 y^2}\right) = 0 = \left(2y - \frac{2}{y^3 x^2}\right)$$

that is, $x^4y^2 - 1 = 0$ and $x^2y^4 - 1 = 0$. From the first equation we get $y^2 = 1/x^4$, and, substituting this into the second equation, we obtain

$$\frac{x^2}{x^8} = 1 = \frac{1}{x^6}$$

Thus $x = \pm 1$ and $y = \pm 1$, and it therefore follows that f has four critical points, namely, $(1, 1)$, $(1, -1)$, $(-1, 1)$, and $(-1, -1)$. To determine whether these are local minima, local maxima, or saddle points we apply Theorem 5:

$$\frac{\partial^2 f}{\partial x^2} = 2 + \frac{6}{x^4 y^2}, \qquad \frac{\partial^2 f}{\partial y^2} = 2 + \frac{6}{x^2 y^4}, \qquad \frac{\partial^2 f}{\partial y\, \partial x} = \frac{4}{x^3 y^3}$$

so

$$\frac{\partial^2 f}{\partial x^2}(a, b) = \frac{\partial^2 f}{\partial y^2}(a, b) = 8$$

where (a, b) is any one of the above four critical points, and $(\partial^2 f/\partial x\, \partial y)(a, b) = \pm 4$.

We see that in any of the above cases $D = 64 - 16 = 48 > 0$ and $(\partial^2 f/\partial x^2)(a, b) > 0$, so each critical point is a local minimum, and these are all the local minima for f.

Finally, note that $d_*^2(a, b) = 3$ for all these critical points and so the points on the surface that are closest to $(0, 0, 0)$ are $(1, 1, 1)$, $(1, -1, -1)$, $(-1, 1, -1)$, and $(-1, -1, 1)$ with $d_* = \sqrt{3}$ at these points. Thus $d_* \geq \sqrt{3}$ and is equal to $\sqrt{3}$ when $(x, y) = (\pm 1, \pm 1)$.

EXAMPLE 8. Analyze the behavior of $z = x^5 y + x y^5 + xy$ at its critical points.

The first partial derivatives are

$$\frac{\partial z}{\partial x} = 5x^4 y + y^5 + y = y(5x^4 + y^4 + 1)$$

and

$$\frac{\partial z}{\partial y} = x(5y^4 + x^4 + 1)$$

The terms $5x^4 + y^4 + 1$ and $5y^4 + x^4 + 1$ are always greater than or equal to 1 so it follows that the only critical point is $(0, 0)$.

The second partials are

$$\frac{\partial^2 z}{\partial x^2} = 20x^3 y, \qquad \frac{\partial^2 z}{\partial y^2} = 20xy^3$$

and

$$\frac{\partial^2 z}{\partial x\, \partial y} = 5x^4 + 5y^4 + 1$$

Thus at $(0, 0)$, $D = -1$, so $(0, 0)$ is a non-degenerate saddle point and the graph of z near $(0, 0)$ looks like Figure 4.2.2.

EXERCISES

In Exercises 1 to 10, find the critical points of the given function and then determine whether they are local maxima, local minima, or saddle points.

1. $f(x, y) = x^2 - y^2 + xy$
2. $f(x, y) = x^2 + y^2 - xy$
3. $f(x, y) = x^2 + y^2 + 2xy$
4. $f(x, y) = x^2 + y^2 + 3xy$
5. $f(x, y) = e^{1+x^2-y^2}$
6. $f(x, y) = x^2 - 3xy + 5x - 2y + 6y^2 + 8$
7. $f(x, y) = 3x^2 + 2xy + 2x + y^2 + y + 4$
8. $f(x, y) = \sin(x^2 + y^2)$ (consider only the critical point $(0, 0)$)
9. $f(x, y) = \cos(x^2 + y^2)$ (consider only the critical points $(0, 0)$, $(\sqrt{\pi/2}, \sqrt{\pi/2})$, and $(0, \sqrt{\pi})$)

*10. $f(x, y) = \log(2 + \sin xy)$

11. An examination of the function $f: \mathbb{R}^2 \to \mathbb{R}$, $(x, y) \mapsto (y - 3x^2)(y - x^2)$, will give an idea of the difficulty of finding conditions that guarantee that a critical point is a relative extremum when Theorem 5 fails. Show that
    (a) the origin is a critical point of f;
    (b) f has a relative minimum at $(0, 0)$ on every straight line through $(0, 0)$, that is, if $g(t) = (at, bt)$, then $f \circ g: \mathbb{R} \to \mathbb{R}$ has a relative minimum at 0, for every choice of a and b;
    (c) the origin is not a relative minimum of f.

12. Let $f(x, y) = Ax^2 + E$. What are the critical points of f? Are they local maxima or local minima?

13. Let $f(x, y) = x^2 - 2xy + y^2$. Here $D = 0$. Can you say if the critical points are local minima, local maxima, or saddle points?

*14. Show that if (x_0, y_0) is a critical point of a C^3 function $f(x, y)$ and $D < 0$ then there are points (x, y) near (x_0, y_0) at which $f(x, y) > f(x_0, y_0)$ and, similarly, points for which $f(x, y) < f(x_0, y_0)$.

15. Determine the nature of the critical points of the function
$$f(x, y, z) = x^2 + y^2 + z^2 + xy$$

In Exercises 16 to 20 let D be the unit disc in $\mathbb{R}^2$, that is, $\{(x, y) | x^2 + y^2 \leq 1\}$ and $\partial D = \{(x, y) | x^2 + y^2 = 1\}$.

*16. Let u be a function on D that is "strictly subharmonic", that is, $\nabla^2 u = (\partial^2 u/\partial x^2) + (\partial^2 u/\partial y^2) > 0$. Show that u cannot have a maximum point in $D \backslash \partial D$ (the set of points in D but not in ∂D).

*17. Let u be a harmonic function, that is, $\nabla^2 u = 0$. Show that if u achieves its maximum value in $D \backslash \partial D$ it also achieves it on ∂D. This is sometimes called the "weak maximum principle" for harmonic functions. (HINT: Consider $\nabla^2(u + \varepsilon e^x)$, $\varepsilon > 0$. This exercise requires the following fact, which is proved in more advanced texts: Given a sequence $\{\mathbf{p}_n\}$, $n = 1, 2, \ldots$ in a closed bounded set A in $\mathbb{R}^2$ or $\mathbb{R}^3$, there exists a point $\mathbf{q}$ such that every neighborhood of $\mathbf{q}$ contains at least one member of $\{\mathbf{p}_n\}$. A set A is *closed* when $\partial A \subset A$ and is *bounded* when it lies in some ball.)

*18. Define the notion of a strict superharmonic function u on D. Show that u cannot have a minimum in $D\backslash\partial D$.

*19. Let u be harmonic in D as in Exercise 17. Show that if u achieves its minimum value in $D\backslash\partial D$ it also achieves it on ∂D. This is sometimes called the "weak minimum principle" for harmonic functions.

*20. Let $\phi\colon \partial D \to \mathbf{R}$ be continuous and let T be a solution on D to $\nabla^2 T = 0$, $T = \phi$ on ∂D.
(a) Use Exercises 16 to 19 to show that such a solution, if it exists, must be unique.
(b) Suppose that $T(x, y)$ represents a temperature function that is independent of time with ϕ representing the temperature of a circular plate at its boundary. Can you give a physical interpretation of (a)?

4.3 CONSTRAINED EXTREMA AND LAGRANGE MULTIPLIERS

Often in problems we want to maximize a function subject to certain *constraints* or *side conditions.* Such situations arise, for example, in economics. Suppose we are selling two kinds of goods, say I and II; let x and y represent the quantity of each sold. Then let $f(x, y)$ represent the profit we earn when x amount of I and y amount of II are sold. But our production is controlled by our capital, so we are constrained to work subject to a relation, say $g(x, y) = c$. Thus we want to maximize $f(x, y)$ among those (x, y) satisfying $g(x, y) = c$. We call the condition $g(x, y) = c$ the constraint in the problem.

The purpose of this section is to develop some methods for handling this and similar problems.

THEOREM 6. *Let $f\colon U \subset \mathbb{R}^n \to \mathbb{R}$ and $g\colon U \subset \mathbb{R}^n \to \mathbb{R}$ be given smooth functions. Let $\mathbf{x}_0 \in U$, $g(\mathbf{x}_0) = c$, and let S be the level set for g with value c (recall that this is the set of points $\mathbf{x} \in \mathbb{R}^n$ with $g(\mathbf{x}) = c$). Assume $\nabla g(\mathbf{x}_0) \neq \mathbf{0}$.*

If $f|S$, which denotes f restricted to S, has a maximum or minimum at $\mathbf{x}_0$, then there is a real number λ such that

$$\nabla f(\mathbf{x}_0) = \lambda \nabla g(\mathbf{x}_0) \tag{1}$$

Proof. Actually, we do not have enough machinery to give a thorough proof, but we can provide the essential points. (The additional technicalities needed are given in Section 4.4.)

Recall that for $n = 3$ the tangent space or tangent plane of S at $\mathbf{x}_0$ is defined as the space orthogonal to $\nabla g(\mathbf{x}_0)$ (see Section 2.5), and for arbitrary n we can give exactly the same definition for the tangent space of S at $\mathbf{x}_0$. This definition can be motivated by considering tangents to paths $\boldsymbol{\sigma}(t)$ that lie in S, as follows: if $\boldsymbol{\sigma}(t)$ is a path in S and $\boldsymbol{\sigma}(0) = \mathbf{x}_0$, then $\boldsymbol{\sigma}'(0)$ is a tangent vector to S at $\mathbf{x}_0$; but

$$\frac{d}{dt}g(\boldsymbol{\sigma}(t)) = \frac{d}{dt}c = 0$$

and on the other hand, by the Chain Rule,

$$\frac{d}{dt}g(\boldsymbol{\sigma}(t))\bigg|_{t=0} = \nabla g(\mathbf{x}_0)\cdot\boldsymbol{\sigma}'(0)$$

so $\nabla g(\mathbf{x}_0)\cdot\boldsymbol{\sigma}'(0) = 0$; that is, $\boldsymbol{\sigma}'(0)$ is orthogonal to $\nabla g(\mathbf{x}_0)$.

If $f|S$ has a maximum at $\mathbf{x}_0$, then certainly $f(\boldsymbol{\sigma}(t))$ has a maximum at $t = 0$. By one-variable calculus, $df(\boldsymbol{\sigma}(t))/dt|_{t=0} = 0$. Hence by the Chain Rule

$$0 = \frac{d}{dt}f(\boldsymbol{\sigma}(t))\bigg|_{t=0} = \nabla f(\mathbf{x}_0)\cdot\boldsymbol{\sigma}'(0)$$

Thus $\nabla f(\mathbf{x}_0)$ is perpendicular to the tangent of every curve in S and so is also perpendicular to the tangent space of S at $\mathbf{x}_0$. Since the space perpendicular to this tangent space is a line, $\nabla f(\mathbf{x}_0)$ and $\nabla g(\mathbf{x}_0)$ are parallel. Since $\nabla g(\mathbf{x}_0) \neq \mathbf{0}$ it follows that $\nabla f(\mathbf{x}_0)$ is a multiple of $\nabla g(\mathbf{x}_0)$, which is exactly the conclusion of the theorem. ■

Let us extract from this proof the geometry of the situation. We can formulate things in the following way.

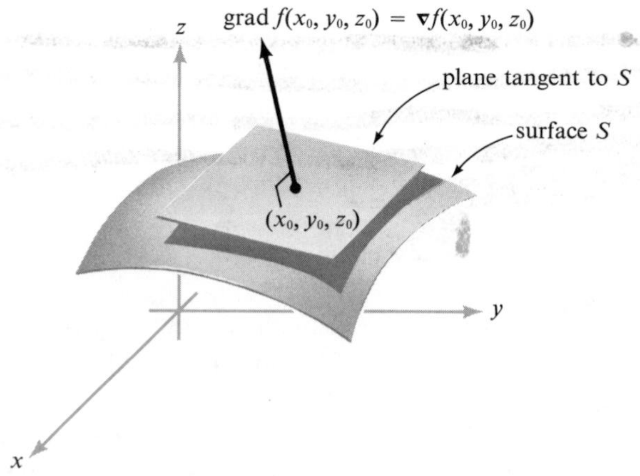

FIGURE 4.3.1
The geometry of constrained extrema.

COROLLARY. *If f, when constrained to a surface S, has a maximum or minimum at $\mathbf{x}_0$, then $\nabla f(\mathbf{x}_0)$ is perpendicular to S at $\mathbf{x}_0$ (see Figure 4.3.1).*

These results tell us that to find the constrained extrema of f we must look among those $\mathbf{x}_0$ satisfying the conclusions of the theorem or of the corollary. We shall give several illustrations of how to use each.

When the method of Theorem 6 is used we must look for a point $\mathbf{x}_0$ and a constant λ, called a *Lagrange multiplier*, such that $\nabla f(\mathbf{x}_0) = \lambda\, \nabla g(\mathbf{x}_0)$. This method is more analytical in nature than the method of the corollary to Theorem 6, which is more geometrical.

Equation (1) says that the partial derivatives of f are proportional to those of g. Finding such points $\mathbf{x}_0$ at which this occurs means solving the simultaneous equations

$$\left.\begin{aligned} \frac{\partial f}{\partial x_1}(x_1, \ldots, x_n) &= \lambda \frac{\partial g}{\partial x_1}(x_1, \ldots, x_n) \\ \frac{\partial f}{\partial x_2}(x_1, \ldots, x_n) &= \lambda \frac{\partial g}{\partial x_2}(x_1, \ldots, x_n) \\ &\vdots \\ \frac{\partial f}{\partial x_n}(x_1, \ldots, x_n) &= \lambda \frac{\partial g}{\partial x_n}(x_1, \ldots, x_n) \\ g(x_1, \ldots, x_n) &= c \end{aligned}\right\} \tag{2}$$

for $x_1, \ldots, x_n$ and λ.

Another way of looking at these equations is as follows: Think of λ as an additional variable and form the auxiliary function $h(x_1, \ldots, x_n, \lambda) = f(x_1, \ldots, x_n) - \lambda[g(x_1, \ldots, x_n) - c]$. Theorem 6 then says that to find the extreme points of $f|S$, we should examine the critical points of h. These are found by solving the equations

$$\left.\begin{aligned} 0 &= \frac{\partial h}{\partial x_1} = \frac{\partial f}{\partial x_1} - \lambda \frac{\partial g}{\partial x_1} \\ &\vdots \\ 0 &= \frac{\partial h}{\partial x_n} = \frac{\partial f}{\partial x_n} - \lambda \frac{\partial g}{\partial x_n} \\ 0 &= \frac{\partial h}{\partial \lambda} = g(x_1, \ldots, x_n) - c \end{aligned}\right\} \tag{3}$$

which are the same as equations (2) above.

Second derivative tests for maxima and minima analogous to those in Section 4.2 will be given in Theorem 7 below. However, in many problems it is possible to distinguish between maxima and minima by geometric means. Since this is usually simpler we consider examples of the latter type first.

EXAMPLE 1. Let $S \subset \mathbb{R}^2$ be a line through $(-1, 0)$ inclined at 45°, and let $f\colon \mathbb{R}^2 \to \mathbb{R}, (x, y) \mapsto x^2 + y^2$. Then $S = \{(x, y) | y - x - 1 = 0\}$, so here we set $g(x, y) = y - x - 1$ and $c = 0$. We have $\nabla g(x, y) = -\mathbf{i} + \mathbf{j} \neq \mathbf{0}$. The relative extrema of $f|S$ must be found among the points at which ∇f is orthogonal to S, that is, inclined at $-45°$. But $\nabla f(x, y) = (2x, 2y)$, which has the desired slope only when $x = -y$, or when (x, y) lies on the line L through the origin inclined at $-45°$. This can occur in the set S only for the single point at which L and S intersect (see Figure 4.3.2). Reference to the level curves of f indicates that this point, $(-\frac{1}{2}, \frac{1}{2})$, is a relative minimum of $f|S$ (but not of f).

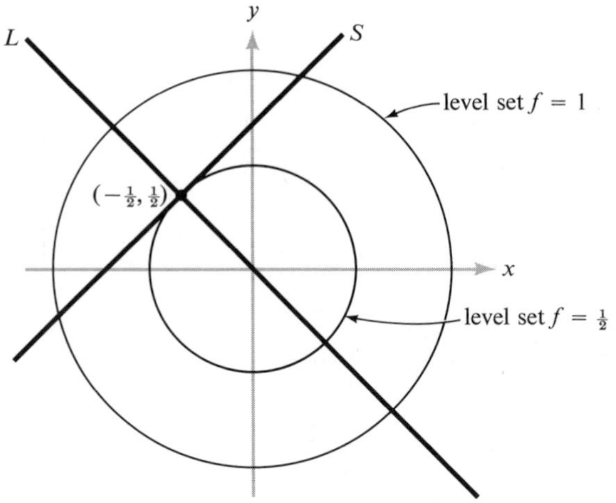

FIGURE 4.3.2
The geometry associated with finding the extrema of $f(x, y) = x^2 + y^2$ restricted to $S = \{(x, y) | y - x - 1 = 0\}$.

EXAMPLE 2. Let $f\colon \mathbb{R}^2 \to \mathbb{R}, (x, y) \mapsto x^2 - y^2$, and let S be the circle of radius 1 around the origin. Thus S is the level curve for g with value 1, where $g\colon \mathbb{R}^2 \to \mathbb{R}, (x, y) \mapsto x^2 + y^2$. As both of these functions have been studied in previous examples, we know their level curves, and these are shown in Figure 4.3.3. Clearly, the gradient of f is orthogonal to S at the four points $(0, \pm 1)$ and $(\pm 1, 0)$, which are relative minima and maxima, respectively, of $f|S$.

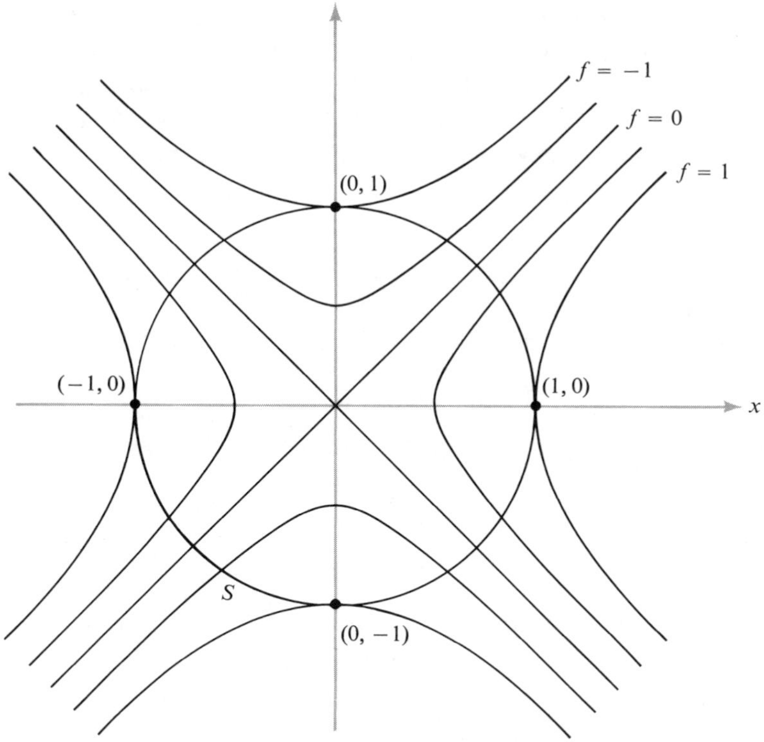

FIGURE 4.3.3
The geometry associated with the problem of finding the extrema of $x^2 - y^2$ *on* $S = \{(x, y) | x^2 + y^2 = 1\}$.

Let us also do this problem analytically by the method of Lagrange multipliers. Clearly

$$\nabla f(x, y) = \left(\frac{\partial f}{\partial x}, \frac{\partial f}{\partial y}\right) = (2x, -2y)$$

and

$$\nabla g(x, y) = (2x, 2y)$$

Note that $\nabla g(x, y) \neq 0$ if $x^2 + y^2 = 1$. Thus, according to Theorem 6, we must find a λ such that

$$(2x, -2y) = \lambda(2x, 2y)$$

and

$$(x, y) \in S \quad \text{that is} \quad x^2 + y^2 = 1$$

These conditions yield three equations, which can be solved for the three unknowns x, y, and λ. From $2x = \lambda 2x$ we conclude that either $x = 0$ or

$\lambda = 1$. If $x = 0$ then $y = \pm 1$ and $-2y = \lambda 2y$ implies $\lambda = -1$. If $\lambda = 1$, then $y = 0$ and $x = \pm 1$. Thus we get the points $(0, \pm 1)$ and $(\pm 1, 0)$, as before. As we have mentioned, this method only locates potential extrema; whether they are maxima, minima, or neither must be determined by other means, such as geometric arguments or the second derivative test given below.*

EXAMPLE 3. Maximize $f(x, y, z) = x + z$ subject to the constraint $x^2 + y^2 + z^2 = 1$.

Again we use Theorem 6. We seek λ and (x, y, z) such that

$$1 = 2x\lambda$$

$$0 = 2y\lambda$$

$$1 = 2z\lambda$$

and

$$x^2 + y^2 + z^2 = 1$$

From the first or third equation, we see that $\lambda \neq 0$. Thus, from the second equation, we get $y = 0$. From the first and third equations, $x = z$, and so from the fourth, $x = \pm 1/\sqrt{2} = z$. Hence our points are $(1/\sqrt{2}, 0, 1/\sqrt{2})$ and $(-1/\sqrt{2}, 0, -1/\sqrt{2})$. Comparing the values of f at these points, we can see that the first point yields the maximum of f and the second the minimum.

EXAMPLE 4. Find the largest volume a rectangular box can have subject to the constraint that the surface area be fixed at 10 square meters.

Here, if x, y, z are the lengths of the sides the volume is $f(x, y, z) = xyz$. The constraint is $2(xy + xz + yz) = 10$; that is, $xy + xz + yz = 5$. Thus our conditions are

$$yz = \lambda(y + z)$$

$$xz = \lambda(x + z)$$

$$xy = \lambda(y + x)$$

$$xy + xz + yz = 5$$

First of all, $x \neq 0$, for $x = 0$ implies $yz = 5$ and $0 = \lambda z$, so $\lambda = 0$ and $yz = 0$. Similarly, $y \neq 0$, $z \neq 0$, $x + y \neq 0$, etc. Elimination of λ from the first two equations gives $yz/(y + z) = xz/(x + z)$, which gives $x = y$; similarly, $y = z$. Substituting these values into the last equation, we obtain

* In these examples, $\nabla g(\mathbf{x}_0) \neq 0$ on the surface S, as required by Theorem 6. If $\nabla g(\mathbf{x}_0)$ were zero for some $\mathbf{x}_0$ on S, then it would have to be included among the possible extrema.

$3x^2 = 5$, or $x = \sqrt{5/3}$. Thus $x = y = z = \sqrt{5/3}$, and $xyz = (5/3)^{3/2}$. This is the solution; it should be geometrically clear that the maximum occurs when $x = y = z$.

A more substantial application of Lagrange multipliers to economics is given in Section 4.5.

Some general guidelines may be useful for problems such as these. First of all, if the surface S is bounded (like an ellipsoid) then f must have a maximum and a minimum on S.* In particular, if f has only two points satisfying the conditions of Theorem 6 or its corollary, then one must be a maximum and one must be a minimum. However, if there are more than two such points, some can also be saddle points. Also, if S is not bounded (e.g., a hyperboloid) then f need not have any maxima nor minima.

If a surface S is defined by a number of constraints

$$\left.\begin{aligned} g_1(x_1, \ldots, x_n) &= c_1 \\ g_2(x_1, \ldots, x_n) &= c_2 \\ &\vdots \\ g_k(x_1, \ldots, x_n) &= c_k \end{aligned}\right\} \tag{4}$$

then Theorem 6 may be generalized as follows: *If f has a maximum or minimum at $\mathbf{x}_0$ on S, there must exist constants $\lambda_1, \ldots, \lambda_k$ such that*†

$$\nabla f(\mathbf{x}_0) = \lambda_1 \nabla g_1(\mathbf{x}_0) + \cdots + \lambda_k \nabla g_k(\mathbf{x}_0) \tag{5}$$

This case may be proved by generalizing the method used to prove Theorem 6. We leave the argument to the reader. Let us give an example of how this more general formulation may be used.

EXAMPLE 5. Find the extreme points of $f(x, y, z) = x + y + z$ subject to the conditions $x^2 + y^2 = 2$ and $x + z = 1$.

Here there are two constraints

$$g_1(x, y, z) = x^2 + y^2 - 2 = 0$$

$$g_2(x, y, z) = x + z - 1 = 0$$

Thus we must find x, y, z, λ_1, and λ_2 such that

$$\nabla f(x, y, z) = \lambda_1 \nabla g_1(x, y, z) + \lambda_2 \nabla g_2(x, y, z)$$

* This is proved in more advanced courses. See, for example, J. Marsden, *Elementary Classical Analysis*, W. H. Freeman and Company, San Francisco, 1974, Chapter 4.

† As with the hypothesis $\nabla g(\mathbf{x}_0) \neq 0$ in Theorem 6, here one must assume that the determinants of the matrix whose columns are $\nabla g_1(\mathbf{x}_0), \ldots, \nabla g_k(\mathbf{x}_0)$ is nonzero; that is, these vectors are linearly independent.

and

$$g_1(x, y, z) = 0$$

$$g_2(x, y, z) = 0$$

that is, computing the gradients and equating components,

$$1 = \lambda_1 \cdot 2x + \lambda_2 \cdot 1$$

$$1 = \lambda_1 \cdot 2y + \lambda_2 \cdot 0$$

$$1 = \lambda_1 \cdot 0 + \lambda_2 \cdot 1$$

and

$$x^2 + y^2 = 2$$

$$x + z = 1$$

These are five equations for x, y, z, λ_1, and λ_2. From the third, $\lambda_2 = 1$ and so $2x\lambda_1 = 0$, $2y\lambda_1 = 1$. Since the second implies $\lambda_1 \neq 0$, we have $x = 0$. Thus $y = \pm\sqrt{2}$ and $z = 1$. Hence our points are $(0, \pm\sqrt{2}, 1)$. By inspection one can show that $(0, \sqrt{2}, 1)$ gives a maximum, and $(0, -\sqrt{2}, 1)$ a minimum.

OPTIONAL

A SECOND DERIVATIVE TEST FOR CONSTRAINED EXTREMA

In Section 4.2, we developed a second derivative test for extrema of functions of several variables by looking at the second-degree term in the Taylor series of f. If the Hessian matrix of second partial derivatives was either positive or negative definite at a critical point of f, then the sign of the contribution these terms made to f was independent of the direction in which we moved away from the critical point. We could conclude therefore that we were at a relative maximum or minimum respectively.

In this section, however, we are not interested in all values of f but only in those obtained by restricting f to some set S that is the level set of another function g. The situation is complicated, first, because the constrained extrema of f need not come at critical points of f and, second, because the variable is only allowed to move in the set S. Nevertheless, a second derivative test can be given in terms of what is called a *bordered Hessian*. We will show how this comes about for the case of a function $f(x, y)$ of two variables subject to the constraint $g(x, y) = c$.

According to the remarks following Theorem 6, the constrained extrema of f are found by looking at the critical points of the auxiliary function

OPTIONAL (*Continued*)

$h(x, y, \lambda) = f(x, y) - \lambda(g(x, y) - c)$. Suppose (x_0, y_0, λ) is such a point and let $\mathbf{v}_0 = (x_0, y_0)$. That is,

$$\left.\frac{\partial f}{\partial x}\right|_{\mathbf{v}_0} = \lambda \left.\frac{\partial g}{\partial x}\right|_{\mathbf{v}_0}$$

$$\left.\frac{\partial f}{\partial y}\right|_{\mathbf{v}_0} = \lambda \left.\frac{\partial g}{\partial y}\right|_{\mathbf{v}_0}$$

$$g(x_0, y_0) = c$$

In a sense this is a one-variable problem. If the function g is at all reasonable, then the set S defined by $g(x, y) = c$ is a curve and we are interested in how f varies as we move along this curve. If we can solve the equation $g(x, y) = c$ for one variable in terms of the other, then we can make this explicit and use the one-variable second derivative test. If $\partial g/\partial y|_{\mathbf{v}_0} \neq 0$, then the curve S is not vertical at $\mathbf{v}_0$ and it is reasonable that we can solve for y as a function of x in a neighborhood of x_0. We will, in fact, prove this in Section 4.4. (If $\partial g/\partial x|_{\mathbf{v}_0} \neq 0$, we can solve for x as a function of y.)

Suppose S is the graph of $y = \phi(x)$. Then $f|S$ can be written as a function of one variable, $f(x, y) = f(x, \phi(x))$. The Chain Rule gives

$$\left.\begin{aligned} \frac{df}{dx} &= \frac{\partial f}{\partial x} + \frac{\partial f}{\partial y}\frac{d\phi}{dx} \\ &\text{and} \\ \frac{d^2 f}{dx^2} &= \frac{\partial^2 f}{\partial x^2} + 2\frac{\partial^2 f}{\partial x\,\partial y}\frac{d\phi}{dx} + \frac{\partial^2 f}{\partial y^2}\left(\frac{d\phi}{dx}\right)^2 + \frac{\partial f}{\partial y}\frac{d^2\phi}{dx^2} \end{aligned}\right\} \quad (6)$$

The relation $g(x, \phi(x)) = c$ can be used to find $d\phi/dx$ and $d^2\phi/dx^2$. Differentiating both sides of $g(x, \phi(x)) = c$ with respect to x gives

$$\frac{\partial g}{\partial x} + \frac{\partial g}{\partial y}\frac{d\phi}{dx} = 0$$

and

$$\frac{\partial^2 g}{\partial x^2} + 2\frac{\partial^2 g}{\partial x\,\partial y}\frac{d\phi}{dx} + \frac{\partial^2 g}{\partial y^2}\left(\frac{d\phi}{dx}\right)^2 + \frac{\partial g}{\partial y}\frac{d^2\phi}{dx^2} = 0$$

so that

$$\left.\begin{aligned} \frac{d\phi}{dx} &= -\frac{\partial g/\partial x}{\partial g/\partial y} \\ &\text{and} \\ \frac{d^2\phi}{dx^2} &= -\frac{1}{\partial g/\partial y}\left[\frac{\partial^2 g}{\partial x^2} - 2\frac{\partial^2 g}{\partial x\,\partial y}\frac{\partial g/\partial x}{\partial g/\partial y} + \frac{\partial^2 g}{\partial y^2}\left(\frac{\partial g/\partial x}{\partial g/\partial y}\right)^2\right] \end{aligned}\right\} \quad (7)$$

OPTIONAL (*Continued*)

Substituting (7) into (6) gives

$$\left.\begin{aligned}
&\frac{df}{dx} = \frac{\partial f}{\partial x} - \frac{\partial f/\partial y}{\partial g/\partial y}\frac{\partial g}{\partial x} \\
&\text{and} \\
&\frac{d^2f}{dx^2} = \frac{1}{(\partial g/\partial y)^2}\left\{\left[\frac{\partial^2 f}{\partial x^2} - \frac{\partial f/\partial y}{\partial g/\partial y}\frac{\partial^2 g}{\partial x^2}\right]\left(\frac{\partial g}{\partial y}\right)^2\right. \\
&\qquad - 2\left[\frac{\partial^2 f}{\partial x\,\partial y} - \frac{\partial f/\partial y}{\partial g/\partial y}\frac{\partial^2 g}{\partial x\,\partial y}\right]\frac{\partial g}{\partial x}\frac{\partial g}{\partial y} \\
&\qquad \left. + \left[\frac{\partial^2 f}{\partial y^2} - \frac{\partial f/\partial y}{\partial g/\partial y}\frac{\partial^2 g}{\partial y^2}\right]\left(\frac{\partial g}{\partial x}\right)^2\right\}
\end{aligned}\right\} \tag{8}$$

At $\mathbf{v}_0$, we know that $\partial f/\partial y = \lambda\, \partial g/\partial y$ and $\partial f/\partial x = \lambda\, \partial g/\partial x$ so (8) becomes

$$\left.\begin{aligned}
&\left.\frac{df}{dx}\right|_{x_0} = \left.\frac{\partial f}{\partial x}\right|_{x_0} - \lambda\left.\frac{\partial g}{\partial x}\right|_{x_0} = 0 \\
&\text{and} \\
&\left.\frac{d^2f}{dx^2}\right|_{x_0} = \frac{1}{(\partial g/\partial y)^2}\left\{\frac{\partial^2 h}{\partial x^2}\left(\frac{\partial g}{\partial y}\right)^2 - 2\frac{\partial^2 h}{\partial x\,\partial y}\frac{\partial g}{\partial x}\frac{\partial g}{\partial y} + \frac{\partial^2 h}{\partial y^2}\left(\frac{\partial g}{\partial x}\right)^2\right\} \\
&\qquad = -\frac{1}{(\partial g/\partial y)^2}\begin{vmatrix} 0 & \dfrac{\partial g}{\partial x} & \dfrac{\partial g}{\partial y} \\ \dfrac{\partial g}{\partial x} & \dfrac{\partial^2 h}{\partial x^2} & \dfrac{\partial^2 h}{\partial x\,\partial y} \\ \dfrac{\partial g}{\partial y} & \dfrac{\partial^2 h}{\partial x\,\partial y} & \dfrac{\partial^2 h}{\partial y^2} \end{vmatrix}
\end{aligned}\right\} \tag{9}$$

where the quantities are evaluated at x_0 and h is the auxilary function introduced above. This 3×3 determinant is called a *bordered Hessian*, and its sign is opposite that of d^2f/dx^2. Therefore if it is negative we must be at a local minimum. If it is positive we are at a local maximum and if it is zero the test is inconclusive. This reasoning leads to the following test.

THEOREM 7. *Let $f\colon U \subset \mathbb{R}^2 \to \mathbb{R}$ and $g\colon U \subset \mathbb{R}^2 \to \mathbb{R}$ be smooth functions. Let $\mathbf{v}_0 \in U$, $g(\mathbf{v}_0) = c$, and S be the level curve for g with value c. Assume $\nabla g(\mathbf{v}_0) \neq 0$ and that there is a real number λ such that $\nabla f(\mathbf{v}_0) = \lambda\, \nabla g(\mathbf{v}_0)$.*

OPTIONAL (*Continued*)

Form the auxiliary function $h = f - \lambda g$ and the bordered Hessian determinant

$$|\bar{H}| = \begin{vmatrix} 0 & \dfrac{\partial g}{\partial x} & \dfrac{\partial g}{\partial y} \\ \dfrac{\partial g}{\partial x} & \dfrac{\partial^2 h}{\partial x^2} & \dfrac{\partial^2 h}{\partial x\, \partial y} \\ \dfrac{\partial g}{\partial y} & \dfrac{\partial^2 h}{\partial x\, \partial y} & \dfrac{\partial^2 h}{\partial y^2} \end{vmatrix} \quad \textit{evaluated at } \mathbf{v}_0$$

(*i*) *If $|\bar{H}| > 0$, then $\mathbf{v}_0$ is a local maximum point for $f|_S$.*
(*ii*) *If $|\bar{H}| < 0$, then $\mathbf{v}_0$ is a local minimum point for $f|_S$.*
(*iii*) *If $|\bar{H}| = 0$, the test is inconclusive and $\mathbf{v}_0$ may be a minimum, a maximum, or neither.*

EXAMPLE 6. Find extreme points of $f(x, y) = (x - y)^n$ subject to the constraint $x^2 + y^2 = 1$ $(n \geq 1)$.

We set the first derivatives of the auxiliary function $h(x, y, \lambda) = (x - y)^n - \lambda(x^2 + y^2 - 1)$ equal to 0:

$$\begin{aligned} n(x - y)^{n-1} - 2\lambda x &= 0 \\ -n(x - y)^{n-1} - 2\lambda y &= 0 \\ -(x^2 + y^2 - 1) \qquad &= 0 \end{aligned}$$

From the first two equations we see that $\lambda(x + y) = 0$. If $\lambda = 0$, then $x = y = \pm\sqrt{2}/2$. If $\lambda \neq 0$, then $x = -y$. The four critical points and their corresponding values of $f(x, y)$ are represented in Figure 4.3.4 and listed below:

(*A*)	$x = \sqrt{2}/2$	$y = \sqrt{2}/2$	$\lambda = 0$	$f(x, y) = 0$
(*B*)	$x = \sqrt{2}/2$	$y = -\sqrt{2}/2$	$\lambda = n(\sqrt{2})^{n-2}$	$f(x, y) = (\sqrt{2})^n$
(*C*)	$x = -\sqrt{2}/2$	$y = -\sqrt{2}/2$	$\lambda = 0$	$f(x, y) = 0$
(*D*)	$x = -\sqrt{2}/2$	$y = +\sqrt{2}/2$	$\lambda = (-1)^{n-2}n(\sqrt{2})^{n-2}$	$f(x, y) = (-\sqrt{2})^n$

By inspection we see that if n is even, then A and C are minimum points and B and D are maxima. If n is odd, then B is a maximum point, D is a minimum, and A and D are neither. Let us see if Theorem 7 is consistent with these observations.

OPTIONAL (*Continued*)

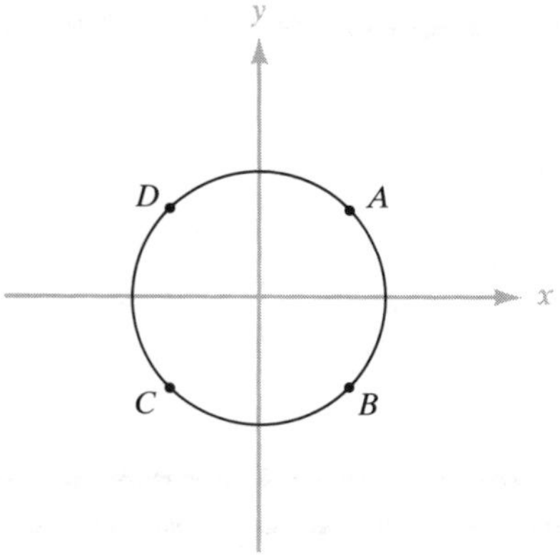

FIGURE 4.3.4
The four critical points in Example 6.

The bordered Hessian matrix is

$$|\bar{H}| = \begin{vmatrix} 0 & 2x & 2y \\ 2x & n(n-1)(x-y)^{n-2} - 2\lambda & -n(n-1)(x-y)^{n-2} \\ 2y & -n(n-1)(x-y)^{n-2} & n(n-1)(x-y)^{n-2} - 2\lambda \end{vmatrix}$$

$$= -4n(n-1)(x-y)^{n-2}(x+y)^2 + 8\lambda(x^2 - y^2)$$

If $n = 1$ or if $n \geq 3$, $|\bar{H}| = 0$ at A, B, C, and D. If $n = 2$, then $|\bar{H}| = 0$ at B and D and -16 at A and C. Thus the second derivative test picks up the minima at A and C but is inconclusive in testing the maxima at B and D for $n = 2$. It is also inconclusive for all other values of n.

Just as in the unconstrained case, there is also a second derivative test for functions of more than two variables. If we are to find extreme points for $f(x_1, \ldots, x_n)$ subject to a single constraint $g(x_1, \ldots, x_n) = c$, we first form the bordered Hessian for the auxiliary function $h(x_1, \ldots, x_n) = f(x_1, \ldots, x_n) - \lambda(g(x_1, \ldots, x_n) - c)$ as follows:

$$\begin{vmatrix} 0 & \partial g/\partial x_1 & \partial g/\partial x_2 & \cdots & \partial g/\partial x_n \\ \partial g/\partial x_1 & \partial^2 h/\partial x_1^2 & \partial^2 h/\partial x_1\,\partial x_2 & \cdots & \partial^2 h/\partial x_1\,\partial x_n \\ \partial g/\partial x_2 & \partial^2 h/\partial x_1\,\partial x_2 & \partial^2 h/\partial x_2^2 & & \vdots \\ \vdots & \vdots & \vdots & & \\ \partial g/\partial x_n & \partial^2 h/\partial x_1\,\partial x_n & \partial^2 h/\partial x_2\,\partial x_n & \cdots & \partial^2 h/\partial x_n^2 \end{vmatrix}$$

OPTIONAL (*Continued*)

Second, we examine the determinants of the diagonal submatrices of order ≥ 3 at the critical points of h. If they are all negative

$$\begin{vmatrix} 0 & \dfrac{\partial g}{\partial x_1} & \dfrac{\partial g}{\partial x_2} \\ \dfrac{\partial g}{\partial x_1} & \dfrac{\partial^2 h}{\partial x_1^2} & \dfrac{\partial^2 h}{\partial x_1\, \partial x_2} \\ \dfrac{\partial g}{\partial x_2} & \dfrac{\partial^2 h}{\partial x_1\, \partial x_2} & \dfrac{\partial^2 h}{\partial x_2^2} \end{vmatrix} < 0, \quad \begin{vmatrix} 0 & \dfrac{\partial g}{\partial x_1} & \dfrac{\partial g}{\partial x_2} & \dfrac{\partial g}{\partial x_3} \\ \dfrac{\partial g}{\partial x_1} & \dfrac{\partial^2 h}{\partial x_1^2} & \dfrac{\partial^2 h}{\partial x_1\, \partial x_2} & \dfrac{\partial^2 h}{\partial x_1\, \partial x_3} \\ \dfrac{\partial g}{\partial x_2} & \dfrac{\partial^2 h}{\partial x_1\, \partial x_2} & \dfrac{\partial^2 h}{\partial x_2^2} & \dfrac{\partial^2 h}{\partial x_2\, \partial x_3} \\ \dfrac{\partial g}{\partial x_3} & \dfrac{\partial^2 h}{\partial x_1\, \partial x_3} & \dfrac{\partial^2 h}{\partial x_2\, \partial x_3} & \dfrac{\partial^2 h}{\partial x_3^2} \end{vmatrix} < 0, \dots,$$

then we are at local minimum of $f\,|\,S$. If they start out positive and alternate in sign (i.e., >0, <0, >0, $<0, \dots$), then we are at a local maximum.*

EXAMPLE 7. Study the local extreme points of $f(x, y, z) = xyz$ on the surface of the unit sphere $x^2 + y^2 + z^2 = 1$ using the second derivative test. Setting the partial derivatives of the auxiliary function $h(x, y, z, \lambda) = xyz - \lambda(x^2 + y^2 + z^2 - 1)$ equal to zero gives

$$\begin{aligned} yz &= 2\lambda x \\ xz &= 2\lambda y \\ xy &= 2\lambda z \\ x^2 + y^2 + z^2 &= 1 \end{aligned}$$

Thus $3xyz = 2\lambda(x^2 + y^2 + z^2) = 2\lambda$. If $\lambda = 0$, the solutions are $(x, y, z, \lambda) = (1, 0, 0, 0)$, $(0, 1, 0, 0)$, and $(0, 0, 1, 0)$. If $\lambda \neq 0$, then we have $2\lambda = 3xyz = 6\lambda z^2$ and so $z^2 = \frac{1}{3}$. Similarly, $x^2 = y^2 = \frac{1}{3}$. Thus the solutions are given by $\lambda = 3xyz = \pm\sqrt{3}/6$. The critical points of h and the corresponding values of f are given in Table 4.1. From it, we see that points E, F, G, and K are minima. Points D, H, I, and J are maxima. To see if this is in accord with the second derivative test, we need to consider two determinants. First we

* For a detailed discussion, see C. Caratheodory, *Calculus of Variations and Partial Differential Equations*, Holden-Day, San Francisco, 1965, or Y. Murata, *Mathematics for Stability and Optimization of Economic Ssytems*, Academic Press, New York, 1977, pp. 263–271.

OPTIONAL (*Continued*)

Table 4.1 The Critical Points of h and Corresponding Values of f

	x	y	z	λ	$f(x, y, z)$
A	1	0	0	0	0
B	0	1	0	0	0
C	0	0	1	0	0
D	$\sqrt{3}/3$	$\sqrt{3}/3$	$\sqrt{3}/3$	$\sqrt{3}/6$	$\sqrt{3}/9$
E	$-\sqrt{3}/3$	$\sqrt{3}/3$	$\sqrt{3}/3$	$-\sqrt{3}/6$	$-\sqrt{3}/9$
F	$\sqrt{3}/3$	$-\sqrt{3}/3$	$\sqrt{3}/3$	$-\sqrt{3}/6$	$-\sqrt{3}/9$
G	$\sqrt{3}/3$	$\sqrt{3}/3$	$-\sqrt{3}/3$	$-\sqrt{3}/6$	$-\sqrt{3}/9$
H	$\sqrt{3}/3$	$-\sqrt{3}/3$	$-\sqrt{3}/3$	$\sqrt{3}/6$	$\sqrt{3}/9$
I	$-\sqrt{3}/3$	$\sqrt{3}/3$	$-\sqrt{3}/3$	$\sqrt{3}/6$	$\sqrt{3}/9$
J	$-\sqrt{3}/3$	$-\sqrt{3}/3$	$\sqrt{3}/3$	$\sqrt{3}/6$	$\sqrt{3}/9$
K	$-\sqrt{3}/3$	$-\sqrt{3}/3$	$-\sqrt{3}/3$	$-\sqrt{3}/6$	$-\sqrt{3}/9$

look at the following:

$$|\bar{H}_2| = \begin{vmatrix} 0 & \partial g/\partial x & \partial g/\partial x \\ \partial g/\partial x & \partial^2 h/\partial x^2 & \partial^2 h/\partial x\partial y \\ \partial g/\partial y & \partial^2 h/\partial x\partial y & \partial^2 h/\partial y^2 \end{vmatrix} = \begin{vmatrix} 0 & 2x & 2y \\ 2x & -2\lambda & z \\ 2y & z & -2\lambda \end{vmatrix} = 16\lambda(x^2 + y^2)$$

Observe that $\operatorname{sign}(|\bar{H}_2|) = \operatorname{sign} \lambda = \operatorname{sign}(xyz)$, where $\operatorname{sign} \alpha = +1$ if $\alpha > 0$ and -1 if $\alpha < 0$. Second, we consider

$$|\bar{H}_3| = \begin{vmatrix} & \partial g/\partial x & \partial g/\partial y & \partial g/\partial z \\ \partial g/\partial x & \partial^2 h/\partial x^2 & \partial^2 h/\partial x\partial y & \partial^2 h/\partial x\partial z \\ \partial g/\partial y & \partial^2 h/\partial x\partial y & \partial^2 h/\partial y^2 & \partial^2 h/\partial y\partial z \\ \partial g/\partial z & \partial^2 h/\partial x\partial z & \partial^2 h/\partial y\partial z & \partial^2 h/\partial z^2 \end{vmatrix} = \begin{vmatrix} 0 & 2x & 2y & 2z \\ 2x & -2\lambda & z & y \\ 2y & z & -2\lambda & x \\ 2z & y & x & -2\lambda \end{vmatrix}$$

which works out to be $+4$ at points A, B, and C and $-16/3$ at the other 8 points. At E, F, G, and K, we have $|\bar{H}_2| < 0$ *and* $|\bar{H}_3| < 0$ so the test indicates there are local minima. At D, H, I, and J we have $|\bar{H}_2| > 0$ and $|\bar{H}_3| < 0$ so the test says these are local maxima. At A, B, and C the second derivative test is inconclusive.

EXERCISES

In Exercises 1 to 5 find the extrema of f subject to the stated constraints.

1. $f(x, y, z) = x - y + z$, $x^2 + y^2 + z^2 = 2$
2. $f(x, y) = x - y$, $x^2 - y^2 = 2$
3. $f(x, y) = x$, $x^2 + 2y^2 = 3$
4. $f(x, y, z) = x + y + z$, $x^2 - y^2 = 1$, $2x + z = 1$
5. $f(x, y) = 3x + 2y$, $2x^2 + 3y^2 = 3$

Find the relative extrema of $f|S$ in Exercises 6 to 9.

6. $f: \mathbb{R}^2 \to \mathbb{R}, (x, y) \mapsto x^2 + y^2, S = \{(x, 2) | x \in \mathbb{R}\}$
7. $f: \mathbb{R}^2 \to \mathbb{R}, (x, y) \mapsto x^2 + y^2, S = \{(x, y) | y \geq 2\}$
8. $f: \mathbb{R}^2 \to \mathbb{R}, (x, y) \mapsto x^2 - y^2, S = \{(x, \cos x) | x \in \mathbb{R}\}$
9. $f: \mathbb{R}^3 \to \mathbb{R}, (x, y, z) \mapsto x^2 + y^2 + z^2, S = \{(x, y, z) | z \geq 2 + x^2 + y^2\}$

10. A rectangular box with no top is to have a surface area of 16 square meters. Find the dimensions that maximize its volume.

11. Design a cylindrical can (with a lid) to contain 1 liter of water, using the minimum amount of metal.

12. Show that solutions of equation (5) are in one-to-one correspondence with the critical points of $h(x_1, \ldots, x_n, \lambda_1, \ldots, \lambda_k) = f(x_1, \ldots, x_n) - \lambda_1[g_1(x_1, \ldots, x_n) - c_1] - \cdots - \lambda_k[g_k(x_1, \ldots, x_n) - c_k]$.

*13. Let A be a nonzero symmetric 3×3 matrix. Thus its entries satisfy $a_{ij} = a_{ji}$. Consider the function $f(\mathbf{x}) = \frac{1}{2}(A\mathbf{x}) \cdot \mathbf{x}$.
  (a) What is ∇f?
  (b) Consider the restriction of f to the unit sphere $S = \{(x, y, z) | x^2 + y^2 + z^2 = 1\}$ in $\mathbb{R}^3$. Assume that f must have a maximum and a minimum on S (see the remarks on p. 223). Show that there must be an $\mathbf{x} \in S$ and a $\lambda \neq 0$ such that $A\mathbf{x} = \lambda\mathbf{x}$. (The $\mathbf{x}$ is called an *eigenvector* while the λ is called an *eigenvalue*.)

*14. Suppose now that A in the function f defined in Exercise 13 is not necessarily symmetric.
  (a) What is ∇f?
  (b) Can one conclude the existence of an eigenvector and eigenvalue as in Exercise 13?

15. Apply the second derivative test to study the nature of the extrema in Exercises 1 and 5.

16. A light ray travels from point A to point B crossing a boundary between two media (see Figure 4.3.5). In the first medium its speed is v_1 and in the second it is v_2. Show that the trip is made in minimum time when *Snell's law* holds:

$$\frac{\sin \theta_1}{\sin \theta_2} = \frac{v_1}{v_2}$$

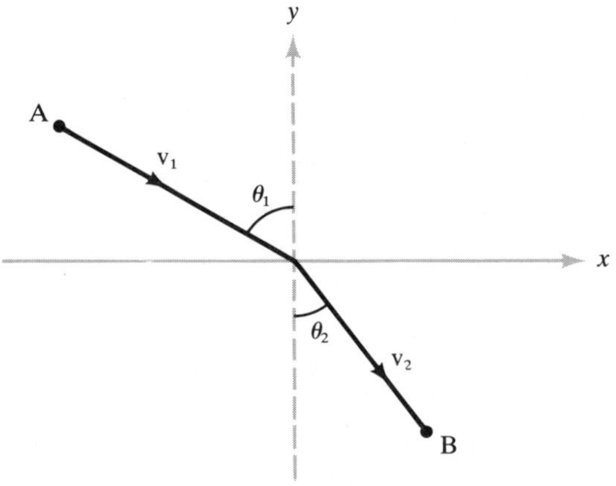

FIGURE 4.3.5
Snell's law of refraction.

17. A parcel delivery service requires that the dimensions of a rectangular box be such that the length plus twice the width plus twice the height be no more than 108 inches. ($l + 2w + 2h \leq 108$). What is the volume of the largest volume box they will deliver?

OPTIONAL

4.4 THE IMPLICIT FUNCTION THEOREM

This section begins with a statement and proof of a special version of the Implicit Function Theorem. This version is appropriate for a study of surfaces and, in particular, allows us to complete the proof of Theorem 6 in Section 4.3. Upon proving this theorem, we state, without proof, the general Implicit (and attendant Inverse) Function Theorem. This discussion will be useful later when we study the Change of Variables Theorem. However, the topics covered in this section are not essential for an understanding of the principal results and applications of this book.

THEOREM 8. *(Special Implicit Function Theorem). Let $F\colon \mathbb{R}^{n+1} \to \mathbb{R}$ be a given function having continuous partial derivatives. Let points in $\mathbb{R}^{n+1}$ be denoted $(\mathbf{x}, z)$, where $\mathbf{x} \in \mathbb{R}^n$ and $z \in \mathbb{R}$. Suppose that $(\mathbf{x}_0, z_0)$ is a point satisfying*

$$F(\mathbf{x}_0, z_0) = 0 \quad \textit{and} \quad \frac{\partial F}{\partial z}(\mathbf{x}_0, z_0) \neq 0$$

OPTIONAL (*Continued*)

Then there is a ball U containing $\mathbf{x}_0$ in $\mathbb{R}^n$ and a neighborhood V of z_0 in $\mathbb{R}$ such that there is a unique function $z = g(\mathbf{x})$ defined for $\mathbf{x}$ in U and z in V that satisfies

$$F(\mathbf{x}, g(\mathbf{x})) = 0$$

Moreover, $z = g(\mathbf{x})$ is continuously differentiable with the derivative given by

$$Dg(\mathbf{x}) = -\frac{1}{\dfrac{\partial F}{\partial z}(\mathbf{x}, g(\mathbf{x}))} \cdot D_{\mathbf{x}}F(\mathbf{x}, z)$$

where $D_{\mathbf{x}}F$ denotes the derivative of F with respect to the variable $\mathbf{x}$, $D_{\mathbf{x}}F = [\partial F/\partial x_1, \dots, \partial F/\partial x_n]$; that is,

$$\frac{\partial g}{\partial x_i} = -\frac{\partial F}{\partial x_i}\bigg/\frac{\partial F}{\partial z}, \qquad i = 1, \dots, n \tag{1}$$

Proof. We shall prove the case $n = 2$, so $F: \mathbb{R}^3 \to \mathbb{R}$. The case for all n is similar but the notation must be modified. We write $\mathbf{x} = (x, y)$ and $\mathbf{x}_0 = (x_0, y_0)$. Since $(\partial F/\partial z)(x_0, y_0, z_0) \neq 0$, it is either positive or negative. Suppose for definiteness that it is positive. By continuity, we can find numbers $a > 0$ and $b > 0$ such that if $\|\mathbf{x} - \mathbf{x}_0\| < a$ and $\|z - z_0\| < a$, then $(\partial F/\partial z)(\mathbf{x}, z) > b$. We can also assume that the other partials are bounded by a number M in this region, that is, $|(\partial F/\partial x)(\mathbf{x}, z)| \leq M$ and $|(\partial F/\partial y)(\mathbf{x}, z)| \leq M$. This also follows from continuity. We can now write

$$\begin{aligned} F(\mathbf{x}, z) = F(\mathbf{x}, z) - F(\mathbf{x}_0, z_0) &= [F(\mathbf{x}, z) - F(\mathbf{x}_0, z)] \\ &\quad + [F(\mathbf{x}_0, z) - F(\mathbf{x}_0, z_0)] \end{aligned} \tag{2}$$

Consider the function

$$h(t) = F(t\mathbf{x} + (1 - t)\mathbf{x}_0, z)$$

for fixed $\mathbf{x}$ and z. By the Mean Value Theorem, there is a number θ between 0 and 1 such that

$$h(1) - h(0) = h'(\theta)$$

that is,

$$F(\mathbf{x}, z) - F(\mathbf{x}_0, z) = D_{\mathbf{x}}F(\theta\mathbf{x} + (1 - \theta)\mathbf{x}_0, z)(\mathbf{x} - \mathbf{x}_0)$$

Substitution of this formula into (2) along with a similar formula for the second term of the equation gives

$$\begin{aligned} F(\mathbf{x}, z) &= D_{\mathbf{x}}F(\theta\mathbf{x} + (1 - \theta)\mathbf{x}_0, z)(\mathbf{x} - \mathbf{x}_0) \\ &\quad + \frac{\partial F}{\partial z}(\mathbf{x}_0, \phi z + (1 - \phi)z_0)(z - z_0) \end{aligned} \tag{3}$$

OPTIONAL (*Continued*)

where ϕ is between 0 and 1. Let a_0 satisfy $0 < a_0 < a$ and choose $\delta > 0$ such that $\delta < a_0$ and $\delta < ba_0/2M$. Then if $\|\mathbf{x} - \mathbf{x}_0\| < \delta$, both $|x - x_0|$ and $|y - y_0|$ are less than δ, so the absolute value of each of the two terms in

$$D_{\mathbf{x}}F(\theta\mathbf{x} + (1 - \theta)\mathbf{x}_0, z)(\mathbf{x} - \mathbf{x}_0) = \frac{\partial F}{\partial x}(\theta\mathbf{x} + (1 - \theta)\mathbf{x}_0, z)(x - x_0)$$

$$+ \frac{\partial F}{\partial y}(\theta\mathbf{x} + (1 - \theta)\mathbf{x}_0, z)(y - y_0)$$

is less than $M \cdot \delta < M\ (ba_0/2M) = ba_0/2$. Thus, $\|\mathbf{x} - \mathbf{x}_0\| < \delta$ implies $|D_{\mathbf{x}}F(\theta\mathbf{x} + (1 - \theta)\mathbf{x}_0, z)(\mathbf{x} - \mathbf{x}_0)| < ba_0$. Therefore, from (3) and the choice of b, $\|\mathbf{x} - \mathbf{x}_0\| < \delta$ implies that

$$F(\mathbf{x}, z_0 + a_0) > 0 \quad \text{and} \quad F(\mathbf{x}, z_0 - a_0) < 0$$

(The inequalities are reversed if $(\partial F/\partial z)(\mathbf{x}_0, z_0) < 0$.) Thus, by the intermediate value theorem applied to $F(\mathbf{x}, z)$ as a function of z, for each $\mathbf{x}$ there is a z between $z_0 - a_0$ and $z_0 + a_0$ such that $F(\mathbf{x}, z) = 0$. This z is unique since, by elementary calculus, a function with a positive derivative is increasing and thus can have no more than one zero.

Let U be the open ball of radius δ and center $\mathbf{x}_0$ in $\mathbb{R}^n$ and let V be the open interval on $\mathbb{R}$ from $z_0 - a_0$ to $z_0 + a_0$. We have proved that if $\mathbf{x}$ is confined to U, there is a unique z in V such that $F(\mathbf{x}, z) = 0$. This defines the function $z = g(\mathbf{x}) = g(x, y)$ required by the theorem. We leave it to the reader to prove from this construction that $z = g(x, y)$ is a continuous function.

It remains to establish the continuous differentiability of $z = g(\mathbf{x})$. From (3) and since $F(\mathbf{x}, z) = 0$ and $z_0 = g(\mathbf{x}_0)$, we have

$$g(\mathbf{x}) - g(\mathbf{x}_0) = -\frac{D_{\mathbf{x}}F(\theta\mathbf{x} + (1 - \theta)\mathbf{x}_0, z)(\mathbf{x} - \mathbf{x}_0)}{\dfrac{\partial F}{\partial z}(\mathbf{x}_0, \phi z + (1 - \phi)z_0)}$$

If we let $\mathbf{x} = (x_0 + h, y_0)$, then this equation becomes

$$\frac{g(x_0 + h, y_0) - g(x_0, y_0)}{h} = -\frac{\dfrac{\partial F}{\partial x}(\theta\mathbf{x} + (1 - \theta)\mathbf{x}_0, z)}{\dfrac{\partial F}{\partial z}(\mathbf{x}_0, \phi z + (1 - \phi)z_0)}$$

OPTIONAL (*Continued*)

As $h \to 0$, it follows that $x \to x_0$ and $z \to z_0$, so we get

$$\frac{\partial g}{\partial x}(x_0, y_0) = \lim_{h \to 0} \frac{g(x_0 + h, y_0) - g(x_0, y_0)}{h} = -\frac{\dfrac{\partial F}{\partial x}(\mathbf{x}_0, z)}{\dfrac{\partial F}{\partial z}(\mathbf{x}_0, z)}$$

The formula

$$\frac{\partial g}{\partial y}(\mathbf{x}_0, y_0) = -\frac{\dfrac{\partial F}{\partial y}(\mathbf{x}_0, z)}{\dfrac{\partial F}{\partial z}(\mathbf{x}_0, z)}$$

is proved in the same way. This derivation holds at any point (x, y) in U by the same argument, so we have proved (1). Since the right hand side of (1) is continuous, we have proved the theorem. ■

Once it is known that $z = g(\mathbf{x})$ exists and is differentiable, formula (1) may be checked by implicit differentiation; that is, the chain rule applied to $F(\mathbf{x}, g(\mathbf{x})) = 0$ gives

$$D_{\mathbf{x}}F(\mathbf{x}, g(\mathbf{x})) + \frac{\partial F}{\partial z}(\mathbf{x}, g(\mathbf{x})) \cdot Dg(\mathbf{x}) = 0$$

which is equivalent to (1).

EXAMPLE 1. In using Theorem 8, it is important to recognize the necessity of taking sufficiently small neighborhoods U and V. For example, consider the equation

$$x^2 + z^2 - 1 = 0$$

that is, $F(x, z) = x^2 + z^2 - 1$, with $n = 1$. Here $(\partial F/\partial z)(x, z) = 2z$, so Theorem 8 applies to a point (x_0, z_0) satisfying $x_0^2 + z_0^2 - 1 = 0$ and $z_0 \neq 0$. Thus near such points, z is a unique function of x. This function is $z = \sqrt{1 - x^2}$ if $z_0 > 0$ and $z = -\sqrt{1 - x^2}$ if $z_0 < 0$. Note that z is defined only for $|x| < 1$ (U must not be too big) and z is unique only if it is near z_0 (V must not be too big). These facts and the nonexistence of $\partial z/\partial x$ at $z_0 = 0$ are of course clear from the fact that $x^2 + z^2 = 1$ defines a circle in the xz-plane (Figure 4.4.1).

OPTIONAL (*Continued*)

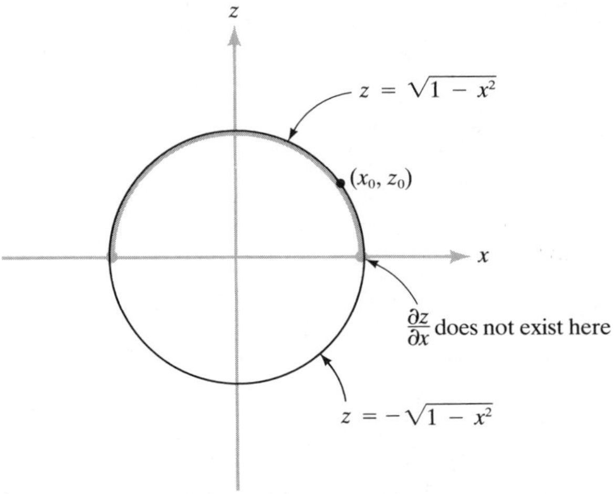

FIGURE 4.4.1
It is necessary to take small neighborhoods in the implicit function theorem.

Let us apply Theorem 8 to the study of surfaces. As in Theorem 6, we are concerned with the level set of a function $g: U \subset \mathbb{R}^n \to \mathbb{R}$, that is, with the surface S consisting of the set of $\mathbf{x}$ satisfying $g(\mathbf{x}) = c_0$ where $c_0 = g(\mathbf{x}_0)$ and where $\mathbf{x}_0$ is given. Let us take $n = 3$ for concreteness. Thus we are dealing with the level surface of a function $g(x, y, z)$ through a given point (x_0, y_0, z_0). As in Theorem 6, let us assume that $\nabla g(x_0, y_0, z_0) \neq 0$. This means that at least one of the partials of g is nonzero. For definiteness, suppose $(\partial g/\partial z)(x_0, y_0, z_0) \neq 0$. By applying Theorem 8 to the function $(x, y, z) \mapsto g(x, y, z) - c_0$, we know there is a unique function $z = k(x, y)$ satisfying $g(x, y, k(x, y)) = c_0$ for (x, y) near (x_0, y_0) and z near z_0. Thus, near z_0 the surface S is the graph of the function k. Since k is continuously differentiable, this surface has a tangent plane at (x_0, y_0, z_0) defined by

$$z = z_0 + \frac{\partial k}{\partial x}(x_0, y_0)(x - x_0) + \frac{\partial k}{\partial y}(x_0, y_0)(y - y_0) \tag{4}$$

But by (1)

$$\frac{\partial k}{\partial x}(x_0, y_0) = -\frac{\dfrac{\partial g}{\partial x}(x_0, y_0, z_0)}{\dfrac{\partial g}{\partial z}(x_0, y_0, z_0)} \quad \text{and} \quad \frac{\partial k}{\partial y}(x_0, y_0) = -\frac{\dfrac{\partial g}{\partial y}(x_0, y_0, z_0)}{\dfrac{\partial g}{\partial z}(x_0, y_0, z_0)}$$

OPTIONAL (*Continued*)

Substituting these two equations into equation (4) gives this equivalent description:

$$0 = (z - z_0)\frac{\partial g}{\partial z}(x_0, y_0, z_0) + (x - x_0)\frac{\partial g}{\partial x}(x_0, y_0, z_0) + (y - y_0)\frac{\partial g}{\partial y}(x_0, y_0, z_0)$$

that is,

$$(x - x_0, y - y_0, z - z_0) \cdot \nabla g(x_0, y_0, z_0) = 0$$

Thus the tangent plane to the level surface of g is the orthogonal complement to $\nabla g(x_0, y_0, z_0)$ through the point (x_0, y_0, z_0). This agrees with the definition on p. 132.

We are now ready to complete the proof of Theorem 6 (see partial proof on p. 217) and must show that every vector tangent to S at (x_0, y_0, z_0) is tangent to a curve in S. By Theorem 8, we need only show this for a graph of the form $z = k(x, y)$. However, if $\mathbf{v} = (x - x_0, y - y_0, z - z_0)$ is tangent to the graph (i.e., it satisfies (4)), then $\mathbf{v}$ is tangent to the curve in S given by

$$\mathbf{c}(t) = (x_0 + t(x - x_0), y_0 + t(y - y_0), k(x_0 + t(x - x_0)))$$

at $t = 0$. This can be checked using the Chain Rule. (See Figure 4.4.2.) We have now completed the technical details of Theorem 6.

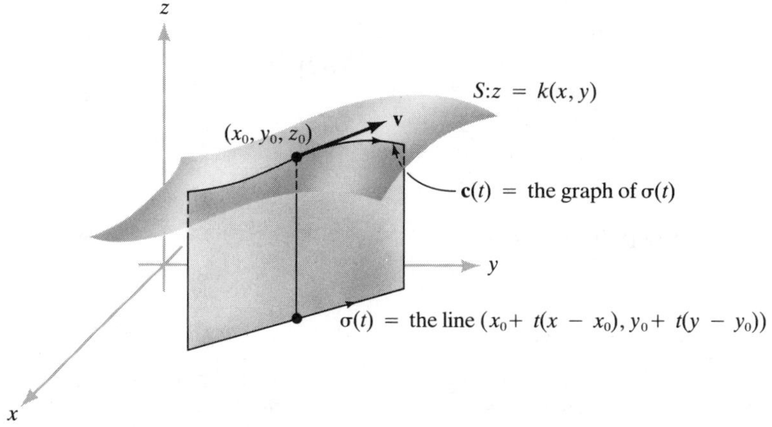

FIGURE 4.4.2
The construction of a curve $\mathbf{c}(t)$ *in the surface* S *whose tangent vector is* $\mathbf{v}$.

OPTIONAL (*Continued*)

EXAMPLE 2. Near what points may the surface

$$x^3 + 3y^2 + 8xz^2 - 3z^3y = 1$$

be represented as a graph of a differentiable function $z = k(x, y)$?

Here we take $F(x, y, z) = x^3 + 3y^2 + 8xz^2 - 3z^3y - 1$ and attempt to solve $F(x, y, z) = 0$ for z as a function of (x, y). By Theorem 8, this may be done near a point (x_0, y_0, z_0) if $(\partial F/\partial z)(x_0, y_0, z_0) \neq 0$, that is, if

$$z_0(16x_0 - 9z_0y_0) \neq 0$$

that is,

$$z_0 \neq 0 \quad \text{and} \quad 16x_0 \neq 9z_0y_0$$

Next we shall state, without proof, the General Implicit Function Theorem.* Instead of attempting to solve one equation for one variable, we attempt to solve m equations for m variables $z_1, \ldots, z_m$:

$$\begin{aligned} F_1(x_1, \ldots, x_n, z_1, \ldots, z_m) &= 0 \\ F_2(x_1, \ldots, x_n, z_1, \ldots, z_m) &= 0 \\ \vdots \qquad\qquad \vdots \qquad\qquad & \ \ \vdots \\ F_m(x_1, \ldots, x_n, z_1, \ldots, z_m) &= 0 \end{aligned} \tag{5}$$

In Theorem 8 we had the condition $\partial F/\partial z \neq 0$. The condition appropriate to the General Implicit Function Theorem is that $\Delta \neq 0$, where Δ is the determinant of the $m \times m$ matrix

$$\begin{bmatrix} \dfrac{\partial F_1}{\partial z_1} & \cdots & \dfrac{\partial F_1}{\partial z_m} \\ \vdots & & \vdots \\ \dfrac{\partial F_m}{\partial z_1} & \cdots & \dfrac{\partial F_m}{\partial z_m} \end{bmatrix}$$

evaluated at the point $(\mathbf{x}_0, \mathbf{z}_0)$ whose neighborhood we are interested in.

* For three different proofs of the general case, consult:

(a) E. Goursat, *A Course in Mathematical Analysis*, I, Dover, New York, 1959, p. 45. (This proof derives the general theorem by successive application of Theorem 8.)

(b) T. M. Apostol, *Mathematical Analysis*, 2nd ed., Addison Wesley, Reading, Mass., 1974.

(c) J. E. Marsden, *Elementary Classical Analysis*, W. H. Freeman and Company, San Francisco, 1974.

Of these sources, the last two use more sophisticated ideas that are usually not covered until a junior level course in analysis. The first however, is easily understood by the reader who has some knowledge of linear algebra.

OPTIONAL (*Continued*)

If $\Delta \neq 0$, *then, near the point* $\mathbf{x}_0, \mathbf{z}_0$, (5) *defines unique functions*

$$z_i = k_i(x_1, \ldots, x_n) \qquad (i = 1, \ldots, m)$$

Their derivatives may be computed by implicit differentiation.

EXAMPLE 3. Show that near the point $(x, y, u, v) = (1, 1, 1, 1)$ we can solve

$$xu + yvu^2 = 2$$

$$xu^3 + y^2v^4 = 2$$

uniquely for u and v as functions of x and y. Compute $(\partial u/\partial x)(1, 1)$.

To check solvability, we form the equations

$$F_1(x, y, u, v) = xu + yvu^2 - 2$$

$$F_2(x, y, u, v) = xu^3 + y^2v^4 - 2$$

and the determinant

$$\Delta = \begin{vmatrix} \dfrac{\partial F_1}{\partial u} & \dfrac{\partial F_1}{\partial v} \\ \dfrac{\partial F_2}{\partial u} & \dfrac{\partial F_2}{\partial v} \end{vmatrix} \quad \text{at} \quad (1, 1, 1, 1)$$

$$= \begin{vmatrix} x + 2yuv & yu^2 \\ 3u^2x & 4y^2v^3 \end{vmatrix} \quad \text{at} \quad (1, 1, 1, 1)$$

$$= \begin{vmatrix} 3 & 1 \\ 3 & 4 \end{vmatrix} = 9$$

Since $\Delta \neq 0$, solvability is assured by the General Implicit Function Theorem. To find $\partial u/\partial x$, we implicitly differentiate the given equations in x using the Chain Rule:

$$x\frac{\partial u}{\partial x} + u + y\frac{\partial v}{\partial x}u^2 + 2yvu\frac{\partial u}{\partial x} = 0$$

$$3xu^2\frac{\partial u}{\partial x} + u^3 + 4y^2v^3\frac{\partial v}{\partial x} = 0$$

Setting $(x, y, u, v) = (1, 1, 1, 1)$ gives

$$3\frac{\partial u}{\partial x} + \frac{\partial v}{\partial x} = -1$$

$$3\frac{\partial u}{\partial x} + 4\frac{\partial v}{\partial x} = -1$$

OPTIONAL (*Continued*)

Solving for $\partial u/\partial x$ by multiplying the first equation by 4 and subtracting gives $\partial u/\partial x = -\frac{1}{3}$.

A special case of the General Implicit Function Theorem is the *Inverse Function Theorem.* Here we attempt to solve the n equations

$$\left.\begin{aligned} f_1(x_1, \dots, x_n) &= y_1 \\ &\cdots \\ f_n(x_1, \dots, x_n) &= y_n \end{aligned}\right\} \tag{6}$$

for $x_1, \dots, x_n$ as functions of $y_1, \dots, y_n$; that is, we are trying to invert the system (6). This is analogous to forming the inverses of functions like $\sin x = y$ and $e^x = y$, with which the reader should be familiar from elementary calculus. Now, however, we are concerned with functions of several variables. The question of solvability is answered by the General Implicit Function Theorem applied to the functions $y_i - f_i(x_1, \dots, x_n)$ with the unknowns $x_1, \dots, x_n$ (called $z_1, \dots, z_n$ above). The condition is $\Delta \neq 0$ where Δ is the determinant of the matrix $Df(\mathbf{x}_0)$, $f = (f_1, \dots, f_n)$ and where $\mathbf{x}_0$ is a point whose neighborhood we are interested in. The quantity Δ is usually denoted by $\partial(f_1, \dots, f_n)/\partial(x_1, \dots, x_n)$, $\partial(y_1, \dots, y_n)/\partial(x_1, \dots, x_n)$, or $Jf(\mathbf{x}_0)$ and is called the *Jacobian determinant* of f. Explicitly,

$$\left.\frac{\partial(f_1, \dots, f_n)}{\partial(x_1, \dots, x_n)}\right|_{\mathbf{x}=\mathbf{x}_0} = Jf(\mathbf{x}_0) = \begin{vmatrix} \dfrac{\partial f_1}{\partial x_1}(\mathbf{x}_0) & \cdots & \dfrac{\partial f_1}{\partial x_n}(\mathbf{x}_0) \\ \vdots & & \vdots \\ \dfrac{\partial f_n}{\partial x_1}(\mathbf{x}_0) & \cdots & \dfrac{\partial f_n}{\partial x_n}(\mathbf{x}_0) \end{vmatrix} \tag{7}$$

The Jacobian determinant will play an important role in our later work on integration (see Section 5.8). The following theorem summarizes this discussion:

THEOREM 9 (*Inverse Function Theorem*). *Let $U \subset \mathbb{R}^n$ be open and let $f_1: U \to \mathbb{R}, \dots$ and $f_n: U \to \mathbb{R}$ have continuous partial derivatives. Consider the equations in (6) near a given solution $\mathbf{x}_0$, $\mathbf{y}_0$. If $[\partial(f_1, \dots, f_n)]/[\partial(x_1, \dots, x_n)] = Jf(\mathbf{x}_0)$ (defined by (7)) is nonzero, then (6) can be solved uniquely as $\mathbf{x} = \mathbf{g}(\mathbf{y})$ for $\mathbf{x}$ near $\mathbf{x}_0$ and $\mathbf{y}$ near $\mathbf{y}_0$. Moreover, the function $\mathbf{g}$ has continuous partial derivatives.*

EXAMPLE 4. Consider the equations

$$x^4 + y^4/x = u(x, y), \qquad \sin x + \cos y = v(x, y)$$

Near which points (x, y) can we solve for x, y in terms of u, v?

OPTIONAL (*Continued*)

Here the functions are $u(x, y) = f_1(x, y) = (x^4 + y^4)/x$ and $v(x, y) = f_2(x, y) = \sin x + \cos y$. We want to know the points near which we can solve for x, y as functions of u and v. According to the inverse function theorem we must first compute $\partial(f_1, f_2)/\partial(x, y)$. We take the domain of $f = (f_1, f_2)$ to be $U = \{(x, y) \in \mathbb{R}^2 \mid x \neq 0\}$. Now

$$\frac{\partial(f_1, f_2)}{\partial(x, y)} = \begin{vmatrix} \dfrac{\partial f_1}{\partial x} & \dfrac{\partial f_1}{\partial y} \\ \dfrac{\partial f_2}{\partial x} & \dfrac{\partial f_2}{\partial y} \end{vmatrix} = \begin{vmatrix} \dfrac{3x^4 - y^4}{x^2} & \dfrac{4y^3}{x} \\ \cos x & -\sin y \end{vmatrix}$$

$$= \frac{(\sin y)}{x^2}(y^4 - 3x^4) - \frac{4y^3}{x}\cos x$$

Therefore, at points where this does not vanish we can solve for x, y in terms of u and v. In other words, we can solve for x, y near those x, y for which $x \neq 0$ and $(\sin y)(y^4 - 3x^4) \neq 4xy^3 \cos x$. Such conditions generally cannot be solved explicitly. For example, if $x_0 = \pi/2$, $y_0 = \pi/2$, we can solve for x, y near x_0, y_0, because there, $\partial(f_1\ f_2)/\partial(x, y) \neq 0$.

EXERCISES

1. Let $F(x, y) = 0$ define a curve in the xy-plane through the point (x_0, y_0). Assume that $(\partial F/\partial y)(x_0, y_0) \neq 0$. Show that this curve can be locally represented by the graph of a function $y = g(x)$. Show that (i) the line orthogonal to $\nabla F(x_0, y_0)$ agrees with (ii) the tangent line to the graph of $y = g(x)$.

2. Show that $xy + z + 3xz^5 = 4$ is solvable for z as a function of (x, y) near $(1, 0, 1)$. Compute $\partial z/\partial x$ and $\partial z/\partial y$ at $(1, 0)$.

3. (a) Check directly (i.e., without using Theorem 8) where we can solve the equation $F(x, y) = y^2 + y + 3x + 1 = 0$ for y in terms of x.

   (b) Check that your answer in (a) agrees with the answer you expect from the Implicit Function Theorem. Compute dy/dx.

4. Discuss the solvability in the system

$$3x + 2y + z^2 + u + v^2 = 0$$
$$4x + 3y + z + u^2 + v + w + 2 = 0$$
$$x + z + w + u^2 + 2 = 0$$

   for u, v, w in terms of x, y, z near $x = y = z = 0$, $u = v = 0$, and $w = -2$.

5. Discuss the solvability of

$$y + x + uv = 0$$
$$uxy + v = 0$$

   for u, v in terms of x, y near $x = y = u = v = 0$ and check directly.

OPTIONAL (*Continued*)

6. Investigate whether or not the system

$$u(x, y, z) = x + xyz$$
$$v(x, y, z) = y + xy$$
$$w(x, y, z) = z + 2x + 3z^2$$

can be solved for x, y, z in terms of u, v, w near $(0, 0, 0)$.

7. Consider $f(x, y) = ((x^2 - y^2)/(x^2 + y^2), xy/(x^2 + y^2))$. Does this map of $\mathbb{R}^2\backslash(0, 0)$ to $\mathbb{R}^2$ have a local inverse near $(x, y) = (0, 1)$?

8. (a) Define $x\colon \mathbb{R}^2 \to \mathbb{R}$ by $x(r, \theta) = r\cos\theta$ and define $y\colon \mathbb{R}^2 \to \mathbb{R}$ by $y(r, \theta) = r\sin\theta$. Show that

$$\left.\frac{\partial(x, y)}{\partial(r, \theta)}\right|_{(r_0, \theta_0)} = r_0$$

(b) When can we form a smooth inverse function $r(x, y)$, $\theta(x, y)$? Check directly and with the Inverse Function Theorem.

(c) Consider the following transformations for spherical coordinates (see Section 1.4):

$$x(r, \phi, \theta) = \rho\sin\phi\cos\theta$$
$$y(r, \phi, \theta) = \rho\sin\phi\sin\theta$$
$$z(r, \phi, \theta) = \rho\cos\phi$$

Show that

$$\frac{\partial(x, y, z)}{\partial(\rho, \phi, \theta)} = \rho^2\sin\phi$$

(d) When can we solve for (ρ, ϕ, θ) in terms of (x, y, z)?

9. Let (x_0, y_0, z_0) be a point of the locus defined by $z^2 + xy - a = 0$, $z^2 + x^2 - y^2 - b = 0$.
   (a) Under what conditions may the part of the locus near (x_0, y_0, z_0) be represented in the form $x = f(z)$, $y = g(z)$?
   (b) Compute $f'(z)$ and $g'(z)$.

10. Is it possible to solve

$$xy^2 + xzu + yv^2 = 3$$
$$u^3yz + 2xv - u^2v^2 = 2$$

for $u(x, y, z)$, $v(x, y, z)$ near $(x, y, z) = (1, 1, 1)$, $(u, v) = (1, 1)$? Compute $\partial v/\partial y$ at $(1, 1, 1)$.

11. The problem of factoring a polynomial $x^n + a_{n-1}x^{n-1} + \cdots + a_0$ into linear factors is, in a sense, an "inverse function" problem. The coefficients a_i are known functions of the n roots r_j. We would like to find the roots as functions of the coefficients in some region. With $n = 3$, apply the inverse function theorem to this problem and state what it tells you about the possibility of doing this.

4.5 SOME APPLICATIONS

In this section we shall give some applications of the mathematical methods that we have developed in the preceding sections. The applications are made to mechanics, geometry, and economics, beginning with mechanics. The student should consult the instructor concerning which of these examples should be studied.

Let $\mathbf{F}$ denote a force field defined on a certain domain U of $\mathbb{R}^3$. Thus $\mathbf{F}\colon U \to \mathbb{R}^3$ is a given vector field. Let us agree that a particle (with mass m) is to move along a path $\boldsymbol{\sigma}(t)$ in such a way that Newton's law holds: Mass $\times$ acceleration $=$ force; that is, the path $\boldsymbol{\sigma}(t)$ is to satisfy the equation

$$m\boldsymbol{\sigma}''(t) = \mathbf{F}(\boldsymbol{\sigma}(t)) \tag{1}$$

If $\mathbf{F}$ is a potential field with potential V, that is, if $\mathbf{F} = -\operatorname{grad} V$, then

$$\tfrac{1}{2}m\|\boldsymbol{\sigma}'(t)\|^2 + V(\boldsymbol{\sigma}(t)) = \text{constant} \tag{2}$$

(The first term is called the *kinetic energy*.) Indeed, by differentiating with the Chain Rule,

$$\begin{aligned}\frac{d}{dt}\left\{\frac{1}{2}m\|\boldsymbol{\sigma}'(t)\|^2 + V(\boldsymbol{\sigma}(t))\right\} &= m\boldsymbol{\sigma}'(t)\cdot\boldsymbol{\sigma}''(t) + \operatorname{grad} V(\boldsymbol{\sigma}(t))\cdot\boldsymbol{\sigma}'(t)\\ &= [m\boldsymbol{\sigma}''(t) + \operatorname{grad} V(\boldsymbol{\sigma}(t))]\cdot\boldsymbol{\sigma}'(t) = 0\end{aligned}$$

since $m\boldsymbol{\sigma}''(t) = -\operatorname{grad} V(\boldsymbol{\sigma}(t))$. This proves (2).

DEFINITION. *A point $\mathbf{x}_0 \in U$ is called a* ***position of equilibrium*** *if the force at that point is zero: $\mathbf{F}(\mathbf{x}_0) = \mathbf{0}$. A point $\mathbf{x}_0$ that is a position of equilibrium is said to be* ***stable*** *if for every $\rho > 0$ and $\varepsilon > 0$, we can choose numbers $\rho_0 > 0$ and $\varepsilon_0 > 0$ such that a material point situated anywhere at a distance less than ρ_0 from $\mathbf{x}_0$, after initially receiving kinetic energy in amount less than ε_0, will forever remain a distance from $\mathbf{x}_0$ less than ρ and posses kinetic energy less than ε (see Figure 4.5.1).*

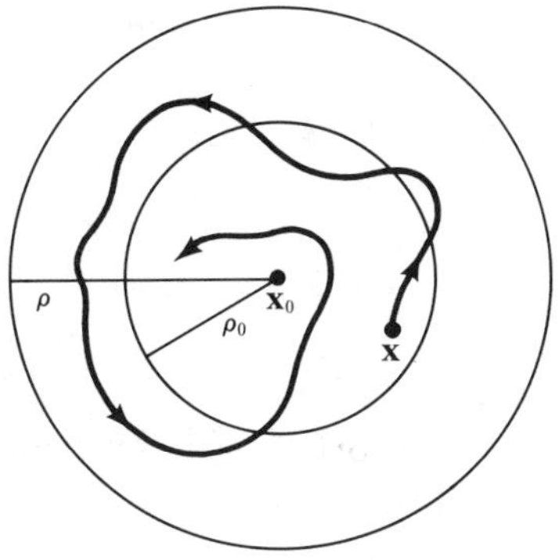

FIGURE 4.5.1
Motion near a stable point $\mathbf{x}_0$.

Thus if we have a position of equilibrium, stability at $\mathbf{x}_0$ means that a slowly moving particle near $\mathbf{x}_0$ will always remain near $\mathbf{x}_0$ and keep moving slowly. If we have an unstable equilibrium point $\mathbf{x}_0$, then $\boldsymbol{\sigma}(t) = \mathbf{x}_0$ solves the equation $m\boldsymbol{\sigma}''(t) = \mathbf{F}(\boldsymbol{\sigma}(t))$, but nearby solutions may move away from $\mathbf{x}_0$ as time progresses. For example a pencil balancing on its tip illustrates an unstable configuration, whereas a ball hanging on a spring illustrates a stable equilibrium.

THEOREM 10.

(*i*) *Critical points of a potential are the positions of equilibrium.*

(*ii*) *In a potential field, a point $\mathbf{x}_0$ at which the potential takes a strict local minimum is a position of stable equilibrium.*

Recall that a function f is said to have a *strict* local minimum at the point $\mathbf{x}_0$ if there exists a neighborhood U of $\mathbf{x}_0$ such that $f(x) > f(\mathbf{x}_0)$ for all $\mathbf{x}$ in U other than $\mathbf{x}_0$.

Proof. The first assertion (*i*) is quite obvious from the definition $\mathbf{F} = -\text{grad } V$; equilibrium points $\mathbf{x}_0$ are exactly critical points of V, at which $\nabla V(\mathbf{x}_0) = \mathbf{0}$.

To prove (*ii*) we shall make use of the law of conservation of energy, equation (2). We have

$$\tfrac{1}{2}m\|\boldsymbol{\sigma}'(t)\|^2 + V(\boldsymbol{\sigma}(t)) = \tfrac{1}{2}m\|\boldsymbol{\sigma}'(0)\|^2 + V(\boldsymbol{\sigma}(0))$$

We shall argue slightly informally to amplify and illuminate the central ideas involved. Let us choose a small neighborhood of $\mathbf{x}_0$ and start our particle with a small kinetic energy. As t increases, the particle moves away from $\mathbf{x}_0$ on a path $\boldsymbol{\sigma}(t)$ and $V(\boldsymbol{\sigma}(t))$ increases (since $V(\boldsymbol{\sigma}(0))$ is a strict minimum), so the kinetic energy must decrease. If the initial kinetic energy is sufficiently small, then in order for the particle to escape from our neighborhood of $\mathbf{x}_0$, outside of which V has increased by a definite amount, the kinetic energy would have to become negative (which is impossible). Thus the particle cannot escape the neighborhood. ■

EXAMPLE 1. Find the points that are positions of equilibrium, and determine whether or not they are stable, if the force field $\mathbf{F} = F_x\mathbf{i} + F_y\mathbf{j} + F_z\mathbf{k}$ is given by $F_x = -k^2x$, $F_y = -k^2y$, $F_z = -k^2z\,(k \neq 0)$.*

The field $\mathbf{F}$ is a potential field with potential $V = \frac{1}{2}k^2(x^2 + y^2 + z^2)$. The only critical point of V is the origin. The Hessian of V at the origin is $HV(0, 0, 0)(h_1, h_2, h_3) = \frac{1}{2}k^2(h_1^2 + h_2^2 + h_3^2)$, which is positive definite. It

* The force field in this example is that governing the motion of a three-dimensional harmonic oscillator.

follows that the origin is a strict minimum of V. Thus, by (*i*) and (*ii*) of Theorem 10, we have shown that the origin is a position of stable equilibrium.

Let a material point in a potential field V be constrained to remain on the level surface S given by the equation $\phi(x, y, z) = 0$, with grad $\phi \neq \mathbf{0}$. If in formula (1) we replace $\mathbf{F}$ by the component of $\mathbf{F}$ parallel to S, we ensure that the particle will remain on S.† By analogy with Theorem 10, we have:

THEOREM 11.

(*i*) *If at a point P on the surface S the potential $V|S$ has an extreme value, then the point P is a position of equilibrium on the surface.*

(*ii*) *If a point $P \in S$ is a strict local minimum of the potential $V|S$, then the point P is a position of stable equilibrium.*

The proof of this theorem will be omitted. It is similar to the proof of Theorem 10, with the additional fact that the equation of motion uses only the component of $\mathbf{F}$ along the surface.‡

EXAMPLE 2. Let $\mathbf{F}$ be the gravitational field near the surface of the Earth, that is, $\mathbf{F} = (F_x, F_y, F_z)$ where $F_x = 0$, $F_y = 0$, and $F_z = -mg$. What are the positions of equilibrium, if a material point with mass m is constrained to the sphere $\phi(x, y, z) = x^2 + y^2 + z^2 - r^2 = 0\,(r > 0)$? Which of these are stable?

Notice that $\mathbf{F}$ is a potential field with $V = mgz$. Using the method of Lagrange multipliers introduced in the preceding section to locate the possible extrema, we have the equations

$$\nabla V = \lambda\,\nabla\phi$$
$$\phi = 0$$

or in terms of components

$$0 = 2\lambda x$$
$$0 = 2\lambda y$$
$$mg = 2\lambda z$$
$$x^2 + y^2 + z^2 - r^2 = 0$$

† If $\phi(x, y, z) = x^2 + y^2 + z^2 - r^2$, the particle is constrained to move on a sphere; for instance, it may be whirling on a string. The part subtracted from $\mathbf{F}$ to make it parallel to S is normal to S and is called the *centripetal force*.

‡ These ideas can be applied to quite a number of interesting physical situations, such as molecular vibrations. The stability of such systems is an important question. For further information consult the physics literature (e.g., H. Goldstein, *Classical Mechanics*, Chapter 10, Addison-Wesley, Reading, Mass., 1950), and the mathematics literature (e.g., M. Hirsch and S. Smale, *Differential Equations, Dynamical Systems and Linear Algebra*, Academic Press, New York, 1974).

and the solution of the above simultaneous equations is $x = 0$, $y = 0$, $z = \pm r$, $\lambda = \pm mg/2r$. By Theorem 11 it follows that the points $P_1 = (0, 0, -r)$ and $P_2 = (0, 0, r)$ are positions of equilibrium. By observation of the potential function $V = mgz$ and by Theorem 11(*ii*) it follows that P_1 is a strict minimum and hence a stable point, whereas P_2 is not. This conclusion should be physically obvious.

Next we turn to a geometric application.

EXAMPLE 3. Suppose we have a curve defined by the equation

$$\phi(x, y) = Ax^2 + 2Bxy + Cy^2 - 1 = 0$$

Find the maximum and minimum distance of the curve to the origin. These are called the lengths of the semi-major and the semi-minor axis.

The problem is equivalent to finding the extreme values of $f(x, y) = x^2 + y^2$ subject to the constraining condition $\phi(x, y) = 0$. Using the Lagrange multiplier method, we have the following equations:

$$2x + \lambda(2Ax + 2By) = 0 \tag{1}$$

$$2y + \lambda(2Bx + 2Cy) = 0 \tag{2}$$

$$Ax^2 + 2Bxy + Cy^2 = 1 \tag{3}$$

Adding (1) $\times\, x$ to (2) $\times\, y$, we obtain $2(x^2 + y^2) + 2\lambda(Ax^2 + 2Bxy + Cy^2) = 0$. By (3), it follows that $x^2 + y^2 + \lambda = 0$. Let $t = -1/\lambda = 1/(x^2 + y^2)$ ($\lambda = 0$ is impossible since $(0, 0)$ is not on the curve $\phi(x, y) = 0$). Then (1) and (2) can be written as follows:

$$\begin{aligned} 2(A - t)x + 2By &= 0 \\ 2Bx + 2(C - t)y &= 0 \end{aligned} \tag{4}$$

If these two equations are to have a nontrivial solution (remember $(x, y) = (0, 0)$ is not on our curve and so is not a solution), it follows from a theorem of linear algebra that their determinant vanishes:*

$$\begin{vmatrix} A - t & B \\ B & C - t \end{vmatrix} = 0$$

Since this equation is quadratic in t, there are two solutions, say t_1 and t_2. Since $-\lambda = x^2 + y^2$, we have $\sqrt{x^2 + y^2} = \sqrt{-\lambda}$. Now $\sqrt{x^2 + y^2}$ is the distance from the point (x, y) to the origin. Therefore, if (x_1, y_1) and (x_2, y_2) denote the nontrivial solutions to (4) corresponding to t_1 and t_2, we have that $\sqrt{x_2^2 + y_2^2} = 1/\sqrt{t_2}$ and $\sqrt{x_1^2 + y_1^2} = 1/\sqrt{t_1}$. Consequently, if $t_1 > t_2$ the lengths of the semi-minor and semi-major axes are $1/\sqrt{t_1}$ and $1/\sqrt{t_2}$,

* The matrix of coefficients of the equations cannot have an inverse, because this would imply that the solution is zero. From Section 1.4 we know that a matrix that does not have an inverse has determinant zero.

respectively. If the curve is an ellipse, both t_1 and t_2 are real and positive. What happens with a hyperbola or a parabola?

Finally we discuss an application to economics.

EXAMPLE 4. Suppose that the output of a manufacturing firm is a quantity Q of a certain product, where Q is a function $f(K, L)$, where K is the amount of capital equipment (or investment) and where L is the amount of labor used. If the price of labor is p, the price of capital is q, and the firm can spend no more than B dollars, how can we find the amount of capital and labor to maximize the output Q?

We would expect that if the amount of capital or labor is increased, then the output Q should also increase; that is,

$$\frac{\partial Q}{\partial K} \geq 0 \quad \text{and} \quad \frac{\partial Q}{\partial L} \geq 0$$

We also expect that as more labor is added to a given amount of capital equipment, we get less additional output for our effort; that is,

$$\frac{\partial^2 Q}{\partial L^2} < 0$$

Similarly,

$$\frac{\partial^2 Q}{\partial K^2} < 0$$

With these assumptions on Q, it is reasonable to expect the level curves of output (called *isoquants*) $Q(K, L) = c$ to look something like the curves sketched in Figure 4.5.2, with $c_1 < c_2 < c_3$.

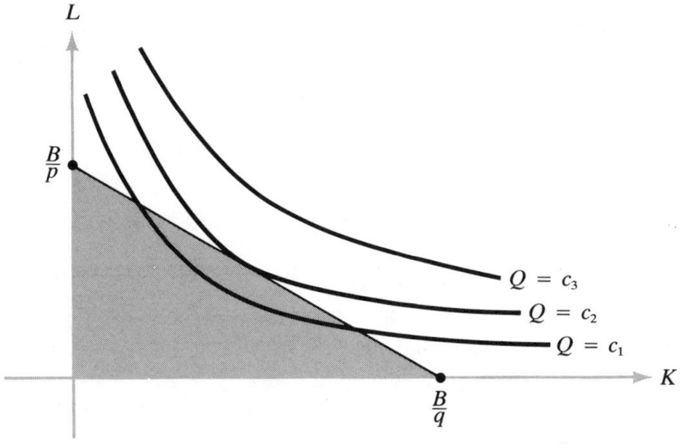

FIGURE 4.5.2
What is the largest value of Q in the shaded triangle?

We can interpret the convexity of the isoquants as follows: As one moves to the right along a given isoquant, it takes more and more capital to replace a unit of labor and still produce the same output. The budget constraint means that we must stay inside the triangle bounded by the axes and the line $pL + qK = B$. Geometrically, it is clear that we produce the most by spending all our money in such a way as to pick the isoquant that just touches, but does not cross, the budget line.

Since the maximum point lies on the boundary of our domain, to find it we apply the method of Lagrange multipliers. To maximize $Q = f(K, L)$ subject to the constraint $pL + qK = B$, we look for critical points of the auxiliary function,

$$h(K, L, \lambda) = f(K, L) - \lambda(pL + qK - B)$$

Thus we want

$$\frac{\partial Q}{\partial K} = \lambda q, \qquad \frac{\partial Q}{\partial L} = \lambda p \quad \text{and} \quad pL + qK = B$$

These are the conditions we must meet in order to maximize output. (The reader is asked to work out a specific case in Exercise 7.)

In this example, λ represents something interesting. Let $k = qK$ and $l = qL$, so that k is the dollar value of the capital used and l is the dollar value of the labor used. Then the first two equations become

$$\frac{\partial Q}{\partial k} = \frac{l}{q}\frac{\partial Q}{\partial K} = \lambda = \frac{l}{p}\frac{\partial Q}{\partial L} = \frac{\partial Q}{\partial l}$$

Thus at the optimum production point the marginal change in output per dollar's worth of additional capital investment is equal to the marginal change of output per dollar's worth of additional labor, and λ is this common value. At the optimum point, the exchange of a dollar's worth of capital for a dollar's worth of labor does not change the output. Away from the optimum point the marginal outputs are different, and one exchange or the other will increase the output.

EXERCISES

1. Let a particle move in a potential field in $\mathbb{R}^2$ given by $V(x, y) = 3x^2 + 2xy + 2x + y^2 + y + 4$. Find the stable equilibrium points, if any.
2. Let a particle move in a potential field in $\mathbb{R}^2$ given by $V(x, y) = x^2 - 2xy + y^2 + y^3 + x^4$. Is $(0, 0)$ a position of stable equilibrium?
3. Let a particle be constrained to move on the sphere $x^2 + y^2 + z^2 = 1$, subject to gravitational forces (as in Example 2) as well as an additional potential $V(x, y, z) = x + y$. Find the stable equilibrium points, if any.

4. Attempt to formulate a definition and a theorem saying that if a potential has a maximum at $\mathbf{x}_0$, then $\mathbf{x}_0$ is a position of unstable equilibrium. Watch out for pitfalls in your argument.
5. Try to find the extrema of $xy + yz$ among points satisfying $xz = 1$.
6. Answer the question posed in the last line of Example 3.
7. Carry out the analysis of Example 4 for the production function $Q(K, L) = AK^{\alpha}L^{1-\alpha}$ where A and α are positive constants, and $0 < \alpha < 1$. This is called a Cobb-Douglas production function and is sometimes used as a simple model for the national economy. Q is then the aggregate output of the economy for a given input of capital and labor.
8. A firm uses wool and cotton fiber to produce cloth. The amount of cloth produced is given by $Q(x, y) = xy - x - y + 1$ where x is the number of pounds of wool, y the number of pounds of cotton, $x > 1$, and $y > 1$. If wool costs p dollars per pound, cotton q dollars per pound, and the firm can spend B dollars on material, what should the ratio of cotton and wool be to produce the most cloth?

REVIEW EXERCISES FOR CHAPTER 4

1. Analyze the behavior of the following functions at the indicated points.
   (a) $z = x^2 - y^2 + 3xy, \quad (x, y) = (0, 0)$
   (b) $z = x^2 + y^2 + Cxy, \quad (x, y) = (0, 0)$
   [Your answer will depend on the constant C.]
2. Find the equation of the plane tangent to the surface at the indicated point.
   (a) $z = 2x^2 + y^2, \quad (x, y) = (0, 0)$
   (b) $z = x^2 - 3y^2 + x, \quad (x, y) = (1, 0)$
   (c) $z = x + 2y, \quad (x, y) = (3, 2)$
3. Find the equation of the plane tangent to the surface S given by the graph of
   (a) $f(x, y) = \sqrt{x^2 + y^2} + (x^2 + y^2)$ at $(1, 0, 2)$; and
   (b) $f(x, y) = \sqrt{x^2 + 2xy - y^2 + 1}$ at $(1, 1, \sqrt{3})$.
4. Find and classify the extreme values (if any) of the functions on $\mathbb{R}^2$ defined by the following expressions.
   (a) $y^2 - x^3$
   (b) $(x - 1)^2 + (x - y)^2$
   (c) $x^2 + xy^2 + y^4$
5. (a) Find the minimum distance from the origin in $\mathbb{R}^3$ to the surface $z = \sqrt{x^2 - 1}$.
   (b) Repeat (a) for the surface $z = 6xy + 7$.
6. Let $f: \mathbb{R}^3 \to \mathbb{R}$ be C^1 and let $z: \mathbb{R}^2 \to \mathbb{R}$ be C^1. Consider the composite function

$$h(x, y) = f(x, y, z(x, y))$$

Show that

$$\frac{\partial h}{\partial x} = \frac{\partial f}{\partial x} + \frac{\partial f}{\partial z}\frac{\partial z}{\partial x}$$

and

$$\frac{\partial h}{\partial y} = \frac{\partial f}{\partial y} + \frac{\partial f}{\partial z}\frac{\partial z}{\partial y}$$

7. (a) Let $f(x, y, z) = (xyz)^2$; compute ∇f.
   (b) Let $\mathbf{F}(x, y, z) = x^2y\mathbf{i} + y^2z\mathbf{j} + z^2y\mathbf{k}$; compute $\nabla \times \mathbf{F}$.
   Compute $\nabla \cdot \mathbf{F}$ and $\nabla \times \mathbf{F}$ for the following vector fields.
   (c) $\mathbf{F} = 2x\mathbf{i} + 3y\mathbf{j} + (4z + x^3)\mathbf{k}$
   (d) $\mathbf{F} = x^2\mathbf{i} + y^2\mathbf{j} + z^2\mathbf{k}$
   (e) $\mathbf{F} = (x + 3y)\mathbf{i} + (y + e^z)\mathbf{j} + (z + \cos x)\mathbf{k}$

8. Find the first few terms in the Taylor expansion of $f(x, y) = e^{xy} \cos x$ about $x = 0, y = 0$.

9. Find the extreme value of $z = xy$, subject to the condition $x + y = 1$.

10. Find the extreme values of $z = \cos^2 x + \cos^2 y$ subject to the condition $x + y = \pi/4$.

11. Find the points on the surface $z^2 - xy = 1$ nearest to the origin.

12. (a) Let $(\partial f/\partial x)(x, y) = (\partial f/\partial y)(x, y) = 0$ for every (x, y) in an open disc D. Show that $f(x, y)$ is a constant in D.
    (b) Use the result of (a) to show that if $\partial f/\partial x = \partial g/\partial x$ and $\partial f/\partial y = \partial g/\partial y$ in D, then f and g differ by a constant.

13. Find the shortest distance from the point $(0, b)$ to the parabola $x^2 - 4y = 0$. Solve this problem using the Lagrange multiplier and also without using Lagrange's method.

14. Solve the following geometric problems by Lagrange's method.
    (a) Find the shortest distance from the point (a_1, a_2, a_3) in $\mathbb{R}^3$ to the plane whose equation is given by $b_1x_1 + b_2x_2 + b_3x_3 + b_0 = 0$, where $(b_1, b_2, b_3) \neq (0, 0, 0)$.
    (b) Find the point on the line of intersection of the two planes $a_1x_1 + a_2x_2 + a_3x_3 + a_0 = 0$ and $b_1x_1 + b_2x_2 + b_3x_3 + b_0 = 0$ that is nearest to the origin.
    (c) Show that the volume of the largest rectangular parallelepiped that can be inscribed in the ellipsoid

$$\frac{x^2}{a^2} + \frac{y^2}{b^2} + \frac{z^2}{c^2} = 1$$

    is $8abc/3\sqrt{3}$.

15. A particle moves in a potential $V(x, y) = x^3 - y^2 + x^2 + 3xy$. Determine if $(0, 0)$ is a stable equilibrium point.

16. Study the nature of the function $f(x, y) = x^3 - 3xy^2$ near $(0, 0)$. Show that the point $(0, 0)$ is a degenerate critical point, that is, $D = 0$. This surface is called a "monkey saddle."

17. Find and sketch the maximum of $f(x, y) = xy$ on the curve $(x + 1)^2 + y^2 = 1$.

18. Find the maximum and minimum of $f(x, y) = xy - y + x - 1$ on the set $x^2 + y^2 \leq 2$.

*19. (In order to work this exercise the reader should be familiar with the technique of diagonalizing a 2×2 matrix.) Let $a(x)$, $b(x)$, and $c(x)$ be three continuous functions defined on $U \cup \partial U$ where U is an open set and ∂U denotes its set of boundary points (see Section 2.2). Use the notation of Lemma 2 in Section 4.2, and assume that for each $x \in U \cup \partial U$ the quadratic form defined by the matrix

$$\begin{bmatrix} a & b \\ b & c \end{bmatrix}$$

is positive definite. For a C^2 function v on $U \cup \partial U$ we define a differential operator L by $Lv = a(\partial^2 v/\partial x^2) + 2b(\partial^2 v/\partial x\,\partial y) + c(\partial^2 v/\partial y^2)$. With this positive definite condition such an operator is called "elliptic." A function v is called *strictly subharmonic relative to* L if $Lv > 0$. Show that a strictly subharmonic function cannot have a maximum point in U.

*20. A function v is said to be in the kernel of the operator L described in Exercise 19 if $Lv = 0$ on $U \cup \partial U$. Arguing as in Exercise 17 of Section 4.2, show that if v achieves its maximum on U it also achieves it on ∂U. This is the weak maximum principle for elliptic operators.

*21. Let L be an elliptic differential operator as in exercises 19 and 20.
(a) Define the notion of strict superharmonic function.
(b) Show that such functions cannot achieve a minimum on U.
(c) If v is as in Exercise 20, show that if v achieves its minimum on U it also achieves it on ∂U.

*22. Let $f\colon \mathbb{R} \to \mathbb{R}$ be C^1 and

$$u = f(x)$$

$$v = -y + xf(x)$$

If $f'(x_0) \neq 0$ show that this transformation of $\mathbb{R}^2$ to $\mathbb{R}^2$ is invertible near (x_0, y) and its inverse is given by

$$x = f^{-1}(u)$$

$$y = -v + uf^{-1}(u)$$

*23. Show that the equations

$$x^2 - y^2 - u^3 + v^2 + 4 = 0$$

$$2xy + y^2 - 2u^2 + 3v^4 + 8 = 0$$

determine functions $u(x, y)$ and $v(x, y)$ near $x = 2$ and $y = -1$ such that $u(2, -1) = 2$ and $v(2, -1) = 1$. Compute $\partial u/\partial x$ at $(2, -1)$.

*24. Show that there exist positive numbers p and q such that there are unique functions u and v from $]-1 - p, -1 + p[$ into $]1 - q, 1 + q[$ satisfying

$$xe^{u(x)} + u(x)e^{v(x)} = 0 = xe^{v(x)} + v(x)e^{u(x)}$$

for all $x \in\,]-1 - p, -1 + p[$ and $u(-1) = 1 = v(-1)$.

25. The Baraboo, Wisconsin plant of International Widget Co., Inc. uses aluminum, iron, and magnesium to produce high quality widgets. The quantity of widgets that may be produced using x tons of aluminum, y tons of iron, and z tons of magnesium is $Q(x, y, z) = xyz$. The cost of raw materials is aluminum, \$6 per ton; iron, \$4 per ton; and magnesium, \$8 per ton. How many tons each of aluminum, iron, and magnesium should be used to manufacture 1,000 widgets at the lowest possible cost? (HINT: Find an extreme value for what function subject to what constraint?)

The following *method of least squares* should be applied to Exercises 26 to 31.

It often happens that the theory behind an experiment indicates that the experimental data should lie approximately along a straight line of the form $y = mx + b$. The actual results, of course, never match the theory exactly. We are then faced with the problem of finding the straight line that best fits some set of experimental data $(x_1, y_1), \ldots, (x_n, y_n)$ as in Figure 4.R.1. If we guess at a straight line $y = mx + b$ to fit the data, each point will deviate vertically from the line by an amount $d_i = y_i - (mx_i + b)$.

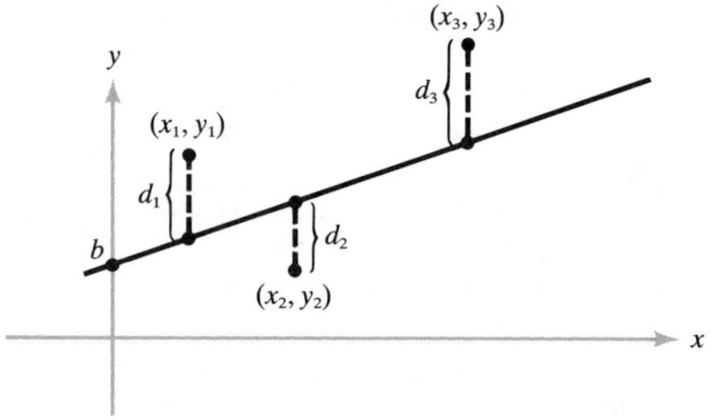

FIGURE 4.R.1
The method of least squares tries to find a straight line that best approximates a set of data.

We would like to choose m and b in such a way as to make the total effect of these deviations as small as possible. However, since some are negative and some positive, we could get a lot of cancellations and still have a pretty bad fit. This leads us to suspect that a better measure of the total error might be the sum of the squares of these deviations. So we are led to the problem of finding m and b that minimize the function

$$s = f(m, b) = d_1^2 + d_2^2 + \cdots + d_n^2 = \sum_{i=1}^{n} (y_i - mx_i - b)^2$$

where $x_1, \ldots, x_n$ and $y_1, \ldots, y_n$ are given data.

26. For each set of three data points, plot the points, write down the function $f(m, b)$ above, find m and b to give the best straight-line fit according to method of least squares, and plot the straight line.
    (a) $(x_1, y_1) = (1, 1)$
        $(x_2, y_2) = (2, 3)$
        $(x_3, y_3) = (4, 3)$
    (b) $(x_1, y_1) = (0, 0)$
        $(x_2, y_2) = (1, 2)$
        $(x_3, y_3) = (2, 3)$
27. Show that if only two data points (x_1, y_1) and (x_2, y_2) are given, this method method produces the line through (x_1, y_1) and (x_2, y_2).
28. Show that the equations for a critical point, $\partial s/\partial b = 0$ and $\partial s/\partial m = 0$, are equivalent to

$$m(\textstyle\sum x_i) + nb = (\sum y_i) \quad \text{and} \quad m(\sum x_i^2) + b(\sum x_i) = (\sum x_i y_i)$$

    where all the sums run from $i = 1$ to $i = n$.
29. If $y = mx + b$ is the best-fitting straight line to the data points $(x_1, y_1), \ldots, (x_n, y_n)$ according to the least squares method, show that

$$\sum_{i=1}^{n} (y_i - mx_i - b) = 0$$

    That is, the positive and negative deviations cancel (see Exercise 28).

30. Use the second derivative test to show that the critical point of f actually does produce a minimum.

31. Use the method of least squares to find the straight line that best fits the points (0, 1), (1, 3), (2, 2), (3, 4), and (4, 5). Plot the points and line.*

* The method of least squares may be varied and generalized in a number of ways. The basic idea can be applied to equations of more complicated curves than the straight line. This might be done to find, for example, the parabola that best fits a given set of data points. These ideas also form part of the basis for the development of the science of cybernetics by Norbert Wiener. Another version of the data is the following problem of least squares approximation: Given a function f defined and integrable on an interval $[a, b]$, find a polynomial P of degree $\leq n$ such that the mean-square error

$$\int_a^b |f(x) - P(x)|^2 \, dx$$

is as small as possible.

CHAPTER 5

INTEGRATION

The order of the operations "putting your socks on" and "putting your shoes on" is of practical importance.

JOHN ADDISON

In this chapter we shall study the integration of real-valued functions of several variables; we begin by considering integrals of functions of two variables, or *double integrals*, as they are called. The double integral has a basic geometric interpretation as volume, and can be defined rigorously as a limit of approximating sums. We shall present several techniques for evaluating double integrals, consider some applications, and then discuss improper integrals. Finally, we shall introduce integrals of functions of three variables, or *triple integrals*.

5.1 INTRODUCTION

In this introductory section, we shall briefly discuss some of the geometric aspects of the double integral, deferring a more rigorous discussion in terms of Riemann sums until Section 5.2.

Let us consider, then, a continuous function of two variables $f: R \subset \mathbb{R}^2 \to \mathbb{R}$ whose domain R is a rectangle with sides parallel to the coordinate axes. The rectangle R can be described in terms of the two closed intervals $[a, b]$ and $[c, d]$, representing the sides of R along the x- and y-axes respec-

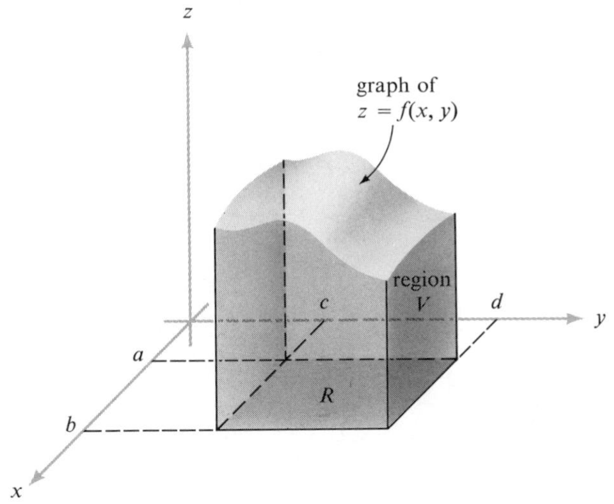

FIGURE 5.1.1
The region V in space is bounded by the graph of f, the rectangle R, and the four vertical sides indicated.

tively, as in Figure 5.1.1. In this case, we say that R is the *Cartesian product* of $[a, b]$ and $[c, d]$ and write $R = [a, b] \times [c, d]$.

Let us assume that $f(x, y) \geq 0$ on R. The graph of $z = f(x, y)$ is then a surface lying above the rectangle R. This surface, the rectangle R, and the four planes $x = a$, $x = b$, $y = c$, and $y = d$ form the boundary of a region V in space (see Figure 5.1.1). The problem of how to rigorously define the volume of V has to be faced, and we shall solve it by the classical method of exhaustion, or, in more modern terms, the method of Riemann sums, in Section 5.2. However in order to gain an intuitive grasp of this method, let us provisionally assume that the volume of a region has been defined. This is not unreasonable since we feel intuitively that every region has a volume. Then the volume of the region above R and under the graph of f is called the *(double) integral* of f over R and is denoted by

$$\int_R f, \qquad \int_R f(x, y)\, dA, \qquad \int_R f(x, y)\, dx\, dy, \quad \text{or} \quad \iint\limits_R f(x, y)\, dx\, dy$$

EXAMPLE 1. (a) If $f(x, y) = k$, where k is a positive constant, then $\int_R f(x, y)\, dA = k(b - a)(d - c)$, since the integral is equal to the volume of a rectangular box with base R and height k.

(b) If $f(x, y) = 1 - x$ and $R = [0, 1] \times [0, 1]$, then $\int_R f(x, y)\, dA = \frac{1}{2}$ since the integral is equal to the volume of the triangular solid shown in Figure 5.1.2.

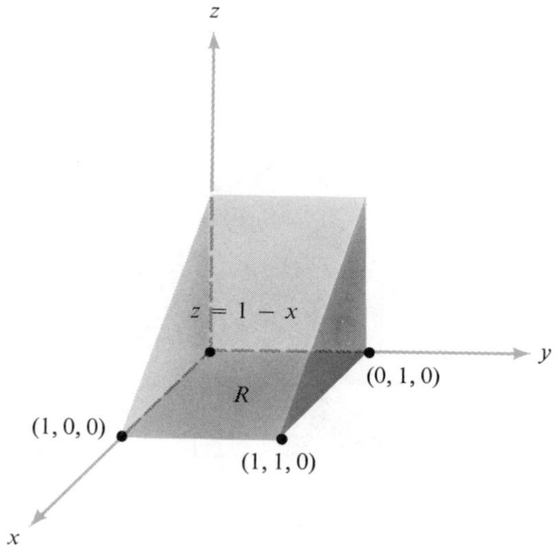

FIGURE 5.1.2
Volume under the graph $z = 1 - x$ over $R = [0, 1] \times [0, 1]$.

EXAMPLE 2. Suppose $z = f(x, y) = x^2 + y^2$ and $R = [-1, 1] \times [0, 1]$. Then the integral $\int_R f = \int_R (x^2 + y^2)\, dx\, dy$ is equal to the volume of the solid sketched in Figure 5.1.3. We shall compute this integral in Example 3.

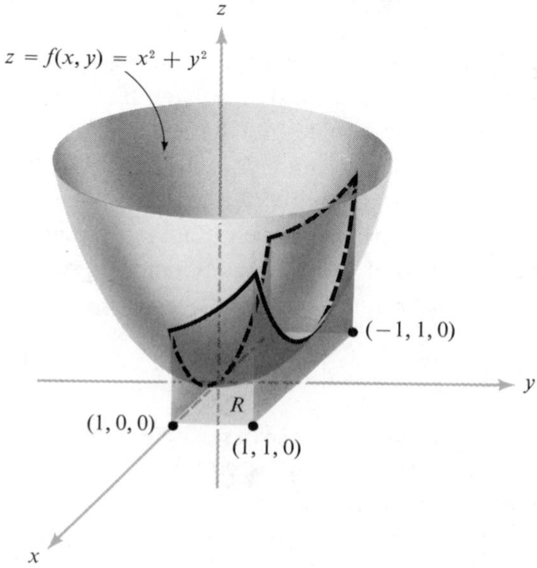

FIGURE 5.1.3
Volume under $z = x^2 + y^2$ over $R = [-1, 1] \times [0, 1]$.

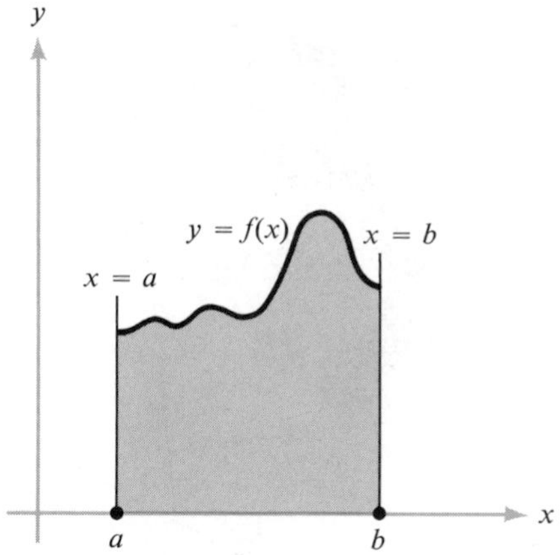

FIGURE 5.1.4
Area under the graph of a nonnegative continuous function f from $x = a$ to $x = b$ is $\int_a^b f(x)\,dx$.

These ideas are similar to the idea of a single integral $\int_a^b f(x)\,dx$, which represents the area under the graph of f if f is ≥ 0 and, say, continuous; see Figure 5.1.4.* We should also recall that $\int_a^b f(x)\,dx$ can be rigorously defined, without recourse to the area concept, as a limit of Riemann sums. Thus we can approximate $\int_a^b f(x)\,dx$ by choosing a partition $a = x_0 < x_1 < \cdots < x_n = b$ of $[a, b]$, selecting points $c_i \in [x_i, x_{i+1}]$, and forming the Riemann sum

$$\sum_{i=0}^{n-1} f(c_i)(x_{i+1} - x_i) \approx \int_a^b f(x)\,dx$$

(see Figure 5.1.5). We shall examine the analogous process for double integrals in the next section.

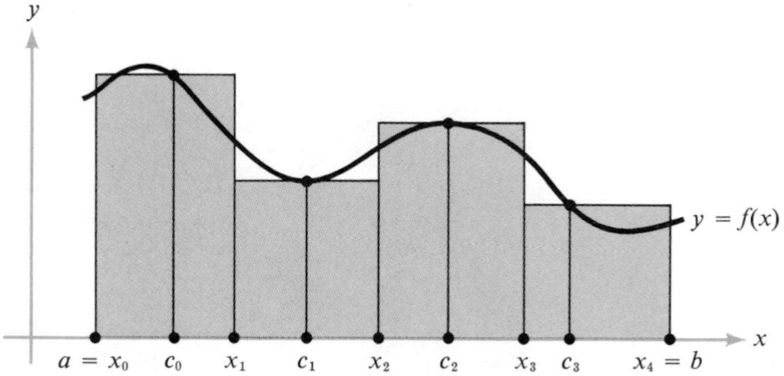

FIGURE 5.1.5
The sum of the areas of the shaded rectangles is a Riemann sum approximating the area under f from $x = a$ to $x = b$.

* Readers not already familiar with this idea should review the appropriate sections of their introductory calculus text.

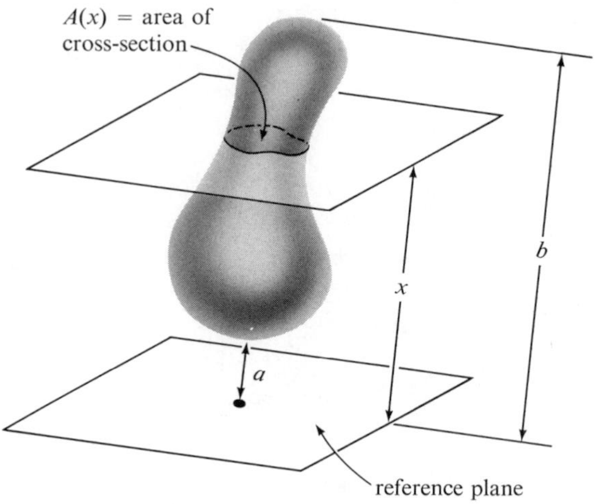

FIGURE 5.1.6

A solid body with cross-sectional area $A(x)$ at distance x from a reference plane.

There is a method for computing volumes known as *Cavalieri's Principle.* Suppose we have a solid body and we let $A(x)$ denote its cross-sectional area measured at a distance x from a reference plane (Figure 5.1.6). According to Cavalieri's Principle, the volume of the body is given by

$$\text{volume} = \int_a^b A(x)\,dx$$

where a and b are the minimum and maximum distances from the reference plane. This can be made intuitively clear. If we partition $[a, b]$ into $x_0 = a < x_1 < \cdots < x_n = b$, then an approximating Riemann sum for the above integral is

$$\sum_{i=0}^{n-1} A(c_i)(x_{i+1} - x_i)$$

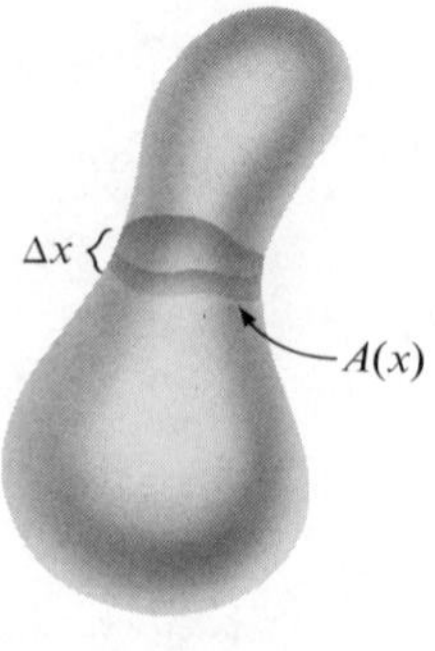

FIGURE 5.1.7

Volume of a slab with cross-sectional area $A(x)$ and thickness Δx equals $A(x)\,\Delta x$. The total volume of the body is $\int_a^b A(x)\,dx$.

But this sum also approximates the volume of the body, since $A(x)\,\Delta x$ is the volume of a slab with cross-sectional area $A(x)$ and thickness Δx (Figure 5.1.7). Therefore it is reasonable to accept the above formula for the volume. A more careful justification of this method is given below.

HISTORICAL NOTE

Bonaventura Cavalieri (1598–1647) was a pupil of Galileo and a professor in Bologna. His investigations into area and volume were important building blocks of the foundations of calculus. Although his methods were criticized by his contemporaries, similar ideas had been used by Archimedes in antiquity, and were later taken up by the "fathers" of calculus, Newton and Leibniz.

We now use Cavalieri's Principle to help us evaluate double integrals.

Consider the solid region under a graph $z = f(x, y)$ defined on the region $[a, b] \times [c, d]$, where f is continuous and greater than zero. There are two natural cross-sectional area functions: one obtained by using cutting planes perpendicular to the x-axis, and the other obtained by using cutting planes perpendicular to the y-axis. The cross section determined by a cutting plane $x = x_0$, of the first sort, is the plane region under the graph of $z = f(x_0, y)$ from $y = c$ to $y = d$ (Figure 5.1.8). When we fix $x = x_0$ we have the function

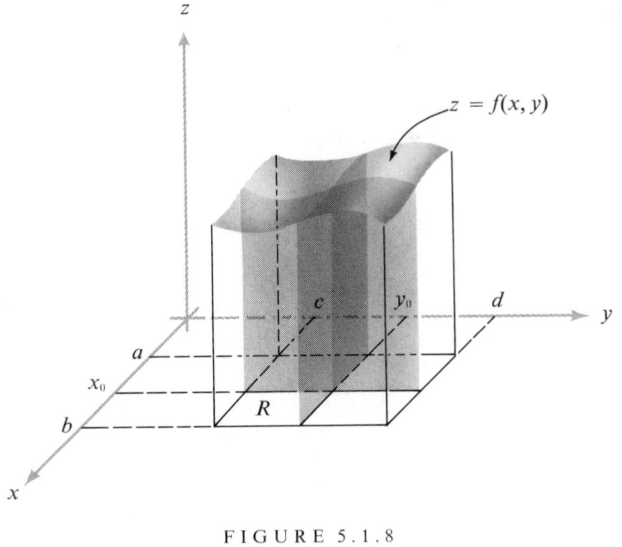

FIGURE 5.1.8

Two different cross sections sweeping out the volume under $z = f(x, y)$.

$y \mapsto f(x_0, y)$, which is continuous on $[c, d]$. The cross-sectional area $A(x_0)$ is, therefore, equal to the integral $\int_c^d f(x_0, y)\, dy$. Thus the cross-sectional area function A has domain $[a, b]$, and $A: x \mapsto \int_c^d f(x, y)\, dy$. By Cavalieri's Principle, the volume V of the region under $z = f(x, y)$ must be equal to

$$V = \int_a^b A(x)\, dx = \int_a^b \left[\int_c^d f(x, y)\, dy \right] dx$$

The integral $\int_a^b [\int_c^d f(x, y)\, dy]\, dx$ is known as an *iterated integral*, because it is obtained by integrating with respect to y and then integrating the result with respect to x. Since $\int_R f(x, y)\, dA$ is equal to the volume V,

$$\int_R f(x, y)\, dA = \int_a^b \left[\int_c^d f(x, y)\, dy \right] dx \tag{1}$$

If we reverse the roles of x and y in the above discussion and use cutting planes perpendicular to the y-axis, we obtain

$$\int_R f(x, y)\, dA = \int_c^d \left[\int_a^b f(x, y)\, dx \right] dy \tag{2}$$

The expression on the right of (2) is the iterated integral obtained by integrating with respect to x and then integrating the result with respect to y.

Thus, if our intuition about volumes is correct, formulas (1) and (2) ought to be valid. This is in fact true when the concepts we are discussing are defined rigorously, and is known as Fubini's Theorem. We shall give a proof of this theorem in the next section.

As the following examples illustrate, the notion of the iterated integral and equations (1) and (2) provide a powerful method for computing the double integral of a function of two variables.

EXAMPLE 3. Let $z = f(x, y) = x^2 + y^2$ and let $R = [-1, 1] \times [0, 1]$. Let us evaluate the integral $\int_R (x^2 + y^2)\, dx\, dy$. By equation (2), we have

$$\int_R (x^2 + y^2)\, dx\, dy = \int_0^1 \left[\int_{-1}^1 (x^2 + y^2)\, dx \right] dy$$

To find $\int_{-1}^1 (x^2 + y^2)\, dx$, we treat y as a constant and integrate with respect to x. Since $x \mapsto x^3/3 + y^2 x$ is an antiderivative of $x \mapsto x^2 + y^2$, we can integrate, using methods of elementary calculus, to obtain

$$\int_{-1}^1 (x^2 + y^2)\, dx = \left[\frac{x^3}{3} + y^2 x \right]_{x=-1}^1 = \frac{2}{3} + 2y^2$$

Next we integrate $y \mapsto \frac{2}{3} + 2y^2$ with respect to y from 0 to 1, to obtain

$$\int_0^1 \left(\frac{2}{3} + 2y^2 \right) dy = \left[\frac{2}{3} y + \frac{2}{3} y^3 \right]_{y=0}^1 = \frac{4}{3}$$

Hence the volume of the solid in Figure 5.1.3 is $\frac{4}{3}$. For completeness, let us evaluate $\int_R (x^2 + y^2)\,dx\,dy$ using (1)—that is, integrating with respect to y and then with respect to x. We have

$$\int_R (x^2 + y^2)\,dx\,dy = \int_{-1}^{1}\left[\int_0^1 (x^2 + y^2)\,dy\right]dx$$

Treating x as a constant in the y-integration, we obtain

$$\int_0^1 (x^2 + y^2)\,dy = \left[x^2 y + \frac{y^3}{3}\right]_{y=0}^{1} = x^2 + \frac{1}{3}$$

Next we evaluate $\int_{-1}^{1} (x^2 + \frac{1}{3})\,dx$ to obtain

$$\int_{-1}^{1}\left(x^2 + \frac{1}{3}\right)dx = \left[\frac{x^3}{3} + \frac{x}{3}\right]_{x=-1}^{1} = \frac{4}{3}$$

which agrees with our previous answer.

EXAMPLE 4. Compute $\int_s \cos x \sin y\,dx\,dy$ where S is the square $[0, \pi/2] \times [0, \pi/2]$ (see Figure 5.1.9). We have by equation (2)

$$\int_S \cos x \sin y\,dx\,dy = \int_0^{\pi/2}\left[\int_0^{\pi/2} \cos x \sin y\,dx\right]dy$$

$$= \int_0^{\pi/2} \sin y\,dy = 1$$

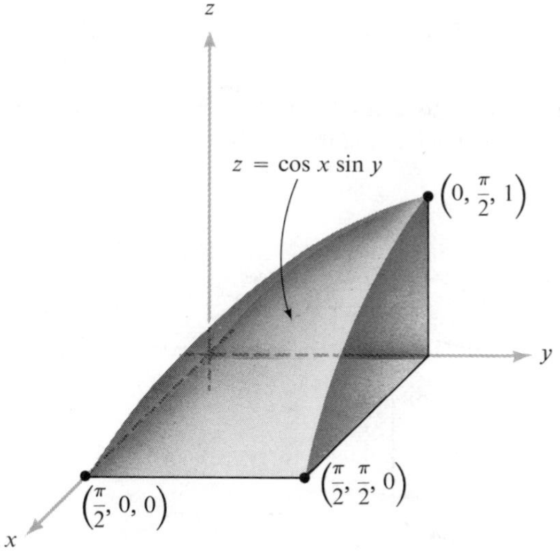

FIGURE 5.1.9
Volume under $z = \cos x \sin y$ *over the rectangle* $[0, \pi/2] \times [0, \pi/2]$.

In the next section, we shall use Riemann sums to rigoriously define the double integral for a wide class of functions of two variables without recourse to the notion of volume. Although we shall drop the requirement that $f(x, y) \geq 0$, equations (1) and (2) will remain valid. Therefore, the iterated integral will again provide the key to computing the double integral. In Section 5.3 we shall treat double integrals over regions more general than rectangles.

Finally, we remark that it is common to delete the brackets in iterated integrals such as (1) and (2) above and write

$$\int_a^b \int_c^d f(x, y)\, dy\, dx = \int_a^b \left[\int_c^d f(x, y)\, dy \right] dx$$

and

$$\int_c^d \int_a^b f(x, y)\, dx\, dy = \int_c^d \left[\int_a^b f(x, y)\, dx \right] dy$$

EXERCISES

1. Evaluate the following iterated integrals.
   (a) $\int_{-1}^{1} \int_0^1 (x^4 y + y^2)\, dy\, dx$ (b) $\int_0^{\pi/2} \int_0^1 (y \cos x + 2)\, dy\, dx$
   (c) $\int_0^1 \int_0^1 (xye^{x+y})\, dy\, dx$ (d) $\int_{-1}^{0} \int_1^2 (-x \log y)\, dy\, dx$
2. Evaluate the integrals in Exercise 1 by integrating with respect to x and then with respect to y.
3. Use Cavalieri's formula to show that the volumes of two cylinders with the same base and height are equal (see Figure 5.1.10).

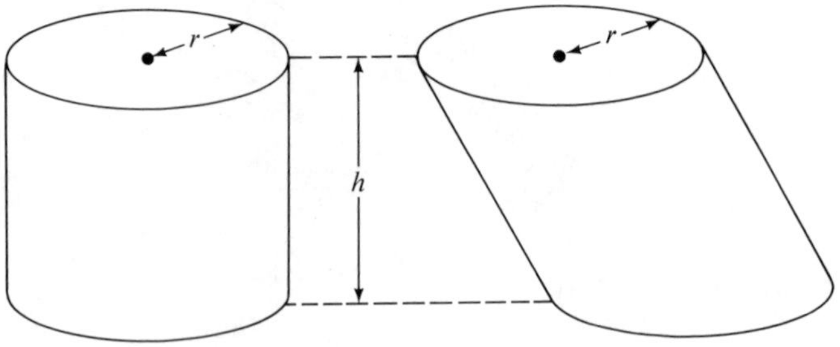

FIGURE 5.1.10
Two cylinders with same base and height have the same volume.

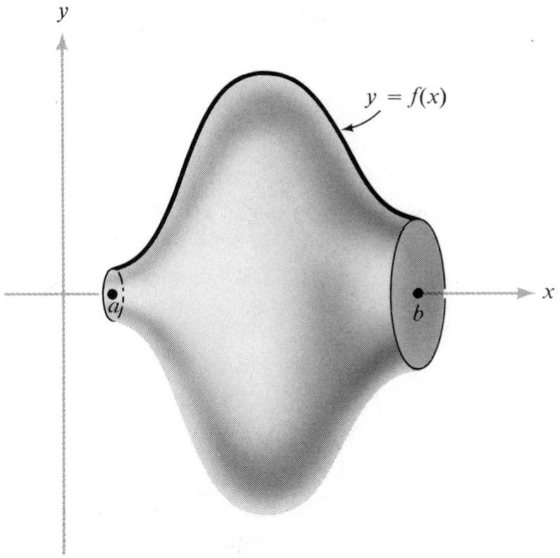

FIGURE 5.1.11
This solid of revolution has volume $\pi \int_a^b [f(x)]^2\,dx$.

4. (a) Argue that the volume of revolution shown in Figure 5.1.11 is

$$\pi \int_a^b [f(x)]^2\,dx$$

(b) Show that the volume of the region obtained by rotating the graph of the parabola $y = -x^2 + 2x + 3$, $-1 \le x \le 3$, about the x-axis is $512\pi/15$ (see Figure 5.1.12).

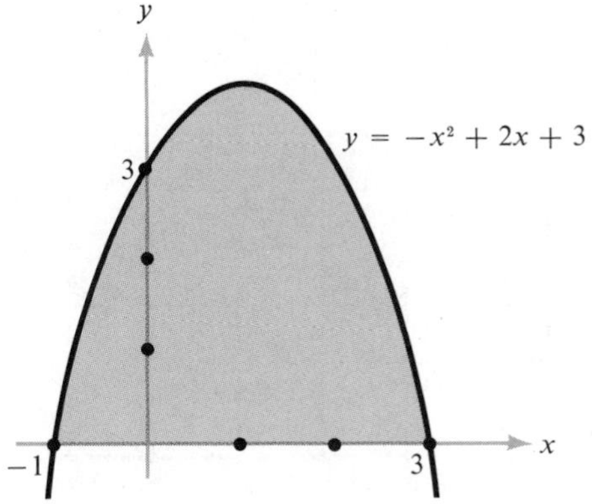

FIGURE 5.1.12
Region between the graph of $y = -x^2 + 2x + 3$ and the x-axis.

5. Evaluate the following double integrals where R is the rectangle $[0, 2] \times [-1, 0]$.
   (a) $\int_R (x^2y^2 + x)\, dy\, dx$
   (b) $\int_R (|y| \cos \frac{1}{4}\pi x)\, dy\, dx$
   (c) $\int_R (-xe^x \sin \frac{1}{2}\pi y)\, dy\, dx$
6. Find the volume bounded by the graph of $f(x, y) = 1 + 2x + 3y$, the rectangle $[1, 2] \times [0, 1]$, and the four vertical sides of the rectangle R as in Figure 5.1.1.
7. Repeat Exercise 6 for the surface $f(x, y) = x^4 + y^2$ and the rectangle $[-1, 1] \times [-3, -2]$.

5.2 THE DOUBLE INTEGRAL OVER A RECTANGLE

We are now ready to give a rigorous definition of the double integral as the limit of a sequence of sums. This will then be used to *define* the volume of the region under the graph of a function $f(x, y)$. We shall not require that $f(x, y) \geq 0$, but if $f(x, y)$ assumes negative values we shall not interpret the integral as a volume, just as we make a similar reservation for the area under the graph of a function of one variable. In addition, we shall discuss some of the fundamental algebraic properties of the double integral and prove Fubini's Theorem, which states that the double integral can be calculated as an iterated integral. To begin, let us establish some notation for partitions and sums.

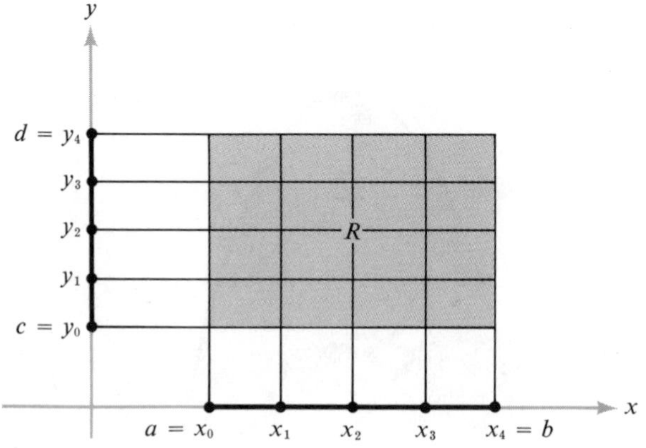

FIGURE 5.2.1
A regular partition of a rectangle R, with $n = 4$.

Consider a rectangle $R \subset \mathbb{R}^2$ that is the Cartesian product $R = [a, b] \times [c, d]$. By the *regular partition* of R of order n we mean the two collections of $n + 1$ equally spaced points $\{x_j\}_{j=0}^n$ *and* $\{y_k\}_{k=0}^n$ with

$$a = x_0 < x_1 < \cdots < x_n = b, \qquad c = y_0 < y_1 < \cdots < y_n = d$$

and

$$x_{j+1} - x_j = \frac{b-a}{n}, \qquad y_{k+1} - y_k = \frac{d-c}{n}$$

(see Figure 5.2.1).

Let R_{jk} be the rectangle $[x_j, x_{j+1}] \times [y_k, y_{k+1}]$, and let c_{jk} be any point in R_{jk}. Suppose $f: R \to \mathbb{R}$ is a bounded real-valued function. Form the sum

$$S_n = \sum_{j,\,k=0}^{n-1} f(c_{jk})\,\Delta x\,\Delta y = \sum_{j,\,k=0}^{n-1} f(c_{jk})\,\Delta A \tag{1}$$

where

$$\Delta x = x_{j+1} - x_j = \frac{b-a}{n}, \qquad \Delta y = y_{k+1} - y_k = \frac{d-c}{n}$$

and

$$\Delta A = \Delta x\,\Delta y$$

This sum is taken over all j's and k's from 0 to $n - 1$, so there are n^2 terms. A sum of this type is called a *Riemann sum* for f.

DEFINITION. *If the sequence $\{S_n\}$ converges to a limit S as $n \to \infty$, with the limit S being the same for any choice of points c_{jk} in the rectangles R_{jk}, then we say that f is* ***integrable*** *over R and we write*

$$\int_R f, \qquad \int_R f(x, y)\,dA, \qquad \int_R f(x, y)\,dx\,dy, \quad \text{or} \quad \iint_R f(x, y)\,dx\,dy$$

for the limit S.

Thus we can rewrite integrability in the following way:

$$\lim_{n\to\infty} \sum_{j,\,k=0}^{n-1} f(c_{jk})\,\Delta x\,\Delta y = \int_R f$$

for any choice of $c_{jk} \in R_{jk}$.

The proof of the following basic theorem is not difficult, but, because it requires some technicalities that are not essential to the main text, it is presented in Appendix B.

THEOREM 1. *Any continuous function defined on a rectangle R is integrable.*

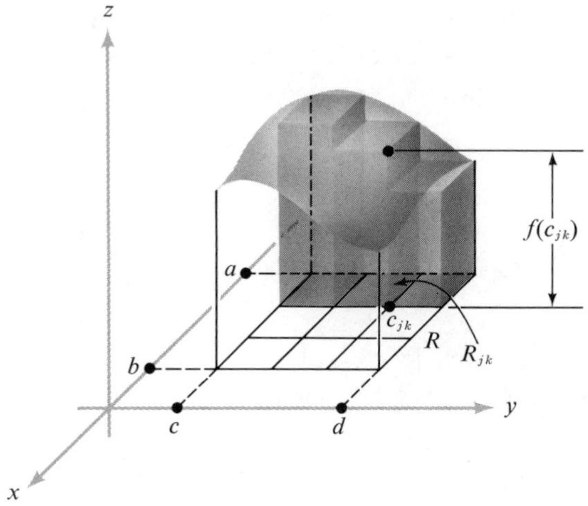

FIGURE 5.2.2
The sum of inscribed boxes approximates the volume under the graph of $z = f(x, y)$.

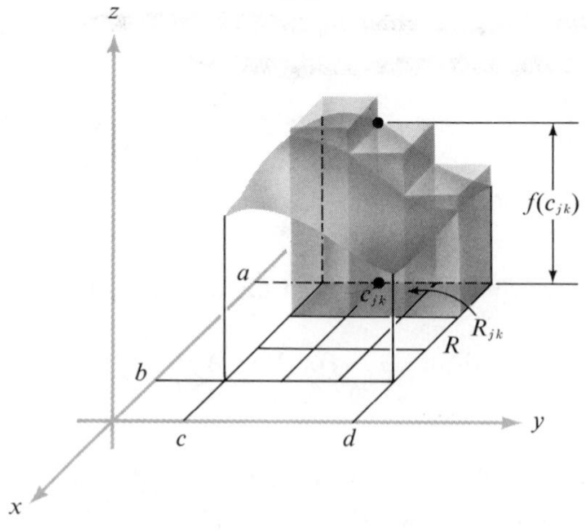

FIGURE 5.2.3
The volume of circumscribed boxes also approximates the volume under $z = f(x, y)$.

If $f(x, y) \geq 0$, the existence of $\lim_{n \to \infty} S_n$ has a straightforward geometric meaning. Consider the graph of $z = f(x, y)$ as the top of a solid whose base is the rectangle R. If we take each c_{jk} to be a point where $f(x, y)$ has its minimum value* on R_{jk}, then $f(c_{jk})\,\Delta x\,\Delta y$ represents the volume of a rectangular box with base R_{jk}. The sum $\sum_{j,k=0}^{n-1} f(c_{jk})\,\Delta x\,\Delta y$ equals the volume of an inscribed solid, part of which is shown in Figure 5.2.2. Similarly, if c_{jk} is a point where $f(x, y)$ has its maximum on R_{jk}, then the sum $\sum_{j,k=0}^{n-1} f(c_{jk})\,\Delta x\,\Delta y$ is equal to the volume of a circumscribed solid (see Figure 5.2.3). Therefore, if $\lim_{n \to \infty} S_n$ exists and is independent of $c_{jk} \in R_{jk}$, it follows that the volumes of the inscribed and circumscribed solids approach the same limit as $n \to \infty$. It is therefore reasonable to call this limit the exact volume of the solid under the graph of f. Thus the method of Riemann sums supports the concepts introduced on an intuitive basis in Section 5.1.

There is a theorem guaranteeing the existence of the integral of certain discontinuous functions as well. We shall need this result in the next section in order to discuss the integral of functions over regions more general than rectangles.

We shall be interested in functions whose discontinuities comprise curves in the xy-plane. Figure 5.2.4 shows two functions defined on a rectangle R

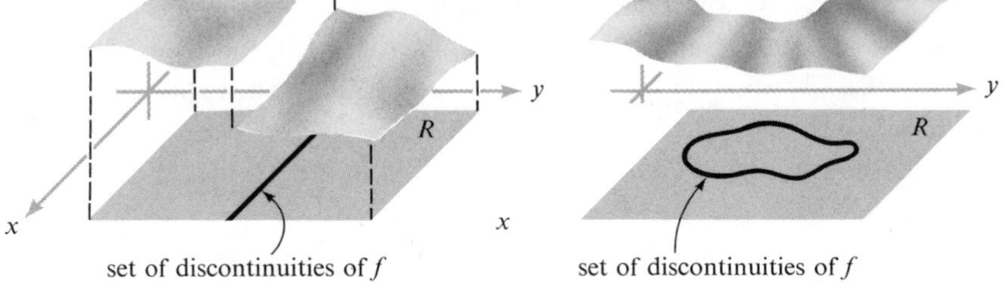

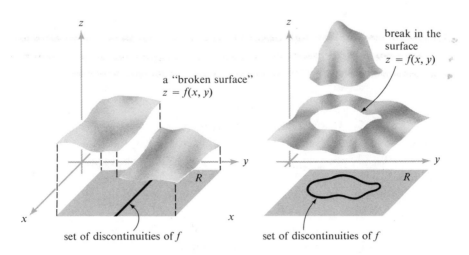

FIGURE 5.2.4
What the graphs of discontinuous functions of two variables might look like.

* Such c_{jk} exist by virtue of the continuity of f on R, but we shall not prove this fact.

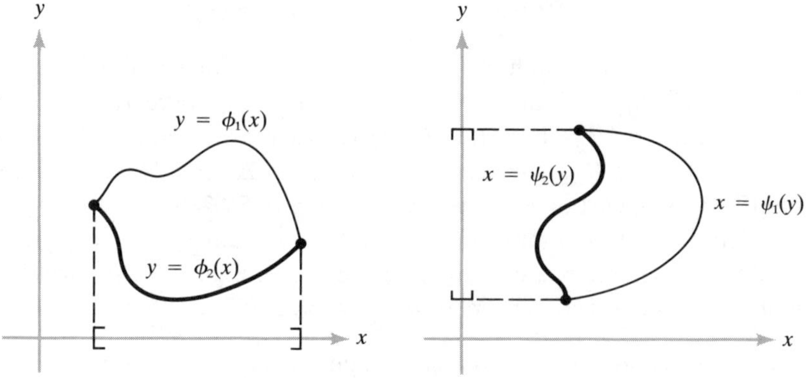

FIGURE 5.2.5
Curves in the plane written as graphs.

whose discontinuities lie along curves. In other words, f is continuous at each point that is in R but not on the curve. Curves we will find useful will be graphs of functions $y = \phi(x)$, $a \leq x \leq b$ or $x = \psi(y)$, $c \leq y \leq d$ or finite unions of such graphs. Some examples are shown in Figure 5.2.5. The next theorem provides an important criterion for determining if a function is integrable. The proof is discussed in Appendix B.

THEOREM 2. *Let $f: R \to \mathbb{R}$ be a bounded real-valued function on the rectangle R, and suppose that the set of points where f is discontinuous comprise a finite union of graphs of continuous functions. Then f is integrable over R.*

Recall that a function is *bounded* if there is a number $M > 0$ such that $-M \leq f(x, y) \leq M$ for all (x, y) in the domain of f. A continuous function on a closed rectangle is always bounded, but, for example, $f(x, y) = 1/x$ on $]0, 1] \times [0, 1]$ is not bounded, because $1/x$ becomes arbitrarily large for x near 0.

Using Theorem 2 and the remarks preceding it, we see that the functions sketched in Figure 5.2.4 are integrable over R, since these functions are bounded and continuous except on graphs of continuous functions.

Geometrically, Theorem 2 implies that if a nonnegative function f is not "too badly behaved," then the volumes of the circumscribed and inscribed solids will approximate the "true" volume under its graph (see Figure 5.2.6).

From the definition of the integral as a limit of sums and the limit theorems, we can deduce some fundamental properties of the integral $\int_R f(x, y)\, dA$; these properties are essentially the same as for the integral of a real-valued function of a single variable.

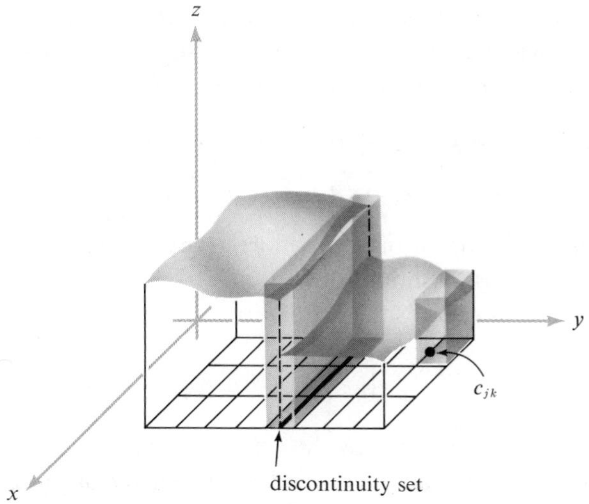

FIGURE 5.2.6
Graph of a discontinuous function and two circumscribing boxes.

Let f and g be integrable functions on the rectangle R, and let c be a constant. Then $f + g$ and cf are integrable, and

(*i*) (*linearity*)
$\int_R [f(x, y) + g(x, y)]\, dA = \int_R f(x, y)\, dA + \int_R g(x, y)\, dA$

(*ii*) (*homogeneity*) $\int_R cf(x, y)\, dA = c \int_R f(x, y)\, dA$

(*iii*) (*monotonicity*) If $f(x, y) \geq g(x, y)$, then

$$\int_R f(x, y)\, dA \geq \int_R g(x, y)\, dA$$

(*iv*) (*additivity*) If R_i, $i = 1, \ldots, m$, are pairwise disjoint rectangles such that f is integrable over each R_i and if $Q = R_1 \cup R_2 \cup \cdots \cup R_m$ is a rectangle, then a bounded function $f: Q \to \mathbb{R}$ is integrable over Q and

$$\int_Q f(x, y)\, dA = \sum_{i=1}^{m} \int_{R_i} f(x, y)\, dA$$

Properties (*i*) and (*ii*) are a consequence of the definition of the integral as a limit of a sum and the following facts for convergent sequences $\{S_n\}$ and $\{T_n\}$, which are proved in texts for one-variable calculus:

$$\operatorname*{limit}_{n\to\infty} (T_n + S_n) = \operatorname*{limit}_{n\to\infty} T_n + \operatorname*{limit}_{n\to\infty} S_n$$

$$\operatorname*{limit}_{n\to\infty} (cS_n) = c \operatorname*{limit}_{n\to\infty} S_n$$

To demonstrate monotonicity we first observe that if $h(x, y) \geq 0$ and $\{S_n\}$ is a sequence of Riemann sums that converges to $\int_R h(x, y)\, dA$, then $S_n \geq 0$ for all n, so that $\int_R h(x, y)\, dA = \lim_{n\to\infty} S_n \geq 0$. If $f(x, y) \geq g(x, y)$ for all $(x, y) \in R$, then $(f - g)(x, y) \geq 0$ for all (x, y) and using (*i*) and (*ii*), we have

$$\int_R f(x, y)\, dA - \int_R g(x, y)\, dA = \int_R [f(x, y) - g(x, y)]\, dA \geq 0$$

This proves (*iii*). The proof of (*iv*) is more technical and a special case is proved in Appendix B. It should be intuitively obvious.

Another important result is the inequality

$$\left|\int_R f\right| \leq \int_R |f| \tag{2}$$

To see why (2) is true, note that, by the definition of absolute value, $-|f| \leq f \leq |f|$; so from the monotonicity and homogeneity of integration (with $c = -1$)

$$-\int_R |f| \leq \int_R f \leq \int_R |f|$$

which is equivalent to formula (2).

Although we have noted the integrability of a variety of functions, we have not yet established rigorously a general method of computing integrals. In the case of one variable we avoid computing $\int_a^b f(x)\, dx$ from its definition as a limit of a sum by using the Fundamental Theorem of Integral Calculus. Let us recall this important theorem tells us that

if f is continuous, then

$$\int_a^b f(x)\, dx = F(b) - F(a)$$

where F is an antiderivative of f; that is, $F' = f$.

This technique will not work as stated for functions $f(x, y)$ of two variables. However, as we indicated in Section 5.1, we can often reduce a double integral over a rectangle to iterated single integrals; the Fundamental Theorem then applies to these single integrals. Fubini's Theorem, which was mentioned in the last section, establishes this reduction to iterated integrals rigorously, by using Riemann sums. As we saw in Section 5.1 this reduction

$$\int_R f(x, y)\, dA = \int_a^b \left[\int_c^d f(x, y)\, dy\right] dx$$

$$= \int_c^d \left[\int_a^b f(x, y)\, dx\right] dy$$

is a consequence of Cavalieri's Principle, at least if $f(x, y) \geq 0$. In terms of Riemann sums, it corresponds to the following equality

$$\begin{aligned}\sum_{j,\,k=0}^{n-1} f(c_{jk})\,\Delta x\,\Delta y &= \sum_{j=0}^{n-1}\left(\sum_{k=0}^{n-1} f(c_{jk})\,\Delta y\right)\Delta x \\ &= \sum_{k=0}^{n-1}\left(\sum_{j=0}^{n-1} f(c_{jk})\,\Delta x\right)\Delta y\end{aligned}$$

which may be proved more generally as follows: *Let* $[a_{jk}]$ *be an* $n \times n$ *matrix,* $0 \leq j \leq n-1$, $0 \leq k \leq n-1$. *Let* $\sum_{j,\,k=0}^{n-1} a_{jk}$ *be the sum of the* n^2 *matrix entries Then*

$$\sum_{j,\,k=0}^{n-1} a_{jk} = \sum_{j=0}^{n-1}\left(\sum_{k=0}^{n-1} a_{jk}\right) = \sum_{k=0}^{n-1}\left(\sum_{j=0}^{n-1} a_{jk}\right) \tag{3}$$

In the first equality, the right-hand side represents summing the matrix entries by rows:

$$\left(\begin{array}{ccccccc} a_{00} & a_{01} & a_{02} & \cdots & & & a_{0(n-1)} \\ \vdots & & & & & & \vdots \\ a_{j0} & a_{j1} & & \cdots & a_{jk} & \cdots & a_{j(n-1)} \\ \vdots & & & & & & \vdots \\ a_{(n-1)0} & a_{(n-1)1} & & \cdots & a_{(n-1)k} & & a_{(n-1)(n-1)} \end{array}\right) \begin{array}{l} \to \sum_{k=0}^{n-1} a_{0k} \\ \quad\vdots \\ \to \sum_{k=0}^{n-1} a_{jk} \\ \quad\vdots \\ \to \sum_{k=0}^{n-1} a_{(n-1)k} \\ \hline \sum_{j=0}^{n-1}\left(\sum_{k=0}^{n-1} a_{jk}\right) \end{array}$$

Clearly this is equal to $\sum_{j,\,k=0}^{n-1} a_{jk}$, that is, the sum of all the a_{jk}'s. Similarly $\sum_{k=0}^{n-1} (\sum_{j=0}^{n-1} a_{jk})$ represents a summing of the matrix entries by columns. This establishes (3) and makes the reduction to iterated integrals quite plausible if we remember that integrals can be approximated by the corresponding Riemann sums. The actual proof of Fubini's Theorem exploits this idea.

Before we proceed to the proof, it may be hepful to recall how Cavalieri's Principle makes plausible the formula

$$\int_R f(x, y)\,dA = \int_a^b \left[\int_c^d f(x, y)\,dy\right] dx = \int_c^d \left[\int_a^b f(x, y)\,dx\right] dy \tag{4}$$

If we slice up the volume under the graph of f into slabs parallel to the y-axis, then we can see that the total volume under the graph is approximately

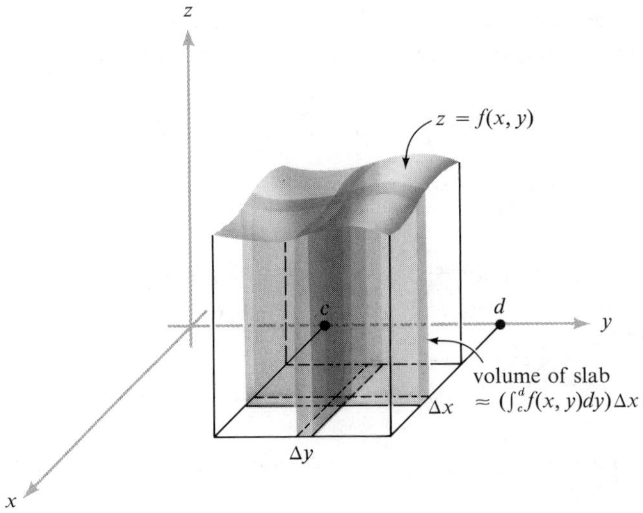

FIGURE 5.2.7
Geometric interpretation of the iterated integral.

equal to the sum of the quantities $[\int_c^d f(x, y)\, dy]\, \Delta x$; that is, we have $\int_R f(x, y)\, dA = \int_a^b [\int_c^d f(x, y)\, dy]\, dx$. Similarly, the second equality above is proved by slicing the volume into slabs parallel to the x-axis (see Figure 5.2.7).

THEOREM 3 (FUBINI'S THEOREM). *Let f be a continuous function with domain a rectangle $R = [a, b] \times [c, d]$. Then*

$$\int_a^b \int_c^d f(x, y)\, dy\, dx = \int_c^d \int_a^b f(x, y)\, dx\, dy = \int_R f(x, y)\, dA \qquad (4')$$

OPTIONAL

Proof. We shall first show that

$$\int_a^b \int_c^d f(x, y)\, dy\, dx = \int_R f(x, y)\, dA$$

Let $c = y_0 < y_1 < \cdots < y_n = d$ be a partition of $[c, d]$ into n equal parts. Define

$$F(x) = \int_c^d f(x, y)\, dy$$

then

$$F(x) = \sum_{k=0}^{n-1} \int_{y_k}^{y_{k+1}} f(x, y)\, dy$$

OPTIONAL (*Continued*)

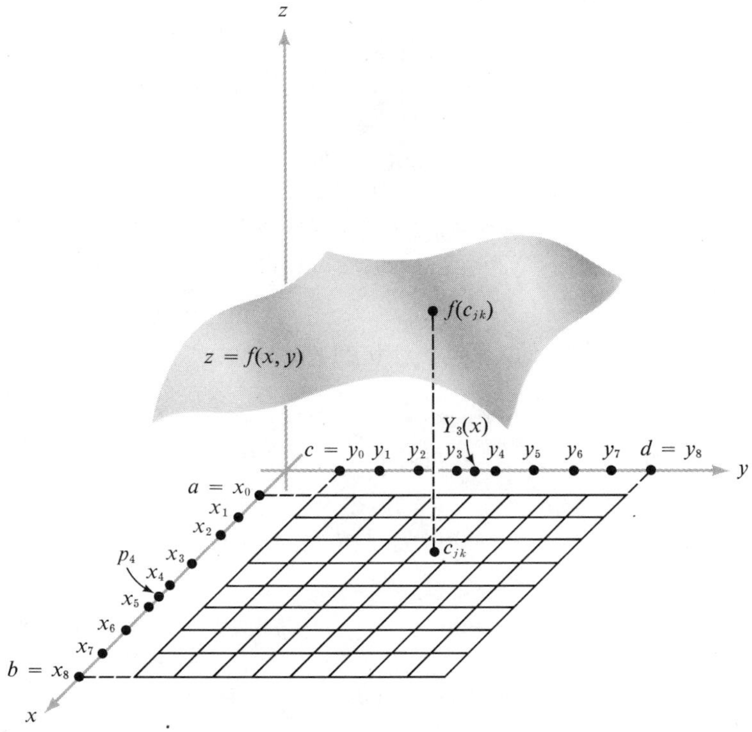

FIGURE 5.2.8

The notation needed in the proof of Fubini's Theorem; $n = 8$.

Using the integral version of the Mean Value Theorem,* for each fixed x and for each k we have (see Figure 5.2.8)

$$\int_{y_k}^{y_{k+1}} f(x, y)\, dy = f(x, Y_k(x))(y_{k+1} - y_k)$$

where the point $Y_k(x)$ belongs to $[y_k, y_{k+1}]$ and may depend on x and n. We have thus shown that

$$F(x) = \sum_{k=0}^{n-1} f(x, Y_k(x))(y_{k+1} - y_k)$$

* This states that if $g(x)$ is continuous on $[a, b]$, $\int_a^b g(x)\, dx = g(c)(b - a)$ for some point $c \in [a, b]$. The more general second mean value theorem was proved in Section 4.1 (see p. 203).

OPTIONAL (*Continued*)

Now by the definition of the integral in one variable as a limit of Riemann sums

$$\int_a^b F(x)\,dx = \int_a^b \left[\int_c^d f(x, y)\,dy\right] dx$$

$$= \lim_{n\to\infty} \sum_{j=0}^{n-1} F(p_j)(x_{j+1} - x_j)$$

where $a = x_0 < x_1 < \cdots < x_n = b$ is a partition of the interval $[a, b]$ into n equal parts and p_j is any point in $[x_j, x_{j+1}]$. Setting $c_{jk} = (p_j, Y_k(p_j)) \in R_{jk}$, we have (substituting p_j for x above)

$$F(p_j) = \sum_{k=0}^{n-1} f(c_{jk})(y_{k+1} - y_k)$$

Therefore

$$\int_a^b \int_c^d f(x, y)\,dy\,dx = \int_a^b F(x)\,dx$$

$$= \lim_{n\to\infty} \sum_{j=0}^{n-1} F(p_j)(x_{j+1} - x_j)$$

$$= \lim_{n\to\infty} \sum_{j=0}^{n-1} \sum_{k=0}^{n-1} f(c_{jk})(y_{k+1} - y_k)(x_{j+1} - x_j)$$

$$= \int_R f(x, y)\,dA$$

Thus we have proved that

$$\int_a^b \int_c^d f(x, y)\,dy\,dx = \int_R f(x, y)\,dA$$

By the same reasoning we can show that

$$\int_c^d \int_a^b f(x, y)\,dx\,dy = \int_R f(x, y)\,dA$$

These two conclusions are exactly what we wanted to prove. ∎

Fubini's Theorem can be generalized to the case where f is not necessarily continuous. Although we shall not present a proof, we state here this more general version.

THEOREM 3′ (FUBINI'S THEOREM). *Let f be a bounded function with domain a rectangle $R = [a, b] \times [c, d]$, and suppose the discontinuities of f form a finite union of graphs of continuous functions. If*

$$\int_c^d f(x, y)\,dy \quad \textit{exists for each} \quad x \in [a, b]$$

then

$$\int_a^b \left[\int_c^d f(x, y)\, dy\right] dx \quad \textit{exists}$$

and

$$\int_a^b \int_c^d f(x, y)\, dy\, dx = \int_R f(x, y)\, dA$$

Similarly, if

$$\int_a^b f(x, y)\, dx \quad \textit{exists for each} \quad y \in [c, d]$$

then

$$\int_c^d \left[\int_a^b f(x, y)\, dx\right] dy \quad \textit{exists}$$

and

$$\int_c^d \int_a^b f(x, y)\, dx\, dy = \int_R f(x, y)\, dA$$

Thus, if all these conditions hold simultaneously,

$$\int_a^b \int_c^d f(x, y)\, dy\, dx = \int_c^d \int_a^b f(x, y)\, dx\, dy = \int_R f(x, y)\, dA$$

The assumptions made for this version of Fubini's Theorem are more complicated than those we made in Theorem 3. They are necessary because if f is not continuous everywhere, for example, there is no guarantee that $\int_c^d f(x, y)\, dy$ will exist for each x.

EXAMPLE 3. Compute $\int_R (x^2 + y)\, dA$ where R is the square $[0, 1] \times [0, 1]$. By Fubini's Theorem,

$$\int_R (x^2 + y)\, dA = \int_0^1 \int_0^1 (x^2 + y)\, dx\, dy = \int_0^1 \left[\int_0^1 (x^2 + y)\, dx\right] dy$$

By the Fundamental Theorem of Integral Calculus the x-integration may be performed:

$$\int_0^1 (x^2 + y)\, dx = \left[\frac{x^3}{3} + yx\right]_{x=0}^1 = \frac{1}{3} + y$$

Thus

$$\int_R (x^2 + y)\, dA = \int_0^1 \left[\frac{1}{3} + y\right] dy = \left[\frac{1}{3}y + \frac{y^2}{2}\right]_0^1 = \frac{5}{6}$$

What we have done is hold y fixed, integrate with respect to x, and then evaluate the result between the given limits for the x-variable. Next we integrated the remaining function (of y alone) with respect to y to obtain the final answer.

A consequence of Fubini's Theorem is that interchanging the order of integration in the iterated integrals does not change the answer. Let us verify this for the above example. We have

$$\int_0^1 \int_0^1 (x^2 + y)\,dy\,dx = \int_0^1 \left[x^2y + \frac{y^2}{2}\right]_{y=0}^1 dx = \int_0^1 \left[x^2 + \frac{1}{2}\right] dx$$
$$= \left[\frac{x^3}{3} + \frac{x}{2}\right]_0^1 = \frac{5}{6}$$

We have seen that when $f(x, y) \geq 0$ on $R = [a, b] \times [c, d]$, the integral $\int_R f(x, y)\,dA$ can be interpreted as a volume. If the function also takes on negative values, then the double integral can be thought of as the sum of all volumes lying between the surface $z = f(x, y)$ and the plane $z = 0$, bounded by the planes $x = a$, $x = b$, $y = c$ and $y = d$; here the volumes above $z = 0$ are counted as positive and those below as negative. However, Fubini's Theorem as stated remains valid in the case where $f(x, y)$ is negative or changes sign on R; that is, there is no restriction on the sign of f in the hypotheses of the theorem.

EXAMPLE 4. Let R be the rectangle $[-2, 1] \times [0, 1]$ and let $f(x, y) = y(x^3 - 12x)$; $f(x, y)$ takes on both positive and negative values on R. Evaluate the integral $\int_R f(x, y)\,dx\,dy = \int_R y(x^3 - 12x)\,dx\,dy$.

By Fubini's Theorem, we may write

$$\int_R y(x^3 - 12x)\,dx\,dy = \int_0^1 \left[\int_{-2}^1 y(x^3 - 12x)\,dx\right] dy = \frac{57}{4}\int_0^1 y\,dy = \frac{57}{8}$$

Alternatively, integrating first with respect to y, we find

$$\int_R y(x^3 - 12x)\,dy\,dx = \int_{-2}^1 \left[\int_0^1 (x^3 - 12x)y\,dy\right] dx$$
$$= \frac{1}{2}\int_{-2}^1 (x^3 - 12x)\,dx$$
$$= \frac{1}{2}\left[\frac{x^4}{4} - 6x^2\right]_{-2}^1 = \frac{57}{8}$$

HISTORICAL NOTE

Although Theorem 3 on the equality of iterated integrals is named after the Italian mathematician Guido Fubini (1879–1943), who proved a very general result of this type in 1907, it was certainly known to Cauchy and his contemporaries that equality held for continuous functions. Cauchy was the first to show that equality need not hold when f is unbounded, and, somewhat later, examples were also found of bounded functions where equality does not hold.

EXERCISES

1. Evaluate each of the following integrals if $R = [0, 1] \times [0, 1]$.
   (a) $\int_R (x^3 + y^2)\, dA$
   (b) $\int_R ye^{xy}\, dA$
   (c) $\int_R (xy)^2 \cos x^3\, dA$
2. Evaluate each of the following integrals if $R = [0, 1] \times [0, 1]$.
   (a) $\int_R (x^m y^n)\, dx\, dy$
   (b) $\int_R (ax + by + c)\, dx\, dy$
   (c) $\int_R \sin(x + y)\, dx\, dy$
3. Let f be continuous, $f \geq 0$, on the rectangle R. If $\int_R f\, dA = 0$, prove that $f = 0$ on R.
4. Compute the volume of the solid bounded by the xz-plane, the yz-plane, the xy-plane, the planes $x = 1$ and $y = 1$, and the surface $z = x^2 + y^4$.
5. Let f be continuous on $[a, b]$ and g continuous on $[c, d]$. Show that

$$\int_R [f(x)g(y)]\, dx\, dy = \left[\int_a^b f(x)\, dx\right]\left[\int_c^d g(y)\, dy\right]$$

   where $R = [a, b] \times [c, d]$.
6. Compute the volume of the solid bounded by $z = \sin y, 0 \leq y \leq \pi/2, 0 \leq x \leq 1$, and the xy-plane.
7. Compute the volume of the solid bounded by $z = x^2 + y$, the rectangle $R = [0, 1] \times [1, 2]$, and the "vertical sides" of R.
8. Let f be continuous on $R = [a, b] \times [c, d]$; for $a < x < b$, $c < y < d$, define

$$F(x, y) = \int_a^x \int_c^y f(u, v)\, dv\, du$$

   Show that $\partial^2 F/\partial x\, \partial y = \partial^2 F/\partial y\, \partial x = f(x, y)$. Use this example to discuss the relationship between Fubini's Theorem and the equality of mixed partial derivatives (see Section 2.6).

*9. Let $f\colon [0, 1] \times [0, 1] \to \mathbb{R}$ be defined by

$$f(x, y) = \begin{cases} 1 & x \text{ rational} \\ 2y & x \text{ irrational} \end{cases}$$

   Show that the iterated integral $\int_0^1 \left[\int_0^1 f(x, y)\, dy\right] dx$ exists but f is not integrable.

*10. Express $\int_R \cosh xy\, dx\, dy$ as a convergent series, where $R = [0, 1] \times [0, 1]$.

*11. Although Fubini's Theorem holds for most functions met in practice, one must still exercise some caution. It certainly does not hold for all functions. For example, one could divide the unit square into infinitely many rectangles of the form $[1/(m + 1), 1/m] \times [1/(n + 1), 1/n]$, as in Figure 5.2.9. Define f in such a way that the volume under the graph of f over each rectangle takes on values according to

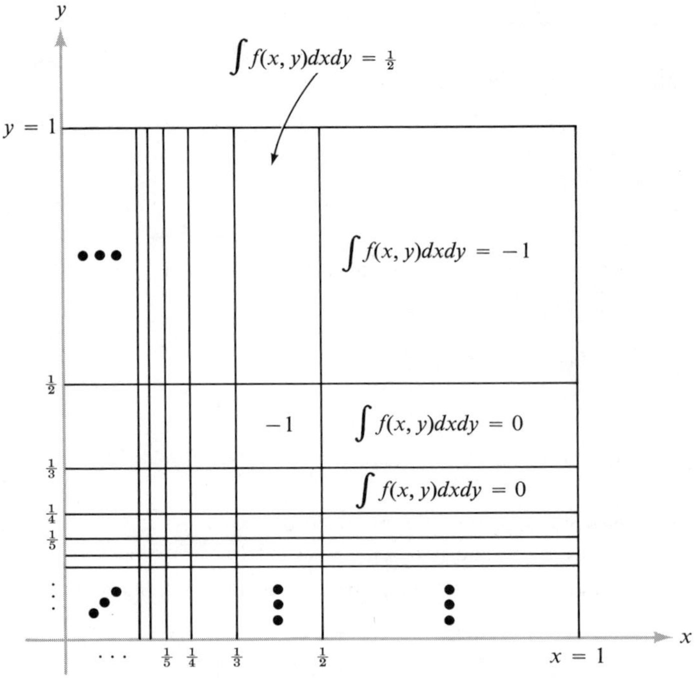

FIGURE 5.2.9

Construction of a function that does not satisfy Fubini's Theorem (Exercise 11).

the following table:

$\cdots$	$\frac{1}{32}$	$\frac{1}{16}$	$\frac{1}{8}$	$\frac{1}{4}$	$\frac{1}{2}$	-1
$\cdots$	$\frac{1}{16}$	$\frac{1}{8}$	$\frac{1}{4}$	$\frac{1}{2}$	-1	0
$\cdots$	$\frac{1}{8}$	$\frac{1}{4}$	$\frac{1}{2}$	-1	0	0
$\cdots$	$\frac{1}{4}$	$\frac{1}{2}$	-1	0	0	0
	$\vdots$	$\vdots$	$\vdots$	$\vdots$	$\vdots$	$\vdots$

Define f to be zero at $(0, 0)$. Each row adds to zero, so adding rows and then columns gives a result of zero. On the other hand the columns add to

$$\cdots \quad -\tfrac{1}{32} \quad -\tfrac{1}{16} \quad -\tfrac{1}{8} \quad -\tfrac{1}{4} \quad -\tfrac{1}{2} \quad -1$$

so adding columns and then rows gives a result of -2. Why does Fubini's Theorem not hold for this function?

5.3 THE DOUBLE INTEGRAL OVER MORE GENERAL REGIONS

Our goal in this section is twofold; first we wish to define the integral $\int_D f(x, y)\, dA$ on regions D more general than rectangles, and second, we want to develop a technique for evaluating this type of integral. To accomplish this, we shall define three special types of subsets of the xy-plane, and then extend the notion of the double integral to them.

Suppose we are given two continuous real-valued functions $\phi_1: [a, b] \to \mathbb{R}$, $\phi_2: [a, b] \to \mathbb{R}$ that satisfy $\phi_2(t) \le \phi_1(t)$ for all $t \in [a, b]$. Let D be the set of all points (x, y) such that

$$x \in [a, b], \qquad \phi_2(x) \le y \le \phi_1(x)$$

This region D is said to be of *type* 1. Figure 5.3.1 shows various examples of regions of type 1. The curves and straight line segments that bound the region taken together constitute the *boundary* of D, denoted ∂D.

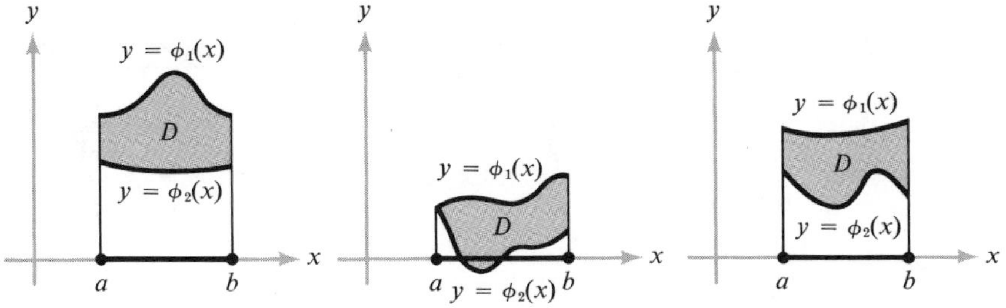

FIGURE 5.3.1
Some regions of type 1.

We say that a region D is of *type* 2 if there are continuous functions $\phi_1, \phi_2: [c, d] \to \mathbb{R}$ such that D is the set of points (x, y) satisfying

$$y \in [c, d], \qquad \phi_2(y) \le x \le \phi_1(y)$$

where $\phi_2(t) \le \phi_1(t)$, $t \in [c, d]$. Again, the curves that bound the region D constitute its boundary ∂D. Some examples of type 2 regions are shown in Figure 5.3.2.

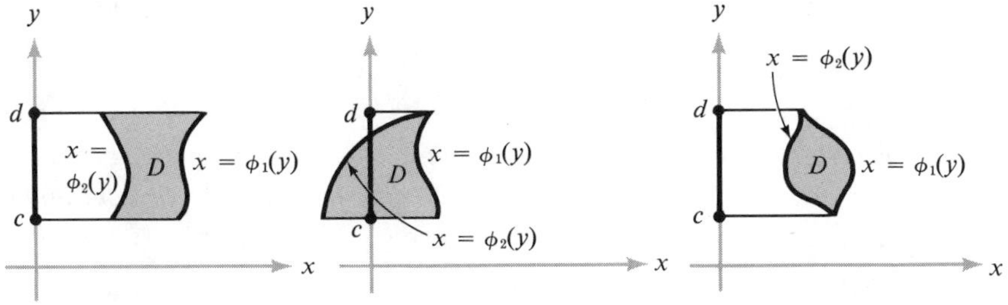

FIGURE 5.3.2
Some regions of type 2.

Finally, a region of *type* 3 is one that is both type 1 and type 2: That is, the region can be described both as a region of type 1 and as a region of type 2. An example of a type 3 region is the unit disc (Figure 5.3.3).

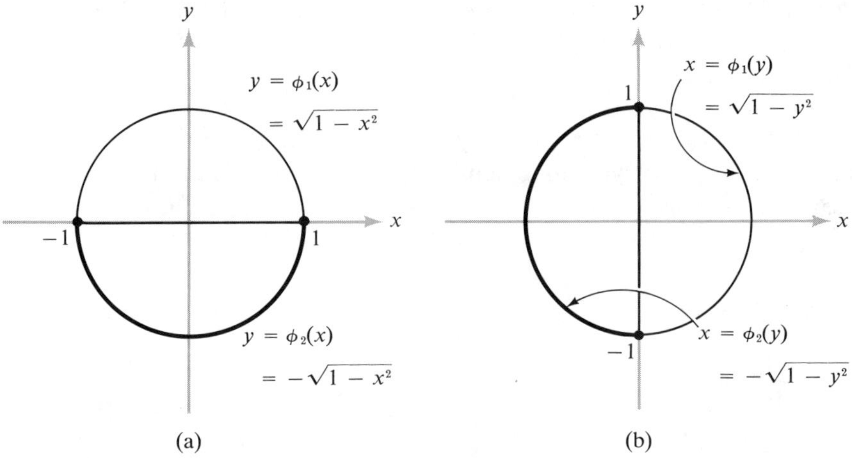

FIGURE 5.3.3
The unit disc, a region of type 3: (a) as a type 1 region, and (b) as a type 2 region.

Sometimes we shall refer to regions of types 1, 2, and 3 as *elementary regions.* Note that the boundary ∂D of an elementary region is the type of set allowed as the discontinuities of a function in Theorem 2.

DEFINITION. *If D is an elementary region in the plane, we can find a rectangle R that contains D. Assume that we have chosen such an R. Given $f: D \to \mathbb{R}$, where f is continuous (and hence bounded), we would like to define $\int_D f(x, y)\, dA$, the* **integral of f over the set** *D. To do this we "extend" f to a function f^* defined on all of R by*

$$f^*(x, y) = \begin{cases} f(x, y) & (x, y) \in D \\ 0 & (x, y) \notin D \quad \text{and} \quad (x, y) \in R \end{cases}$$

Now f^* is bounded (since f is) and continuous except possibly on the boundary of D (see Figure 5.3.4). The boundary of D consists of graphs of continuous functions, so f^* is integrable over R by Theorem 2, Section 5.2. Therefore we can define

$$\int_D f(x, y)\, dA = \int_R f^*(x, y)\, dA$$

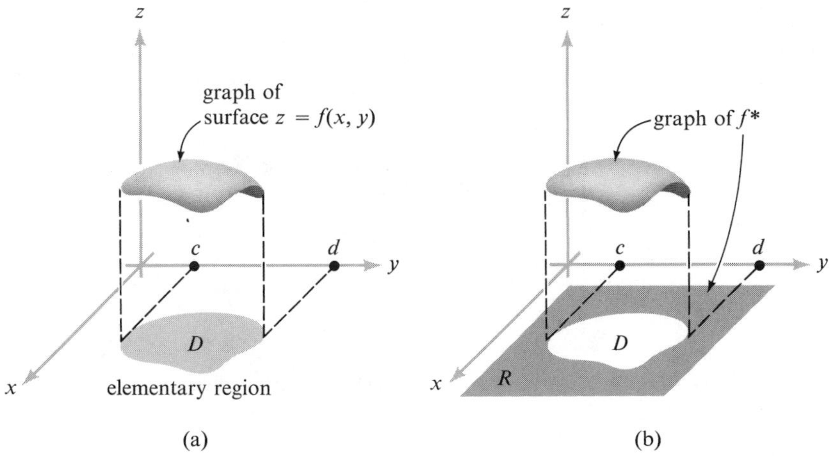

FIGURE 5.3.4

(a) Graph of surface $z = f(x, y)$ over elementary region D. (b) Shaded region shows graph of $z = f^*(x, y)$ on some rectangle R containing D. From this picture we see that boundary points of D may be points of discontinuity of f^*, since the graph of $z = f^*(x, y)$ can be broken at these points.

When $f(x, y) \geq 0$ on D, we can interpret the integral $\int_D f(x, y)\, dA$ as the volume of the three-dimensional region between the graph of f and D, as is evident from Figure 5.3.4.

We have defined $\int_D f(x, y)\, dx\, dy$ by choosing a rectangle R that encloses D. It should be intuitively clear that the value of $\int_D f(x, y)\, dx\, dy$ does not depend on the particular R we select; we shall demonstrate this fact at the end of this section.

If $R = [a, b] \times [c, d]$ is a rectangle containing D, we can use the results on iterated integrals in Section 5.2 to obtain

$$\int_D f(x, y)\, dA = \int_R f^*(x, y)\, dA = \int_a^b \int_c^d f^*(x, y)\, dy\, dx$$
$$= \int_c^d \int_a^b f^*(x, y)\, dx\, dy$$

where f^* equals f in D and zero outside D, as above. Assume D is a region of type 1 determined by functions $\phi_1\colon [a, b] \to \mathbb{R}$ and $\phi_2\colon [a, b] \to \mathbb{R}$. Consider the iterated integral

$$\int_a^b \int_c^d f^*(x, y)\, dy\, dx$$

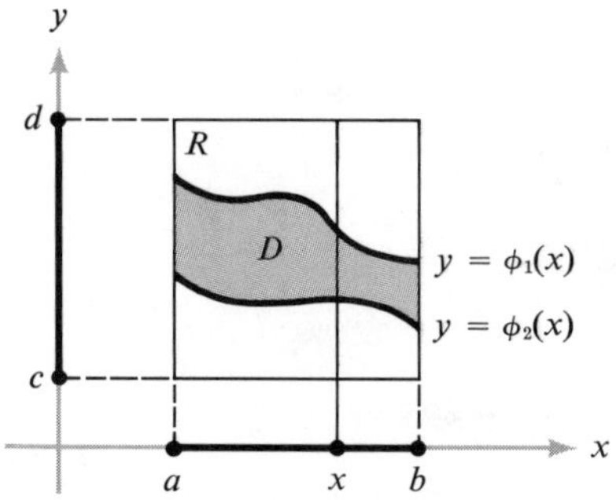

FIGURE 5.3.5
The region between two graphs—a type 1 region.

and, in particular, the inner integral $\int_c^d f^*(x, y)\,dy$ for some fixed x (Figure 5.3.5). Since by definition $f^*(x, y) = 0$ if $y > \phi_1(x)$ or $y < \phi_2(x)$, we obtain

$$\int_c^d f^*(x, y)\,dy = \int_{\phi_2(x)}^{\phi_1(x)} f^*(x, y)\,dy = \int_{\phi_2(x)}^{\phi_1(x)} f(x, y)\,dy$$

We summarize what we have obtained in the following.

THEOREM 4. *If D is a region of type 1, as shown in Figure 5.3.5, then*

$$\int_D f(x, y)\,dA = \int_a^b \int_{\phi_2(x)}^{\phi_1(x)} f(x, y)\,dy\,dx \tag{1}$$

In the case $f(x, y) = 1$ for all $(x, y) \in D$, $\int_D f(x, y)\,dA$ is the area of D. We can check this for formula (1) as follows:

$$\int_a^b \int_{\phi_2(x)}^{\phi_1(x)} f(x, y)\,dy\,dx = \int_a^b [\phi_1(x) - \phi_2(x)]\,dx = A(D)$$

which is the formula for the area of D learned in elementary calculus.

EXAMPLE 1. Find $\int_T (x^3y + \cos x)\,dA$, where T is the triangle consisting of all points (x, y) such that $0 \le x \le \pi/2$, $0 \le y \le x$. Referring to Figure 5.3.6 and formula (1), we have

$$\begin{aligned}\int_T (x^3y + \cos x)\,dA &= \int_0^{\pi/2}\int_0^x (x^3y + \cos x)\,dy\,dx \\ &= \int_0^{\pi/2}\left[\frac{x^3y^2}{2} + y\cos x\right]_{y=0}^{x} dx = \int_0^{\pi/2}\left(\frac{x^5}{2} + x\cos x\right) dx \\ &= \left[\frac{x^6}{12}\right]_0^{\pi/2} + \int_0^{\pi/2} (x\cos x)\,dx = \frac{\pi^6}{(12)(64)} + [x\sin x + \cos x]_0^{\pi/2} \\ &= \frac{\pi^6}{768} + \frac{\pi}{2} - 1\end{aligned}$$

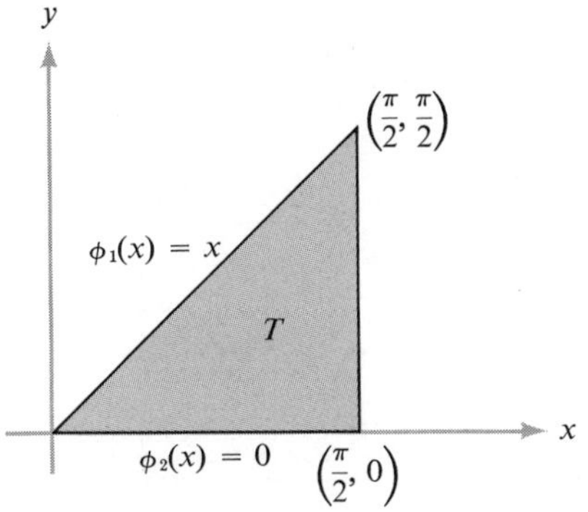

FIGURE 5.3.6

A triangle T represented as a region of type 1.

In the next example we shall use formula (1) to find the volume of a solid whose base is a nonrectangular region D.

EXAMPLE 2. Find the volume of the tetrahedron bounded by the planes $y = 0$, $z = 0$, $x = 0$, and the plane $y - x + z = 1$ (Figure 5.3.7).

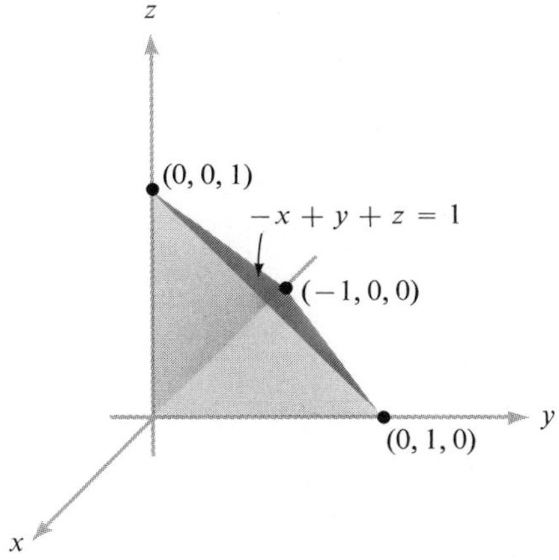

FIGURE 5.3.7

A tetrahedron bounded by the planes $y = 0$, $z = 0$, $x = 0$, and $y - x + z = 1$.

We first note that the given tetrahedron has a triangular base D whose points (x, y) satisfy $-1 \le x \le 0$ and $0 \le y \le 1 + x$; hence D is a region of type 1. (In fact, D is type 3; see Figure 5.3.8.)

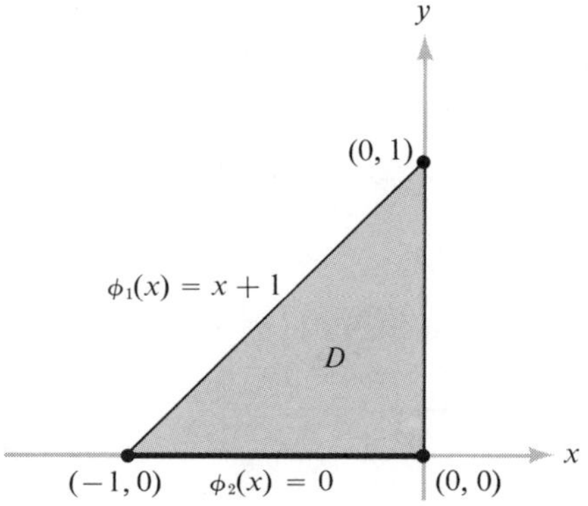

FIGURE 5.3.8

The base of the tetrahedron in Figure 5.3.7 represented as a region of type 1.

For any point (x, y) in D, the height of the surface z above (x, y) is $1 - y + x$. Thus, the volume we seek is given by the integral

$$\int_D (1 - y + x)\, dA$$

Using formula (1) with $\phi_1(x) = x + 1$ and $\phi_2(x) = 0$, we have

$$\begin{aligned}\int_D (1 - y + x)\, dA &= \int_{-1}^{0} \int_{0}^{1+x} (1 - y + x)\, dy\, dx \\ &= \int_{-1}^{0} \left[(1 + x)y - \frac{y^2}{2}\right]_{y=0}^{1+x} dx \\ &= \int_{-1}^{0} \left[\frac{(1 + x)^2}{2}\right] dx \\ &= \left[\frac{(1 + x)^3}{6}\right]_{-1}^{0} = \frac{1}{6}\end{aligned}$$

EXAMPLE 3. Let D be a region of type 1. Describe its area $A(D)$ as a limit of Riemann sums.

If we recall the definition, $A(D) = \int_D dx\, dy$ is the integral over a containing rectangle R of the function $f = 1$. A Riemann sum S_n for this integral is obtained by dividing R into subrectangles and forming the sum $S_n = \sum_{j,k=0}^{n-1} f^*(c_{jk})\, \Delta x\, \Delta y$ as in (1), Section 5.2. Now $f^*(c_{jk})$ is 1 or 0 depending on whether or not c_{jk} is in D. Consider those subrectangles R_{jk} that meet D, and choose c_{jk} in $D \cap R_{jk}$. Thus S_n is the sum of the areas of the subrectangles that meet D and $A(D)$ is the limit of these as $n \to \infty$. Thus $A(D)$ is the limit of the areas of the rectangles "circumscribing" D. The reader should draw a figure to accompany this discussion.

The methods for treating regions of type 2 are entirely analogous. Specifically, we have the following.

THEOREM 4′. *If D is the set of points* (x, y) *such that* $y \in [c, d]$ *and* $\phi_2(y) \leq x \leq \phi_1(y)$, *and if f is continuous on D, then*

$$\int_D f(x, y)\, dA = \int_c^d \left[\int_{\phi_2(y)}^{\phi_1(y)} f(x, y)\, dx \right] dy \tag{2}$$

To find the area of D we substitute $f = 1$ in formula (2); this yields

$$\int_D dA = \int_c^d (\phi_1(y) - \phi_2(y))\, dy$$

Note again that this result for area agrees with the results of single-variable calculus for the area of a region between two curves.

Either the method for type 1 or the method for type 2 regions can be used for integrals over regions of type 3.

It also follows from formulas (1) and (2) that $\int_D f\, dA$ is independent of the choice of the rectangle R enclosing D used in the definition of $\int_D f\, dA$. To see this let us consider the case when D is of type 1. Then formula (1) holds; moreover, on the right side of this formula R does not appear, and thus $\int_D f\, dA$ is independent of R.

EXERCISES

1. Evaluate the following iterated integrals and draw the regions D determined by the limits. State whether the regions are of type 1, type 2, or both.

   (a) $\int_0^1 \int_0^{x^2} dy\, dx$ (b) $\int_1^2 \int_{2x}^{3x+1} dy\, dx$

   (c) $\int_0^1 \int_1^{e^x} (x + y)\, dy\, dx$ (d) $\int_0^1 \int_{x^3}^{x^2} y\, dy\, dx$

2. Repeat Exercise 1 for the following iterated integrals.

   (a) $\int_{-3}^2 \int_0^{y^2} (x^2 + y)\, dx\, dy$

   (b) $\int_{-1}^1 \int_{-2|x|}^{|x|} e^{x+y}\, dy\, dx$

   (c) $\int_0^1 \int_0^{(1-x^2)^{1/2}} dy\, dx$

   (d) $\int_0^{\pi/2} \int_0^{\cos x} y \sin x\, dy\, dx$

   (e) $\int_0^1 \int_{y^2}^{y} (x^n + y^m)\, dx\, dy, \quad m, n > 0$

   (f) $\int_{-1}^0 \int_0^{2(1-x^2)^{1/2}} x\, dy\, dx$

3. Use double integrals to compute the area of a circle of radius r.

4. Using double integrals, determine the area of an ellipse with semiaxes of length a and b.

5. What is the volume of a barn that has a rectangular base 20 ft by 40 ft, and vertical walls 30 ft high at the front (which we assume is on the 20-ft side of the barn) and 40 ft high at the rear? The barn has a flat roof.

6. Let D be the region bounded by the positive x- and y-axes and the line $3x + 4y = 10$. Compute

$$\int_D (x^2 + y^2)\, dA$$

7. Let D be the region bounded by the y-axis and the parabola $x = -4y^2 + 3$. Compute

$$\int_D x^3 y\, dx\, dy$$

8. Evaluate $\int_0^1 \int_0^{x^2} (x^2 + xy - y^2)\, dy\, dx$. Describe this iterated integral as a double integral over a certain region D.

9. Let D be the region given as the set of (x, y) where $1 \le x^2 + y^2 \le 2$ and $y \ge 0$. Is D an elementary region? Evaluate $\int_D f(x, y)\, dA$ where $f(x, y) = 1 + xy$.

10. Find the area enclosed by one period of the sine function $\sin x$, for $0 \le x \le 2\pi$ and the x-axis.

11. Find the volume of the region inside the surface $z = x^2 + y^2$ and between $z = 0$ and $z = 10$.

12. Compute the volume of a cone of base radius r and height h.

13. Evaluate $\int_D y\, dA$ where D is the set of points (x, y) such that $0 \le 2x/\pi \le y$, $y \le \sin x$.

14. From Exercise 5, Section 5.2, we know that $\int_a^b \int_c^d f(x)g(y)\, dy\, dx = (\int_a^b f(x)\, dx) \times (\int_c^d g(y)\, dy)$. Is this true if we integrate $f(x)g(y)$ over any region D (for example, a region of type 1)?

15. Let D be a region given as the set of (x, y) with $-\phi(x) \le y \le \phi(x)$ and $a \le x \le b$, where ϕ is a nonnegative continuous function on the interval $[a, b]$. Let $f(x, y)$ be a function on D such that $f(x, y) = -f(x, -y)$ for all $(x, y) \in D$. Argue that $\int_D f(x, y)\, dA = 0$.

16. Use the methods of this section to show that the area of the parallelogram D determined by vectors $\mathbf{a}$ and $\mathbf{b}$ is $|a_1 b_2 - a_2 b_1|$, where $\mathbf{a} = a_1\mathbf{i} + a_2\mathbf{j}$, $\mathbf{b} = b_1\mathbf{i} + b_2\mathbf{j}$.

17. Describe the area $A(D)$ of a region as a limit of areas of inscribed rectangles, as in Example 3.

5.4 CHANGING THE ORDER OF INTEGRATION

Suppose that D is a region of type 3. Thus, being of types 1 and 2, it can be given as the set of points (x, y) such that

$$a \le x \le b, \qquad \phi_2(x) \le y \le \phi_1(x)$$

and also as the set of points (x, y) such that

$$c \le y \le d, \qquad \psi_2(y) \le x \le \psi_1(y)$$

Hence we have the formulas

$$\int_D f(x, y)\, dA = \int_a^b \int_{\phi_2(x)}^{\phi_1(x)} f(x, y)\, dy\, dx$$

$$= \int_c^d \int_{\psi_2(y)}^{\psi_1(y)} f(x, y)\, dx\, dy$$

If we are to compute one of the iterated integrals above, we may do so by evaluating the other iterated integral; this technique is called *changing the order of integration.* It is often useful to make such a change when evaluating iterated integrals, since one of the iterated integrals may be more difficult to compute than the other.

EXAMPLE 1. By changing the order of integration, evaluate

$$\int_0^a \int_0^{(a^2-x^2)^{1/2}} (a^2 - y^2)^{1/2}\, dy\, dx$$

Note that x varies between 0 and a, and for fixed x, $0 \le y \le (a^2 - x^2)^{1/2}$. Thus the iterated integral is equivalent to the double integral

$$\int_D (a^2 - y^2)^{1/2}\, dy\, dx$$

where D is the set of points (x, y) such that $0 \le x \le a$ and $0 \le y \le (a^2 - x^2)^{1/2}$. But this is the representation of one quarter (the positive quadrant portion) of the disc of radius a; hence D can also be described as the set of points (x, y) satisfying

$$0 \le y \le a, \qquad 0 \le x \le (a^2 - y^2)^{1/2}$$

(see Figure 5.4.1). Thus

$$\begin{aligned}\int_0^a \int_0^{(a^2-x^2)^{1/2}} (a^2 - y^2)^{1/2}\, dy\, dx &= \int_0^a \int_0^{(a^2-y^2)^{1/2}} (a^2 - y^2)^{1/2}\, dx\, dy \\ &= \int_0^a \left[x(a^2 - y^2)^{1/2}\right]_{x=0}^{(a^2-y^2)^{1/2}} dy \\ &= \int_0^a (a^2 - y^2)\, dy = \left[a^2 y - \frac{y^3}{3}\right]_0^a = \frac{2a^3}{3}\end{aligned}$$

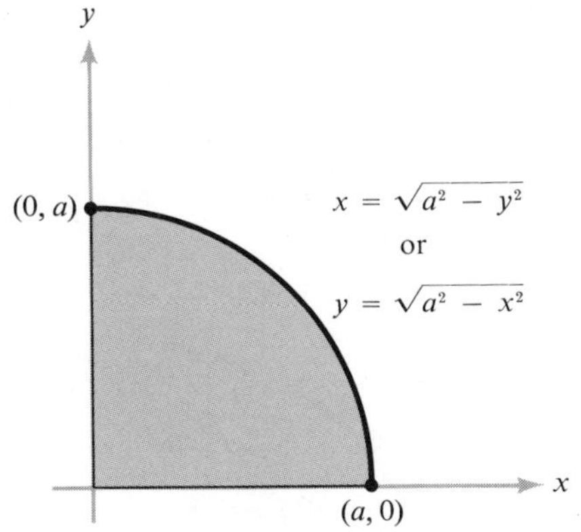

FIGURE 5.4.1
The positive-quadrant portion of a disc of radius a.

We could have evaluated the initial iterated integral directly but, as the reader can easily verify for himself, changing the order of integration makes the problem much simpler computationally.

The next example shows that it may even be "impossible" to evaluate an iterated integral and yet be possible to evaluate the iterated integral obtained by changing the order of integration.

EXAMPLE 2. Evaluate

$$\int_1^2 \int_0^{\log x} (x - 1)\sqrt{1 + e^{2y}}\, dy\, dx$$

First observe that we could not compute this integral in the order given by using the Fundamental Theorem. However, the integral is equal to $\int_D (x - 1)\sqrt{1 + e^{2y}}\, dA$, where D is the set of (x, y) such that

$$1 \le x \le 2 \quad \text{and} \quad 0 \le y \le \log x$$

The region D is of type 3 (see Figure 5.4.2) and can therefore be described by

$$0 \le y \le \log 2 \quad \text{and} \quad e^y \le x \le 2$$

Thus the given iterated integral is equal to

$$\begin{aligned}\int_0^{\log 2} \int_{e^y}^2 (x - 1)\sqrt{1 + e^{2y}}\, dx\, dy &= \int_0^{\log 2} \sqrt{1 + e^{2y}} \left[\int_{e^y}^2 (x - 1)\, dx\right] dy \\ &= \int_0^{\log 2} \sqrt{1 + e^{2y}} \left[\frac{x^2}{2} - x\right]_{e^y}^2 dy \\ &= -\int_0^{\log 2} \left(\frac{e^{2y}}{2} - e^y\right)\sqrt{1 + e^{2y}}\, dy \\ &= -\frac{1}{2}\int_0^{\log 2} e^{2y}\sqrt{1 + e^{2y}}\, dy \\ &\quad + \int_0^{\log 2} e^y\sqrt{1 + e^{2y}}\, dy \qquad (1)\end{aligned}$$

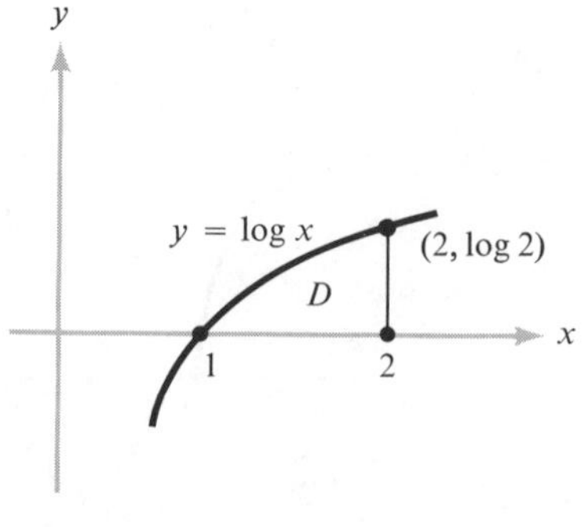

FIGURE 5.4.2
D is the region of integration for Example 2.

In the first integral in expression (1) we substitute $u = e^{2y}$ and in the second, $v = e^y$. Hence we obtain

$$-\frac{1}{4}\int_1^4 \sqrt{1+u}\,du + \int_1^2 \sqrt{1+v^2}\,dv \tag{2}$$

Both integrals in (2) are easily found with techniques of calculus (or by consulting the table of integrals at the back of the book). For the first integral we get

$$\frac{1}{4}\int_1^4 \sqrt{1+u}\,du = \left[\tfrac{1}{6}(1+u)^{3/2}\right]_1^4 = \tfrac{1}{6}[(1+4)^{3/2} - 2^{3/2}]$$

$$= \tfrac{1}{6}[5^{3/2} - 2^{3/2}] \tag{3}$$

The second integral is (see formula 43 on p. 524)

$$\int_1^2 \sqrt{1+v^2}\,dv = \tfrac{1}{2}\left[v\sqrt{1+v^2} + \log(\sqrt{1+v^2}+v)\right]_1^2$$

$$= \tfrac{1}{2}[2\sqrt{5} + \log(\sqrt{5}+2)] - \tfrac{1}{2}[\sqrt{2} + \log(\sqrt{2}+1)] \tag{4}$$

Finally, we subtract (3) from (4) to obtain the answer

$$\frac{1}{2}\left(2\sqrt{5} - \sqrt{2} + \log\frac{\sqrt{5}+2}{\sqrt{2}+1}\right) - \frac{1}{6}[5^{3/2} - 2^{3/2}]$$

To conclude this section we mention an important analog of the Mean Value Theorem of integral calculus.

THEOREM 5 (MEAN VALUE THEOREM FOR DOUBLE INTEGRALS). *Suppose $f\colon D \to \mathbb{R}$ is continuous and D is an elementary region. Then for some point (x_0, y_0) in D we have*

$$\int_D f(x, y)\,dA = f(x_0, y_0)\cdot A(D)$$

where $A(D)$ denotes the area of D.

Proof. We cannot prove this theorem with complete rigor because it requires some concepts about continuous functions not proved in this course, but we can sketch the main ideas that underlie the proof (see the proof of a one-variable version on p. 203).

Since f is continuous on D it has a maximum value M and a minimum value m (proved in advanced calculus). Thus

$$m \le f(x, y) \le M \tag{5}$$

for all $(x, y) \in D$. Furthermore, $f(x_1, y_1) = m$ and $f(x_2, y_2) = M$ for some pairs (x_1, y_1) and (x_2, y_2) in D. From inequality (5) it follows that

$$mA(D) = \int_D m\,dA \le \int_D f(x, y)\,dA \le \int_D M\,dA = MA(D)$$

Therefore dividing through by $A(D)$ we get

$$m \leq \frac{1}{A(D)} \int_D f(x, y)\, dA \leq M \tag{6}$$

Since a continuous function on D takes on every value between its maximum and minimum values (this is the Intermediate Value Theorem proved in advanced calculus), and the number $(1/A(D)) \int_D f(x, y)\, dA$ is by inequality (6) between these values, there must be a point $(x_0, y_0) \in D$ with

$$f(x_0, y_0) = \frac{1}{A(D)} \int_D f(x, y)\, dA$$

But this is precisely the conclusion of Theorem 5. ■

EXERCISES

1. In the following integrals, change the order of integration, sketch the corresponding regions, and evaluate the integral both ways.
   (a) $\int_0^1 \int_x^1 xy\, dy\, dx$
   (b) $\int_0^{\pi/2} \int_0^{\cos\theta} \cos\theta\, dr\, d\theta$
   (c) $\int_0^1 \int_1^{2-y} (x + y)^2\, dx\, dy$
   (d) $\int_a^b \int_a^y f(x, y)\, dx\, dy$ (Express your answer in terms of antiderivatives.)
2. Find
   (a) $\int_{-1}^1 \int_{|y|}^1 (x + y)^2\, dx\, dy$ and (b) $\int_{-3}^3 \int_{-\sqrt{(9-y^2)}}^{\sqrt{(9-y^2)}} x^2\, dx\, dy.$
3. If $f(x, y) = e^{\sin(x+y)}$ and $D = [-\pi, \pi] \times [-\pi, \pi]$ show that
$$\frac{1}{e} \leq \frac{1}{4\pi^2} \int_D f(x, y)\, dA \leq e$$
4. Prove that
$$2 \int_a^b \int_x^b f(x)f(y)\, dy\, dx = \left(\int_a^b f(x)\, dx\right)^2$$
5. Compute the volume of an ellipsoid with semiaxes a, b, and c. (HINT: Use symmetry and first find the volume of one half of the ellipsoid.)
6. Compute $\int_D f(x, y)\, dA$ where $f(x, y) = y^2\sqrt{x}$ and D is the set of (x, y) where $x > 0$, $y > x^2$, $y < 10 - x^2$.
7. Find the volume of the region determined by $x^2 + y^2 + z^2 \leq 10$, $z \geq 2$.
8. Evaluate $\iint_D e^{x-y}\, dx\, dy$ where D is the interior of the triangle with vertices (0, 0), (1, 3), and (2, 2).

9. Evaluate $\iint_D y^3(x^2 + y^2)^{-3/2}\, dx\, dy$ where D is the region determined by the conditions $\frac{1}{2} \le y \le 1$ and $x^2 + y^2 \le 1$.
10. Given that the double integral $\iint_D f(x, y)\, dx\, dy$ of a positive continuous function f equals the iterated integral $\int_0^1 [\int_{x^2}^{x} f(x, y)\, dy]\, dx$, sketch the region D and interchange the order of integration.
11. Given that the double integral $\iint_D f(x, y)\, dx\, dy$ of a positive continuous function f equals the iterated integral $\int_0^1 [\int_y^{\sqrt{2-y^2}} f(x, y)]\, dx\, dy$, sketch the region D and interchange the order of integration.

OPTIONAL

5.5 IMPROPER INTEGRALS

The notion of an improper integral arises in considering certain integrals such as the area of a hemisphere (this is explained in Example 2, Section 6.4). Thus, for completeness of the theory of integration, we must consider such integrals.

In the previous sections we defined the notion of the integral for functions of two variables, and stated criteria guaranteeing that f was indeed integrable over a set D. Recall that one of the hypotheses of Theorem 2 (Section 5.2) was that f be bounded. The following example shows how the sum S_n may fail to converge if f is not bounded.

Let R be the unit square $[0, 1] \times [0, 1]$ and let $f\colon R \to \mathbb{R}$ be defined by

$$f(x, y) = \begin{cases} \dfrac{1}{\sqrt{x}} & x \neq 0 \\ 0 & x = 0 \end{cases}$$

Clearly, f is not bounded on R since, as x gets close to zero, f gets arbitrarily large. Let R_{ij} be a regular partition of R and form the sum (1) of Section 5.2

$$S_n = \sum_{i=0}^{n-1} \sum_{j=0}^{n-1} f(c_{ij})\, \Delta x\, \Delta y$$

Let R_{11} be the subrectangle that contains (0, 0) (see Figure 5.5.1) and choose some $c_{11} \in R_{11}$. For a fixed n, we can make S_n as large as we please by picking c_{11} closer and closer to (0, 0); hence $\lim_{n\to\infty} S_n$ cannot be *independent* of the choice of the c_{ij}.

OPTIONAL (*Continued*)

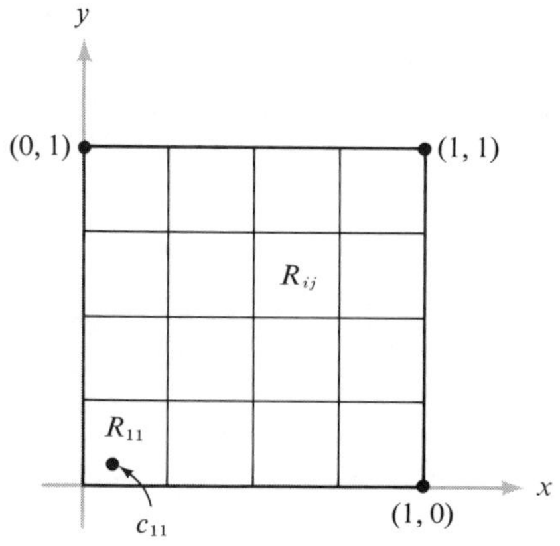

FIGURE 5.5.1
Location of R_{11} *in a partition of* $[0, 1] \times [0, 1]$.

However, let us formally evaluate the iterated integral of f, following the rules for integrating a function of a single variable. We have

$$\int_0^1 \int_0^1 f(x, y)\, dx\, dy = \int_0^1 \int_0^1 \frac{dx}{\sqrt{x}}\, dy = \int_0^1 [2\sqrt{x}]_0^1\, dy$$

$$= \int_0^1 2\, dy = 2$$

Moreover, if we reverse the order of integration we also obtain

$$\int_0^1 \int_0^1 \frac{dy}{\sqrt{x}}\, dx = 2$$

So in some sense this function is integrable. The question is, in what sense?

Recall from one-variable calculus how the improper integral $\int_0^1 dx/\sqrt{x}$ is treated: $1/\sqrt{x}$ is unbounded on the interval $]0, 1]$, yet $\lim_{\delta \to 0} \int_\delta^1 (dx/\sqrt{x}) = 2$, and we *define* $\int_0^1 (dx/\sqrt{x})$ to be this limit. Similarly, for the two-variable case we shall allow the function to be unbounded at certain points on the boundary of its domain and define the improper integral through a limiting process.

More specifically, suppose the region D is of type 1 and $f: D \to \mathbb{R}$ is continuous and bounded except at certain points on the boundary. Assume that D is described by $a \le x \le b$, $\phi_2(x) \le y \le \phi_1(x)$. Choose numbers

OPTIONAL (*Continued*)

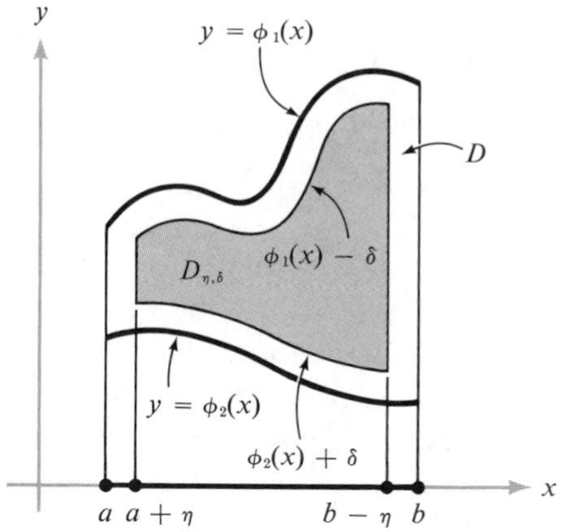

FIGURE 5.5.2
A shrunken domain $D_{\eta,\delta}$ for improper integrals.

$\delta, \eta > 0$ such that $D_{\eta,\delta}$ is the subset of D consisting of points (x, y) with $a + \eta \le x \le b - \eta$, $\phi_2(x) + \delta \le y \le \phi_1(x) - \delta$ (Figure 5.5.2), where η and δ are chosen small enough so that $D_{\eta,\delta} \subset D$. (If either $\phi_2(a) = \phi_1(a)$ or $\phi_2(b) = \phi_1(b)$ we must modify this slightly because in this case $D_{\eta,\delta}$ may not be a subset of D (see Example 2).) Since f is continuous and bounded on $D_{\eta,\delta}$ the integral $\int_{D_{\eta,\delta}} f$ exists. We can now ask what happens as the region $D_{\eta,\delta}$ expands to fill the region D, that is, as $(\eta, \delta) \to (0, 0)$.

If

$$\lim_{(\eta,\delta)\to(0,0)} \int_{D_{\eta,\delta}} f\,dA$$

exists we define $\int_D f$ to be equal to this limit and say that it is the *improper integral* of f over D. This definition is exactly analogous to the definition of improper integral for a function of one variable.

Since f is integrable over $D_{\eta,\delta}$ we can apply Fubini's Theorem to obtain

$$\int_{D_{\eta,\delta}} f\,dA = \int_{a+\eta}^{b-\eta} \int_{\phi_2(x)+\delta}^{\phi_1(x)-\delta} f(x, y)\,dy\,dx \tag{1}$$

So if f is integrable over D,

$$\int_D f\,dA = \lim_{(\eta,\delta)\to(0,0)} \int_{a+\eta}^{b-\eta} \int_{\phi_2(x)+\delta}^{\phi_1(x)-\delta} f(x, y)\,dy\,dx$$

OPTIONAL (*Continued*)

It may be convenient to work with the iterated limits

$$\lim_{\eta \to 0} \int_{a+\eta}^{b-\eta} \left[\lim_{\delta \to 0} \int_{\phi_2(x)+\delta}^{\phi_1(x)-\delta} f(x, y)\, dy \right] dx \tag{2}$$

if these limits exist. If the limits do exist, then we denote (2) by

$$\int_a^b \int_{\phi_2(x)}^{\phi_1(x)} f(x, y)\, dy\, dx$$

and call it the *iterated improper integral* of f over D. Using more advanced techniques it is possible to show that if $|f|$ is integrable, then if the iterated improper integral exists, it equals $\int_D f\, dA$; that is, formula (2) may be used to evaluate the improper integral. Also, if $f \geq 0$, the existence of the limits (2) implies the existence of the double limit defining $\int_D f\, dA$; so (2) equals $\int_D f\, dA$ in this case. The definition for D a region of type 2 is analogous.

Finally, let us consider the case where D is a region of type 3 and f is unbounded at points on ∂D. For example, suppose D is the set of points (x, y) with

$$a \leq x \leq b, \qquad \phi_2(x) \leq y \leq \phi_1(x)$$

and is also the set of points (x, y) with

$$c \leq y \leq d, \qquad \psi_2(y) \leq x \leq \psi_1(y)$$

If $|f|$ is integrable and

$$\int_c^d \int_{\psi_2(y)}^{\psi_1(y)} f(x, y)\, dx\, dy \quad \text{and} \quad \int_a^b \int_{\phi_2(x)}^{\phi_1(x)} f(x, y)\, dy\, dx$$

exist, then it can be shown that both iterated integrals are equal and their common value is $\int_D f\, dA$. This is Fubini's Theorem for improper integrals.

EXAMPLE 1. Evaluate $\int_D f(x, y)\, dy\, dx$ where $f(x, y) = 1/\sqrt{1 - x^2 - y^2}$ and D is the unit disc $x^2 + y^2 \leq 1$.

We can describe D as the set of points (x, y) with $-1 \leq x \leq 1$, $-\sqrt{1 - x^2} \leq y \leq \sqrt{1 - x^2}$. Now since ∂D is the set of points (x, y) with $x^2 + y^2 = 1$, f is undefined at every point on ∂D, since at such points the denominator of f is 0. We calculate the iterated improper integrals and obtain

$$\begin{aligned}
\int_{-1}^1 \int_{-\sqrt{(1-x^2)}}^{\sqrt{(1-x^2)}} \frac{dy\, dx}{\sqrt{1 - x^2 - y^2}} &= \int_{-1}^1 \left[\sin^{-1}\left(\frac{y}{\sqrt{1 - x^2}} \right) \right]_{-\sqrt{(1-x^2)}}^{\sqrt{(1-x^2)}} dx \\
&= \int_{-1}^1 [\sin^{-1}(1) - \sin^{-1}(-1)]\, dx \\
&= \pi \int_{-1}^1 dx = 2\pi
\end{aligned}$$

OPTIONAL (*Continued*)

In this example we used the fact, stated above, that the iterated improper integral is equal to the improper integral of $1/(\sqrt{1 - x^2 - y^2})$ over the unit disc.

EXAMPLE 2. Let $f(x, y) = 1/(x - y)$ and let D be the set of (x, y) with $0 \le x \le 1$ and $0 \le y \le x$. Show that f is not integrable over D.

Since the denominator of f is zero on the line $y = x$, f is unbounded on part of the boundary of D. Let $0 < \eta < 1$ and $0 < \delta < \eta$ and let $D_{\eta, \delta}$ be the set of (x, y) with $\eta \le x \le 1 - \eta$ and $\delta \le y \le x - \delta$ (Figure 5.5.3). We choose $\delta < \eta$ to guarantee that $D_{\eta, \delta}$ is contained in D. Consider

$$\begin{aligned}\int_{D_{\eta,\delta}} f\, dA &= \int_{0+\eta}^{1-\eta} \int_{\delta}^{x-\delta} \frac{1}{x - y}\, dy\, dx \\ &= \int_{\eta}^{1-\eta} [-\log(x - y)]_{\delta}^{x-\delta}\, dx \\ &= \int_{\eta}^{1-\eta} [-\log(\delta) + \log(x - \delta)]\, dx \\ &= [-\log \delta] \int_{\eta}^{1-\eta} dx + \int_{\eta}^{1-\eta} \log(x - \delta)\, dx \\ &= -(1 - 2\eta)\log \delta + [(x - \delta)\log(x - \delta) - (x - \delta)]_{\eta}^{1-\eta}\end{aligned}$$

In the last step we used the fact that $\int \log u\, du = u \log u - u$. Continuing the above set of equalities, we have

$$\begin{aligned}&= -(1 - 2\eta)\log \delta + (1 - \eta - \delta)\log(1 - \eta - \delta) \\ &\quad - (1 - \eta - \delta) - (\eta - \delta)\log(\eta - \delta) + (\eta - \delta)\end{aligned}$$

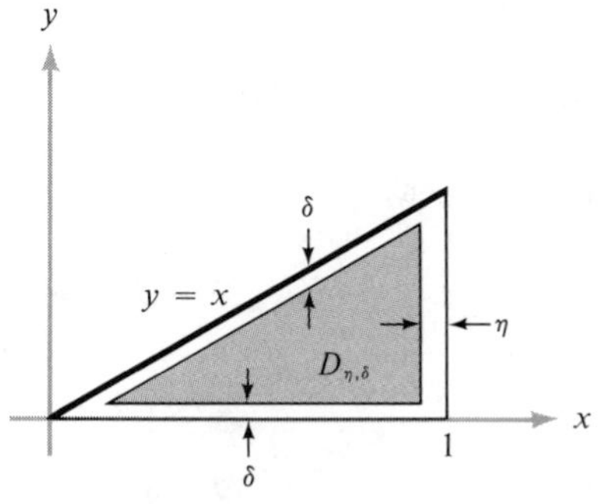

FIGURE 5.5.3
The shrunken domain $D_{\eta, \delta}$ for a triangular domain D.

OPTIONAL (*Continued*)

As $(\eta, \delta) \to (0, 0)$ the second term converges to $1 \log 1 = 0$, and the third and fifth terms converge to 1 and 0, respectively. Let $v = \eta - \delta$. Since $v \log v \to 0$ as $v \to 0$ (a limit established in calculus) we see that the fourth term goes to zero as $(\eta, \delta) \to (0, 0)$. It is the first term which will give us trouble. Now

$$-(1 - 2\eta)\log \delta = -\log \delta + 2\eta \log \delta \tag{3}$$

and it is not hard to see that this does not converge as $(\eta, \delta) \to (0, 0)$. For example, let $\eta = 2\delta$; then (3) becomes

$$-\log \delta + 4\delta \log \delta$$

As before, $4\delta \log \delta \to 0$ as $\delta \to 0$ but $-\log \delta \to +\infty$ as $\delta \to 0$, which shows that (3) does not converge. Hence $\lim_{(\eta, \delta) \to (0, 0)} \int_{D_{\eta,\delta}} f\, dA$ does not exist and so f is not integrable.

It is important to consider improper integrals because they do arise from natural problems. For example, as we shall see later, one of the formulas for computing the surface area of a hemisphere forces us to consider the improper integral of Example 1.

EXERCISES

1. Evaluate the following integrals if they exist.

   (a) $\int_D \frac{1}{\sqrt{xy}}\, dA, \quad D = [0, 1] \times [0, 1]$

   (b) $\int_D \frac{1}{\sqrt{|x - y|}}\, dx\, dy, \quad D = [0, 1] \times [0, 1]$ (HINT: Divide D into two pieces.)

   (c) $\int_D y/x\, dx\, dy$, D bounded by $x = 1$, $x = y$, and $x = 2y$

   (d) $\int_0^1 \int_0^{e^y} \log x\, dx\, dy$

2. (a) Discuss how you would define $\int_D f\, dA$ if D is an unbounded region, for example, the set of (x, y) such that $a \le x < \infty$ and $\phi_2(x) \le y \le \phi_1(x)$, where $\phi_2 \le \phi_1$ are given (Figure 5.5.4).

   (b) Evaluate $\int_D xye^{-(x^2 + y^2)}\, dx\, dy$ if $x \ge 0$, $0 \le y \le 1$.

OPTIONAL (*Continued*)

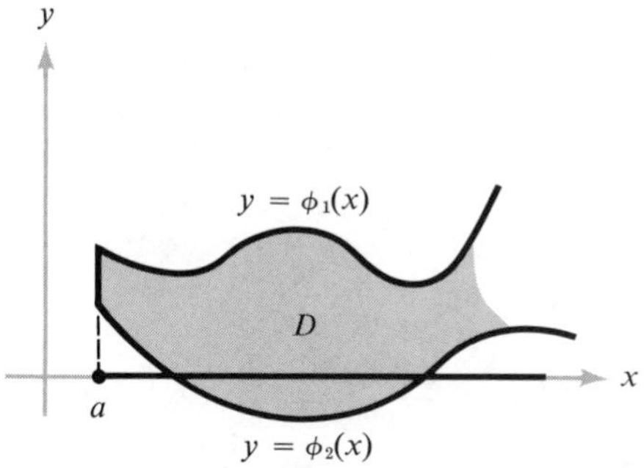

FIGURE 5.5.4
An unbounded region D.

3. Using Exercise 2, integrate e^{-xy} for $x \geq 0, 1 \leq y \leq 2$ in two ways (assume Fubini's Theorem) to show that

$$\int_0^{\infty} \frac{e^{-x} - e^{-2x}}{x}\, dx = \log 2$$

4. Show that the integral $\int_0^1 \int_0^a (x/\sqrt{a^2 - y^2})\, dy\, dx$ exists, and compute its value.

*5. Discuss whether the integral

$$\int_D \frac{x + y}{x^2 + 2xy + y^2}\, dx\, dy$$

exists where $D = [0, 1] \times [0, 1]$. If it exists compute its value.

6. Let f be a non-negative function that may be unbounded and discontinuous on the boundary of an elementary region D. Let g be a similar function such that $f(x, y) \leq g(x, y)$ whenever both are defined. Suppose $\int_D g(x, y)\, dA$ exists. Argue informally that this implies the existence of $\int_D f(x, y)\, dA$.

7. Use Exercise 6 to show that

$$\int_D \frac{\sin^2(x - y)}{\sqrt{1 - x^2 - y^2}}\, dy\, dx$$

exists where D is the unit disc $x^2 + y^2 \leq 1$.

8. Let f be as in Exercise 6 and let g be a function such that $0 \leq g(x, y) \leq f(x, y)$ whenever both are defined. Suppose that $\int_D g(x, y)\, dA$ does not exist. Argue informally that $\int_D f(x, y)\, dA$ cannot exist.

OPTIONAL (*Continued*)

9. Use Exercise 8 to show that

$$\int_D \frac{e^{x^2+y^2}}{x-y}\,dy\,dx$$

does not exist, where D is the set of (x, y) with $0 \le x \le 1$ and $0 \le y \le x$.

5.6 THE TRIPLE INTEGRAL

Given a continuous function $f\colon C \to \mathbb{R}$, where C is some rectangular parallelepiped in $\mathbb{R}^3$, we can define the integral of f over C as a limit of sums just as we did for a function of two variables. Briefly, we partition the three sides of C into n equal parts and form the sum

$$S_n = \sum_{i=0}^{n-1} \sum_{j=0}^{n-1} \sum_{k=0}^{n-1} f(c_{ijk})\,\Delta V$$

where $c_{ijk} \in C_{ijk}$, the ijkth rectangular parallelepiped (or box) in the partition of C, and ΔV is the volume of C_{ijk} (see Figure 5.6.1).

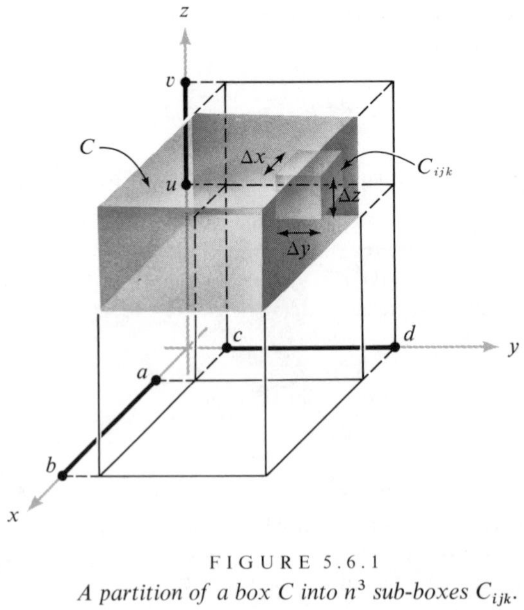

FIGURE 5.6.1
A partition of a box C into n^3 sub-boxes C_{ijk}.

DEFINITION. *Let f be a bounded function of three variables defined on C. If $\lim_{n\to\infty} S_n$ exists and the limit is independent of the points c_{ijk}, we call the limit of S_n the* **triple integral** *(or simply the integral) of f over C and denote it by*

$$\int_C f\,dV, \qquad \int_C f(x, y, z)\,dV, \qquad \int_C f(x, y, z)\,dx\,dy\,dz,$$

$$\text{or} \quad \iiint_C f(x, y, z)\,dx\,dy\,dz$$

As before, one can prove that continuous functions defined on C are integrable. Moreover, bounded functions whose discontinuities are confined to graphs of continuous functions (such as $x = \alpha(y, z)$, $y = \beta(x, z)$, or $z = \gamma(x, z)$) are integrable. This is the analog of Theorem 2 of Section 5.2.

Suppose the rectangular parallelepiped C is the Cartesian product $[a, b] \times [c, d] \times [u, v]$. Then by analogy with functions of two variables, there are various iterated integrals we can consider, namely

$$\int_u^v \int_c^d \int_a^b f(x, y, z)\,dx\,dy\,dz, \qquad \int_u^v \int_a^b \int_c^d f(x, y, z)\,dy\,dx\,dz,$$

$$\text{and} \quad \int_a^b \int_u^v \int_c^d f(x, y, z)\,dy\,dz\,dx, \text{ etc.}$$

The order of dx, dy, and dz indicates how the integration is carried out. For example, the first integral above stands for

$$\int_u^v \left[\int_c^d \left(\int_a^b f(x, y, z)\,dx\right) dy\right] dz$$

As in the two-variable case, Fubini's Theorem is valid: If f is continuous, then the six possible iterated integrals are all equal. In other words, a triple integral may be reduced to a threefold iterated integration.

To complete the analogy with the double integral, consider the problem of evaluating triple integrals over more general bounded sets $W \subset \mathbb{R}^3$ (that is, those sets that can be contained in some box). Given $f\colon W \to \mathbb{R}$, extend f to a function f^* that agrees with f on W and is zero outside of W. If B is a box containing W and ∂W consists of graphs of finitely many continuous functions, f^* will be integrable and we define

$$\int_W f(x, y, z)\,dV = \int_B f^*(x, y, z)\,dV$$

As in the two-dimensional case, this integral is independent of the choice of B.

As in the two-variable case, we shall restrict our attention to regions of special types. A region W is of ***type*** 1 if it can be described as the set of all (x, y, z) such that

$$a \le x \le b, \qquad \phi_2(x) \le y \le \phi_1(x) \quad \text{and} \quad \gamma_2(x, y) \le z \le \gamma_1(x, y) \quad (1)$$

In this definition, $\gamma_i\colon D \to \mathbb{R}$, $i = 1, 2$ are continuous functions, D is a region of type 1, and $\gamma_1(x, y) = \gamma_2(x, y)$ implies $(x, y) \in \partial D$. The last condition means that the surfaces $z = \gamma_1(x, y)$ and $z = \gamma_2(x, y)$, if they intersect at all, do so only for $(x, y) \in \partial D$.

A three-dimensional region will also be said to be of type 1 if it can be expressed as the set of all (x, y, z) such that

$$c \le y \le d, \qquad \psi_2(y) \le x \le \psi_1(y) \quad \text{and} \quad \gamma_2(x, y) \le z \le \gamma_1(x, y) \quad (2)$$

where $\gamma_i\colon D \to \mathbb{R}$ are as above and D is a two-dimensional region of type 2. Figure 5.6.2 shows two regions of type 1 that are described by conditions (1) and (2) respectively.

A region W is of ***type*** 2 if it can be expressed in the form (1) or (2) with the roles of x and z interchanged, and W is of ***type*** 3 if it can be expressed in the form (1) or (2) with y and z interchanged. A region W that is of type 1, 2, and 3 is said to be of ***type*** 4 (Figure 5.6.3). An example of a type 4 region is the ball of radius r, $x^2 + y^2 + z^2 \le r^2$.

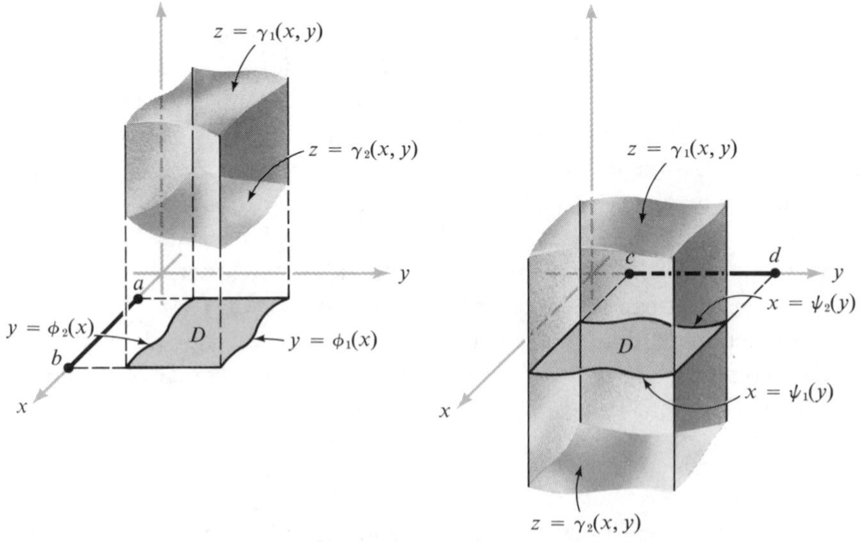

FIGURE 5.6.2
Some regions of type 1 in space.

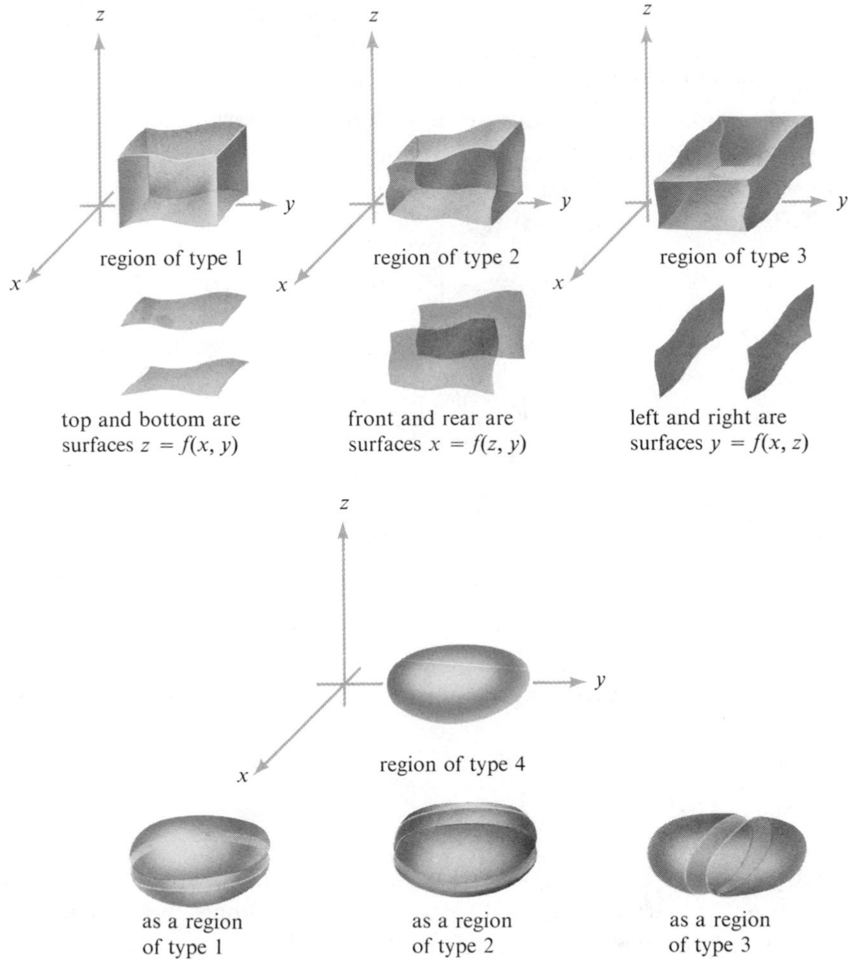

FIGURE 5.6.3
The four possible types of regions in space.

Suppose W is of type 1. Then either

$$\int_W f(x, y, z)\, dV = \int_a^b \int_{\phi_2(x)}^{\phi_1(x)} \int_{\gamma_2(x,y)}^{\gamma_1(x,y)} f(x, y, z)\, dz\, dy\, dx$$

$$= \int_D \left[\int_{\gamma_2(x,y)}^{\gamma_1(x,y)} f(x, y, z)\, dz \right] dy\, dx \tag{3}$$

or

$$\int_W f(x, y, z)\, dV = \int_c^d \int_{\psi_2(y)}^{\psi_1(y)} \int_{\gamma_2(x,y)}^{\gamma_1(x,y)} f(x, y, z)\, dz\, dx\, dy$$

$$= \int_D \left[\int_{\gamma_2(x,y)}^{\gamma_1(x,y)} f(x, y, z)\, dz \right] dx\, dy \tag{4}$$

according to whether W is defined by (1) or by (2). The proofs of (3) and (4) by Fubini's Theorem are the same as for the two-dimensional case. The student should sketch, or at least try to visualize, the figures associated with (3) and (4). Once all the terms are understood, the formulas are easily remembered. It may help to recall the intuition Cavalieri's Principle provided of Fubini's Theorem.

If $f(x, y, z) = 1$ for all $(x, y, z) \in W$, then we obtain

$$\int_W f(x, y, z)\, dV = \int_W 1\, dV = \text{volume}\,(W)$$

In case W is of type 1 and formula (3) is applicable, we get the formula

$$\begin{aligned}\text{volume}\,(W) &= \int_a^b \int_{\phi_2(x)}^{\phi_1(x)} \int_{\gamma_2(x, y)}^{\gamma_1(x, y)} dz\, dy\, dx \\ &= \int_a^b \int_{\phi_2(x)}^{\phi_1(x)} [\gamma_1(x, y) - \gamma_2(x, y)]\, dy\, dx\end{aligned}$$

Can you see how to prove this formula from Cavalieri's Principle?

EXAMPLE 1. Verify the formula for the volume of a ball: $\int_W dV = \frac{4}{3}\pi$, where W is the unit ball $x^2 + y^2 + z^2 \leq 1$.

The region W is of type 1; we can describe it as the set of (x, y, z) satisfying

$$-1 \leq x \leq 1, \qquad -\sqrt{1 - x^2} \leq y \leq \sqrt{1 - x^2},$$

and

$$-\sqrt{1 - x^2 - y^2} \leq z \leq \sqrt{1 - x^2 - y^2}$$

(see Figure 5.6.4). Describing W this way is often the most difficult step in the evaluation of a triple integral. Once this has been done appropriately, it remains only to evaluate the given triple integral by using an equivalent iterated integral. In this case we may apply (3) to obtain

$$\int_W dV = \int_{-1}^{1} \int_{-(1-x^2)^{1/2}}^{(1-x^2)^{1/2}} \int_{-(1-x^2-y^2)^{1/2}}^{(1-x^2-y^2)^{1/2}} dz\, dy\, dx$$

Holding y and x fixed and integrating with respect to z yields

$$\int_{-1}^{1} \int_{-(1-x^2)^{1/2}}^{(1-x^2)^{1/2}} \left[z \Big|_{-(1-x^2-y^2)^{1/2}}^{(1-x^2-y^2)^{1/2}} \right] dy\, dx$$

$$= 2 \int_{-1}^{1} \left[\int_{-(1-x^2)^{1/2}}^{(1-x^2)^{1/2}} (1 - x^2 - y^2)^{1/2}\, dy \right] dx$$

Now since x is fixed in the dy-integral, this integral can be expressed as $\int_{-a}^{a} (a^2 - y^2)^{1/2}\, dy$, where $a = (1 - x^2)^{1/2}$. This integral represents the area

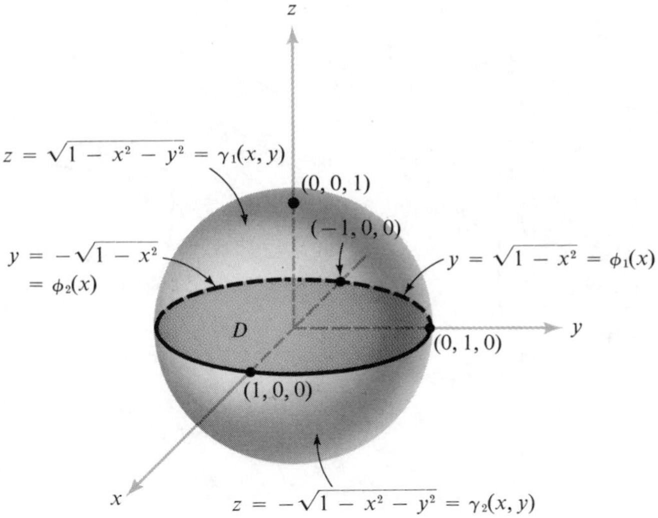

FIGURE 5.6.4
The unit ball expressed as a region of type 1.

of a semicircular region of radius a, so that

$$\int_{-a}^{a} (a^2 - y^2)^{1/2}\, dy = \frac{a^2}{2}\pi$$

(Of course we could have evaluated the integral directly by using the table of integrals in Appendix C, but this trick saves quite a bit of effort.) Thus

$$\int_{-(1-x^2)^{1/2}}^{(1-x^2)^{1/2}} (1 - x^2 - y^2)^{1/2}\, dy = \frac{1 - x^2}{2}\pi$$

and so

$$\begin{aligned} 2\int_{-1}^{1}\int_{-(1-x^2)^{1/2}}^{(1-x^2)^{1/2}} (1 - x^2 - y^2)^{1/2}\, dy\, dx &= 2\int_{-1}^{1} \pi \frac{1 - x^2}{2}\, dx \\ &= \pi\int_{-1}^{1} (1 - x^2)\, dx \\ &= \pi\left[x - \frac{x^3}{3}\right]_{-1}^{1} = \frac{4}{3}\pi \end{aligned}$$

EXAMPLE 2. Let W be the region bounded by the planes $x = 0$, $y = 0$, $z = 2$ and the surface $z = x^2 + y^2$, $x \geq 0$, $y \geq 0$. Compute $\int_W x\, dx\, dy\, dz$.

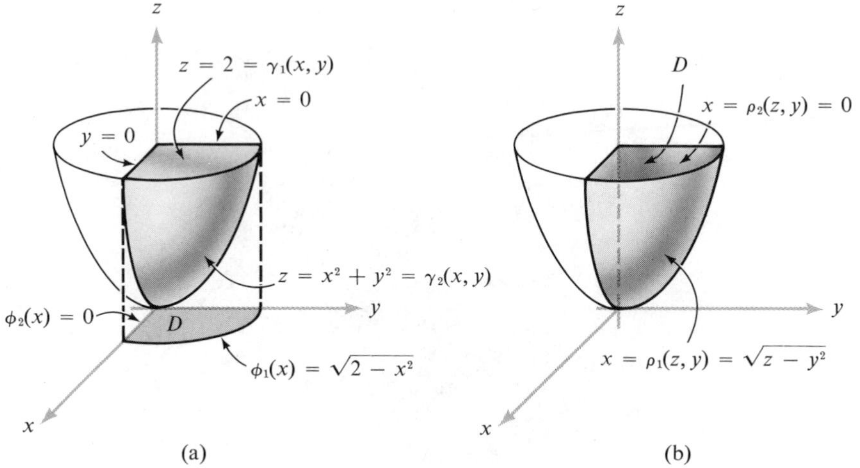

FIGURE 5.6.5

(a) A region of type 1 in $\mathbb{R}^3$ over a domain D. (b) Here D, denoted by the shaded area, is the set of (z, y) with $0 \leq z \leq 2$ and $0 \leq y \leq \sqrt{z}$.

The region W is sketched in Figure 5.6.5(a). To write this as a region of type 1, let $\gamma_1(x, y) = 2$, $\gamma_2(x, y) = x^2 + y^2$, $\phi_1(x) = \sqrt{2 - x^2}$, $\phi_2(x) = 0$, $a = 0$, and $b = \sqrt{2}$. Thus, by formula (3), p. 301,

$$\begin{aligned}
\int_W x\,dx\,dy\,dz &= \int_0^{\sqrt{2}} \left[\int_0^{(2-x^2)^{1/2}} \left(\int_{x^2+y^2}^{2} x\,dz \right) dy \right] dx \\
&= \int_0^{\sqrt{2}} \int_0^{(2-x^2)^{1/2}} x(2 - x^2 - y^2)\,dy\,dx \\
&= \int_0^{\sqrt{2}} x \left[(2 - x^2)^{3/2} - \frac{(2 - x^2)^{3/2}}{3} \right] dx \\
&= \int_0^{\sqrt{2}} \frac{2x}{3}(2 - x^2)^{3/2}\,dx = \left. \frac{-2(2 - x^2)^{5/2}}{15} \right|_0^{\sqrt{2}} \\
&= 2 \cdot \frac{2^{5/2}}{15} = \frac{8\sqrt{2}}{15}
\end{aligned}$$

We can also evaluate the integral by writing W as a region of type 2. We see that W can be expressed as the set of (x, y, z) with $\rho_2(z, y) = 0 \leq x \leq (z - y^2)^{1/2} = \rho_1(z, y)$ and $(z, y) \in D$, where D is the subset of the yz-plane with $0 \leq z \leq 2$ and $0 \leq y \leq z^{1/2}$ (see Figure 5.6.5(b)).

Therefore

$$\int_W x\,dx\,dy\,dz = \int_D \left[\int_{\rho_2(z,y)}^{\rho_1(z,y)} x\,dx\right] dy\,dz$$

$$= \int_0^2 \left[\int_0^{z^{1/2}} \left(\int_0^{(z-y^2)^{1/2}} x\,dx\right) dy\right] dz$$

$$= \int_0^2 \int_0^{z^{1/2}} \left(\frac{z-y^2}{2}\right) dy\,dz = \frac{1}{2}\int_0^2 \left(z^{3/2} - \frac{z^{3/2}}{3}\right) dz$$

$$= \frac{1}{2}\int_0^2 \frac{2}{3} z^{3/2}\,dz = \left[\frac{2}{15} z^{5/2}\right]_0^2 = \frac{2}{15}\cdot 2^{5/2} = \frac{8\sqrt{2}}{15}.$$

EXERCISES

1. Evaluate $\int_W x^2\,dV$ where $W = [0, 1] \times [0, 1] \times [0, 1]$.
2. Evaluate $\int_W e^{-xy} y\,dV$ where $W = [0, 1] \times [0, 1] \times [0, 1]$.
3. Evaluate $\int_W x^2 \cos z\,dV$ where W is the region bounded by the planes $z = 0$, $z = \pi$, $y = 0$, $y = \pi$, $x = 0$, and $x + y = 1$.
4. Find the volume of the region bounded by $z = x^2 + 3y^2$ and $z = 9 - x^2$.
5. Evaluate $\int_0^1 \int_0^{2x} \int_{x^2+y^2}^{x+y} dz\,dy\,dx$ and sketch the region of integration.
6. Find the volume of the solid bounded by the surfaces $x^2 + 2y^2 = 2$, $z = 0$, and $x + y + 2z = 2$.
7. Find the volume of the solid of revolution $z^2 \geq x^2 + y^2$ within the surface $x^2 + y^2 + z^2 = 1$.
8. Change the order of integration in

$$\int_0^1 \int_0^x \int_0^y f(x, y, z)\,dz\,dy\,dx$$

to obtain five other forms of the answer. Sketch the region.

*9. Let f be continuous and let B_ε be the ball of radius ε centered at the point (x_0, y_0, z_0). Let $|B_\varepsilon|$ be the volume of B_ε. Prove that

$$\operatorname*{limit}_{\varepsilon \to 0} \frac{1}{|B_\varepsilon|} \int_{B_\varepsilon} f(x, y, z)\,dV = f(x_0, y_0, z_0)$$

10. Find the volume of the region bounded by the surfaces $z = x^2 + y^2$ and $z = 10 - x^2 - 2y^2$. Sketch.
11. Let W be a bounded set whose boundary consists of graphs of continuous functions. Assume W is symmetric in the xy-plane: $(x, y, z) \in W$ implies $(x, y, -z) \in W$. Suppose f is a bounded continuous function on W and $f(x, y, z) = -f(x, y, -z)$. Prove that $\int_W f(x, y, z)\,dV = 0$.
12. Use the result of Exercise 11 to prove that $\int_W (1 + x + y)\,dV = 4\pi/3$, where W, the unit ball, is the set of (x, y, z) with $x^2 + y^2 + z^2 \leq 1$.

13. Evaluate $\iiint_S xyz\, dx\, dy\, dz$ where S is the region determined by the conditions $x \geq 0$, $y \geq 0$, $z \geq 0$, and $x^2 + y^2 + z^2 \leq 1$.
14. Let B be the region determined by the conditions $0 \leq x \leq 1$, $0 \leq y \leq 1$, and $0 \leq z \leq xy$.
    (a) Find the volume of B.
    (b) Evaluate $\iiint\limits_B x\, dx\, dy\, dz$.
    (c) Evaluate $\iiint\limits_B y\, dx\, dy\, dz$.
    (d) Evaluate $\iiint\limits_B z\, dx\, dy\, dz$.
    (e) Evaluate $\iiint\limits_B xy\, dx\, dy\, dz$.
15. For each of the following regions W, find appropriate limits $\phi_1(x)$, $\phi_2(x)$, $\gamma_1(x, y)$, and $\gamma_2(x, y)$, and write the triple integral over the region W as an iterated integral in the form

$$\iiint\limits_W f\, dV = \int_a^b \left[\int_{\phi_2(x)}^{\phi_1(x)} \left[\int_{\gamma_2(x, y)}^{\gamma_1(x, y)} f(x, y, z)\, dz \right] dy \right] dx$$

    (a) $W = \{(x, y, z) | \sqrt{x^2 + y^2} \leq z \leq 1\}$
    (b) $W = \{(x, y, z) | \frac{1}{2} \leq z \leq 1 \text{ and } x^2 + y^2 + z^2 \leq 1\}$
16. Let B be the region bounded by the planes $x = 0$, $y = 0$, $z = 0$, $x + y = 1$, and $z = x + y$.
    (a) Find the volume of B.
    (b) Evaluate $\iiint\limits_B x\, dx\, dy\, dz$.
    (c) Evaluate $\iiint\limits_B y\, dx\, dy\, dz$.

5.7 THE GEOMETRY OF MAPS FROM $\mathbb{R}^2$ TO $\mathbb{R}^2$

In Chapter 3 we studied vector fields on $\mathbb{R}^2$ and on $\mathbb{R}^3$. Now we wish to investigate these from a somewhat different point of view. We shall be interested in what maps from $\mathbb{R}^2$ to $\mathbb{R}^2$ and $\mathbb{R}^3$ to $\mathbb{R}^3$ do to subsets of these spaces. This geometric understanding will be useful in the text section when we discuss the Change of Variables formula for multiple integrals.

Let D^* be a subset of $\mathbb{R}^2$; suppose we consider a continuously differentiable map $T: D^* \to \mathbb{R}^2$. Then T takes points in D^* to points in $\mathbb{R}^2$. We denote this image set of points by D or by $T(D^*)$; hence, $D = T(D^*)$ is the set of all points $(x, y) \in \mathbb{R}^2$ such that

$$(x, y) = T(x^*, y^*) \quad \text{for some} \quad (x^*, y^*) \in D^*$$

One way to understand the geometry of the map T is to see how it *deforms* or changes D^*. For example, Figure 5.7.1 illustrates a map T that takes a slightly twisted region into a disc.

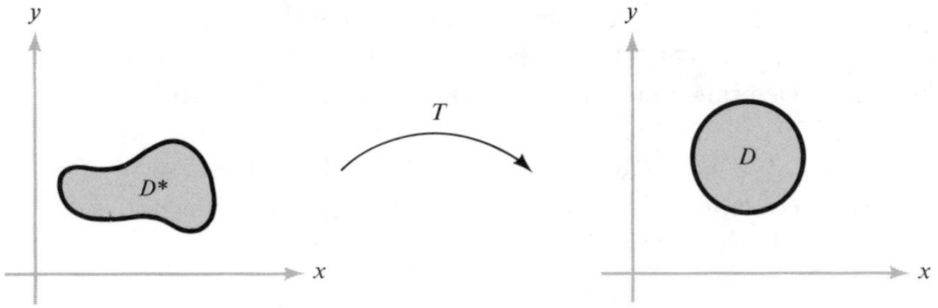

FIGURE 5.7.1
A function T from a domain D to a domain D.*

EXAMPLE 1. Let $D^* \subset \mathbb{R}^2$ be the rectangle $D^* = [0, 1] \times [0, 2\pi]$. Then all points in D^* are of the form (r, θ) where $0 \leq \theta \leq 2\pi, 0 \leq r \leq 1$. Let T be defined by $T(r, \theta) = (r \cos \theta, r \sin \theta)$. Find the image set D.

We set $(x, y) = (r \cos \theta, r \sin \theta)$. Since $x^2 + y^2 = r^2 \cos^2 \theta + r^2 \sin^2 \theta = r^2 \leq 1$, the set of points $(x, y) \in \mathbb{R}^2$ such that $(x, y) \in D$ has the property that $x^2 + y^2 \leq 1$, and so D is contained in the unit disc. In addition, any point (x, y) in the unit disc can be written as $(r \cos \theta, r \sin \theta)$ for $0 \leq r \leq 1$ and $0 \leq \theta \leq 2\pi$. Thus D is the unit disc (see Figure 5.7.2).

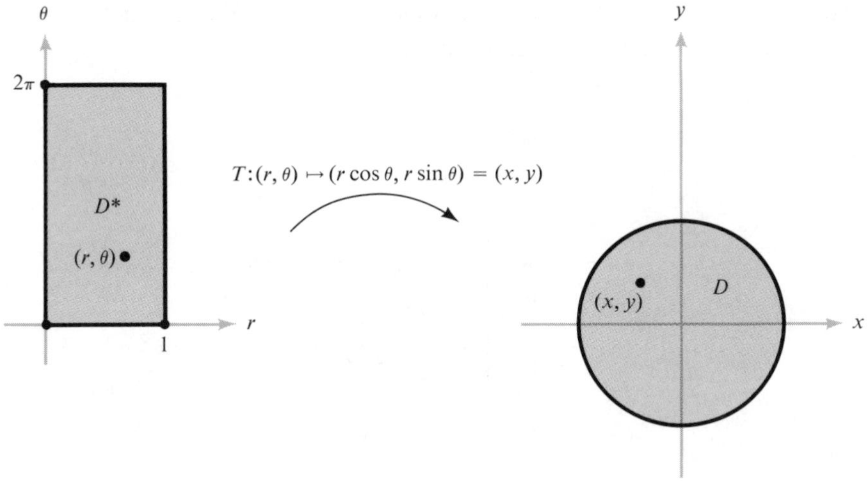

FIGURE 5.7.2
T gives a change of variables to polar coordinates. The unit circle is the image of a rectangle.

EXAMPLE 2. Let T be defined by $T(x, y) = ((x + y)/2, (x - y)/2)$. Let $D^* = [-1, 1] \times [-1, 1] \subset \mathbb{R}^2$ be a square with side of length 2 centered at the origin. Determine the image D obtained by applying T to D^*.

Let us first determine the effect of T on the line $\boldsymbol{\sigma}_1(t) = (t, 1)$, where $-1 \le t \le 1$. We have $T(\boldsymbol{\sigma}_1(t)) = ((t + 1)/2, (t - 1)/2)$. The map $t \mapsto T(\boldsymbol{\sigma}_1(t))$ is a parametrization of the line $y = x - 1$, $0 \le x \le 1$, since $(t - 1)/2 = (t + 1)/2 - 1$ (see Figure 5.7.3). Let

$$\boldsymbol{\sigma}_2(t) = (1, t), \qquad -1 \le t \le 1$$
$$\boldsymbol{\sigma}_3(t) = (t, -1), \qquad -1 \le t \le 1$$
$$\boldsymbol{\sigma}_4(t) = (-1, t), \qquad -1 \le t \le 1$$

be parametrizations of the other edges of the square D^*. Using the same argument as above, we see that $T \circ \boldsymbol{\sigma}_2$ is a parametrization of the line $y = 1 - x, 0 \le x \le 1$, $T \circ \boldsymbol{\sigma}_3$ the line $y = x + 1$, $-1 \le x \le 0$, and $T \circ \boldsymbol{\sigma}_4$ the line $y = -x - 1$, $-1 \le x \le 0$. By this time it seems reasonable to guess that T "flips" the square D^* over and takes it to the square D whose vertices are $(1, 0), (0, 1), (-1, 0), (0, -1)$ (Figure 5.7.4). To prove that this is indeed the case, let $-1 \le \alpha \le 1$ and let L_α (Figure 5.7.3) be a fixed line parametrized by $\boldsymbol{\sigma}(t) = (\alpha, t)$, $-1 \le t \le 1$; then $T(\boldsymbol{\sigma}(t)) = ((\alpha + t)/2, (\alpha - t)/2)$ is a parametrization of the line $y = -x + \alpha$, $(\alpha - 1)/2 \le x \le (\alpha + 1)/2$. This line begins, for $t = -1$, at the point $((\alpha - 1)/2, (1 + \alpha)/2)$ and ends up at the point $((1 + \alpha)/2, (\alpha - 1)/2)$; as is easily checked, these points lie on the lines $T \circ \boldsymbol{\sigma}_3$ and $T \circ \boldsymbol{\sigma}_1$, respectively. Thus as α varies between -1 and 1, L_α sweeps out the square D^* while $T(L_\alpha)$ sweeps out the square D determined by the vertices $(-1, 0)$, $(0, 1)$, $(1, 0)$, and $(0, -1)$.

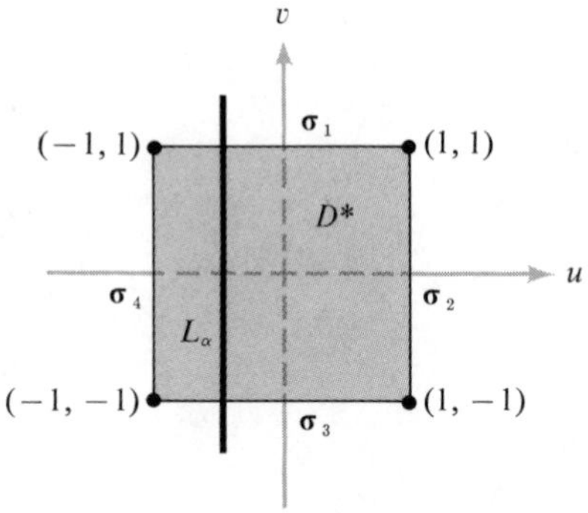

FIGURE 5.7.3
Domain for the transformation T of Example 2.

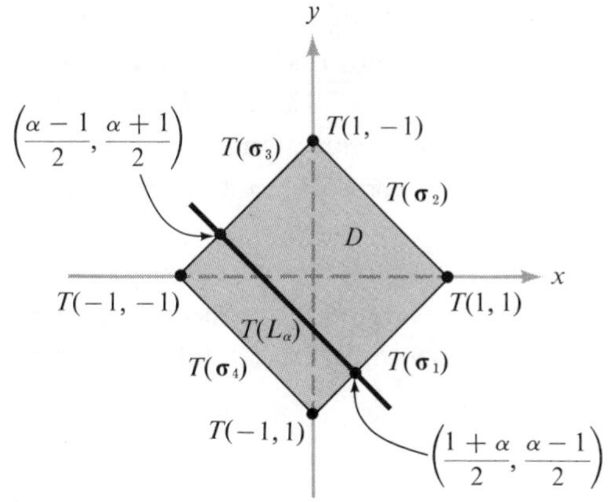

FIGURE 5.7.4
The effect of T on the region D.*

The following theorem supplies an easier way to describe the image $T(D^*)$.

THEOREM 6. *Let A be a* 2×2 *matrix with* $\det A \neq 0$ *and let T be the linear mapping of* $\mathbb{R}^2$ *to* $\mathbb{R}^2$ *given by* $T(x) = Ax$ *(matrix multiplication); see Example 4, Section 2.4. Then T transforms parallelograms into parallelograms and vertices into vertices. Moreover, if* $T(D^*)$ *is a parallelogram,* D^* *must be a parallelogram.*

The proof of Theorem 6 is left as an exercise at the end of this section (see Exercise 8). This theorem simplifies the result of Example 2 because we need only find the vertices of $T(D^*)$ and then connect them by straight lines.

Although we cannot visualize the graph of a function $T: \mathbb{R}^2 \to \mathbb{R}^2$, it does help to consider how the function deforms subsets. However, simply looking at these deformations does not give us a complete picture of the behavior of T. We may characterize T further using the notion of a one-to-one correspondence.

DEFINITION. *The function T is* ***one-to-one*** *on* D^* *if* $T(u, v) = T(u', v')$ *for* (u, v) *and* $(u', v') \in D^*$ *implies that* $u = u'$ *and* $v = v'$.

Geometrically, this statement means that two different points of D^* do not get sent into the same point of D by T. For example, the function $T(x, y) = (x^2 + y^2, y^4)$ is not one-to-one because $T(1, -1) = (2, 1) = T(1, 1)$ yet $(1, -1) \neq (1, 1)$. In other words, a function is one-to-one when it does not collapse two different points together.

EXAMPLE 3. Consider the function $T: \mathbb{R}^2 \to \mathbb{R}^2$ of Example 1, $T(r, \theta) = (r \cos \theta, r \sin \theta)$. Show that T is not one-to-one if its domain is all of $\mathbb{R}^2$.

If $\theta_1 \neq \theta_2$, then $T(0, \theta_1) = T(0, \theta_2)$ and so T cannot be one-to-one. This observation implies that if L is the side of the rectangle $D^* = [0, 1] \times [0, 2\pi]$ (Figure 5.7.5), where $0 \leq \theta \leq 2\pi$ and $r = 0$, then T maps all of L into a single point, the center of the unit disc D.

However, if we consider the set $S^* =]0, 1] \times]0, 2\pi]$ then $T: S^* \to S$ is one-to-one (see Exercise 1). Evidently, in determining whether a function is one-to-one, the domain chosen must be carefully considered.

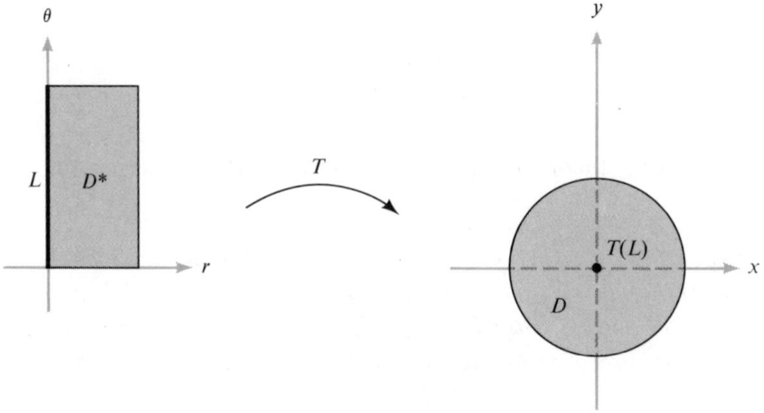

FIGURE 5.7.5
The polar-coordinate transformation T takes the line L to the point (0, 0).

EXAMPLE 4. Show that the function $T: \mathbb{R}^2 \to \mathbb{R}^2$ of Example 2 is one-to-one.

Suppose $T(x, y) = T(x', y')$; then

$$((x + y)/2, (x - y)/2) = ((x' + y')/2, (x' - y')/2)$$

and we have

$$x + y = x' + y'$$

$$x - y = x' - y'$$

Adding, we have

$$2x = 2x'$$

Thus $x = x'$ and hence $y = y'$, which shows that T is one-to-one (with domain all of $\mathbb{R}^2$). Actually, since T is linear and $T(x) = Ax$, where A is a 2×2 matrix, it suffices to show that $\det A \neq 0$ (see Exercise 6).

In Examples 1 and 2 we have been determining the image $D = T(D^*)$ of a region D^* under a mapping T. What will be of interest to us in the next section is, in part, the inverse problem; namely, given D and a one-to-one mapping T of $\mathbb{R}^2$ to $\mathbb{R}^2$, find D^* such that $T(D^*) = D$.

Before we examine this question in more detail we introduce the notion of "onto."

DEFINITION. *The function T is **onto** D if for every point $(x, y) \in D$ there exists a (u, v) in the domain of T such that $T(u, v) = (x, y)$.*

Thus if T is onto, we can solve the equation $T(u,v) = (x, y)$ for every $(x, y) \in D$. If T is, in addition, one-to-one, this solution is unique.

For *linear* mappings T of $\mathbb{R}^2$ to $\mathbb{R}^2$ (or $\mathbb{R}^n$ to $\mathbb{R}^n$) it turns out that one-to-one and onto are equivalent notions (see exercises 6 and 7).

If we are given a region D and a mapping T, the determination of a region D^* such that $T(D^*) = D$ will only be possible when for every $(x, y) \in D$ there is a (u, v) in the domain of T such that $T(u, v) = (x, y)$ (i.e., T must be onto D). The next example shows that this cannot always be done.

EXAMPLE 5. Let $T: \mathbb{R}^2 \to \mathbb{R}^2$ be given by $T(u, v) = (u, 0)$. Let $D = [0, 1] \times [0, 1]$. Since T takes all of $\mathbb{R}^2$ to one axis it is impossible to find a D^* such that $T(D^*) = D$.

EXAMPLE 6. Let T be defined as in Example 2 and let D be the square whose vertices are $(1, 0), (0, 1), (-1, 0), (0, -1)$. Find a D^* with $T(D^*) = D$.

Since T is linear and $T(x) = Ax$, where A is a 2×2 matrix with $\det A \neq 0$ we know that $T: \mathbb{R}^2 \to \mathbb{R}^2$ is onto (see Exercises 6 and 7), and thus D^* can be found. By Theorem 5, D^* must be a parallelogram. In order to find D^* it suffices to find the four points that are mapped onto the vertices of D, then, by connecting these points, we will have found D^*. For the vertex $(1, 0)$ of D we must solve $T(x, y) = (1, 0) = ((x + y)/2, (x - y)/2)$ so $(x + y)/2 = 1$, $(x - y)/2 = 0$. Thus $(x, y) = (1, 1)$ is a vertex of D^*. Solving for the other vertices we find that $D^* = [-1, 1] \times [-1, 1]$.

EXAMPLE 7. Let D be the region in the first quadrant lying between the arcs of the circles $x^2 + y^2 = a^2$, $x^2 + y^2 = b^2$, $(0 < a < b)$ (see Figure 5.7.6). These circles have equations $r = a$ and $r = b$ in polar coordinates. Let T be the polar coordinate transformation given by $T(r, \theta) = (r \cos\theta, r \sin\theta) = (x, y)$. Let us find D^* such that $T(D^*) = D$.

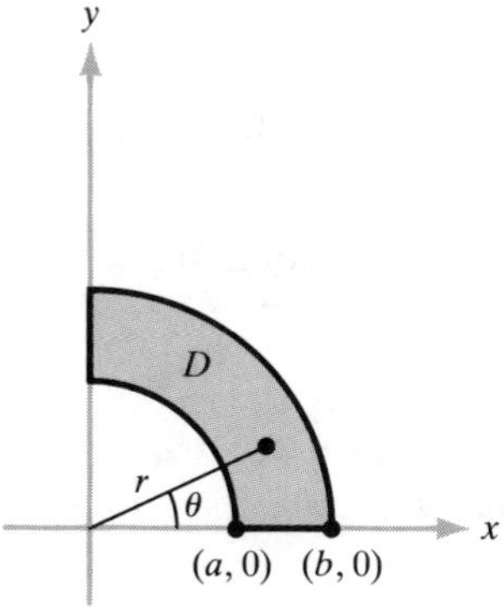

FIGURE 5.7.6
We seek a region D^* in the θr-plane whose image under the polar coordinate mapping is D.

In the region D, $a^2 \leq x^2 + y^2 \leq b^2$, and, since $r^2 = x^2 + y^2$, we see that $a \leq r \leq b$. Clearly for this region θ varies between $0 \leq \theta \leq \pi/2$. Thus if $D^* = [a, b] \times [0, \pi/2]$ it is not hard to see that $T(D^*) = D$ and T is one-to-one.

OPTIONAL

A REMARK ON THE INVERSE FUNCTION THEOREM AND MAPS FROM $\mathbb{R}^2$ TO $\mathbb{R}^2$.

The Inverse Function Theorem discussed in Section 4.4 is relevant to the material here. It states that if the determinant of $DT(u_0, v_0)$ (which is the matrix of partial derivatives of T evaluated at (u_0, v_0)) is not zero, then for (u, v) near (u_0, v_0) and (x, y) near $(x_0, y_0) = T(u_0, v_0)$, the equation $T(u, v) = (x, y)$ can be *uniquely solved* for (u, v) as functions of (x, y). In particular, by uniqueness, T is one-to-one near (u_0, v_0); also, T is onto a neighborhood of (x_0, y_0) since $T(u, v) = (x, y)$ is solvable for (u, v) if (x, y) is near (x_0, y_0). If D^* and D are elementary regions and $T: D^* \to D$ has the property that the determinant of $DT(u, v)$ is not zero for any (u, v) and T maps D^* onto D, then one can prove that T is one-to-one on D^*. (This proof is beyond the scope of this text.)

EXERCISES

1. Let $S^* =]0, 1] \times]0, 2\pi]$ and define $T(r, \theta) = (r \cos \theta, r \sin \theta)$. Determine the image set S. Show that T is one-to-one on S^*.

2. Define

$$T(x^*, y^*) = \left(\frac{x^* - y^*}{\sqrt{2}}, \frac{x^* + y^*}{\sqrt{2}}\right)$$

Show that T rotates the unit square, $D^* = [0, 1] \times [0, 1]$.

3. Let $D^* = [0, 1] \times [0, 1]$ and define T on D^* by $T(u, v) = (-u^2 + 4u, v)$. Find D. Is T one-to-one?

4. Let D^* be the parallelogram bounded by the lines $y = 3x - 4$, $y = 3x$, $y = \frac{1}{2}x$, and $y = \frac{1}{2}(x + 4)$. Let $D = [0, 1] \times [0, 1]$. Find a T such that D is the image of D^* under T.

5. Let $T: \mathbb{R}^3 \to \mathbb{R}^3$ be defined by $(\rho, \phi, \theta) \mapsto (x, y, z)$, where

$$x = \rho \sin \phi \cos \theta, \qquad y = \rho \sin \phi \sin \theta, \qquad z = \rho \cos \phi$$

Let D^* be the set of points (ρ, ϕ, θ) such that $\phi \in [0, \pi]$, $\theta \in [0, 2\pi]$, $\rho \in [0, 1]$. Find $D = T(D^*)$. Is T one-to-one? If not, can we eliminate some subset of D^* (as we did, in Exercise 1, to D^* in Example 1) so that, on the remainder, T will be one-to-one?

In Exercises 6 and 7 let $T(x) = Ax$, where A is a 2×2 matrix.

6. Show T is one-to-one if and only if the determinant of A is not zero.

7. Show $\det A \neq 0$ if and only if T is onto.

8. Suppose $T: \mathbb{R}^2 \to \mathbb{R}^2$ is linear; $T(x) = Ax$, where A is a 2×2 matrix. Show that if $\det A \neq 0$, T takes parallelograms onto parallelograms. (HINT: The general parallelogram in $\mathbb{R}^2$ can be described by the set of points $\mathbf{q} = \mathbf{p} + \lambda\mathbf{v} + \mu\mathbf{w}$ for $\lambda, \mu \in [0, 1]$ where $\mathbf{p}, \mathbf{v}, \mathbf{w}$ are vectors in $\mathbb{R}^2$ with $\mathbf{v}$ not a scalar multiple of $\mathbf{w}$.)

9. Suppose $T: \mathbb{R}^2 \to \mathbb{R}^2$ is as in Exercise 8 and that $T(P^*) = P$ is a parallelogram. Show that P^* is a parallelogram.

5.8 THE CHANGE OF VARIABLES THEOREM

Given two regions D and D^* of type 1 or 2 in $\mathbb{R}^2$, a differentiable map T on D^* with image D, that is, $T(D^*) = D$, and any real-valued integrable function $f: D \to \mathbb{R}$, we would like to express $\int_D f(x, y)\, dA$ as an integral over D^* of the composite function $f \circ T$. In this section we shall see how to do this.

Assume the region D^* is a subset of $\mathbb{R}^2$ of type 1 with the coordinate variables designated by (u, v). Furthermore, assume that D is a subset of type 1 of the xy-plane. The map T is given by two coordinate functions

$$T(u, v) = (x(u, v), y(u, v)) \quad \text{for} \quad (u, v) \in D^*$$

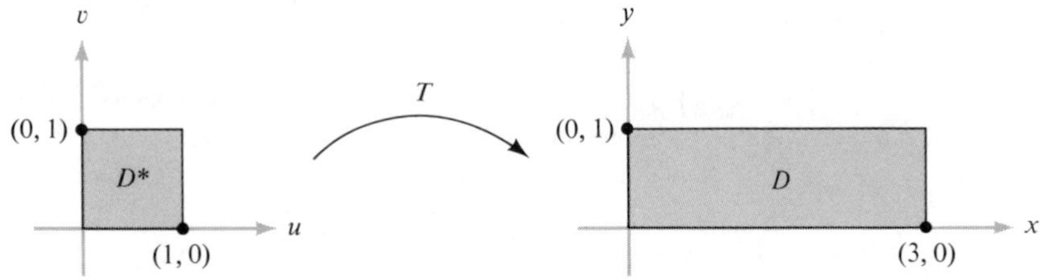

FIGURE 5.8.1
The map $T: (u, v) \mapsto (-u^2 + 4u, v)$ takes the square D^* onto the rectangle D.

Now, as a first guess one might conjecture that

$$\int_D f(x, y)\, dx\, dy \stackrel{?}{=} \int_{D^*} f(x(u, v), y(u, v))\, du\, dv \tag{1}$$

where $f \circ T(u, v) = f(x(u, v), y(u, v))$ is the composite function defined on D^*. However, if we consider the function $f: D \to \mathbb{R}^2$ where $f(x, y) = 1$, then equation (1) would imply

$$A(D) = \int_D dx\, dy \stackrel{?}{=} \int_{D^*} du\, dv = A(D^*) \tag{2}$$

It is easy to see that (2) will hold only for a few special cases and not for a general map T. For example, define T by $T(u, v) = (-u^2 + 4u, v)$. Restrict T to the unit square $D^* = [0, 1] \times [0, 1]$ in the uv-plane (see Figure 5.8.1). Then, as in Exercise 3, Section 5.7, T takes D^* onto $D = [0, 3] \times [0, 1]$. Clearly $A(D) \neq A(D^*)$, and so formula (2) is not valid.

What is needed here is a measure of how a transformation $T: \mathbb{R}^2 \to \mathbb{R}^2$ distorts the area of a region. This is given by the *Jacobian determinant*, which is defined as follows.

DEFINITION. *Let $T: D^* \subset \mathbb{R}^2 \to \mathbb{R}^2$ be a C^1 transformation given by $x = x(u, v)$ and $y = y(u, v)$. The **Jacobian** of T, written $\partial(x, y)/\partial(u, v)$, is the determinant of the derivative matrix $DT(x, y)$ of T:*

$$\frac{\partial(x, y)}{\partial(u, v)} = \begin{vmatrix} \dfrac{\partial x}{\partial u} & \dfrac{\partial x}{\partial v} \\ \dfrac{\partial y}{\partial u} & \dfrac{\partial y}{\partial v} \end{vmatrix}$$

EXAMPLE 1. The function from $\mathbb{R}^2$ to $\mathbb{R}^2$ that transforms polar coordinates into Cartesian coordinates is given by

$$x = r \cos \theta, \qquad y = r \sin \theta$$

and its Jacobian is

$$\frac{\partial(x, y)}{\partial(r, \theta)} = \begin{vmatrix} \cos\theta & -r\sin\theta \\ \sin\theta & r\cos\theta \end{vmatrix} = r(\cos^2\theta + \sin^2\theta) = r$$

If we make suitable restrictions on the function T, we can show that the area of $D = T(D^*)$ is obtained by integrating the absolute value of the Jacobian $\partial(x, y)/\partial(u, v)$ over D^*; that is, we have the equations

$$A(D) = \int_D dx\, dy = \int_{D^*} \left|\frac{\partial(x, y)}{\partial(u, v)}\right| du\, dv \tag{3}$$

To illustrate, from Example 1 in Section 5.7 take $T\colon D^* \to D$, where $D = T(D^*)$ is the set of (x, y) with $x^2 + y^2 \leq 1$ and $D^* = [0, 1] \times [0, 2\pi]$, and $T(r, \theta) = (r\cos\theta, r\sin\theta)$. By formula (3)

$$A(D) = \int_{D^*} \left|\frac{\partial(x, y)}{\partial(r, \theta)}\right| dr\, d\theta = \int_{D^*} r\, dr\, d\theta \tag{4}$$

(here r and θ play the role of u and v). From the above computation it follows that

$$\int_{D^*} r\, dr\, d\theta = \int_0^{2\pi}\int_0^1 r\, dr\, d\theta = \int_0^{2\pi}\left[\frac{r^2}{2}\bigg|_0^1\right] d\theta = \frac{1}{2}\int_0^{2\pi} d\theta = \pi$$

is the area of D, confirming formula (3) in this case. In fact, we may recall from first-year calculus that (4) is the correct formula for the area of a region in polar coordinates.

It is not easy to prove rigorously assertion (3), that the Jacobian determinant is a measure of how a transformation distorts area. However, looked at in the proper way, it becomes quite plausible.

Recall that $A(D) = \int_D dx\, dy$ was obtained by dividing up D into little rectangles, summing their areas, and then taking the limit of this sum as the size of the subrectangles tended to zero (see Example 3, Section 5.3). The problem is that T may map rectangles into regions whose area is not easy to compute. The solution is to approximate these images by simpler regions whose area we can compute. A useful tool for doing this is the derivative of T, which we know (from Chapter 2) gives the best linear approximation to T.

Consider a small rectangle D^* in the uv-plane as shown in Figure 5.8.2. Let T' denote the derivative of T evaluated at (u_0, v_0), so T' is a 2×2 matrix. From our work in Chapter 2 (see p. 110) we know that a good approximation to $T(u, v)$ is given by

$$T(u_0, v_0) + T'\begin{pmatrix}\Delta u \\ \Delta v\end{pmatrix}$$

where $\Delta u = u - u_0$ and $\Delta v = v - v_0$. But this mapping takes D^* into a parallelogram with vertex at $T(u_0, v_0)$ and with adjacent sides given by the

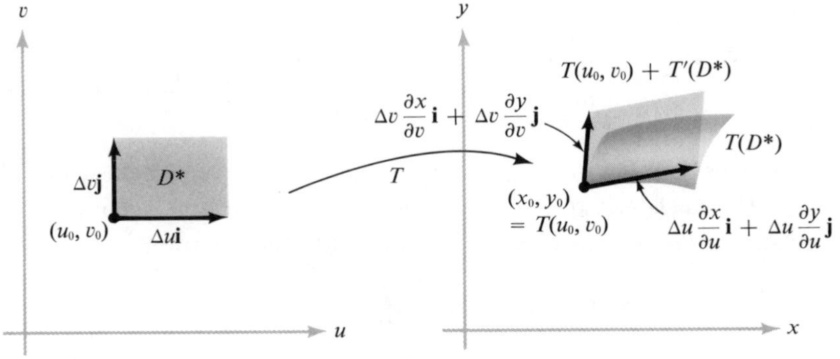

FIGURE 5.8.2
Effect of the transformation T on a small rectangle D^*.

vectors

$$T'(\Delta u\mathbf{i}) = \begin{bmatrix} \frac{\partial x}{\partial u} & \frac{\partial x}{\partial v} \\ \frac{\partial y}{\partial u} & \frac{\partial y}{\partial v} \end{bmatrix} \begin{bmatrix} \Delta u \\ 0 \end{bmatrix} = \Delta u \begin{bmatrix} \frac{\partial x}{\partial u} \\ \frac{\partial y}{\partial u} \end{bmatrix} = \Delta u \mathbf{T}_u$$

$$T'(\Delta v\mathbf{j}) = \begin{bmatrix} \frac{\partial x}{\partial u} & \frac{\partial x}{\partial v} \\ \frac{\partial y}{\partial u} & \frac{\partial y}{\partial v} \end{bmatrix} \begin{bmatrix} 0 \\ \Delta v \end{bmatrix} = \Delta v \begin{bmatrix} \frac{\partial x}{\partial v} \\ \frac{\partial y}{\partial v} \end{bmatrix} = \Delta v \mathbf{T}_v$$

where

$$\mathbf{T}_u = \frac{\partial x}{\partial u}\mathbf{i} + \frac{\partial y}{\partial u}\mathbf{j}$$

$$\mathbf{T}_v = \frac{\partial x}{\partial v}\mathbf{i} + \frac{\partial y}{\partial v}\mathbf{j}$$

are evaluated at (u_0, v_0).

We know from Section 1.3 that the area of the parallelogram with sides equal to the vectors $a\mathbf{i} + b\mathbf{j}$ and $c\mathbf{i} + d\mathbf{j}$ is equal to the absolute value of the determinant

$$\begin{vmatrix} a & b \\ c & d \end{vmatrix} = \begin{vmatrix} a & c \\ b & d \end{vmatrix}$$

Thus the area of $T(D^*)$ is approximately equal to the absolute value of

$$\begin{vmatrix} \frac{\partial x}{\partial u}\Delta u & \frac{\partial x}{\partial v}\Delta v \\ \frac{\partial y}{\partial u}\Delta u & \frac{\partial y}{\partial v}\Delta v \end{vmatrix} = \begin{vmatrix} \frac{\partial x}{\partial u} & \frac{\partial x}{\partial v} \\ \frac{\partial y}{\partial u} & \frac{\partial y}{\partial v} \end{vmatrix} \Delta u\, \Delta v = \frac{\partial(x, y)}{\partial(u, v)} \Delta u\, \Delta v$$

evaluated at (u_0, v_0). But the absolute value of this is just $|\partial(x, y)/\partial(u, v)|\, \Delta u\, \Delta v$.

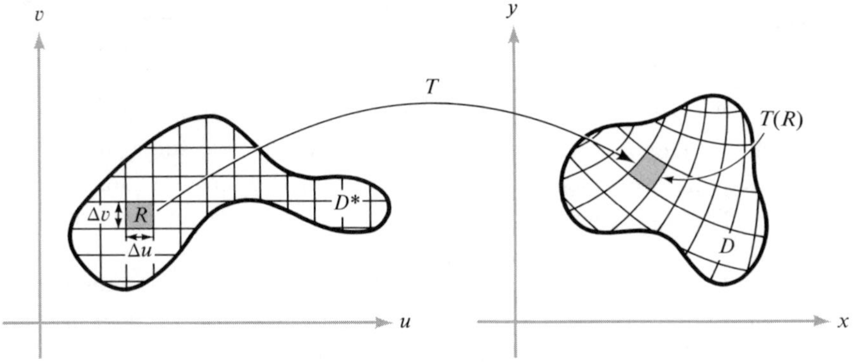

FIGURE 5.8.3

The area of the little rectangle R is $\Delta u\,\Delta v$. The area of $T(R)$ is approximately $|\partial(x, y)/\partial(u, v)|\,\Delta u\,\Delta v$.

This fact and a partitioning argument should make formula (3) plausible. Indeed, if we partition D^* into small rectangles with sides of length Δu and Δv, the images of these rectangles are approximated by parallelograms with sides $\mathbf{T}_u\,\Delta u$ and $\mathbf{T}_v\,\Delta v$ and hence with area $|\partial(x, y)/\partial(u, v)|\,\Delta u\,\Delta v$. Thus the area of D^* is approximately $\sum \Delta u\,\Delta v$, where the sum is taken over all the rectangles R inside D^* (see Figure 5.8.3). Hence the area of $T(D^*)$ is approximately the sum $\sum |\partial(x, y)/\partial(u, v)|\,\Delta u\,\Delta v$. In the limit, this sum becomes

$$\int_{D^*} \left|\frac{\partial(x, y)}{\partial(u, v)}\right| du\,dv.$$

Let us give another argument for the special case (4) of formula (3), that is, the case of polar coordinates. Consider a region D in the xy-plane and a grid corresponding to a partition of the r and θ variables (Figure 5.8.4). The area of the shaded region shown is approximately $(\Delta r)(r_{jk}\,\Delta\theta)$, since the arc

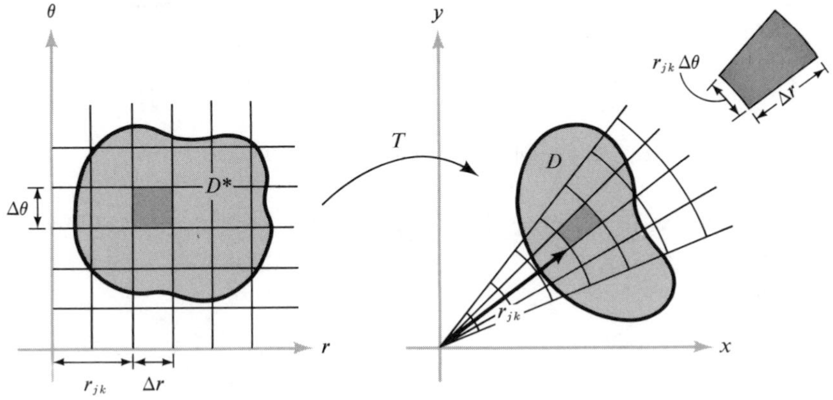

FIGURE 5.8.4

D^* gets mapped to D under the polar-coordinate mapping T.

length of a segment of a circle of radius R subtending an angle ϕ is $R\phi$. The total area is then the limit of $\sum r_{jk}\,\Delta r\,\Delta\theta$; that is, $\int_{D^*} r\,dr\,d\theta$. The key idea is thus that the jkth "polar rectangle" in the grid has area approximately equal to $r_{jk}\,\Delta r\,\Delta\theta$. (For n large, the jkth polar rectangle will look like a rectangle with sides of length $r_{jk}\,\Delta\theta$ and Δr.) This should provide some insight into why we say the "area element $\Delta x\,\Delta y$" is transformed into the "area element $r\,\Delta r\,\Delta\theta$." The following example explicates these ideas for a special case.

EXAMPLE 2. Let the elementary region D in the xy-plane be bounded by the graph of a polar equation $r = f(\theta)$ where $\theta_0 \le \theta \le \theta_1$ and $f(\theta) \ge 0$ (see Figure 5.8.5). In the $r\theta$-plane we consider the type-2 region D^* where $\theta_0 \le \theta \le \theta_1$ and $0 \le r \le f(\theta)$. Under the transformation $x = r\cos\theta$, $y = r\sin\theta$, the region D^* is carried onto the region D. We have

$$\begin{aligned} A(D) = \int_D dx\,dy &= \int_{D^*} \left|\frac{\partial(x, y)}{\partial(r, \theta)}\right| dr\,d\theta \\ &= \int_{D^*} r\,dr\,d\theta = \int_{\theta_0}^{\theta_1}\left[\int_0^{f(\theta)} r\,dr\right] d\theta \\ &= \int_{\theta_0}^{\theta_1}\left[\frac{r^2}{2}\right]_0^{f(\theta)} d\theta = \int_{\theta_0}^{\theta_1} \frac{[f(\theta)]^2}{2}\,d\theta \end{aligned}$$

This formula for $A(D)$ may be familiar from one-variable calculus.

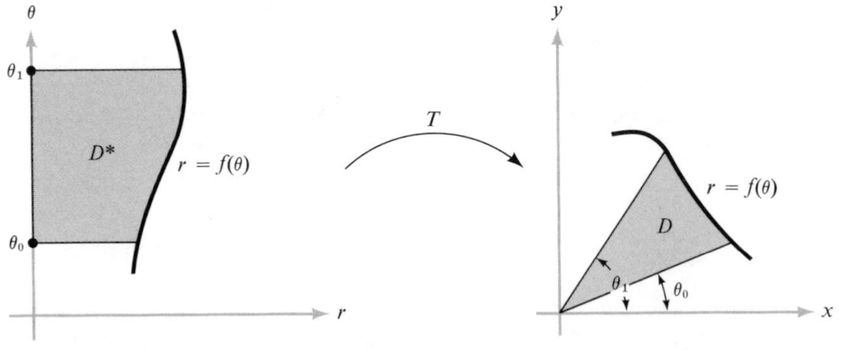

FIGURE 5.8.5
Effect on the region D^* of the polar coordinate mapping.

Before stating the change of variables formula, which is the culmination of the above discussion, let us recall the corresponding result from one-variable calculus:

$$\int_a^b f(x(u))\frac{dx}{du}\,du = \int_{x(a)}^{x(b)} f(x)\,dx \tag{5}$$

where f is continuous and $u \mapsto x(u)$ is continuously differentiable on $[a, b]$.

Proof. Let F be an antiderivative of f; that is, $F' = f$, which is possible by the Fundamental Theorem of Calculus. The right-hand side of (5) becomes

$$\int_{x(a)}^{x(b)} f(x)\, dx = F(x(b)) - F(x(a))$$

To evaluate the left hand side of (5), let $G(u) = F(x(u))$. By the Chain Rule, $G'(u) = F'(x(u))x'(u) = f(x(u))x'(u)$. Hence, again by the Fundamental Theorem,

$$\int_a^b f(x(u))x'(u)\, du = \int_a^b G'(u)\, du = G(b) - G(a) = F(x(b)) - F(x(a))$$

as required. ■

Suppose now that the C^1 function $u \mapsto x(u)$ is one-to-one on $[a, b]$. Thus, we must have either $dx/du \geq 0$ on $[a, b]$ or $dx/du \leq 0$ on $[a, b]$. (If dx/du is positive and then negative, the function $x = x(u)$ rises and then falls, and thus is not one-to-one; a similar statement applies if dx/du is negative then positive.) Let I^* denote the interval $[a, b]$, and let I denote the closed interval with endpoints $x(a)$ and $x(b)$. (Thus, $I = [x(a), x(b)]$ if $u \mapsto x(u)$ is increasing and $I = [x(b), x(a)]$ if $u \mapsto x(u)$ is decreasing.) With these notations we can rewrite formula (5) as

$$\int_{I^*} f(x(u)) \left|\frac{dx}{du}\right| du = \int_I f(x)\, dx$$

This is the formula that generalizes to double integrals: I^* becomes D^*, I becomes D, and $|dx/du|$ is replaced by $|\partial(x, y)/\partial(u, v)|$.

Let us state the result formally (the technical proof is omitted).

THEOREM 7 (CHANGE OF VARIABLES FOR DOUBLE INTEGRALS). *Let D and D^* be elementary regions in the plane and let $T: D^* \to D$ be C^1; suppose that T is one-to-one on D^*. Furthermore, suppose that $D = T(D^*)$. Then for any integrable function $f: D \to \mathbb{R}$, we have*

$$\int_D f(x, y)\, dx\, dy = \int_{D^*} f(x(u, v), y(u, v)) \left|\frac{\partial(x, y)}{\partial(u, v)}\right| du\, dv \qquad (6)$$

One of the purposes of the Change of Variables Theorem is to supply a method by which some double integrals can be simplified. One might encounter an integral $\int_D f\, dA$ on which either the integrand f or the region D is complicated and for which direct evaluation is difficult. Therefore, a T is chosen so that the integral is easier to evaluate with the new integrand $f \circ T$ or with the new region D^* (defined by $T(D^*) = D$). Unfortunately, the problem may actually become more complicated if T is not selected carefully.

EXAMPLE 3. Let P be the parallelogram bounded by $y = 2x$, $y = 2x - 2$, $y = x$, and $y = x + 1$ (see Figure 5.8.6). Evaluate $\int_P xy\, dx\, dy$ by making the

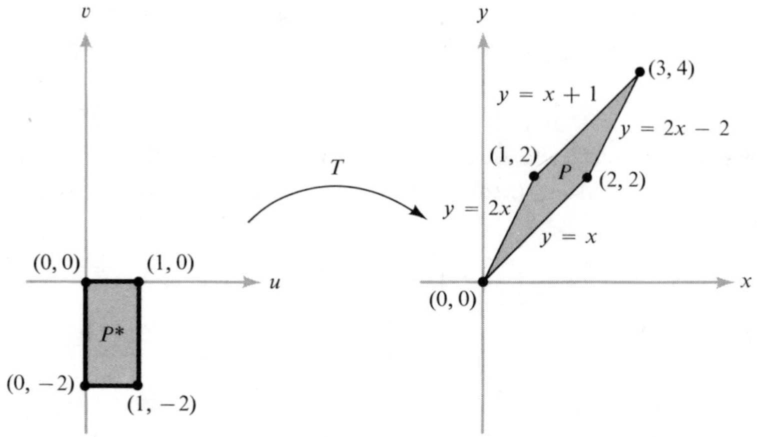

FIGURE 5.8.6
The effect of $T(u, v) = (u - v, 2u - v)$ on the rectangle P^*.

change of variables

$$x = u - v, \qquad y = 2u - v$$

that is, $T(u, v) = (u - v, 2u - v)$.

The transformation T is one-to-one (see Exercise 6, Section 5.7) and is designed so that it takes the *rectangle* P^* bounded by $v = 0$, $v = -2$, $u = 0$, $u = 1$ onto P. The use of T simplifies the region of integration from P to P^*. Moreover,

$$\left|\frac{\partial(x, y)}{\partial(u, v)}\right| = \left|\det\begin{bmatrix} 1 & -1 \\ 2 & -1 \end{bmatrix}\right| = 1$$

Therefore

$$\begin{aligned}
\int_P xy\, dx\, dy &= \int_{P^*} (u - v)(2u - v)\, du\, dv \\
&= \int_{-2}^{0}\int_0^1 (2u^2 - 3vu + v^2)\, du\, dv \\
&= \int_{-2}^{0}\left[\frac{2}{3}u^3 - \frac{3u^2v}{2} + v^2u\right]_0^1 dv \\
&= \int_{-2}^{0}\left[\frac{2}{3} - \frac{3}{2}v + v^2\right] dv \\
&= \left[\frac{2}{3}v - \frac{3}{4}v^2 + \frac{v^3}{3}\right]_{-2}^{0} = -\left[\frac{2}{3}(-2) - 3 - \frac{8}{3}\right] \\
&= -\left[-\frac{12}{3} - 3\right] = 7
\end{aligned}$$

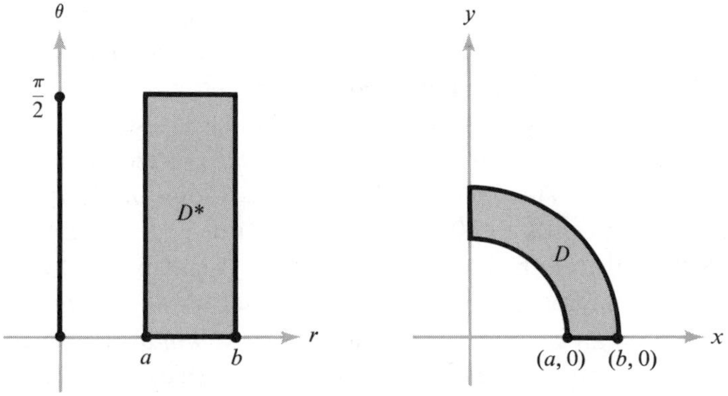

FIGURE 5.8.7

The polar-coordinate mapping takes a rectangle D^* onto part of an annulus D.

EXAMPLE 4. Evaluate $\int_D \log(x^2 + y^2)\, dx\, dy$, where D is the region in the first quadrant lying between the arcs of the circles

$$x^2 + y^2 = a^2, \qquad x^2 + y^2 = b^2 \qquad (0 < a < b)$$

(see Figure 5.8.7). These circles have the simple equations $r = a$ and $r = b$ in polar coordinates. Moreover, $r^2 = x^2 + y^2$ appears in the integrand. Thus, a change to polar coordinates, will simplify both the integrand and the region of integration. From Example 7, Section 5.7, the polar coordinate transformation

$$x = r\cos\theta, \qquad y = r\sin\theta$$

sends the rectangle D^* given by $a \le r \le b$, $0 \le \theta \le \frac{1}{2}\pi$ onto the region D. This transformation is one-to-one on D^* and so, by Theorem 7, we have

$$\int_D \log(x^2 + y^2)\, dx\, dy = \int_{D^*} \log r^2 \left|\frac{\partial(x, y)}{\partial(r, \theta)}\right| dr\, d\theta$$

Now $|\partial(x, y)/\partial(r, \theta)| = r$, as we have seen before; hence the right-hand integral becomes

$$\int_a^b \int_0^{\pi/2} r \log r^2\, d\theta\, dr = \frac{\pi}{2}\int_a^b r\log r^2\, dr = \frac{\pi}{2}\int_a^b 2r\log r\, dr$$

Applying integration by parts, or using the formula

$$\int x\log x\, dx = \frac{x^2}{2}\log x - \frac{x^2}{4}$$

from the table of integrals in Appendix C, we obtain the result

$$\frac{\pi}{2}\int_a^b 2r\log r\, dr = \frac{\pi}{2}\left(b^2\log b - a^2\log a - \frac{1}{2}(b^2 - a^2)\right)$$

Suppose we consider the rectangle D^* defined by $0 \le \theta \le 2\pi$, $0 \le r \le a$ in the $r\theta$-plane. Then the transformation T given by $T(r, \theta) = (r\cos\theta, r\sin\theta)$ takes D^* onto the disc D with equation $x^2 + y^2 \le a^2$ in the xy-plane. This transformation represents the change from Cartesian coordinates to polar coordinates. However, T does not satisfy the requirements of the Change of Variables Theorem since it is not one-to-one on D^*: In particular, T sends all points with $r = 0$ to $(0, 0)$ (see Figure 5.8.8 and Example 3 of Section 5.7). However, the Change of Variables Theorem is still valid in this case. Basically, the reason for this is that the set of points where T is not one-to-one is on an edge of D^*, which is the graph of a smooth curve and therefore, for the purpose of integration, can be neglected. In summary, the formula

$$\int_D f(x, y)\, dx\, dy = \int_{D^*} f(r\cos\theta, r\sin\theta) r\, dr\, d\theta \qquad (7)$$

is valid when T sends D^* onto D in a one-to-one fashion except possibly for points on the boundary of D^*. Example 2 provided a simple example of this for the case where $f(x, y)$ is constantly 1. Now we shall consider a more challenging example.

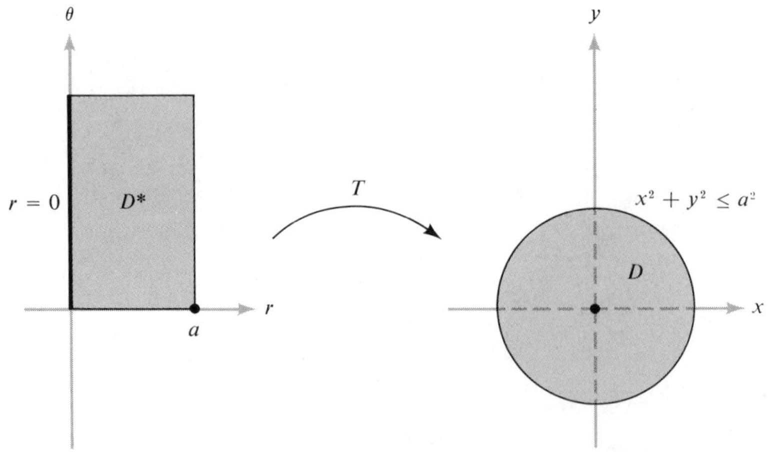

FIGURE 5.8.8

If the rectangle stands against the θ-axis between 0 and 2π, the annulus becomes a disc.

EXAMPLE 5. Evaluate $\int_R \sqrt{x^2 + y^2}\, dx\, dy$ where $R = [0, 1] \times [0, 1]$.

This double integral is equal to the volume of the three-dimensional region shown in Figure 5.8.9. As it stands this integral is difficult to evaluate. Since the integrand is a simple function of $r^2 = x^2 + y^2$ we again try a polar

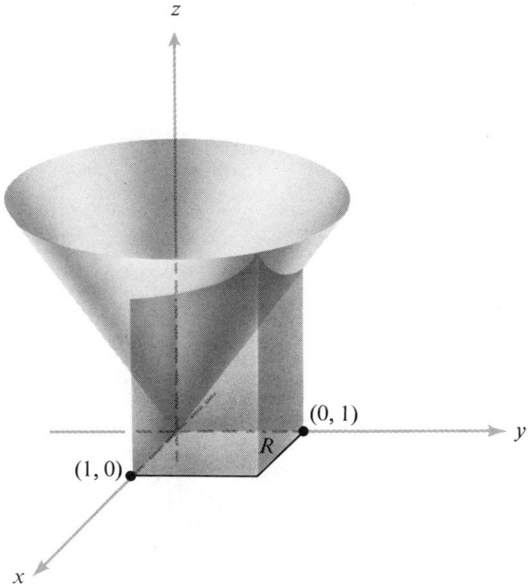

FIGURE 5.8.9
Volume of the region under $z = \sqrt{x^2 + y^2}$ *and over* $R = [0, 1] \times [0, 1]$.

coordinate change of variables. This will result in a simplification of the integrand but, unfortunately, not in the domain of integration. However, the simplification is sufficient to enable us to evaluate the integral. To apply Theorem 7 with polar coordinates, we refer to Figure 5.8.10. The reader can verify that R is the image under $T(r, \theta) = (r \cos \theta, r \sin \theta)$ of the region $D^* = D_1^* \cup D_2^*$ where for D_1^* we have $0 \le \theta \le \frac{1}{4}\pi$ and $0 \le r \le \sec \theta$; for D_2^* we have $\frac{1}{4}\pi \le \theta \le \frac{1}{2}\pi$, $0 \le r \le \csc \theta$. The transformation T sends D_1^* onto a triangle T_1 and D_2^* onto a triangle T_2. The transformation T is one-to-one except when $r = 0$, so we can apply Theorem 7. From the symmetry of $z = \sqrt{x^2 + y^2}$ on R, we can see that

$$\int_R \sqrt{x^2 + y^2}\, dx\, dy = 2 \int_{T_1} \sqrt{x^2 + y^2}\, dx\, dy$$

Applying formula (7), we obtain

$$\int_{T_1} \sqrt{x^2 + y^2}\, dx\, dy = \int_{D_1^*} \sqrt{r^2} r\, dr\, d\theta = \int_{D_1^*} r^2\, dr\, d\theta$$

Next we use iterated integration to obtain

$$\int_{D_1^*} r^2\, dr\, d\theta = \int_0^{\pi/4} \left[\int_0^{\sec \theta} r^2\, dr \right] d\theta = \frac{1}{3} \int_0^{\pi/4} \sec^3 \theta\, d\theta$$

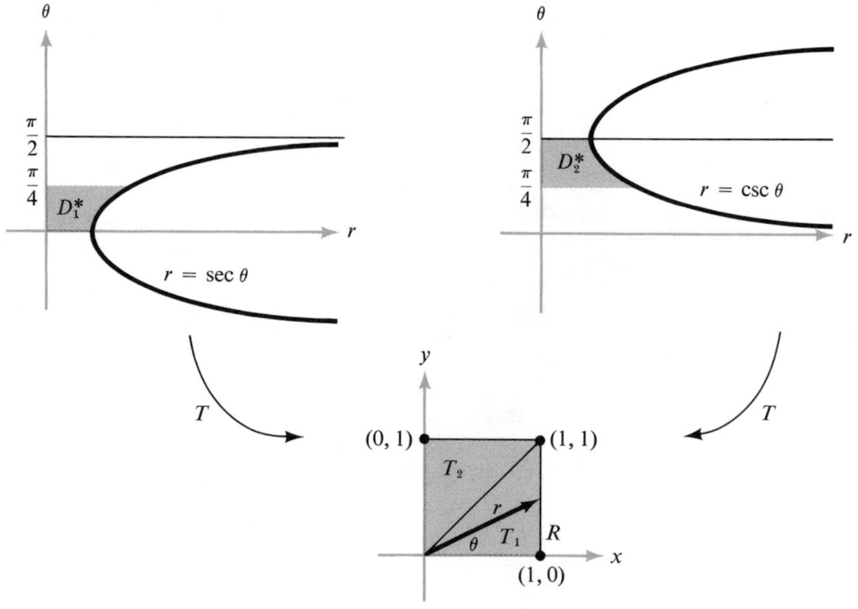

FIGURE 5.8.10
The polar-coordinate transformation takes D_1^* to the triangle T_1 and D_2^* to T_2.

Consulting a table of integrals (see the back of the book) to find $\int \sec^3 x\, dx$, we have

$$\int_0^{\pi/4} \sec^3 \theta\, d\theta = \left[\frac{\sec\theta\tan\theta}{2}\right]_0^{\pi/4} + \frac{1}{2}\int_0^{\pi/4} \sec\theta\, d\theta$$

$$= \frac{\sqrt{2}}{2} + \frac{1}{2}\int_0^{\pi/4} \sec\theta\, d\theta$$

Consulting the table again for $\int \sec x\, dx$, we find

$$\frac{1}{2}\int_0^{\pi/4} \sec\theta\, d\theta = \frac{1}{2}[\log|\sec\theta + \tan\theta|]_0^{\pi/4} = \frac{1}{2}\log(1+\sqrt{2})$$

Combining these results and recalling the factor $\frac{1}{3}$, we obtain

$$\int_{D_1^*} r^2\, dr\, d\theta = \frac{1}{3}\left(\frac{\sqrt{2}}{2} + \frac{1}{2}\log(1+\sqrt{2})\right) = \frac{1}{6}(\sqrt{2} + \log(1+\sqrt{2}))$$

Multiplying by 2, we obtain the answer

$$\int_R \sqrt{x^2+y^2}\, dx\, dy = \frac{1}{3}(\sqrt{2} + \log(1+\sqrt{2}))$$

There is also a change of variables formula for triple integrals, which we state below. First we must define the Jacobian of a transformation from $\mathbb{R}^3$ to $\mathbb{R}^3$—it is a simple extension of the two-variable case.

DEFINITION. *Let $T: W \subset \mathbb{R}^3 \to \mathbb{R}^3$ be a C^1 function defined by $x = x(u, v, w)$, $y = y(u, v, w)$, $z = z(u, v, w)$. Then the **Jacobian** of T, written $\partial(x, y, z)/\partial(u, v, w)$, is the determinant*

$$\begin{vmatrix} \dfrac{\partial x}{\partial u} & \dfrac{\partial x}{\partial v} & \dfrac{\partial x}{\partial w} \\[2ex] \dfrac{\partial y}{\partial u} & \dfrac{\partial y}{\partial v} & \dfrac{\partial y}{\partial w} \\[2ex] \dfrac{\partial z}{\partial u} & \dfrac{\partial z}{\partial v} & \dfrac{\partial z}{\partial w} \end{vmatrix}$$

The absolute value of this determinant is equal to the volume of the parallelepiped determined by the vectors

$$\mathbf{T}_u = \frac{\partial x}{\partial u}\mathbf{i} + \frac{\partial y}{\partial u}\mathbf{j} + \frac{\partial z}{\partial u}\mathbf{k}$$

$$\mathbf{T}_v = \frac{\partial x}{\partial v}\mathbf{i} + \frac{\partial y}{\partial v}\mathbf{j} + \frac{\partial z}{\partial v}\mathbf{k}$$

$$\mathbf{T}_w = \frac{\partial x}{\partial w}\mathbf{i} + \frac{\partial y}{\partial w}\mathbf{j} + \frac{\partial z}{\partial w}\mathbf{k}$$

(see Section 1.3). Just as in the two-variable case, the Jacobian measures how the transformation T distorts its domain. Hence for volume (triple) integrals, the change of variables formula takes the following form:

$$\int_D f(x, y, z)\, dx\, dy\, dz$$

$$= \int_{D^*} f(x(u, v, w), y(u, v, w), z(u, v, w)) \left| \frac{\partial(x, y, z)}{\partial(u, v, w)} \right| du\, dv\, dw \qquad (8)$$

where D^* is an elementary region in uvw-space corresponding to D in xyz-space, under a coordinate change $T: (u, v, w) \mapsto (x(u, v, w), y(u, v, w), z(u, v, w))$, provided T is C^1 and one-to-one except on a set that is the union of graphs of functions of two variables.

Let us apply formula (8) to cylindrical and spherical coordinates (see Section 1.4). First we compute the Jacobian for the map defining the change to cylindrical coordinates. Since

$$x = r\cos\theta, \qquad y = r\sin\theta, \qquad z = z$$

we have

$$\frac{\partial(x, y, z)}{\partial(r, \theta, z)} = \begin{vmatrix} \cos\theta & -r\sin\theta & 0 \\ \sin\theta & r\cos\theta & 0 \\ 0 & 0 & 1 \end{vmatrix} = r$$

Thus, we obtain the formula

$$\int_D f(x, y, z)\, dx\, dy\, dz = \int_{D^*} f(r\cos\theta, r\sin\theta, z) r\, dr\, d\theta\, dz \tag{9}$$

Next let us consider the spherical coordinate system. Recall that it is given by

$$x = \rho\sin\phi\cos\theta, \qquad y = \rho\sin\phi\sin\theta, \qquad z = \rho\cos\phi$$

Therefore we have

$$\frac{\partial(x, y, z)}{\partial(\rho, \theta, \phi)} = \begin{vmatrix} \sin\phi\cos\theta & -\rho\sin\phi\sin\theta & \rho\cos\phi\cos\theta \\ \sin\phi\sin\theta & \rho\sin\phi\cos\theta & \rho\cos\phi\sin\theta \\ \cos\phi & 0 & -\rho\sin\phi \end{vmatrix}$$

Expanding along the last row, we get

$$\begin{aligned} \frac{\partial(x, y, z)}{\partial(\rho, \theta, \phi)} &= \cos\phi \begin{vmatrix} -\rho\sin\phi\sin\theta & \rho\cos\phi\cos\theta \\ \rho\sin\phi\cos\theta & \rho\cos\phi\sin\theta \end{vmatrix} \\ &\quad -\rho\sin\phi \begin{vmatrix} \sin\phi\cos\theta & -\rho\sin\phi\sin\theta \\ \sin\phi\sin\theta & \rho\sin\phi\cos\theta \end{vmatrix} \\ &= -\rho^2\cos^2\phi\sin\phi\sin^2\theta - \rho^2\cos^2\phi\sin\phi\cos^2\theta \\ &\quad - \rho^2\sin^3\phi\cos^2\theta - \rho^2\sin^3\phi\sin^2\theta \\ &= -\rho^2\cos^2\phi\sin\phi - \rho^2\sin^3\phi \\ &= -\rho^2\sin\phi \end{aligned}$$

Thus we arrive at the formula

$$\int_D f(x, y, z)\, dx\, dy\, dz$$

$$= \int_{D^*} f(\rho\sin\phi\cos\theta, \rho\sin\phi\sin\theta, \rho\cos\phi)\rho^2\sin\phi\, d\rho\, d\theta\, d\phi \tag{10}$$

In order to prove the validity of formula (10), one must show that the transformation S on the set D^* is one-to-one except on a set that is the union of finitely many graphs of continuous functions. We shall leave this verification as an exercise (see Exercise 18).

EXAMPLE 6. Evaluate $\int_D \exp(x^2 + y^2 + z^2)^{3/2}\, dV$ where D is the unit ball in $\mathbb{R}^3$.

First note that we cannot easily integrate this function using iterated integrals (try it!). Hence, let us try a change of variables. The transformation into spherical coordinates seems appropriate, since then the entire quantity $x^2 + y^2 + z^2$ can be replaced by one variable, namely, ρ^2. If D^* is the region such that

$$0 \leq \rho \leq 1, \qquad 0 \leq \theta \leq 2\pi, \qquad 0 \leq \phi \leq \pi$$

we may apply formula (10) and write

$$\int_D \exp(x^2 + y^2 + z^2)^{3/2}\, dV = \int_{D^*} \rho^2 e^{\rho^3} \sin\phi\, d\rho d\theta d\phi$$

This integral equals the iterated integral

$$\begin{aligned}\int_0^1 \int_0^\pi \int_0^{2\pi} e^{\rho^3}\rho^2 \sin\phi\, d\theta\, d\phi\, d\rho &= 2\pi \int_0^1 \int_0^\pi e^{\rho^3}\rho^2 \sin\phi\, d\phi\, d\rho \\ &= -2\pi \int_0^1 \rho^2 e^{\rho^3}[\cos\phi]_0^\pi\, d\rho \\ &= 4\pi \int_0^1 e^{\rho^3}\rho^2\, d\rho = \tfrac{4}{3}\pi \int_0^1 e^{\rho^3}(3\rho^2)\, d\rho \\ &= [\tfrac{4}{3}\pi e^{\rho^3}]_0^1 = \tfrac{4}{3}\pi(e - 1)\end{aligned}$$

EXAMPLE 7. Let D be the ball of radius R and center $(0, 0, 0)$ in $\mathbb{R}^3$. Then $\int_D dx\, dy\, dz$ is the volume of D. This integral may be evaluated by reducing it to iterated integrals (Example 1, Section 5.6) or by regarding D as a volume of revolution, but let us evaluate it here by using spherical coordinates. We get

$$\begin{aligned}\int_D dx\, dy\, dz &= \int_0^\pi \int_0^{2\pi} \int_0^R \rho^2 \sin\phi\, d\rho\, d\theta\, d\phi \\ &= \frac{R^3}{3} \int_0^\pi \int_0^{2\pi} \sin\phi\, d\theta\, d\phi = \frac{2\pi R^3}{3} \int_0^\pi \sin\phi\, d\phi \\ &= \frac{2\pi R^3}{3}\{-(\cos(\pi) - \cos(0))\} = \frac{4\pi R^3}{3}\end{aligned}$$

which is the familiar formula for the volume of a solid sphere.

EXERCISES

1. Let D be the unit circle. Evaluate

$$\int_D \exp(x^2 + y^2)\, dx\, dy$$

by making a change of variables to polar coordinates.

2. Let D be the region $0 \leq y \leq x$ and $0 \leq x \leq 1$. Evaluate

$$\int_D (x + y)\, dx\, dy$$

by making the change of variables $x = u + v$, $y = u - v$. Check your answer by evaluating the integral directly by using an iterated integral.

3. Let $T(u, v) = (x(u, v), y(u, v))$ be the mapping defined by $T(u, v) = (4u, 2u + 3v)$. Let D^* be the rectangle $[0, 1] \times [1, 2]$. Find $D = T(D^*)$ and evaluate

   (a) $\int_D xy\, dx\, dy$ (b) $\int_D (x - y)\, dx\, dy$

   by making a change of variables to evaluate them as integrals over D^*.

4. Repeat Exercise 3 for $T(u, v) = (u, v(1 + u))$.

5. Evaluate

$$\int_D \frac{dx\, dy}{\sqrt{1 + x + 2y}}$$

   where $D = [0, 1] \times [0, 1]$, by setting $T(u, v) = (u, v/2)$ and evaluating an integral over D^*, where $T(D^*) = D$.

6. Define $T(u, v) = (u^2 - v^2, 2uv)$. Let D^* be the set of (u, v) with $u^2 + v^2 \leq 1$, $u \geq 0$, $v \geq 0$. Find $T(D^*) = D$. Evaluate $\int_D dx\, dy$.

7. Let $T(u, v)$ be as in Exercise 6. By making this change of variables evaluate

$$\int_D \frac{dx\, dy}{\sqrt{x^2 + y^2}}$$

8. Let D^* be a region of type 1 in the uv-plane bounded by

$$v = g(u), \qquad v = h(u) \leq g(u)$$

   for $a \leq u \leq b$. Let $T: \mathbb{R}^2 \to \mathbb{R}^2$ be the transformation given by

$$x = u, \qquad y = \psi(u, v)$$

   where ψ is C^1 and $\partial\psi/\partial v$ is never zero. Assume $T(D^*) = D$ is a region of type 1; show that if $f: D \to \mathbb{R}$ is continuous then

$$\int_D f(x, y)\, dx\, dy = \int_{D^*} f(u, \psi(u, v)) \left|\frac{\partial\psi}{\partial v}\right| du\, dv$$

9. Find the area inside the curve $r = 1 + \sin\theta$.

10. (a) Express $\int_0^1 \int_0^{x^2} xy\, dy\, dx$ as an integral over the triangle D^*, which is the set of (u, v) where $0 \leq u \leq 1, 0 \leq v \leq u$. (HINT: Find a one-to-one mapping T of D^* onto the given region of integration.)

    (b) Evaluate this integral directly and over D^*.

*11. Let D be the region bounded by $x^{3/2} + y^{3/2} = a^{3/2}$, for $x \geq 0, y \geq 0$, and the coordinate axes $x = 0, y = 0$. Express $\int_D f(x, y)\, dx\, dy$ as an integral over the triangle D^*, which is the set of points $0 \leq u \leq a, 0 \leq v \leq a - u$.

12. Let D be the unit disc. Express $\int_D (1 + x^2 + y^2)^{3/2}\, dx\, dy$ as an integral over the rectangle $[0, 1] \times [0, 2\pi]$ and evaluate.

13. Using polar coordinates, find the area bounded by the *lemniscate* $(x^2 + y^2)^2 = 2a^2(x^2 - y^2)$.

14. Redo Exercise 11 of Section 5.3 using a change of variables and compare the effort involved in each method.

*15. A change of variables can help to find the value of an improper integral over the unbounded region $\mathbb{R}^2$. Evaluate

$$\int_{\mathbb{R}^2} e^{-x^2-y^2}\, dx\, dy$$

by changing to polar coordinates. Could you have evaluated this integral directly (see Exercise 2, Section 5.5)?

16. Let $T: \mathbb{R}^3 \to \mathbb{R}^3$ be defined by

$$T(u, v, w) = (u \cos v \cos w, u \sin v \cos w, u \sin w)$$

(a) Show that T is onto the unit sphere; that is, every (x, y, z) with $x^2 + y^2 + z^2 = 1$ can be written as $(x, y, z) = T(u, v, w)$ for some (u, v, w).

(b) Show that T is not one-to-one.

17. Determine the equations of the following curves and surfaces in spherical and cylindrical coordinates.

(a) $\dfrac{x^2}{a^2} + \dfrac{y^2}{b^2} + \dfrac{z^2}{c^2} = 1$ (b) $z^2 = x^2 + y^2$

(c) the line $y = x = z$ (d) $z = \tan^{-1}\dfrac{y}{x}, x^2 + y^2 = 1$

18. Show that the spherical change-of-coordinate mapping $S(\rho, \theta, \phi) = (\rho \sin \phi \cos \theta, \rho \sin \phi \sin \theta, \rho \cos \phi)$ is one-to-one except on a set that is a union of finitely many graphs of continuous functions.

19. Describe the surface $\sqrt{x^2 + y^2 + z^2} = \rho = \theta$. What is the Cartesian representation of this surface?

20. Describe the surface $r = z \cos \theta$ in rectangular coordinates.

21. Let D be the unit ball. Evaluate

$$\int_D \frac{dx\, dy\, dz}{\sqrt{2 + x^2 + y^2 + z^2}}$$

by making the appropriate change of variables.

22. Evaluate $\iint_A 1/[(x^2 + y^2)^2]\, dx\, dy$ where A is determined by the conditions $x^2 + y^2 \leq 1$ and $x + y \geq 1$.

*23. Let D be the first octant of the ball $x^2 + y^2 + z^2 \leq a^2$, where $x \geq 0, y \geq 0, z \geq 0$. Evaluate the improper integral

$$\int_D (x^2 + y^2 + z^2)^{1/4}/\sqrt{z + (x^2 + y^2 + z^2)^2}\, dx\, dy\, dz$$

by changing variables.

24. Evaluate $\iint_D x^2\, dx\, dy$ where D is determined by the conditions $0 \leq x \leq y$ and $x^2 + y^2 \leq 1$.

*25. Let D be the unbounded region defined as the set of (x, y, z) with $x^2 + y^2 + z^2 \geq 1$. By making a change of variables evaluate the improper integral

$$\int_D \frac{dx\, dy\, dz}{(x^2 + y^2 + z^2)^2}$$

26. Evaluate the following by using cylindrical coordinates.
    (a) $\iiint_B z\,dx\,dy\,dz$ where B is the region within the cylinder $x^2 + y^2 = 1$ above the xy-plane and below the cone $z = (x^2 + y^2)^{1/2}$
    (b) $\iiint_D (x^2 + y^2 + z^2)^{-1/2}\,dx\,dy\,dz$ where D is the region determined by the conditions $\frac{1}{2} \le z \le 1$ and $x^2 + y^2 + z^2 \le 1$
27. Evaluate $\iint_B (x + y)\,dx\,dy$ where B is the rectangle in the xy-plane with vertices at (0, 1), (1, 0), (3, 4), and (4, 3).
28. Evaluate $\iint_D (x + y)\,dx\,dy$ where D is the square with vertices at (0, 0), (1, 2), (3, 1), and (2, −1).
29. Let E be the ellipsoid $(x^2/a^2) + (y^2/b^2) + (z^2/c^2) \le 1$, where a, b, and c are positive.
    (a) Find the volume of E.
    (b) Evaluate $\iiint_E [(x^2/a^2) + (y^2/b^2) + (z^2/c^2)]\,dx\,dy\,dz$. (HINT: Change variables and then use spherical coordinates.)

REVIEW EXERCISES FOR CHAPTER 5

1. Evaluate each of the following integrals:
   (a) $\int_0^3 \int_{-x^2+1}^{x^2+1} xy\,dy\,dx$
   (b) $\int_0^1 \int_{\sqrt{x}}^1 (x + y)^2\,dy\,dx$
   (c) $\int_0^1 \int_{e^x}^{e^{2x}} x \ln y\,dy\,dx$
2. Reverse the order of integration of the integrals in Exercise 1 and evaluate.
3. Find the volume between the surfaces $x^2 + y^2 = z$ and $x^2 + y^2 + z^2 = 2$.
4. Find the volume enclosed by the cone $x^2 + y^2 = z^2$ and the plane $2z - y - 2 = 0$.

*5. Use the ideas in Exercise 2, Section 5.5 to evaluate $\int_{\mathbb{R}^2} f(x, y)\,dx\,dy$ where $f(x, y) = 1/(1 + x^2 + y^2)^{3/2}$. (HINT: You may assume that changing variables and Fubini's Theorem are valid for improper integrals.)

*6. In Exercise 2, Section 5.5, we discussed integrals over unbounded regions. Use the polar change of coordinates to show $\int_{-\infty}^{\infty} e^{-x^2}\,dx = \sqrt{\pi}$. (HINT: use Fubini's Theorem (you may assume its validity) to show that

$$\left(\int_{-\infty}^{\infty} e^{-x^2}\,dx\right)^2 = \int_{-\infty}^{\infty}\int_{-\infty}^{\infty} e^{-x^2-y^2}\,dx\,dy$$

and use Exercise 15, Section 5.8.)

*7. Find $\int_{\mathbb{R}^3} f(x, y, z)\,dx\,dy\,dz$ where $f(x, y, z) = \exp(-(x^2 + y^2 + z^2)^{3/2})$.

*8. Find $\int_{\mathbb{R}^3} f(x, y, \dot{z})\,dx\,dy\,dz$ where

$$f(x, y, z) = \frac{1}{(1 + (x^2 + y^2 + z^2)^{3/2})^{3/2}}$$

9. A cylindrical hole of diameter 1 is bored through a sphere of radius 2. Assuming that the axis of the cylinder is the same as the axis of the sphere, find the volume of the solid that remains.

10. Let C_1 and C_2 be two cylinders of infinite extent, of diameter 2, and with axes on the x- and y-axes respectively. Find the volume of $C_1 \cap C_2$.

*11. Suppose D is the unbounded region on $\mathbb{R}^2$ given by the set of (x, y) with $0 \le x < \infty$, $0 \le y \le x$. Let $f(x, y) = x^{-3/2}e^{y-x}$. Does the improper integral $\int_D f(x, y)\, dx\, dy$ exist?

12. Write the iterated integral $\int_0^1 \int_1^{1-x} \int_x^1 f(x, y, z)\, dz\, dy\, dx$ as an integral over a region in $\mathbb{R}^3$ and then rewrite it in five other possible orders of integration.

13. Evaluate each of the following iterated integrals.

(a) $\int_0^\infty \int_0^y xe^{-y^3}\, dx\, dy$

(b) $\int_0^1 \int_y^{y^3} e^{x/y}\, dx\, dy$

(c) $\int_0^{\pi/2} \int_0^{(\arcsin y)/y} y \cos xy\, dx\, dy$

14. Evaluate each of the following iterated integrals.

(a) $\int_0^1 \int_0^z \int_0^y xy^2z^3\, dx\, dy\, dz$

(b) $\displaystyle\int_0^1 \int_0^y \int_0^{x/\sqrt{3}} \frac{x}{x^2 + z^2}\, dz\, dx\, dy$

(c) $\int_1^2 \int_1^z \int_{1/y}^2 yz^2\, dx\, dy\, dz$

15. Find the volume bounded by $x/a + y/b + z/c = 1$ and the coordinate planes.

16. Find the volume determined by $z \le 6 - x^2 - y^2$ and $z \ge \sqrt{x^2 + y^2}$.

17. (a) Let $\mathbf{F} = 3x^2y\mathbf{i} + zx\mathbf{j} + e^{xy}\mathbf{k}$. Compute $\nabla \cdot \mathbf{F}$ and $\nabla \times \mathbf{F}$.
(b) It is always true that $\nabla \times (\nabla \times \mathbf{F}) = \mathbf{0}$?

18. In (a) to (d) below, make the indicated change of variables. (Do not evaluate.)

(a) $\int_0^1 \int_{-1}^1 \int_{-\sqrt{(1-y^2)}}^{\sqrt{(1-y^2)}} (x^2 + y^2)^{1/2}\, dx\, dy\, dz$, cylindrical coordinates

(b) $\int_{-1}^1 \int_{-\sqrt{(1-y^2)}}^{\sqrt{(1-y^2)}} \int_{-\sqrt{(4-x^2-y^2)}}^{\sqrt{(4-x^2-y^2)}} xyz\, dz\, dx\, dy$, cylindrical coordinates

(c) $\int_{-\sqrt{2}}^{\sqrt{2}} \int_{-\sqrt{(2-y^2)}}^{\sqrt{(2-y^2)}} \int_{\sqrt{(x^2+y^2)}}^{\sqrt{(4-x^2-y^2)}} z^2\, dz\, dx\, dy$, spherical coordinates

(d) $\int_0^1 \int_0^{\pi/4} \int_0^{2\pi} \rho^3 \sin 2\phi\, d\theta\, d\phi\, d\rho$, rectangular coordinates

19. Evaluate $\iint_B (x^4 + 2x^2y^2 + y^4)\, dx\, dy$, where B is the portion of the disc of radius 2 (centered at $(0, 0)$) in the first quadrant.

20. Change the order of integration and evaluate

$$\int_0^2 \int_{y/2}^1 (x + y)^2\, dx\, dy$$

21. Evaluate $\iint_B e^{-x^2-y^2}\, dx\, dy$, where B consists of those (x, y) satisfying $x^2 + y^2 \le 1$ and $y \le 0$. Discuss the geometric meaning of your answer.

22. Change the order of integration and evaluate

$$\int_0^1 \int_{y^{1/2}}^1 (x^2 + y^3x)\, dx\, dy$$

23. Let D be the region in the xy-plane inside the unit circle $x^2 + y^2 = 1$. Evaluate $\iint_D f(x, y)\, dx\, dy$ in each of the following cases.
(a) $f(x, y) = xy$ (b) $f(x, y) = x^2y^2$ (c) $f(x, y) = x^3y^3$

24. Evaluate $\iiint_D (x^2 + y^2 + z^2)\, xyz\, dx\, dy\, dz$ over each of the following regions.
   (a) the sphere $D = \{(x, y, z) | x^2 + y^2 + z^2 \le R^2\}$
   (b) the hemisphere $D = \{(x, y, z) | x^2 + y^2 + z^2 \le R^2 \text{ and } r \ge 0\}$
   (c) the octant $D = \{(x, y, z) | x \ge 0,\ y \ge 0,\ z \ge 0, \text{ and } z^2 + y^2 + z^2 \le R^2\}$
25. Let C be the cone shaped region $\{(x, y, z) | \sqrt{x^2 + y^2} \le z \le 1\}$ and evaluate $\iiint_C (1 + \sqrt{x^2 + y^2})\, dx\, dy\, dz$.
26. Let ρ, θ, ϕ be spherical coordinates in $\mathbb{R}^3$ and suppose that a surface surrounding the origin is described by a continuous function $\rho = f(\theta, \phi)$. Show that the volume enclosed by the surface is

$$V = \frac{1}{3} \int_0^{2\pi} \int_0^{\pi} (f(\theta, \phi))^3 \sin \phi \, d\phi \, d\theta$$

Assume that (1) $f(0, \phi) = f(2\pi, \phi)$, (2) $f(\theta, \phi) > 0$ for $0 \le \phi \le \pi$ and $0 \le \theta < 2\pi$, and (3) $f(\theta, 0)$ and $f(\theta, \pi)$ are constants.
27. Evaluate $\iint_B \exp[(y - x)/(y + x)]\, dx\, dy$ where B is the interior of the triangle with vertices at (0, 0), (0, 1), and (1, 0).
28. Let E be the solid ellipsoid, $E = \{(x, y, z) | (x^2/a^2) + (y^2/b^2) + (z^2/c^2) \le 1\}$ where $a > 0$, $b > 0$, and $c > 0$. Evaluate

$$\iiint xyz\, dx\, dy\, dz$$

   (a) over the whole ellipsoid; and
   (b) over that part of it in the first quadrant:

$$x \ge 0,\ y \ge 0, \quad \text{and} \quad z \ge 0, \qquad \frac{x^2}{a^2} + \frac{y^2}{b^2} + \frac{z^2}{c^2} \le 1$$

   (see Exercise 29 of Section 5.8).
29. Let B be the region in the first quadrant bounded by the curves $xy = 1$, $xy = 3$, $x^2 - y^2 = 1$, and $x^2 - y^2 = 4$. Evaluate $\iint_B (x^2 + y^2)\, dx\, dy$.

Exercises 30, 31, and 32 use the following: If the *density* of an object near a point (x, y, z) is given by a function $\rho(x, y, z)$, then the *total mass* of the object is given by

$$M = \iiint_V \rho(x, y, z)\, dx\, dy\, dz$$

where V is the region of space occupied by the object. (Notice that the units used for measuring volume in the density should agree with those of the region of integration.)

30. Suppose the density of a solid sphere of radius R is given by $(1 + d^3)^{-1}$ where d is the distance to the center of the sphere. Find the total mass of the sphere.
31. The density of the material of a spherical shell whose inner radius is 1 meter and whose outer radius is 2 meters is $0.4\, d^2$ g/cm^3, where d is the distance to the center of the sphere in meters. Find the total mass of the shell.
32. If the shell in Exercise 31 were dropped in a large tank of pure water, would it float? What if the shell leaked? (Assume that the density of water is exactly 1 g/cm^3.)

Exercises 33, 34, and 35 are based on the following definition: The *average value* of a function f over a region W is defined by

$$\bar{f} = \frac{\displaystyle\iiint_W f(x, y, z)\, dx\, dy\, dz}{\displaystyle\iiint_W dx\, dy\, dz}$$

33. The temperature at points in the cube $C = \{(x, y, z) | -1 \le x \le 1, -1 \le y \le 1,$ and $-1 \le z \le 1\}$ is $32\, d^2$ where d is the distance to the origin.
    (a) What is the average temperature?
    (b) At what points of the cube is the temperature equal to the average temperature?

34. Suppose W is a *path connected region.* That is, given any two points of W there is a continuous path joining them. If f is a continuous function on W, use the intermediate value theorem to show that there is at least one point in W at which the value of f is equal to the average of f over W. (Compare this to the Mean Value Theorem for Double Integrals given as Theorem 5.) What happens if W is not connected?

35. The *center of mass* of a body is the point whose coordinates are the average coordinates of the body weighted according to mass density. That is, if a body occupies the region V with density $\rho(x, y, z)$, the center of mass is the point $(\bar{x}, \bar{y}, \bar{z})$ where

$$\bar{x} = \frac{1}{M} \iiint_V x\rho(x, y, z)\, dx\, dy\, dz$$

$$\bar{y} = \frac{1}{M} \iiint_V y\rho(x, y, z)\, dx\, dy\, dz$$

$$\bar{z} = \frac{1}{M} \iiint_V z\rho(x, y, z)\, dx\, dy\, dz$$

and M is the total mass

$$M = \iiint_V \rho(x, y, z)\, dx\, dy\, dz$$

Find the center of mass of the solid hemisphere

$$V = \{(x, y, z) | x^2 + y^2 + z^2 \le 1 \text{ and } z \ge 0\}$$

if the density is constant.

CHAPTER 6

INTEGRALS OVER PATHS AND SURFACES

I hold in fact: (1) That small portions of space are of a nature analogous to little hills on a surface which is on the average flat. (2) That this property of being curved or distorted is continually passed on from one portion of space to another after the manner of a wave. (3) That this variation of curvature of space is really what happens in that phenomenon which we call the motion of matter whether ponderable or ethereal. (4) That in this physical world nothing else takes place but this variation, subject, possibly, to the law of continuity.

W. K. CLIFFORD (1870)

In Chapter 5 we studied integration over regions in $\mathbb{R}^2$ and $\mathbb{R}^3$. For example, we learned how to evaluate integrals like

$$\int_D f(x, y)\, dA$$

where D is a region in $\mathbb{R}^2$. In this chapter we shall discuss integration over paths and surfaces. This is basic to an understanding of Chapter 7. Indeed, that chapter will relate our results on vector differential calculus (Chapter 3) and vector integral calculus (this chapter) by proving the profound theorems of Green, Gauss, and Stokes. In that chapter we shall also examine some significant physical applications.

6.1 THE PATH INTEGRAL

In this section we shall introduce the concept of a path integral; this is one of the several ways in which integrals of functions of one variable can be generalized to functions of several variables. Besides those in Chapter 5, there are other generalizations, to be discussed in later sections.

Suppose we are given a scalar function $f\colon \mathbb{R}^3 \to \mathbb{R}$, so f sends points in $\mathbb{R}^3$ to real numbers. It can be extremely useful to be able to integrate the function f along a path $\boldsymbol{\sigma}\colon I = [a, b] \to \mathbb{R}^3$, where $\boldsymbol{\sigma}(t) = (x(t), y(t), z(t))$. To motivate this notion, let us suppose that the image of $\boldsymbol{\sigma}$ represents a wire. If $f(x, y, z)$ denotes the mass density at (x, y, z), we might want to know the total mass of the wire. If $f(x, y, z)$ indicates temperature, we might want to know the average temperature along the wire. Both types of problems require integrating $f(x, y, z)$ over $\boldsymbol{\sigma}$.

DEFINITION. *The **path integral** or the **integral of** $f(x, y, z)$ **along the path** $\boldsymbol{\sigma}$ is defined when $\boldsymbol{\sigma}\colon I = [a, b] \to \mathbb{R}^3$ is C^1 and when the composite function $t \mapsto f(x(t), y(t), z(t))$ is continuous on I. We define this integral by the equation*

$$\int_{\boldsymbol{\sigma}} f\, ds = \int_a^b f(x(t), y(t), z(t))\|\boldsymbol{\sigma}'(t)\|\, dt$$

Sometimes $\int_{\boldsymbol{\sigma}} f\, ds$ is denoted $\int_{\boldsymbol{\sigma}} f(x, y, z)\, ds$. Notice that another way to write the definition is

$$\int_{\boldsymbol{\sigma}} f\, ds = \int_a^b f(\boldsymbol{\sigma}(t))\|\boldsymbol{\sigma}'(t)\|\, dt$$

If $\boldsymbol{\sigma}(t)$ is only piecewise C^1 or $f(\boldsymbol{\sigma}(t))$ is piecewise continuous, we can still form $\int_{\boldsymbol{\sigma}} f\, ds$ by breaking $[a, b]$ into pieces over which $f(\boldsymbol{\sigma}(t))\|\boldsymbol{\sigma}'(t)\|$ is continuous, and summing the integrals over the pieces.

Note first that we recover the definition of the arc length of $\boldsymbol{\sigma}$ when $f = 1$ (see Section 3.2), and second that f need only be defined on the image curve C of $\boldsymbol{\sigma}$ and not necessarily on the whole space in order for the above definition to make sense.

EXAMPLE 1. Let $\boldsymbol{\sigma}$ be the helix $\boldsymbol{\sigma}\colon [0, 2\pi] \to \mathbb{R}^3$, $t \mapsto (\cos t, \sin t, t)$ (see Figure 3.1.8), and let $f(x, y, z) = x^2 + y^2 + z^2$. To evaluate the integral $\int_{\boldsymbol{\sigma}} f(x, y, z)\, ds$, we first find

$$\begin{aligned}\|\boldsymbol{\sigma}'(t)\| &= \sqrt{\left[\frac{d(\cos t)}{dt}\right]^2 + \left[\frac{d(\sin t)}{dt}\right]^2 + \left[\frac{dt}{dt}\right]^2}\\ &= \sqrt{\sin^2 t + \cos^2 t + 1} = \sqrt{2}\end{aligned}$$

We substitute for x, y, and z to obtain

$$f(x, y, z) = x^2 + y^2 + z^2 = \cos^2 t + \sin^2 t + t^2 = 1 + t^2$$

along $\boldsymbol{\sigma}$. This yields

$$\int_{\boldsymbol{\sigma}} f(x, y, z)\, ds = \int_0^{2\pi} (1 + t^2)\sqrt{2}\, dt = \sqrt{2}\left[t + \frac{t^3}{3}\right]_0^{2\pi}$$

$$= \frac{2\sqrt{2}\pi}{3}(3 + 4\pi^2)$$

If we think of the helix as a wire and $f(x, y, z) = x^2 + y^2 + z^2$ as the mass density, then the total mass of the wire is $(2\sqrt{2}\pi)(3 + 4\pi^2)/3$.

To motivate the definition of the path integral we shall consider "Riemann-like" sums S_N in the same general way as we have done before to define arc length. For simplicity, let $\boldsymbol{\sigma}$ be C^1 on I. Subdivide the interval $I = [a, b]$ by means of a partition

$$a = t_0 < t_1 < \cdots < t_N = b$$

This leads to a decomposition of $\boldsymbol{\sigma}$ into paths $\boldsymbol{\sigma}_i$ (Figure 6.1.1) defined on $[t_i, t_{i+1}]$ for $0 \le i \le N - 1$. Denote the arc length of $\boldsymbol{\sigma}_i$ by Δs_i; thus

$$\Delta s_i = \int_{t_i}^{t_{i+1}} \|\boldsymbol{\sigma}'(t)\|\, dt$$

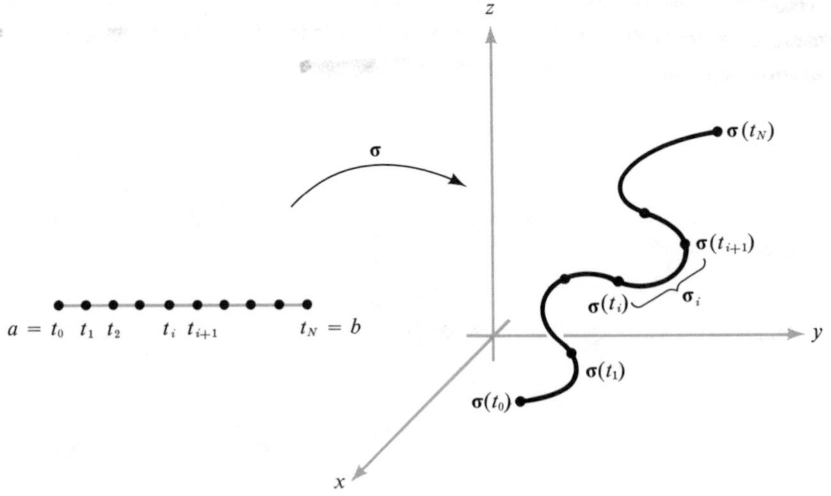

FIGURE 6.1.1
Breaking $\boldsymbol{\sigma}$ into smaller paths $\boldsymbol{\sigma}_i$.

When N is large, the arc length Δs_i is small and $f(x, y, z)$ is approximately constant for points on $\boldsymbol{\sigma}_i$. We consider the sums

$$S_N = \sum_{i=0}^{N-1} f(x_i, y_i, z_i)\, \Delta s_i$$

where $(x_i, y_i, z_i) = \boldsymbol{\sigma}(t)$ for some $t \in [t_i, t_{i+1}]$. These sums are basically Riemann sums, and from their theory it can be shown that

$$\operatorname*{limit}_{N \to \infty} S_N = \int_I f(x(t), y(t), z(t)) \|\boldsymbol{\sigma}'(t)\|\, dt = \int_{\boldsymbol{\sigma}} f(x, y, z)\, ds$$

Thus the path integral can be expressed as a limit of Riemann sums.

An important special case of the path integral occurs when the path $\boldsymbol{\sigma}$ describes a plane curve. Let us examine this case in detail. Suppose then that all points $\boldsymbol{\sigma}(t)$ lie in the xy-plane. Let f be a real-valued function of two variables. The path integral of f along $\boldsymbol{\sigma}$ is

$$\int_{\boldsymbol{\sigma}} f(x, y)\, ds = \int_a^b f(x(t), y(t)) \sqrt{x'(t)^2 + y'(t)^2}\, dt$$

When $f(x, y) \geq 0$, this integral has a natural geometric interpretation as the "area of a fence." We can construct a "fence" with base the image of $\boldsymbol{\sigma}$ and with height $f(x, y)$ at (x, y) (Figure 6.1.2). If $\boldsymbol{\sigma}$ winds only once around the image of $\boldsymbol{\sigma}$, the integral $\int_{\boldsymbol{\sigma}} f(x, y)\, ds$ represents the area of a side of this fence. Readers should try to justify this interpretation, for themselves, using an argument like the one used to justify the arc-length formula.

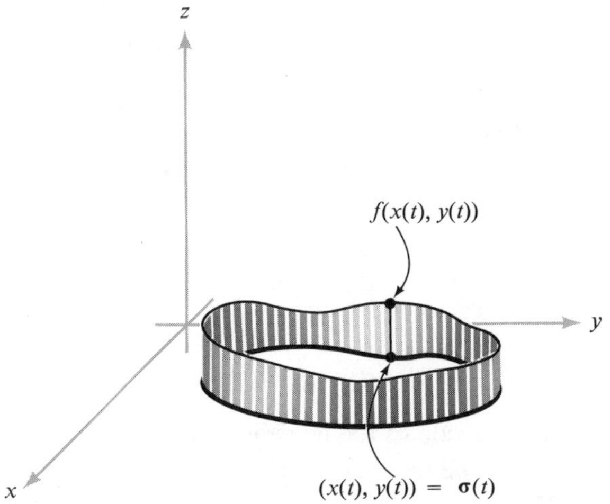

FIGURE 6.1.2
The path integral as the area of a fence.

EXAMPLE 2. Tom Sawyer's aunt has asked him to whitewash both sides of the old fence shown in Figure 6.1.3. Tom estimates that for each 25 square feet of whitewashing he lets someome do for him, the willing victim will pay 5 cents. How much can Tom hope to earn, assuming his aunt will provide whitewash free of charge?

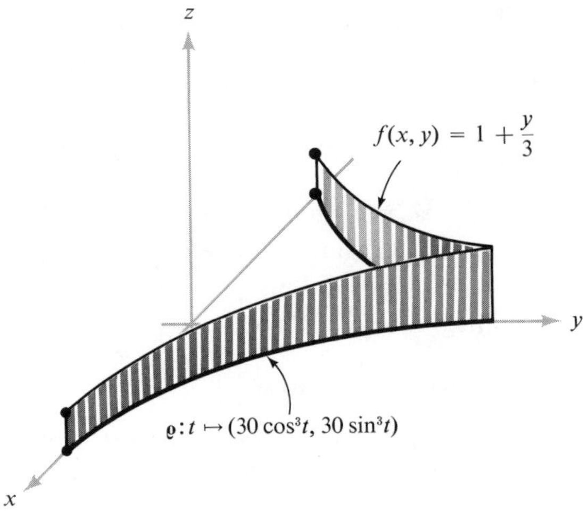

FIGURE 6.1.3
Tom Sawyer's fence.

From Figure 6.1.3, the base of the fence in the first quadrant is the path $\mathbf{\rho}: [0, \pi/2] \to \mathbb{R}^2$, $t \mapsto (30\cos^3 t, 30\sin^3 t)$, and the height of the fence at (x, y) is $f(x, y) = 1 + y/3$. The area of one side of half of the fence is equal to the integral $\int_{\mathbf{\rho}} f(x, y)\, ds = \int_{\mathbf{\rho}} (1 + y/3)\, ds$. Since $\mathbf{\rho}'(t) = (-90\cos^2 t \sin t,$ $90\sin^2 t \cos t)$, we have $\|\mathbf{\rho}'(t)\| = 90 \sin t \cos t$. Thus the integral is

$$\begin{aligned}\int_{\mathbf{\rho}} \left(1 + \frac{y}{3}\right) ds &= \int_0^{\pi/2} \left(1 + \frac{30\sin^3 t}{3}\right) 90 \sin t \cos t\, dt \\ &= 90 \int_0^{\pi/2} (\sin t + 10 \sin^4 t) \cos t\, dt \\ &= 90 \left[\frac{\sin^2 t}{2} + 2\sin^5 t\right]_0^{\pi/2} \\ &= 90(\tfrac{1}{2} + 2) = 225\end{aligned}$$

Hence, the area of one side of the fence is 450 square feet. Since both sides are to be whitewashed we must multiply by 2 to find the total area, which is 900 square feet. Dividing by 25 and then multiplying by 5, we find that Tom could realize as much as \$1.80 for the job.

We shall see many further applications of the path integral in Chapter 7, when we study vector analysis.

This concludes our study of integration of *scalar* functions over paths. In the next section we shall turn our attention to the integration of *vector fields* over paths.

EXERCISES

1. Let $f(x, y, z) = y$ and $\boldsymbol{\sigma}(t) = (0, 0, t)$, $0 \leq t \leq 1$. Prove that $\int_{\boldsymbol{\sigma}} f\, ds = 0$.
2. Evaluate the following path integrals $\int_{\boldsymbol{\sigma}} f(x, y, z)\, ds$, where
   (a) $f(x, y, z) = x + y + z$ and $\boldsymbol{\sigma}: t \mapsto (\sin t, \cos t, t)$, $t \in [0, 2\pi]$;
   (b) $f(x, y, z) = \cos z$, $\boldsymbol{\sigma}$ as in (a); and
   (c) $f(x, y, z) = x \cos z$, $\boldsymbol{\sigma}: t \mapsto t\mathbf{i} + t^2\mathbf{j}$, $t \in [0, 1]$.
3. Evaluate the following path integrals $\int_{\boldsymbol{\sigma}} f(x, y, z)\, ds$, where
   (a) $f(x, y, z) = \exp\sqrt{z}$, and $\boldsymbol{\sigma}: t \mapsto (1, 2, t^2)$, $t \in [0, 1]$;
   (b) $f(x, y, z) = yz$, and $\boldsymbol{\sigma}: t \mapsto (t, 3t, 2t)$, $t \in [1, 3]$; and
   (c) $f(x, y, z) = (x + y)/(y + z)$, and $\boldsymbol{\sigma}: t \mapsto (t, \frac{2}{3}t^{3/2}, t)$, $t \in [1, 2]$.
4. (a) Show that the path integral of $f(x, y)$ along a path given in polar coordinates by $r = r(\theta)$, $\theta_1 \leq \theta \leq \theta_2$, is

$$\int_{\theta_1}^{\theta_2} f(r \cos \theta, r \sin \theta) \sqrt{r^2 + \left(\frac{dr}{d\theta}\right)^2}\, d\theta$$

   (b) Compute the arc length of $r = 1 + \cos \theta$, $0 \leq \theta \leq 2\pi$.
5. Let $f: \mathbb{R}^3 \backslash \{xz\text{-plane}\} \to \mathbb{R}$ be defined by $f(x, y, z) = 1/y^3$. Evaluate $\int_{\boldsymbol{\sigma}} f(x, y, z)\, ds$ where $\boldsymbol{\sigma}: [1, e] \to \mathbb{R}^3$ is given by $\boldsymbol{\sigma}(t) = (\log t)\mathbf{i} + t\mathbf{j} + 2\mathbf{k}$.
6. Write the following limit as a path integral of $f(x, y, z) = xy$ over some path $\boldsymbol{\sigma}$ on $[0, 1]$ and evaluate

$$\lim_{N \to \infty} \sum_{i=1}^{N-1} (t_i^*)^2 \cdot (t_{i+1}^2 - t_i^2)$$

   (here $t_1, \ldots, t_N$ is a partition of $[0, 1]$ and $t_i \leq t_i^* \leq t_{i+1}$).
7. Let $f(x, y) = 2x - y$, $x = t^4$, $y = t^4$, $-1 \leq t \leq 1$.
   (a) Compute the integral of f along this path and interpret the answer geometrically.
   (b) Evaluate the arc-length function $s(t)$ and redo (a) in terms of s (you may wish to consult Exercise 2, Section 3.2).

Exercises 8 to 12 are concerned with the application of the path integral to the problem of defining the average value of a scalar function along a path. We define the number

$$\frac{\int_{\boldsymbol{\sigma}} f(x, y, z)\, ds}{l(\boldsymbol{\sigma})}$$

as the *average value* of f along $\boldsymbol{\sigma}$. Here $l(\boldsymbol{\sigma})$ is the length of the path:

$$l(\boldsymbol{\sigma}) = \int_{\boldsymbol{\sigma}} \|\boldsymbol{\sigma}'(t)\| \, dt$$

(This is analogous to the average of a function over a region as defined in Review Exercise 33 in Chapter 5.)

8. (a) Justify the formula $[\int_{\boldsymbol{\sigma}} f(x, y, z)\, ds / l(\boldsymbol{\sigma})]$ for the average value of f along $\boldsymbol{\sigma}$ using Riemann sums.
   (b) Show that the average value of f along $\boldsymbol{\sigma}$ in Example 1 is $(1 + \frac{4}{3}\pi^2)$.
   (c) In exercises 2(a) and 2(b) above, find the average value of f over the given curves.
9. Find the average y-coordinate of the points on the semicircle parametrized by $\boldsymbol{\rho}\colon [0, \pi] \to \mathbb{R}^3$, $\theta \mapsto (0, a \sin \theta, a \cos \theta)$; $a > 0$.
10. Suppose the semicircle in Exercise 9 is made of a wire with a uniform density of 2 grams per unit length.
    (a) What is the total mass of the wire?
    (b) Where is the center of mass of this configuration of wire? (Consult Review Exercise 35 in Chapter 5).
11. Let $\boldsymbol{\sigma}$ be the path given by $\boldsymbol{\sigma}(t) = (t^2, t, 3)$ for $t \in [0, 1]$.
    (a) Find $l(\boldsymbol{\sigma})$, the length of the path.
    (b) Find the average y-coordinate along the path $\boldsymbol{\sigma}$.
12. If $f\colon [a, b] \to \mathbb{R}$ is piecewise continuously differentiable, let the *length of the graph* of f on $[a, b]$ be defined as the length of the path $t \mapsto (t, f(t))$ for $t \in [a, b]$.
    (a) Show that the length of the graph of f on $[a, b]$ is

$$\int_a^b \sqrt{1 + (f'(x))^2}\, dx$$

    (b) Find the length of the graph of $y = \log x$ from $x = 1$ to $x = 2$.
13. Find the mass of a wire that follows the intersection of the sphere $x^2 + y^2 + z^2 = 1$ and the plane $x + y + z = 0$ if the density at (x, y, z) is given by $\rho(x, y, z) = x^2$ grams per unit length of wire.
14. Evaluate $\int_{\boldsymbol{\sigma}} f\, ds$ where $f(x, y, z) = z$ and $\boldsymbol{\sigma}(t) = (t \cos t, t \sin t, t)$ for $0 \le t \le t_0$.

6.2 LINE INTEGRALS

If **F** is a force field in space, then a test particle (for example, a small unit charge in an electric force field or a unit mass in a gravitational field) will experience the force **F**. Suppose the particle moves along the image of a path $\boldsymbol{\sigma}$ while being acted upon by **F**. One of the fundamental concepts in physics is the *work done* by **F** on the particle as it traces out the path $\boldsymbol{\sigma}$. If $\boldsymbol{\sigma}$ is a straight-line displacement given by the vector **d** and **F** is a constant force, then the work done by **F** in moving the particle along the path is $\mathbf{F} \cdot \mathbf{d}$.

$$\mathbf{F} \cdot \mathbf{d} = (\text{force}) \times (\text{displacement in direction of force})$$

More generally, if the path is curved we can imagine that it is made up of a succession of infinitesimal straight-line displacements or that it is approximated by a finite number of straight-line displacements. Then (as in our derivation of the formulas for arc length, in Section 3.2, and the path integral, in Section 6.1) we are led to the following formula for the work done by the force field $\mathbf{F}$ on a particle moving along a path $\boldsymbol{\sigma}: [a, b] \to \mathbb{R}^3$:

$$\text{work done by } \mathbf{F} = \int_a^b \mathbf{F}(\boldsymbol{\sigma}(t)) \cdot \boldsymbol{\sigma}'(t)\, dt$$

Without giving a full proof, we can justify this derivation in some detail. If t ranges over a small interval t to $t + \Delta t$ the particle moves from $\boldsymbol{\sigma}(t)$ to $\boldsymbol{\sigma}(t + \Delta t)$, a vector displacement of $\Delta\mathbf{s} = \boldsymbol{\sigma}(t + \Delta t) - \boldsymbol{\sigma}(t)$ (see Figure 6.2.1).

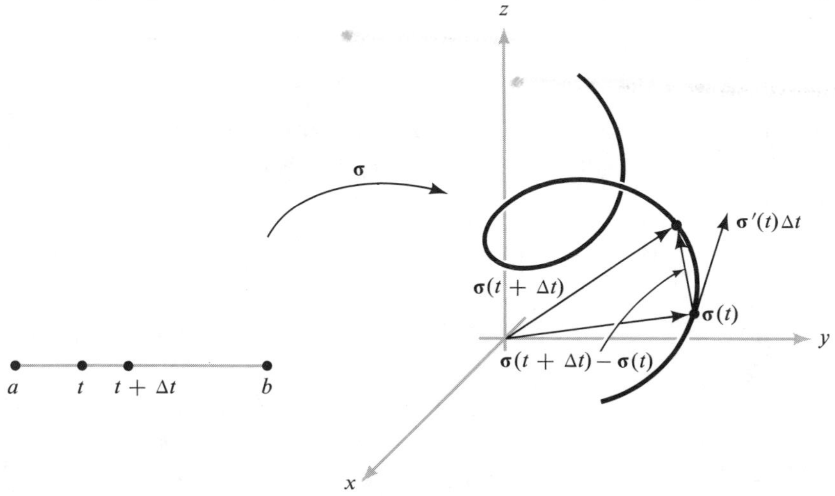

FIGURE 6.2.1
For small Δt, $\Delta\mathbf{s} = \boldsymbol{\sigma}(t + \Delta t) - \boldsymbol{\sigma}(t) \approx \boldsymbol{\sigma}'(t)\, \Delta t$.

Now from the definition of the derivative we have the approximation $\Delta\mathbf{s} \approx \boldsymbol{\sigma}'(t)\, \Delta t$. The work done in going from $\boldsymbol{\sigma}(t)$ to $\boldsymbol{\sigma}(t + \Delta t)$ is therefore approximately

$$\mathbf{F}(\boldsymbol{\sigma}(t)) \cdot \Delta\mathbf{s} \approx \mathbf{F}(\boldsymbol{\sigma}(t)) \cdot \boldsymbol{\sigma}'(t)\, \Delta t$$

If we subdivide the interval $[a, b]$ into n equal parts $a = t_0 < t_1 < \cdots < t_n = b$, with $\Delta t = t_{i+1} - t_i$, then the work done by $\mathbf{F}$ is approximately

$$\sum_{i=0}^{n-1} \mathbf{F}(\boldsymbol{\sigma}(t_i)) \cdot \Delta\mathbf{s} \approx \sum_{i=0}^{n-1} \mathbf{F}(\boldsymbol{\sigma}(t_i)) \cdot \boldsymbol{\sigma}'(t_i)\, \Delta t$$

As $n \to \infty$ this approximation becomes better and better, so it is reasonable to take as our definition of work the limit of the above sum as $n \to \infty$. But this limit is given by the integral

$$\int_a^b \mathbf{F}(\boldsymbol{\sigma}(t)) \cdot \boldsymbol{\sigma}'(t)\, dt$$

This fundamental physical notion of work leads us to make the following mathematical definition.

DEFINITION. *Let* $\mathbf{F}$ *be a vector field on* $\mathbb{R}^3$, *continuous on the* C^1 *path* $\boldsymbol{\sigma}\colon [a, b] \to \mathbb{R}^3$. *We define* $\int_{\boldsymbol{\sigma}} \mathbf{F} \cdot d\mathbf{s}$, *the* ***line integral*** *of* $\mathbf{F}$ *along* $\boldsymbol{\sigma}$, *by the formula*

$$\int_{\boldsymbol{\sigma}} \mathbf{F} \cdot d\mathbf{s} = \int_a^b \mathbf{F}(\boldsymbol{\sigma}(t)) \cdot \boldsymbol{\sigma}'(t)\, dt$$

that is, we integrate the dot product of $\mathbf{F}$ *with* $\boldsymbol{\sigma}'$ *over the interval* $[a, b]$.

As is the case with scalar functions, we can also define $\int_{\boldsymbol{\sigma}} \mathbf{F} \cdot d\mathbf{s}$ *if*

$$\mathbf{F}(\boldsymbol{\sigma}(t)) \cdot \boldsymbol{\sigma}'(t)$$

is only piecewise continuous.

For paths $\boldsymbol{\sigma}$ that satisfy $\boldsymbol{\sigma}'(t) \neq \mathbf{0}$, there is another useful formula for the line integral. Namely, if $\mathbf{T}(t) = \boldsymbol{\sigma}'(t)/\|\boldsymbol{\sigma}'(t)\|$ denotes the unit tangent vector, we have

$$\begin{aligned}\int_{\boldsymbol{\sigma}} \mathbf{F} \cdot d\mathbf{s} &= \int_a^b \mathbf{F}(\boldsymbol{\sigma}(t)) \cdot \boldsymbol{\sigma}'(t)\, dt && \text{(by definition)}\\ &= \int_a^b \left[\mathbf{F}(\boldsymbol{\sigma}(t)) \cdot \frac{\boldsymbol{\sigma}'(t)}{\|\boldsymbol{\sigma}'(t)\|}\right] \|\boldsymbol{\sigma}'(t)\|\, dt && (\text{cancelling } \|\boldsymbol{\sigma}'(t)\|)\\ &= \int_a^b [\mathbf{F}(\boldsymbol{\sigma}(t)) \cdot \mathbf{T}(t)] \|\boldsymbol{\sigma}'(t)\|\, dt && (1)\end{aligned}$$

This formula says that $\int_{\boldsymbol{\sigma}} \mathbf{F} \cdot d\mathbf{s}$ is equal to something that looks like the path integral of the tangential component $\mathbf{F}(\boldsymbol{\sigma}(t)) \cdot \mathbf{T}(t)$ of $\mathbf{F}$ along $\boldsymbol{\sigma}$. In fact, we can consider the last part of (1) as the path integral of a function f along $\boldsymbol{\sigma}$, although technically it is not always such an expression. If $\boldsymbol{\sigma}$ does not intersect itself (i.e., if $\boldsymbol{\sigma}(t_1) = \boldsymbol{\sigma}(t_2)$ implies $t_1 = t_2$) then each point P on C (the image curve of $\boldsymbol{\sigma}$) can be written uniquely as $\boldsymbol{\sigma}(t)$ for some t. If we define $f(P) = f(\boldsymbol{\sigma}(t)) = \mathbf{F}(\boldsymbol{\sigma}(t)) \cdot \mathbf{T}(t)$, f is a function on C and by definition its path integral along $\boldsymbol{\sigma}$ is given by (1) and there is no difficulty. Although if $\boldsymbol{\sigma}$ intersects itself we cannot define f as a function on C as above (why?), in this case it is still useful to think of the right side of (1) as a path integral. To compute a line integral in any particular case, one can either use the original definition or integrate the tangential component of $\mathbf{F}$ along $\boldsymbol{\sigma}$, as prescribed by formula (1), whichever is easier or more appropriate.

EXAMPLE 1. Let $\boldsymbol{\sigma}(t) = (\sin t, \cos t, t)$, with $0 \leq t \leq 2\pi$. Let $\mathbf{F}(x, y, z) = x\mathbf{i} + y\mathbf{j} + z\mathbf{k}$. Then $\mathbf{F}(\boldsymbol{\sigma}(t)) = \mathbf{F}(\sin t, \cos t, t) = (\sin t)\mathbf{i} + (\cos t)\mathbf{j} + t\mathbf{k}$, and $\boldsymbol{\sigma}'(t) = (\cos t)\mathbf{i} - (\sin t)\mathbf{j} + \mathbf{k}$. Therefore,

$$\mathbf{F}(\boldsymbol{\sigma}(t)) \cdot \boldsymbol{\sigma}'(t) = \sin t \cos t - \cos t \sin t + t = t$$

and so

$$\int_{\boldsymbol{\sigma}} \mathbf{F} \cdot d\mathbf{s} = \int_0^{2\pi} t\, dt = 2\pi^2$$

Another common way of writing line integrals is

$$\int_{\boldsymbol{\sigma}} \mathbf{F} \cdot d\mathbf{s} = \int_{\boldsymbol{\sigma}} F_1\, dx + F_2\, dy + F_3\, dz$$

where F_1, F_2, and F_3 are the components of the vector field $\mathbf{F}$. We call the expression $F_1\, dx + F_2\, dy + F_3\, dz$ a *differential form.** By *definition* the integral of a differential form is

$$\int_{\boldsymbol{\sigma}} F_1\, dx + F_2\, dy + F_3\, dz = \int_a^b \left(F_1 \frac{dx}{dt} + F_2 \frac{dy}{dt} + F_3 \frac{dz}{dt} \right) dt = \int_{\boldsymbol{\sigma}} \mathbf{F} \cdot d\mathbf{s}$$

EXAMPLE 2. Evaluate $\int_{\boldsymbol{\sigma}} x^2\, dx + xy\, dy + dz$, where $\boldsymbol{\sigma}\colon [0, 1] \to \mathbb{R}^3$ is given by $\boldsymbol{\sigma}(t) = (t, t^2, 1) = (x(t), y(t), z(t))$.

We compute $dx/dt = 1$, $dy/dt = 2t$, $dz/dt = 0$; therefore,

$$\begin{aligned} \int_{\boldsymbol{\sigma}} x^2\, dx + xy\, dy + dz &= \int_0^1 \left([x(t)]^2 \frac{dx}{dt} + [x(t)y(t)] \frac{dy}{dt} \right) dt \\ &= \int_0^1 (t^2 + 2t^4)\, dt \\ &= \left[\frac{1}{3} t^3 + \frac{2}{5} t^5 \right]_0^1 = \frac{11}{15} \end{aligned}$$

EXAMPLE 3. Evaluate $\int_{\boldsymbol{\sigma}} \cos z\, dx + e^x\, dy + e^y\, dz$, where $\boldsymbol{\sigma}(t) = (1, t, e^t)$ and $0 \le t \le 2$.

We compute $dx/dt = 0$, $dy/dt = 1$, $dz/dt = e^t$, and so

$$\begin{aligned} \int_{\boldsymbol{\sigma}} \cos z\, dx + e^x\, dy + e^y\, dz &= \int_0^2 (0 + e + e^{2t})\, dt \\ &= \left[et + \tfrac{1}{2} e^{2t} \right]_0^2 = 2e + \tfrac{1}{2} e^4 - \tfrac{1}{2} \end{aligned}$$

EXAMPLE 4. Let $\boldsymbol{\sigma}$ be the path

$$x = \cos^3 \theta, \qquad y = \sin^3 \theta, \qquad z = \theta, \qquad 0 \le \theta \le \frac{7\pi}{2}$$

(see Figure 6.2.2). Evaluate the integral $\int_{\boldsymbol{\sigma}} (\sin z\, dx + \cos z\, dy - (xy)^{1/3}\, dz)$.

* See Section 7.6 for a brief discussion of the general theory of differential forms.

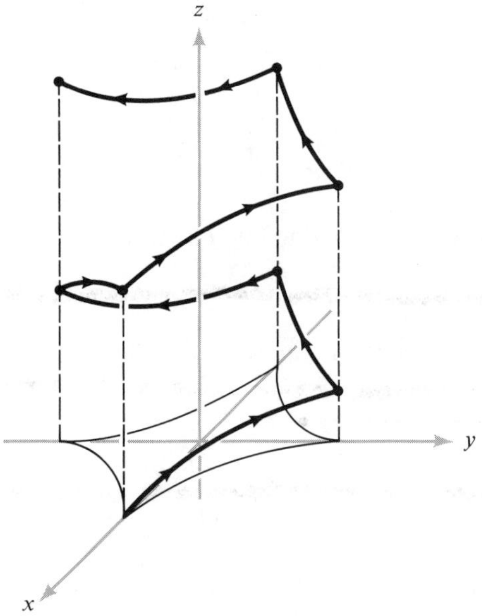

FIGURE 6.2.2
The image of the path $x = \cos^3 \theta$, $y = \sin^3 \theta$, $z = \theta$; $0 \le \theta \le 7\pi/2$.

In this case we have

$$\frac{dx}{d\theta} = -3\cos^2\theta \sin\theta, \qquad \frac{dy}{d\theta} = 3\sin^2\theta\cos\theta, \qquad \frac{dz}{d\theta} = 1$$

and so the integral is

$$\begin{aligned}\int_\sigma \sin z\, dx + \cos z\, dy - (xy)^{1/3}\, dz &= \int_0^{7\pi/2} (-3\cos^2\theta\sin^2\theta + 3\sin^2\theta\cos^2\theta - \cos\theta\sin\theta)\, d\theta \\ &= -\int_0^{7\pi/2} \cos\theta\sin\theta\, d\theta = -\left[\tfrac{1}{2}\sin^2\theta\right]_0^{7\pi/2} = -\tfrac{1}{2}\end{aligned}$$

EXAMPLE 5. Suppose $\mathbf{F}$ is the vector force field $\mathbf{F}(x, y, z) = x^3\mathbf{i} + y\mathbf{j} + z\mathbf{k}$. We may parametrize a circle of radius a in the yz-plane by setting

$$x = 0, \qquad y = a\cos\theta, \qquad z = a\sin\theta, \qquad 0 \le \theta \le 2\pi$$

Since the force field $\mathbf{F}$ is normal to the circle at every point on the circle, $\mathbf{F}$ will not do any work on a particle moving along the circle (Figure 6.2.3). The work done by $\mathbf{F}$ must therefore be 0. We can verify this by a direct

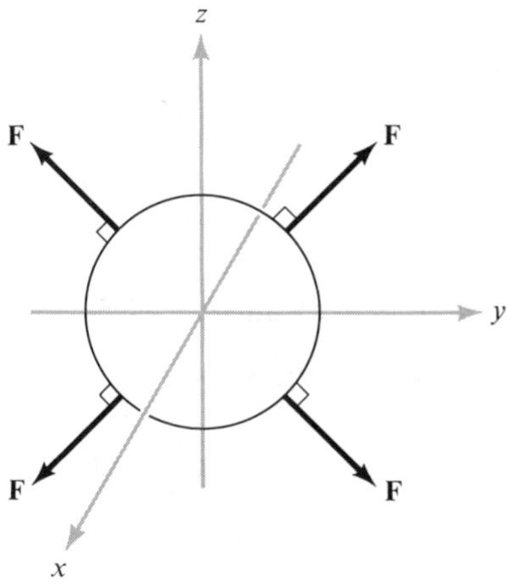

FIGURE 6.2.3
A vector field **F** *normal to a circle in the yz-plane.*

computation:

$$W = \int_{\boldsymbol{\sigma}} \mathbf{F} \cdot d\mathbf{s} = \int_{\boldsymbol{\sigma}} x^3\, dx + y\, dy + z\, dz$$
$$= \int_0^{2\pi} (0 - a^2 \cos\theta \sin\theta + a^2 \cos\theta \sin\theta)\, d\theta = 0$$

as inferred.

EXAMPLE 6. If we consider the field and curve of Example 4, we see that the work done by the field is $-\frac{1}{2}$, a negative quantity. This means that the field impedes movement along that path.

We have seen that the line integral $\int_{\boldsymbol{\sigma}} \mathbf{F} \cdot d\mathbf{s}$ depends not only on the field **F** but also on the path $\boldsymbol{\sigma}\colon [a, b] \to \mathbb{R}^3$. In general, if $\boldsymbol{\sigma}$ and $\boldsymbol{\rho}$ are two different paths in $\mathbb{R}^3$, $\int_{\boldsymbol{\sigma}} \mathbf{F} \cdot d\mathbf{s} \neq \int_{\boldsymbol{\rho}} \mathbf{F} \cdot d\mathbf{s}$. On the other hand, we shall see that it is true that $\int_{\boldsymbol{\sigma}} \mathbf{F} \cdot d\mathbf{s} = \pm\int_{\boldsymbol{\rho}} \mathbf{F} \cdot d\mathbf{s}$ for every vector field **F** if $\boldsymbol{\rho}$ is what we call a *reparametrization of* $\boldsymbol{\sigma}$.

DEFINITION. *Let $h\colon I \to I_1$ be a C^1 real-valued function that is a one-to-one map of an interval $I = [a, b]$ onto another interval $I_1 = [a_1, b_1]$. Let $\boldsymbol{\sigma}\colon I_1 \to \mathbb{R}^2$ be a piecewise C^1 path. Then we call the composition*

$$\boldsymbol{\rho} = \boldsymbol{\sigma} \circ h\colon I \to \mathbb{R}^3$$

a ***reparametrization*** *of* $\boldsymbol{\sigma}$.

This means that $\boldsymbol{\rho}(t) = \boldsymbol{\sigma}(h(t))$, so h changes the variable; alternatively, one can think of h as changing the speed at which a point moves along the path. Indeed, observe that $\boldsymbol{\rho}'(t) = \boldsymbol{\sigma}'(h(t))h'(t)$, so the length of the velocity vector for $\boldsymbol{\sigma}$ is multiplied by the scalar factor $|h'(t)|$.

It is implicit in the definition that h must carry endpoints to endpoints; that is, either $h(a) = a_1$ and $h(b) = b_1$, or $h(a) = b_1$ and $h(b) = a_1$. We thus distinguish two types of reparametrizations. If $\boldsymbol{\sigma} \circ h$ is a reparametrization of $\boldsymbol{\sigma}$ then either

$$\boldsymbol{\sigma} \circ h(a) = \boldsymbol{\sigma}(a_1) \quad \text{and} \quad \boldsymbol{\sigma} \circ h(b) = \boldsymbol{\sigma}(b_1)$$

or

$$\boldsymbol{\sigma} \circ h(a) = \boldsymbol{\sigma}(b_1) \quad \text{and} \quad \boldsymbol{\sigma} \circ h(b) = \boldsymbol{\sigma}(a_1)$$

In the first case, the reparametrization is called *orientation preserving*, and a particle tracing the path $\boldsymbol{\sigma} \circ h$ moves in the same direction as a particle tracing $\boldsymbol{\sigma}$. In the second case, the reparametrization is called *orientation reversing*, and a particle tracing the path $\boldsymbol{\sigma} \circ h$ moves in the opposite direction to that of a particle tracing $\boldsymbol{\sigma}$ (Figure 6.2.4).

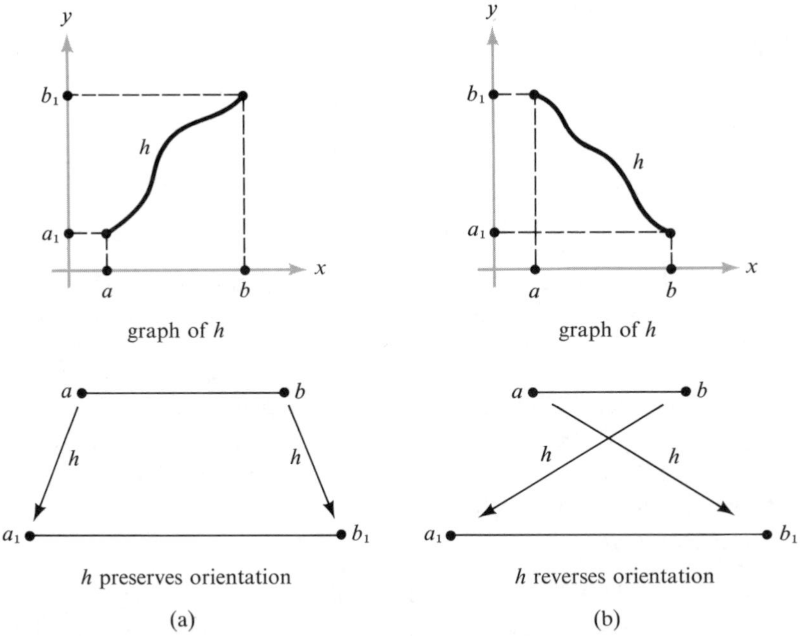

FIGURE 6.2.4
Illustrating an orientation-preserving reparameterization (a) and an orientation-reversing reparametrization (b).

For example, if C is the image of a path $\boldsymbol{\sigma}$ as shown in Figure 6.2.5, that is, $C = \boldsymbol{\sigma}([a_1, b_1])$, and if h is orientation preserving, then $\boldsymbol{\sigma} \circ h(t)$ will go from $\boldsymbol{\sigma}(a_1)$ to $\boldsymbol{\sigma}(b_1)$ as t goes from a to b; and if h is orientation reversing, $\boldsymbol{\sigma} \circ h(t)$ will go from $\boldsymbol{\sigma}(b_1)$ to $\boldsymbol{\sigma}(a_1)$ as t goes from a to b.

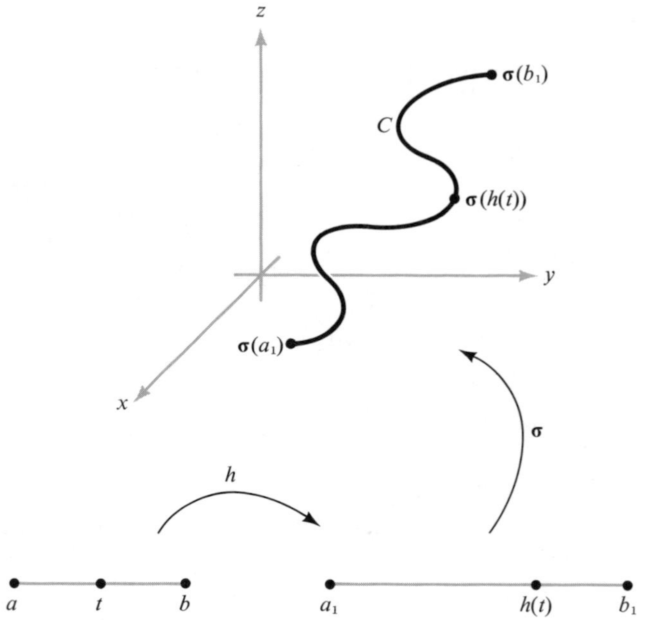

FIGURE 6.2.5
The path $\boldsymbol{\rho} = \boldsymbol{\sigma} \circ h$ is a reparametrization of $\boldsymbol{\sigma}$.

EXAMPLE 7. Let $\boldsymbol{\sigma}: [a, b] \to \mathbb{R}^3$ be any piecewise C^1 path. Then:

(a) The path $\boldsymbol{\sigma}_{\rm op}: [a, b] \to \mathbb{R}^3$, $t \mapsto \boldsymbol{\sigma}(a + b - t)$, is a reparametrization of $\boldsymbol{\sigma}$ corresponding to the map $h: [a, b] \to [a, b]$, $t \mapsto a + b - t$; we call $\boldsymbol{\sigma}_{\rm op}$ the *opposite path* to $\boldsymbol{\sigma}$. This reparametrization is orientation reversing.

(b) The path $\boldsymbol{\rho}: [0, 1] \to \mathbb{R}^3$, $t \mapsto \boldsymbol{\sigma}(a + (b - a)t)$, is an orientation-preserving reparametrization of $\boldsymbol{\sigma}$ corresponding to a change of coordinates $h: [0, 1] \to [a, b]$, $t \mapsto a + (b - a)t$.

THEOREM 1. *Let* $\mathbf{F}$ *be a vector field continuous on the* C^1 *path* $\boldsymbol{\sigma}: [a_1, b_1] \to \mathbb{R}^3$, *and let* $\boldsymbol{\rho}: [a, b] \to \mathbb{R}^3$ *be a reparametrization of* $\boldsymbol{\sigma}$. *If* $\boldsymbol{\rho}$ *is orientation preserving, then*

$$\int_{\boldsymbol{\rho}} \mathbf{F} \cdot d\mathbf{s} = \int_{\boldsymbol{\sigma}} \mathbf{F} \cdot d\mathbf{s}$$

while if $\boldsymbol{\rho}$ *is orientation reversing, then*

$$\int_{\boldsymbol{\rho}} \mathbf{F} \cdot d\mathbf{s} = -\int_{\boldsymbol{\sigma}} \mathbf{F} \cdot d\mathbf{s}$$

Proof. By hypothesis, we have a map h such that $\boldsymbol{\rho} = \boldsymbol{\sigma} \circ h$. By the Chain Rule

$$\boldsymbol{\rho}'(t) = \boldsymbol{\sigma}'(h(t))h'(t)$$

so we have

$$\int_{\boldsymbol{\rho}} \mathbf{F} \cdot d\mathbf{s} = \int_a^b [\mathbf{F}(\boldsymbol{\sigma}(h(t))) \cdot \boldsymbol{\sigma}'(h(t))]h'(t)\, dt$$

By changing variables with $s = h(t)$ (see p. 318), this becomes

$$\int_{h(a)}^{h(b)} \mathbf{F}(\boldsymbol{\sigma}(s)) \cdot \boldsymbol{\sigma}'(s)\, ds$$

$$= \begin{cases} \displaystyle\int_{a_1}^{b_1} \mathbf{F}(\boldsymbol{\sigma}(s)) \cdot \boldsymbol{\sigma}'(s)\, ds = \int_{\boldsymbol{\sigma}} \mathbf{F} \cdot d\mathbf{s} & \text{if } \boldsymbol{\rho} \text{ is orientation preserving} \\[2ex] \displaystyle\int_{b_1}^{a_1} \mathbf{F}(\boldsymbol{\sigma}(s)) \cdot \boldsymbol{\sigma}'(s)\, ds = -\int_{\boldsymbol{\sigma}} \mathbf{F} \cdot d\mathbf{s} & \text{if } \boldsymbol{\rho} \text{ is orientation reversing} \end{cases}$$ ■

Theorem 1 also holds for piecewise C^1 paths, as may be seen by breaking up the intervals into segments on which the paths are C^1 and summing the integrals over the several intervals.

Thus, if it is convenient to reparametrize a path when evaluating an integral, Theorem 1 assures us that the value of the integral will not be affected, except possibly for the sign, depending on the orientation.

EXAMPLE 8. Let $\mathbf{F}(x, y, z) = yz\mathbf{i} + xz\mathbf{j} + xy\mathbf{k}$ and $\boldsymbol{\sigma}: [-5, 10] \to \mathbb{R}^3$, $t \mapsto (t, t^2, t^3)$. Evaluate $\int_{\boldsymbol{\sigma}} \mathbf{F} \cdot d\mathbf{s}$ and $\int_{\boldsymbol{\sigma}_{\text{op}}} \mathbf{F} \cdot d\mathbf{s}$.

For $\boldsymbol{\sigma}$ we have $dx/dt = 1$, $dy/dt = 2t$, $dz/dt = 3t^2$, and $\mathbf{F}(\boldsymbol{\sigma}(t)) = t^5\mathbf{i} + t^4\mathbf{j} + t^3\mathbf{k}$. Therefore

$$\int_{\boldsymbol{\sigma}} \mathbf{F} \cdot d\mathbf{s} = \int_{-5}^{10} \left(F_1 \frac{dx}{dt} + F_2 \frac{dy}{dt} + F_3 \frac{dz}{dt} \right) dt$$

$$= \int_{-5}^{10} (t^5 + 2t^5 + 3t^5)\, dt = [t^6]_{-5}^{10} = 984{,}375$$

On the other hand, for

$$\boldsymbol{\sigma}_{\text{op}}: [-5, 10] \to \mathbb{R}^3, t \mapsto \boldsymbol{\sigma}(5 - t) = (5 - t, (5 - t)^2, (5 - t)^3)$$

we have $dx/dt = -1$, $dy/dt = -10 + 2t = -2(5 - t)$, $dz/dt = -75 + 30t - 3t^2 = -3(5 - t)^2$, and $\mathbf{F}(\boldsymbol{\sigma}_{\text{op}}(t)) = (5 - t)^5\mathbf{i} + (5 - t)^4\mathbf{j} + (5 - t)^3\mathbf{k}$. Therefore

$$\int_{\boldsymbol{\sigma}_{\text{op}}} \mathbf{F} \cdot d\mathbf{s} = \int_{-5}^{10} (-(5 - t)^5 - 2(5 - t)^5 - 3(5 - t)^5)\, dt$$

$$= [(5 - t)^6]_{-5}^{10} = -984{,}375$$

We are interested in reparametrizations because if the image of a particular $\boldsymbol{\sigma}$ can be represented in many ways, we want to be sure that integrals over this image do not depend on the particular parametrization. For example, for some problems the unit circle may be conveniently represented by the map $\boldsymbol{\rho}$ given by

$$x(t) = \cos 2t, \qquad y(t) = \sin 2t, \qquad 0 \le t \le \pi$$

Theorem 1 guarantees that any integral computed for this representation will be the same as when we represent the circle by the map $\boldsymbol{\sigma}$ given by

$$x(t) = \cos t, \qquad y(t) = \sin t, \qquad 0 \le t \le 2\pi$$

since $\boldsymbol{\rho} = \boldsymbol{\sigma} \circ h$, where $h(t) = 2t$, and thus $\boldsymbol{\rho}$ is a reparametrization of $\boldsymbol{\sigma}$. However, notice that the map $\boldsymbol{\gamma}$ given by

$$x(t) = \cos t, \qquad y(t) = \sin t, \qquad 0 \le t \le 4\pi$$

is not a reparametrization of $\boldsymbol{\sigma}$. Although it traces out the same image (the circle), it does so twice. (Why does this imply that $\boldsymbol{\gamma}$ is not a reparametrization of $\boldsymbol{\sigma}$?)

The line integral is an *oriented integral*, in that a change of sign occurs (as we have seen in Theorem 1) if the orientation of the curve is reversed. The path integral does not have this property. This follows from the fact that changing t to $-t$(reversing orientation) just changes the sign of $\boldsymbol{\sigma}'(t)$, not its length. This is one of the differences between line and path integrals. The following theorem, proved by the same method as Theorem 1, shows that path integrals are unchanged under reparametrizations.

THEOREM 2. *Let $\boldsymbol{\sigma}$ be piecewise C^1, f a continuous (real-valued) function on the image of $\boldsymbol{\sigma}$, and let $\boldsymbol{\rho}$ be any reparametrization of $\boldsymbol{\sigma}$. Then*

$$\int_{\boldsymbol{\sigma}} f(x, y, z)\, ds = \int_{\boldsymbol{\rho}} f(x, y, z)\, ds \tag{2}$$

We next consider a simple but often very useful technique for evaluating line integrals. Recall that a vector field $\mathbf{F}$ is a *gradient vector field* if $\mathbf{F} = \nabla f$ for some real-valued function f. Thus

$$\mathbf{F} = \frac{\partial f}{\partial x}\mathbf{i} + \frac{\partial f}{\partial y}\mathbf{j} + \frac{\partial f}{\partial z}\mathbf{k}$$

Suppose G, $g\colon [a, b] \to \mathbb{R}$ are real-valued continuous functions with $G' = g$. Then by the Fundamental Theorem of Calculus

$$\int_a^b g(x)\, dx = G(b) - G(a)$$

Thus the value of the integral of g depends only on the value of G at the endpoints of the interval $[a, b]$. Since ∇f represents the derivative of f, one can ask whether $\int_{\boldsymbol{\sigma}} \nabla f \cdot d\mathbf{s}$ is completely determined by the value of f at the endpoints $\boldsymbol{\sigma}(a)$ and $\boldsymbol{\sigma}(b)$. The answer is contained in the following generalization of the Fundamental Theorem of Calculus.

THEOREM 3. *Suppose that* $f: \mathbb{R}^3 \to \mathbb{R}$ *is* C^1 *and that* $\boldsymbol{\sigma}: [a, b] \to \mathbb{R}^3$ *is a piecewise* C^1 *path. Then*

$$\int_{\boldsymbol{\sigma}} \nabla f \cdot d\mathbf{s} = f(\boldsymbol{\sigma}(b)) - f(\boldsymbol{\sigma}(a))$$

Proof. We apply the Chain Rule to the composite function

$$F: t \mapsto f(\boldsymbol{\sigma}(t))$$

to obtain

$$F'(t) = (f \circ \boldsymbol{\sigma})'(t) = \nabla f(\boldsymbol{\sigma}(t)) \cdot \boldsymbol{\sigma}'(t)$$

The function F is a real function of the variable t, and so, by the Fundamental Theorem of Calculus, we have

$$\int_a^b F'(t)\, dt = F(b) - F(a) = f(\boldsymbol{\sigma}(b)) - f(\boldsymbol{\sigma}(a))$$

Therefore,

$$\begin{aligned}\int_{\boldsymbol{\sigma}} \nabla f \cdot d\mathbf{s} = \int_a^b \nabla f(\boldsymbol{\sigma}(t)) \cdot \boldsymbol{\sigma}'(t)\, dt &= \int_a^b F'(t)\, dt = F(b) - F(a) \\ &= f(\boldsymbol{\sigma}(b)) - f(\boldsymbol{\sigma}(a)) \quad \blacksquare\end{aligned}$$

EXAMPLE 9. Let $\boldsymbol{\sigma}$ be the path $\boldsymbol{\sigma}(t) = (t^4/4, \sin^3(t\pi/2), 0)$, $t \in [0, 1]$. Evaluate

$$\int_{\boldsymbol{\sigma}} y\, dx + x\, dy$$

(which means $\int_{\boldsymbol{\sigma}} y\, dx + x\, dy + 0\, dz$).

We recognize $y\, dx + x\, dy$, or equivalently, the vector field $y\mathbf{i} + x\mathbf{j} + 0\mathbf{k}$, as the gradient of the function $f(x, y, z) = xy$. Thus

$$\int_{\boldsymbol{\sigma}} y\, dx + x\, dy = f(\boldsymbol{\sigma}(1)) - f(\boldsymbol{\sigma}(0)) = \tfrac{1}{4} \cdot 1 - 0 = \tfrac{1}{4}$$

Obviously, if one can recognize the integrand as a gradient, then evaluation of the integral becomes much easier. For example, the reader should try to work out the above integral directly. In one-variable calculus, every integral is, in principle, obtainable by finding an antiderivative. For vector fields, however, this is not always true, because a vector field need not always be a gradient. This point will be examined in detail in Section 7.3.

We have seen how to define path integrals (integrals of scalar functions) and line integrals (integrals of vector functions) over parametrized curves. We have also seen that our work is simplified if we make a judicious choice of parametrization.

Since these integrals are independent of the parametrization (except possibly for the sign), it seems natural to try to write out the theory in a way that is independent of the parametrization, and that is thereby more "geometric." We do this briefly and somewhat informally in the following discussion.

DEFINITION. *We define a **simple curve** to be the image of a piecewise C^1 map $\boldsymbol{\sigma}: I \to \mathbb{R}^3$ that is one-to-one on an interval I. Thus a simple curve is one that does not intersect itself (Figure 6.2.6). If $I = [a, b]$, we call $\boldsymbol{\sigma}(a)$ and $\boldsymbol{\sigma}(b)$ the **endpoints** of the curve. Each simple curve C has two orientations or directions associated with it. If P and Q are the endpoints of the curve, then we can consider C either as directed from P to Q or from Q to P. The simple curve C together with a sense of direction is called an **oriented simple curve** or **directed simple curve** (Figure 6.2.7).*

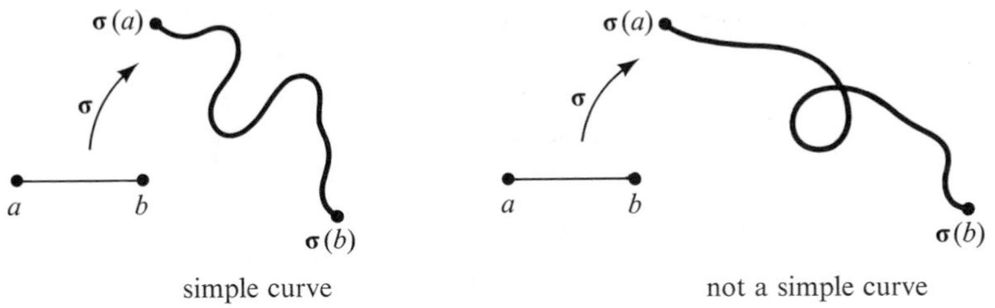

FIGURE 6.2.6
A simple curve that has no self intersections is shown on the left. On the right we have a curve with a self intersection that is not, therefore, simple.

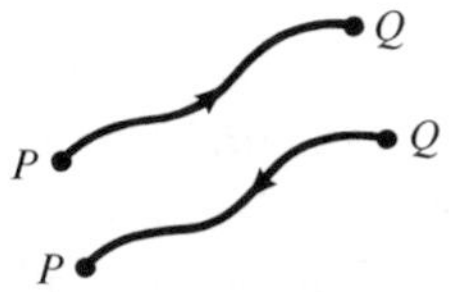

FIGURE 6.2.7
There are two possible senses of direction on a curve joining P and Q.

EXAMPLE 10. If $I = [a, b]$ is a closed interval on the x-axis, then I, as a curve, has two orientations: one corresponding to motion from a to b (left to right) and the other corresponding to motion from b to a (right to left). If f is a real function continuous on I, then denoting I with the first orientation by I^+, and I with the second orientation by I^-, we have

$$\int_{I^+} f(x)\, dx = \int_a^b f(x)\, dx = -\int_b^a f(x)\, dx = -\int_{I^-} f(x)\, dx$$

DEFINITION. *By a **simple closed curve** we mean the image of a piecewise C^1 map $\boldsymbol{\sigma}: [a, b] \to \mathbb{R}^3$ that is one-to-one on $[a, b[$ and satisfies $\boldsymbol{\sigma}(a) = \boldsymbol{\sigma}(b)$ (Figure 6.2.8). If $\boldsymbol{\sigma}$ satisfies the condition $\boldsymbol{\sigma}(a) = \boldsymbol{\sigma}(b)$ but is not necessarily one-to-one on $[a, b[$, we call its image a **closed curve**. Simple closed curves have two orientations, corresponding to the two possible directions of motion along the curve (Figure 6.2.9).*

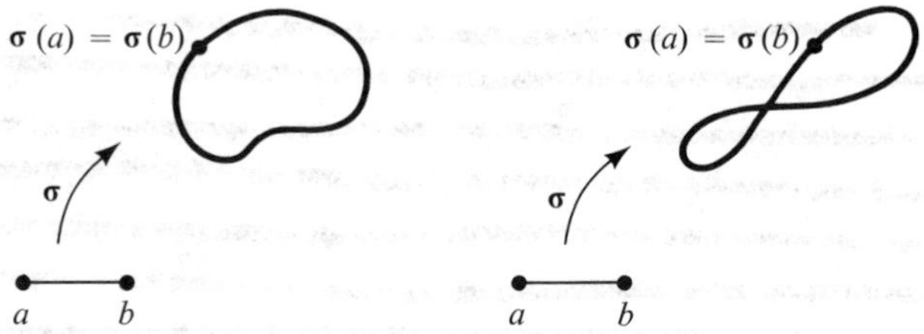

FIGURE 6.2.8
A simple closed curve (left) and a closed curve that is not simple (right).

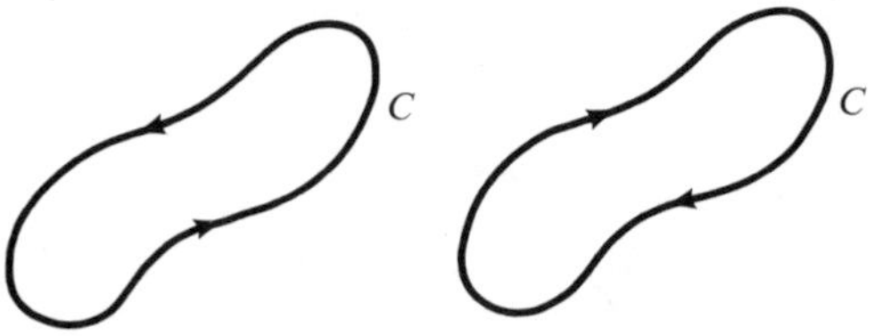

FIGURE 6.2.9
Two possible orientations for a simple closed curve C.

Given a simple curve or a simple closed curve C, we can write $\int_C f\,ds$ and (*if C is oriented*) $\int_C \mathbf{F}\cdot d\mathbf{s}$ unambiguously, by virtue of theorems 1 and 2. (We have not proved that any two one-to-one paths $\boldsymbol{\sigma}$ and $\boldsymbol{\eta}$ with the same image must be reparametrizations of each other, but this technical point will be omitted here.) The point we want to make here is that *while a curve must be parametrized to make integration along it tractable, it is not necessary to include the parametrization in our notation for the integral.*

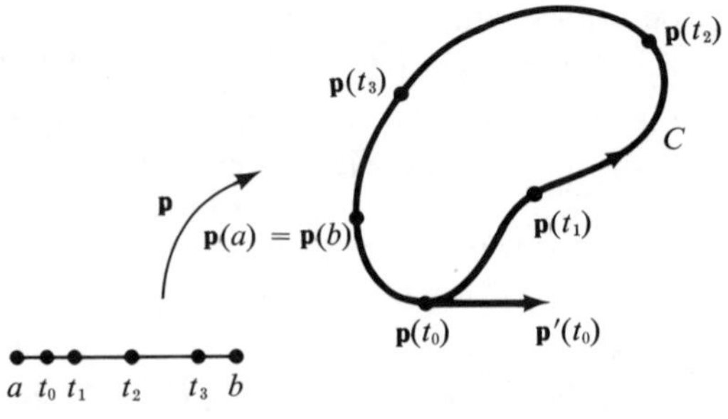

FIGURE 6.2.10
As t goes from a to b, $\boldsymbol{\rho}(t)$ moves around the curve C in some fixed direction.

A given simple closed curve can be parametrized in many different ways. Figure 6.2.10 shows C represented as the image of a map $\boldsymbol{\rho}$, with $\boldsymbol{\rho}(t)$ progressing in the correct direction around an oriented curve C as t ranges from a to b. Note that $\boldsymbol{\rho}'(t_0)$ points in this direction also. The speed with which we traverse C may vary from parametrization to parametrization, but the integral will not, according to theorems 1 and 2.

The following precaution should be noted in regard to these remarks. It is possible to have two mappings $\boldsymbol{\sigma}$ and $\boldsymbol{\eta}$ with the same image and inducing the same orientation on the image, such that

$$\int_{\boldsymbol{\sigma}} \mathbf{F} \cdot d\mathbf{s} \neq \int_{\boldsymbol{\eta}} \mathbf{F} \cdot d\mathbf{s}$$

For an example, let $\boldsymbol{\sigma}(t) = (\cos t, \sin t, 0)$, $\boldsymbol{\eta}(t) = (\cos 2t, \sin 2t, 0)$, $0 \leq t \leq 2\pi$ with $\mathbf{F}(x, y, z) = (y, 0, 0)$. Then

$$\int_{\boldsymbol{\sigma}} \mathbf{F} \cdot d\mathbf{s} = \int_0^{2\pi} F_1(\boldsymbol{\sigma}(t)) \frac{dx}{dt}\, dt$$

(the terms containing F_2 and F_3 are zero)

$$= -\int_0^{2\pi} \sin^2 t\, dt = -\pi$$

But $\int_{\boldsymbol{\eta}} \mathbf{F} \cdot d\mathbf{s} = -2\int_0^{2\pi} \sin^2 2t\, dt = -2\pi$. Clearly $\boldsymbol{\sigma}$ and $\boldsymbol{\eta}$ have the same image, namely, the unit circle in the xy-plane, and, moreover, they traverse the unit circle in the same direction; yet $\int_{\boldsymbol{\sigma}} \mathbf{F} \cdot d\mathbf{s} \neq \int_{\boldsymbol{\eta}} \mathbf{F} \cdot d\mathbf{s}$. The reason for this is that $\boldsymbol{\sigma}$ is one-to-one but $\boldsymbol{\eta}$ is not ($\boldsymbol{\eta}$ traverses the unit circle twice in a counterclockwise direction); therefore $\boldsymbol{\eta}$ is not a parametrization of the unit circle as a simple closed curve.

If C is an oriented simple curve or an oriented simple closed curve we therefore may define

$$\int_C \mathbf{F} \cdot d\mathbf{s} = \int_{\boldsymbol{\sigma}} \mathbf{F} \cdot d\mathbf{s} \quad \text{and} \quad \int_C f\, ds = \int_{\boldsymbol{\sigma}} f\, ds \tag{3}$$

where $\boldsymbol{\sigma}$ is any orientation-preserving parametrization of C. As we have mentioned, these integrals do not depend on the choice of $\boldsymbol{\sigma}$ as long as $\boldsymbol{\sigma}$ is one-to-one (except possibly at the endpoints). If $\mathbf{F} = P\mathbf{i} + Q\mathbf{j} + R\mathbf{k}$ is a vector field, then in differential-form notation we write

$$\int_C \mathbf{F} \cdot d\mathbf{s} = \int_C P\, dx + Q\, dy + R\, dz$$

If C^- is the same curve as C, but with the opposite orientation, then

$$\int_C \mathbf{F} \cdot d\mathbf{s} = -\int_{C^-} \mathbf{F} \cdot d\mathbf{s}$$

If C is a (oriented) curve that is made up of several (oriented) component curves C_i, $i = 1, \ldots, k$, as in Figure 6.2.11, then we shall write $C = C_1 + C_2 + \cdots + C_K$. Since we can parametrize C by parametrizing the pieces

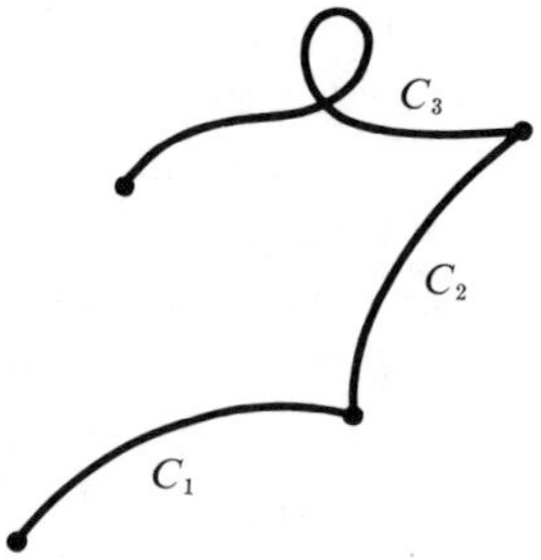

FIGURE 6.2.11
A curve can be made up of several components.

$C_1, \ldots, C_K$ separately, we get

$$\int_C \mathbf{F} \cdot d\mathbf{s} = \int_{C_1} \mathbf{F} \cdot d\mathbf{s} + \int_{C_2} \mathbf{F} \cdot d\mathbf{s} + \cdots + \int_{C_k} \mathbf{F} \cdot d\mathbf{s} \tag{4}$$

One reason for writing a curve as a sum of components is that it may be easier to parametrize the components C_i individually than it is to parametrize C as a whole. If that is the case, formula (4) provides a convenient way of evaluating $\int_C \mathbf{F} \cdot d\mathbf{s}$.

EXAMPLE 11. Consider C, the perimeter of the unit square in $\mathbb{R}^2$, oriented in the counterclockwise sense (see Figure 6.2.12). Evaluate the line integral $\int_C x^2\, dx + xy\, dy$.

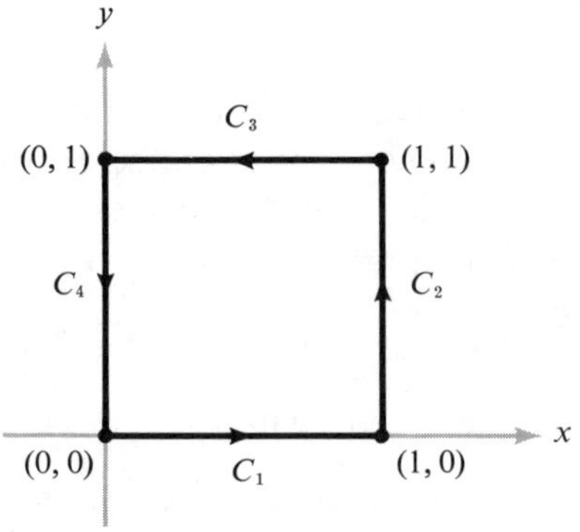

FIGURE 6.2.12
The perimeter of the unit square, parametrized in four pieces.

We may evaluate the integral by using any parametrization of C that induces the given orientation. For example:

$$\boldsymbol{\sigma}\colon [0,4] \to \mathbb{R}^2, \qquad t \mapsto \begin{cases} (t, 0) & 0 \le t \le 1 \\ (1, t-1) & 1 \le t \le 2 \\ (3-t, 1) & 2 \le t \le 3 \\ (0, 4-t) & 3 \le t \le 4 \end{cases}$$

Then

$$\begin{aligned} \int_C x^2\,dx + xy\,dy &= \int_0^1 (t^2 + 0)\,dt + \int_1^2 (0 + (t-1))\,dt \\ &\quad + \int_2^3 (-(3-t)^2 + 0)\,dt + \int_3^4 (0+0)\,dt \\ &= \tfrac{1}{3} + \tfrac{1}{2} + (-\tfrac{1}{3}) + 0 = \tfrac{1}{2} \end{aligned}$$

Now let us re-evaluate this line integral, using formula (4) and parametrizing the C_i separately. Notice that the curve $C = C_1 + C_2 + C_3 + C_4$, where C_i are the oriented curves pictured in Figure 6.2.12. These can be parametrized as follows:

$$C_1\colon \boldsymbol{\sigma}_1(t) = (t, 0),\ 0 \le t \le 1$$

$$C_2\colon \boldsymbol{\sigma}_2(t) = (1, t),\ 0 \le t \le 1$$

$$C_3\colon \boldsymbol{\sigma}_3(t) = (1 - t, 1),\ 0 \le t \le 1$$

$$C_4\colon \boldsymbol{\sigma}_4(t) = (0, 1 - t),\ 0 \le t \le 1$$

So

$$\int_{C_1} x^2\,dx + xy\,dy = \int_0^1 t\,dt = \tfrac{1}{3}$$

$$\int_{C_2} x^2\,dx + xy\,dy = \int_0^1 t\,dt = \tfrac{1}{2}$$

$$\int_{C_3} x^2\,dx + xy\,dy = \int_0^1 -(1-t)^2\,dt = -\tfrac{1}{3}$$

$$\int_{C_4} x^2\,dx + xy\,dy = \int_0^1 0\,dt = 0$$

Thus

$$\int_C x^2\,dx + xy\,dy = \tfrac{1}{3} + \tfrac{1}{2} - \tfrac{1}{3} + 0 = \tfrac{1}{2}$$

as before

EXAMPLE 12. An interesting application of the line integral is the mathematical formulation of Ampère's law, which relates electric currents to their magnetic effects.* Suppose $\mathbf{H}$ denotes a magnetic field in $\mathbb{R}^3$, and let C be a

* The discovery that electric currents produce magnetic effects was made by Oersted about 1820. See any elementary physics text for discussions of the physical basis of these ideas.

closed oriented curve in $\mathbb{R}^3$. Ampère's law states that (in appropriate physical units)

$$\int_C \mathbf{H} \cdot d\mathbf{s} = I$$

where I is the net current that passes through any surface bounded by C (see Figure 6.2.13).

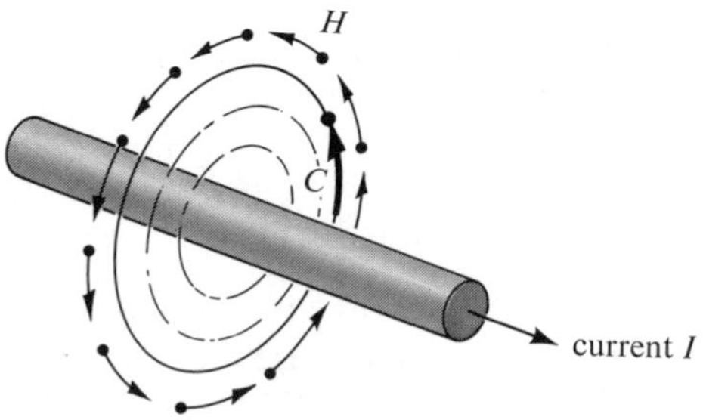

FIGURE 6.2.13

The magnetic field $\mathbf{H}$ *surrounding a wire carrying a current* I *satisfies Ampere's law:* $\int_C \mathbf{H} \cdot d\mathbf{s} = I$.

Finally, let us mention that the line integral has another important physical meaning, specifically, the interpretation of $\int_C \mathbf{V} \cdot d\mathbf{s}$ where $\mathbf{V}$ is the velocity field of a fluid. We shall discuss this interpretation in Section 7.2. Thus, a wide variety of physical concepts, from the notion of work to electromagnetic fields and the motions of fluids, can be analyzed with the help of line integrals.

EXERCISES

1. Let $\mathbf{F}(x, y, z) = x\mathbf{i} + y\mathbf{j} + z\mathbf{k}$. Evaluate the integral of $\mathbf{F}$ along each of the following paths.
   (a) $\boldsymbol{\sigma}(t) = (t, t, t), \quad 0 \le t \le 1$
   (b) $\boldsymbol{\sigma}(t) = (\cos t, \sin t, 0), \quad 0 \le t \le 2\pi$
2. Evaluate each of the following integrals.
   (a) $\int_{\boldsymbol{\sigma}} x\,dy - y\,dx, \quad \boldsymbol{\sigma}(t) = (\cos t, \sin t), \quad 0 \le t \le 2\pi$
   (b) $\int_{\boldsymbol{\sigma}} x\,dx + y\,dy, \quad \boldsymbol{\sigma}(t) = (\cos \pi t, \sin \pi t), \quad 0 \le t \le 2$
   (c) $\int_{\boldsymbol{\sigma}} yz\,dx + xz\,dy + xy\,dz$, where $\boldsymbol{\sigma}$ consists of straight line segments joining $(1, 0, 0)$ to $(0, 1, 0)$ to $(0, 0, 1)$
3. Consider the force $\mathbf{F}(x, y, z) = x\mathbf{i} + y\mathbf{j} + z\mathbf{k}$. Compute the work done in moving a particle along the parabola $y = x^2$, $z = 0$, from $x = -1$ to $x = 2$.

4. Let $\boldsymbol{\sigma}$ be a smooth path.
   (a) Suppose $\mathbf{F}$ is perpendicular to $\boldsymbol{\sigma}'(t)$ at $\boldsymbol{\sigma}(t)$. Show that

$$\int_{\boldsymbol{\sigma}} \mathbf{F} \cdot d\mathbf{s} = 0$$

   (b) If $\mathbf{F}$ is parallel to $\boldsymbol{\sigma}'(t)$ at $\boldsymbol{\sigma}(t)$, show that

$$\int_{\boldsymbol{\sigma}} \mathbf{F} \cdot d\mathbf{s} = \int_{\boldsymbol{\sigma}} \|\mathbf{F}\|\, ds$$

   (By parallel to $\boldsymbol{\sigma}'(t)$ we mean that $\mathbf{F}(\boldsymbol{\sigma}(t)) = \lambda(t)\boldsymbol{\sigma}'(t)$ where $\lambda(t) > 0$.)

5. Suppose $\boldsymbol{\sigma}$ has length l, and $\|\mathbf{F}\| \leq M$. Then prove

$$\left| \int_{\boldsymbol{\sigma}} \mathbf{F} \cdot d\mathbf{s} \right| \leq Ml$$

6. Evaluate $\int_{\boldsymbol{\sigma}} \mathbf{F} \cdot d\mathbf{s}$ where $\mathbf{F}(x, y, z) = y\mathbf{i} + 2x\mathbf{j} + y\mathbf{k}$ and $\boldsymbol{\sigma}(t) = t\mathbf{i} + t^2\mathbf{j} + t^3\mathbf{k}$, $0 \leq t \leq 1$.

7. Evaluate $\int_{\boldsymbol{\sigma}} y\, dx + (3y^3 - x)\, dy + z\, dz$ for each of the paths $\boldsymbol{\sigma}(t) = (t, t^n, 0)$, $0 \leq t \leq 1, n = 1, 2, 3, \ldots$

8. This exercise refers to Example 12. Let L be a very long wire, a planar section of which (with the plane perpendicular to the wire) is shown in Figure 6.2.14. Suppose this plane is the xy-plane. Experiments show that $\mathbf{H}$ is tangent to every circle in the xy-plane whose center is the axis of L, and that the magnitude of $\mathbf{H}$ is constant on every such circle C. Thus $\mathbf{H} = H\mathbf{T}$, where $\mathbf{T}$ is a unit tangent vector to C and H is some scalar. Using this information, show that $H = I/2\pi r$, where r is the radius of circle C, and I is the current flowing in the wire.

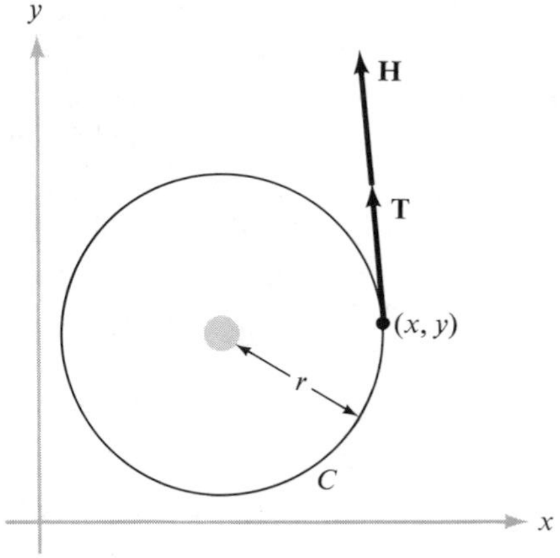

FIGURE 6.2.14
A planar section of a long wire and a curve C about the wire.

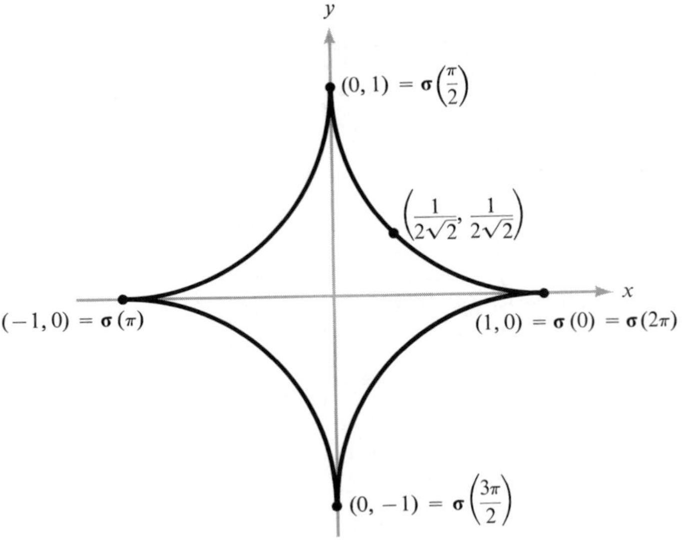

FIGURE 6.2.15
The hypocycloid $\boldsymbol{\sigma}(t) = (\cos^3 t, \sin^3 t)$ *(Exercise 9).*

9. The image of $t \mapsto (\cos^3 t, \sin^3 t), 0 \le t \le 2\pi$, in the plane is shown in Figure 6.2.15. Evaluate the integral of the vector field $\mathbf{F}(x, y) = x\mathbf{i} + y\mathbf{j}$ around this curve.

10. Suppose $\boldsymbol{\sigma}, \boldsymbol{\psi}$ are two paths with the same endpoints, and $\mathbf{F}$ is a vector field. Show that $\int_{\boldsymbol{\sigma}} \mathbf{F} \cdot d\mathbf{s} = \int_{\boldsymbol{\psi}} \mathbf{F} \cdot d\mathbf{s}$ is equivalent to $\int_C \mathbf{F} \cdot d\mathbf{s} = 0$, where C is the closed curve obtained by first moving around $\boldsymbol{\sigma}$ and then moving around $\boldsymbol{\psi}$ in the opposite direction.

11. Let $\boldsymbol{\sigma}(t)$ be a path and $\mathbf{T}$ the unit tangent vector. What is $\int_{\boldsymbol{\sigma}} \mathbf{T} \cdot d\mathbf{s}$?

12. Let $\mathbf{F} = (z^3 + 2xy)\mathbf{i} + x^2\mathbf{j} + 3xz^2\mathbf{k}$. Show that the integral of $\mathbf{F}$ around the circumference of the unit square is zero.

13. Using the path in Exercise 9, observe that a C^1 map $\boldsymbol{\sigma}: [a, b] \to \mathbb{R}^3$ can have an image that does not "look smooth." Do you think this could happen if $\boldsymbol{\sigma}'(t)$ were always nonzero?

14. What is the value of the integral of any gradient field around a closed curve C?

15. Evaluate $\int_C 2xyz\, dx + x^2z\, dy + x^2y\, dz$, where C is an oriented simple curve connecting (1, 1, 1) to (1, 2, 4).

16. Prove Theorem 2.

*17. Let $\boldsymbol{\sigma}: [a, b] \to \mathbb{R}^3$ be a path such that $\boldsymbol{\sigma}'(t) \ne \mathbf{0}$. When this condition holds, $\boldsymbol{\sigma}$ is said to be *regular*. Let the function f be defined by $f(x) = \int_a^x \|\boldsymbol{\sigma}'(t)\|\, dt$.
(a) What is df/dx?
(b) Using (a) prove that $f: [a, b] \to [0, L]$, where L is the length of $\boldsymbol{\sigma}$, has a differentiable inverse $g: [0, L] \to [a, b]$ satisfying $f \circ g(s) = s$, $g \circ f(x) = x$. (You may use the one-variable Inverse Function Theorem of Section 4.4.)

(c) Compute dg/ds.

(d) Recall that a path $s \mapsto \boldsymbol{\rho}(s)$ is unit speed or parametrized by arc length (see Exercises 6 to 11, Section 3.2) if $\|\boldsymbol{\rho}'(x)\| = 1$. Show that the reparametrization of $\boldsymbol{\sigma}$ given by $\boldsymbol{\rho}(s) = \boldsymbol{\sigma} \circ g(s)$ is unit speed. Conclude that any regular path can be reparametrized by arc length. (Thus, for example, the Frenet formulas of Section 3.2 can be applied to the reparametrization $\boldsymbol{\rho}$.)

6.3 PARAMETRIZED SURFACES

In Sections 6.1 and 6.2 we studied integrals of scalar and vector functions along curves. Now we shall turn to integrals over surfaces. Let us begin by studying the geometry of surfaces themselves.

We are already used to one kind of surface, namely, the graph of a function $f(x, y)$. Graphs were extensively studied in Chapter 2, and we know how to compute their tangent planes. However, it would be unduly limiting to restrict ourselves to this case. For example, many surfaces arise as level surfaces of functions. Suppose our surface S is the set of points (x, y, z) where $x - z + z^3 = 0$. Here S is a sheet that (relative to the xy-plane) doubles back on itself (see Figure 6.3.1). Obviously, we want to call S a surface, since it is

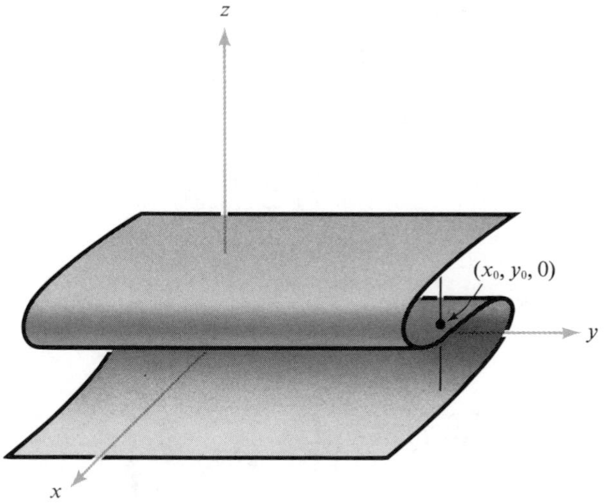

FIGURE 6.3.1
A surface that is not the graph of a function $z = f(x, y)$.

just a plane with a wrinkle. However, S is *not* the graph of some function $z = f(x, y)$, because this means that for each $(x_0, y_0) \in \mathbb{R}^2$ there must be one z_0 with $(x_0, y_0, z_0) \in S$. As Figure 6.3.1 illustrates, this condition is violated.

Another example is the torus or surface of a doughnut, which is depicted in Figure 6.3.2. Anyone would call a torus a surface; yet, by the same reason-

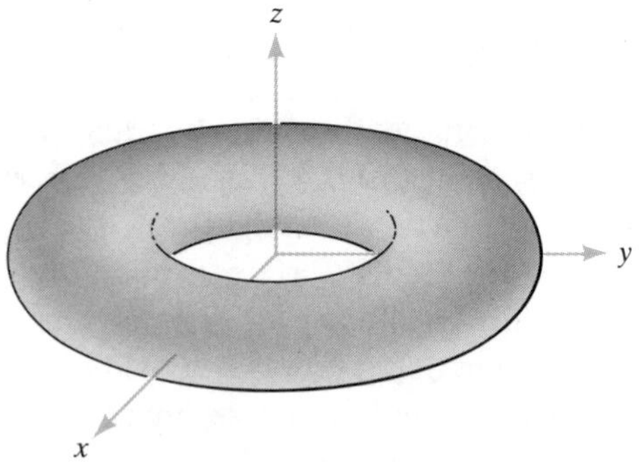

FIGURE 6.3.2
The torus is not the graph of a function of the form $z = f(x, y)$.

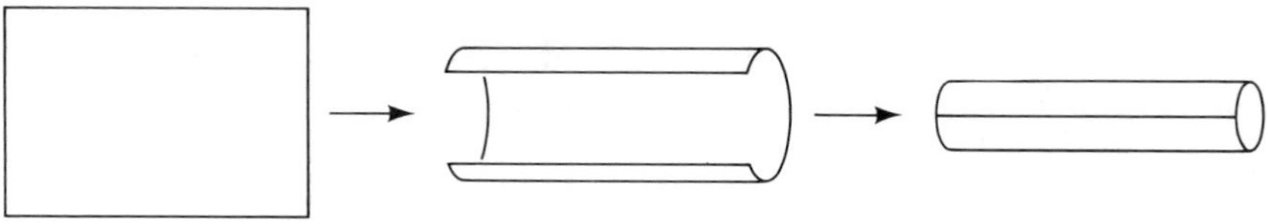

FIGURE 6.3.3
The first step in obtaining a torus from a rectangle: making a cylinder.

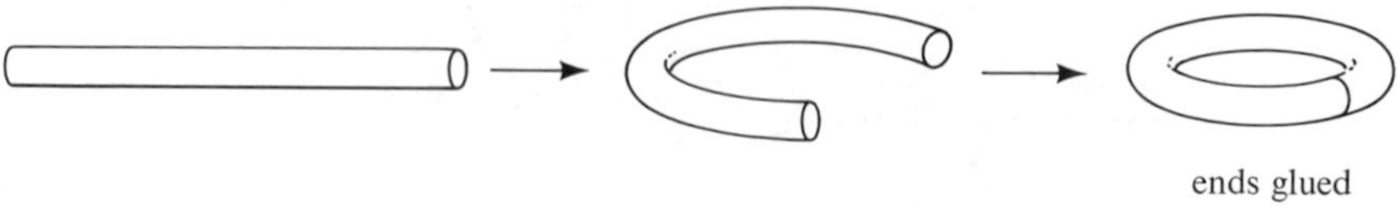

FIGURE 6.3.4
Bending the cylinder and gluing the ends to get a doughnut.

ing as above, a torus cannot be the graph of a differentiable function of two variables. These observations encourage us to extend our definition of a surface. The motivation for the definition that follows is partly that a surface can be thought of as being obtained from the plane by "rolling," "bending," and "pushing." For example, to get a torus we take a portion of the plane and roll it (like a cigarette—see Figure 6.3.3), then take the two "ends" and bring them together until they meet (Figure 6.3.4).

With surfaces, just as with curves, we want to distinguish a map (a parametrization) from its image (a geometric object).

DEFINITION. *A **parametrized surface** is a function* $\mathbf{\Phi}: D \subset \mathbb{R}^2 \to \mathbb{R}^3$, *where* D *is some domain in* $\mathbb{R}^2$. *The **surface** S corresponding to the function* $\mathbf{\Phi}$ *is its image:* $S = \mathbf{\Phi}(D)$. *We can write*

$$\mathbf{\Phi}(u, v) = (x(u, v), y(u, v), z(u, v))$$

If $\mathbf{\Phi}$ *is differentiable or* C^1 *(which is the same as saying that* $x(u, v)$, $y(u, v)$, *and* $z(u, v)$ *are differentiable or* C^1 *functions of* (u, v)*—see Chapter 2), we call* S *a **differentiable or*** C^1 ***surface**.*

We can think of $\mathbf{\Phi}$ as twisting or bending the region D in the plane to yield the surface S (see Figure 6.3.5). Thus each point (u, v) in D becomes a label for a point $(x(u, v), y(u, v), z(u, v))$ on S.

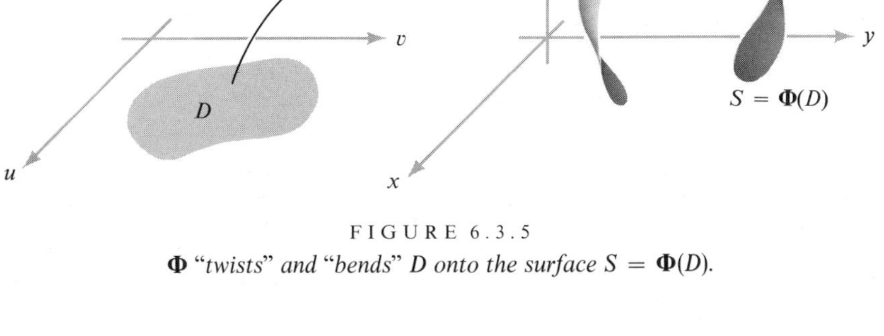

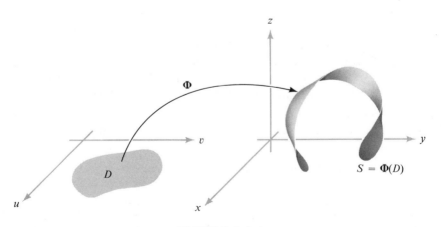

FIGURE 6.3.5
$\mathbf{\Phi}$ *"twists" and "bends"* D *onto the surface* $S = \mathbf{\Phi}(D)$.

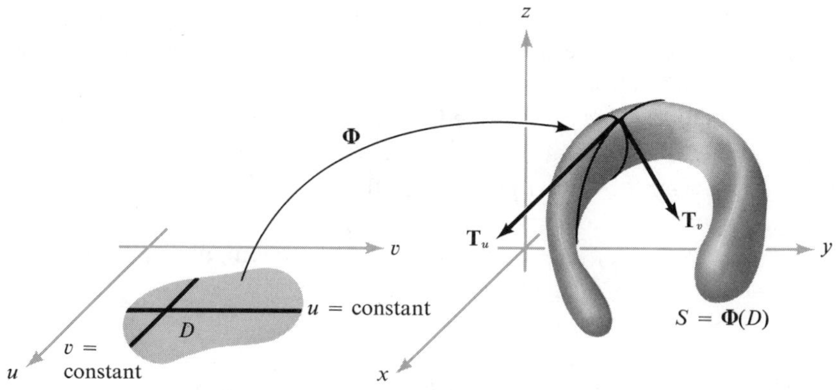

FIGURE 6.3.6
The tangent vectors $\mathbf{T}_u$ *and* $\mathbf{T}_v$ *that are tangent to curves on a surface S, and hence tangent to S.*

Suppose that $\mathbf{\Phi}$ is differentiable at $(u_0, v_0) \in \mathbb{R}^2$. Fixing u at u_0, we get a map $\mathbb{R} \to \mathbb{R}^3$ given by $t \mapsto \mathbf{\Phi}(u_0, t)$, whose image is a curve on the surface (Figure 6.3.6). From chapters 2 and 3 we know that the tangent vector to this curve at the point $\mathbf{\Phi}(u_0, v_0)$ is given by

$$\mathbf{T}_v = \frac{\partial x}{\partial v}(u_0, v_0)\mathbf{i} + \frac{\partial y}{\partial v}(u_0, v_0)\mathbf{j} + \frac{\partial z}{\partial v}(u_0, v_0)\mathbf{k}$$

Similarly, if we fix v and consider the curve $t \mapsto \mathbf{\Phi}(t, v_0)$ we obtain the tangent vector to this curve at $\mathbf{\Phi}(u_0, v_0)$, given by

$$\mathbf{T}_u = \frac{\partial x}{\partial u}(u_0, v_0)\mathbf{i} + \frac{\partial y}{\partial u}(u_0, v_0)\mathbf{j} + \frac{\partial z}{\partial u}(u_0, v_0)\mathbf{k}$$

Since the vectors $\mathbf{T}_u$ and $\mathbf{T}_v$ are tangent to two curves on the surface at $\mathbf{\Phi}(u_0, v_0)$, they ought to determine the plane tangent to the surface at this point; that is, $\mathbf{T}_u \times \mathbf{T}_v$ ought to be normal to the surface.

We say that the surface S is *smooth** at $\mathbf{\Phi}(u_0, v_0)$ if $\mathbf{T}_u \times \mathbf{T}_v \neq \mathbf{0}$ at (u_0, v_0); the surface is smooth if it is smooth at all points $\mathbf{\Phi}(u_0, v_0) \in S$. The nonzero vector $\mathbf{T}_u \times \mathbf{T}_v$ is *normal* to S (recall that the vector product of $\mathbf{T}_u$ and $\mathbf{T}_v$ is perpendicular to the plane spanned by $\mathbf{T}_u$ and $\mathbf{T}_v$); the fact that it is nonzero ensures that there will be a tangent plane. Intuitively, a smooth surface has no "corners."†

* Strictly speaking, smoothness depends on the parametrization $\mathbf{\Phi}$ and not just on its image S. Therefore this terminology is somewhat imprecise; however, it is descriptive and should not cause confusion. See Exercise 9.

† In Section 4.4 we showed that level surfaces $f(x, y, z) = 0$ were in fact graphs of functions of two variables in some neighborhood of a point (x_0, y_0, z_0) satisfying $\nabla f(x_0, y_0, z_0) \neq \mathbf{0}$. This united two concepts of surface. Again using the Implicit Function Theorem it is likewise possible to show that the image of a parametrized surface $\mathbf{\Phi}$ in the neighborhood of a point (u_0, y_0) where $\mathbf{T}_u \times \mathbf{T}_v \neq \mathbf{0}$ is also the graph of a function of two variables. Thus all definitions of surface are consistent. (See Exercise 10.)

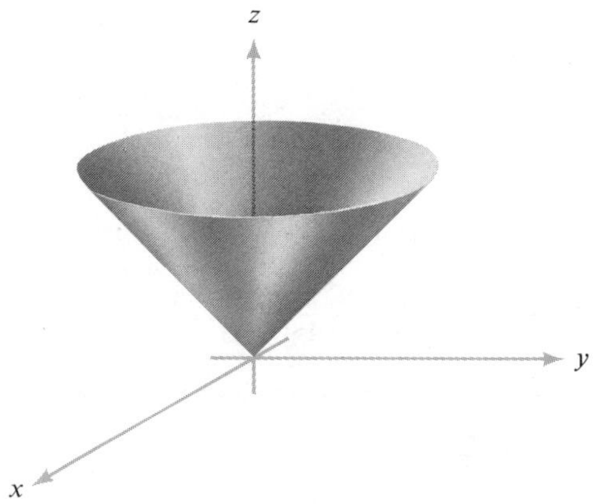

FIGURE 6.3.7
The surface $z = \sqrt{x^2 + y^2}$ is a cone.

EXAMPLE 1. Consider the surface given by the equations

$$x = u \cos v, \qquad y = u \sin v, \qquad z = u, \qquad u \geq 0$$

These equations describe the surface $z = \sqrt{x^2 + y^2}$ (square the equations for x, y, and z to check this), which is shown in Figure 6.3.7. This surface is a cone with a "point" at (0, 0, 0); it is a differentiable surface because each component function is differentiable. However, the surface is not smooth at (0, 0, 0). To see this, compute $\mathbf{T}_u$ and $\mathbf{T}_v$ at $(0, 0) \in \mathbb{R}^2$:

$$\begin{aligned}\mathbf{T}_u &= \frac{\partial x}{\partial u}(0, 0)\mathbf{i} + \frac{\partial y}{\partial u}(0, 0)\mathbf{j} + \frac{\partial z}{\partial u}(0, 0)\mathbf{k} \\ &= (\cos 0)\mathbf{i} + (\sin 0)\mathbf{j} + \mathbf{k} = \mathbf{i} + \mathbf{k}\end{aligned}$$

and similarly

$$\mathbf{T}_v = 0(-\sin 0)\mathbf{i} + 0(\cos 0)\mathbf{j} + 0\mathbf{k} = \mathbf{0}$$

Thus $\mathbf{T}_u \times \mathbf{T}_v = \mathbf{0}$, and so, by definition, the surface is not smooth at (0, 0, 0).

Let us summarize our conclusions in a formal definition:

DEFINITION. *If a parametrized surface $\mathbf{\Phi}: D \subset \mathbb{R}^2 \to \mathbb{R}^3$ is* ***smooth*** *at $\mathbf{\Phi}(u_0, v_0)$, that is, if $\mathbf{T}_u \times \mathbf{T}_v \neq \mathbf{0}$ at (u_0, v_0), we define the* ***tangent plane*** *of the surface at $\mathbf{\Phi}(u_0, v_0)$ to be the plane determined by the vectors $\mathbf{T}_u$ and $\mathbf{T}_v$. Thus $\mathbf{n} = \mathbf{T}_u \times \mathbf{T}_v$ is a normal vector, and an equation of the tangent plane at (x_0, y_0, z_0) on the surface is given by*

$$(x - x_0, y - y_0, z - z_0) \cdot \mathbf{n} = 0 \tag{1}$$

where $\mathbf{n}$ *is evaluated at* (u_0, v_0). *If* $\mathbf{n} = (n_1, n_2, n_3) = n_1\mathbf{i} + n_2\mathbf{j} + n_3\mathbf{k}$ *then* (1) *becomes*

$$n_1(x - x_0) + n_2(y - y_0) + n_3(z - z_0) = 0 \tag{1'}$$

EXAMPLE 2. Let $\mathbf{\Phi}: \mathbb{R}^2 \to \mathbb{R}^3$ be given by

$$x = u \cos v, \qquad y = u \sin v, \qquad z = u^2 + v^2$$

Then

$$\mathbf{T}_u = (\cos v)\mathbf{i} + (\sin v)\mathbf{j} + 2u\mathbf{k}$$

$$\mathbf{T}_v = -u(\sin v)\mathbf{i} + u(\cos v)\mathbf{j} + 2v\mathbf{k}$$

and the tangent plane at $\mathbf{\Phi}(u, v)$ is the set of vectors through $\mathbf{\Phi}(u, v)$ perpendicular to

$$\mathbf{T}_u \times \mathbf{T}_v = (-2u^2 \cos v + 2v \sin v, -2u^2 \sin v - 2v \cos v, u)$$

if this vector is nonzero. Since $\mathbf{T}_u \times \mathbf{T}_v$ is equal to $\mathbf{0}$ at $(u, v) = (0, 0)$, there is no tangent plane at $\mathbf{\Phi}(0, 0) = (0, 0, 0)$. However, we can find an equation of the tangent plane at all the other points, where $\mathbf{T}_u \times \mathbf{T}_v \neq \mathbf{0}$. For instance, let us find the equation of the plane tangent to the surface under consideration at the point $\mathbf{\Phi}(1, 0) = (1, 0, 1)$. At this point

$$\mathbf{n} = \mathbf{T}_u \times \mathbf{T}_v = (-2, 0, 1) = -2\mathbf{i} + \mathbf{k}$$

Since we have the vector $\mathbf{n}$ normal to the surface and a point $(1, 0, 1)$ on the surface, we can use formula (1′) to obtain an equation of the tangent plane:

$$-2(x - 1) + (z - 1) = 0$$

that is

$$z = 2x - 1$$

EXAMPLE 3. Suppose a surface S is the graph of a differentiable function $g: \mathbb{R}^2 \to \mathbb{R}$. Then the surface is smooth at all points $(u_0, v_0, g(u_0, v_0)) \in \mathbb{R}^3$.

To show this, we write S in parametric form as follows:

$$x = u, \qquad y = v, \qquad z = g(u, v)$$

which is the same as $z = g(x, y)$. Then

$$\mathbf{T}_u = \mathbf{i} + \frac{\partial g}{\partial u}(u_0, v_0)\mathbf{k}$$

$$\mathbf{T}_v = \mathbf{j} + \frac{\partial g}{\partial v}(u_0, v_0)\mathbf{k}$$

and for $(u_0, v_0) \in \mathbb{R}^2$,

$$\mathbf{n} = \mathbf{T}_u \times \mathbf{T}_v = -\frac{\partial g}{\partial u}(u_0, v_0)\mathbf{i} - \frac{\partial g}{\partial v}(u_0, v_0)\mathbf{j} + \mathbf{k} \neq \mathbf{0} \tag{2}$$

This is nonzero because the coefficient of $\mathbf{k}$ is 1; consequently the parametrization $(u, v) \mapsto (u, v, g(u, v))$ is smooth at all points. Moreover the tangent plane at $(x_0, y_0, z_0) = (u_0, v_0, g(u_0, v_0))$ is given, by formula (1), as

$$(x - x_0, y - y_0, z - z_0) \cdot \left(-\frac{\partial g}{\partial u}, -\frac{\partial g}{\partial v}, 1\right) = 0$$

where the partial derivatives are evaluated at (u_0, v_0). Remembering that $x = u$ and $y = v$, we can write this as

$$z - z_0 = \left(\frac{\partial g}{\partial x}\right) \cdot (x - x_0) + \left(\frac{\partial g}{\partial y}\right) \cdot (y - y_0) \tag{3}$$

where $\partial g/\partial x$ and $\partial g/\partial y$ are evaluated at (x_0, y_0).

This example also shows that the definition of the tangent plane for parametrized surfaces agrees with the one for surfaces obtained as graphs, since (3) is the same formula we derived (in Chapter 2) for the plane tangent to S at $(x_0, y_0, z_0) \in S$ (see p. 110).

It is also useful to consider piecewise smooth surfaces, that is, surfaces composed of a certain number of images of smooth parametrized surfaces. For example, the surface of a cube in $\mathbb{R}^3$ is such a surface. These surfaces are considered in Section 6.4.

EXERCISES

In Exercises 1 to 3, find an equation for the plane tangent to the given surface at the specified point.

1. $x = 2u, \quad y = u^2 + v, \quad z = v^2$, at $(0, 1, 1)$
2. $x = u^2 - v^2, \quad y = u + v, \quad z = u^2 + 4v$, at $(-\frac{1}{4}, \frac{1}{2}, 2)$
3. $x = u^2, \quad y = u \sin e^v, \quad z = \frac{1}{3}u \cos e^v$, at $(13, -2, 1)$
4. Are the surfaces in Exercises 1 and 2 smooth?
5. Find an expression for a unit vector normal to the surface

$$x = \cos v \sin u, \qquad y = \sin v \sin u, \qquad z = \cos u$$

for $u \in [0, \pi]$ and $v \in [0, 2\pi]$. Identify this surface.

6. Repeat Exercise 5 for the surface

$$x = (2 - \cos v)\cos u, \qquad y = (2 - \cos v)\sin u, \qquad z = \sin v$$

for $-\pi \le u \le \pi$, $-\pi \le v \le \pi$. Is this surface smooth?

7. (a) Develop a formula for the plane tangent to the surface $x = h(y, z)$.
   (b) Obtain a similar formula for $y = k(x, z)$.
8. Find the equation of the plane tangent to each surface at the indicated point.
   (a) $x = u^2, y = v^2, z = u^2 + v^2, u = 1, v = 1$
   (b) $z = 3x^2 + 8xy, x = 1, y = 0$
   (c) $x^3 + 3xy + z^2 = 2, x = 1, y = \frac{1}{3}, z = 0$

*9. Consider the surfaces $\mathbf{\Phi}_1(u, v) = (u, v, 0)$ and $\mathbf{\Phi}_2(u, v) = (u^3, v^3, 0)$.
(a) Show that the image of $\mathbf{\Phi}_1$ and $\mathbf{\Phi}_2$ is the xy-plane.
(b) Show that $\mathbf{\Phi}_1$ describes a smooth surface, yet $\mathbf{\Phi}_2$ does not. Conclude that the notion of smoothness of a surface S depends on the existence of at least one smooth parametrization for S.
(c) Prove that the tangent plane of S is well defined independent of the smooth parametrization (you will need to use the inverse function theorem from Section 4.4).
(d) After these remarks, do you think you can find a smooth parametrization of the cone in Figure 6.3.7?

*10. Let $\mathbf{\Phi}$ be a smooth surface; that is, $\mathbf{\Phi}$ is C^1 and $\mathbf{T}_u \times \mathbf{T}_v \neq \mathbf{0}$ at (u_0, v_0).
(a) Use the Implicit Function Theorem (Section 4.4) to show that the image of $\mathbf{\Phi}$ near (u_0, v_0) is the graph of a C^1 function of two variables, say $z = f(x, y)$. (This will hold if the z-component of $\mathbf{T}_u \times \mathbf{T}_v$ is nonzero.)
(b) Show that the tangent plane at $\mathbf{\Phi}(u_0, v_0)$ defined by the plane spanned by $\mathbf{T}_u$ and $\mathbf{T}_v$ coincides with the tangent plane of the graph of $z = f(x, y)$ at this point.

11. Consider the surface in $\mathbb{R}^3$ parametrized by

$$\mathbf{\Phi}(r, \theta) = (r \cos \theta, r \sin \theta, \theta), \qquad 0 \leq r \leq 1 \quad \text{and} \quad 0 \leq \theta \leq 4\pi$$

(a) Sketch and describe the surface.
(b) Find an expression for a unit normal to the surface.
(c) Find an equation for the plane tangent to the surface at the point (x_0, y_0, z_0).
(d) If (x_0, y_0, z_0) is a point on the surface, show that the horizontal line segment of unit length from the z-axis through (x_0, y_0, z_0) is contained in the surface and in the plane tangent to the surface at (x_0, y_0, z_0).

12. Given a sphere of radius 2 centered at the origin, find the equation for the plane that is tangent to it at the point $(1, 1, \sqrt{2})$ by considering the sphere as:
(a) a surface parametrized by $\mathbf{\Phi}(\theta, \phi) = (2 \cos \theta \sin \phi, 2 \sin \theta \sin \phi, 2 \cos \phi)$;
(b) a level surface of $f(x, y, z) = x^2 + y^2 + z^2$; and
(c) the graph of $g(x, y) = \sqrt{4 - x^2 - y^2}$.

13. (a) Find a parametrization for the hyperboloid $x^2 + y^2 - z^2 = 1$
(b) Find an expression for a unit normal to this surface.
(c) Find an equation for the plane tangent to the surface at $(x_0, y_0, 0)$, where $x_0^2 + y_0^2 = 1$.
(d) Show that the lines $(x_0, y_0, 0) + t(-y_0, x_0, 1)$ and $(x_0, y_0, 0) + t(y_0, -x_0, 1)$ lie in the surface *and* in the tangent plane found in (c).

*14. A parametrized surface is describe by a differentiable function $\mathbf{\Phi}: \mathbb{R}^2 \to \mathbb{R}^3$. According to Chapter 2, the derivative should give a linear approximation that supplies a representation of the tangent plane. This exercise demonstrates that this is indeed the case.
(a) Show that the range of the linear transformation $D\mathbf{\Phi}(u_0, v_0)$ is the plane spanned by $\mathbf{T}_u$ and $\mathbf{T}_v$. ($\mathbf{T}_u$ and $\mathbf{T}_v$ are evaluated at (u_0, v_0).)
(b) Show that $\mathbf{w} \perp \mathbf{T}_u \times \mathbf{T}_v$ if and only if $\mathbf{w}$ is in the range of $D\mathbf{\Phi}(u_0, v_0)$.
(c) Show that the tangent plane as defined in this section is the same as the "parametrized plane"

$$(u, v) \mapsto \mathbf{\Phi}(u_0, v_0) + D\mathbf{\Phi}(u_0, v_0)\begin{pmatrix} u - u_0 \\ u - v_0 \end{pmatrix}$$

6.4 AREA OF A SURFACE

Before proceeding to general surface integrals, let us first consider the problem of computing the area of a surface, just as we considered the problem of finding the arc length of a curve before discussing path integrals.

Our object here is to derive a formula for the area of a surface. There are various ways of obtaining such formulas. In order to avoid certain difficulties involved in deriving surface area from a limiting process involving Riemann sums, we shall take a simpler route and *define* surface area as a double integral. Then we shall present an argument for the plausibility of this definition.

In Section 6.3 we defined a parametrized surface S to be the *image* of a function $\mathbf{\Phi}: D \subset \mathbb{R}^2 \to \mathbb{R}^3$, written as $\mathbf{\Phi}(u, v) = (x(u, v), y(u, v), z(u, v))$. The map $\mathbf{\Phi}$ was called the parametrization of S. Then S was said to be smooth at $\mathbf{\Phi}(u, v) \in S$ if $\mathbf{T}_u \times \mathbf{T}_v \neq \mathbf{0}$, where

$$\mathbf{T}_u = \frac{\partial x}{\partial u}(u, v)\mathbf{i} + \frac{\partial y}{\partial u}(u, v)\mathbf{j} + \frac{\partial z}{\partial u}(u, v)\mathbf{k}$$

and

$$\mathbf{T}_v = \frac{\partial x}{\partial v}(u, v)\mathbf{i} + \frac{\partial y}{\partial v}(u, v)\mathbf{j} + \frac{\partial z}{\partial v}(u, v)\mathbf{k}$$

Recall that a smooth surface (loosely speaking) is one that has no corners or breaks.

In the rest of this chapter and in the next one we shall consider only piecewise smooth surfaces that are unions of images of parametrized surfaces $\mathbf{\Phi}_i: D_i \to \mathbb{R}^3$ for which:

(*i*) D_i is an elementary region in the plane;

(*ii*) $\mathbf{\Phi}_i$ is C^1 and one-to-one, except possibly on the boundary of D_i; and

(*iii*) S_i, the image of $\mathbf{\Phi}_i$ is smooth, except possibly at a finite number of points.

DEFINITION. *We define the* ***surface area**** $A(S)$ *of a parametrized surface by*

$$A(S) = \int_D \|\mathbf{T}_u \times \mathbf{T}_v\| \, du \, dv \tag{1}$$

where $\|\mathbf{T}_u \times \mathbf{T}_v\|$ *is the norm of* $\mathbf{T}_u \times \mathbf{T}_v$. *If S is a union of surfaces S_i, its area is the sum of the areas of the S_i.*

As the reader can easily verify, we have

$$\|\mathbf{T}_u \times \mathbf{T}_v\| = \sqrt{\left[\frac{\partial(x, y)}{\partial(u, v)}\right]^2 + \left[\frac{\partial(y, z)}{\partial(u, v)}\right]^2 + \left[\frac{\partial(x, z)}{\partial(u, v)}\right]^2} \tag{2}$$

* As we have not yet discussed the independence of parametrization, it may seem that $A(S)$ depends on the parametrization $\mathbf{\Phi}$. We shall discuss independence of parametrization in Section 6.6; the use of this notation here should not cause confusion.

where

$$\frac{\partial(x, y)}{\partial(u, v)} = \begin{vmatrix} \frac{\partial x}{\partial u} & \frac{\partial x}{\partial v} \\ \frac{\partial y}{\partial u} & \frac{\partial y}{\partial v} \end{vmatrix}$$

and so on. Thus formula (1) becomes

$$A(S) = \int_D \sqrt{\left[\frac{\partial(x, y)}{\partial(u, v)}\right]^2 + \left[\frac{\partial(y, z)}{\partial(u, v)}\right]^2 + \left[\frac{\partial(x, z)}{\partial(u, v)}\right]^2}\, du\, dv \tag{3}$$

We can also justify this definition by analyzing the integral $\int_D \|\mathbf{T}_u \times \mathbf{T}_v\| \, du\, dv$ in terms of Riemann sums. For simplicity, suppose D is a rectangle; consider the nth regular partition of D, and let R_{ij} be the ijth rectangle in the partition, with vertices (u_i, v_j), (u_{i+1}, v_j), (u_i, v_{j+1}), and (u_{i+1}, v_{j+1}), $0 \le i \le n-1$, $0 \le j \le n-1$. Denote the values of $\mathbf{T}_u$ and $\mathbf{T}_v$ at (u_i, v_j) by $\mathbf{T}_{u_i}$ and $\mathbf{T}_{v_j}$. We can think of the vectors $\Delta u \mathbf{T}_{u_i}$ and $\Delta v \mathbf{T}_{v_j}$ as tangent to the surface at $\mathbf{\Phi}(u_i, v_j) = (x_{ij}, y_{ij}, z_{ij})$ where $\Delta u = u_{i+1} - u_i$, $\Delta v = v_{j+1} - v_j$. Then these vectors form a parallelogram P_{ij} that lies in the plane tangent to the surface at (x_{ij}, y_{ij}, z_{ij}) (see Figure 6.4.1). We thus have a "patchwork cover" of the surface by the P_{ij}. For n large, the area of P_{ij} is a good approximation to the area of $\mathbf{\Phi}(R_{ij})$. Since the area of the parallelogram spanned by two vectors $\mathbf{v}_1$ and $\mathbf{v}_2$ is $\|\mathbf{v}_1 \times \mathbf{v}_2\|$ (see Chapter 1), we see that

$$A(P_{ij}) = \|\Delta u \mathbf{T}_{u_i} \times \Delta v \mathbf{T}_{v_j}\| = \|\mathbf{T}_{u_i} \times \mathbf{T}_{v_j}\| \,\Delta u\, \Delta v$$

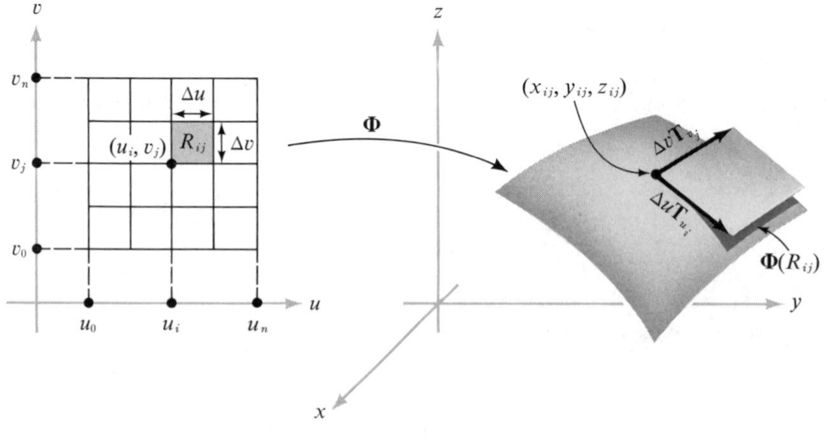

FIGURE 6.4.1

$\|\mathbf{T}_{u_i} \times \mathbf{T}_{v_j}\| \,\Delta u\, \Delta v$ is equal to the area of a parallelogram that approximates the area of a patch on a surface $S = \mathbf{\Phi}(D)$.

Therefore, the area of the cover is

$$A_n = \sum_{i=0}^{n-1} \sum_{j=0}^{n-1} A(P_{ij}) = \sum_{i=0}^{n-1} \sum_{j=0}^{n-1} \|\mathbf{T}_{u_i} \times \mathbf{T}_{v_j}\| \, \Delta u \, \Delta v$$

As $n \to \infty$, the sums A_n converge to the integral

$$\int_D \|\mathbf{T}_u \times \mathbf{T}_v\| \, du \, dv$$

Since A_n should approximate the surface area better and better as $n \to \infty$, we are led to formula (1) for a reasonable definition of $A(S)$.

EXAMPLE 1. Let D be the region determined by $0 \le \theta \le 2\pi, 0 \le r \le 1$ and let the function $\mathbf{\Phi}: D \to \mathbb{R}^3$ defined by

$$x = r \cos \theta, \qquad y = r \sin \theta, \qquad z = r$$

be a parametrization of a cone S. We compute

$$\frac{\partial(x, y)}{\partial(r, \theta)} = \begin{vmatrix} \cos \theta & -r \sin \theta \\ \sin \theta & r \cos \theta \end{vmatrix} = r$$

$$\frac{\partial(y, z)}{\partial(r, \theta)} = \begin{vmatrix} \sin \theta & r \cos \theta \\ 1 & 0 \end{vmatrix} = -r \cos \theta$$

and

$$\frac{\partial(x, z)}{\partial(r, \theta)} = \begin{vmatrix} \cos \theta & -r \sin \theta \\ 1 & 0 \end{vmatrix} = r \sin \theta$$

so the area integrand is

$$\begin{aligned} \|\mathbf{T}_r \times \mathbf{T}_\theta\| &= \sqrt{r^2 + r^2 \cos^2 \theta + r^2 \sin^2 \theta} \\ &= r\sqrt{2} \end{aligned}$$

Clearly, $\|\mathbf{T}_r \times \mathbf{T}_\theta\|$ vanishes for $r = 0$, but $\mathbf{\Phi}(0, \theta) = (0, 0, 0)$ for any θ. Thus $(0, 0, 0)$ is the only point where the surface is not smooth. We have

$$\begin{aligned} \int_D \|\mathbf{T}_r \times \mathbf{T}_\theta\| \, dr \, d\theta &= \int_0^{2\pi} \int_0^1 \sqrt{2} r \, dr \, d\theta \\ &= \int_0^{2\pi} \tfrac{1}{2}\sqrt{2} \, d\theta \\ &= \sqrt{2}\pi \end{aligned}$$

To confirm that this is the area of $\mathbf{\Phi}(D)$ we must verify that $\mathbf{\Phi}$ is one-to-one (for points not on the boundary of D). Let D^0 be the set of (r, θ) with $0 < r < 1$ and $0 < \theta < 2\pi$. Hence, D^0 is D without its boundary. To see that $\mathbf{\Phi}: D^0 \to \mathbb{R}^3$ is one-to-one, assume that $\mathbf{\Phi}(r, \theta) = \mathbf{\Phi}(r', \theta')$ for (r, θ) and $(r', \theta') \in D^0$. Then

$$r \cos \theta = r' \cos \theta', \qquad r \sin \theta = r' \sin \theta', \qquad r = r'$$

From these equations it follows that $\cos\theta = \cos\theta'$ and $\sin\theta = \sin\theta'$. Thus either $\theta = \theta'$ or $\theta = \theta' + 2\pi n$. But the second case is impossible, since both θ and θ' belong to the open interval $]0, 2\pi[$ and thus cannot be 2π radians apart. This proves that off the boundary $\mathbf{\Phi}$ is one-to-one. (Is $\mathbf{\Phi}: D \to \mathbb{R}^3$ one-to-one?)

In future examples we shall not usually verify that the parametrization is one-to-one when it is intuitively clear.

EXAMPLE 2. A helicoid is defined by $\mathbf{\Phi}: D \to \mathbb{R}^3$ where

$$x = r\cos\theta, \qquad y = r\sin\theta, \qquad z = \theta$$

and D is the region where $0 \le \theta \le 2\pi$ and $0 \le r \le 1$ (Figure 6.4.2). We compute $\partial(x, y)/\partial(r, \theta) = r$ as before, and

$$\frac{\partial(y, z)}{\partial(r, \theta)} = \begin{vmatrix} \sin\theta & r\cos\theta \\ 0 & 1 \end{vmatrix} = \sin\theta$$

$$\frac{\partial(x, z)}{\partial(r, \theta)} = \begin{vmatrix} \cos\theta & -r\sin\theta \\ 0 & 1 \end{vmatrix} = \cos\theta$$

The area integrand is therefore $\sqrt{r^2 + 1}$, which never vanishes, so the surface is smooth. The area of the helicoid is

$$\int_D \|\mathbf{T}_r \times \mathbf{T}_\theta\|\, dr\, d\theta = \int_0^{2\pi}\int_0^1 \sqrt{r^2+1}\, dr\, d\theta = 2\pi \int_0^1 \sqrt{r^2+1}\, dr$$

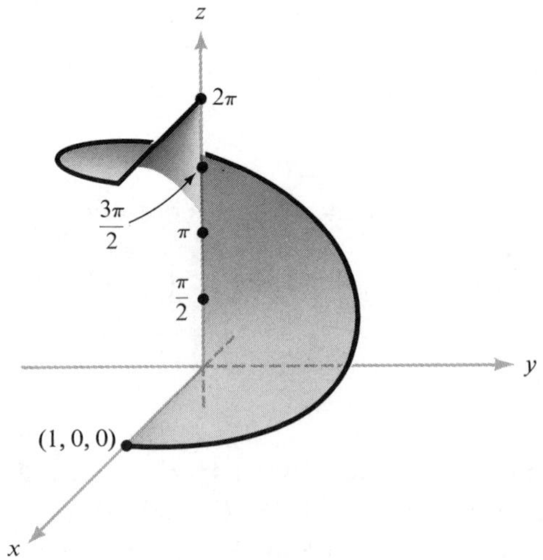

FIGURE 6.4.2
The helicoid $x = r\cos\theta$, $y = r\sin\theta$, $z = \theta$.

After a little computation (using the integral tables in Appendix C), we find that this integral is equal to

$$\pi(\sqrt{2} + \log(1 + \sqrt{2}))$$

A surface S given in the form $z = f(x, y)$, where $(x, y) \in D$, admits the parametrization

$$x = u, \qquad y = v, \qquad z = f(u, v)$$

for $(u, v) \in D$. When f is C^1, this parametrization is smooth, and the formula for surface area reduces to

$$A(S) = \int_D \left(\sqrt{\left(\frac{\partial f}{\partial x}\right)^2 + \left(\frac{\partial f}{\partial y}\right)^2 + 1}\right) dA \tag{4}$$

after applying the formulas

$$\mathbf{T}_u = \mathbf{i} + \frac{\partial f}{\partial u}\mathbf{k}$$

$$\mathbf{T}_v = \mathbf{j} + \frac{\partial f}{\partial v}\mathbf{k}$$

and

$$\mathbf{T}_u \times \mathbf{T}_v = -\frac{\partial f}{\partial u}\mathbf{i} - \frac{\partial f}{\partial v}\mathbf{j} + \mathbf{k} = -\frac{\partial f}{\partial x}\mathbf{i} - \frac{\partial f}{\partial y}\mathbf{j} + \mathbf{k}$$

as noted in Example 2 of Section 6.3.

OPTIONAL

SURFACE AREA AND IMPROPER INTEGRALS

In formula (4) we have assumed that $\partial f/\partial x$ and $\partial f/\partial y$ are continuous (and hence, bounded) functions on D. However, it is important to consider areas of surfaces for which either $(\partial f/\partial x)(x_0, y_0)$ or $(\partial f/\partial y)(x_0, y_0)$ gets arbitrarily large as (x_0, y_0) approaches the boundary of D. For example, consider the hemisphere

$$z = \sqrt{1 - x^2 - y^2}$$

where D is the region $x^2 + y^2 \leq 1$ (see Figure 6.4.3). We have

$$\frac{\partial f}{\partial x} = \frac{-x}{\sqrt{1 - x^2 - y^2}}, \qquad \frac{\partial f}{\partial y} = \frac{-y}{\sqrt{1 - x^2 - y^2}} \tag{5}$$

OPTIONAL (*Continued*)

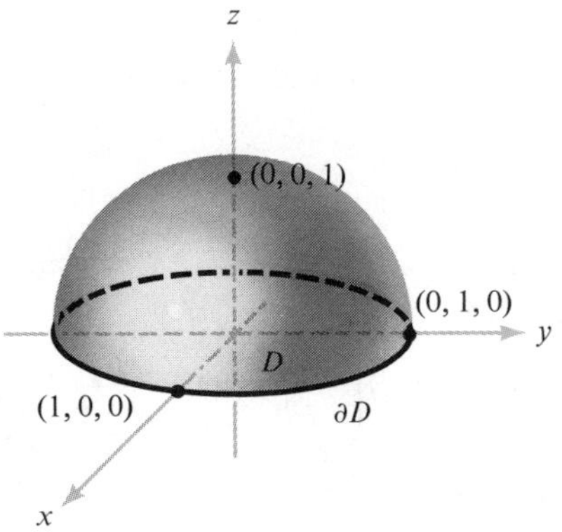

FIGURE 6.4.3
The hemisphere $z = \sqrt{1 - x^2 - y^2}$.

The boundary of D is the unit circle $x^2 + y^2 = 1$, so, as (x, y) gets close to ∂D, the value of $x^2 + y^2$ approaches 1. Hence, the denominators in (5) go to zero.

We want to be able to deal with cases such as these, so we define the area $A(S)$ of a surface S described by $z = f(x, y)$ over a region D, where f is differentiable with possible discontinuities of $\partial f/\partial x$ and $\partial f/\partial y$ on ∂D, as

$$A(S) = \int_D \left(\sqrt{\left(\frac{\partial f}{\partial x}\right)^2 + \left(\frac{\partial f}{\partial y}\right)^2 + 1} \right) dA$$

whenever $\sqrt{(\partial f/\partial x)^2 + (\partial f/\partial y)^2 + 1}$ is integrable over D, *even if the integral is improper*; in fact, this is one of the reasons we introduced the notion of improper integral in Chapter 5.

EXAMPLE 3. Compute the area of the surface of the sphere S described by $x^2 + y^2 + z^2 = 1$.

We shall compute the area of the upper hemisphere S^+, where

$$x^2 + y^2 + z^2 = 1, \qquad z \geq 0$$

and then multiply our result by two. Hence we have

$$f(x, y) = \sqrt{1 - x^2 - y^2}, \qquad x^2 + y^2 \leq 1$$

OPTIONAL (*Continued*)

Let D be the region $x^2 + y^2 \leq 1$. Then applying formula (4) and calculations (5) above we get

$$A(S^+) = \int_D \left(\sqrt{\left(\frac{\partial f}{\partial x}\right)^2 + \left(\frac{\partial f}{\partial y}\right)^2 + 1}\right) dA$$

$$= \int_D \left(\sqrt{\frac{x^2}{1 - x^2 - y^2} + \frac{y^2}{1 - x^2 - y^2} + 1}\right) dA$$

$$= \int_D \frac{1}{\sqrt{1 - x^2 - y^2}}\, dy\, dx$$

which is an improper integral. However we may apply Fubini's Theorem in this case to obtain the iterated improper integral

$$\int_{-1}^{1} \int_{-(1-x^2)^{1/2}}^{(1-x^2)^{1/2}} \frac{1}{\sqrt{1 - x^2 - y^2}}\, dy\, dx = \int_{-1}^{1} \left[\sin^{-1} \frac{y}{(1 - x^2)^{1/2}}\right]_{-(1-x^2)^{1/2}}^{(1-x^2)^{1/2}} dx$$

$$= \int_{-1}^{1} \left[\frac{\pi}{2} + \frac{\pi}{2}\right] dx = \int_{-1}^{1} \pi\, dx = 2\pi$$

Thus, the area of the entire sphere is 4π. For another way of computing this area without improper integrals, see Exercise 1.

In most books on one-variable calculus, it is shown that the lateral surface area generated by revolving the graph of a function $y = f(x)$ about the x-axis is given by

$$A_1 = 2\pi \int_a^b |f(x)|\sqrt{1 + [f'(x)]^2}\, dx \tag{6}$$

If the graph is revolved about the y-axis, we have

$$A_2 = 2\pi \int_a^b |x|\sqrt{1 + [f'(x)]^2}\, dx \tag{7}$$

We shall derive (6) by using the methods developed above; one can obtain (7) in a similar fashion (Exercise 10).

To derive formula (6) from formula (3), we must give a parametrization of S. Define the parametrization by

$$x = u, \qquad y = f(u)\cos v, \qquad z = f(u)\sin v$$

over the region D given by

$$a \leq u \leq b, \qquad 0 \leq v \leq 2\pi$$

This is indeed a parametrization of S, because for fixed u

$$(u, f(u)\cos v, f(u)\sin v)$$

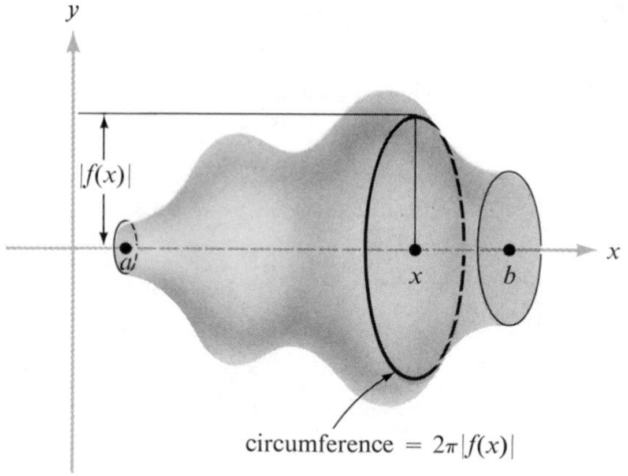

FIGURE 6.4.4
The curve $y = f(x)$ rotated about the x-axis.

traces out a circle of radius $|f(u)|$ with the center $(u, 0, 0)$ (Figure 6.4.4). Now

$$\frac{\partial(x, y)}{\partial(u, v)} = -f(u)\sin v, \qquad \frac{\partial(y, z)}{\partial(u, v)} = f(u)f'(u), \qquad \frac{\partial(x, z)}{\partial(u, v)} = f(u)\cos v,$$

so by formula (3)

$$\begin{aligned} A(S) &= \int_D \left(\sqrt{\left[\frac{\partial(x, y)}{\partial(u, v)}\right]^2 + \left[\frac{\partial(y, z)}{\partial(u, v)}\right]^2 + \left[\frac{\partial(x, z)}{\partial(u, v)}\right]^2}\right) du\, dv \\ &= \int_D \left(\sqrt{[f(u)]^2 \sin^2 v + [f(u)]^2[f'(u)]^2 + [f(u)]^2 \cos^2 v}\right) du\, dv \\ &= \int_D |f(u)|\sqrt{1 + [f'(u)]^2}\, du\, dv \\ &= \int_a^b \int_0^{2\pi} |f(u)|\sqrt{1 + [f'(u)]^2}\, dv\, du \\ &= 2\pi \int_a^b |f(u)|\sqrt{1 + [f'(u)]^2}\, du \end{aligned}$$

which is formula (6).

If S is the surface of revolution then $2\pi|f(x)|$ is the circumference of the vertical cross section to S at the point x (Figure 6.4.4). Observe that we can write

$$2\pi \int_a^b |f(x)|\sqrt{1 + [f'(x)]^2}\, dx = \int_{\boldsymbol{\sigma}} 2\pi|f(x)|\, ds$$

where the expression on the right is the integral of $2\pi|f(x)|$ along the path $\boldsymbol{\sigma}: [a, b] \to \mathbb{R}^2$, $t \mapsto (t, f(t))$. Therefore, the lateral surface area of a solid of revolution is obtained by integrating the cross-sectional circumference along the path determined by the given function.

HISTORICAL NOTE

Recall that the Calculus was invented (or discovered?) by Isaac Newton (1647–1727) about 1669 and Gottfried Wilhelm Leibniz (1646–1716) about 1684. In the beginning of the eighteenth century, mathematicians were interested in the problem of finding paths of shortest length on a surface by using the methods of the Calculus. At this time surfaces were regarded as boundaries of solids that were defined by inequalities (the ball $x^2+y^2+z^2 \leq 1$ is bounded by the sphere $x^2 + y^2 + z^2 = 1$).

Christian Huygens (1629–1695) was the first person since Archimedes to give results on the areas of special surfaces beyond the sphere, and he obtained the areas of portions of surfaces of revolution, such as the paraboloid and hyperboloid.

The brilliant and prolific mathematician Leonhard Euler (1707–1783) presented the first fundamental work on the theory of surfaces in 1760 with "Recherches sur la courbure des surfaces," and it may have been in this work that a surface was formally defined for the first time as the graph of $z = f(x, y)$. Euler was interested in studying the curvature of surfaces, and in 1771 he introduced the notion of the parametric surfaces that are described in this section.

After the rapid development of the Calculus in the early eighteenth century, formulas for the lengths of curves and areas of surfaces were developed. Although we do not know when the area formulas presented in this section first appeared, they were certainly common by the end of the eighteenth century. The underlying concepts of the length of a curve and the area of a surface were understood intuitively before this time, and the use of calculus formulas to compute areas was considered a great achievement.

Augustin-Louis Cauchy (1789–1857) was the first to take the step of defining the quantities of length and surface areas by integrals as we have done in this book. The question of defining surface area independent of integrals was taken up somewhat later, but this posed many difficult problems that were not properly resolved until this century.

We end this section by describing the fascinating classical problem of Plateau, which has enjoyed a long history in mathematics. The Belgian physicist Joseph Plateau (1801–1883) carried out many experiments from 1830 to 1869 on surface tension and capillary phenomena, experiments that had enormous impact at the time and were repeated by notable nineteenth-century physicists, such as Michael Faraday (1791–1867).

If a wire is dipped into a soap or glycerine solution, then one usually withdraws a soap film spanning the wire. Some examples are given in Figure 6.4.5, although readers might like to perform the experiment for themselves.

HISTORICAL NOTE (*Continued*)

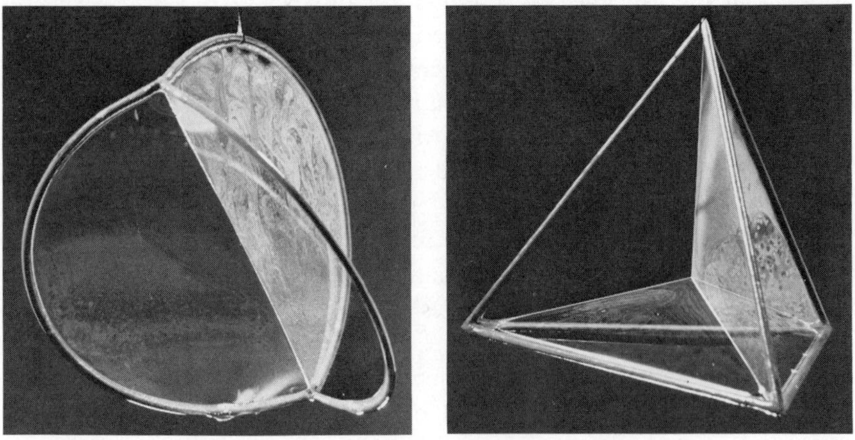

FIGURE 6.4.5
Two soap films spanning wires. (Fritz Goro.)

Plateau raised the mathematical question: For a given boundary (wire), how does one prove the existence of such a surface (soap film) and how many surfaces can there be? The underlying physical principle is that nature tends to minimize area; that is, the surface that forms should be a surface of least area among all possible surfaces that have the given curve as their boundary.

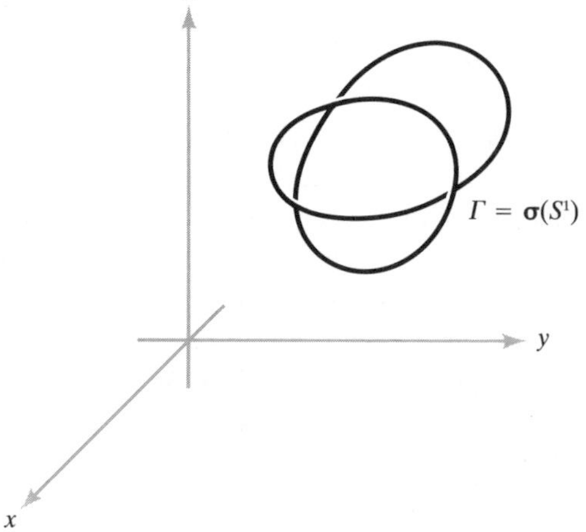

FIGURE 6.4.6
Γ *is to be the boundary of a soap film.*

HISTORICAL NOTE (*Continued*)

Plateau formulated the problem in a special way. Let $D \subset \mathbb{R}^2$ be the unit disc defined by $\{(x, y) | x^2 + y^2 \leq 1\}$; let $S^1 = \partial D$ be its boundary. Furthermore suppose that $\boldsymbol{\sigma}: [0, 2\pi] \to \mathbb{R}^3$ is a simple closed curve with $\Gamma = \boldsymbol{\sigma}([0, 2\pi])$ its image representing a wire in $\mathbb{R}^3$ (Figure 6.4.6).

Let $\mathscr{S}$ be the set of all maps $\boldsymbol{\Phi}: D \to \mathbb{R}^3$ such that $\boldsymbol{\Phi}(\partial D) = \Gamma$, $\boldsymbol{\Phi}$ is C^1, and $\boldsymbol{\Phi}$ is one-to-one on ∂D. Each $\boldsymbol{\Phi} \in \mathscr{S}$ represents a parametric C^1 surface "spanning" the wire Γ. For each $\boldsymbol{\Phi} \in \mathscr{S}$ we have the area of the image surface, $A(\boldsymbol{\Phi}) = \int_D \|\mathbf{T}_u \times \mathbf{T}_v\| \, du \, dv$. Thus area is a function that assigns to each parametric surface its area. Plateau asked whether A has a minimum on $\mathscr{S}$; that is, does there exist a $\boldsymbol{\Phi}_0$ such that $A(\boldsymbol{\Phi}_0) \leq A(\boldsymbol{\Phi})$ for all $\boldsymbol{\Phi} \in \mathscr{S}$? Unfortunately, the methods of this book are not adequate to solve this problem. We can tackle questions of finding minima of real-valued functions of several variables but in no way can the set $\mathscr{S}$ be thought of as a region in $\mathbb{R}^n$ for *any n*! The set $\mathscr{S}$ is really a space of functions, and the problem of finding minima of functions like A on such sets is part of a subject called the "Calculus of Variations," a subject that is almost as old as the Calculus itself. It is also a discipline that is intimately connected with partial differential equations.

Plateau showed that if a minimum

$$\boldsymbol{\Phi}_0(u, v) = (x(u, v), y(u, v), z(u, v))$$

existed at all it would have to satisfy (after suitable normalizations) the partial differential equations

(*i*) $\nabla^2 \boldsymbol{\Phi}_0 = 0$

(*ii*) $\dfrac{\partial \boldsymbol{\Phi}_0}{\partial u} \cdot \dfrac{\partial \boldsymbol{\Phi}_0}{\partial v} = 0$

(*iii*) $\left\| \dfrac{\partial \boldsymbol{\Phi}_0}{\partial u} \right\| = \left\| \dfrac{\partial \boldsymbol{\Phi}_0}{\partial v} \right\|$

where $\|\mathbf{w}\|$ denotes the "norm" or length of the vector $\mathbf{w}$.

For well over seventy years mathematicians such as Riemann, Wierstrass, H. A. Schwarz, Darboux, and Lebesgue puzzled over the challenge posed by Plateau. In 1931 the question was finally settled when Jessie Douglas showed that such a $\boldsymbol{\Phi}_0$ existed. However, many questions about soap films remain unsolved, and this area of research is still active today.

EXERCISES

1. Find the surface area of the unit sphere S represented parametrically by $\boldsymbol{\Phi}: D \to S \subset \mathbb{R}^3$ where D is the rectangle $0 \leq \theta \leq 2\pi$, $0 \leq \phi \leq \pi$ and $\boldsymbol{\Phi}$ is given by the

equations

$$x = \cos\theta \sin\phi, \qquad y = \sin\theta \sin\phi, \qquad z = \cos\phi$$

Note that we can represent the entire sphere parametrically, but we cannot represent it in the form $z = f(x, y)$. Compare with Example 3.

2. In Exercise 1, what happens if we allow ϕ to vary from $-\pi/2$ to $\pi/2$; from 0 to 2π? Why do we obtain different answers?

3. Find the area of the helicoid in Example 2 if the domain D is $0 \le r \le 1$ and $0 \le \theta \le 3\pi$.

4. The torus T can be represented parametrically by the function $\mathbf{\Phi}: D \to \mathbb{R}^3$, where $\mathbf{\Phi}$ is given by the coordinate functions $x = (R + \cos\phi)\cos\theta$, $y = (R + \cos\phi)\sin\theta$, $z = \sin\phi$; D is the rectangle $[0, 2\pi] \times [0, 2\pi]$, that is, $0 \le \theta, \phi \le 2\pi$; and $R > 1$ is fixed (see Figure 6.4.7). Show that $A(T) = (2\pi)^2 R$, first by using formula (3) and then by using formula (7).

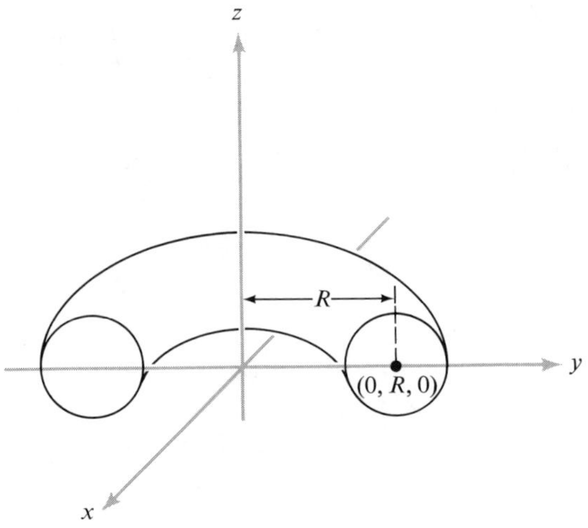

FIGURE 6.4.7

A cross section of a torus.

5. Let $\mathbf{\Phi}(u, v) = (u - v, u + v, uv)$ and let D be the unit disc in the uv-plane. Find the area of $\mathbf{\Phi}(D)$.

6. Find the area of the portion of the unit sphere that is cut out by the cone $x^2 + y^2 = z^2$, $z \ge 0$ (see Exercise 1).

*7. The cylinder $x^2 + y^2 = x$ divides the unit sphere S into two regions S_1 and S_2, where S_1 is inside the cylinder and S_2 outside. Find the ratio of areas $A(S_2)/A(S_1)$.

*8. Suppose a surface S that is the graph of a function $z = f(x, y)$, $(x, y) \in D \subset \mathbb{R}^2$ can also be described as the set of $(x, y, z) \in \mathbb{R}^3$ with $F(x, y, z) = 0$ (a level surface). Derive a formula for $A(S)$ that involves only F.

9. Represent the ellipsoid E:

$$\frac{x^2}{a^2} + \frac{y^2}{b^2} + \frac{z^2}{c^2} = 1$$

parametrically and write out the integral for its surface area $A(E)$ (do not evaluate the integral).

10. Let the curve $y = f(x)$, $a \le x \le b$ be rotated about the y-axis. Show that the area of the surface swept out is

$$A = 2\pi \int_a^b \sqrt{1 + [f'(x)]^2}\,|x|\,dx$$

11. Find the area of the surface obtained by rotating the curve $y = x^2$, $0 \le x \le 1$, about the y-axis.

12. Use formula (4) to compute the surface area of the cone in Example 1.

13. Find the area of the surface defined by $x + y + z = 1$, $x^2 + 2y^2 \le 1$.

14. Show that for the vectors $\mathbf{T}_u$ and $\mathbf{T}_v$ we have the formula

$$\|\mathbf{T}_u \times \mathbf{T}_v\| = \sqrt{\left[\frac{\partial(x, y)}{\partial(u, v)}\right]^2 + \left[\frac{\partial(y, z)}{\partial(u, v)}\right]^2 + \left[\frac{\partial(x, z)}{\partial(u, v)}\right]^2}$$

15. Compute the area of the surface given by

$$x = r\cos\theta, \qquad y = 2r\cos\theta, \qquad z = \theta, \qquad 0 \le r \le 1, \qquad 0 \le \theta \le 2\pi$$

Sketch.

16. Prove *Pappus' Theorem*: Let $\boldsymbol{\sigma}\colon [a, b] \to \mathbb{R}^2$ be a C^1 path whose image lies in the right half plane and is a simple closed curve. The area of the lateral surface generated by rotating the image of $\boldsymbol{\sigma}$ about the y-axis is equal to $2\pi\bar{x}l(\boldsymbol{\sigma})$ where $\bar{x}$ is the average value of x coordinates of points on $\boldsymbol{\sigma}$, and $l(\boldsymbol{\sigma})$ is the length of $\boldsymbol{\sigma}$. (See Exercises 8–12, Section 6-1, for a discussion of average values.)

17. Take a thin length of wire and bend it into a simple closed curve. Fill a small bowl with water, add some dishwashing liquid, and stir thoroughly. Dip the wire into this solution and withdraw slowly. Write a few sentences on your observations.

6.5 INTEGRALS OF SCALAR FUNCTIONS OVER SURFACES

Now we are ready to define the integral of a *scalar* function f *over a surface* S. This concept is a natural generalization of the area of a surface, which corresponds to the integral over S of the scalar function $f(x, y, z) = 1$. This is quite analogous to considering the path integral as a generalization of arc length. In the next section we shall deal with the integral of a *vector* function $\mathbf{F}$ over a surface. These concepts will play a crucial role in the vector analysis treated in the final chapter.

Let us start with a surface S parametrized by $\mathbf{\Phi}: D \to S \subset \mathbb{R}^3$, $\mathbf{\Phi}(u, v) = (x(u, v), y(u, v), z(u, v))$.

DEFINITION. *If $f(x, y, z)$ is a real-valued continuous function defined on S, we define the **integral of f over S** to be*

$$\int_S f(x, y, z)\, dS = \int_S f\, dS = \int_D f(\mathbf{\Phi}(u, v))\|\mathbf{T}_u \times \mathbf{T}_v\|\, du\, dv \qquad (1)$$

where $\mathbf{T}_u$ and $\mathbf{T}_v$ have the same meaning as in Section 6.3. Written out, equation (1) becomes

$$\int_S f\, dS =$$

$$\int_D f(x(u, v), y(u, v), z(u, v)) \sqrt{\left[\frac{\partial(x, y)}{\partial(u, v)}\right]^2 + \left[\frac{\partial(y, z)}{\partial(u, v)}\right]^2 + \left[\frac{\partial(x, z)}{\partial(u, v)}\right]^2}\, du\, dv \quad (2)$$

Thus if f is identically 1, we recover the area formula (3) of Section 6.4. Like surface area, the surface integral is independent of the particular parametrization used. This will be discussed in Section 6.6.

We can gain some intuitive knowledge about this integral by considering it as a limit of sums. Let D be a rectangle partitioned into n^2 rectangles R_{ij}. Let $S_{ij} = \mathbf{\Phi}(R_{ij})$ be the portion of the surface $\mathbf{\Phi}(D)$ corresponding to R_{ij} (see Figure (6.5.1), and let $A(S_{ij})$ be the area of this portion of the surface.

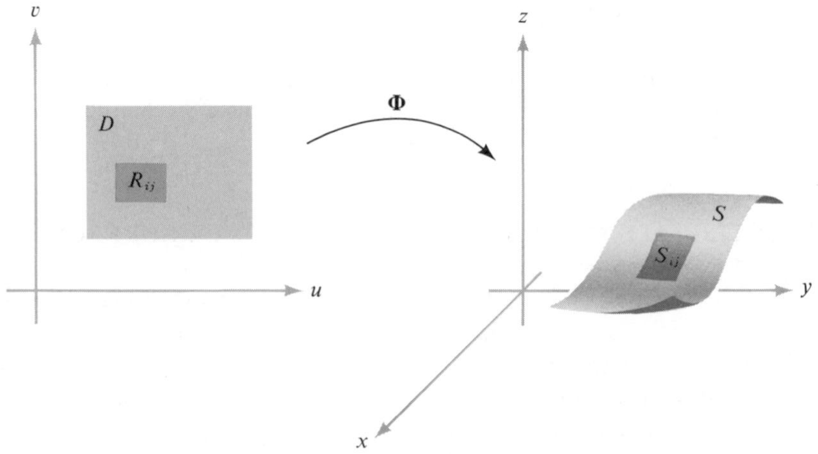

FIGURE 6.5.1

$\mathbf{\Phi}$ *takes a portion* R_{ij} *of* D *to a portion of* S.

For large n, f will be approximately constant on S_{ij}, and we form the sum

$$S_n = \sum_{i=0}^{n-1} \sum_{j=0}^{n-1} f(\mathbf{\Phi}(u_i, v_j)) A(S_{ij}) \tag{3}$$

where $(u_i, v_j) \in R_{ij}$. But from Section 6.4 we have a formula for $A(S_{ij})$:

$$A(S_{ij}) = \int_{R_{ij}} \|\mathbf{T}_u \times \mathbf{T}_v\| \, du \, dv$$

which, by the Mean Value Theorem for Integrals (see p. 289), equals $\|\mathbf{T}_{u_i^*} \times \mathbf{T}_{v_j^*}\| \, \Delta u \, \Delta v$ for some point (u_i^*, v_j^*) in R_{ij}. Hence our sum becomes

$$S_n = \sum_{i=0}^{n-1} \sum_{j=0}^{n-1} f(\mathbf{\Phi}(u_i, v_j)) \|\mathbf{T}_{u_i^*} \times \mathbf{T}_{v_j^*}\| \, \Delta u \, \Delta v$$

which is an approximating sum for the last integral in (1). Therefore, $\lim_{n \to \infty} S_n = \int_S f \, dS$. Note that each term in the sum (3) is the value of f at some point $\mathbf{\Phi}(u_i, v_j)$ times the area of S_{ij}. Compare this with the Riemann-sum interpretation of the path integral in Section 6.1.

If S is a union of parametrized surfaces S_i, $i = 1, \ldots, N$, that do not intersect except possibly along curves defining their boundaries, then the integral of f over S is defined by

$$\int_S f \, dS = \sum_{i=1}^{N} \int_{S_i} f \, dS$$

as we should expect. For example, the integral over the surface of a cube may be expressed as the sum of the integrals over the six sides.

EXAMPLE 1. Suppose a helicoid is described as in Example 2, Section 6.4, and let f be given by $f(x, y, z) = \sqrt{x^2 + y^2 + 1}$. As before

$$\frac{\partial(x, y)}{\partial(r, \theta)} = r, \qquad \frac{\partial(y, z)}{\partial(r, \theta)} = \sin \theta, \qquad \frac{\partial(x, z)}{\partial(r, \theta)} = \cos \theta$$

Also, $f(r \cos \theta, r \sin \theta, \theta) = \sqrt{r^2 + 1}$. Therefore

$$\begin{aligned} \int_S f(x, y, z) \, dS &= \int_D f(\mathbf{\Phi}(r, \theta)) \|\mathbf{T}_r \times \mathbf{T}_\theta\| \, dr \, d\theta \\ &= \int_0^{2\pi} \int_0^1 \sqrt{r^2 + 1} \sqrt{r^2 + 1} \, dr \, d\theta \\ &= \int_0^{2\pi} \tfrac{4}{3} \, d\theta = \tfrac{8}{3}\pi \end{aligned}$$

Suppose S is the graph of a C^1 function $z = g(x, y)$. Then we can parametrize S by

$$x = u, \qquad y = v, \qquad z = g(u, v)$$

In this case

$$\|\mathbf{T}_u \times \mathbf{T}_v\| = \sqrt{1 + \left(\frac{\partial g}{\partial u}\right)^2 + \left(\frac{\partial g}{\partial v}\right)^2}$$

and so

$$\int_S f(x, y, z)\,dS = \int_D f(x, y, g(x, y))\sqrt{1 + \left(\frac{\partial g}{\partial x}\right)^2 + \left(\frac{\partial g}{\partial y}\right)^2}\,dx\,dy \qquad (4)$$

EXAMPLE 2. Let S be the surface defined by $z = x^2 + y$, where D is the region $0 \le x \le 1$, $-1 \le y \le 1$. Evaluate $\int_S x\,dS$.

If we let $z = g(x, y) = x^2 + y$, formula (4) gives

$$\begin{aligned}\int_S x\,dS &= \int_D x\sqrt{1 + \left(\frac{\partial g}{\partial x}\right)^2 + \left(\frac{\partial g}{\partial y}\right)^2}\,dx\,dy \\ &= \int_{-1}^{1}\int_0^1 x\sqrt{1 + 4x^2 + 1}\,dx\,dy \\ &= \frac{1}{8}\int_{-1}^{1}\left[\int_0^1 [2 + 4x^2]^{1/2}(8x\,dx)\right]dy \\ &= \frac{2}{3}\cdot\frac{1}{8}\int_{-1}^{1}[(2 + 4x^2)^{3/2}]_0^1\,dy \\ &= \frac{1}{12}\int_{-1}^{1}[6^{3/2} - 2^{3/2}]\,dy = \tfrac{1}{6}[6^{3/2} - 2^{3/2}] \\ &= \sqrt{6} - \frac{\sqrt{2}}{3} = \sqrt{2}\left(\sqrt{3} - \frac{1}{3}\right)\end{aligned}$$

EXAMPLE 3. Evaluate $\int_S z^2\,dS$ where S is the unit sphere $x^2 + y^2 + z^2 = 1$.

For this problem it is convenient to represent the sphere parametrically by the equations $x = \cos\theta\sin\phi$, $y = \sin\theta\sin\phi$, $z = \cos\phi$, over the region D in the $\theta\phi$-plane given by $0 \le \phi \le \pi$, $0 \le \theta \le 2\pi$. From equation (1) we get

$$\int_S z^2\,dS = \int_D (\cos\phi)^2\|\mathbf{T}_\theta \times \mathbf{T}_\phi\|\,d\theta\,d\phi$$

Now a little computation [use formula (2) of Section 6.4] shows that

$$\|\mathbf{T}_\theta \times \mathbf{T}_\phi\| = |\sin\phi|$$

so

$$\begin{aligned}\int_S z^2\,dS &= \int_0^{2\pi}\int_0^{\pi}\cos^2\phi|\sin\phi|\,d\phi\,d\theta \\ &= \int_0^{2\pi}\int_0^{\pi}\cos^2\phi\sin\phi\,d\phi\,d\theta = \frac{1}{3}\int_0^{2\pi}[-\cos^3\phi]_0^{\pi}\,d\theta \\ &= \frac{2}{3}\int_0^{2\pi}d\theta = \frac{4\pi}{3}\end{aligned}$$

which completes our evaluation of the surface integral.

We shall now develop a formula for surface integrals when the surface can be represented as a graph. To do so, we let S be the graph of $z = g(x, y)$ and consider formula (4); we wish to interpret this result geometrically. We claim that

$$\int_S f(x, y, z)\, dS = \int_D \frac{f(x, y, g(x, y))}{\cos\theta}\, dx\, dy \tag{5}$$

where θ is the angle the normal to the surface makes with the unit vector $\mathbf{k}$ at the point $(x, y, g(x, y))$ (see Figure 6.5.2).

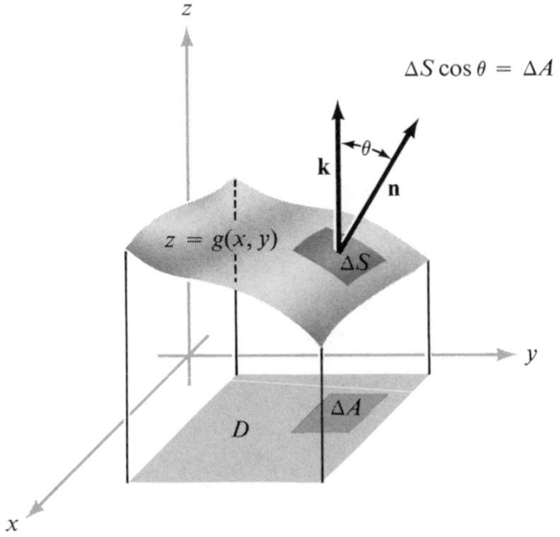

FIGURE 6.5.2
The area of a patch of area ΔS over a patch ΔA is $\Delta S = \Delta A/\cos\theta$ where θ is the angle the normal **n** *makes with* **k**.

Since $\phi(x, y, z) = z - g(x, y) = 0$, a normal vector is $\nabla\phi$; that is,

$$\mathbf{n} = -(\partial g/\partial x)\mathbf{i} - (\partial g/\partial y)\mathbf{j} + \mathbf{k}$$

(see Example 3 of Section 6.3, or recall that the normal to a surface $g(x, y, z) =$ constant is ∇g.) Thus

$$\cos\theta = \frac{\mathbf{n}\cdot\mathbf{k}}{\|\mathbf{n}\|} = 1\Bigg/\sqrt{\left(\frac{\partial g}{\partial x}\right)^2 + \left(\frac{\partial g}{\partial y}\right)^2 + 1}$$

Substitution of this formula into (4) gives (5).

The result is, in fact, obvious geometrically, for if a small rectangle in the xy-plane has area ΔA then the area of the portion above it on the surface is $\Delta S = \Delta A/\cos\theta$ (Figure 6.5.2). This approach can help us to remember formula (5) and to apply it in problems.

EXAMPLE 4. Compute $\int_S x\, dS$ where S is the triangle with vertices (1, 0, 0), (0, 1, 0), (0, 0, 1) (see Figure 6.5.3).

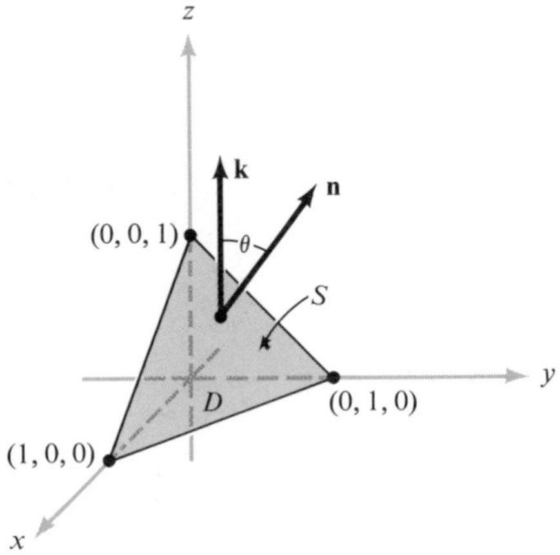

FIGURE 6.5.3

In computing a specific surface integral, one finds a formula for the normal **n** *and computes the angle* θ *in preparation for formula (5).*

This surface is the plane described by the equation $x + y + z = 1$. Since the surface is a plane, the angle θ is constant and a unit normal vector is $\mathbf{n} = (1/\sqrt{3}, 1/\sqrt{3}, 1/\sqrt{3})$. Thus, $\cos\theta = 1/\sqrt{3}$, and by (5)

$$\int_S x\, dS = \sqrt{3}\int_D x\, dx\, dy$$

where D is the domain in the xy-plane. But

$$\sqrt{3}\int_D x\, dx\, dy = \sqrt{3}\int_0^1\int_0^{1-x} x\, dy\, dx = \sqrt{3}\int_0^1 x(1-x)\, dx = \frac{\sqrt{3}}{6}$$

Integrals of functions over surfaces are useful for computing the mass of a surface when the mass density function m is known. The total mass of a surface with mass density m is given by

$$M(S) = \int_S m(x, y, z)\, dS \tag{6}$$

EXAMPLE 5. Let $\mathbf{\Phi}: D \to \mathbb{R}^3$ be the parametrization of the helicoid $S = \mathbf{\Phi}(D)$ of Example 2 of Section 6.4. Recall that $\mathbf{\Phi}(r, \theta) = (r\cos\theta, r\sin\theta, \theta)$, $0 \le \theta \le 2\pi, 0 \le r \le 1$. Suppose S has a mass density at $(x, y, z) \in S$ equal to twice the distance of (x, y, z) to the central axis (see Figure 6.4.2), namely $m(x, y, z) = 2\sqrt{x^2 + y^2} = 2r$, in the cylindrical coordinate system. Find the total mass of the surface.

Applying formula (6),

$$M(S) = \int_S 2\sqrt{x^2 + y^2}\, dS = \int_S 2r\, dS = \int_D 2r\|\mathbf{T}_r \times \mathbf{T}_\theta\|\, dr\, d\theta$$

From Example 2 of Section 6.4 we see that $\|\mathbf{T}_r \times \mathbf{T}_\theta\| = \sqrt{1 + r^2}$. Thus

$$\begin{aligned} M(S) &= \int_D 2r\sqrt{1 + r^2}\, dr\, d\theta = \int_0^{2\pi} \int_0^1 2r\sqrt{1 + r^2}\, dr\, d\theta \\ &= \int_0^{2\pi} \left[\tfrac{2}{3}(1 + r^2)^{3/2}\right]_0^1 d\theta = \int_0^{2\pi} \tfrac{2}{3}[2^{3/2} - 1]\, d\theta \\ &= \frac{4\pi}{3}[2^{3/2} - 1] \end{aligned}$$

EXERCISES

1. Compute $\int_S xy\, dS$ where S is the surface of the tetrahedron with sides $z = 0$, $y = 0$, $x + z = 1$, and $x = y$.
2. Let $\mathbf{\Phi}: D \subset \mathbb{R}^2 \to \mathbb{R}^3$ be a parametrization of a surface S defined by

$$x = x(u, v), \qquad y = y(u, v), \qquad z = z(u, v)$$

   (a) Let

$$\frac{\partial \mathbf{\Phi}}{\partial u} = \left(\frac{\partial x}{\partial u}, \frac{\partial y}{\partial u}, \frac{\partial z}{\partial u}\right), \qquad \frac{\partial \mathbf{\Phi}}{\partial v} = \left(\frac{\partial x}{\partial v}, \frac{\partial y}{\partial v}, \frac{\partial z}{\partial v}\right)$$

   that is, $\partial\mathbf{\Phi}/\partial u = \mathbf{T}_u$ and $\partial\mathbf{\Phi}/\partial v = \mathbf{T}_v$, and set

$$E = \left\|\frac{\partial \mathbf{\Phi}}{\partial u}\right\|^2, \qquad F = \frac{\partial \mathbf{\Phi}}{\partial u} \cdot \frac{\partial \mathbf{\Phi}}{\partial v}, \qquad G = \left\|\frac{\partial \mathbf{\Phi}}{\partial v}\right\|^2$$

   Show that the surface area of S is $\int_D \sqrt{EG - F^2}\, du\, dv$. In this notation, how can we express $\int_S f\, dS$ for a general function f?

   (b) What does the formula become if the vectors $\partial\mathbf{\Phi}/\partial u$ and $\partial\mathbf{\Phi}/\partial v$ are orthogonal?

   (c) Use parts (a) and (b) to compute the surface area of a sphere of radius a.
3. Evaluate $\int_S z\, dS$, where S is the upper hemisphere of radius a, that is, the set of (x, y, z) with $z = \sqrt{a^2 - x^2 - y^2}$.
4. Evaluate $\int_S (x + y + z)\, dS$, where S is the boundary of the unit ball B; that is, S is the set of (x, y, z) with $x^2 + y^2 + z^2 = 1$. (Hint: Use the symmetry of the problem.)
5. Evaluate $\int_S xyz\, dS$, where S is the triangle with vertices $(1, 0, 0)$, $(0, 2, 0)$, and $(0, 1, 1)$.
6. Let a surface S be defined implicitly by $F(x, y, z) = 0$ for (x, y) in a domain D of $\mathbb{R}^2$. Show that

$$\int_S \left|\frac{\partial F}{\partial z}\right| dS = \int_D \sqrt{\left[\frac{\partial F}{\partial x}\right]^2 + \left[\frac{\partial F}{\partial y}\right]^2 + \left[\frac{\partial F}{\partial z}\right]^2}\, dx\, dy$$

   Compare with Exercise 8 of Section 6.4.
7. Evaluate $\int_S z\, dS$, where S is the surface $z = x^2 + y^2$, $x^2 + y^2 \le 1$.

8. Evaluate $\int_S z^2\, dS$, where S is the boundary of the cube $C = [-1, 1] \times [-1, 1] \times [-1, 1]$. (HINT: Do each face separately and add the results.)

9. Find the mass of a spherical surface S of radius R such that at each point $(x, y, z) \in S$ the mass density is equal to the distance of (x, y, z) to some fixed point $(x_0, y_0, z_0) \in S$.

10. A metallic surface S is in the shape of a hemisphere $z = \sqrt{R^2 - x^2 - y^2}$, $0 \leq x^2 + y^2 \leq R^2$. The mass density at $(x, y, z) \in S$ is given by $m(x, y, z) = x^2 + y^2$. Find the total mass of S.

11. Let S be the sphere of radius R.
   (a) Argue by symmetry that

$$\int_S x^2\, dS = \int_S y^2\, dS = \int_S z^2\, dS$$

   (b) Use this fact and some clever thinking to evaluate, with very little computation, the integral

$$\int_S x^2\, dS$$

   (c) Does this help in Exercise 10?

12. (a) Use Riemann sums to justify the formula

$$\frac{1}{A(S)} \int_S f(x, y, z)\, dS$$

   for the *average value* of f over the surface S.
   (b) In Example 3 of this section, show that the average of $f(x, y, z) = z^2$ is $\frac{1}{3}$.
   (c) Define the *center of gravity* $(\bar{x}, \bar{y}, \bar{z})$ of a surface S to be such that $\bar{x}$, $\bar{y}$, and $\bar{z}$ are the average values of the x, y, and z coordinates on S. Show that the center of gravity of the triangle in Example 4 of this section is $(\frac{1}{3}, \frac{1}{3}, \frac{1}{3})$.

13. Find the x-, y-, and z-coordinates of the center of gravity of the octant of the sphere of radius R determined by $x \geq 0$, $y \geq 0$, $z \geq 0$. (HINT: Write this octant as a parametrized surface—see Example 3 of this section and Exercise 12.)

14. Find the z-coordinate of the center of gravity (the average z-coordinate) of the surface of a hemisphere ($z \leq 0$) of radius (see Exercise 12). Argue by symmetry that the average x- and y-coordinates are both zero.

*15. Let *Dirichlet's functional* for a parametrized surface $\mathbf{\Phi}: D \to \mathbb{R}^3$ be defined by

$$J(\mathbf{\Phi}) = \frac{1}{2} \int_D \left(\left\| \frac{\partial \mathbf{\Phi}}{\partial u} \right\|^2 + \left\| \frac{\partial \mathbf{\Phi}}{\partial v} \right\|^2 \right) du\, dv$$

Using Exercise 15, Section 1.5, argue that the area $A(\mathbf{\Phi}) \leq J(\mathbf{\Phi})$ and equality holds if

(a) $\left\| \dfrac{\partial \mathbf{\Phi}}{\partial u} \right\|^2 = \left\| \dfrac{\partial \mathbf{\Phi}}{\partial v} \right\|^2$; and

(b) $\dfrac{\partial \mathbf{\Phi}}{\partial u} \cdot \dfrac{\partial \mathbf{\Phi}}{\partial v} = 0.$

Compare these equations with Exercise 2 and the remarks at the end of Section 6.4. A parametrization $\mathbf{\Phi}$ that satisfies (a) and (b) is said to be *conformal*.

HISTORICAL NOTE

Dirichlet's functional played a major role in the mathematics of the nineteenth century. The mathematician Georg Friedrich Bernhard Riemann (1826–1866) used it to develop his complex function theory and to give a proof of the famous Riemann mapping theorem. Today it is still used extensively as a tool in the study of partial differential equations.

*16. Let $D \subset \mathbb{R}^2$ and $\mathbf{\Phi}: D \to \mathbb{R}^2$ be a smooth function $\mathbf{\Phi}(u, v) = (x(u, v), y(u, v))$ satisfying (a) and (b) of Exercise 15. Show that x and y satisfy the *Cauchy-Riemann equations* $\partial x/\partial u = \partial y/\partial v$, $\partial x/\partial v = -\partial y/\partial u$ or the *conjugate Cauchy-Riemann equations* $\partial x/\partial u = -\partial y/\partial v$, $\partial x/\partial v = \partial y/\partial u$. Conclude that $\nabla^2 \mathbf{\Phi} = 0$ (i.e., each component of $\mathbf{\Phi}$ is harmonic).

17. (a) Compute the area of the portion of the cone $x^2 + y^2 = z^2$ with $z \geq 0$ that is inside the sphere $x^2 + y^2 + z^2 = 2rz$, $r > 0$.
(b) What is the area of that portion of the sphere that is inside the cone?

*18. Let S be a sphere of radius r and $\mathbf{p}$ be a point inside or outside the sphere (but not on it). Show that

$$\int_S \frac{1}{\|\mathbf{x} - \mathbf{p}\|}\, dS = \begin{cases} 4\pi r & \text{if } \mathbf{p} \text{ is inside } S \\ 4\pi r^2/d & \text{if } \mathbf{p} \text{ is outside } S \end{cases}$$

where d is the distance from $\mathbf{p}$ to the center of the sphere.

19. Find the surface area of that part of the cylinder $x^2 + z^2 = a^2$ that is inside the cylinder $x^2 + y^2 = 2ay$ and also in the positive octant ($x \geq 0$, $y \geq 0$, $z \geq 0$). Assume $a > 0$.

6.6 SURFACE INTEGRALS OF VECTOR FUNCTIONS

In this section we shall turn our attention to integrals of *vector* functions over surfaces. The definition we give here is a natural extension of that for scalar functions discussed in Section 6.5.

DEFINITION. *Let* $\mathbf{F}$ *be a vector field defined on* S*, the image of a parametrized surface* $\mathbf{\Phi}$*. The* ***surface integral*** *of* $\mathbf{F}$ *over* $\mathbf{\Phi}$*, denoted by*

$$\int_{\mathbf{\Phi}} \mathbf{F} \cdot d\mathbf{S}, \quad \textit{or} \quad \textit{sometimes by} \iint_{\mathbf{\Phi}} \mathbf{F} \cdot d\mathbf{S}$$

is defined by

$$\int_{\mathbf{\Phi}} \mathbf{F} \cdot d\mathbf{S} = \int_D \mathbf{F} \cdot (\mathbf{T}_u \times \mathbf{T}_v)\, du\, dv$$

where $\mathbf{T}_u$ *and* $\mathbf{T}_v$ *are defined as on p. 362 (see Figure 6.6.1).*

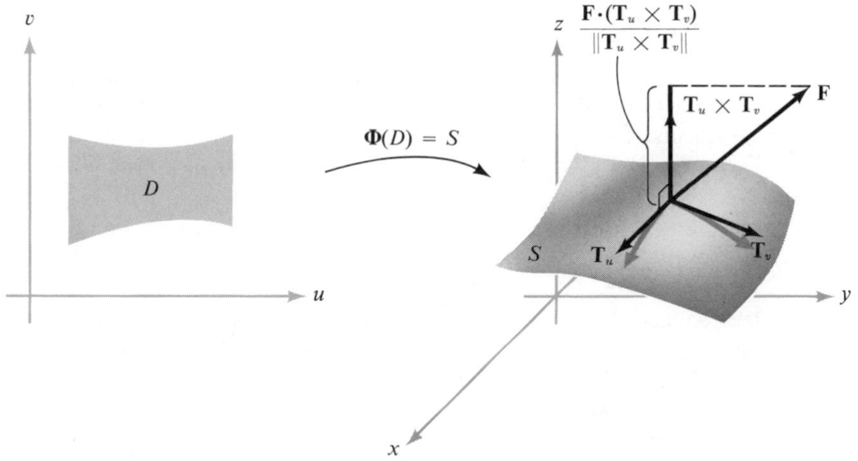

FIGURE 6.6.1
The geometric significance of $\mathbf{F}\cdot(\mathbf{T}_u \times \mathbf{T}_v)$.

EXAMPLE 1. Let D be the rectangle in the $\theta\phi$-plane defined by

$$0 \le \theta \le 2\pi, \qquad 0 \le \phi \le \pi$$

and let the surface S be defined by the parametrization $\mathbf{\Phi}: D \to \mathbb{R}^3$ given by

$$x = \cos\theta \sin\phi, \qquad y = \sin\theta\sin\phi, \qquad z = \cos\phi$$

Thus θ and ϕ are the angles of spherical coordinates (see p. 44), and S is the unit sphere parametrized by $\mathbf{\Phi}$. Let $\mathbf{r}$ be the position vector $\mathbf{r}(x, y, z) = x\mathbf{i} + y\mathbf{j} + z\mathbf{k}$. We compute $\int_{\mathbf{\Phi}} \mathbf{r}\cdot d\mathbf{S}$ as follows. First we find

$$\mathbf{T}_\theta = (-\sin\phi\sin\theta)\mathbf{i} + (\sin\phi\cos\theta)\mathbf{j}$$

$$\mathbf{T}_\phi = (\cos\theta\cos\phi)\mathbf{i} + (\sin\theta\cos\phi)\mathbf{j} - (\sin\phi)\mathbf{k}$$

and hence

$$\mathbf{T}_\theta \times \mathbf{T}_\phi = (-\sin^2\phi\cos\theta)\mathbf{i} - (\sin^2\phi\sin\theta)\mathbf{j} - (\sin\phi\cos\phi)\mathbf{k}$$

Then we evaluate

$$\begin{aligned}\mathbf{r}\cdot(\mathbf{T}_\theta \times \mathbf{T}_\phi) &= (x\mathbf{i} + y\mathbf{j} + z\mathbf{k})\cdot(\mathbf{T}_\theta \times \mathbf{T}_\phi)\\ &= [(\cos\theta\sin\phi)\mathbf{i} + (\sin\theta\sin\phi)\mathbf{j} + (\cos\phi)\mathbf{k}]\\ &\quad\cdot(-\sin\phi)[(\sin\phi\cos\theta)\mathbf{i} + (\sin\phi\sin\theta)\mathbf{j} + (\cos\phi)\mathbf{k}]\\ &= (-\sin\phi)(\sin^2\phi\cos^2\theta + \sin^2\phi\sin^2\theta + \cos^2\phi)\\ &= -\sin\phi\end{aligned}$$

Thus

$$\int_{\mathbf{\Phi}} \mathbf{r} \cdot d\mathbf{S} = \int_D -\sin\phi \, d\phi \, d\theta = \int_0^{2\pi} (-2) \, d\theta = -4\pi$$

An analogy can be drawn between the surface integral $\int_{\mathbf{\Phi}} \mathbf{F} \cdot d\mathbf{S}$ and the line integral $\int_{\boldsymbol{\sigma}} \mathbf{F} \cdot d\mathbf{s}$. Recall that the line integral is an oriented integral. We needed the notion of orientation of a curve to extend the definition of $\int_{\boldsymbol{\sigma}} \mathbf{F} \cdot d\mathbf{s}$ to line integrals $\int_C \mathbf{F} \cdot d\mathbf{s}$ over oriented curves. We should like to extend the definition of $\int_{\mathbf{\Phi}} \mathbf{F} \cdot d\mathbf{S}$ to surfaces in a similar fashion; that is, given a surface S parametrized by a mapping $\mathbf{\Phi}$, we want to define $\int_S \mathbf{F} \cdot d\mathbf{S} = \int_{\mathbf{\Phi}} \mathbf{F} \cdot d\mathbf{S}$ and show that it is independent of the parametrization, except possibly for the sign. In order to accomplish this, we need the notion of orientation of a surface.

DEFINITION. *An* ***oriented surface*** *is a two-sided surface with one side specified as the* ***outside*** *or* ***positive side****; we call the other side the* ***inside*** *or* ***negative side***.* *At each point $(x, y, z) \in S$ there are two unit normal vectors $\mathbf{n}_1$ and $\mathbf{n}_2$, where $\mathbf{n}_1 = -\mathbf{n}_2$ (see Figure 6.6.2). Each of these two normals can be associated with one side of the surface. Thus to specify a side of a surface S, at each point we choose a unit normal vector $\mathbf{n}$ that points away from the positive side of S at that point.*

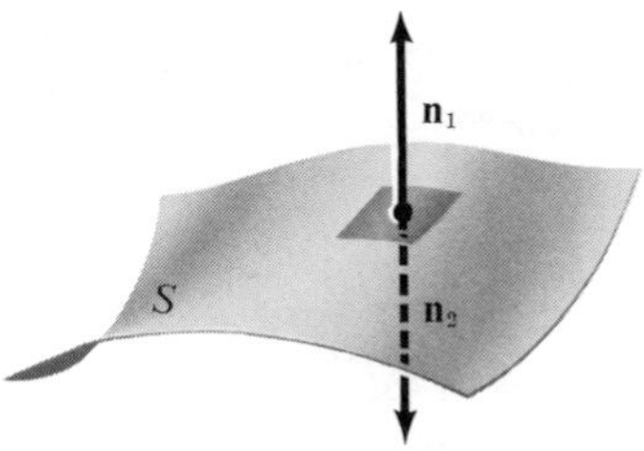

FIGURE 6.6.2
The two possible unit normals to a surface at a point.

This definition assumes that our surface does have two sides. We should give an example of a surface with only one side. The first known example of such a surface was the Möbius strip (named after the Dutch mathematician A. F. Möbius, who, along with the mathematician J. B. Listing, discovered it in 1858). Pictures of such a surface are given in Figures 6.6.3 and 6.6.4.

* We use the phrase "side" in an intuitive sense. This concept can be developed rigorously. Also, the choice of the side to be named the "outside" is often dictated by the surface itself, as, for example, is the case with a sphere. In other cases the naming is somewhat arbitrary (see the piece of surface depicted in Figure 6.6.2).

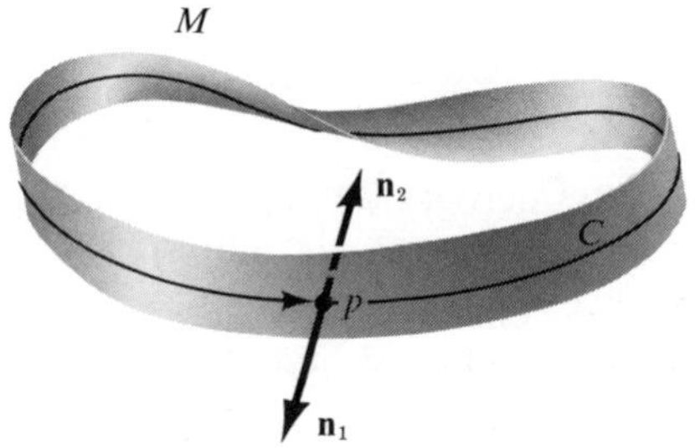

FIGURE 6.6.3
The Möbius strip: slide $\mathbf{n}_2$ *around C once; when* $\mathbf{n}_2$ *returns to its initial point,* $\mathbf{n}_2$ *will coincide with* $\mathbf{n}_1 = -\mathbf{n}_2$.

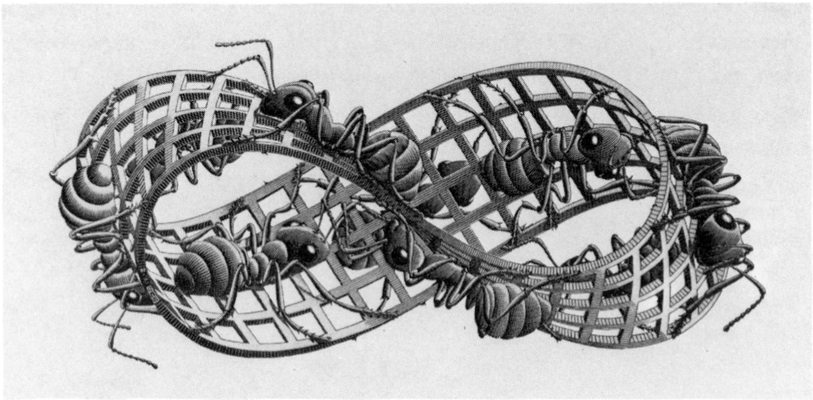

FIGURE 6.6.4
Ants walking on a Möbius strip. (Moebius Strip II 1963, by M. C. Escher. Escher Foundation, Haags Gemeentemuseum, The Hague.)

At each point of M there are two unit normals $\mathbf{n}_1$ and $\mathbf{n}_2$. However, $\mathbf{n}_1$ doesn't determine a unique side of M, and neither does $\mathbf{n}_2$. To see this intuitively, we can slide $\mathbf{n}_2$ around the closed curve C (Figure 6.6.3). When $\mathbf{n}_2$ returns to a fixed point p on C it will coincide with $\mathbf{n}_1$, showing that both $\mathbf{n}_1$ and $\mathbf{n}_2$ point away from the same side of M and, consequently, that M has only one side.

Figure 6.6.4 is a Möbius strip as drawn by the well-known twentieth-century mathematician and artist M. C. Escher. It depicts ants crawling along the Möbius band. After one trip around the band (without crossing an edge) they end up on the "opposite side" of the surface.

Let $\mathbf{\Phi}: D \to \mathbb{R}^3$ be a parametrization of an oriented surface S and suppose S is smooth at $\mathbf{\Phi}(u_0, v_0)$, $(u_0, v_0) \in D$; that is, the unit normal vector $(\mathbf{T}_{u_0} \times \mathbf{T}_{v_0})/\|\mathbf{T}_{u_0} \times \mathbf{T}_{v_0}\|$ is defined. If $\mathbf{n}(\mathbf{\Phi}(u_0, v_0))$ denotes the unit normal to S at $\mathbf{\Phi}(u_0, v_0)$ pointing to the positive side of S at that point, it follows that $(\mathbf{T}_{u_0} \times \mathbf{T}_{v_0})/\|\mathbf{T}_{u_0} \times \mathbf{T}_{v_0}\| = \pm\mathbf{n}(\mathbf{\Phi}(u_0, v_0))$. The parametrization $\mathbf{\Phi}$ is said to be *orientation preserving* if $(\mathbf{T}_u \times \mathbf{T}_v)/\|\mathbf{T}_u \times \mathbf{T}_v\| = \mathbf{n}(\mathbf{\Phi}(u, v))$ at all $(u, v) \in D$ for which S is smooth at $\mathbf{\Phi}(u, v)$. In other words, $\mathbf{\Phi}$ is orientation preserving if the vector $\mathbf{T}_u \times \mathbf{T}_v$ points away from the outside of the surface. If $\mathbf{T}_u \times \mathbf{T}_v$ points away from the inside of the surface at all $(u, v) \in D$ for which S is smooth at $\mathbf{\Phi}(u, v)$, then $\mathbf{\Phi}$ is said to be *orientation reversing*. Using the above notation, this condition corresponds to $(\mathbf{T}_u \times \mathbf{T}_v)/\|\mathbf{T}_u \times \mathbf{T}_v\| = -\mathbf{n}(\mathbf{\Phi}(u, v))$.

EXAMPLE 2. We can give the unit sphere $x^2 + y^2 + z^2 = 1$ in $\mathbb{R}^3$ (Figure 6.6.5) an orientation by selecting the unit vector $\mathbf{n}(x\ y, z) = \mathbf{r}$, where $\mathbf{r} = x\mathbf{i} + y\mathbf{j} + z\mathbf{k}$, which points away from the outside of the surface. This choice corresponds to our intuitive notion of outside for the sphere.

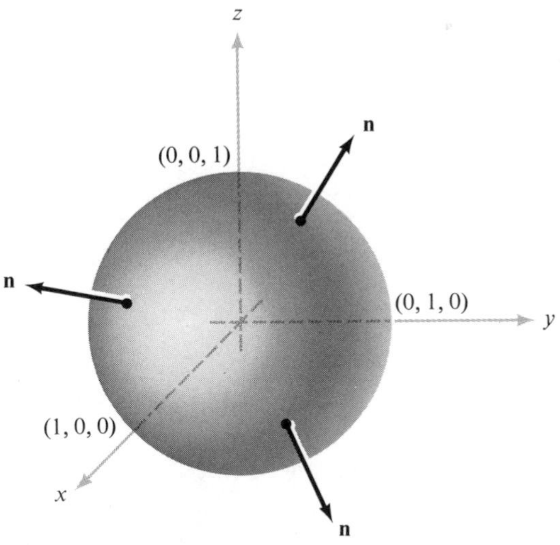

FIGURE 6.6.5
The unit sphere oriented by its outward normal **n**.

Now that the sphere S is an oriented surface, consider the parametrization $\mathbf{\Phi}$ of S given in Example 1. The cross product of the tangent vectors $\mathbf{T}_\theta$ and $\mathbf{T}_\phi$—that is, a normal to S—is given by

$$(-\sin\phi)[(\cos\theta\sin\phi)\mathbf{i} + (\sin\theta\sin\phi)\mathbf{j} + (\cos\phi)\mathbf{k}] = -\mathbf{r}\sin\phi$$

Since $-\sin\phi \le 0$ for $0 \le \phi \le \pi$, this normal vector points inward from the sphere. Thus the given parametrization $\mathbf{\Phi}$ is orientation reversing.

EXAMPLE 3. Let S be a surface described by $z = f(x, y)$. There are two unit normal vectors to S at $(x_0, y_0, f(x_0, y_0))$, namely $\pm\mathbf{n}$, where

$$\mathbf{n} = \frac{-\dfrac{\partial f}{\partial x}(x_0, y_0)\mathbf{i} - \dfrac{\partial f}{\partial y}(x_0, y_0)\mathbf{j} + \mathbf{k}}{\sqrt{\left(\dfrac{\partial f}{\partial x}(x_0, y_0)\right)^2 + \left(\dfrac{\partial f}{\partial y}(x_0, y_0)\right)^2 + 1}}$$

We can orient all such surfaces* by taking the positive side of S to be the side away from which $\mathbf{n}$ points (Figure 6.6.6). Thus the positive side of such a surface is determined by the unit normal $\mathbf{n}$ with positive $\mathbf{k}$ component. If we parametrize this surface by $\mathbf{\Phi}(u, v) = (u, v, f(u, v))$, then $\mathbf{\Phi}$ will be orientation preserving.

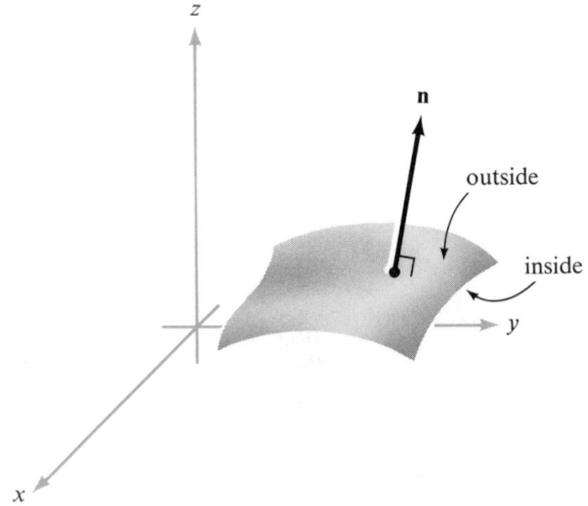

FIGURE 6.6.6

n *points away from the outside of the surface.*

* If we had given a rigorous definition of orientation we could use this argument to show all surfaces $z = f(x, y)$ are in fact orientable; that is, they "have two sides."

We now state without proof a theorem showing that the integral over an oriented surface is independent of the parametrization. The proof of this theorem is analogous to that of Theorem 1 (Section 6.2); the heart of the proof is in the change of variables formula—this time applied to double integrals.

THEOREM 4. *Let S be an oriented surface and let $\mathbf{\Phi}_1$ and $\mathbf{\Phi}_2$ be two smooth orientation-preserving parametrizations, with $\mathbf{F}$ a continuous vector field defined on S. Then*

$$\int_{\mathbf{\Phi}_1} \mathbf{F} \cdot d\mathbf{S} = \int_{\mathbf{\Phi}_2} \mathbf{F} \cdot d\mathbf{S}$$

If $\mathbf{\Phi}_1$ is orientation preserving and $\mathbf{\Phi}_2$ orientation reversing, then

$$\int_{\mathbf{\Phi}_1} \mathbf{F} \cdot d\mathbf{S} = -\int_{\mathbf{\Phi}_2} \mathbf{F} \cdot d\mathbf{S}$$

If f is a real-valued continuous function defined on S, and if $\mathbf{\Phi}_1$ and $\mathbf{\Phi}_2$ are parametrizations of S, then

$$\int_{\mathbf{\Phi}_1} f\, dS = \int_{\mathbf{\Phi}_2} f\, dS$$

If $f = 1$ we obtain

$$A(S) = \int_{\mathbf{\Phi}_1} dS = \int_{\mathbf{\Phi}_2} dS$$

thus showing that area is independent of parametrization.

We can thus unambiguously use the notation

$$\int_S \mathbf{F} \cdot d\mathbf{S} = \int_{\mathbf{\Phi}} \mathbf{F} \cdot d\mathbf{S}$$

(or a sum of such integrals if S is a union of parametrized surfaces that intersect only along their boundary curves) where $\mathbf{\Phi}$ is an orientation-preserving parametrization. Theorem 4 guarantees that the value of the integral does not depend on the selection of $\mathbf{\Phi}$.

Recall from formula (1) of Section 6.2 that a line integral $\int_{\boldsymbol{\sigma}} \mathbf{F} \cdot d\mathbf{s}$ can be thought of as the path integral of the tangential component of $\mathbf{F}$ along $\boldsymbol{\sigma}$ (although for the case in which $\boldsymbol{\sigma}$ intersects itself, the integral obtained is technically not a path integral). A similar situation will hold for surface integrals in our setting, since we are assuming (see Section 6.4) that the mappings $\mathbf{\Phi}$ defining the surface S are one-to-one except perhaps on the boundary of D, which can be ignored for the purposes of integration. Thus in defining integrals over surfaces we assume in this book that the surfaces are nonintersecting.

For an oriented smooth surface S and an orientation-preserving parametrization $\mathbf{\Phi}$ of S we can express $\int_S \mathbf{F} \cdot d\mathbf{S}$ as an integral of a real-valued function f over the surface. Let $\mathbf{n} = (\mathbf{T}_u \times \mathbf{T}_v)/\|\mathbf{T}_u \times \mathbf{T}_v\|$ be the unit normal pointing to the outside of S. Then

$$\begin{aligned}\int_S \mathbf{F}\cdot d\mathbf{S} = \int_{\mathbf{\Phi}} \mathbf{F}\cdot d\mathbf{S} &= \int_D \mathbf{F}\cdot(\mathbf{T}_u \times \mathbf{T}_v)\,du\,dv \\ &= \int_D \mathbf{F}\cdot\left(\frac{\mathbf{T}_u \times \mathbf{T}_v}{\|\mathbf{T}_u \times \mathbf{T}_v\|}\right)\|\mathbf{T}_u \times \mathbf{T}_v\|\,du\,dv \\ &= \int_D (\mathbf{F}\cdot\mathbf{n})\|\mathbf{T}_u \times \mathbf{T}_v\|\,du\,dv = \int_S (\mathbf{F}\cdot\mathbf{n})\,dS = \int_S f\,dS\end{aligned}$$

where $f = \mathbf{F}\cdot\mathbf{n}$. We have thus proved the following theorem.

THEOREM 5. $\int_S \mathbf{F}\cdot d\mathbf{S}$, *the surface integral of* $\mathbf{F}$ *over* S, *is equal to the integral of the normal component of* $\mathbf{F}$ *over the surface. In short,*

$$\int_S \mathbf{F}\cdot d\mathbf{S} = \int_S \mathbf{F}\cdot\mathbf{n}\,dS$$

This observation can often save a great deal of computational effort, as Example 4 below demonstrates.

The geometric and physical significance of the surface integral can be understood by expressing it as a limit of Riemann sums. For simplicity, we assume D is a rectangle. Fix a parametrization $\mathbf{\Phi}$ of S that preserves orientation and partition the region D into n^2 pieces D_{ij}, $0 \le i \le n-1$, $0 \le j \le n-1$. We let Δu denote the length of the horizontal side of D_{ij} and Δv denote the length of the vertical side of D_{ij}. Let (u, v) be a point in D_{ij} and

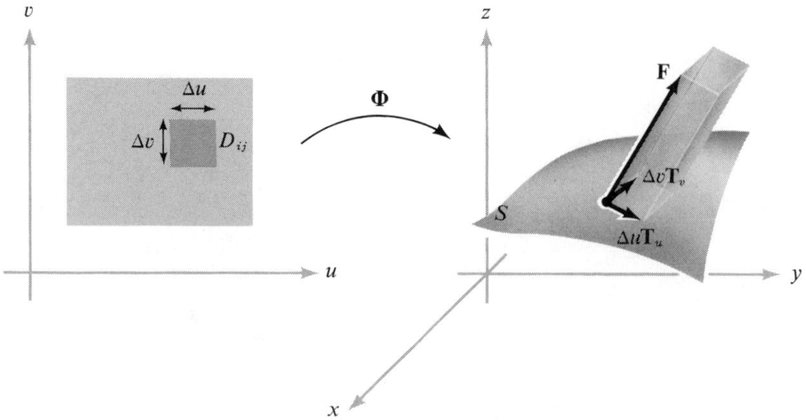

FIGURE 6.6.7

$\mathbf{F}\cdot(\mathbf{T}_u \times \mathbf{T}_v) > 0$ *when the parallelepiped formed by* $\Delta v\mathbf{T}_v$, $\Delta u\mathbf{T}_u$, *and* $\mathbf{F}$ *lies to the "outside" of the surface* S.

$(x, y, z) = \mathbf{\Phi}(u, v)$, the corresponding point on the surface. We consider the parallelogram with sides $\Delta u\mathbf{T}_u$ and $\Delta v\mathbf{T}_v$ lying in the plane tangent to S at (x, y, z) and the parallelepiped formed by $\mathbf{F}$, $\Delta u\mathbf{T}_u$, and $\Delta v\mathbf{T}_v$. The volume of the parallelepiped is the absolute value of the triple product

$$\mathbf{F}\cdot(\Delta u\mathbf{T}_u \times \Delta v\mathbf{T}_v) = \mathbf{F}\cdot(\mathbf{T}_u \times \mathbf{T}_v)\,\Delta u\,\Delta v$$

The vector $\mathbf{T}_u \times \mathbf{T}_v$ is normal to the surface at (x, y, z) and points away from the outside of the surface. Thus the number $\mathbf{F}\cdot(\mathbf{T}_u \times \mathbf{T}_v)$ is positive when the parallelepiped lies on the outside of the surface (Figure 6.6.7).

In general, the parallelepiped lies on that side of the surface away from which $\mathbf{F}$ is pointing. If we think of $\mathbf{F}$ as the velocity field of a fluid, $\mathbf{F}(x, y, z)$ is pointing in the direction in which fluid is moving across the surface near (x, y, z). Moreover, the number

$$|\mathbf{F}\cdot(\mathbf{T}_u\,\Delta u \times \mathbf{T}_v\,\Delta v)|$$

measures the amount of fluid that passes through the tangent parallelogram per unit time. Since the sign of $\mathbf{F}\cdot(\Delta u\mathbf{T}_u \times \Delta v\mathbf{T}_v)$ is positive if the vector $\mathbf{F}$ is pointing outward at (x, y, z) and negative if $\mathbf{F}$ is pointing inward, the sum $\sum_{i,j}\mathbf{F}\cdot(\mathbf{T}_u \times \mathbf{T}_v)\,\Delta u\,\Delta v$ is an approximate measure of the net quantity of fluid to flow outward across the surface per unit time. (Remember that "outward" or "inward" depends on our choice of parametrization. Figure 6.6.8 illustrates $\mathbf{F}$ directed outward and inward, given $\mathbf{T}_u$ and $\mathbf{T}_v$.) Hence, *the integral* $\int_S \mathbf{F}\cdot d\mathbf{S}$ *is the net quantity of fluid to flow across the surface per unit time, that is, the rate of fluid flow.* Therefore this integral is also called the *flux* of $\mathbf{F}$ across the surface.

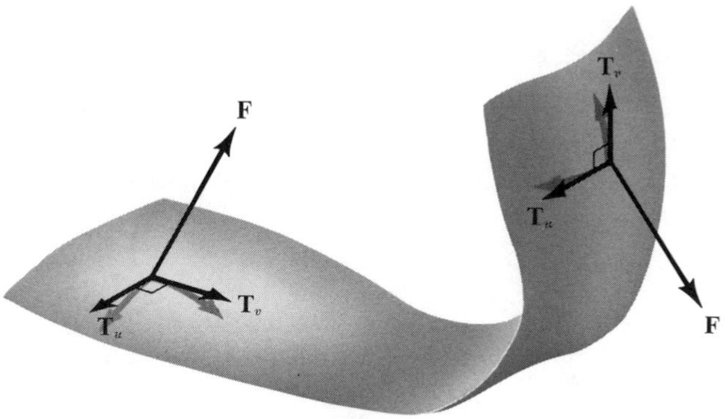

FIGURE 6.6.8
(left) When $\mathbf{F}\cdot(\mathbf{T}_u \times \mathbf{T}_v) > 0$, $\mathbf{F}$ *points outward; (right) when* $\mathbf{F}\cdot(\mathbf{T}_u \times \mathbf{T}_v) < 0$, $\mathbf{F}$ *points inward.*

In the case where $\mathbf{F}$ represents electric or magnetic fields, $\int_S \mathbf{F} \cdot d\mathbf{S}$ is also commonly known as the flux. The reader may be familiar with physical laws (such as Faraday's law) that relate flux of a vector field to a circulation (or current) in a bounding loop. This is the historical and physical basis of Stokes' Theorem, which we shall meet in Section 7.2. The corresponding principle in fluid mechanics is called Kelvin's Circulation Theorem.

Surface integrals also apply to the study of heat flow. Let $T(x, y, z)$ be the temperature at a point $(x, y, z) \in W \subset \mathbb{R}^3$, where W is some region and T is a C^1 function. Then

$$\nabla T = \frac{\partial T}{\partial x}\mathbf{i} + \frac{\partial T}{\partial y}\mathbf{j} + \frac{\partial T}{\partial z}\mathbf{k}$$

represents the temperature gradient, and heat "flows" with the vector field $-k\,\nabla T = \mathbf{F}$, where k is a positive constant (see Section 7.5). Therefore $\int_S \mathbf{F} \cdot d\mathbf{S}$ is the total rate of heat flow or flux across the surface S.

EXAMPLE 4. Suppose a temperature function is given as $T(x, y, z) = x^2 + y^2 + z^2$, and let S be the unit sphere $x^2 + y^2 + z^2 = 1$ oriented with the outward normal (see Example 2). Find the heat flux across the surface S if $k = 1$.

We have

$$\mathbf{F} = -\nabla T(x, y, z) = -2x\mathbf{i} - 2y\mathbf{j} - 2z\mathbf{k}$$

On S, $\mathbf{n}(x, y, z) = x\mathbf{i} + y\mathbf{j} + z\mathbf{k}$ is the unit "outward" normal to S at (x, y, z), and $f(x, y, z) = \mathbf{F} \cdot \mathbf{n} = -2x^2 - 2y^2 - 2z^2 = -2$ is the normal component of $\mathbf{F}$. From Theorem 5 we can see that the surface integral of $\mathbf{F}$ is equal to the integral of its normal component $f = \mathbf{F} \cdot \mathbf{n}$ over S. Thus $\int_S \mathbf{F} \cdot d\mathbf{S} = \int_S f\,dS = -2\int_S dS = -2A(S) = -2(4\pi) = -8\pi$. The flux of heat is directed towards the center of the sphere (Why towards?). Consequently, our observation that $\int_S \mathbf{F} \cdot d\mathbf{S} = \int_S f\,dS$ has saved us considerable computational time.

In this example $\mathbf{F}(x, y, z) = -2x\mathbf{i} - 2y\mathbf{j} - 2z\mathbf{k}$ could also represent a magnetic field, in which case $\int_S \mathbf{F} \cdot d\mathbf{S} = -8\pi$ would be the magnetic flux across S.

EXAMPLE 5. There is an important physical law, due to the great mathematician and physicist K. F. Gauss, that relates the flux of an electric field $\mathbf{E}$ over a "closed" surface S (for example, a sphere or an ellipsoid) to the net charge Q enclosed by the surface, namely

$$\int_S \mathbf{E} \cdot d\mathbf{S} = Q \qquad (1)$$

(see Figure 6.6.9). Gauss' law will be discussed in detail in Chapter 7. This law is analogous to Ampère's law (see Example 12, Section 6.2).

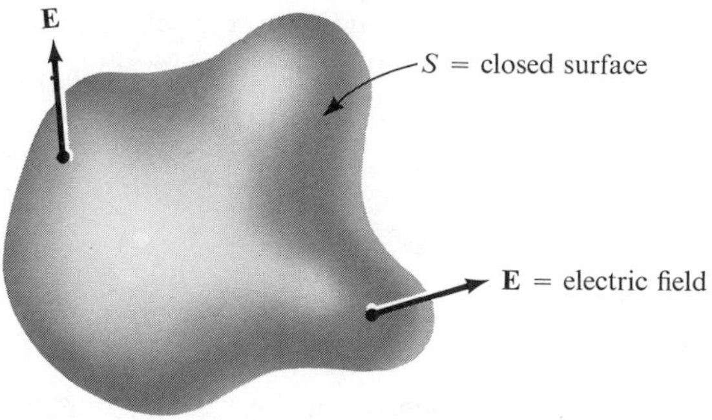

FIGURE 6.6.9
Gauss' Law: $\int_S \mathbf{E} \cdot d\mathbf{S} = Q$, *where Q is the net charge inside S.*

Suppose that $\mathbf{E} = E\mathbf{n}$; that is, $\mathbf{E}$ is a constant scalar multiple of the unit normal to S. Then Gauss' law (1) becomes

$$\int_S \mathbf{E} \cdot d\mathbf{S} = \int_S E\, dS = E \int_S dS = Q$$

Hence

$$E = \frac{Q}{A(S)} \tag{2}$$

In the case where S is the sphere of radius R, (2) becomes

$$E = \frac{Q}{4\pi R^2} \tag{3}$$

(see Figure 6.6.10).

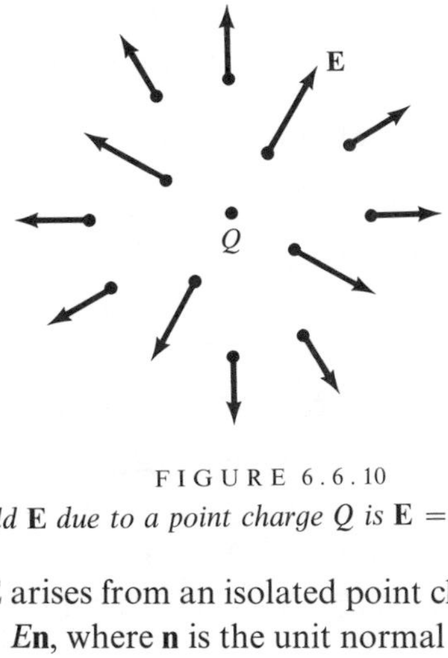

FIGURE 6.6.10
The field $\mathbf{E}$ *due to a point charge Q is* $\mathbf{E} = Q\mathbf{n}/4\pi R^2$.

Now suppose that $\mathbf{E}$ arises from an isolated point charge Q. From symmetry it follows that $\mathbf{E} = E\mathbf{n}$, where $\mathbf{n}$ is the unit normal to any sphere centered at Q. Hence (3) holds. Consider a second point charge Q_0 located at a distance R from Q. The force $\mathbf{F}$ that acts on this second charge Q_0 is given by

$$\mathbf{F} = \mathbf{E}Q_0 = EQ_0\mathbf{n} = \frac{QQ_0}{4\pi R^2}\mathbf{n}$$

If F is the magnitude of $\mathbf{F}$ we have

$$F = \frac{QQ_0}{4\pi R^2}$$

which is the well-known *Coulomb's law* for the force between two point charges.*

An important kind of surface, as we know from the preceding section, is the graph of a function. Let us work out the surface-integral formulas for this case. Consider the surface S described by $z = f(x, y)$, $(x, y) \in D$, where S is oriented so that

$$\mathbf{n} = \frac{-\dfrac{\partial f}{\partial x}\mathbf{i} - \dfrac{\partial f}{\partial y}\mathbf{j} + \mathbf{k}}{\sqrt{\left(\dfrac{\partial f}{\partial x}\right)^2 + \left(\dfrac{\partial f}{\partial y}\right)^2 + 1}}$$

points outward. We have seen that we can parametrize S by $\mathbf{\Phi}\colon D \to \mathbb{R}^3$ given by $\mathbf{\Phi}(x, y) = (x, y, f(x, y))$. In this case, $\int_S \mathbf{F} \cdot d\mathbf{S}$ can be written in a particularly simple form. We have

$$\mathbf{T}_x = \mathbf{i} + \frac{\partial f}{\partial x}\mathbf{k}, \qquad \mathbf{T}_y = \mathbf{j} + \frac{\partial f}{\partial y}\mathbf{k}$$

Thus $\mathbf{T}_x \times \mathbf{T}_y = -(\partial f/\partial x)\mathbf{i} - (\partial f/\partial y)\mathbf{j} + \mathbf{k}$. If $\mathbf{F} = F_1\mathbf{i} + F_2\mathbf{j} + F_3\mathbf{k}$ is a continuous vector field then we get the formula

$$\begin{aligned}\int_S \mathbf{F} \cdot d\mathbf{S} &= \int_D \mathbf{F} \cdot (\mathbf{T}_x \times \mathbf{T}_y)\, dx\, dy \\ &= \int_D \left[F_1\left(-\frac{\partial f}{\partial x}\right) + F_2\left(-\frac{\partial f}{\partial y}\right) + F_3\right] dx\, dy \qquad (4)\end{aligned}$$

EXAMPLE 6. The equations

$$z = 12, \qquad x^2 + y^2 \le 25$$

describe a disc of radius 5 lying in the plane $z = 12$. Suppose $\mathbf{r}$ is the vector field

$$\mathbf{r}(x, y, z) = x\mathbf{i} + y\mathbf{j} + z\mathbf{k}$$

* Sometimes one sees the formula $F = (1/4\pi\varepsilon_0)QQ_0/R^2$. The extra constant ε_0 appears when the MKS units are used for measuring charge. We are using CGS, or Gaussian units.

Then $\int_S \mathbf{r} \cdot d\mathbf{S}$ is easily computed; we have $\partial z/\partial x = \partial z/\partial y = 0$ since $z = 12$ is constant on the disc, so

$$\mathbf{r}(x, y, z) \cdot (\mathbf{T}_x \times \mathbf{T}_y) = \mathbf{r}(x, y, z) \cdot (\mathbf{i} \times \mathbf{j}) = \mathbf{r}(x, y, z) \cdot \mathbf{k} = z$$

and the integral becomes

$$\int_S \mathbf{r} \cdot d\mathbf{S} = \int_D z \, dx \, dy = \int_D 12 \, dx \, dy = 12(\text{area of } D) = 300\pi$$

A second solution: since the disc is parallel to the xy-plane, the outward unit normal is $\mathbf{k}$. So $\mathbf{n}(x, y, z) = \mathbf{k}$ and $\mathbf{r} \cdot \mathbf{n} = z$. However, $\|\mathbf{T}_x \times \mathbf{T}_y\| = \|\mathbf{k}\| = 1$, so we know from the discussion following Theorem 4 (p. 393) that

$$\int_S \mathbf{r} \cdot d\mathbf{S} = \int_S \mathbf{r} \cdot \mathbf{n} \, dS = \int_S z \, dS = \int_D 12 \, dx \, dy = 300\pi$$

Alternatively, we may solve this problem by using formula (4) directly, with $f(x, y) = 12$ and D the disc $x^2 + y^2 \leq 25$:

$$\int_S \mathbf{r} \cdot d\mathbf{S} = \int_D [x \cdot 0 + y \cdot 0 + 12] \, dx \, dy = 12(\text{Area of } D) = 300\pi$$

EXERCISES

1. Let the temperature of a point in $\mathbb{R}^3$ be given by $T(x, y, z) = 3x^2 + 3z^2$. Compute the heat flux across the surface $x^2 + z^2 = 2, 0 \leq y \leq 2$, if $k = 1$.
2. Compute the heat flux across the unit sphere S if $T(x, y, z) = x$ (see Example 4). Can you interpret your answer physically?
3. Let S be the closed surface that consists of the hemisphere $x^2 + y^2 + z^2 = 1$, $z \geq 0$, and its base $x^2 + y^2 \leq 1$, $z = 0$. Let $\mathbf{E}$ be the electric field defined by $\mathbf{E}(x, y, z) = 2x\mathbf{i} + 2y\mathbf{j} + 2z\mathbf{k}$. Find the electric flux across S. (HINT: Break S into two pieces S_1 and S_2 and evaluate $\int_{S_1} \mathbf{E} \cdot d\mathbf{S}$ and $\int_{S_2} \mathbf{E} \cdot d\mathbf{S}$ separately.)
4. Let the velocity field of a fluid be described by $\mathbf{F} = \sqrt{y}\mathbf{j}$ (measured in meters/second). Compute how many cubic meters of fluid per second is crossing the surface $x^2 + z^2 = y, 0 \leq y \leq 1$.
5. Evaluate $\int_S (\nabla \times \mathbf{F}) \cdot d\mathbf{S}$, where S is the surface $x^2 + y^2 + 3z^2 = 1$, $z \leq 0$, and $\mathbf{F} = y\mathbf{i} - x\mathbf{j} + zx^3y^2\mathbf{k}$.
6. Evaluate $\int_S (\nabla \times \mathbf{F}) \cdot d\mathbf{S}$ where $\mathbf{F} = (x^2 + y - 4)\mathbf{i} + 3xy\mathbf{j} + (2xz + z^2)\mathbf{k}$ and S is the surface $x^2 + y^2 + z^2 = 16$, $z \geq 0$.
7. Let S be the surface of the unit sphere. Let $\mathbf{F}$ be a vector field and F_r its radial component. Prove that

$$\int_S \mathbf{F} \cdot d\mathbf{S} = \int_{\theta=0}^{2\pi} \int_{\phi=0}^{\pi} F_r \sin\phi \, d\phi \, d\theta$$

What is the formula for real-valued functions f?

*8. Prove the Mean Value Theorem for surface integrals: If $\mathbf{F}$ is a continuous vector field, then

$$\int_S \mathbf{F} \cdot \mathbf{n} \, dS = [\mathbf{F}(Q) \cdot \mathbf{n}(Q)]A(S)$$

for some $Q \in S$, where $A(S)$ is the area of S. (HINT: Prove it for real functions first, by reducing the problem to one of a double integral: Show that if $g \geq 0$, then

$$\int_D fg\, dA = f(Q) \int_D g\, dA$$

for some $Q \in D$ (do it by considering $(\int_D fg\, dA)/(\int_D g\, dA)$ and using the Intermediate Value Theorem).)

9. Work out a formula like that in Exercise 7 for integration over the surface of a cylinder.

10. Let S be a surface in $\mathbb{R}^3$ that is actually a subset D of the xy-plane. Show that the integral of a scalar function $f(x, y, z)$ over S reduces to the double integral of $f(x, y, z)$ over D. What does the surface integral of a vector field over S become?

11. Let the velocity field of a fluid be described by $\mathbf{F} = \mathbf{i} + x\mathbf{j} + z\mathbf{k}$ (measured in meters/second). Compute how many cubic meters of fluid per second are crossing the surface described by $x^2 + y^2 + z^2 = 1, z \geq 0$.

12. (a) A uniform fluid that flows vertically downward (heavy rain) is described by the vector field $\mathbf{F}(x, y, z) = (0, 0, -1)$. Find the total flux through the cone $z = (x^2 + y^2)^{1/2}; x^2 + y^2 \leq 1$.
    (b) The rain is driven sideways by a strong wind so that it falls at a 45° angle, and it is described by $\mathbf{F}(x, y, z) = -(\sqrt{2}/2, 0, \sqrt{2}/2)$. Now what is the flux through the cone?

13. For $a > 0, b > 0, c > 0$, let S be the upper half ellipsoid

$$S = \left\{(x, y, z) \middle| \frac{x^2}{a^2} + \frac{y^2}{b^2} + \frac{z^2}{c^2} = 1, z \geq 0\right\}$$

with orientation determined by the upward normal. Compute $\int_S \mathbf{F} \cdot d\mathbf{S}$ where $\mathbf{F}(x, y, z) = (x^3, 0, 0)$.

14. If S is the upper hemisphere $\{(x, y, z) | x^2 + y^2 + z^2 = 1, z \geq 0\}$ oriented by the normal pointing out of the sphere, compute $\int_S \mathbf{F} \cdot d\mathbf{S}$ for (a) and (b).
    (a) $\mathbf{F}(x, y, z) = x\mathbf{i} + y\mathbf{j}$
    (b) $\mathbf{F}(x, y, z) = y\mathbf{i} - x\mathbf{j}$
    (c) For each of the vector fields above, compute $\int_s (\nabla \times \mathbf{F}) \cdot d\mathbf{S}$ and $\int_C \mathbf{F} \cdot d\mathbf{s}$ where C is the unit circle in the xy-plane traversed in the counterclockwise direction (as viewed from the positive z-axis). (Notice that C is the boundary of S. The phenomenon illustrated here will be studied more thoroughly in the next chapter, using Stokes' Theorem.)

REVIEW EXERCISES FOR CHAPTER 6

1. Integrate $f(x, y, z) = xyz$ along the following paths.
    (a) $\boldsymbol{\sigma}(t) = (e^t \cos t, e^t \sin t, 3), 0 \leq t \leq 2\pi$
    (b) $\boldsymbol{\sigma}(t) = (\cos t, \sin t, t), 0 \leq t \leq 2\pi$
    (c) $\boldsymbol{\sigma}(t) = \frac{3}{2}t^2\mathbf{i} + 2t^2\mathbf{j} + t\mathbf{k}, 0 \leq t \leq 1$
    (d) $\boldsymbol{\sigma}(t) = t\mathbf{i} + (1/\sqrt{2})t^2\mathbf{j} + (1/3)t^3\mathbf{k}, 0 \leq t \leq 1$
2. Compute the integral of f along the path $\boldsymbol{\sigma}$ in each of the following cases.
    (a) $f(x, y, z) = x + y + yz; \boldsymbol{\sigma}(t) = (\sin t, \cos t, t), 0 \leq t \leq 2\pi$
    (b) $f(x, y, z) = x + \cos^2 z; \boldsymbol{\sigma}(t) = (\sin t, \cos t, t), 0 \leq t \leq 2\pi$
    (c) $f(x, y, z) = x + y + z; \boldsymbol{\sigma}(t) = (t, t^2, \frac{2}{3}t^3), 0 \leq t \leq 1$

3. Compute each of the following line integrals
(a) $\int_C (\sin \pi x)\, dy - (\cos \pi y)\, dz$ where C is the triangle whose vertices are $(1, 0, 0)$, $(0, 1, 0)$, and $(0, 0, 1)$ in that order
(b) $\int_C (\sin z)\, dx + (\cos z)\, dy - (xy)^{1/3}\, dz$ where C is the path parametrized by $\boldsymbol{\sigma}(\theta) = (\cos^3 \theta, \sin^3 \theta, \theta)$, $0 \leq \theta \leq 7\pi/2$

4. If $\mathbf{F}(\mathbf{x})$ is orthogonal to $\boldsymbol{\sigma}'(t)$ at each point on the curve $\mathbf{x} = \boldsymbol{\sigma}(t)$, what can you say about $\int_{\boldsymbol{\sigma}} \mathbf{F} \cdot d\mathbf{s}$?

5. Find the work done by the force $\mathbf{F}(x, y) = (x^2 - y^2)\mathbf{i} + 2xy\mathbf{j}$ in moving a particle counterclockwise around the square with corners $(0, 0)$, $(a, 0)$, (a, a), $(0, a)$, $a > 0$.

6. A ring in the shape of the curve $x^2 + y^2 = a^2$ is formed of thin wire weighing $|x| + |y|$ grams per unit length at (x, y). Find the mass of the ring.

7. Find a parametrization for each of the following surfaces.
(a) $x^2 + y^2 + z^2 - 4x - 6y = 12$
(b) $2x^2 + y^2 + z^2 - 8x = 1$
(c) $4x^2 + 9y^2 - 2z^2 = 8$

8. Find the area of the surface defined by $\boldsymbol{\Phi}: (u, v) \mapsto (x, y, z)$ where

$$x = h(u, v) = u + v, \qquad y = g(u, v) = u, \qquad z = f(u, v) = v$$

$0 \leq u \leq 1, 0 \leq v \leq 1$. Sketch.

9. Write a formula for the surface area of $\boldsymbol{\Phi}: (r, \theta) \mapsto (x, y, z)$ where

$$x = r \cos \theta, \qquad y = 2r \sin \theta, \qquad z = r$$

$0 \leq r \leq 1, 0 \leq \theta \leq 2\pi$. Sketch.

10. Suppose $z = f(x, y)$, and $(\partial f/\partial x)^2 + (\partial f/\partial y)^2 = c$, $c > 0$. Show that the area of the graph of f lying over a region D in the xy-plane is $\sqrt{1 + c}$ times the area of D.

11. Compute the integral of $f(x, y, z) = x^2 + y^2 + z^2$ over the surface in Exercise 8.

12. Find $\int_S f\, dS$ in each of the following cases.
(a) $f(x, y, z) = x$; S is the part of the plane $x + y + z = 1$ in the positive octant $x \geq 0, y \geq 0, z \geq 0$
(b) $f(x, y, z) = x^2$; S is the part of the plane $x = z$ inside the cylinder $x^2 + y^2 = 1$
(c) $f(x, y, z) = x$; S is the part of the cylinder $x^2 + y^2 = 2x$ with $0 \leq z \leq \sqrt{x^2 + y^2}$

13. Compute the integral of $f(x, y, z) = xyz$ over the rectangle with vertices $(1, 0, 1)$, $(2, 0, 0)$, $(1, 1, 1)$, and $(2, 1, 0)$.

14. Compute the integral of $x + y$ over the surface of the unit sphere.

15. Compute the integral of x over the triangle with vertices $(1, 1, 1)$, $(2, 1, 1)$, and $(2, 0, 3)$.

16. A paraboloid of revolution S is parametrized by $\boldsymbol{\Phi}(u, v) = (u \cos v, u \sin v, u^2)$, $0 \leq u \leq 1, 0 \leq v \leq 2\pi$.
(a) Find an equation in x, y, and z describing the surface.
(b) What are the geometric meanings of the parameters u and v?
(c) Find a unit vector orthogonal to the surface at $\boldsymbol{\Phi}(u, v)$
(d) Find the equation for the tangent plane at $\boldsymbol{\Phi}(u_0, v_0) = (1, 1, 2)$ and express your answer in the following two ways:
  (*i*) parametrized by u and v; and
  (*ii*) in terms of x, y, and z.
(e) Find the area of S.

17. Let $f(x, y, z) = xe^y \cos \pi z$.
    (a) Compute $\mathbf{F} = \nabla f$.
    (b) Evaluate $\int_{\mathbf{c}} \mathbf{F} \cdot d\mathbf{s}$ where $\mathbf{c}(t) = (3\cos^4 t, 5\sin^7 t, 0)$, $0 \leq t \leq \pi$.

18. Let $\mathbf{F}(x, y, z) = x\mathbf{i} + y\mathbf{j} + z\mathbf{k}$. Evaluate $\int_S \mathbf{F} \cdot d\mathbf{S}$ where S is the upper hemisphere of the unit sphere $x^2 + y^2 + z^2 = 1$.

19. Let $\mathbf{F}(x, y, z) = x\mathbf{i} + y\mathbf{j} + z\mathbf{k}$. Evaluate $\int_{\mathbf{c}} \mathbf{F} \cdot d\mathbf{s}$ where $\mathbf{c}(t) = (e^t, t, t^2)$, $0 \leq t \leq 1$.

20. Let $\mathbf{F} = \nabla f$ for a given scalar function. Let $\mathbf{c}(t)$ be a closed curve, that is, $\mathbf{c}(b) = \mathbf{c}(a)$. Show that $\int_{\mathbf{c}} \mathbf{F} \cdot d\mathbf{s} = 0$.

21. Consider the surface $\mathbf{\Phi}(u, v) = (u^2 \cos v, u^2 \sin v, u)$. Compute the unit normal at $u = 1, v = 0$. Compute the equation of the tangent plane at this point.

22. Let S be the part of the cone $z^2 = x^2 + y^2$ with z between 1 and 2 oriented by the normal pointing out of the cone. Compute $\int_S \mathbf{F} \cdot d\mathbf{S}$ where $\mathbf{F}(x, y, z) = (x^2, y^2, z^2)$.

23. Let $\mathbf{F} = x\mathbf{i} + x^2\mathbf{j} + yz\mathbf{k}$ represent the velocity field of a fluid (velocity measured in meters/second). Compute how many cubic meters of fluid per second are crossing the xy-plane through the square $0 \leq x \leq 1, 0 \leq y \leq 1$.

24. Let $\mathbf{F} = x^3\mathbf{i} + y^3\mathbf{j} + z^3\mathbf{k}$. Compute (a) div $\mathbf{F} = \nabla \cdot \mathbf{F}$ and (b) $\nabla \times \mathbf{F}$.

25. Let S be a surface and $\mathbf{c}$ a closed curve bounding S. Verify the equality

$$\int_S (\nabla \times \mathbf{F}) \cdot d\mathbf{S} = \int_{\mathbf{c}} \mathbf{F} \cdot d\mathbf{s}$$

if $\mathbf{F}$ is a gradient field (use Exercise 20).

26. Calculate $\int_S \mathbf{F} \cdot d\mathbf{S}$ where $\mathbf{F}(x, y, z) = (x, y, -y)$ and S is the cylindrical surface defined by $x^2 + y^2 = 1$, $0 \leq z \leq 1$, with normal pointing out of the cylinder.

27. Let S be the portion of the cylinder $x^2 + y^2 = 4$ between the planes $z = 0$ and $z = x + 3$. Compute the following.

    (a) $\int_S x^2\, dS$ (b) $\int_S y^2\, dS$ (c) $\int_S z^2\, dS$

28. Let Γ be the curve of intersection of the plane $z = ax + by$ with the cylinder $x^2 + y^2 = 1$. Find all values of the real numbers a and b such that $a^2 + b^2 = 1$ and

$$\int_\Gamma y\,dx + (z - x)\,dy - y\,dz = 0$$

29. A circular helix that lies on the cylinder $x^2 + y^2 = R^2$ with pitch p may be described parametrically by

$$x = R\cos\theta, \qquad y = R\sin\theta, \qquad z = p\theta, \qquad \theta \geq 0$$

A particle slides under the action of gravity (which acts parallel to the z-axis) without friction along the helix. If the particle starts out at the height $z_0 > 0$, then when it reaches the height z, $0 \leq z < z_0$, along the helix, its speed is given by

$$\frac{ds}{dt} = \sqrt{(z_0 - z)2g}$$

where s is arc length along the helix, g is the constant of gravity, and t is time.
    (a) Find the length of the part of the helix between the planes $z = z_0$ and $z = z_1$, $0 \leq z_1 < z_0$.
    (b) Compute the time T_0 it takes the particle to reach the plane $z = 0$.

CHAPTER 7

VECTOR ANALYSIS

All the theory of the motion of fluids has just been reduced to the solution of analytic formulas.

L. EULER

We are now prepared to tie together the vector differential calculus (see Chapter 3) and the vector integral calculus (see Chapter 6). This will be done by means of the important theorems of Green, Gauss, and Stokes. We shall also point out some of the physical applications of these theorems to the study of electricity and magnetism, hydrodynamics, heat conduction, and differential equations (the last through a brief introduction to potential theory).

HISTORICAL NOTE

Many of these basic theorems had their origins in physics. For example, Green's Theorem, discovered about 1828, arose in connection with potential theory (this includes gravitational and electrical potentials). Gauss' Theorem—the Divergence Theorem—arose in connection with electrostatics (this theorem should actually be jointly credited to the Russian mathematician Ostrogradsky). Stokes' Theorem was first suggested in a letter to Stokes from the physicist Lord Kelvin in 1850 and was used by Stokes on the examination for the Smith Prize in 1854.

7.1 GREEN'S THEOREM

Green's Theorem relates a line integral along a closed curve C in the plane $\mathbb{R}^2$ to a double integral over the region enclosed by C. This important result will be generalized in the following sections to curves and surfaces in $\mathbb{R}^3$. We shall be referring to line integrals around curves that are the boundaries of elementary regions of type 1, 2 or 3 (see Section 5.3). To understand the ideas in this section you may need to refer to Section 6.2.

A simple closed curve C that is the boundary of a region of type 1, 2, or 3 has two orientations—counterclockwise (positive) and clockwise (negative). We denote C with the counterclockwise orientation as C^+, and with the clockwise orientation as C^- (Figure 7.1.1).

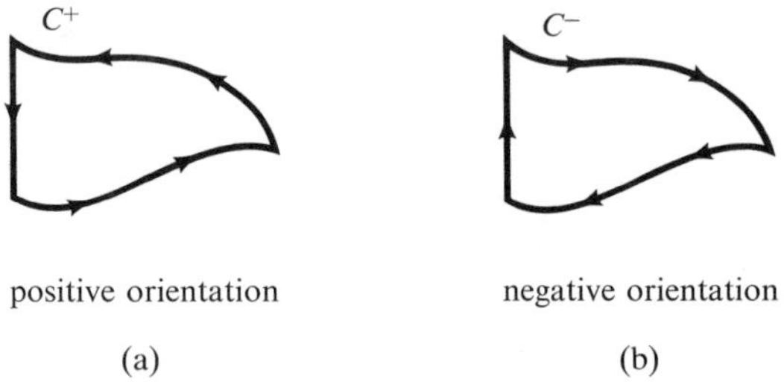

FIGURE 7.1.1
Positive orientation of C (a) and negative orientation of C (b).

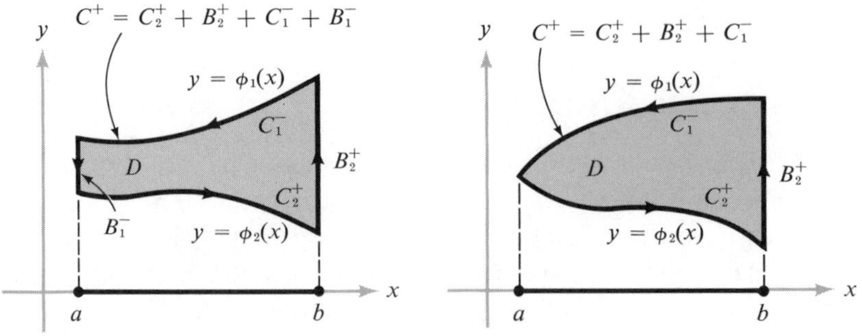

FIGURE 7.1.2
Two examples showing how to break the positively oriented boundary of a region D of type 1 into oriented components.

The boundary C of a region of type 1 can be decomposed into top and bottom portions, C_1 and C_2, and (if applicable) left and right vertical portions, B_1 and B_2. Then we write, following Figure 7.1.2,

$$C^+ = C_2^+ + B_2^+ + C_1^- + B_1^-$$

where the pluses denote the curves oriented in the direction of left to right or bottom to top, and the minuses denote the curves oriented from right to left or from top to bottom.

We may make a similar decomposition of the boundary of a region of type 2 into left and right portions, and upper and lower horizontal portions (if applicable) (Figure 7.1.3).

The boundary of a region of type 3 has two decompositions—one into upper and lower halves, the other into left and right halves.

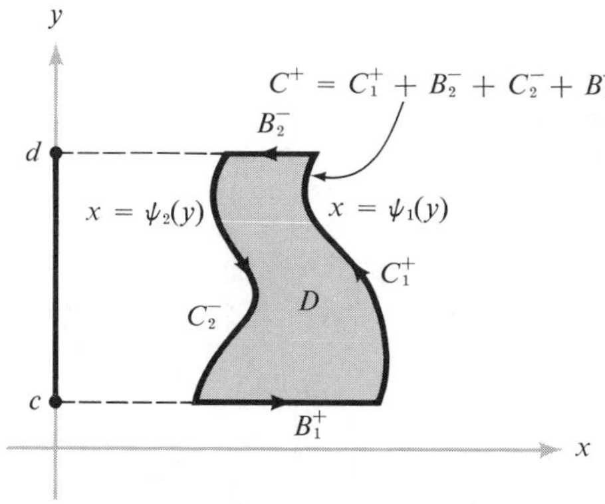

FIGURE 7.1.3
An example showing how to break the positively oriented boundary of a region D of type 2 into oriented components.

We shall now prove two lemmas in preparation for Green's Theorem.

LEMMA 1. *Let D be a region of type 1 and let C be its boundary. Suppose $P: D \to \mathbb{R}$ is C^1. Then*

$$\int_{C^+} P\,dx = -\int_D \frac{\partial P}{\partial y}\,dx\,dy$$

(The left-hand side denotes the line integral $\int_{C^+} P\,dx + Q\,dy + R\,dz$ where $Q = 0$ and $R = 0$.)

Proof. Suppose D is described by

$$a \le x \le b, \qquad \phi_2(x) \le y \le \phi_1(x)$$

We decompose C^+ by writing $C^+ = C_2^+ + B_2^+ + C_1^- + B_1^-$ (see Figure 7.1.2). By Fubini's Theorem, we may evaluate the double integral as an iterated integral (see p. 282)

$$\begin{aligned}\int_D \frac{\partial P}{\partial y}(x, y)\, dx\, dy &= \int_a^b \int_{\phi_2(x)}^{\phi_1(x)} \frac{\partial P}{\partial y}(x, y)\, dy\, dx \\ &= \int_a^b [P(x, \phi_1(x)) - P(x, \phi_2(x))]\, dx\end{aligned}$$

However, since C_2^+ can be parametrized by $x \mapsto (x, \phi_2(x))$, $a \le x \le b$, and C_1^+ can be parametrized by $x \mapsto (x, \phi_1(x))$, $a \le x \le b$, we have

$$\int_a^b P(x, \phi_2(x))\, dx = \int_{C_2^+} P(x, y)\, dx$$

and

$$\int_a^b P(x, \phi_1(x))\, dx = \int_{C_1^+} P(x, y)\, dx$$

Thus by reversing orientations

$$-\int_a^b P(x, \phi_1(x))\, dx = \int_{C_1^-} P(x, y)\, dx$$

Hence

$$\int_D \frac{\partial P}{\partial y}\, dx\, dy = -\int_{C_2^+} P\, dx - \int_{C_1^-} P\, dx$$

Since x is constant on B_2^+ and B_1^- we have

$$\int_{B_2^+} P\, dx = 0 = \int_{B_1^-} P\, dx$$

and so

$$\begin{aligned}\int_{C^+} P\, dx &= \int_{C_2^+} P\, dx + \int_{B_2^+} P\, dx + \int_{C_1^-} P\, dx + \int_{B_1^-} P\, dx \\ &= \int_{C_2^+} P\, dx + \int_{C_1^-} P\, dx\end{aligned}$$

Thus

$$\int_D \frac{\partial P}{\partial y}\, dx\, dy = -\int_{C_2^+} P\, dx - \int_{C^-} P\, dx = -\int_{C^+} P\, dx \quad \blacksquare$$

We now prove the analogous lemma with the roles of x and y interchanged.

LEMMA 2. *Let D be a region of type 2 with boundary C. Then if $Q: D \to \mathbb{R}$ is C^1,*

$$\int_{C^+} Q\, dy = \int_D \frac{\partial Q}{\partial x}\, dx\, dy$$

The negative sign does not occur here because reversing the role of x and y corresponds to a change of orientation for the plane.

Proof. Suppose D is given by

$$\psi_2(y) \le x \le \psi_1(y), \qquad c \le y \le d$$

Using the notation of Figure 7.1.3 we have

$$\int_{C^+} Q\,dy = \int_{C_2^- + B_1^+ + C_1^+ + B_2^-} Q\,dy = \int_{C_1^+} Q\,dy + \int_{C_2^-} Q\,dy$$

where C_1^+ is the curve parametrized by $y \mapsto (\psi_1(y), y)$, $c \le y \le d$, and C_2^+ is the curve $y \mapsto (\psi_2(y), y)$, $c \le y \le d$. Applying Fubini's Theorem, we obtain

$$\begin{aligned}\int_D \frac{\partial Q}{\partial x}\,dx\,dy &= \int_c^d \int_{\psi_2(y)}^{\psi_1(y)} \frac{\partial Q}{\partial x}\,dx\,dy \\ &= \int_c^d [Q(\psi_1(y), y) - Q(\psi_2(y), y)]\,dy \\ &= \int_{C_1^+} Q\,dy - \int_{C_2^+} Q\,dy = \int_{C_1^+} Q\,dy + \int_{C_2^-} Q\,dy \\ &= \int_{C^+} Q\,dy \quad \blacksquare\end{aligned}$$

Adding the results of Lemmas 1 and 2 proves the following important theorem.

THEOREM 1 (GREEN'S THEOREM). *Let D be a region of type 3 and let C be its boundary. Suppose $P\colon D \to \mathbb{R}$ and $Q\colon D \to \mathbb{R}$ are C^1. Then*

$$\int_{C^+} P\,dx + Q\,dy = \int_D \left(\frac{\partial Q}{\partial x} - \frac{\partial P}{\partial y}\right) dx\,dy$$

The correct (positive) orientation for the boundary curves of region D can be remembered by the following device: "*If you walk along the curve C with the correct orientation, the region D will be on your left.*" See Figure 7.1.4.

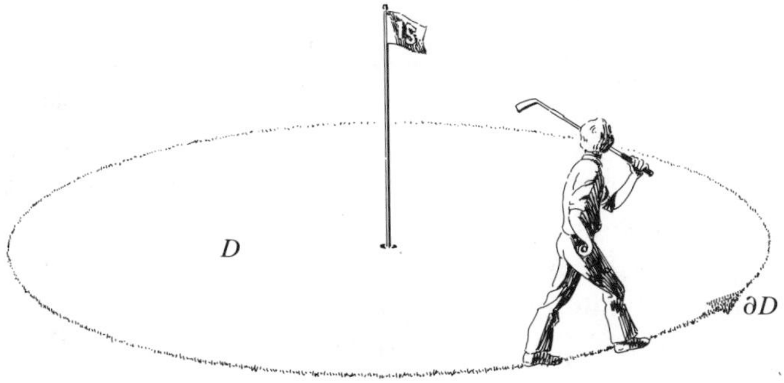

FIGURE 7.1.4
The correct orientation for the boundary of a region D.

Green's Theorem actually applies to any "decent" region in $\mathbb{R}^2$. In Exercise 8 we indicate a generalization of Green's theorem for regions that are not type 3, but that can be broken up into pieces, each of which is of type 3. An example is shown in Figure 7.1.5. The region D is an annulus; its boundary consists of two curves $C = C_1 + C_2$ with the indicated orientations. (Note that for the inner region the correct orientation is *clockwise*; the device on p. 407 still works for remembering the orientation.) If Theorem 1 is applied to each of the regions D_1, D_2, D_3, and D_4 and the results summed, the equality of Green's Theorem will be obtained for D and its boundary curve C.

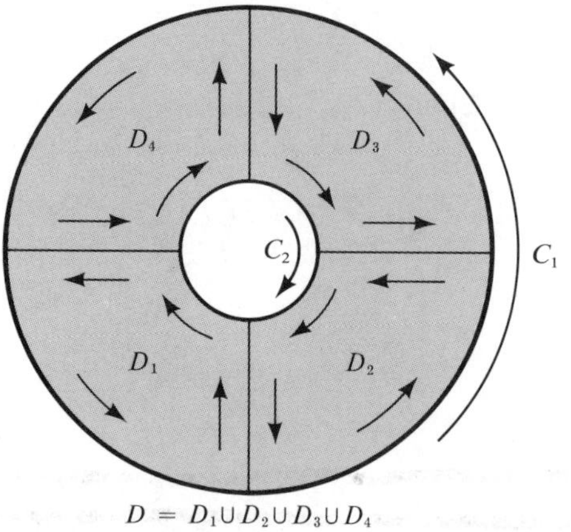

FIGURE 7.1.5
Green's Theorem applies to $D = D_1 \cup D_2 \cup D_3 \cup D_4$.

Let us use the notation ∂D for the oriented curve C^+, that is, the boundary curve of D oriented in the correct sense as described by the device on p. 407. Then we may write Green's Theorem as

$$\int_{\partial D} P\,dx + Q\,dy = \int_D \left(\frac{\partial Q}{\partial x} - \frac{\partial P}{\partial y} \right) dx\,dy$$

Green's Theorem is very useful because it relates a line integral around the boundary of a region to an area integral over the interior of the region, and in many cases it is easier to evaluate the line integral than the area integral. For example, if we know that P vanishes on the boundary, we can immediately conclude that $\int_D (\partial P/\partial y)\,dx\,dy = 0$ even though $\partial P/\partial y$ need not vanish on the interior. (Can you construct such a P on the unit square?)

EXAMPLE 1. Verify Green's Theorem for $P(x, y) = x$ and $Q(x, y) = xy$ where D is the unit disc $x^2 + y^2 \leq 1$.

We can evaluate both sides in Green's Theorem directly. The boundary of D is the unit circle parametrized by $x = \cos t$, $y = \sin t$, $0 \le t \le 2\pi$, so

$$\begin{aligned}\int_{\partial D} P\,dx + Q\,dy &= \int_0^{2\pi} [(\cos t)(-\sin t) + \cos t \sin t \cos t]\,dt\\ &= \left[\frac{\cos^2 t}{2}\right]_0^{2\pi} + \left[-\frac{\cos^3 t}{3}\right]_0^{2\pi} = 0\end{aligned}$$

On the other hand

$$\int_D \left(\frac{\partial Q}{\partial x} - \frac{\partial P}{\partial y}\right) dx\,dy = \int_D y\,dx\,dy$$

which is zero also. Thus Green's Theorem is verified in this case.

We can use Green's Theorem to obtain a formula for the area of a region bounded by a simple closed curve.

THEOREM 2. *If C is a simple closed curve that bounds a region to which Green's Theorem applies, then the area of the region D bounded by C is*

$$A = \frac{1}{2}\int_{\partial D} x\,dy - y\,dx$$

Proof. Let $P(x, y) = -y$, $Q(x, y) = x$; then by Green's Theorem we have

$$\begin{aligned}\frac{1}{2}\int_{\partial D} x\,dy - y\,dx &= \frac{1}{2}\int_D \left(\frac{\partial x}{\partial x} - \frac{\partial(-y)}{\partial y}\right) dx\,dy\\ &= \int_D dx\,dy = A \quad \blacksquare\end{aligned}$$

EXAMPLE 2. The area of the region enclosed by the hypocycloid $x^{2/3} + y^{2/3} = a^{2/3}$ can be computed using the parametrization

$$x = a\cos^3\theta, \qquad y = a\sin^3\theta, \qquad 0 \le \theta \le 2\pi$$

(see Figure 7.1.6). Thus

$$\begin{aligned}A &= \frac{1}{2}\int_{\partial D} x\,dy - y\,dx\\ &= \frac{1}{2}\int_0^{2\pi} [(a\cos^3\theta)(3a\sin^2\theta\cos\theta) - (a\sin^3\theta)(-3a\cos^2\theta\sin\theta)]\,d\theta\\ &= \frac{3}{2}a^2\int_0^{2\pi} [\sin^2\theta\cos^4\theta + \cos^2\theta\sin^4\theta]\,d\theta = \frac{3}{2}a^2\int_0^{2\pi} \sin^2\theta\cos^2\theta\,d\theta\\ &= \frac{3}{8}a^2\int_0^{2\pi} \sin^2 2\theta\,d\theta = \frac{3}{8}a^2\int_0^{2\pi}\left[\frac{1-\cos 4\theta}{2}\right] d\theta\\ &= \frac{3}{16}a^2\int_0^{2\pi} d\theta - \frac{3}{16}a^2\int_0^{2\pi}\cos 4\theta\,d\theta = \frac{3}{8}\pi a^2\end{aligned}$$

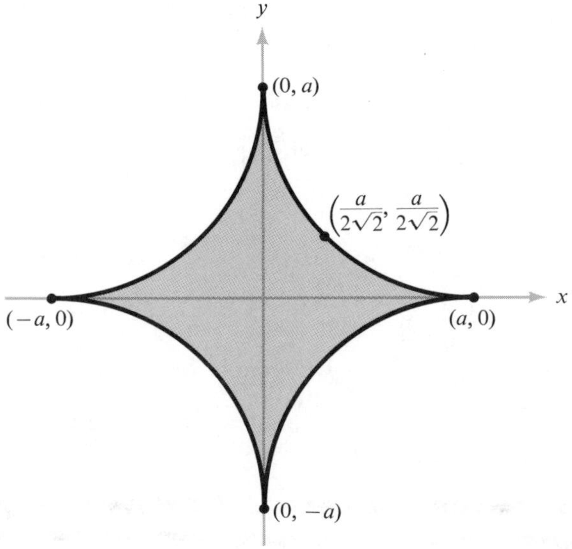

FIGURE 7.1.6
The hypocycloid $x = a\cos^3\theta$, $y = a\sin^3\theta$, $0 \le \theta \le 2\pi$.

The statement of Green's Theorem contained in Theorem 1 is not the form that we shall generalize. We can rewrite the theorem neatly in the language of vector fields.

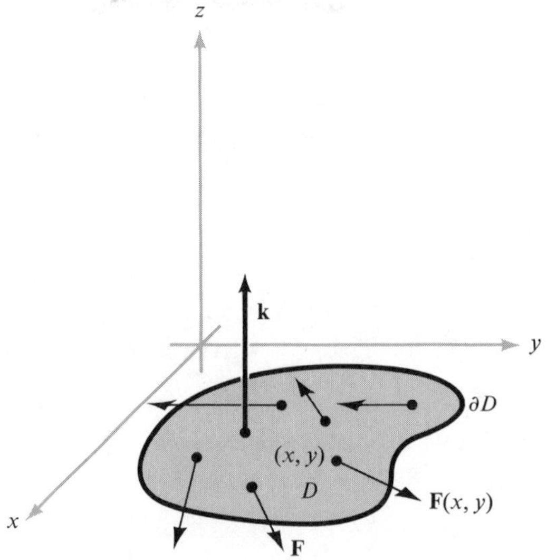

FIGURE 7.1.7
The elements of the vector form of Green's Theorem.

THEOREM 3 (VECTOR FORM OF GREEN'S THEOREM). *Let $D \subset \mathbb{R}^2$ be a region of type 3 and let ∂D be its boundary (oriented counterclockwise). Let $\mathbf{F} = P\mathbf{i} + Q\mathbf{j}$ be a C^1 vector field on D. Then*

$$\int_{\partial D} \mathbf{F} \cdot d\mathbf{s} = \int_D (\text{curl } \mathbf{F}) \cdot \mathbf{k} \, dA = \int_D (\nabla \times \mathbf{F}) \cdot \mathbf{k} \, dA$$

(see Figure 7.1.7).

This result follows easily from Theorem 1 after we interpret the various symbols. We shall ask the reader to supply the details in Exercise 14.

EXAMPLE 3. Let $\mathbf{F} = (xy^2, y + x)$. Integrate $(\nabla \times \mathbf{F}) \cdot \mathbf{k}$ over the region in the first quadrant bounded by the curves $y = x^2$ and $y = x$.

Method 1. Here we compute

$$\nabla \times \mathbf{F} = \left(0, 0, \frac{\partial F_2}{\partial x} - \frac{\partial F_1}{\partial y}\right) = (1 - 2xy)\mathbf{k}.$$

Thus $(\nabla \times \mathbf{F}) \cdot \mathbf{k} = 1 - 2xy$. This can be integrated over the given region D (see Figure 7.1.8) using an iterated integral as follows:

$$\begin{aligned}\iint_D (\nabla \times \mathbf{F}) \cdot \mathbf{k} \, dx \, dy &= \int_0^1 \int_{x^2}^x (1 - 2xy) \, dy \, dx \\ &= \int_0^1 [y - xy^2]_{x^2}^x \, dx \\ &= \int_0^1 [x - x^3 - x^2 + x^5] \, dx = \tfrac{1}{2} - \tfrac{1}{4} - \tfrac{1}{3} + \tfrac{1}{6} = \tfrac{1}{12}\end{aligned}$$

Method 2. Here we use Theorem 3 to obtain

$$\iint_D (\nabla \times \mathbf{F}) \cdot \mathbf{k} \, dx \, dy = \int_{\partial D} \mathbf{F} \cdot d\mathbf{s}$$

The line integral of $\mathbf{F}$ along the curve $y = x$ from left to right is

$$\int_0^1 F_1 \, dx + F_2 \, dy = \int_0^1 (x^3 + 2x) \, dx = \tfrac{1}{4} + 1 = \tfrac{5}{4}$$

Along the curve $y = x^2$ we get

$$\int_0^1 F_1 \, dx + F_2 \, dy = \int_0^1 (x^5 \, dx + (x + x^2)(2x \, dx) = \tfrac{1}{6} + \tfrac{2}{3} + \tfrac{1}{2} = \tfrac{4}{3}$$

Thus, remembering that the integral along $y = x$ is to be taken from right to left, as in Figure 7.1.8,

$$\int_{\partial D} \mathbf{F} \cdot d\mathbf{s} = \tfrac{4}{3} - \tfrac{5}{4} = \tfrac{1}{12}$$

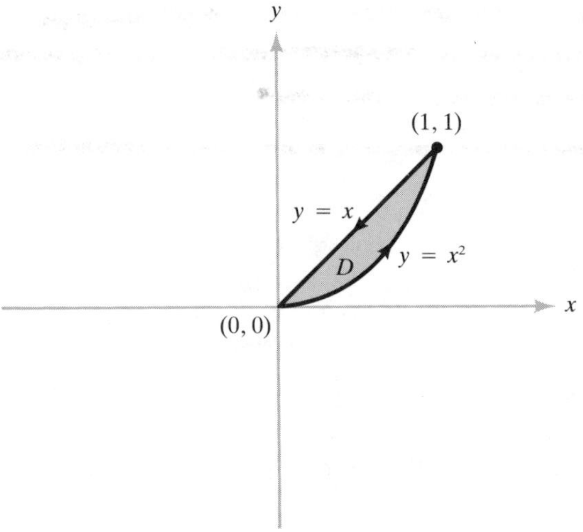

FIGURE 7.1.8
The region bounded by the curves $y = x^2$ and $y = x$.

There is yet another form of Green's Theorem that is capable of being generalized to $\mathbb{R}^3$.

THEOREM 4 (DIVERGENCE THEOREM IN THE PLANE). *Let $D \subset \mathbb{R}^2$ be a region of type 3 and let ∂D be its boundary. Let $\mathbf{n}$ denote the outward unit normal to ∂D, which is given by*

$$\mathbf{n} = (y'(t), -x'(t))/\sqrt{(x'(t))^2 + (y'(t))^2}$$

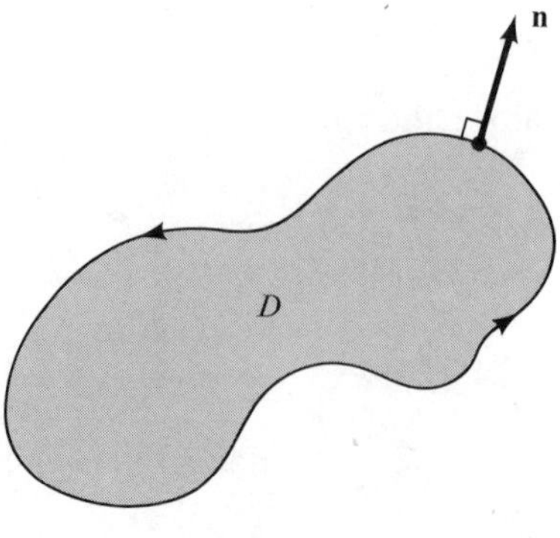

FIGURE 7.1.9
$\mathbf{n}$ *is the outward unit normal to ∂D.*

if $\boldsymbol{\sigma}:[a,b] \to \mathbb{R}^2$, $t \mapsto \boldsymbol{\sigma}(t) = (x(t), y(t))$ *is a positively oriented parametrization of* ∂D *(see Figure 7.1.9).*

Let $\mathbf{F} = P\mathbf{i} + Q\mathbf{j}$ *be a* C^1 *vector field on D. Then*

$$\int_{\partial D} \mathbf{F} \cdot \mathbf{n}\, ds = \int_D \operatorname{div} \mathbf{F}\, dA$$

Proof. Since $\boldsymbol{\sigma}'(t) = (x'(t), y'(t))$ is tangent to ∂D it is clear that $\mathbf{n} \cdot \boldsymbol{\sigma}' = 0$, so $\mathbf{n}$ is normal to the boundary. The sign of $\mathbf{n}$ is chosen to make it correspond to the *outward* (rather than the inward) direction. By definition of the path integral (see Section 6.1)

$$\begin{aligned}\int_{\partial D} \mathbf{F} \cdot \mathbf{n}\, ds &= \int_a^b \left(\frac{P(x(t), y(t))y'(t) - Q(x(t), y(t))x'(t)}{\sqrt{(x'(t))^2 + (y'(t))^2}}\right)\sqrt{(x'(t))^2 + (y'(t))^2}\, dt \\ &= \int_a^b [P(x(t), y(t))\, y'(t) - Q(x(t), y(t))x'(t)]\, dt \\ &= \int_{\partial D} P\, dy - Q\, dx\end{aligned}$$

By Green's Theorem, this equals

$$\int_D \left(\frac{\partial P}{\partial x} + \frac{\partial Q}{\partial y}\right) dx\, dy = \int_D \operatorname{div} \mathbf{F}\, dA \quad \blacksquare$$

EXAMPLE 4. Let $\mathbf{F} = y^3\mathbf{i} + x^5\mathbf{j}$. Compute the integral of the normal com ponent of $\mathbf{F}$ around the unit square.

This can be done by the Divergence Theorem. Indeed,

$$\int_{\partial D} \mathbf{F} \cdot \mathbf{n}\, ds = \int_D \operatorname{div} \mathbf{F}\, dA$$

But $\operatorname{div} \mathbf{F} = 0$, so the integral is zero.

EXERCISES

1. Evaluate $\int_C y\, dx - x\, dy$ where C is the boundary of the square $[-1, 1] \times [-1, 1]$ oriented in the counterclockwise direction (use Green's Theorem).
2. Find the area of the disc D of radius R using Green's Theorem.
3. Verify Green's Theorem for the disc D with center $(0, 0)$ and radius R and the functions:
   (a) $P(x, y) = xy^2$, $Q(x, y) = -yx^2$;
   (b) $P(x, y) = x + y$, $Q(x, y) = y$; and
   (c) $P(x, y) = xy = Q(x, y)$.
4. Using the Divergence Theorem show that $\int_{\partial D} \mathbf{F} \cdot \mathbf{n}\, ds = 0$ where $\mathbf{F}(x, y) = y\mathbf{i} - x\mathbf{j}$ and D is the unit disc. Verify this directly.

5. Find the area bounded by one arc of the cycloid $x = a(\theta - \sin\theta)$, $y = a(1 - \cos\theta)$, $a > 0$, $0 \le \theta \le 2\pi$, and the x-axis (use Green's Theorem).

6. Under the conditions of Green's Theorem, prove that

(a) $$\int_{\partial D} PQ\,dx + PQ\,dy = \int_D \left[Q\left(\frac{\partial P}{\partial x} - \frac{\partial P}{\partial y}\right) + P\left(\frac{\partial Q}{\partial x} - \frac{\partial Q}{\partial y}\right)\right] dx\,dy$$

(b) $$\int_{\partial D} \left(Q\frac{\partial P}{\partial x} - P\frac{\partial Q}{\partial x}\right) dx + \left(P\frac{\partial Q}{\partial y} - Q\frac{\partial P}{\partial y}\right) dy = 2\int_D \left(P\frac{\partial^2 Q}{\partial x \partial y} - Q\frac{\partial^2 P}{\partial x \partial y}\right) dx\,dy$$

7. Evaluate $\int_C (2x^3 - y^3)\,dx + (x^3 + y^3)\,dy$, where C is the unit circle, and verify Green's Theorem for this case.

8. Prove the following generalization of Green's Theorem: Let D be a region in the xy-plane with boundary a finite number of oriented simple closed curves. Suppose that by means of a finite number of line segments parallel to the coordinate axes, D can be decomposed into a finite number of regions D_i of type 3 with the boundary of each D_i oriented counterclockwise (see Figure 7.1.5). Then if P and Q are C^1 on D,

$$\int_D \left(\frac{\partial Q}{\partial x} - \frac{\partial P}{\partial y}\right) dx\,dy = \int_{\partial D} P\,dx + Q\,dy$$

where ∂D is the oriented boundary of D. (HINT: Apply Green's Theorem to each D_i.)

9. Verify Green's Theorem for the integrand of Exercise 7 ($P = 2x^3 - y^3$, $Q = x^3 + y^3$) and the annular region D described by $a \le x^2 + y^2 \le b$, with boundaries oriented as in Figure 7.1.5.

10. Let D be a region for which Green's Theorem holds. Suppose f is harmonic; that is,

$$\frac{\partial^2 f}{\partial x^2} + \frac{\partial^2 f}{\partial y^2} = 0$$

on D. Prove that

$$\int_{\partial D} \frac{\partial f}{\partial y}\,dx - \frac{\partial f}{\partial x}\,dy = 0$$

11. (a) Verify the Divergence Theorem for $\mathbf{F} = x\mathbf{i} + y\mathbf{j}$ and D the unit disc $x^2 + y^2 \le 1$.
(b) Evaluate the integral of the normal component of $2xy\mathbf{i} - y^2\mathbf{j}$ around the ellipse $x^2/a^2 + y^2/b^2 = 1$.

12. Let $P(x, y) = -y/(x^2 + y^2)$, $Q(x, y) = x/(x^2 + y^2)$. Assuming D is the unit disc, investigate why Green's Theorem fails for this P and Q.

13. Use Green's Theorem to evaluate $\int_{C^+} (y^2 + x^3)\,dx + x^4\,dy$, where C^+ is the perimeter of $[0, 1] \times [0, 1]$ in the counterclockwise direction.

14. Verify Theorem 3.

15. Use Theorem 2 to compute the area inside the ellipse $x^2/a^2 + y^2/b^2 = 1$.

CHÚT !!

*16. Use Green's Theorem to prove the change of variables formula in the following special case:

$$\int_D dx\, dy = \int_{D^*} \left|\frac{\partial(x, y)}{\partial(u, v)}\right| du\, dv$$

for a transformation $(u, v) \mapsto (x(u, v), y(u, v))$. Formulate the necessary hypotheses on the functions $x = x(u, v)$ and $y = y(u, v)$ and on $\partial(x, y)/\partial(u, v)$ for your proof.

17. Evaluate $\int_{\boldsymbol{\sigma}} (x^5 - 2xy^3)\, dx - 3x^2y^2\, dy$, where $\boldsymbol{\sigma}$ is the path $\boldsymbol{\sigma}(t) = (t^8, t^{10})$, $0 \le t \le 1$.

18. Prove the identity

$$\int_{\partial D} \phi\, \nabla\phi \cdot \mathbf{n}\, ds = \int_D (\phi\, \nabla^2\phi + \nabla\phi \cdot \nabla\phi)\, dA$$

19. Use Green's Theorem to find the area of one loop of the four-leafed rose $r = 3 \sin 2\theta$. (HINT: $x\, dy - y\, dx = r^2\, d\theta$.)

20. Show that if C is a simple closed curve that bounds a region to which Green's Theorem applies, then the area of the region D bounded by C is

$$A = \int_{\partial D} x\, dy = -\int_{\partial D} y\, dx$$

Show how this implies Theorem 2.

Exercises 21 to 28 illustrate the application of Green's Theorem to partial differential equations. They are particularly concerned with solutions to Laplace's equation, that is, with harmonic functions. (See Section 7.5 for additional results). For these exercises, let D be an open region in $\mathbb{R}^2$ with boundary ∂D. Let $u\colon D \cup \partial D \to \mathbb{R}$ be a continuous function that is C^2 on D. Suppose $\mathbf{p} \in D$ and the *closed* discs $B_\rho = B_\rho(\mathbf{p})$ of radius ρ centered at $\mathbf{p}$ are contained in D for $0 < \rho \le R$. Define $I(\rho)$ by

$$I(\rho) = \frac{1}{\rho} \int_{\partial B_\rho} u\, ds$$

*21. Show that $\lim_{\rho \to 0} I(\rho) = 2\pi u(\mathbf{p})$

*22. Let $\mathbf{n}$ denote the outward unit normal to ∂B_ρ and $\partial u/\partial n = \nabla u \cdot \mathbf{n}$. Show that

$$\int_{\partial B} \frac{\partial u}{\partial n}\, ds = \iint_{B_\rho} \nabla^2 u\, dA$$

*23. Show that $I'(\rho) = \dfrac{1}{\rho} \int_{B_\rho} \nabla^2 u\, dA$

*24. Suppose u satisfies Laplace's equation: $\nabla^2 u = 0$ on D. Use the preceding exercises to show that

$$u(\mathbf{p}) = \frac{1}{2\pi R} \int_{\partial B_R} u\, ds$$

(This expresses the fact that the value of a harmonic function at a point is determined by its value on the circumference of any disc centered about it.)

*25. Use Exercise 24 to show that if u is harmonic (i.e., $\nabla^2 u = 0$), then $u(\mathbf{p})$ can be expressed as an area integral

$$u(\mathbf{p}) = \frac{1}{\pi R^2} \iint_{B_R} u \, dA$$

*26. Suppose u is a harmonic function defined on D (i.e., $\nabla^2 u = 0$ on D) and that u has a local maximum (or minimum) at a point $\mathbf{p}$ in D.
   (a) Show that u must be constant on some disc centered at $\mathbf{p}$. (HINT: Use the results of Exercise 25).
   (b) Suppose that D is path connected (i.e., for any points $\mathbf{p}$ and $\mathbf{q}$ in D there is a continuous path $\boldsymbol{\sigma}: [0, 1] \to D$ such that $\boldsymbol{\sigma}(0) = \mathbf{p}$ and $\boldsymbol{\sigma}(1) = \mathbf{q}$ and that the maximum (or minimum) at $\mathbf{p}$ is absolute. (Thus $u(\mathbf{q}) \le u(\mathbf{p})$ (or $u(\mathbf{q}) \ge u(\mathbf{p})$) for every $\mathbf{q}$ in D.) Show that u must be constant on D.

The result in Exercise 26 is called a *strong maximum* (or *minimum*) *principal* for harmonic functions. Compare this with Exercises 16 to 19 in Section 4.2.

*27. A function is called *subharmonic* on D if $\nabla^2 u \ge 0$ everywhere in D. It is called *superharmonic* if $\nabla^2 u \le 0$.
   (a) Derive a strong maximum principal for subharmonic functions.
   (b) Derive a strong minimum principal for superharmonic functions.

*28. Suppose D is the disc $\{(x, y) | x^2 + y^2 < 1\}$ and C is the circle $\{(x, y) | x^2 + y^2 = 1\}$. In Section 7.5 we shall show that if f is a continuous real-valued function on C, then there is a continuous function u on $D \cup C$ that agrees with f on C and is harmonic on D. That is, f has a harmonic extension to the disc. Assuming this, show the following.
   (a) If q is a nonconstant continuous function on $D \cup C$ that is subharmonic (but not harmonic) on D, then there is a continuous function u on $D \cup C$ that is harmonic on D such that u agrees with q on C and $q < u$ everywhere on D.
   (b) The same assertion holds if "subharmonic" is replaced by "superharmonic" and "$q < u$" by "$q > u$".

7.2 STOKES' THEOREM

Stokes' Theorem relates the line integral of a vector field around a simple closed curve C in $\mathbb{R}^3$ to an integral over a surface S for which C is the boundary. In this regard it is very much like Green's Theorem.

Let us begin by recalling a few facts from Chapter 6. Consider a surface S that is the graph of a function $f(x, y)$, so S is parametrized by

$$\begin{cases} x = u \\ y = v \\ z = f(u, v) = f(x, y) \end{cases}$$

for (u, v) in some domain D. The integral of a vector function $\mathbf{F}$ over S was developed in Section 6.6 as

$$\int_S \mathbf{F} \cdot d\mathbf{S} = \int_D \left[F_1\left(-\frac{\partial z}{\partial x}\right) + F_2\left(-\frac{\partial z}{\partial y}\right) + F_3 \right] dx\, dy \tag{1}$$

where $\mathbf{F} = F_1\mathbf{i} + F_2\mathbf{j} + F_3\mathbf{k}$.

In Section 7.1 we assumed that the regions D under consideration were of type 3; this was an essential requirement in our proof of Green's Theorem but we noted there that the theorem is valid for a wider class of regions. In this section we shall assume that D is a region whose boundary is a simple closed curve and to which Green's Theorem applies. Green's Theorem involves choosing an orientation on the Boundary of D, as was explained in Section 7.1. The choice of orientation that validates Green's Theorem will be called *positive*. Recall that if D is of type 3, then the positive orientation is the counterclockwise one.

Suppose that $\boldsymbol{\sigma}: [a, b] \to \mathbb{R}^2$, $\boldsymbol{\sigma}(t) = (x(t), y(t))$, is a parametrization of ∂D in the positive direction. Then we define the *boundary curve* ∂S to be the oriented simple closed curve that is the image of the mapping $\boldsymbol{\eta}: t \mapsto (x(t), y(t), f(x(t), y(t)))$ with the *orientation* induced by $\boldsymbol{\eta}$ (Figure 7.2.1).

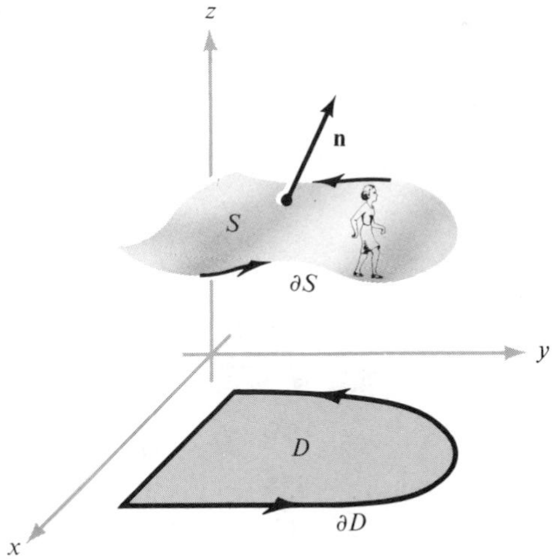

FIGURE 7.2.1
The induced orientation on ∂S; as you walk around the boundary, the surface should be on your left.

To remember this orientation (that is, the positive direction) on ∂S, think of an "observer" walking along the boundary of the surface with the normal being her upright direction; she is moving in the positive direction if the surface is on her left. This orientation on ∂S is often called the *orientation induced by an upward normal* $\mathbf{n}$.

We are now ready to state and prove one of the primary results of this section.

THEOREM 5 (STOKES' THEOREM FOR GRAPHS). *Let S be the oriented surface defined by a C^2 function $z = f(x, y)$, $(x, y) \in D$, and let $\mathbf{F}$ be a C^1 vector field on S. Then if ∂S denotes the oriented boundary curve of S as defined above we have*

$$\int_S \operatorname{curl} \mathbf{F} \cdot d\mathbf{S} = \int_S (\nabla \times \mathbf{F}) \cdot d\mathbf{S} = \int_{\partial S} \mathbf{F} \cdot d\mathbf{s}$$

Remember that $\int_{\partial S} \mathbf{F} \cdot d\mathbf{s}$ is the integral around ∂S of the tangential component of $\mathbf{F}$, while $\int_S \mathbf{G} \cdot d\mathbf{S}$ is the integral over S of $\mathbf{G} \cdot \mathbf{n}$, the normal component of $\mathbf{G}$ (see sections 6.2 and 6.6). Thus Stokes' Theorem says that the integral of *the normal component of the curl of a vector field* $\mathbf{F}$ *over a surface* S *is equal to the integral of the tangential component of* $\mathbf{F}$ *around the boundary* ∂S.

Proof. If $\mathbf{F} = F_1\mathbf{i} + F_2\mathbf{j} + F_3\mathbf{k}$ then

$$\operatorname{curl} \mathbf{F} = \left(\frac{\partial F_3}{\partial y} - \frac{\partial F_2}{\partial z}\right)\mathbf{i} + \left(\frac{\partial F_1}{\partial z} - \frac{\partial F_3}{\partial x}\right)\mathbf{j} + \left(\frac{\partial F_2}{\partial x} - \frac{\partial F_1}{\partial y}\right)\mathbf{k}$$

Therefore we use formula (1) to write

$$\int_S \operatorname{curl} \mathbf{F} \cdot d\mathbf{S} = \int_D \left[\left(\frac{\partial F_3}{\partial y} - \frac{\partial F_2}{\partial z}\right)\left(-\frac{\partial z}{\partial x}\right) + \left(\frac{\partial F_1}{\partial z} - \frac{\partial F_3}{\partial x}\right)\left(-\frac{\partial z}{\partial y}\right) + \left(\frac{\partial F_2}{\partial x} - \frac{\partial F_1}{\partial y}\right)\right] dA \qquad (2)$$

On the other hand,

$$\int_{\partial S} \mathbf{F} \cdot d\mathbf{s} = \int_{\boldsymbol{\eta}} \mathbf{F} \cdot d\mathbf{s} = \int_{\boldsymbol{\eta}} F_1\, dx + F_2\, dy + F_3\, dz$$

where $\boldsymbol{\eta}: [a, b] \to \mathbb{R}^3$, $\boldsymbol{\eta}(t) = (x(t), y(t), f(x(t), y(t)))$ is the orientation-preserving parametrization of the oriented simple closed curve ∂S discussed above. Thus

$$\int_{\partial S} \mathbf{F} \cdot d\mathbf{s} = \int_a^b \left(F_1 \frac{dx}{dt} + F_2 \frac{dy}{dt} + F_3 \frac{dz}{dt}\right) dt \qquad (3)$$

But by the Chain Rule

$$\frac{dz}{dt} = \frac{\partial z}{\partial x}\frac{dx}{dt} + \frac{\partial z}{\partial y}\frac{dy}{dt}$$

Substituting this expression into (3) we obtain

$$\int_{\partial S} \mathbf{F} \cdot d\mathbf{s} = \int_a^b \left[\left(F_1 + F_3 \frac{\partial z}{\partial x} \right) \frac{dx}{dt} + \left(F_2 + F_3 \frac{\partial z}{\partial y} \right) \frac{dy}{dt} \right] dt$$

$$= \int_\sigma \left(F_1 + F_3 \frac{\partial z}{\partial x} \right) dx + \left(F_2 + F_3 \frac{\partial z}{\partial y} \right) dy$$

$$= \int_{\partial D} \left(F_1 + F_3 \frac{\partial z}{\partial x} \right) dx + \left(F_2 + F_3 \frac{\partial z}{\partial y} \right) dy \qquad (4)$$

Applying Green's Theorem to (4) yields (we are assuming that Green's Theorem applies to D)

$$\int_D \left[\frac{\partial (F_2 + F_3\, \partial z / \partial y)}{\partial x} - \frac{\partial (F_1 + F_3\, \partial z / \partial x)}{\partial y} \right] dA$$

Now we use the Chain Rule, remembering that F_1, F_2, and F_3 are functions of x, y, and z is a function of x and y, to obtain

$$\int_D \left[\left(\frac{\partial F_2}{\partial x} + \frac{\partial F_2}{\partial z} \cdot \frac{\partial z}{\partial x} + \frac{\partial F_3}{\partial x} \cdot \frac{\partial z}{\partial y} + \frac{\partial F_3}{\partial z} \cdot \frac{\partial z}{\partial x} \cdot \frac{\partial z}{\partial y} + F_3 \cdot \frac{\partial^2 z}{\partial x\, \partial y} \right) \right.$$

$$\left. - \left(\frac{\partial F_1}{\partial y} + \frac{\partial F_1}{\partial z} \cdot \frac{\partial z}{\partial y} + \frac{\partial F_3}{\partial y} \cdot \frac{\partial z}{\partial x} + \frac{\partial F_3}{\partial z} \cdot \frac{\partial z}{\partial y} \cdot \frac{\partial z}{\partial x} + F_3 \frac{\partial^2 z}{\partial y\, \partial x} \right) \right] dA$$

The last two terms in each parenthesis cancel each other, and we can rearrange terms to obtain the integral (2), which completes the proof. ■

EXAMPLE 1. Let $\mathbf{F} = ye^z\mathbf{i} + xe^z\mathbf{j} + xye^z\mathbf{k}$. Show that the integral of $\mathbf{F}$ around an oriented simple closed curve C that is the boundary of a surface S is 0. (Assume S is the graph of a function, as in Theorem 5.)

Indeed, $\int_C \mathbf{F} \cdot d\mathbf{s} = \int_S (\nabla \times \mathbf{F}) \cdot d\mathbf{S}$ by Stokes' Theorem. But we compute

$$\nabla \times \mathbf{F} = \begin{vmatrix} \mathbf{i} & \mathbf{j} & \mathbf{k} \\ \frac{\partial}{\partial x} & \frac{\partial}{\partial y} & \frac{\partial}{\partial z} \\ ye^z & xe^z & xye^z \end{vmatrix} = \mathbf{0}$$

so $\int_C \mathbf{F} \cdot d\mathbf{s} = 0$.

EXAMPLE 2. Use Stokes' Theorem to evaluate the line integral

$$\int_C -y^3\, dx + x^3\, dy - z^3\, dz$$

where C is the intersection of the cylinder $x^2 + y^2 = 1$ and the plane $x + y + z = 1$, and the orientation on C corresponds to counterclockwise motion in the xy-plane.

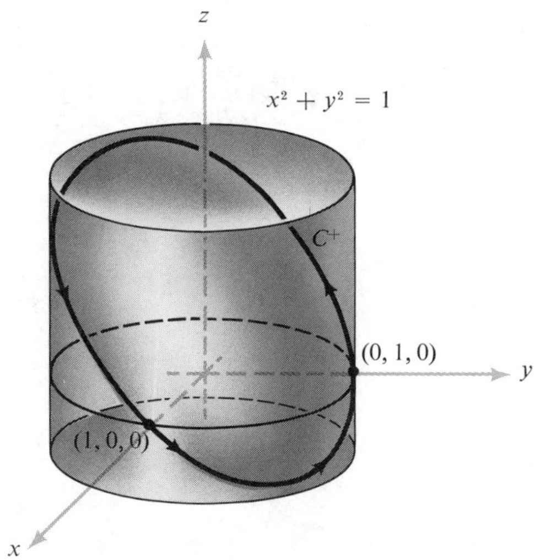

FIGURE 7.2.2
The curve C is the intersection of the cylinder $x^2 + y^2 = 1$ and the plane $x + y + z = 1$.

The curve C bounds the surface S defined by $z = 1 - x - y = f(x, y)$ for (x, y) in $D = \{(x, y) | x^2 + y^2 \leq 1\}$ (Figure 7.2.2). We set $\mathbf{F} = -y^3\mathbf{i} + x^3\mathbf{j} - z^3\mathbf{k}$, which has curl $\nabla \times \mathbf{F} = (3x^2 + 3y^2)\mathbf{k}$. Then by Stokes' Theorem the line integral is equal to the surface integral

$$\int_S (\nabla \times \mathbf{F}) \cdot d\mathbf{S}$$

But $\nabla \times \mathbf{F}$ has only a $\mathbf{k}$ component. Thus by formula (1) we have

$$\int_S (\nabla \times \mathbf{F}) \cdot d\mathbf{S} = \int_D (3x^2 + 3y^2)\, dx\, dy$$

This integral can be evaluated by changing to polar coordinates. Doing this, we get

$$3 \int_D (x^2 + y^2)\, dx\, dy = 3 \int_0^1 \int_0^{2\pi} r^2 \cdot r\, d\theta\, dr = 6\pi \int_0^1 r^3\, dr = \frac{6\pi}{4} = \frac{3\pi}{2}$$

Let us verify this result by directly evaluating the line integral

$$\int_C -y^3\, dx + x^3\, dy - z^3\, dz$$

We can parametrize the curve ∂D by the equations

$$x = \cos t, \qquad y = \sin t, \qquad z = 0, \qquad 0 \leq t \leq 2\pi$$

The curve C is therefore parametrized by the equations

$$x = \cos t, \qquad y = \sin t, \qquad z = 1 - \sin t - \cos t, \qquad 0 \leq t \leq 2\pi$$

Thus

$$\int_C -y^3\,dx + x^3\,dy - z^3\,dz$$

$$= \int_0^{2\pi} [(-\sin^3 t)(-\sin t) + (\cos^3 t)(\cos t)$$

$$- (1 - \sin t - \cos t)^3(-\cos t + \sin t)]\,dt$$

$$= \int_0^{2\pi} (\cos^4 t + \sin^4 t)\,dt - \int_0^{2\pi} (1 - \sin t - \cos t)^3(-\cos t + \sin t)\,dt$$

The second integrand is of the form $u^3\,du$, where $u = 1 - \sin t - \cos t$, and thus the integral is equal to

$$\tfrac{1}{4}[1 - \sin t - \cos t)^4]_0^{2\pi} = 0$$

Hence we are left with $\int_0^{2\pi} (\cos^4 t + \sin^4 t)\,dt$.

This can be evaluated using formulas (18) and (19) of Appendix C. We can also proceed as follows. Using the trigonometric identities

$$\sin^2 t = \frac{1 - \cos 2t}{2}, \qquad \cos^2 t = \frac{1 + \cos 2t}{2}$$

we reduce the above integral to

$$\frac{1}{2}\int_0^{2\pi} (1 + \cos^2 2t)\,dt = \pi + \frac{1}{2}\int_0^{2\pi} \cos^2 2t\,dt$$

Again using the fact that

$$\cos^2 2t = \frac{1 + \cos 4t}{2}$$

we find

$$\pi + \frac{1}{4}\int_0^{2\pi} (1 + \cos 4t)\,dt = \pi + \frac{1}{4}\int_0^{2\pi} dt + \frac{1}{4}\int_0^{2\pi} \cos 4t\,dt$$

$$= \pi + \frac{\pi}{2} + 0 = \frac{3\pi}{2}$$

In order to simplify the proof of Stokes' Theorem above we assumed that the surface S could be described as the graph of a function $z = f(x, y)$, $(x, y) \in D$, where D is some region to which Green's Theorem applies. However, without too much more effort we can obtain a more general theorem for oriented parametrized surfaces S. The main complication is in the definition of ∂S.

Suppose $\mathbf{\Phi}: D \to \mathbb{R}^3$ is a parametrization of a surface S and $\boldsymbol{\sigma}(t) = (u(t), v(t))$ is a parametrization of ∂D. We might be tempted to define ∂S as the curve parametrized by $t \mapsto \boldsymbol{\eta}(t) = \mathbf{\Phi}(u(t), v(t))$. However with this definition ∂S might not be the boundary of S in any reasonable geometric sense.

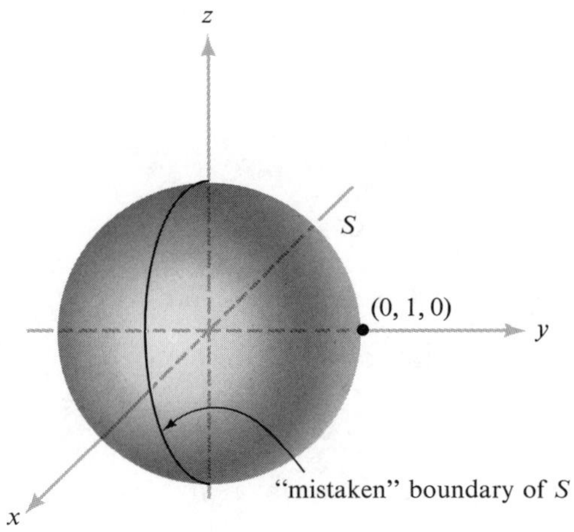

FIGURE 7.2.3
The surface S is a portion of a sphere.

For example, we would conclude that the boundary of the unit sphere S parametrized by spherical coordinates in $\mathbb{R}^3$ is half of the great circle on S lying in the xz-plane, but clearly in a geometric sense S is a smooth surface (no points or cusps) with no boundary or edge at all (see Figure 7.2.3 and Exercise 14). Thus this great circle is in some sense the "mistaken" boundary of S.

We can get around this difficulty by assuming that $\mathbf{\Phi}$ is one-to-one on all of D. Then the image of ∂D under $\mathbf{\Phi}$, namely $\mathbf{\Phi}(\partial D)$ will be the geometric boundary of $S = \mathbf{\Phi}(D)$. If $\boldsymbol{\sigma}(t) = (u(t), v(t))$ is a parametrization of ∂D in the positive direction, we define ∂S to be the oriented simple closed curve that is the image of the mapping $\boldsymbol{\eta}: t \mapsto \mathbf{\Phi}(u(t), v(t))$ with the orientation on ∂S induced by $\boldsymbol{\eta}$ (see Figure 7.2.1).

THEOREM 6 (STOKES' THEOREM FOR PARAMETRIZED SURFACES). *Suppose S is an oriented surface defined by a one-to-one parametrization $\mathbf{\Phi}: D \subset \mathbb{R}^2 \to S$. Let ∂S denote the oriented boundary of S and let $\mathbf{F}$ be a C^1 vector field on S. Then*

$$\int_S (\nabla \times \mathbf{F}) \cdot d\mathbf{S} = \int_{\partial S} \mathbf{F} \cdot d\mathbf{s}$$

This is proved in the same way as Theorem 5.

EXAMPLE 3. Let S be the surface shown in Figure 7.2.4, with the indicated orientation. Let $\mathbf{F} = y\mathbf{i} - x\mathbf{j} + e^{xz}\mathbf{k}$. Evaluate $\int_S (\nabla \times \mathbf{F}) \cdot d\mathbf{S}$.

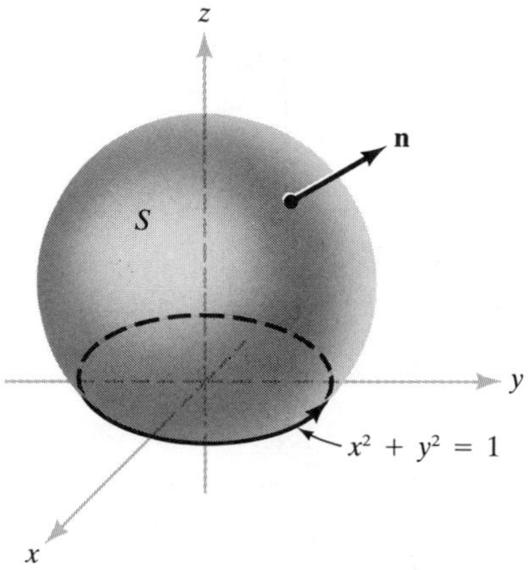

FIGURE 7.2.4
The boundary of a surface S that is parametrized by $\mathbf{\Phi}$: $D \to \mathbb{R}^3$ *is the image of the boundary of D only if* $\mathbf{\Phi}$ *is one-to-one on D.*

This is a parametrized surface and could be parametrized using spherical coordinates based at the center of the sphere. However, we need not explicitly find $\mathbf{\Phi}$ in order to solve this problem. By Theorem 6, $\int_S (\nabla \times \mathbf{F}) \cdot d\mathbf{S} = \int_{\partial S} \mathbf{F} \cdot d\mathbf{s}$, and so if we parametrize ∂S by $x(t) = \cos t$, $y(t) = \sin t$, $0 \le t \le 2\pi$, we determine

$$\int_{\partial S} \mathbf{F} \cdot d\mathbf{s} = \int_0^{2\pi} \left[y \frac{dx}{dt} - x \frac{dy}{dt} \right] dt$$

$$= \int_0^{2\pi} (-\sin^2 t - \cos^2 t)\, dt = -\int_0^{2\pi} dt = -2\pi$$

and therefore $\int_S (\nabla \times \mathbf{F}) \cdot d\mathbf{S} = -2\pi$.

Let us now use Stokes' Theorem to justify the physical interpretation of $\nabla \times \mathbf{F}$ in terms of paddle wheels that was proposed in Chapter 3. Paraphrasing Theorem 6 we have

$$\int_S (\text{curl } \mathbf{F}) \cdot \mathbf{n}\, dS = \int_S \text{curl } \mathbf{F} \cdot d\mathbf{S} = \int_{\partial S} \mathbf{F} \cdot d\mathbf{s} = \int_{\partial S} F_{\mathrm{T}}\, ds$$

where F_{T} is the tangential component of $\mathbf{F}$. This says that the integral of the normal component of the curl of a vector field over an oriented surface S is equal to the line integral of $\mathbf{F}$ along ∂S, which in turn is equal to the path integral of the tangential component of $\mathbf{F}$ over ∂S.

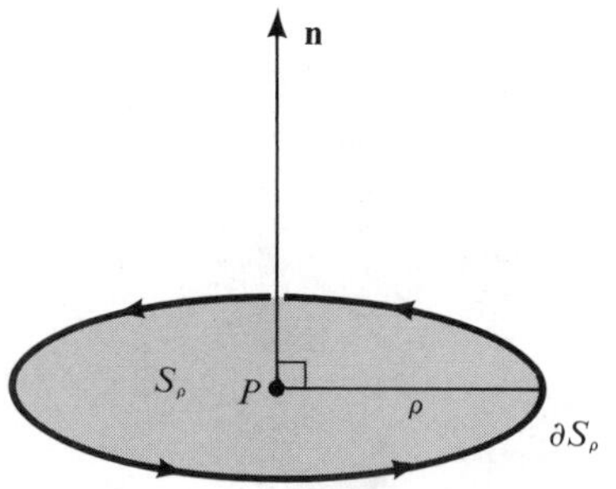

FIGURE 7.2.5
A normal **n** *induces an orientation on the boundary* ∂S_ρ *of the disc* S_ρ.

Suppose **V** represents the velocity vector field of a fluid. Consider a point P and a unit vector **n**. Let S_ρ denote the disc of radius ρ and center P, which is perpendicular to **n**. By Stokes' Theorem

$$\int_{S_\rho} \text{curl}\,\mathbf{V}\cdot d\mathbf{S} = \int_{S_\rho} \text{curl}\,\mathbf{V}\cdot\mathbf{n}\,dS = \int_{\partial S_\rho} \mathbf{V}\cdot d\mathbf{s}$$

where ∂S_ρ has the orientation induced by **n** (see Figure 7.2.5). It is not difficult to show (see Exercise 8, Section 6.6) that there is a point Q in S_ρ such that

$$\int_{S_\rho} \text{curl}\,\mathbf{V}\cdot\mathbf{n}\,dS = (\text{curl}\,\mathbf{V}(Q)\cdot\mathbf{n})A(S_\rho)$$

(this is the Mean Value Theorem for Integrals, proved as on p. 289) where $A(S_\rho) = \pi\rho^2$ is the area of S_ρ, curl $\mathbf{V}(Q)$ is the value of curl **V** at Q, and **n** is evaluated at Q as well. Thus

$$\begin{aligned}\lim_{\rho\to 0}\frac{1}{A(S_\rho)}\int_{\partial S_\rho}\mathbf{V}\cdot d\mathbf{s} &= \lim_{\rho\to 0}\frac{1}{A(S_\rho)}\int_{S_\rho}(\text{curl}\,\mathbf{V})\cdot d\mathbf{S}\\ &= \lim_{\rho\to 0}\text{curl}\,\mathbf{V}(Q)\cdot\mathbf{n}\\ &= \text{curl}\,\mathbf{V}(P)\cdot\mathbf{n}\end{aligned}$$

Thus,*

$$\text{curl}\,\mathbf{V}(P)\cdot\mathbf{n}(P) = \lim_{\rho\to 0}\frac{1}{A(S_\rho)}\int_{\partial S_\rho}\mathbf{V}\cdot d\mathbf{s} \tag{5}$$

Let us pause to consider the physical meaning of $\int_C \mathbf{V}\cdot d\mathbf{s}$ when **V** is the velocity field of a fluid (see p. 171). Suppose, for example, that **V** points in the

* Some physics texts adopt (5) as the *definition* of curl, and use it to "prove" Stokes' Theorem easily. However this raises the danger of circular reasoning, for to show that (5) really defines a vector "curl $\mathbf{V}(P)$" requires Stokes' Theorem, or some similar argument.

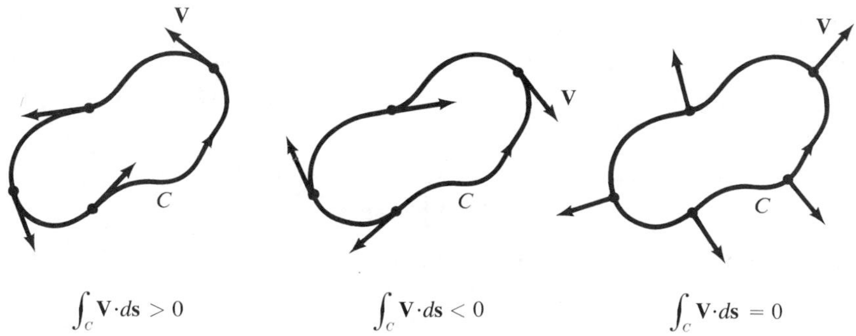

FIGURE 7.2.6
The intuitive meaning of the possible signs of $\int_C \mathbf{V}\cdot d\mathbf{s}$

direction tangent to the oriented curve C (Figure 7.2.6). Then clearly $\int_C \mathbf{V}\cdot d\mathbf{s} > 0$, and particles on C tend to rotate counterclockwise. If $\mathbf{V}$ is pointing in the opposite direction, $\int_C \mathbf{V}\cdot d\mathbf{s} < 0$. If $\mathbf{V}$ is perpendicular to C then particles don't rotate on C at all and $\int_C \mathbf{V}\cdot d\mathbf{s} = 0$. In general $\int_C \mathbf{V}\cdot d\mathbf{s}$, being the integral of the tangential component of $\mathbf{V}$, represents the net amount of turning of the fluid in a counterclockwise direction around C. One therefore refers to $\int_C \mathbf{V}\cdot d\mathbf{s}$ as the *circulation* of $\mathbf{V}$ around C (see Figure 7.2.7).

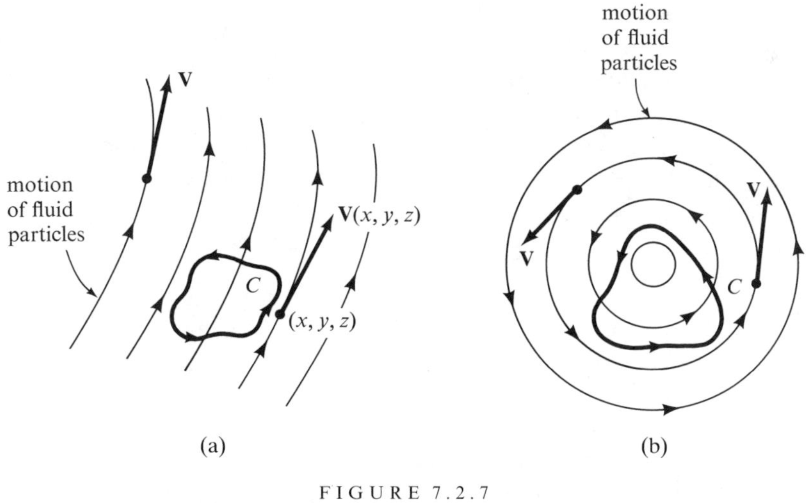

FIGURE 7.2.7
Circulation of a vector field (velocity field of a fluid): (a) circulation about C is zero; (b) nonzero circulation about C ("whirlpool").

These results allow us to see just what curl $\mathbf{V}$ means for the motion of a fluid. The circulation $\int_{\partial S_\rho} \mathbf{V} \cdot d\mathbf{s}$ is the net velocity of the fluid around ∂S_ρ, so curl $\mathbf{V} \cdot \mathbf{n}$ represents the turning or rotating effect of the fluid around the axis $\mathbf{n}$. More precisely, formula (5) states that

curl $\mathbf{V}(P) \cdot \mathbf{n}$ *is the circulation of* $\mathbf{V}$ *per unit area on a surface perpendicular to* $\mathbf{n}$.

Observe that the magnitude of curl $\mathbf{V} \cdot \mathbf{n}$ is maximized when $\mathbf{n} =$ curl $\mathbf{V}/\|\text{curl } \mathbf{V}\|$. Therefore, the rotating effect at P is greatest about the axis parallel to curl $\mathbf{V}/\|\text{curl } \mathbf{V}\|$. Thus curl $\mathbf{V}$ is aptly called the *vorticity vector*.

EXAMPLE 4 (FARADAY'S LAW). A basic law of electromagnetic theory is that if $\mathbf{E}(t, x, y, z)$ and $\mathbf{H}(t, x, y, z)$ represent the electric and magnetic fields at time t, then $\nabla \times \mathbf{E} = -\partial \mathbf{H}/\partial t$, where $\nabla \times \mathbf{E}$ is computed by holding t fixed, and $\partial \mathbf{H}/\partial t$ is computed by holding x, y, and z constant.

Let us use Stokes' Theorem to determine what this means physically. Assume S is a surface to which Stokes' Theorem applies. Then

$$\int_{\partial S} \mathbf{E} \cdot d\mathbf{s} = \int_S (\nabla \times \mathbf{E}) \cdot d\mathbf{S} = -\int_S \frac{\partial \mathbf{H}}{\partial t} \cdot d\mathbf{S}$$

$$= -\frac{\partial}{\partial t} \int_S \mathbf{H} \cdot d\mathbf{S}$$

(The last equality may be justified if $\mathbf{H}$ is C^1.) Thus we obtain

$$\int_{\partial S} \mathbf{E} \cdot d\mathbf{s} = -\frac{\partial}{\partial t} \int_S \mathbf{H} \cdot d\mathbf{S}$$

This equality is known as *Faraday's law*. The quantity $\int_{\partial S} \mathbf{E} \cdot d\mathbf{s}$ represents the "voltage" around ∂S, and if ∂S were a wire, a current would flow in proportion to this voltage. Also $\int_S \mathbf{H} \cdot d\mathbf{S}$ is called the *flux of* $\mathbf{H}$, or the magnetic flux. Thus, Faraday's law says that *the voltage around a loop equals the negative of the rate of change of magnetic flux through the loop.*

EXERCISES

1. Redo Exercise 5 of Section 6.6 using Stokes' Theorem.
2. Redo Exercise 6 of Section 6.6 using Stokes' Theorem.
3. Verify Stokes' Theorem for the upper hemisphere $z = \sqrt{1 - x^2 - y^2}$, $z \geq 0$, and the radial vector field $\mathbf{F}(x, y, z) = x\mathbf{i} + y\mathbf{j} + z\mathbf{k}$.
4. Let S be a surface with boundary ∂S, and suppose $\mathbf{E}$ is an electric field that is perpendicular to ∂S. Show that the induced magnetic flux across S is constant in time. (HINT: Use Faraday's law.)

5. Let S be the capped cylindrical surface shown in Figure 7.2.8. S is the union of two surfaces S_1 and S_2, where S_1 is the set of (x, y, z) with $x^2 + y^2 = 1, 0 \leq z \leq 1$, and S_2 is the set of (x, y, z) with $x^2 + y^2 + (z - 1)^2 = 1$, $z \geq 1$. Set $\mathbf{F}(x, y, z) = (zx + z^2y + x)\mathbf{i} + (z^3yx + y)\mathbf{j} + z^4x^2\mathbf{k}$. Compute $\int_S (\nabla \times \mathbf{F}) \cdot d\mathbf{S}$. (HINT: Stokes' Theorem holds for this surface.)

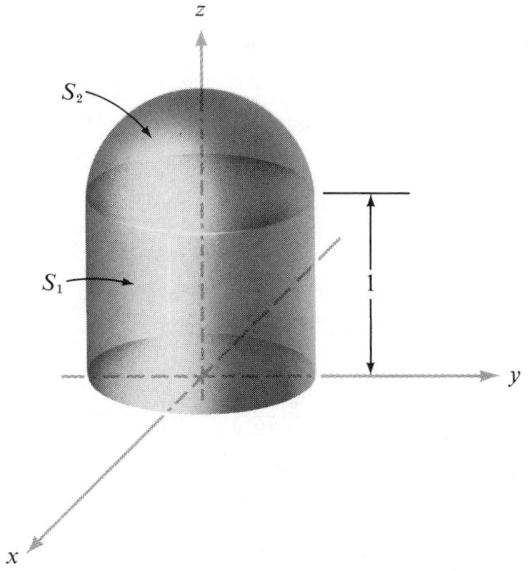

FIGURE 7.2.8
The capped cylinder is the union of S_1 and S_2.

6. Let $\boldsymbol{\sigma}$ consist of straight lines joining (1, 0, 0), (0, 1, 0), and (0, 0, 1) and let S be the triangle with these vertices. Verify Stokes' Theorem directly with $\mathbf{F} = yz\mathbf{i} + xz\mathbf{j} + xy\mathbf{k}$.

7. Prove that Faraday's law implies $\nabla \times \mathbf{E} = -\partial \mathbf{H}/\partial t$.

8. Let S be a surface and let $\mathbf{F}$ be perpendicular to the tangent to the boundary of S. Show that

$$\int_S (\nabla \times \mathbf{F}) \cdot d\mathbf{S} = 0$$

What does this mean physically if $\mathbf{F}$ is an electric field?

9. Consider two surfaces S_1, S_2 with the same boundary ∂S. Describe with sketches how S_1 and S_2 must be oriented to ensure that

$$\int_{S_1} (\nabla \times \mathbf{F}) \cdot d\mathbf{S} = \int_{S_2} (\nabla \times \mathbf{F}) \cdot d\mathbf{S}$$

10. For a surface S and a fixed vector $\mathbf{v}$ prove

$$2 \int_S \mathbf{v} \cdot \mathbf{n}\, dS = \int_{\partial S} (\mathbf{v} \times \mathbf{r}) \cdot d\mathbf{s}$$

where $\mathbf{r}(x, y, z) = (x, y, z)$.

11. Argue informally that if S is a closed surface then

$$\int_S (\nabla \times \mathbf{F}) \cdot d\mathbf{S} = 0$$

(see Exercise 9). (A *closed surface* is one that forms the boundary of a region in space; thus, for example, a sphere is a closed surface.)

12. If C is a closed curve which is the boundary of a surface S and f and g are C^2 functions, show that
(a) $\int_C f\nabla g \cdot d\mathbf{s} = \int_S (\nabla f \times \nabla g) \cdot d\mathbf{S}$ and
(b) $\int_C (f\nabla g + g\nabla f) \cdot d\mathbf{s} = 0$.

13. If C is a closed curve that is the boundary of a surface S and $\mathbf{v}$ is a constant vector, show that

$$\int_C \mathbf{v} \cdot d\mathbf{s} = 0$$

Show that this is true even if C is not the boundary of a surface S.

14. Show that the parametrization $\mathbf{\Phi}: D \to \mathbb{R}^3$, $D = [0, \pi] \times [0, 2\pi]$, $\mathbf{\Phi}(\phi, \theta) = (\cos\theta \sin\phi, \sin\theta \sin\phi, \cos\phi)$, of the unit sphere takes the boundary of D to half of a great circle on S.

15. Verify Theorem 6 for the helicoid $\mathbf{\Phi}(r, \theta) = (r\cos\theta, r\sin\theta, \theta)$, $(r, \theta) \in [0, 1] \times [0, \pi/2]$, and the vector field $\mathbf{F}(x, y, z) = (z, x, y)$.

*16. Prove Theorem 6.

17. Let $\mathbf{F} = x^2\mathbf{i} + (2xy + x)\mathbf{j} + z\mathbf{k}$. Let C be the circle $x^2 + y^2 = 1$ and S the disc $x^2 + y^2 \leq 1$ within the plane $z = 0$. Determine
(a) the flux of $\mathbf{F}$ out of S and
(b) the circulation of $\mathbf{F}$ around C.
(c) Find the flux of $\nabla \times \mathbf{F}$. Verify Stokes' Theorem directly in this case.

*18. Faraday's law relates the line integral of the electric field around a loop C to the surface integral of the rate of change of the magnetic field over a surface S with boundary C. Regarding the equation $\nabla \times \mathbf{E} = -\partial\mathbf{H}/\partial t$ as the basic equation, Faraday's law is a consequence of Stokes' Theorem, as we have seen in Example 4.
Suppose we are given electric and magnetic fields in space that satisfy $\nabla \times \mathbf{E} = -\partial\mathbf{H}/\partial t$. Suppose C is the boundary of the Möbius band shown in Figure 6.6.3 and 6.6.4. Since the Möbius band cannot be oriented, Stokes' Theorem does not apply. What becomes of Faraday's law? What do you guess $\int_C \mathbf{E} \cdot d\mathbf{s}$ equals?

19. Integrate $\nabla \times \mathbf{F}$, $\mathbf{F} = (3y, -xz, yz^2)$ over the portion of the surface $2z = x^2 + y^2$ below the plane $z = 2$ both directly and by using Stokes' Theorem.

7.3 CONSERVATIVE FIELDS

We saw in Section 6.2 that in the case of a gradient force field $\mathbf{F} = \nabla f$, line integrals of $\mathbf{F}$ were evaluated as follows:

$$\int_{\boldsymbol{\sigma}} \mathbf{F} \cdot d\mathbf{s} = f(\boldsymbol{\sigma}(b)) - f(\boldsymbol{\sigma}(a))$$

The value of the integral depends only on the endpoints $\boldsymbol{\sigma}(b)$ and $\boldsymbol{\sigma}(a)$ of the path. In other words, if we used another path with the same endpoints, we would still get the same answer. This leads us to say that the integral is *path independent*.

Gradient fields are very important in physical problems. Usually $V = -f$ represents a potential energy (gravitational, electrical, and so on), and $\mathbf{F}$ represents a force.* Consider the example of a particle of mass m in the field of the earth; in this case one takes f to be GmM/r or $V = -GmM/r$, where G is the gravitational constant, M is the mass of the earth, and r is the distance from the center of the earth. The corresponding force is $\mathbf{F} = (GmM/r^3)\mathbf{r} = (GmM/r^2)\mathbf{n}$, where $\mathbf{n}$ is the unit radial vector. (We shall discuss this case further below.) Note that $\mathbf{F}$ fails to be defined at the one point $r = 0$.

We now wish to characterize those vector fields that can be written as a gradient. Our task is simplified considerably by Stokes' Theorem.

THEOREM 7. *Let* $\mathbf{F}$ *be a* C^1 *vector field defined on* $\mathbb{R}^3$ *except possibly for a finite number of points. The following conditions on* $\mathbf{F}$ *are all equivalent:*

(*i*) *For any oriented simple closed curve* C, $\int_C \mathbf{F}\cdot d\mathbf{s} = 0$.

(*ii*) *For any two oriented simple curves* C_1, C_2 *with the same endpoints,* $\int_{C_1} \mathbf{F}\cdot d\mathbf{s} = \int_{C_2} \mathbf{F}\cdot d\mathbf{s}$.

(*iii*) $\mathbf{F}$ *is the gradient of some function* f*; that is,* $\mathbf{F} = \nabla f$ *(and if* $\mathbf{F}$ *has an exceptional point where it fails to be defined,* f *is also undefined there).*

(*iv*) $\nabla \times \mathbf{F} = \mathbf{0}$.

A vector field satisfying one (and hence, all) of these conditions is called a conservative vector field.†

Proof. We shall establish the following chain of implications, which will prove the theorem:

$$(i) \Rightarrow (ii) \Rightarrow (iii) \Rightarrow (iv) \Rightarrow (i)$$

First we show that condition (*i*) implies condition (*ii*). Suppose $\boldsymbol{\sigma}_1$ and $\boldsymbol{\sigma}_2$ are parametrizations representing C_1 and C_2, with the same endpoints. Construct the closed curve $\boldsymbol{\sigma}$ obtained by first traversing $\boldsymbol{\sigma}_1$, then $-\boldsymbol{\sigma}_2$ (Figure 7.3.1), or symbolically $\boldsymbol{\sigma} = \boldsymbol{\sigma}_1 - \boldsymbol{\sigma}_2$. Assuming $\boldsymbol{\sigma}$ is simple, (*i*) gives

$$\int_{\boldsymbol{\sigma}} \mathbf{F}\cdot d\mathbf{s} = \int_{\boldsymbol{\sigma}_1} \mathbf{F}\cdot d\mathbf{s} - \int_{\boldsymbol{\sigma}_2} \mathbf{F}\cdot d\mathbf{s} = 0$$

so (*ii*) holds. (If $\boldsymbol{\sigma}$ is not simple, an additional argument, omitted here, is needed.)

* If the minus sign is used, then V is decreasing in the direction $\mathbf{F}$. Thus a particle acted on by $\mathbf{F}$ moves in a direction that decreases the potential.

† In the plane $\mathbb{R}^2$ exceptional points are not allowed (see Exercise 12). Theorem 7 can be proved in the same way if $\mathbf{F}$ is defined and C^1 only on an open convex set in $\mathbb{R}^2$ or $\mathbb{R}^3$. (A set D is convex if $P, Q \in D$ implies the line joining P and Q belongs to D.)

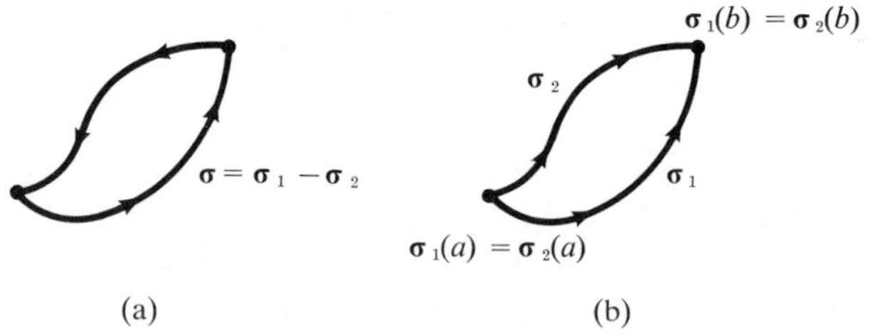

FIGURE 7.3.1
Constructing an oriented simple closed curve $\boldsymbol{\sigma}_1 - \boldsymbol{\sigma}_2$ *(a) from two oriented simple curves (b).*

Next we prove (*ii*) implies (*iii*). Let C be any oriented simple curve joining a point such as (0, 0, 0) to (x, y, z), and suppose C is represented by the parametrization $\boldsymbol{\sigma}$ (if (0, 0, 0) is the exceptional point of $\mathbf{F}$, we can choose a different starting point for $\boldsymbol{\sigma}$ without affecting the argument). Define f to be $\int_{\boldsymbol{\sigma}} \mathbf{F} \cdot d\mathbf{s}$. By hypothesis (*ii*) f is independent of C. We shall show that $\mathbf{F} = \text{grad } f$. Indeed, choose $\boldsymbol{\sigma}$ to be the path shown in Figure 7.3.2, so that

$$f(x, y, z) = \int_0^x F_1(t, 0, 0)\, dt + \int_0^y F_2(x, t, 0)\, dt + \int_0^z F_3(x, y, t)\, dt$$

where $\mathbf{F} = (F_1, F_2, F_3)$. It then follows immediately that $\partial f/\partial z = F_3$. Permuting x, y, and z, we can similarly show $\partial f/\partial x = F_1$ and $\partial f/\partial y = F_2$; that is $\nabla f = \mathbf{F}$. Third, (*iii*) implies (*iv*) because, as proved in Section 3.4,

$$\nabla \times \nabla f = \mathbf{0}$$

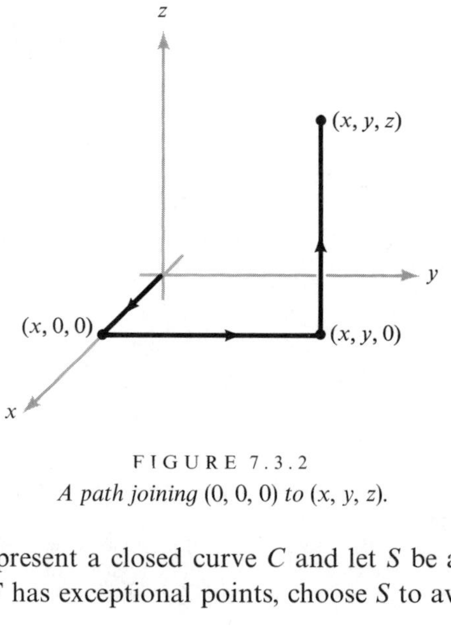

FIGURE 7.3.2
A path joining (0, 0, 0) *to* (x, y, z).

Finally, let $\boldsymbol{\sigma}$ represent a closed curve C and let S be any surface whose boundary is $\boldsymbol{\sigma}$ (if $\mathbf{F}$ has exceptional points, choose S to avoid them). Figure

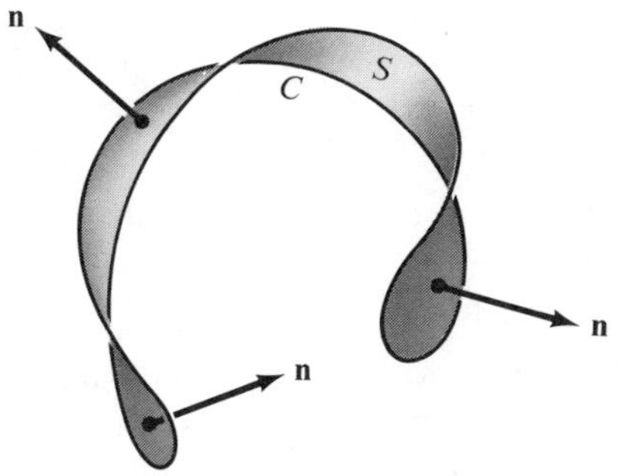

FIGURE 7.3.3
A surface S spanning a curve C.

7.3.3 indicates that we can probably always find such a surface; however, a formal proof of this would require the development of many more sophisticated mathematical ideas than we can present here. By Stokes' Theorem

$$\int_C \mathbf{F} \cdot d\mathbf{s} = \int_\sigma \mathbf{F} \cdot d\mathbf{s} = \int_S (\nabla \times \mathbf{F}) \cdot \mathbf{n}\, dS = \int_S (\text{curl } \mathbf{F}) \cdot \mathbf{n}\, dS$$

Since $\nabla \times \mathbf{F} = \mathbf{0}$, this integral vanishes, so that $(iv) \Rightarrow (i)$. ■

There are several useful physical interpretations of $\int_C \mathbf{F} \cdot d\mathbf{s}$. We have already seen that one is the work done by $\mathbf{F}$ in moving a particle along C. A second interpretation is the notion of circulation, which we encountered at the end of the last section. In this case we think of $\mathbf{F}$ as the velocity field of a fluid; that is, to each point P in space, $\mathbf{F}$ assigns the velocity vector of the fluid at P. Take C to be a closed curve, and let $\Delta\mathbf{s}$ be a small directed chord of C. Then $\mathbf{F} \cdot \Delta\mathbf{s}$ is approximately the tangential component of $\mathbf{F}$ times $\|\Delta\mathbf{s}\|$. The integral $\int_C \mathbf{F} \cdot d\mathbf{s}$ is the net tangential component around C. This means that a small paddle wheel placed in the fluid would rotate if the circulation of the fluid were nonzero, $\int_C \mathbf{F} \cdot d\mathbf{s} \neq 0$ (see Figure 7.3.4). Thus we often

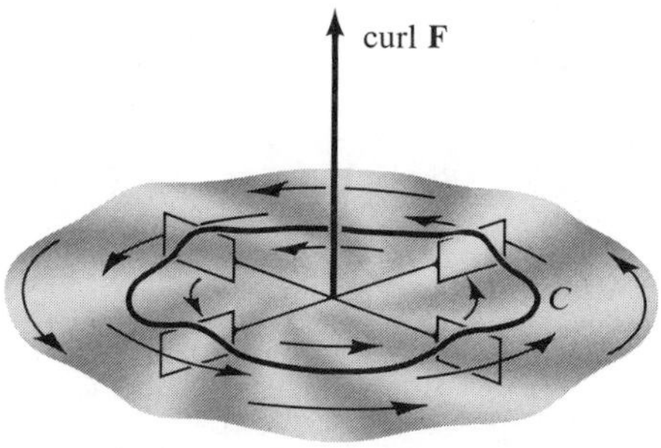

FIGURE 7.3.4
$\int_C \mathbf{F} \cdot d\mathbf{S} \neq 0$ *implies that a paddle wheel in a fluid with velocity field* $\mathbf{F}$ *will rotate around its axis.*

speak of the line integral

$$\int_C \mathbf{F} \cdot d\mathbf{s}$$

as being the *circulation* of **F** around C.

There is a similar interpretation in electromagnetic theory: If **F** represents an electric field, then a current will flow around a loop C if $\int_C \mathbf{F} \cdot d\mathbf{s} \neq 0$.

By Theorem 7, a field **F** has no circulation if and only if curl $\mathbf{F} = \nabla \times \mathbf{F} = \mathbf{0}$. Hence, a vector field **F** with curl $\mathbf{F} = \mathbf{0}$ is called *irrotational*. We have therefore proved that a vector field in $\mathbb{R}^3$ is irrotational if and only if it is a gradient field for some function, that is, if and only if $\mathbf{F} = \nabla f$. The function f is called a *potential* for **F**.

EXAMPLE 1. Consider the vector field **F** on $\mathbb{R}^3$ defined by

$$\mathbf{F}(x, y, z) = y\mathbf{i} + (z \cos yz + x)\mathbf{j} + (y \cos yz)\mathbf{k}$$

Show that **F** is irrotational and find a scalar potential for **F**.

We compute $\nabla \times \mathbf{F}$:

$$\begin{aligned}
\nabla \times \mathbf{F} &= \begin{vmatrix} \mathbf{i} & \mathbf{j} & \mathbf{k} \\ \dfrac{\partial}{\partial x} & \dfrac{\partial}{\partial y} & \dfrac{\partial}{\partial z} \\ y & x + z \cos yz & y \cos yz \end{vmatrix} \\
&= (\cos yz - yz \sin yz - \cos yz + yz \sin yz)\mathbf{i} \\
&\quad + (0 - 0)\mathbf{j} + (1 - 1)\mathbf{k} \\
&= 0\mathbf{i} + 0\mathbf{j} + 0\mathbf{k} = \mathbf{0}
\end{aligned}$$

so **F** is irrotational. We can find a scalar potential in several ways.

Method 1. By the technique used to prove that *(ii)* implies *(iii)* in Theorem 7, we may set

$$\begin{aligned}
f(x, y, z) &= \int_0^x F_1(t, 0, 0)\, dt + \int_0^y F_2(x, t, 0)\, dt + \int_0^z F_3(x, y, t)\, dt \\
&= \int_0^x 0\, dt + \int_0^y x\, dt + \int_0^z y \cos yt\, dt \\
&= 0 + xy + \sin yz = xy + \sin yz
\end{aligned}$$

One easily verifies that $\nabla f = \mathbf{F}$ as required:

$$\nabla f = \frac{\partial f}{\partial x}\mathbf{i} + \frac{\partial f}{\partial y}\mathbf{j} + \frac{\partial f}{\partial z}\mathbf{k} = y\mathbf{i} + (x + z \cos yz)\mathbf{j} + (y \cos yz)\mathbf{k}$$

Method 2. Because we know that f exists, we know that we can solve the equations

$$\frac{\partial f}{\partial x} = y$$

$$\frac{\partial f}{\partial y} = x + z \cos yz$$

$$\frac{\partial f}{\partial z} = y \cos yz$$

for $f(x, y, z)$. These are equivalent to the simultaneous equations
(a) $f(x, y, z) = xy + h_1(y, z)$
(b) $f(x, y, z) = \sin yz + xy + h_2(x, z)$
(c) $f(x, y, z) = \sin yz + h_3(x, y)$
for functions h_1, h_2, h_3 independent of x, y, and z (respectively). When $h_1(y, z) = \sin yz$, $h_2(x, z) = 0$, and $h_3(x, y) = xy$ the three equations agree, and so yield at potential for $\mathbf{F}$. However, we have only guessed at the values of h_1, h_2, and h_3. To derive f more systematically we note that since $f(x, y, z) = xy + h_1(y, z)$ and $\partial f/\partial z = y \cos yz$, we find that

$$\frac{\partial h_1(y, z)}{\partial z} = y \cos yz$$

or

$$h_1(y, z) = \int y \cos yz \, dz + g(y) = \sin yz + g(y)$$

Therefore, plugging this back into (a) we get

$$f(x, y, z) = xy + \sin yz + g(y)$$

but by (b)

$$g(y) = h_2(x, z)$$

Since the right side of this equation is a function of x and z, and the left side is a function of y alone, we conclude that they must equal some constant C. Thus

$$f(x, y, z) = xy + \sin yz + C$$

and we have determined f up to a constant.

EXAMPLE 2. A mass M at the origin in $\mathbb{R}^3$ exerts a force on a unit mass located at $\mathbf{r} = (x, y, z)$ with magnitude GM/r^2 and directed toward the origin. Here G is the gravitational constant, which depends on the units of measurement, and $r = \|\mathbf{r}\| = \sqrt{x^2 + y^2 + z^2}$. If we remember that $-\mathbf{r}/r$

is a unit vector directed towards the origin, then we can write the force field as

$$\mathbf{F}(x, y, z) = -\frac{GM\mathbf{r}}{r^3}$$

We shall show that $\mathbf{F}$ is irrotational and we shall find a scalar potential for $\mathbf{F}$. Notice that $\mathbf{F}$ is not defined at the origin, but Theorem 7 still applies since it allows an exceptional point.

First let us verify that $\nabla \times \mathbf{F} = \mathbf{0}$. Referring to Formula 11 of the table in Section 3.5, we get

$$\nabla \times \mathbf{F} = -GM\left\{\nabla\left(\frac{1}{r^3}\right) \times \mathbf{r} + \frac{1}{r^3}\nabla \times \mathbf{r}\right\}$$

But $\nabla(1/r^3) = -3\mathbf{r}/r^5$ (see Exercise 8, Section 3.5) so the first term vanishes, since $\mathbf{r} \times \mathbf{r} = \mathbf{0}$. The second term vanishes because

$$\nabla \times \mathbf{r} = \begin{vmatrix} \mathbf{i} & \mathbf{j} & \mathbf{k} \\ \dfrac{\partial}{\partial x} & \dfrac{\partial}{\partial y} & \dfrac{\partial}{\partial z} \\ x & y & z \end{vmatrix} = \left(\frac{\partial z}{\partial y} - \frac{\partial y}{\partial z}\right)\mathbf{i} + \left(\frac{\partial x}{\partial z} - \frac{\partial z}{\partial x}\right)\mathbf{j} + \left(\frac{\partial y}{\partial x} - \frac{\partial x}{\partial y}\right)\mathbf{k} = \mathbf{0}$$

Hence $\nabla \times \mathbf{F} = \mathbf{0}$ (for $\mathbf{r} \neq \mathbf{0}$).

If we recall the formula $\nabla(r^n) = nr^{n-2}\mathbf{r}$ (Exercise 8, Section 3.5) then we can read off a scalar potential for $\mathbf{F}$ by inspection. We have $\mathbf{F} = -\nabla\phi$, where $\phi(x, y, z) = -GM/r$ is called the *gravitational potential energy*.

By Theorem 3 of Section 6.2, the work done by $\mathbf{F}$ in moving a unit mass particle from a point P_1 to P_2 is given by

$$\phi(P_1) - \phi(P_2) = GM\left(\frac{1}{r_2} - \frac{1}{r_1}\right)$$

where r_1 is the radial distance of P_1 from the origin, with r_2 similarly defined.

By the same proof, Theorem 7 is also true for smooth vector fields $\mathbf{F}$ on $\mathbb{R}^2$. However, in this case $\mathbf{F}$ cannot have any exceptional points; that is, $\mathbf{F}$ must be smooth everywhere (see Exercise 12).

If $\mathbf{F} = P\mathbf{i} + Q\mathbf{j}$ then

$$\nabla \times \mathbf{F} = \left(\frac{\partial Q}{\partial x} - \frac{\partial P}{\partial y}\right)\mathbf{k}$$

and so the condition $\nabla \times \mathbf{F} = \mathbf{0}$ reduces to

$$\frac{\partial P}{\partial y} = \frac{\partial Q}{\partial x}$$

Thus we have:

COROLLARY. *If* **F** *is a* C^1 *vector field on* $\mathbb{R}^2$ *of the form* $P\mathbf{i} + Q\mathbf{j}$ *with* $\partial P/\partial y = \partial Q/\partial x$, *then* $\mathbf{F} = \nabla f$ *for some* f *on* $\mathbb{R}^2$.

EXAMPLE 3. (a) Determine whether the vector field

$$\mathbf{F} = e^{xy}\mathbf{i} + e^{x+y}\mathbf{j}$$

is a gradient field.

Here $P(x, y) = e^{xy}$ and $Q(x, y) = e^{x+y}$, so we compute

$$\frac{\partial P}{\partial y} = xe^{xy}, \qquad \frac{\partial Q}{\partial x} = e^{x+y}$$

These are not equal so **F** cannot have a potential function.

(b) Repeat part (a) for

$$\mathbf{F} = (2x \cos y)\mathbf{i} - (x^2 \sin y)\mathbf{j}$$

In this case, we find

$$\frac{\partial P}{\partial y} = -2x \sin y = \frac{\partial Q}{\partial x}$$

and so **F** has a potential function f. To compute f we solve the equations

$$\frac{\partial f}{\partial x} = 2x \cos y, \qquad \frac{\partial f}{\partial y} = -x^2 \sin y$$

Thus

$$f(x, y) = x^2 \cos y + h_1(y)$$

and

$$f(x, y) = x^2 \cos y + h_2(x)$$

If $h_1 = h_2 = 0$ both equations are satisfied, and we find that $f(x, y) = x^2 \cos y$ is a potential for **F**.

EXAMPLE 4. Let $\boldsymbol{\sigma}: [1, 2] \to \mathbb{R}^2$ be given by

$$x = e^{t-1}, \qquad y = \sin(\pi/t)$$

Compute the integral

$$\int_{\boldsymbol{\sigma}} \mathbf{F} \cdot d\mathbf{s} = \int_{\boldsymbol{\sigma}} 2x \cos y \, dx - x^2 \sin y \, dy$$

where $\mathbf{F} = (2x \cos y)\mathbf{i} - (x^2 \sin y)\mathbf{j}$.

We have $\boldsymbol{\sigma}(1) = (1, 0)$ and $\boldsymbol{\sigma}(2) = (e, 1)$. Since $\partial(2x \cos y)/\partial y = \partial(-x^2 \sin y)/\partial x$, **F** is irrotational and hence a gradient vector field (as we saw in Example 3). Thus by Theorem 7 we can replace $\boldsymbol{\sigma}$ by any piecewise C^1 curve having the same endpoints, in particular by the polygonal path from

(1, 0) to $(e, 0)$ to $(e, 1)$. So the line integral must be equal to

$$\int_{\boldsymbol{\sigma}} \mathbf{F} \cdot d\mathbf{s} = \int_1^e 2t \cos 0 \, dt + \int_0^1 -e^2 \sin t \, dt$$
$$= (e^2 - 1) + e^2(\cos 1 - 1) = e^2 \cos 1 - 1$$

Alternatively, using Theorem 3 of Section 6.2 we have

$$\int_{\boldsymbol{\sigma}} 2x \cos y \, dx - x^2 \sin y \, dy = \int_{\boldsymbol{\sigma}} \nabla f \cdot d\mathbf{s}$$
$$= f(\boldsymbol{\sigma}(2)) - f(\boldsymbol{\sigma}(1)) = e^2 \cos 1 - 1$$

since $f(x, y) = x^2 \cos y$ is a potential function for $\mathbf{F}$.

Evidently this technique is simpler than computing the integral directly.

We conclude this section with a theorem that is quite similar in spirit to Theorem 7. Theorem 7 was motivated partly as a converse to the result that curl $\nabla f = \mathbf{0}$ for any C^1 function $f\colon \mathbb{R}^3 \to \mathbb{R}$—or, if curl $\mathbf{F} = \mathbf{0}$ then $\mathbf{F} = \nabla f$. We also know (Formula 10 in the table in Section 3.5) that div(curl $\mathbf{G}$) = 0 for any C^2 vector field $\mathbf{G}$. We may ask about the converse statement: If div $\mathbf{F} = 0$, is $\mathbf{F}$ the curl of a vector field $\mathbf{G}$? The following theorem answers this in the affirmative.

THEOREM 8. *If* $\mathbf{F}$ *is a* C^1 *vector field on* $\mathbb{R}^3$ *with* div $\mathbf{F} = 0$, *then there exists a* C^1 *vector field* $\mathbf{G}$ *with* $\mathbf{F} =$ curl $\mathbf{G}$.

The proof is outlined in Exercise, 14. We should warn the reader at this point that, unlike the $\mathbf{F}$ in Theorem 7, the vector field $\mathbf{F}$ in Theorem 8 is not allowed to have an exceptional point. For example, the gravitational force field $\mathbf{F} = -(GM\mathbf{r}/r^3)$ has the property that div $\mathbf{F} = 0$, and yet there is no $\mathbf{G}$ for which $\mathbf{F} =$ curl $\mathbf{G}$ (see Exercise 21). Theorem 8 does not apply because the gravitational force field $\mathbf{F}$ is not defined at $\mathbf{0} \in \mathbb{R}^3$.

EXERCISES

1. Show that any two potential functions for a vector field differ at most by a constant.
2. (a) Let $\mathbf{F}(x, y) = (xy, y^2)$ and let $\boldsymbol{\sigma}$ be the path $y = 2x^2$ joining (0, 0) to (1, 2) in $\mathbb{R}^2$. Evaluate $\int_{\boldsymbol{\sigma}} \mathbf{F} \cdot d\mathbf{s}$.
   (b) Does the integral in (a) depend on the path joining (0, 0) to (1, 2)?
3. Let $\mathbf{F}(x, y, z) = (2xyz + \sin x)\mathbf{i} + x^2z\mathbf{j} + x^2y\mathbf{k}$. Find a function f such that $\mathbf{F} = \nabla f$.
4. Evaluate $\int_{\boldsymbol{\sigma}} \mathbf{F} \cdot d\mathbf{s}$, where $\boldsymbol{\sigma}(t) = (\cos^5 t, \sin^3 t, t^4)$, $0 \le t \le \pi$, and $\mathbf{F}$ is as in Exercise 3.

5. What is the work done by the force $\mathbf{F} = -\mathbf{r}/\|\mathbf{r}\|^3$ in moving a particle from a point $\mathbf{r}_0 \in \mathbb{R}^3$ "to ∞", where $\mathbf{r}(x, y, z) = (x, y, z)$?

6. In Exercise 5, show that $\mathbf{F} = \nabla(1/r)$, $r \neq 0$, $r = \|\mathbf{r}\|$. In what sense is the integral of $\mathbf{F}$ independent of path?

7. Let $\mathbf{F}(x, y, z) = xy\mathbf{i} + y\mathbf{j} + z\mathbf{k}$. Can there exist a function f such that $\mathbf{F} = \nabla f$?

*8. Let $\mathbf{F} = F_1\mathbf{i} + F_2\mathbf{j} + F_3\mathbf{k}$ and suppose each F_k satisfies the homogeneity condition

$$F_k(tx, ty, tz) = tF_k(x, y, z), \qquad k = 1, 2, 3$$

Suppose also $\nabla \times \mathbf{F} = \mathbf{0}$. Prove that $\mathbf{F} = \nabla f$ where

$$2f(x, y, z) = xF_1(x, y, z) + yF_2(x, y, z) + zF_3(x, y, z)$$

(HINT: You might use Review Exercise 21, Chapter 2.)

9. Let $\mathbf{F}(x, y, z) = (e^x \sin y)\mathbf{i} + (e^x \cos y)\mathbf{j} + z^2\mathbf{k}$. Evaluate the integral $\int_{\boldsymbol{\sigma}} \mathbf{F} \cdot d\mathbf{s}$, where $\boldsymbol{\sigma}(t) = (\sqrt{t}, t^3, \exp\sqrt{t})$, $0 \leq t \leq 1$.

10. Let a fluid have the velocity field $\mathbf{F}(x, y, z) = xy\mathbf{i} + yz\mathbf{j} + xz\mathbf{k}$. What is the circulation around the unit circle? Interpret your answer.

11. The mass of the earth is approximately 6×10^{27} g and that of the sun is 330,000 times as much. The gravitational constant is 6.7×10^{-8} meters^2/second^2 gram. The distance of the earth from the sun is about 1.5×10^{12} cm. Compute, approximately, the work necessary to increase the distance of the earth from the sun by 1 cm.

12. (a) Show that $\int_C (x\,dy - y\,dx)/(x^2 + y^2) = 2\pi$, where C is the unit circle.
(b) Conclude that the associated vector field $(-y/(x^2 + y^2))\mathbf{i} + (x/(x^2 + y^2))\mathbf{j}$ is not a conservative field.
(c) Show, however, that $\partial P/\partial y = \partial Q/\partial x$. Does this contradict the corollary to Theorem 7? If not, why not?

13. Determine which of the following vector fields $\mathbf{F}$ in the plane is the gradient of a scalar function f. If such an f exists, find it.
(a) $\mathbf{F}(x, y) = x\mathbf{i} + y\mathbf{j}$
(b) $\mathbf{F}(x, y) = xy\mathbf{i} + xy\mathbf{j}$
(c) $\mathbf{F}(x, y) = (x^2 + y^2)\mathbf{i} + 2xy\mathbf{j}$

14. Prove Theorem 8. (HINT: Define $\mathbf{G} = G_1\mathbf{i} + G_2\mathbf{j} + G_3\mathbf{k}$ by

$$G_1(x, y, z) = \int_0^z F_2(x, y, t)\,dt - \int_0^y F_3(x, t, 0)\,dt$$

$$G_2(x, y, z) = -\int_0^z F_1(x, y, t)\,dt$$

and $G_3(x, y, z) = 0$.)

15. Is each of the following vector fields the curl of some other vector field? If so, find the vector field.
(a) $\mathbf{F} = x\mathbf{i} + y\mathbf{j} + z\mathbf{k}$
(b) $\mathbf{F} = (x^2 + 1)\mathbf{i} + (z - 2xy)\mathbf{j} + y\mathbf{k}$

16. Let $\mathbf{F} = xz\mathbf{i} - yz\mathbf{j} + y\mathbf{k}$. Verify that $\nabla \cdot \mathbf{F} = 0$. Find a $\mathbf{G}$ such that $\mathbf{F} = \nabla \times \mathbf{G}$.

17. Let $\mathbf{F} = (x \cos y)\mathbf{i} - (\sin y)\mathbf{j} + (\sin x)\mathbf{k}$. Find a $\mathbf{G}$ such that $\mathbf{F} = \nabla \times \mathbf{G}$.

18. By using different paths from (0, 0, 0) to (x, y, z), show that the function f defined in the proof of Theorem 7 for "(*ii*) implies (*iii*)" satisfies $\partial f/\partial x = F_1$ and $\partial f/\partial y = F_2$.

19. Let $\mathbf{F}$ be a vector field on $\mathbb{R}^3$ given by $\mathbf{F} = -y\mathbf{i} + x\mathbf{j}$.
   (a) Show that $\mathbf{F}$ is rotational, that is, $\mathbf{F}$ is not irrotational.
   (b) Suppose $\mathbf{F}$ represents the velocity vector field of a fluid. Show that if we place a cork in this fluid it will revolve in a plane parallel to the xy-plane, in a circular trajectory about the z-axis.
   (c) In what direction does the cork revolve?

*20. Let $\mathbf{G}$ be the vector field on $\mathbb{R}^3\backslash\{z\text{-axis}\}$ defined by

$$\mathbf{G} = \frac{-y}{x^2 + y^2}\mathbf{i} + \frac{x}{x^2 + y^2}\mathbf{j}$$

   (a) Show that $\mathbf{G}$ is irrotational.
   (b) Show that 19(b) holds for $\mathbf{G}$ also.
   (c) How can we resolve the fact that the trajectories of $\mathbf{F}$ and $\mathbf{G}$ are both the same (circular about the z-axis) yet $\mathbf{F}$ is rotational and $\mathbf{G}$ is not? (HINT: The property of being rotational is a local condition, that is, a property of the fluid in the neighborhood of a point.)

*21. Let $\mathbf{F} = -(GM\mathbf{r}/r^3)$ be the gravitational force field defined on $\mathbb{R}^3\backslash\{\mathbf{0}\}$.
   (a) Show that div $\mathbf{F} = 0$.
   (b) Show that $\mathbf{F} \neq$ curl $\mathbf{G}$ for any C^1 vector field $\mathbf{G}$ on $\mathbb{R}^3\backslash\{\mathbf{0}\}$.

7.4 GAUSS' THEOREM

Gauss' Theorem states that the flux of a vector field out of a closed surface equals the integral of the divergence of that vector field over the volume enclosed by the surface. The result parallels Stokes' Theorem and Green's Theorem in that it relates an integral over a closed geometrical object (curve or surface) to an integral over a contained region (surface or volume).

We shall begin by asking the reader to review the various regions in space that were introduced when we considered the volume integral; these regions are illustrated in Figure 5.6.3. As that figure indicates, the boundary of a region of type 1, 2, or 3 in $\mathbb{R}^3$ is a surface made up of a finite number (at most six, at least two) of surfaces that can be described as graphs of functions from $\mathbb{R}^2$ to $\mathbb{R}$. This kind of surface is called a *closed surface*; it has no boundary. The surfaces $S_1, S_1, \ldots, S_N$ composing such a closed surface are called its *faces*.

EXAMPLE 1. The cube in Figure 7.4.1 is a region of type 4 (recall that this means it is simultaneously of types 1, 2, 3), with six squares composing its boundary. The sphere is the boundary of a solid ball, which is also a region of type 4.

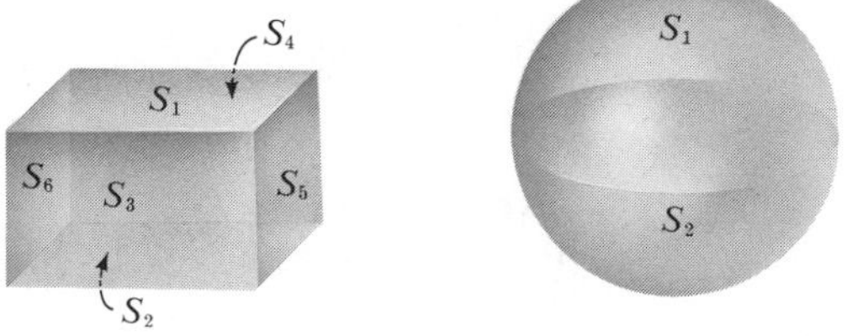

FIGURE 7.4.1
Regions of type 4 (a) and the surfaces S_i (b) composing their boundaries.

Closed surfaces can be oriented in two ways. The outward orientation makes the normal point outward into space, and the inward orientation makes the normal point into the bounded region (Figure 7.4.2).

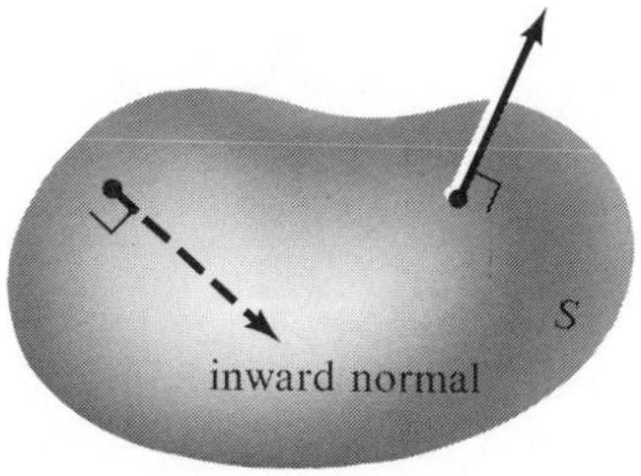

FIGURE 7.4.2
Two possible orientations for a closed surface.

Suppose S is a closed surface oriented in one of these two ways and $\mathbf{F}$ is a vector field on S. Then, as we defined above (p. 393),

$$\int_S \mathbf{F} \cdot d\mathbf{S} = \sum_i \int_{S_i} \mathbf{F} \cdot d\mathbf{S}$$

If S is given the outward orientation, the integral $\int_S \mathbf{F} \cdot d\mathbf{S}$ measures the total flux of $\mathbf{F}$ outwards across S. That is, if we think of $\mathbf{F}$ as the velocity field of a fluid, $\int_S \mathbf{F} \cdot d\mathbf{S}$ indicates the amount of fluid leaving the region bounded by S per unit time. If S is given the inward orientation, the integral $\int_S \mathbf{F} \cdot d\mathbf{S}$ measures the total flux of $\mathbf{F}$ inwards across S.

We recall another common way of writing these surface integrals, a way that explicitly specifies the orientation of S. Let the orientation of S be given by a unit normal vector $\mathbf{n}(x, y, z)$ at each point of S. Then we have the oriented integral

$$\int_S \mathbf{F} \cdot d\mathbf{S} = \int_S (\mathbf{F} \cdot \mathbf{n})\, dS$$

that is, the integral of the normal component of **F** over S. In the remainder of this section, if S is a closed surface enclosing a region Ω, we adopt the convention that $S = \partial\Omega$ is given the outward orientation, with outward unit normal $\mathbf{n}(x, y, z)$ at each point $(x, y, z) \in S$. Furthermore, we denote the surface with the opposite (inward) orientation by $\partial\Omega_{\text{op}}$. Then the associated unit normal direction for this orientation is $-\mathbf{n}$. Thus

$$\int_{\partial\Omega} \mathbf{F} \cdot d\mathbf{S} = \int_S (\mathbf{F} \cdot \mathbf{n})\, dS = -\int_S (\mathbf{F} \cdot (-\mathbf{n}))\, dS = -\int_{\partial\Omega_{\text{op}}} \mathbf{F} \cdot d\mathbf{S}$$

EXAMPLE 2. The unit cube Ω given by

$$0 \le x \le 1, \qquad 0 \le y \le 1, \qquad 0 \le z \le 1$$

is a region in space of type 4 (see Figure 7.4.3). We write the faces as

$$\begin{aligned}
&S_1: z = 1, && 0 \le x \le 1, && 0 \le y \le 1\\
&S_2: z = 0, && 0 \le x \le 1, && 0 \le y \le 1\\
&S_3: x = 1, && 0 \le y \le 1, && 0 \le z \le 1\\
&S_4: x = 0, && 0 \le y \le 1, && 0 \le z \le 1\\
&S_5: y = 1, && 0 \le x \le 1, && 0 \le z \le 1\\
&S_6: y = 0, && 0 \le x \le 1, && 0 \le z \le 1
\end{aligned}$$

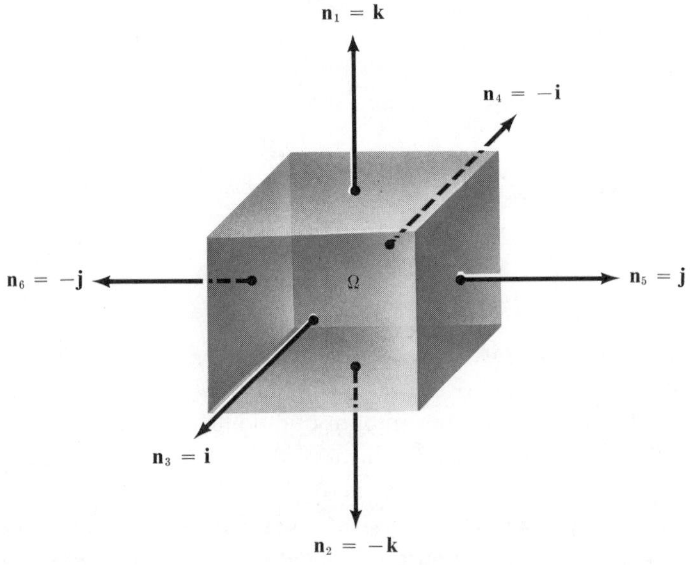

FIGURE 7.4.3
The outward orientation on the cube.

From Figure 7.4.3 we see that

$$\mathbf{n}_1 = \mathbf{k} = -\mathbf{n}_2$$

$$\mathbf{n}_3 = \mathbf{i} = -\mathbf{n}_4$$

$$\mathbf{n}_5 = \mathbf{j} = -\mathbf{n}_6,$$

so for a continuous vector field $\mathbf{F} = F_1\mathbf{i} + F_2\mathbf{j} + F_3\mathbf{k}$

$$\int_{\partial\Omega} \mathbf{F} \cdot d\mathbf{S} = \int_S \mathbf{F} \cdot \mathbf{n}\, dS = \int_{S_1} F_3\, dS - \int_{S_2} F_3\, dS + \int_{S_3} F_1\, dS - \int_{S_4} F_1\, dS + \int_{S_5} F_2\, dS - \int_{S_6} F_2\, dS$$

We have now come to the last of the three central theorems of this chapter. This theorem relates surface integrals to volume integrals; in words, the theorem states that if Ω is a region in $\mathbb{R}^3$, then the flux of a field $\mathbf{F}$ outward across the closed surface $\partial\Omega$ is equal to the integral of div $\mathbf{F}$ over Ω. (See p. 395 for the interpretation of surface integrals in terms of flux.)

THEOREM 9 (GAUSS' DIVERGENCE THEOREM). *Let Ω be a region in space of type 4. Denote by $\partial\Omega$ the oriented closed surface that bounds Ω. Let $\mathbf{F}$ be a smooth vector field defined on Ω. Then*

$$\int_\Omega (\nabla \cdot \mathbf{F})\, dV = \int_{\partial\Omega} \mathbf{F} \cdot d\mathbf{S}$$

or alternatively

$$\int_\Omega (\text{div}\,\mathbf{F})\, dV = \int_{\partial\Omega} (\mathbf{F} \cdot \mathbf{n})\, dS$$

Proof. If $\mathbf{F} = P\mathbf{i} + Q\mathbf{j} + R\mathbf{k}$, then by definition, div $\mathbf{F} = \partial P/\partial x + \partial Q/\partial y + \partial R/\partial z$, so we can write (using additivity of the volume integral)

$$\int_\Omega \text{div}\,\mathbf{F}\, dV = \int_\Omega \frac{\partial P}{\partial x}\, dV + \int_\Omega \frac{\partial Q}{\partial y}\, dV + \int_\Omega \frac{\partial R}{\partial z}\, dV$$

On the other hand, the surface integral in question is

$$\begin{aligned}\int_{\partial\Omega} \mathbf{F} \cdot \mathbf{n}\, dS &= \int_{\partial\Omega} (P\mathbf{i} + Q\mathbf{j} + R\mathbf{k}) \cdot \mathbf{n}\, dS \\ &= \int_{\partial\Omega} P\mathbf{i} \cdot \mathbf{n}\, dS + \int_{\partial\Omega} Q\mathbf{j} \cdot \mathbf{n}\, dS + \int_{\partial\Omega} R\mathbf{k} \cdot \mathbf{n}\, dS\end{aligned}$$

The theorem will follow if we establish the three equalities

$$\int_{\partial\Omega} P\mathbf{i} \cdot \mathbf{n}\, dS = \int_\Omega \frac{\partial P}{\partial x}\, dV \tag{1}$$

$$\int_{\partial\Omega} Q\mathbf{j} \cdot \mathbf{n}\, dS = \int_\Omega \frac{\partial Q}{\partial y}\, dV \tag{2}$$

$$\int_{\partial\Omega} R\mathbf{k} \cdot \mathbf{n}\, dS = \int_\Omega \frac{\partial R}{\partial z}\, dV \tag{3}$$

We shall prove (3); the other two equalities can be proved in an analogous fashion.

Since Ω is a region of type 1 (as well as of types 2 and 3), there exists a pair of functions

$$z = f_1(x, y), \qquad z = f_2(x, y)$$

with common domain an elementary region D in the xy-plane, such that Ω is the set of all points (x, y, z) satisfying

$$f_2(x, y) \le z \le f_1(x, y), \qquad (x, y) \in D$$

By formula (4) of Section 5.6, we have

$$\int_\Omega \frac{\partial R}{\partial z}\, dV = \int_D \left(\int_{z=f_2(x,y)}^{z=f_1(x,y)} \frac{\partial R}{\partial z}\, dz \right) dx\, dy$$

and so

$$\int_\Omega \frac{\partial R}{\partial z}\, dV = \int_D [R(x, y, f_1(x, y)) - R(x, y, f_2(x, y))]\, dx\, dy \tag{4}$$

The boundary of Ω is a closed surface whose top S_1 is the graph of $z = f_1(x, y)$, $(x, y) \in D$, and whose bottom S_2 is the graph of $z = f_2(x, y)$, $(x, y) \in D$. The four other sides of $\partial\Omega$ consists of surfaces S_3, S_4, S_5, and S_6 whose normals are always perpendicular to the z-axis. (See, for example, Figure 7.4.4. Note that some of the other four sides might be absent—for instance, if Ω is a solid ball, and $\partial\Omega$ is a sphere—but this will not affect the argument.) By definition

$$\int_{\partial\Omega} R\mathbf{k} \cdot \mathbf{n}\, dS = \int_{S_1} R\mathbf{k} \cdot \mathbf{n}_1\, dS + \int_{S_2} R\mathbf{k} \cdot \mathbf{n}_2\, dS + \sum_{i=3}^{6} \int_{S_i} R\mathbf{k} \cdot \mathbf{n}_i\, dS$$

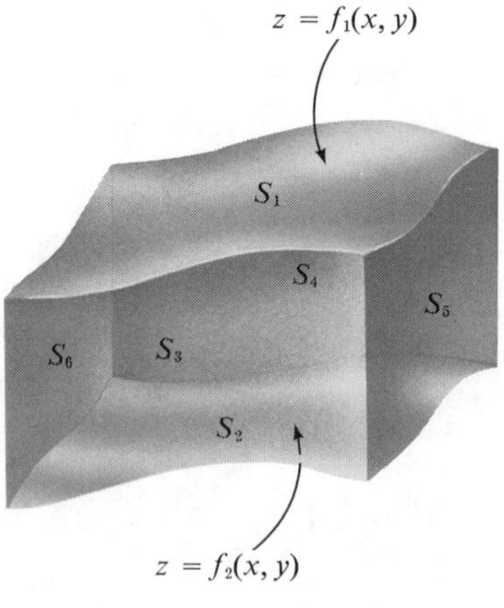

FIGURE 7.4.4

A region Ω of type 1 for which $\int_{\partial\Omega} R\mathbf{k} \cdot \mathbf{n}\, dS = \int_\Omega (\partial R/\partial z)\, dV$. The four sides of $\partial\Omega$ S_3, S_4, S_5, S_6 have normals perpendicular to the z-axis.

Since on each of S_3, S_4, S_5, and S_6 the normal $\mathbf{n}_i$ is perpendicular to $\mathbf{k}$, we have $\mathbf{k} \cdot \mathbf{n} = 0$ along these faces, and so the integral reduces to

$$\int_{\partial\Omega} R\mathbf{k} \cdot \mathbf{n}\, dS = \int_{S_1} R\mathbf{k} \cdot \mathbf{n}_1\, dS + \int_{S_2} R\mathbf{k} \cdot \mathbf{n}_2\, dS \tag{5}$$

The surface S_2 is defined by $z = f_2(x, y)$, so

$$\mathbf{n}_2 = \frac{\dfrac{\partial f_2}{\partial x}\mathbf{i} + \dfrac{\partial f_2}{\partial y}\mathbf{j} - \mathbf{k}}{\sqrt{\left(\dfrac{\partial f_2}{\partial x}\right)^2 + \left(\dfrac{\partial f_2}{\partial y}\right)^2 + 1}}$$

(since S_2 is the bottom portion of Ω, for $\mathbf{n}_2$ to point outward it must have a negative $\mathbf{k}$ component; see Example 2). Thus

$$\mathbf{n}_2 \cdot \mathbf{k} = \frac{-1}{\sqrt{\left(\dfrac{\partial f_2}{\partial x}\right)^2 + \left(\dfrac{\partial f_2}{\partial y}\right)^2 + 1}}$$

and

$$\begin{aligned}\int_{S_2} R(\mathbf{k} \cdot \mathbf{n}_2)\, dS &= \int_D R(x, y, f_2(x, y)) \left(\frac{-1}{\sqrt{\left(\dfrac{\partial f_2}{\partial x}\right)^2 + \left(\dfrac{\partial f_2}{\partial y}\right)^2 + 1}}\right) \times \\ &\qquad \left(\sqrt{\left(\dfrac{\partial f_2}{\partial x}\right)^2 + \left(\dfrac{\partial f_2}{\partial y}\right)^2 + 1}\right) dA \\ &= -\int_D R(x, y, f_2(x, y))\, dx\, dy \end{aligned} \tag{6}$$

This equation also follows from formula (4), Section 6.5.

Similarly, on the top face S_1 we have

$$\mathbf{k} \cdot \mathbf{n}_1 = \frac{1}{\sqrt{\left(\dfrac{\partial f_1}{\partial x}\right)^2 + \left(\dfrac{\partial f_1}{\partial y}\right) + 1}}$$

and so

$$\int_{S_1} R(\mathbf{k} \cdot \mathbf{n}_1)\, dS = \int_D R(x, y, f_1(x, y))\, dx\, dy \tag{7}$$

Substituting (6) and (7) into equation (5) and then comparing with (4) we obtain

$$\int_\Omega \frac{\partial R}{\partial z}\, dV = \int_{\partial\Omega} R(\mathbf{k} \cdot \mathbf{n})\, dS$$

The remaining equalities (1) and (2) can be established in the same way to complete the proof. ■

The reader should note that the proof is similar to that of Green's Theorem. By the procedure used in Exercise 8 of Section 7.1, we can extend Gauss'

Theorem to any region that can be broken up into subregions of type 4. This includes all regions of interest to us here. As an example, consider the region between two closed surfaces, one inside the other. The surface of this region consists of two pieces oriented as shown in Figure 7.4.5. We shall apply the Divergence Theorem to such a region when we prove Gauss' law in Theorem 10 below.

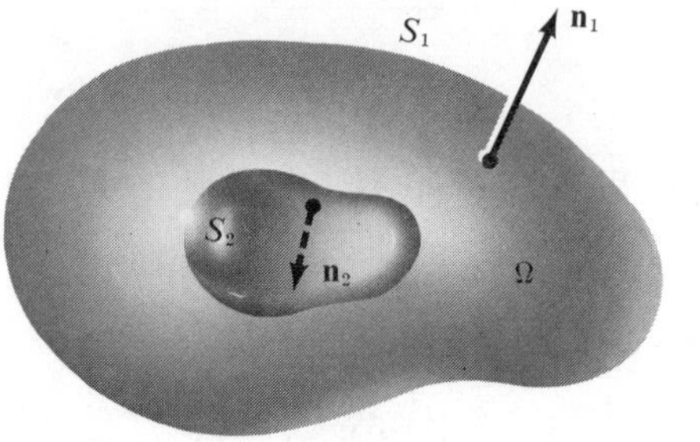

FIGURE 7.4.5
A more general region to which Gauss' Theorem applies.

EXAMPLE 3. Consider $\mathbf{F} = 2x\mathbf{i} + y^2\mathbf{j} + z^2\mathbf{k}$. Let S be the unit sphere $x^2 + y^2 + z^2 = 1$. Evaluate $\int_S \mathbf{F} \cdot \mathbf{n}\, dS$.

By Gauss' Theorem,

$$\int_\Omega (\text{div}\,\mathbf{F})\, dV = \int_S \mathbf{F} \cdot \mathbf{n}\, dS$$

where Ω is the ball bounded by the sphere. The integral on the left is

$$2\int_\Omega (1 + y + z)\, dV = 2\int_\Omega dV + 2\int_\Omega y\, dV + 2\int_\Omega z\, dV$$

By symmetry we can argue that $\int_\Omega y\, dV = \int_\Omega z\, dV = 0$ (for example, see Exercise 11, Section 5.6). Thus

$$2\int_\Omega (1 + y + z)\, dV = 2\int_\Omega dV = \frac{8\pi}{3}$$

(since the unit ball has volume $4\pi/3$; see Example 1, Section 5.6). Readers can convince themselves that direct computation of $\int_S \mathbf{F} \cdot \mathbf{n}\, dS$ proves unwieldy.

EXAMPLE 4. Use the Divergence Theorem to evaluate

$$\int_{\partial W} (x^2 + y + z)\, dS$$

where W is the solid ball $x^2 + y^2 + z^2 \leq 1$.

In order to apply Gauss' Divergence Theorem we must find some vector field

$$\mathbf{F} = F_1\mathbf{i} + F_2\mathbf{j} + F_3\mathbf{k}$$

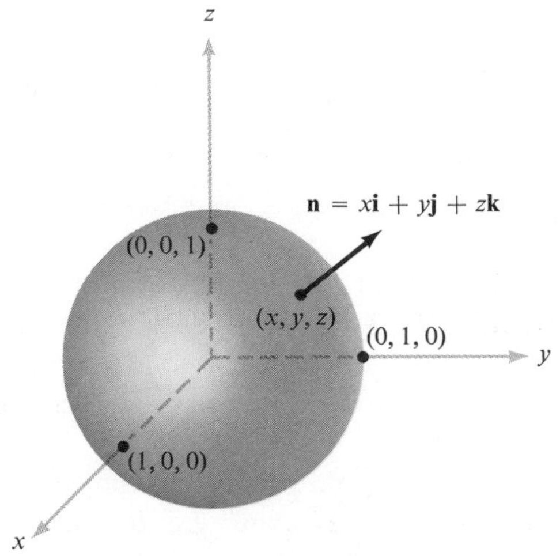

FIGURE 7.4.6
n *is the unit normal to* ∂W*, the boundary of the ball* W.

on W with

$$\mathbf{F}\cdot\mathbf{n} = x^2 + y + z$$

At any point $(x, y, z) \in \partial W$ the outward unit normal $\mathbf{n}$ to ∂W is

$$\mathbf{n} = x\mathbf{i} + y\mathbf{j} + z\mathbf{k}$$

since on ∂W, $x^2 + y^2 + z^2 = 1$, and the radius vector $\mathbf{r} = x\mathbf{i} + y\mathbf{j} + z\mathbf{k}$ is normal to the sphere ∂W (Figure 7.4.6). Therefore, if $\mathbf{F}$ is the desired vector field, then

$$\mathbf{F}\cdot\mathbf{n} = F_1 x + F_2 y + F_3 z$$

We set

$$F_1 x = x^2, \qquad F_2 y = y, \qquad F_3 z = z$$

and solve for F_1, F_2, and F_3 to find that

$$\mathbf{F} = x\mathbf{i} + \mathbf{j} + \mathbf{k}$$

Computing div $\mathbf{F}$ we get

$$\operatorname{div}\mathbf{F} = 1 + 0 + 0 = 1$$

Thus by Gauss' Divergence Theorem

$$\int_{\partial W} (x^2 + y + z)\, dS = \int_W dV = \text{volume}\,(W) = \tfrac{4}{3}\pi$$

The physical meaning of divergence is that at a point P, div $\mathbf{F}(P)$ is the rate of net outward flux at P per unit volume. This follows from Gauss' Theorem and the Mean Value Theorem for Integrals (as well as from the optional discussion in Section 3.4): If Ω_ρ is a ball in $\mathbb{R}^3$ of radius ρ centered at P, then there is a point $Q \in \Omega_\rho$ such that

$$\int_{\partial\Omega_\rho} \mathbf{F} \cdot \mathbf{n}\, dS = \int_{\Omega_\rho} \operatorname{div} \mathbf{F}\, dV = \operatorname{div} \mathbf{F}(Q) \cdot \text{volume}\,(\Omega_\rho)$$

and so

$$\operatorname{div} \mathbf{F}(P) = \underset{\rho\to 0}{\text{limit}} \operatorname{div} \mathbf{F}(Q) = \underset{\rho\to 0}{\text{limit}} \frac{1}{V(\Omega_\rho)} \int_{\partial\Omega_\rho} \mathbf{F} \cdot \mathbf{n}\, dS$$

This is analogous to the limit formulation of curl given at the end of Section 7.2. Thus, if $\operatorname{div} \mathbf{F}(P) > 0$, we consider P to be a *source*, for there is a net outwards flow near P. If $\operatorname{div} \mathbf{F}(P) < 0$, P is called a *sink* for $\mathbf{F}$.

A C^1 vector field $\mathbf{F}$ defined on $\mathbb{R}^3$ is called *divergence free* if $\operatorname{div} \mathbf{F} = 0$. If $\mathbf{F}$ is divergence free, we have $\int_S \mathbf{F} \cdot d\mathbf{S} = 0$ for all closed surfaces S. The converse can also be demonstrated readily using Gauss' Theorem: If $\int_S \mathbf{F} \cdot d\mathbf{S} = 0$ for all closed surfaces S, then $\mathbf{F}$ is divergence free. If $\mathbf{F}$ is divergence free, we thus see that the flux of $\mathbf{F}$ across any closed surface S is 0, so if $\mathbf{F}$ is the velocity field of a fluid, the net amount of fluid to flow out of any region will be 0. Thus, exactly as much fluid must flow into the region as flows out (in unit time). A fluid with this property is therefore called *incompressible*. (Additional justification for this terminology is given in Exercise 18.)

EXAMPLE 5. Evaluate $\int_S \mathbf{F} \cdot d\mathbf{S}$, where $\mathbf{F}(x, y, z) = xy^2\mathbf{i} + x^2y\mathbf{j} + y\mathbf{k}$ and S is the surface of the cylinder $x^2 + y^2 = 1$, $-1 < z < 1$, and $x^2 + y^2 \le 1$ when $z = \pm 1$. Interpret physically.

One can compute this integral directly but, as in many other cases, it is easier to use the Divergence Theorem.

Now S is the boundary of the region Ω given by $x^2 + y^2 \le 1$, $-1 \le z \le 1$. Thus $\int_S \mathbf{F} \cdot d\mathbf{S} = \int_\Omega (\operatorname{div} \mathbf{F})\, dV$. Moreover,

$$\int_\Omega (\operatorname{div} \mathbf{F})\, dV = \int_\Omega (x^2 + y^2)\, dx\, dy\, dz = \int_{-1}^{1} \left(\int_{x^2+y^2\le 1} (x^2 + y^2)\, dx\, dy \right) dz$$

$$= 2 \int_{x^2+y^2\le 1} (x^2 + y^2)\, dx\, dy$$

Before evaluating the double integral, we note that $\int_{\partial\Omega} \mathbf{F} \cdot \mathbf{n}\, dS = 2 \int_{x^2+y^2\le 1} (x^2 + y^2)\, dx\, dy > 0$. This means that $\int_{\partial\Omega} \mathbf{F} \cdot d\mathbf{S}$, the next flux of $\mathbf{F}$ out of the cylinder, is positive, which agrees with the fact that $\operatorname{div} \mathbf{F} = x^2 + y^2 \ge 0$ inside the cylinder.

We change variables to polar coordinates to evaluate the double integral:

$$x = r\cos\theta, \qquad y = r\sin\theta, \qquad 0 \le r \le 1, \qquad 0 \le \theta \le 2\pi$$

Hence, we have $\partial(x, y)/\partial(r, \theta) = r$ and $x^2 + y^2 = r^2$. Thus

$$\int_{x^2+y^2\le 1} (x^2 + y^2)\, dx\, dy = \int_0^{2\pi} \left(\int_0^1 r^3\, dr \right) d\theta = \tfrac{1}{2}\pi$$

Therefore, $\int_\Omega \operatorname{div} \mathbf{F}\, dV = \pi$.

As we remarked above, Gauss' Divergence Theorem can be applied to regions in space more general than those of type 4. To conclude this section, we shall use this observation to prove an important result.

THEOREM 10 (GAUSS' LAW). *Let M be a region in $\mathbb{R}^3$ of type 4. Then if $(0, 0, 0) \notin \partial M$ we have*

$$\int_{\partial M} \frac{\mathbf{r} \cdot \mathbf{n}}{r^3}\, dS = \begin{cases} 4\pi & \text{if } (0, 0, 0) \in M \\ 0 & \text{if } (0, 0, 0) \notin M \end{cases}$$

where

$$\mathbf{r}(x, y, z) = x\mathbf{i} + y\mathbf{j} + z\mathbf{k}$$

and

$$r(x, y, z) = \|\mathbf{r}(x, y, z)\| = \sqrt{x^2 + y^2 + z^2}$$

OPTIONAL

Proof. First suppose $(0, 0, 0) \notin M$. Then $\mathbf{r}/r^3$ is a C^1 vector field on M and ∂M, so by the Divergence Theorem

$$\int_{\partial M} \frac{\mathbf{r} \cdot \mathbf{n}}{r^3}\, dS = \int_M \nabla \cdot \left(\frac{\mathbf{r}}{r^3}\right) dV$$

But $\nabla \cdot (\mathbf{r}/r^3) = 0$ for $r \neq 0$, as the reader can easily verify (see Exercise 8, Section 3.5). Thus

$$\int_{\partial M} \frac{\mathbf{r} \cdot \mathbf{n}}{r^3}\, dS = 0$$

Now let us suppose $(0, 0, 0) \in M$. We can no longer use the above method, because $\mathbf{r}/r^3$ is not smooth on M, due to the singularity at $\mathbf{r} = (0, 0, 0)$. Since $(0, 0, 0) \in M$, there is an $\varepsilon > 0$ such that the ball N of radius ε centered at $(0, 0, 0)$ is contained completely inside M. Now let Ω be the region between M and N. Then Ω has boundary $\partial N \cup \partial M = S$. But the orientation on ∂N induced by the *outward* normal on Ω is the opposite of that obtained from N (see Figure 7.4.7). Now $\nabla \cdot (\mathbf{r}/r^3) = 0$ on Ω, so, by the (generalized) Divergence Theorem,

$$\int_S \frac{\mathbf{r} \cdot \mathbf{n}}{r^3}\, dS = \int_\Omega \nabla \cdot \left(\frac{\mathbf{r}}{r^3}\right) = 0$$

Since

$$\int_S \frac{\mathbf{r} \cdot \mathbf{n}}{r^3}\, dS = \int_{\partial M} \frac{\mathbf{r} \cdot \mathbf{n}}{r^3}\, dS + \int_{\partial N} \frac{\mathbf{r} \cdot \mathbf{n}}{r^3}\, dS$$

where $\mathbf{n}$ is the outward normal to S, we have

$$\int_{\partial M} \frac{\mathbf{r} \cdot \mathbf{n}}{r^3}\, dS = -\int_{\partial N} \frac{\mathbf{r} \cdot \mathbf{n}}{r^3}\, dS$$

OPTIONAL (*Continued*)

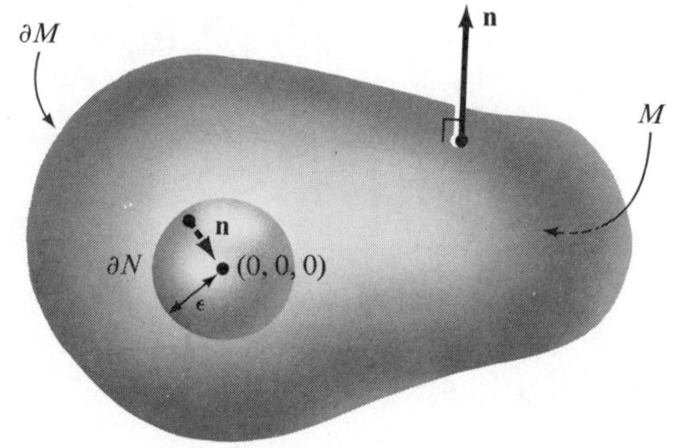

FIGURE 7.4.7
Induced outward orientation on S.

Now on ∂N, $\mathbf{n} = -\mathbf{r}/r$ and $r = \varepsilon$, since ∂N is a sphere of radius ε, so

$$-\int_{\partial N} \frac{\mathbf{r} \cdot \mathbf{n}}{r^3}\, dS = \int_{\partial N} \frac{\varepsilon^2}{\varepsilon^4}\, dS = \frac{1}{\varepsilon^2} \int_{\partial N} dS$$

But $\int_{\partial N} dS = 4\pi\varepsilon^2$, the surface area of the sphere of radius ε. This proves the result.

Gauss' law has the following physical interpretation. The potential due to a point charge Q at (0, 0, 0) is given by

$$\phi(x, y, z) = \frac{Q}{4\pi r} = \frac{Q}{4\pi\sqrt{x^2 + y^2 + z^2}}$$

and the corresponding electric field is

$$\mathbf{E} = -\nabla\phi = \frac{Q}{4\pi}\left(\frac{\mathbf{r}}{r^3}\right)$$

Thus Theorem 10 states that the total electric flux $\int_{\partial M} \mathbf{E} \cdot d\mathbf{S}$ (that is, the flux of $\mathbf{E}$ out of a closed surface ∂M) equals Q if the charge lies inside M and zero otherwise. (A generalization is given in Exercise 10.) Note that even if $(0, 0, 0) \notin M$, $\mathbf{E}$ will still be nonzero on M.

For a continuous charge distribution described by a charge density ρ, the field $\mathbf{E}$ is related to the density ρ by

$$\operatorname{div} \mathbf{E} = \nabla \cdot \mathbf{E} = \rho$$

Thus by Gauss' Theorem,

$$\int_S \mathbf{E} \cdot d\mathbf{S} = \int_\Omega \rho \, dV = Q$$

or the flux out of a surface is equal to the total charge inside.

EXERCISES

1. Let S be a closed surface. Use Gauss' Theorem to show that if $\mathbf{F}$ is a C^2 vector field, then $\int_S (\nabla \times \mathbf{F}) \cdot d\mathbf{S} = 0$. (Compare with Exercise 8 of Section 7.2.)
2. Let $\mathbf{F} = x^3\mathbf{i} + y^3\mathbf{j} + z^3\mathbf{k}$. Evaluate the surface integral of $\mathbf{F}$ over the unit sphere.
3. Evaluate $\int_{\partial\Omega} \mathbf{F} \cdot d\mathbf{S}$, where $\mathbf{F} = x\mathbf{i} + y\mathbf{j} - z\mathbf{k}$ and Ω is the unit cube (in the first octant). Perform the calculation directly and check by using the Divergence Theorem.
4. Repeat Exercise 3 for
   (a) $\mathbf{F} = \mathbf{i} + \mathbf{j} + \mathbf{k}$; and
   (b) $\mathbf{F} = x^2\mathbf{i} + y^2\mathbf{j} + z^2\mathbf{k}$.
5. Let $\mathbf{F} = y\mathbf{i} + z\mathbf{j} + xz\mathbf{k}$. Evaluate $\int_{\partial\Omega} \mathbf{F} \cdot d\mathbf{S}$ for each of the following regions Ω.
   (a) $x^2 + y^2 \leq z \leq 1$
   (b) $x^2 + y^2 \leq z \leq 1$ and $x \geq 0$
   (c) $x^2 + y^2 \leq z \leq 1$ and $x \leq 0$
6. Repeat Exercise 5 for $\mathbf{F} = (x - y)\mathbf{i} + (y - z)\mathbf{j} + (z - x)\mathbf{k}$.
7. Let S be the surface of a region Ω. Show that

$$\int_S \mathbf{r} \cdot \mathbf{n} \, dS = 3 \text{ volume}(\Omega)$$

   Attempt to explain this geometrically. (HINT: Assume $(0, 0, 0) \in \Omega$ and consider the skew cone with its vertex at $(0, 0, 0)$ with base ΔS and altitude $\|\mathbf{r}\|$. Its volume is $\frac{1}{3}(\Delta S) \cdot (\mathbf{r} \cdot \mathbf{n})$.)
8. Evaluate $\int_S \mathbf{F} \cdot \mathbf{n} \, dS$, where $\mathbf{F} = 3xy^2\mathbf{i} + 3x^2y\mathbf{j} + z^3\mathbf{k}$ and S is the surface of the unit sphere.
*9. Show $\int_\Omega (1/r^2) \, dx \, dy \, dz = \int_{\partial\Omega} (\mathbf{r} \cdot \mathbf{n}/r^2) \, dS$ where $\mathbf{r} = x\mathbf{i} + y\mathbf{j} + z\mathbf{k}$.
10. Fix vectors $\mathbf{v}_1, \ldots, \mathbf{v}_k \in \mathbb{R}^3$ and numbers ("charges") $q_1, \ldots, q_k$. Set $\phi(x, y, z) = \sum_{i=1}^k q_i/(4\pi\|\mathbf{r} - \mathbf{v}_i\|)$, where $\mathbf{r} = (x, y, z)$. Show that for a closed surface S, and $\mathbf{E} = -\nabla\phi$,

$$\int_S \mathbf{E} \cdot d\mathbf{S} = Q$$

    where Q is the total charge inside S. (Assume Gauss' law from Theorem 10 and that none of the charges are on S.)
11. Prove *Green's identities*

$$\int_{\partial\Omega} f\nabla g \cdot \mathbf{n} \, dS = \int_\Omega [f\nabla^2 g + \nabla f \cdot \nabla g] \, dV$$

$$\int_{\partial\Omega} (f\nabla g - g\nabla f) \cdot \mathbf{n} \, dS = \int_\Omega (f\nabla^2 g - g\nabla^2 f) \, dV$$

12. Suppose $\mathbf{F}$ satisfies div $\mathbf{F} = 0$ and curl $\mathbf{F} = \mathbf{0}$. Show that we can write $\mathbf{F} = \nabla f$, where $\nabla^2 f = 0$.

*13. Let ρ be a continuous function on $\mathbb{R}^3$ such that $\rho(\mathbf{q}) = 0$ except for $\mathbf{q}$ in some region Ω. Let $\mathbf{q} \in \Omega$ be denoted by $\mathbf{q} = (x, y, z)$. The *potential* of ρ is defined as the function

$$\phi(\mathbf{p}) = \int_\Omega \frac{\rho(\mathbf{q})}{4\pi\|\mathbf{p} - \mathbf{q}\|} dV(\mathbf{q})$$

where $\|\mathbf{p} - \mathbf{q}\|$ is the distance between $\mathbf{p}$ and $\mathbf{q}$.

(a) Using the method of Theorem 10, show that $\int_{\partial W} \nabla\phi \cdot \mathbf{n}\, dS = -\int_W \rho\, dV$ for those regions W that can be partitioned into a finite union of regions of type 4.

(b) Show that ϕ satisfies *Poisson's equation*

$$\nabla^2\phi = -\rho$$

(HINT: Use part (a)).

(Notice that if ρ is a charge density, then the integral defining ϕ may be thought of as the sum of the potentials at $\mathbf{p}$ caused by point charges distributed over Ω according to the density ρ.)

14. Suppose $\mathbf{F}$ is tangent to the closed surface S of a region Ω. Prove that

$$\int_\Omega (\operatorname{div} \mathbf{F})\, dV = 0$$

*15. Use Gauss' law and symmetry to prove that the electric field due to a charge Q evenly spread over the surface of a sphere is the same outside the surface as the field from a point charge Q located at the center of the sphere. What is the field inside the sphere?

*16. Reformulate Exercise 15 in terms of gravitational fields.

17. Show how Gauss' law can be used to solve part (b) of Exercise 21 in Section 7.3.

*18. (*Transport Theorem*). Let $\phi(\mathbf{x}, t)$ be the flow of the vector field $\mathbf{F}$ on $\mathbb{R}^3$ (see Section 3.4), and let $J(\mathbf{x}, t)$ be the Jacobian of the map $\phi_t \colon \mathbf{x} \mapsto \phi(\mathbf{x}, t)$ for t fixed.

(a) Using the proof of Theorem 3, Section 3.4, show that

$$\frac{\partial}{\partial t} J(\mathbf{x}, t) = [\operatorname{div} \mathbf{F}(\mathbf{x})] J(\mathbf{x}, t)$$

(b) Using the Change of Variables Theorem and (a), show that if $f(x, y, z, t)$ is a given function and $\Omega \subset \mathbb{R}^3$ is any region, then

$$\frac{d}{dt} \iiint_{\Omega_t} f(x, y, z, t)\, dx\, dy\, dz = \iiint_{\Omega_t} \left(\frac{Df}{Dt} + f \operatorname{div} \mathbf{F} \right) dx\, dy\, dz$$

(Transport equation)

where $\Omega_t = \phi_t(\Omega)$, which is the region moving with the flow, and where $Df/Dt = \partial f/\partial t + D_{\mathbf{x}} f \cdot \mathbf{F}$ is the material derivative (Exercise 7, Section 3.3).

(c) Taking $f = 1$ in (b), show that the following assertions are equivalent:

(*i*) $\operatorname{div} \mathbf{F} = 0$;

(*ii*) $\operatorname{volume}(\Omega_t) = \operatorname{volume}(\Omega)$; and

(*iii*) $J(\mathbf{x}, t) = 1$.

*19. Let $\phi, J, \mathbf{F}, f$ be as in Exercise 18. Prove the vector form of the Transport Theorem, namely,

$$\frac{d}{dt} \int_{\Omega_t} (f\mathbf{F})\, dx\, dy\, dz = \int_{\Omega_t} \left(\frac{\partial}{\partial t}(f\mathbf{F}) + \mathbf{F} \cdot \nabla(f\mathbf{F}) + (f\mathbf{F}) \operatorname{div} \mathbf{F} \right) dx\, dy\, dz$$

where $\mathbf{F} \cdot \nabla(f\mathbf{F})$ denotes the 3×3 derivative matrix $D(f\mathbf{F})$ operating on the column vector $\mathbf{F}$; in cartesian coordinates, $\mathbf{F} \cdot \nabla\mathbf{G}$ is the vector whose i^{th} component is

$$\sum_{j=1}^{3} F_j \frac{\partial G^i}{\partial x_j}$$

(labeling (x, y, z) as (x_1, x_2, x_3)).

OPTIONAL

7.5 APPLICATIONS TO PHYSICS AND DIFFERENTIAL EQUATIONS*

We can apply the concepts developed in this chapter to the formulation of some physical theories. Let us first discuss an important equation that is referred to as a *conservation* equation. For fluids, it expresses the conservation of mass, and for electromagnetic theory, the conservation of charge. We shall apply the equation to heat conduction and to electromagnetism.

Let $\mathbf{V}(t, x, y, z)$ be a C^1 vector field on $\mathbb{R}^3$ for each t and let $\rho(t, x, y, z)$ be a C^1 real-valued function. By the *law of conservation of mass* for $\mathbf{V}$ and ρ, we shall mean that the condition

$$\frac{d}{dt}\int_{\Omega} \rho \, dV = -\int_{\partial\Omega} \mathbf{J} \cdot \mathbf{n} \, dS$$

holds for all regions Ω in $\mathbb{R}^3$, where $\mathbf{J} = \rho\mathbf{V}$. See Figure 7.5.1.

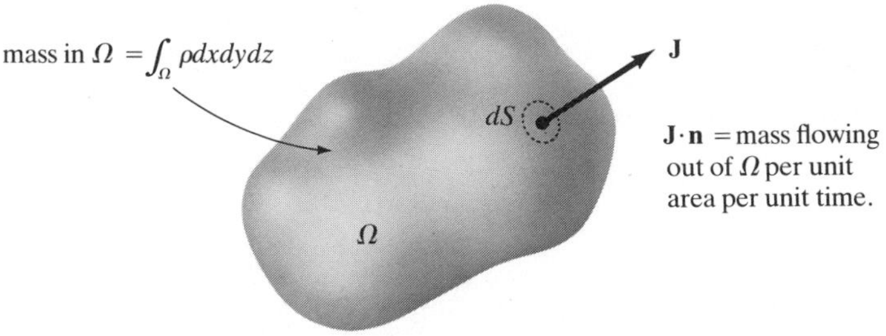

FIGURE 7.5.1

The rate of change of mass in Ω equals the rate at which mass crosses $\partial\Omega$.

* For additional examples the reader may profitably refer to H. M. Schey, *Div, Grad, Curl, and All That*, W. W. Norton, New York, 1973.

OPTIONAL (*Continued*)

If we think of ρ as a mass density (ρ could also be charge density), that is, the mass per unit volume, and $\mathbf{V}$ as the velocity field of a fluid, the condition just says that the rate of change of total mass in Ω equals the rate at which mass flows *into* Ω. Recall that $\int_{\partial\Omega} \mathbf{J} \cdot \mathbf{n}\, dS$ is called the flux of $\mathbf{J}$. We need the following result.

THEOREM 11. *For* $\mathbf{V}$ *and* ρ *defined on* $\mathbb{R}^3$, *the law of conservation of mass for* $\mathbf{V}$ *and* ρ *is equivalent to the condition*

$$\operatorname{div} \mathbf{J} + \frac{\partial \rho}{\partial t} = 0 \tag{1}$$

that is

$$\rho \operatorname{div} \mathbf{V} + \mathbf{V} \cdot \nabla\rho + \frac{\partial \rho}{\partial t} = 0 \tag{1'}$$

NOTE: Here, div $\mathbf{J}$ means that we compute div $\mathbf{J}$ for t held fixed, and $\partial\rho/\partial t$ means we differentiate ρ with respect to t for x, y, z fixed.

Proof. First, observe that $(d/dt)\int_\Omega \rho\, dx\, dy\, dz = \int_\Omega (\partial\rho/\partial t)\, dx\, dy\, dz$, and

$$\int_{\partial\Omega} \mathbf{J} \cdot \mathbf{n}\, dS = \int_\Omega \operatorname{div} \mathbf{J}\, d\mathbf{V}$$

by the Divergence Theorem. Thus conservation of mass is equivalent to the condition

$$\int_\Omega \left(\operatorname{div} \mathbf{J} + \frac{\partial \rho}{\partial t} \right) dx\, dy\, dz = 0$$

Since this is to hold for all regions Ω, this is equivalent to div $\mathbf{J} + \partial\rho/\partial t = 0$. ■

The equation div $\mathbf{J} + \partial\rho/\partial t = 0$ is called the *equation of continuity*. This is not the only equation governing fluid motion and it does not determine the motion of the fluid, but it is just one equation that must hold. Later in this section we shall obtain the additional equations needed to determine the flow. Using the Transport Theorem (Exercise 18, Section 7.4), the reader can check that the equation of continuity is equivalent to the condition

$$\frac{d}{dt} \int_{\Omega_t} \rho\, dx\, dy\, dz = 0$$

OPTIONAL (*Continued*)

which says that the mass of a region moving with the fluid is constant in time. (The notation Ω_t is explained in the same exercise.)

The fluids that the continuity equation governs can be compressible. If $\text{div}\,\mathbf{V} = 0$ (incompressible case) and ρ is constant, equation (1′) follows automatically. But in general, even for incompressible fluids the equation is not automatic, because ρ can depend on (x, y, z) and t. Thus, while $\text{div}\,\mathbf{V} = 0$ may hold, $\text{div}(\rho\mathbf{V}) \neq 0$ may still be true.

Next we discuss *Euler's equation for a perfect fluid.* Consider a homogeneous nonviscous fluid moving in space with a velocity field $\mathbf{V}$. The fluid being "perfect" means that if Ω is any portion of the fluid, forces of pressure act on the boundary of Ω along its normal. We assume that the force per unit area acting on $\partial\Omega$ is $-p\mathbf{n}$, where $p(x, y, z, t)$ is some function called the *pressure*. See Figure 7.5.2. Thus the total force acting on Ω is

$$\mathbf{F}_{\partial\Omega} = \text{Force} = -\int_{\partial\Omega} p\mathbf{n}\, dS$$

This is a *vector* quantity; the ith component of $\mathbf{F}_{\partial\Omega}$ is the integral of the ith

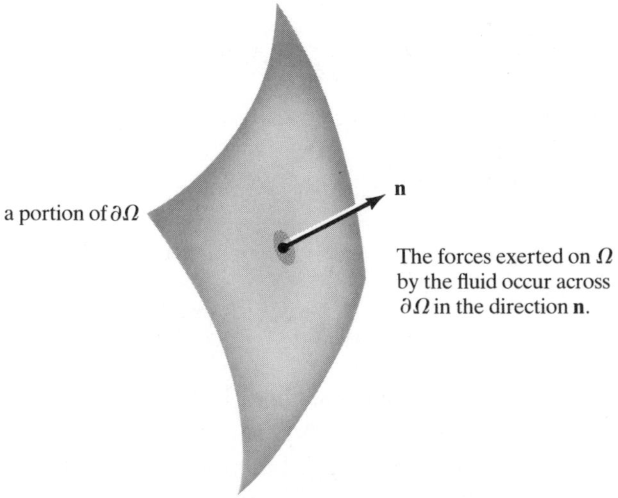
a portion of ∂Ω
n
The forces exerted on Ω
by the fluid occur across
∂Ω in the direction n.

FIGURE 7.5.2
The force acting on $\partial\Omega$ per unit area is $-p\mathbf{n}$.

OPTIONAL (*Continued*)

component of $p\mathbf{n}$ over the surface $\partial\Omega$ (this is therefore the surface integral of a real-valued function). If $\mathbf{e}$ is any fixed vector in space, we have

$$\mathbf{F}_{\partial\Omega} \cdot \mathbf{e} = -\int_{\partial\Omega} p\mathbf{e} \cdot \mathbf{n}\, dS$$

which is the integral of a scalar over $\partial\Omega$.

Notice that, by the divergence theorem and identity 8, p. 190,

$$\begin{aligned} \mathbf{e} \cdot \mathbf{F}_{\partial\Omega} &= -\int_{\Omega} \operatorname{div}(p\mathbf{e})\, dx\, dy\, dz \\ &= -\int_{\Omega} (\operatorname{grad} p) \cdot \mathbf{e}\, dx\, dy\, dz \end{aligned}$$

so

$$\mathbf{F}_{\partial\Omega} = -\int_{\Omega} \nabla p\, dx\, dy\, dz$$

Now we apply Newton's second law to a moving region Ω_t. Here, $\Omega_t = \phi_t(\Omega)$, where $\phi_t(\mathbf{x}) = \phi(\mathbf{x}, t)$ denotes the flow of $\mathbf{V}$. The rate of change of momentum of Ω_t equals the force acting on it:

$$\frac{d}{dt}\int_{\Omega_t} \rho\mathbf{V}\, dx\, dy\, dz = \mathbf{F}_{\partial\Omega_t} = -\int_{\Omega_t} \nabla p\, dx\, dy\, dz$$

We apply the vector form of the Transport Theorem to the left-hand side (Exercise19, Section 7.4) to get

$$\int_{\Omega_t} \left\{\frac{\partial}{\partial t}(\rho\mathbf{V}) + \mathbf{V} \cdot \nabla(\rho\mathbf{V}) + \rho\mathbf{V} \operatorname{div} \mathbf{V}\right\} dx\, dy\, dz = -\int_{\Omega_t} \nabla p\, dx\, dy\, dz$$

Since Ω_t is arbitrary, this is equivalent to

$$\frac{\partial}{\partial t}(\rho\mathbf{V}) + \mathbf{V} \cdot \nabla(\rho\mathbf{V}) + \rho\mathbf{V} \operatorname{div} \mathbf{V} = -\nabla p$$

Simplification using the equation of continuity (1′) gives

$$\rho\left(\frac{\partial\mathbf{V}}{\partial t} + \mathbf{V} \cdot . \nabla\mathbf{V}\right) = -\nabla p \qquad (2)$$

This is Euler's equation for a perfect fluid. For compressible fluids, p is a given function of ρ (for instance, for many gases, $p = A\,\rho^{\gamma}$ for constants A and γ). If the fluid is, on the other hand, incompressible, p is to be determined from the condition $\operatorname{div} \mathbf{V} = 0$. Equations (1) and (2) then govern the motion of the fluid completely. (An example that shows how complex particular solutions of (1) and (2) can be is given on the cover of this book.)

HISTORICAL NOTE

The equations describing the motion of a fluid were first derived by Leonhard Euler in 1755, in a paper entitled "General Principles of the Motion of Fluids." Euler did basic work in mechanics as well as in pure mathematics; he essentially began the subject of analytical mechanics (as opposed to the geometric methods used by Newton). He is responsible for the equations of a rigid body (such as a tumbling satellite) and the formulation of many basic equations of mechanics as minimum principles. Euler wrote the first comprehensive textbook on calculus and contributed to virtually all branches of mathematics. He wrote several books and hundreds of research papers after he became totally blind, and he was working on a new treatise on fluid mechanics at the time of his death in 1783. Euler's equations for a fluid were eventually modified to include viscous effects by Navier and Stokes; the resulting Navier–Stokes' equations are described in virtually every textbook on fluid mechanics. Stokes is, of course, also responsible for Stokes' theorem, one of the main results in this book.

We shall now turn our attention to the *heat equation*, one of the most important equations of applied mathematics. It has been, and remains, one of the prime motivations for the study of partial differential equations.

Let us argue intuitively. If $T(t, x, y, z)$ (a C^2 function) denotes the temperature in a body at time t, then ∇T represents the temperature gradient and heat "flows" with the vector field $-\nabla T = \mathbf{F}$. Note that ∇T points in the direction of increasing T (Chapter 3). Since heat flows from hot to cold, we have inserted a minus sign. The energy density, that is, the energy per unit volume, is $c\rho_0 T$, where c is a constant (specific heat) and ρ_0 is the mass density, assumed constant. (We accept these assertions from elementary physics.) The *energy flux vector* is $\mathbf{J} = \kappa\mathbf{F}$, where κ is a constant called the *conductivity*.

We propose that energy be conserved. Formally, this means that $\mathbf{J}$ and $\rho = c\rho_0 T$ should obey the law of conservation of mass with ρ playing the role of "mass"; that is,

$$\frac{d}{dt}\int_\Omega \rho\, dV = -\int_{\partial\Omega} \mathbf{J}\cdot\mathbf{n}\, dS$$

OPTIONAL (*Continued*)

By Theorem 11 this assertion is equivalent to

$$\operatorname{div} \mathbf{J} + \frac{\partial \rho}{\partial t} = 0$$

But $\operatorname{div} \mathbf{J} = \operatorname{div}(-\kappa \nabla T) = -\kappa \nabla^2 T$. (Recall that $\nabla^2 T = \partial^2 T/\partial x^2 + \partial^2 T/\partial y^2 + \partial^2 T/\partial z^2$ and ∇^2 is the Laplace operator.) Continuing, we have $\partial \rho/\partial t = \partial(c\rho_0 T)/\partial t = c\rho_0(\partial T/\partial t)$. Thus equation (1) becomes, in this case,

$$\frac{\partial T}{\partial t} = \frac{\kappa}{c\rho_0} \nabla^2 T = k \nabla^2 T \tag{3}$$

where $k = \kappa/c\rho_0$ is called the *diffusivity*. Equation (3) is the important heat equation.

Just as (1) and (2) govern the flow of an ideal fluid, (3) governs the conduction of heat, in the following sense. If $T(0, x, y, z)$ is a given initial temperature distribution, then a unique $T(t, x, y, z)$ is determined that satisfies equation (3). In other words the initial condition, at $t = 0$, gives us the result for $t > 0$. Notice that if T does not change with time (steady-state case) then we must have $\nabla^2 T = 0$ (Laplace's equation).

We shall now discuss *Maxwell's equations*, governing electromagnetic fields. The form of these equations depends on the physical units one is employing, and changing units introduces factors like 4π, the velocity of light, and so on. We shall choose the system in which Maxwell's equations are simplest.

Let $\mathbf{E}(t, x, y, z)$ and $\mathbf{H}(t, x, y, z)$ be C^1 functions of (t, x, y, z) that are vector fields for each t. They satisfy (by definition) Maxwell's equations with charge density $\rho(t, x, y, z)$ and current density $\mathbf{J}(t, x, y, z)$ when the following hold:

$$\nabla \cdot \mathbf{E} = \rho \qquad \text{(Gauss' law)} \tag{4}$$

$$\nabla \cdot \mathbf{H} = 0 \qquad \text{(no magnetic sources)} \tag{5}$$

$$\nabla \times \mathbf{E} + \frac{\partial \mathbf{H}}{\partial t} = \mathbf{0} \qquad \text{(Faraday's law)} \tag{6}$$

and

$$\nabla \times \mathbf{H} - \frac{\partial \mathbf{E}}{\partial t} = \mathbf{J} \qquad \text{(Ampère's law)} \tag{7}$$

Of these laws, (4) and (6) were discussed earlier in Sections 7.4 and 7.2 in integral form; historically, they arose in these forms as physically observed

OPTIONAL (*Continued*)

laws. Ampère's law was mentioned for a special case in Example 12, Section 6.2.

Physically, one interprets **E** as the *electric field* and **H** as the *magnetic field*. As time t progresses, these fields interact with each other and with any charges and currents that are present according to the above equations. For example, the propagation of electromagnetic waves in a vacuum is governed by these equations with $\mathbf{J} = \mathbf{0}$ and $\rho = 0$.

Since $\nabla \cdot \mathbf{H} = 0$ we can apply Theorem 8 of Section 7.3 to conclude that $\mathbf{H} = \nabla \times \mathbf{A}$ for some vector field **A**. (We are assuming that **H** is defined on all of $\mathbb{R}^3$ for each time t.) This vector field **A** is not unique, and we can equally well use $\mathbf{A}' = \mathbf{A} + \nabla f$ for any function $f(t, x, y, z)$, since $\nabla \times \nabla f = \mathbf{0}$. (This freedom in the choice of A is called *gauge freedom*.) For any such choice of **A**, we have by (6)

$$\mathbf{0} = \nabla \times \mathbf{E} + \frac{\partial \mathbf{H}}{\partial t} = \nabla \times \mathbf{E} + \frac{\partial}{\partial t} \nabla \times \mathbf{A}$$

$$= \nabla \times \mathbf{E} + \nabla \times \frac{\partial \mathbf{A}}{\partial t} = \nabla \times \left(\mathbf{E} + \frac{\partial \mathbf{A}}{\partial t}\right)$$

Hence applying Theorem 7, Section 7.3, there is a real-valued function ϕ on $\mathbb{R}^3$ such that

$$\mathbf{E} + \frac{\partial \mathbf{A}}{\partial t} = -\nabla \phi$$

Substituting this equation and $\mathbf{H} = \nabla \times \mathbf{A}$ into equation (7) and using the following identity from Table 3.1,

$$\nabla \times (\nabla \times \mathbf{A}) = \nabla(\nabla \cdot \mathbf{A}) - \nabla^2 \mathbf{A}$$

we get

$$\mathbf{J} = \nabla \times \mathbf{H} - \frac{\partial \mathbf{E}}{\partial t} = \nabla \times (\nabla \times \mathbf{A}) - \frac{\partial}{\partial t}\left(-\frac{\partial \mathbf{A}}{\partial t} - \nabla \phi\right)$$

$$= \nabla(\nabla \cdot \mathbf{A}) - \nabla^2 \mathbf{A} + \frac{\partial^2 \mathbf{A}}{\partial t^2} + \frac{\partial}{\partial t}(\nabla \phi)$$

Thus

$$\nabla^2 \mathbf{A} - \frac{\partial^2 \mathbf{A}}{\partial t^2} = -\mathbf{J} + \nabla(\nabla \cdot \mathbf{A}) + \frac{\partial}{\partial t}(\nabla \phi)$$

that is

$$\nabla^2 \mathbf{A} - \frac{\partial^2 \mathbf{A}}{\partial t^2} = -\mathbf{J} + \nabla\left(\nabla \cdot \mathbf{A} + \frac{\partial \phi}{\partial t}\right) \tag{8}$$

OPTIONAL (*Continued*)

Again using the equation $\mathbf{E} + \partial\mathbf{A}/\partial t = -\nabla\phi$ and the equation $\nabla \cdot \mathbf{E} = \rho$, we obtain

$$\rho = \nabla \cdot \mathbf{E} = \nabla \cdot \left(-\nabla\phi - \frac{\partial \mathbf{A}}{\partial t}\right) = -\nabla^2\phi - \frac{\partial(\nabla \cdot \mathbf{A})}{\partial t}$$

that is

$$\nabla^2\phi = -\rho - \frac{\partial(\nabla \cdot \mathbf{A})}{\partial t} \tag{9}$$

Now let us exploit the freedom in our choice of **A**. We impose the "condition"

$$\nabla \cdot \mathbf{A} + \frac{\partial\phi}{\partial t} = 0 \tag{10}$$

We must be sure we can do this. Supposing we have a given $\mathbf{A}_0$ and a corresponding ϕ_0, can we choose a new $\mathbf{A} = \mathbf{A}_0 + \nabla f$ and then a new ϕ such that $\nabla \cdot \mathbf{A} + \partial\phi/\partial t = 0$? With this new **A**, the new ϕ is $\phi_0 - \partial f/\partial t$; we leave verification as an exercise for the reader. Condition (10) on f then becomes

$$0 = \nabla \cdot (\mathbf{A}_0 + \nabla f) + \frac{\partial(\phi_0 - \partial f/\partial t)}{\partial t} = \nabla \cdot \mathbf{A}_0 + \nabla^2 f + \frac{\partial\phi_0}{\partial t} - \frac{\partial^2 f}{\partial t^2}$$

or

$$\nabla^2 f - \frac{\partial^2 f}{\partial t^2} = -\left(\nabla \cdot \mathbf{A}_0 + \frac{\partial\phi_0}{\partial t}\right) \tag{11}$$

Thus to be able to choose **A** and ϕ satisfying $\nabla \cdot \mathbf{A} + \partial\phi/\partial t = 0$, we must be able to solve equation (11) for f. One can indeed do this under general conditions, although we do not prove it here. Equation (11) is called the *inhomogeneous wave equation.*

If we accept that **A** and ϕ can be chosen to satisfy $\nabla \cdot \mathbf{A} + \partial\phi/\partial t = 0$, then the equations (8) and (9) for **A** and ϕ become

$$\nabla^2\mathbf{A} - \frac{\partial^2\mathbf{A}}{\partial t^2} = -\mathbf{J} \tag{8'}$$

$$\nabla^2\phi - \frac{\partial^2\phi}{\partial t^2} = -\rho \tag{9'}$$

Equation (9′) follows from (9) by substituting $-\partial\phi/\partial t$ for $\nabla \cdot \mathbf{A}$. Thus the wave equation appears again.

OPTIONAL (*Continued*)

Conversely, if **A** and ϕ satisfy the equations $\nabla \cdot \mathbf{A} + \partial\phi/\partial t = 0$, $\nabla^2\phi - \partial^2\phi/\partial t^2 = -\rho$, and $\nabla^2\mathbf{A} - \partial^2\mathbf{A}/\partial t^2 = -\mathbf{J}$, then $\mathbf{E} = -\nabla\phi - \partial\mathbf{A}/\partial t$ and $\mathbf{H} = \nabla \times \mathbf{A}$ satisfy Maxwell's equations. *This procedure then reduces Maxwell's equations to a study of the wave equation.**

This is fortunate because the solutions to the wave equation have been well studied (one learns how to solve it in most courses in differential equations). To indicate the wavelike nature of the solutions, for example, observe that for any function f

$$\phi(t, x, y, z) = f(x - t)$$

solves the wave equation $\nabla^2\phi - (\partial^2\phi/\partial t^2) = 0$. This solution just propagates the graph of f like a wave; thus one might conjecture that solutions of Maxwell's equations are wavelike in nature. Historically, this was Maxwell's great achievement, and it soon led to Hertz's discovery of radio waves.

Next we shall show briefly how vector analysis can be used to solve differential equations by a method called "potential theory" or "the Green's-function method." The presentation will be quite informal; the reader may consult the aforementioned references (see preceding footnote) for further information.

Suppose we wish to solve Poisson's equation $\nabla^2 u = \rho$ for $u(x, y, z)$ where $\rho(x, y, z)$ is a given function (this equation arises from Gauss' law if $\mathbf{E} = \nabla u$).

A function $G(\mathbf{x}, \mathbf{y})$ that has the properties

$$G(\mathbf{x}, \mathbf{y}) = G(\mathbf{y}, \mathbf{x}) \text{ and } \nabla^2 G(\mathbf{x}, \mathbf{y}) = \delta(\mathbf{x} - \mathbf{y}) \tag{12}$$

(in this expression **y** is held fixed), that is, which solves the differential equation with ρ replaced by δ, is called the *Green's function* for this differential equation. Here $\delta(\mathbf{x} - \mathbf{y})$ represents the Dirac delta function, "defined" by†

(*i*) $\delta(\mathbf{x} - \mathbf{y}) = 0$ for $\mathbf{x} \neq \mathbf{y}$ and

(*ii*) $\int_{\mathbb{R}^3} \delta(\mathbf{x} - \mathbf{y})\, d\mathbf{y} = 1$.

* There are variations on this procedure. For further details see, for example, G. F. D. Duff and D. Naylor, *Differential Equations of Applied Mathematics*, Wiley, New York, 1966, or books on electromagnetic theory, such as J. D. Jackson, *Classical Electrodynamics*, Wiley, New York, 1962.

† This is not a precise definition; nevertheless, it is enough here to assume that δ is a symbolic expression with the operational property (13). See the references in the preceding footnote for a more careful definition of δ.

OPTIONAL (*Continued*)

It has the following operational property that formally follows from (*i*) and (*ii*): For any continuous function $f(\mathbf{x})$

$$\int_{\mathbb{R}^3} f(\mathbf{y})\delta(\mathbf{x} - \mathbf{y})\, d\mathbf{y} = f(\mathbf{x}) \tag{13}$$

This is sometimes called the *sifting property* of δ.

THEOREM 12. *If* $G(\mathbf{x}, \mathbf{y})$ *satisfies the differential equation* $\nabla^2 u = \rho$ *with* ρ *replaced by* $\delta(\mathbf{x} - \mathbf{y})$, *then*

$$u(\mathbf{x}) = \int_{\mathbb{R}^3} G(\mathbf{x}, \mathbf{y})\rho(\mathbf{y})\, d\mathbf{y} \tag{14}$$

is a solution to $\nabla^2 u = \rho$.

Proof. To see this, note that

$$\begin{aligned}\nabla^2 \int_{\mathbb{R}^3} G(\mathbf{x}, \mathbf{y})\rho(\mathbf{y})\, d\mathbf{y} &= \int_{\mathbb{R}^3} (\nabla^2 G(\mathbf{x}, \mathbf{y}))\rho(\mathbf{y})\, d\mathbf{y} \\ &= \int_{\mathbb{R}^3} \delta(\mathbf{x} - \mathbf{y})\rho(\mathbf{y})\, d\mathbf{y} && \text{by (12)} \\ &= \rho(\mathbf{x}) && \text{by (13)} \quad \blacksquare\end{aligned}$$

The "function" $\rho(\mathbf{x}) = \delta(\mathbf{x})$ represents a unit charge concentrated at a single point (see conditions (*i*) and (*ii*) above). Thus $G(\mathbf{x}, \mathbf{y})$ *represents the potential at* **x** *due to a charge placed at* **y**.

We claim that equation (12) is satisfied if we choose

$$G(\mathbf{x}, \mathbf{y}) = -\frac{1}{4\pi\|\mathbf{x} - \mathbf{y}\|} \tag{15}$$

Clearly $G(\mathbf{x}, \mathbf{y}) = G(\mathbf{y}, \mathbf{x})$. To check the second part of (12), we must verify that $\nabla^2 G(\mathbf{x}, \mathbf{y})$ has the following two formal properties of the δ-function:

(*i*) $\nabla^2 G(\mathbf{x}, \mathbf{y}) = 0$ for $\mathbf{x} \neq \mathbf{y}$; and

(*ii*) $\int_{\mathbb{R}^3} \nabla^2 G(\mathbf{x}, \mathbf{y})\, d\mathbf{y} = 1$

Property (*i*) is true because the gradient of G is

$$\nabla G(\mathbf{x}, \mathbf{y}) = \frac{\mathbf{r}}{4\pi r^3}$$

where $\mathbf{r} = \mathbf{x} - \mathbf{y}$ is the vector from **y** to **x** and $r = \|\mathbf{r}\|$ (see Exercise 8, Section 3.5), and therefore for $r \neq 0$, $\nabla \cdot \nabla G(\mathbf{x}, \mathbf{y}) = 0$ (as in the afore-

OPTIONAL (*Continued*)

mentioned exercise). For (*ii*) let B be a ball about $\mathbf{x}$; by (*i*),

$$\int_{\mathbb{R}^3} \nabla^2 G(\mathbf{x}, \mathbf{y})\, d\mathbf{y} = \int_B \nabla^2 G(\mathbf{x}, \mathbf{y})\, d\mathbf{y}$$

This, in turn, equals

$$\int_{\partial B} \nabla G(\mathbf{x}, \mathbf{y}) \cdot \mathbf{n}\, dS$$

by Gauss' Theorem. Thus, by Theorem 10,

$$\int_{\partial B} \nabla G(\mathbf{x}, \mathbf{y}) \cdot \mathbf{n}\, dS = \int_{\partial B} \frac{\mathbf{r} \cdot \mathbf{n}}{4\pi r^3}\, dS = 1$$

which proves (*ii*).

Thus, *the solution of* $\nabla^2 u = \rho$ *is*

$$u(\mathbf{x}) = \int_{\mathbb{R}^3} \frac{-\rho(\mathbf{y})}{4\pi\|\mathbf{x} - \mathbf{y}\|}\, d\mathbf{y} \tag{16}$$

by Theorem 12.

In two dimensions, one can similarly show that

$$G(\mathbf{x}, \mathbf{y}) = \frac{1}{2\pi} \log\|\mathbf{x} - \mathbf{y}\| \tag{17}$$

so the solution of $\nabla^2 u = \rho$ is

$$u(\mathbf{x}) = \frac{1}{2\pi} \int_{\mathbb{R}^2} \rho(\mathbf{y}) \log\|\mathbf{x} - \mathbf{y}\|\, d\mathbf{y} \tag{18}$$

We now turn to the problem of using Green's functions to solve Poisson's equation in a bounded region with given boundary conditions. To do this, we need Green's first and second identities, which can be obtained from the divergence theorem. We start with the identity

$$\int_V \nabla \cdot \mathbf{F}\, dV = \int_S \mathbf{F} \cdot \mathbf{n}\, dS$$

where V is a region in space, S is its boundary, and $\mathbf{n}$ is the outward unit normal vector at any point on S. Replacing $\mathbf{F}$ by $f\nabla g$, where f and g are scalar functions, we obtain

$$\int_V \nabla f \cdot \nabla g\, dV + \int_V f \nabla^2 g\, dV = \int_S f \frac{\partial g}{\partial n}\, dS \tag{19}$$

OPTIONAL (*Continued*)

where $\partial g/\partial n = \nabla g \cdot \mathbf{n}$. This is *Green's first identity*. If we simply permute f and g, and subtract the result from the above equation, we obtain *Green's second identity*

$$\int_V (f\nabla^2 g - g\nabla^2 f)\, dV = \int_S \left(f\frac{\partial g}{\partial n} - g\frac{\partial f}{\partial n} \right) dS \tag{20}$$

It is this identity that we shall use.

Consider Poisson's equation

$$\nabla^2 u = \rho$$

in some region V, and the corresponding equations for the Green's function

$$G(\mathbf{x}, \mathbf{y}) = G(\mathbf{y}, \mathbf{x}) \text{ and } \nabla^2 G(\mathbf{x}, \mathbf{y}) = \delta(\mathbf{x} - \mathbf{y})$$

Inserting u and G into (20) we obtain

$$\int_V (u\nabla^2 G - G\nabla^2 u)\, dV = \int_S \left(u\frac{\partial G}{\partial n} - G\frac{\partial u}{\partial n} \right) dS$$

Choosing our integration variable to be $\mathbf{y}$ and using $G(\mathbf{x}, \mathbf{y}) = G(\mathbf{y}, \mathbf{x})$, this becomes

$$\int_V [u(\mathbf{y})\delta(\mathbf{x} - \mathbf{y}) - G(\mathbf{x}, \mathbf{y})\rho(\mathbf{y})]\, d\mathbf{y} = \int_S \left(u\frac{\partial G}{\partial n} - G\frac{\partial u}{\partial n} \right) dS$$

that is

$$u(\mathbf{x}) = \int_V G(\mathbf{x}, \mathbf{y})\rho(\mathbf{y})\, d\mathbf{y} + \int_S \left(u\frac{\partial G}{\partial n} - G\frac{\partial u}{\partial n} \right) dS \tag{21}$$

Note that for an unbounded region, this becomes identical to our previous result (14) for all of space. Equation (21) enables us to solve for u in a bounded region where $\rho = 0$ by incorporating the conditions that u must obey on S.

If $\rho = 0$, (21) reduces to

$$u = \int_S \left(u\frac{\partial G}{\partial n} - G\frac{\partial u}{\partial n} \right) dS$$

or fully

$$u(\mathbf{x}) = \int_S \left[u(\mathbf{y})\frac{\partial G}{\partial n}(\mathbf{x}, \mathbf{y}) - G(\mathbf{x}, \mathbf{y})\frac{\partial u}{\partial n}(\mathbf{y}) \right] dS(\mathbf{y}) \tag{22}$$

where u appears on both sides of the equation. The crucial point is that

OPTIONAL (*Continued*)

evaluation of the integral requires only that we know the behavior of u on S. Commonly either u is given on the boundary (*Dirichlet problem*) or $\partial u/\partial n$ is given on the boundary (*Neumann problem*). If we know u on the boundary, we want to make $G\ \partial u/\partial n$ vanish on the boundary so we can evaluate the integral. So if u is given on S we must find a G such that $G(\mathbf{x}, \mathbf{y})$ vanishes whenever $\mathbf{y}$ lies on S. This is called the *Dirichlet Green's function for the region V*. Conversely, if $\partial u/\partial n$ is given on S we must find a G such that $\partial G/\partial n$ vanishes on S. This is the *Neumann Green's function.*

Thus, a Dirichlet Green's function $G(\mathbf{x}, \mathbf{y})$ is defined for $\mathbf{x}$ and $\mathbf{y}$ in the volume V and satisfies these three conditions:

(a) $G(\mathbf{x}, \mathbf{y}) = G(\mathbf{y}, \mathbf{x})$

(b) $\nabla^2 G(\mathbf{x}, \mathbf{y}) = \delta(\mathbf{x} - \mathbf{y})$ and

(c) $G(\mathbf{x}, \mathbf{y}) = 0$ when $\mathbf{y}$ lies on S, the boundary of the region V

(Note that by (a), in (b) and (c) the variables $\mathbf{x}$ and $\mathbf{y}$ can be interchanged without changing the condition).

It is perhaps surprising that *condition* (a) *is actually a consequence of* (b) *and* (c), *provided* (b) *and* (c) *also hold with* $\mathbf{x}$ *and* $\mathbf{y}$ *interchanged.*

To see this, we fix $\mathbf{y}$ and use (20) with $f(\mathbf{x}) = G(\mathbf{x}, \mathbf{y})$ and $g(\mathbf{x}) = G(\mathbf{y}, \mathbf{x})$. By (b), $\nabla^2 f(\mathbf{x}) = \delta(\mathbf{x} - \mathbf{y})$ and $\nabla^2 g(\mathbf{x}) = \delta(\mathbf{x} - \mathbf{y})$, and by (c), f and g vanish on S. Thus, (20) becomes

$$\int_V f(\mathbf{x})\delta(\mathbf{x} - \mathbf{y}) - g(\mathbf{x})\delta(\mathbf{x} - \mathbf{y})\, dV = 0$$

or

$$f(\mathbf{x}) - g(\mathbf{x}) = 0.$$

Thus $G(\mathbf{x}, \mathbf{y}) = G(\mathbf{y}, \mathbf{x})$. This means, in effect, that in examples it is not necessary to check condition (a). (This result is sometimes called the *principle of reciprocity*.)

Doing any particular Dirichlet or Neumann problem thus becomes the task of finding the appropriate Green's function. We shall do this by modifying the Green's function for Laplace's equations on all $\mathbb{R}^2$ or $\mathbb{R}^3$, namely (16) and (17).

As an example, we shall now use the two-dimensional Green's-function method to construct the Dirichlet Green's function for the disc of radius R (see Figure 7.5.3). This will enable us to solve $\nabla^2 u = 0$ (or $\nabla^2 u = \rho$) with u given on the boundary circle.

OPTIONAL (*Continued*)

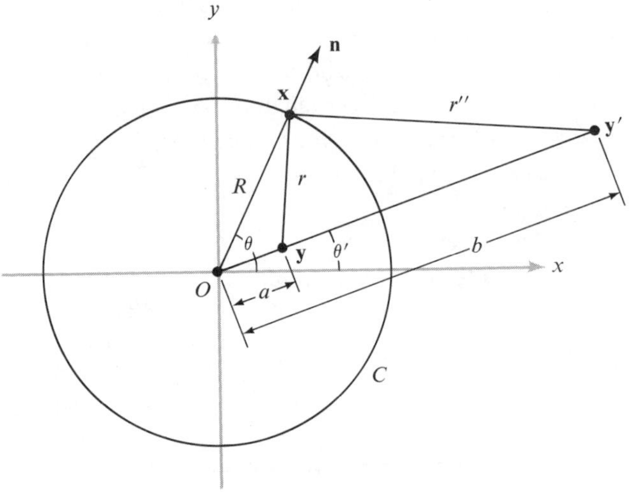

FIGURE 7.5.3

Geometry of the construction of the Green's function for a disc.

In Figure 7.5.3 we have drawn the point **x** on the circumference because that is where we want G to vanish.* The Green's function $G(\mathbf{x}, \mathbf{y})$ that we shall find will, of course, be valid for all **x**, **y** in the disc. The point $\mathbf{y}'$ represents the "reflection" of the point **y** into the region outside the circle, such that $ab = R^2$. Now when $\mathbf{x} \in C$, by the similarity of the triangles $\mathbf{x}O\mathbf{y}$ and $\mathbf{x}O\mathbf{y}'$,

$$\frac{r}{R} = \frac{r''}{b}$$

or

$$r = \frac{r''R}{b} = \frac{r''a}{R}$$

So if we choose our Green's function to be

$$G(\mathbf{x}, \mathbf{y}) = \frac{1}{2\pi}\left(\log r - \log \frac{r''a}{R}\right) \tag{23}$$

we see that G is zero if **x** is on C. Since $r''a/R$ reduces to r when **y** is on C, G also vanishes when **y** is on C. If we can show that G satisfies $\nabla^2 G = \delta(\mathbf{x} - \mathbf{y})$

* According to the procedure above, $G(\mathbf{x}, \mathbf{y})$ is supposed to vanish when either **x** or **y** is on C. We have chosen **x** on C to begin with.

OPTIONAL (*Continued*)

in the circle, then we will have proved that G is indeed the Dirichlet Green's function. From (17) we know that $\nabla^2(\log r)/2\pi = \delta(\mathbf{x} - \mathbf{y})$, so

$$\nabla^2 G(\mathbf{x}, \mathbf{y}) = \delta(\mathbf{x} - \mathbf{y}) - \delta(\mathbf{x} - \mathbf{y}')$$

but $\mathbf{y}'$ is always outside the circle, so $\mathbf{x}$ can never be equal to $\mathbf{y}'$ and so $\delta(\mathbf{x} - \mathbf{y}')$ is always zero. Hence

$$\nabla^2 G(\mathbf{x}, \mathbf{y}) = \delta(\mathbf{x} - \mathbf{y})$$

and thus G is the Dirichlet Green's function for the circle.

Now we shall consider the problem of solving

$$\nabla^2 u = 0$$

in this circle if $u(R, \theta) = f(\theta)$ is the given boundary condition. By (22) we have a solution

$$u = \int_C \left(u \frac{\partial G}{\partial n} - G \frac{\partial u}{\partial n} \right) dS$$

But $G = 0$ on C, so we are left with the integral

$$u = \int_C u \frac{\partial G}{\partial n} \, dS$$

where we can replace u by $f(\theta)$ since the integral is around C. Thus the task of solving the Dirichlet problem in the circle is reduced to finding $\partial G/\partial n$. From (23) we can write

$$\frac{\partial G}{\partial n} = \frac{1}{2\pi} \left(\frac{1}{r} \frac{\partial r}{\partial n} - \frac{1}{r''} \frac{\partial r''}{\partial n} \right)$$

Now

$$\frac{\partial r}{\partial n} = \boldsymbol{\nabla} r \cdot \mathbf{n}$$

and

$$\boldsymbol{\nabla} r = \frac{\mathbf{r}}{r}$$

where $\mathbf{r} = \mathbf{x} - \mathbf{y}$, so

$$\frac{\partial r}{\partial n} = \frac{\mathbf{r} \cdot \mathbf{n}}{r} = \frac{r \cos(nr)}{r} = \cos(nr)$$

where (nr) represents the angle between $\mathbf{n}$ and $\mathbf{r}$. Likewise

$$\frac{\partial r''}{\partial n} = \cos(nr'')$$

OPTIONAL (*Continued*)

Now, in triangle $\mathbf{xy}O$, we have, by the cosine law

$$a^2 = r^2 + R^2 - 2rR\cos(nr)$$

and in triangle $\mathbf{xy}'O$, we get

$$b^2 = (r'')^2 + R^2 - 2r''R\cos(nr'')$$

and so

$$\frac{\partial r}{\partial n} = \cos(nr) = \frac{R^2 + r^2 - a^2}{2rR}$$

and

$$\frac{\partial r''}{\partial n} = \cos(nr'') = \frac{R^2 + (r'')^2 - b^2}{2r''R}$$

Hence

$$\frac{\partial G}{\partial n} = \frac{1}{2\pi}\left[\frac{R^2 + r^2 - a^2}{2r^2R} - \frac{R^2 + (r'')^2 - b^2}{2(r'')^2R}\right]$$

Using the relationship between r and r'' when $\mathbf{x}$ is on C, we get

$$\left.\frac{\partial G}{\partial n}\right|_{\mathbf{x}\in C} = \frac{1}{2\pi}\left(\frac{R^2 - a^2}{Rr^2}\right)$$

Thus the solution can be written as

$$u = \frac{1}{2\pi}\int_C f(\theta)\frac{R^2 - a^2}{Rr^2}\,ds$$

Let us write this in a more explicit and tractable form. First, note that in triangle $\mathbf{xy}O$, we can write

$$r = [a^2 + R^2 - 2aR\cos(\theta - \theta')]^{1/2}$$

where θ and θ' are the polar angles in $\mathbf{x}$- and $\mathbf{y}$-space, respectively. Second, our solution must be valid for all $\mathbf{y}$ in the circle; hence the distance of $\mathbf{y}$ from the origin must now become a variable, which we shall call r'. Finally, note that $ds = R\,d\theta$ on C, so we can write the solution in polar coordinates as

$$u(r', \theta') = \frac{R^2 - (r')^2}{2\pi}\int_0^{2\pi}\frac{f(\theta)\,d\theta}{(r')^2 + R^2 - 2r'R\cos(\theta - \theta')}$$

This is known as *Poisson's formula in two dimensions.** As an exercise, the reader should use this to write down the solution of $\nabla^2 u = \rho$ with u a given function $f(\theta)$ on the boundary.

* There are several ways of deriving this famous formula. For the method of complex variables, see J. Marsden, *Basic Complex Analysis*, W. H. Freeman and Company, San Francisco, 1973, p. 145. For the method of Fourier series, see J. Marsden, *Elementary Classical Analysis*, W. H. Freeman and Company, 1974, p. 466.

OPTIONAL (*Continued*)

EXERCISES

1. (a) Supply the details for the assertion on p. 452 that

$$\frac{d}{dt}\int_{\Omega_t} \rho \, dx\, dy\, dz = 0$$

is equivalent to the law of conservation of mass.

(b) Using (a) and the Change of Variables Theorem show that $\rho(\mathbf{x}, t)$ can be expressed in terms of the Jacobian $J(\mathbf{x}, t)$ of the flow map $\phi(\mathbf{x}, t)$ and $\rho(\mathbf{x}, 0)$ by the equation

$$\rho(\mathbf{x}, t)J(\mathbf{x}, t) = \rho(\mathbf{x}, 0)$$

(c) What can you conclude from (b) for incompressible flow?

2. Let $\mathbf{V}$ be a vector field with flow $\phi(\mathbf{x}, t)$ and let $\mathbf{V}$ and ρ satisfy the law of conservation of mass. Let Ω_t be the region transported with the flow. Prove the following version of the Transport Theorem (see Exercise 18, Section 7.4):

$$\frac{d}{dt}\int_{\Omega_t} \rho f \, dx\, dy\, dz = \int_{\Omega_t} \rho \frac{Df}{Dt}\, dx\, dy\, dz$$

3. (*Bernoulli's law*) (a) Let $\mathbf{V}$, ρ satisfy the law of conservation of mass and equation (2) (Euler's equation for a perfect fluid). Suppose $\mathbf{V}$ is irrotational and hence that $\mathbf{V} = \nabla\phi$ for a function ϕ. Show that if C is a path connecting two points P_1 and P_2, then

$$\left(\frac{\partial\phi}{\partial t} + \frac{1}{2}\|\mathbf{V}\|^2\right)\Bigg|_{P_1}^{P_2} + \int_C \frac{dp}{\rho} = 0$$

(HINT: You will require the identity

$$\mathbf{V}\cdot(\nabla V) = \tfrac{1}{2}\nabla(\|\mathbf{V}\|^2) + (\nabla \times \mathbf{V}) \times \mathbf{V}$$

from Table 3.1, p. 190.)

(b) If in (a), $\mathbf{V}$ is stationary—that is, $\partial\mathbf{V}/\partial t = 0$—and ρ is constant, show that

$$\frac{1}{2}\|\mathbf{V}\|^2 + \frac{p}{\rho}$$

is constant in space. Deduce that, in this situation, "higher pressure is associated with lower fluid speed."

4. Using Exercise 3, show that if ϕ satisfies Laplace's equation $\nabla^2\phi = 0$, then $\mathbf{V} = \nabla\phi$ is a stationary solution to Euler's equation for a perfect *incompressible* fluid.

5. Verify that Maxwell's equations imply the equation of continuity for $\mathbf{J}$ and ρ.

6. Let H denote the upper half space $z \geq 0$. For a point $\mathbf{x} = (x, y, z)$ in H, let $R(\mathbf{x}) = (x, y, -z)$, the reflection of $\mathbf{x}$ in the xy-plane. Let $G(\mathbf{x}, \mathbf{y}) = -1/4\pi\|\mathbf{x} - \mathbf{y}\|$ be the Green's function for all of $\mathbb{R}^3$.

(a) Verify that the function $\tilde{G}$ defined by

$$\tilde{G}(\mathbf{x}, \mathbf{y}) = G(\mathbf{x}, \mathbf{y}) - G(R(\mathbf{x}), \mathbf{y})$$

is the Green's function for the Laplacian in H.

OPTIONAL (*Continued*)

(b) Write down a formula for the solution u of the problem

$$\nabla^2 u = \rho \text{ in } H$$
$$u(x, y, 0) = \phi(x, y)$$

7. (a) With notation as in Figure 7.5.3, show that the Dirichlet problem for the sphere of radius R in three dimensions has Green's function

$$G(\mathbf{x}, \mathbf{y}) = \frac{1}{4\pi}\left(\frac{R}{ar''} - \frac{1}{r}\right)$$

(b) Prove Poisson's formula in three dimensions:

$$u(\mathbf{y}) = \frac{R(R^2 - a^2)}{4\pi} \int_0^{2\pi} \int_0^{\pi} \frac{f(\theta, \phi)\sin\theta \, d\theta \, d\phi}{(R^2 + a^2 - 2Ra\cos\gamma)^{3/2}}$$

where $f(\theta, \phi)$ is the function given on the boundary sphere, $\|\mathbf{y}\| = a$, and γ is the angle between $\mathbf{y}$ and the point $\mathbf{x} = (R\cos\theta\sin\phi, R\sin\theta\sin\phi, R\cos\phi)$ on the surface of the sphere.

Exercises 8 to 14 give some sample applications of vector calculus to *shock waves.**

8. Consider the equation

$$u_t + uu_x = 0$$

for a function $u(x, t)$, $-\infty < x < \infty$, $t \geq 0$ where $u_t = \partial u/\partial t$ and $u_x = \partial u/\partial x$. Let $u(x, 0) = u_0(x)$ be the given value of u at $t = 0$. The curves $(x(s), t(s))$ in the xt-plane defined by

$$\dot{x} = u, \qquad \dot{t} = 1$$

are called *characteristic curves* (the dot $\dot{}$ denotes the derivative with respect to s).

(a) Show that u is constant along each characteristic curve by showing that $\dot{u} = 0$.

(b) Show that the slopes of the characteristic curves are given by $dt/dx = 1/u$, and use it to prove that the characteristic curves are straight lines determined by the initial data.

(c) Suppose that $x_1 < x_2$ and $u_0(x_1) > u_0(x_2) > 0$. Show that the two characteristics through the points $(x_1, 0)$ and $(x_2, 0)$ intersect at a point $P = (\bar{x}, \bar{t})$, with $\bar{t} > 0$. Show that this together with the result in (a) implies that the solution cannot be continuous at P (see Figure 7.5.4).

(d) Calculate $\bar{t}$.

* For additional details, consult A. J. Chorin and J. E. Marsden, *A Mathematical Introduction to Fluid Mechanics*, Springer-Verlag, New York, 1979, and P. D. Lax, "The Formation and Decay of Shock Waves, *Am. Math. Monthly*, 79 (1972): 227–241. We are grateful to Joel Smoller for suggesting this series of exercises.

OPTIONAL (*Continued*)

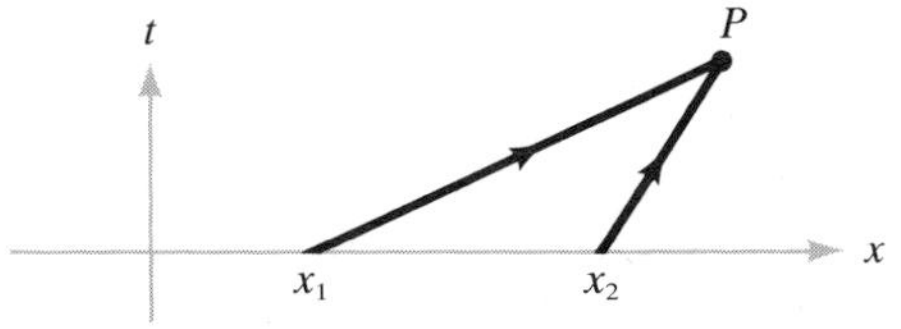

FIGURE 7.5.4
Characteristics of the equation $u_t + uu_x = 0$.

9. Repeat Exercise 8 for the equation

$$u_t + f(u)_x = 0 \tag{24}$$

where $f'' > 0$ and $f'(u_0(x_2)) > 0$. The characteristics are now defined by $\dot{x} = f'(u)$, $\dot{t} = 1$. We call (24) an equation in *divergence form.* (This exercise shows that a continous solution is generally impossible—irrespective of the smoothness of f!)

10. (*Weak solutions*) Since equations of the form in Exercise 9 arise in many physical applications (gas dynamics, magnetohydrodynamics, nonlinear optics (lasers)) and because it would be nice for a solution to exist for all time (t), it is desirable to make sense out of the equation by reinterpreting it when discontinuities develop. To this end, let $\phi = \phi(x, t)$ be a C^1 function. Let D be a rectangle in the xt-plane determined by $-a \le x \le a$ and $0 \le t \le T$, such that $\phi(x, t) = 0$ for $x = \pm a$, $x = T$, and for all (x, t) in the upper half plane outside D. Let u be a "*genuine*" solution of (24).
   (a) Show that

$$\iint_{t \ge 0} (u\phi_t + f(u)\phi_x)\, dx\, dt + \int_{t=0} u_0(x)\phi(x, 0)\, dx = 0 \tag{25}$$

   (HINT: Start with $\iint_D (u_t + f(u)_x)\phi\, dx\, dt = 0$.) Thus, if u is a smooth solution, then (25) holds for all ϕ as above. We call the function u a *weak* solution of (24) if (25) holds for all such ϕ.
   (b) Show that if u is a weak solution, which is C^1 in an open set Ω in the upper half of the xt-plane, then u is a genuine solution of (24) in Ω.

11. (*The jump condition*, which is also known in gas dynamics as the Rankine-Hugoniot condition) The definition of weak solution given in Exercise 10 clearly allows discontinuous solutions. However, the reader shall now determine that not every type of discontinuity is admissible, for there is a connection between the discontinuity curve and the values of the solution on both sides of the discontinuity.
   Let u be a (weak) solution of (24) and suppose Γ is a smooth curve in the xt-plane such that u "jumps" across a curve Γ; that is, u is C^1 except for a jump

OPTIONAL (*Continued*)

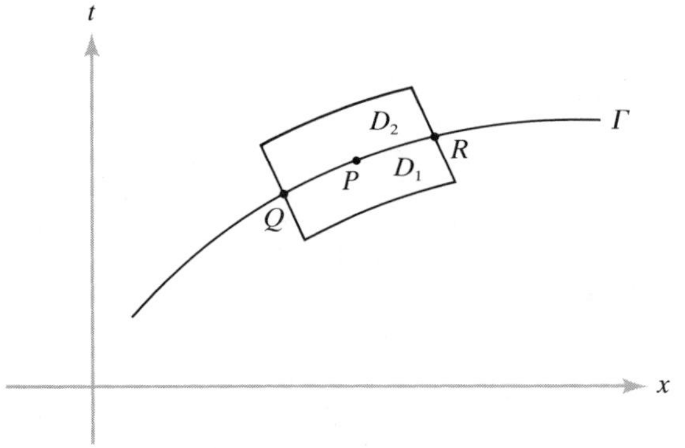

FIGURE 7.5.5

The solution u jumps in value from u_1 to u_2 across Γ.

discontinuity across Γ. We call Γ a *shock wave*. Choose a point $P \in \Gamma$ and construct, near P, a "rectangle" $D = D_1 \cup D_2$, as shown in Figure 7.5.5. Choose ϕ to vanish on D and outside D.

(a) Show that

$$\iint_D (u\phi_t + f(u)\phi_x)\,dx\,dt = 0$$

and

$$\iint_{D_i} (u\phi_t + f(u)\phi_x)\,dx\,dt = \iint_{D_i} [(u\phi)_t + (f(u)\phi)_x]\,dx\,dt$$

(b) Suppose that u jumps in value from u_1 to u_2 across Γ so that when (x, t) approaches a point (x_0, t_0) on Γ from ∂D_i, $u(x, t)$ approaches the value $u_i(x_0, t_0)$. Show that

$$0 = \int_{\partial D_1} \phi\{-u dx + f(u)\,dt\} + \int_{\partial D_2} \phi\{-u dx + f(u)\,dt\}$$

and deduce that

$$0 = \int_\Gamma \phi\{[-u]\,dx + [f(u)]\,dt\}$$

where $[\alpha(u)] = \alpha(u_2) - \alpha(u_1)$ denotes the jump in the quantity $\alpha(u)$ across Γ.

(c) If the curve Γ defines x implicitly as a function of t, and ∂D intersects Γ at $Q = (x(t_1), t_1)$ and $R = (x(t_2), t_2)$, show that

$$0 = \int_Q^R \phi\{[-u]\,dx + [f(u)]\,dt\} = \int_{t_1}^{t_2} \phi\left\{[-u]\frac{dx}{dt} + [f(u)]\right\} dt$$

OPTIONAL (*Continued*)

(d) Show that at the point P on Γ,

$$[u] \cdot s = [f(u)] \tag{26}$$

where $s = dx/dt$ at P. The number s is called the *speed* of the discontinuity. Equation (26) is called the *jump condition*; it is the relation that any discontinuous solution will satisfy.

12. (*Loss of uniqueness*) One drawback of accepting weak solutions is loss of uniqueness. (In gas dynamics, some mathematical solutions are extraneous and rejected on physical grounds. For example, discontinuous solutions of rarefaction shock waves are rejected because they indicate that entropy *decreases* across the discontinuity.)

    Consider the equation

$$u_t + \left(\frac{u^2}{2}\right)_x = 0, \qquad \text{with initial data } u(x, 0) = \begin{cases} -1, & x \geq 0 \\ 1, & x < 0 \end{cases}$$

Show that for every $\alpha \geq 1$, u_α is a weak solution, where u_α is defined by

$$u_\alpha(x, t) = \begin{cases} 1, & x \leq \dfrac{1 - \alpha}{2} t \\ -\alpha, & \dfrac{1 - \alpha}{2} t \leq x \leq 0 \\ \alpha, & 0 \leq x \leq \dfrac{\alpha - 1}{2} t \\ -1, & \dfrac{\alpha - 1}{2} t < x \end{cases}$$

(It can be shown that if $f'' > 0$, uniqueness can be recovered by imposing an additional constraint on the solutions. Thus, there is a unique solution satisfying the "entropy" condition

$$\frac{u(x + a, t) - u(x, t)}{a} \leq \frac{E}{t}$$

for some $E > 0$ and all $a \neq 0$. Hence for fixed t, $u(x, t)$ can only "jump down" as x increases. In our example, this holds only for the solution with $\alpha = 1$.)

13. (The solution of (24) depends on the particular divergence form used.) The equation $u_t + uu_x = 0$ can be written in the two divergence forms

(*i*) $$u_t + (\tfrac{1}{2}u^2)_x = 0$$

(*ii*) $$(\tfrac{1}{2}u^2)_t + (\tfrac{1}{3}u^3)_x = 0$$

Show that a weak solution of (*i*) need not be a weak solution of (*ii*). (HINT: The equations have different jump conditions: In (*i*) $s = \frac{1}{2}(u_2 + u_1)$ while in (*ii*) $s = \frac{2}{3}(u_2^2 + u_1u_2 + u_1^2)/(u_2 + u_1)$.)

OPTIONAL (*Continued*)

14. (*Noninvariance of weak solutions under nonlinear transformation.*) Consider equation (24) where $f'' > 0$.
    (a) Show that the transformation $v = f'(u)$ takes this equation into

$$v_t + vv_x = 0 \tag{27}$$

    (b) Show that the above transformation *does not* necessarily map discontinuous solutions of (24) into discontinuous solutions of (27). (HINT: Check the jump conditions; for (27) $s[v] = \frac{1}{2}[v^2]$ implies $s[f'(u)] = \frac{1}{2}[f'(u)^2]$; for (24), $s[u] = [f(u)]$.)

7.6 DIFFERENTIAL FORMS

The theory of differential forms provides a convenient and elegant way of phrasing Green's, Stokes', and Gauss' theorems. In fact, the use of differential forms shows that these theorems are all manifestations of a single underlying mathematical theory, and provides the necessary language to generalize them to n dimensions. In this section we shall give a very elementary exposition of the theory of forms. Since our primary goal is to show that the theorems of Green, Stokes, and Gauss are all manifestations of the same phenomenon, we shall be satisfied with less than the strongest possible version of these theorems. Moreover, we shall introduce forms in a purely axiomatic and non-constructive manner, thereby avoiding the tremendous amount of formal algebraic preliminaries that is usually required for their construction. Thus to the purist our approach will be far from complete, but to the student it may be comprehensible. We hope that this will motivate some students to delve further into the theory of differential forms.

We shall begin by introducing the notion of a 0-form.

DEFINITION. *Let K be an open set in $\mathbb{R}^3$. A **0-form** on K is a real-valued function $f: K \to \mathbb{R}$. When we differentiate f once, it is assumed to be C^1, and C^2 when we differentiate twice.*

Given two 0-forms f_1 and f_2 on K, we can add them in the usual way to get a new 0-form $f_1 + f_2$ or multiply them to get a 0-form $f_1 f_2$.

OPTIONAL (*Continued*)

EXAMPLE 1. $f_1(x, y, z) = xy + yz$ and $f_2(x, y, z) = y \sin xz$ are 0-forms on $\mathbb{R}^3$.

$$(f_1 + f_2)(x, y, z) = xy + yz + y \sin xz$$

and

$$(f_1 f_2)(x, y, z) = y^2 x \sin xz + y^2 z \sin xz$$

DEFINITION. *The* ***basic*** *1-forms are the expressions dx, dy, and dz. At present we consider these to be only formal symbols. A* ***1-form*** *ω on an open set K is a formal linear combination*

$$\omega = P(x, y, z)\, dx + Q(x, y, z)\, dy + R(x, y, z)\, dz$$

or simply

$$\omega = P\, dx + Q\, dy + R\, dz$$

where P, Q, and R are real-valued functions on K. By the expression $P\, dx$ we mean the 1-form $P\, dx + 0 \cdot dy + 0 \cdot dz$, and similarly for $Q\, dy$ and $R\, dz$. Also the order of $P\, dx$, $Q\, dy$, and $R\, dz$ is immaterial, so

$$P\, dx + Q\, dy + R\, dz = R\, dz + P\, dx + Q\, dy \quad \text{etc.}$$

Given two 1-forms $\omega_1 = P_1\, dx + Q_1\, dy + R_1\, dz$ and $\omega_2 = P_2\, dx + Q_2\, dy + R_2\, dz$, we can add them to get a new 1-form $\omega_1 + \omega_2$ defined by

$$\omega_1 + \omega_2 = (P_1 + P_2)\, dx + (Q_1 + Q_2)\, dy + (R_1 + R_2)\, dz$$

and given a 0-form f, we can form the 1-form $f\omega_1$ defined by

$$f\omega_1 = (fP_1)\, dx + (fQ_1)\, dy + (fR_1)\, dz$$

EXAMPLE 2. Let $\omega_1 = (x + y^2)\, dx + (zy)\, dy + (e^{xyz})\, dz$ and $\omega_2 = \sin y\, dx + \sin x\, dy$ be 1-forms. Then

$$\omega_1 + \omega_2 = (x + y^2 + \sin y)\, dx + (zy + \sin x)\, dy + (e^{xyz})\, dz$$

If $f(x, y, z) = x$ then

$$f\omega_2 = x \sin y\, dx + x \sin x\, dy$$

DEFINITION. *The* ***basic*** *2-forms are the formal expressions $dx\, dy$, $dy\, dz$, and $dz\, dx$. These expressions should be thought of as products of dx and dy, dy and dz, and dz and dx.*

A ***2-form*** *η on K is a formal expression*

$$\eta = F\, dx\, dy + G\, dy\, dz + H\, dz\, dx$$

where F, G, and H are real-valued functions on K. The order of $F\, dx\, dy$, $G\, dy\, dz$, and $H\, dz\, dx$ is immaterial; for example

$$F\, dx\, dy + G\, dy\, dz + H\, dz\, dx = H\, dz\, dx + F\, dx\, dy + G\, dy\, dz \quad \text{etc.}$$

OPTIONAL (*Continued*)

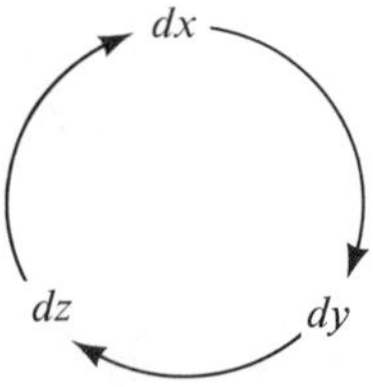

FIGURE 7.6.1
The cyclic order of dx, dy, and dz.

At this point it is useful to note that in a 2-form the basic 1-forms dx, dy, and dz always appear in cyclic pairs (see Figure 7.6.1), that is, $dx\,dy$, $dy\,dz$, and $dz\,dx$.

By analogy with 0-forms and 1-forms we can add two 2-forms

$$\eta_i = F_i\,dx\,dy + G_i\,dy\,dz + H_i\,dz\,dx,$$

$i = 1$ and 2, to obtain a new 2-form

$$\eta_1 + \eta_2 = (F_1 + F_2)\,dx\,dy + (G_1 + G_2)\,dy\,dz + (H_1 + H_2)\,dz\,dx$$

Similarly, if f is a 0-form and η is a 2-form we can take the product

$$f\eta = (fF)\,dx\,dy + (fG)\,dy\,dz + (fH)\,dz\,dx$$

Finally, by the expression $F\,dx\,dy$ we mean the 2-form $F\,dx\,dy + 0 \cdot dy\,dz + 0 \cdot dz\,dx$.

EXAMPLE 3. The expressions

$$\eta_1 = x^2\,dx\,dy + y^3x\,dy\,dz + \sin zy\,dz\,dx$$

and

$$\eta_2 = y\,dy\,dz$$

are 2-forms. Their sum is

$$\eta_1 + \eta_2 = x^2\,dx\,dy + (y^3x + y)\,dy\,dz + \sin zy\,dz\,dx$$

If $f(x, y, z) = xy$, then

$$f\eta_2 = xy^2\,dy\,dz$$

DEFINITION. *A **basic 3-form** is a formal expression $dx\,dy\,dz$ (in cyclic order, Figure 7.6.1). A **3-form** ν on an open set $K \subset \mathbb{R}^3$ is an expression of the form $\nu = f(x, y, z)\,dx\,dy\,dz$, where f is a real-valued function on K.*

OPTIONAL (*Continued*)

We can add two 3-forms and we can multiply them by 0-forms in the obvious way. There seems to be little difference between a 0-form and a 3-form, since both involve a single real-valued function. But we distinguish them for a purpose that will become clear when we multiply and differentiate forms.

EXAMPLE 4. Let $\nu_1 = y\,dx\,dy\,dz$, $\nu_2 = e^{x^2}\,dx\,dy\,dz$, and $f(x, y, z) = xyz$. Then $\nu_1 + \nu_2 = (y + e^{x^2})\,dx\,dy\,dz$ and $f\nu_1 = y^2xz\,dx\,dy\,dz$.

WARNING: Although we can add two 0-forms, two 1-forms, two 2-forms, or two 3-forms, we *never* add a k-form and a j-form if $k \neq j$. For example, we never write

$$f(x, y, z)\,dx\,dy + g(x, y, z)\,dz$$

Now that we have defined these formal objects (forms), one can legitimately ask what they are good for, how they are used, and, perhaps most important, what they mean. The answer to the first question will become clear as we proceed, but we can immediately describe how to use and interpret them.

A real-valued function on a domain K in $\mathbb{R}^3$ is a rule that assigns to each point in K a real number. Differential forms are, in some sense, generalizations of the real-valued functions we have studied in calculus. In fact, 0-forms on an open set K are just functions on K. Thus a 0-form f takes points in K to real numbers.

We should like to interpret differential k-forms (for $k \geq 1$) not as functions on points in K, but as functions on geometric objects such as curves and surfaces. Many of the early Greek geometers viewed lines and curves as being made up of infinitely many points, and planes and surfaces as being made up of infinitely many curves. Consequently there is at least some historical justification for applying this geometric hierarchy to the interpretation of differential forms.

Given an open subset $K \subset \mathbb{R}^3$, we shall distinguish four types of subsets of K (see Figure 7.6.2):

(*i*) points in K,

(*ii*) oriented simple curves and oriented simple closed curves C in K,

(*iii*) oriented surfaces $S \subset K$,

(*iv*) elementary subregions (of types 1 to 4) $R \subset K$.

We shall begin with 1-forms. Let

$$\omega = P(x, y, z)\,dx + Q(x, y, z)\,dy + R(x, y, z)\,dz$$

OPTIONAL (*Continued*)

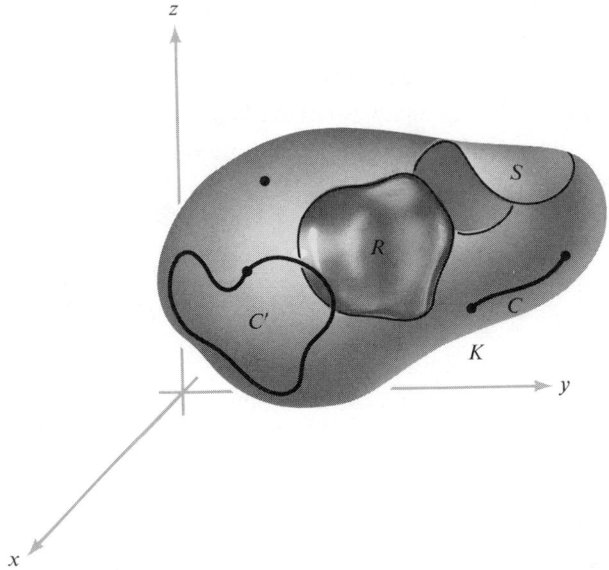

FIGURE 7.6.2
The four geometric types of subsets of an open set $K \subset \mathbb{R}^3$ to which the theory of forms applies.

be a 1-form on K and let C be an oriented simple curve as in Figure 7.6.2. The real number that ω assigns to C is given by the formula

$$\int_C \omega = \int_C P(x, y, z)\, dx + Q(x, y, z)\, dy + R(x, y, z)\, dz \tag{1}$$

Recall (see Section 6.2) that this integral is evaluated as follows. Let $\boldsymbol{\sigma}: [a, b] \to K, \boldsymbol{\sigma}(t) = (x(t), y(t), z(t))$ be an orientation-preserving parametrization of C. Then

$$\int_C \omega = \int_\sigma \omega = \int_a^b \left[P(x(t), y(t), z(t)) \cdot \frac{dx}{dt} + Q(x(t), y(t), z(t)) \cdot \frac{dy}{dt} \right.$$
$$\left. + R(x(t), y(t), z(t)) \cdot \frac{dz}{dt} \right] dt$$

Theorem 1 of Section 6.2 guarantees that $\int_C \omega$ does not depend on the choice of the parametrization $\boldsymbol{\sigma}$.

OPTIONAL (*Continued*)

We can thus interpret a 1-form ω on K as a rule assigning a real number to each curve $C \subset K$; a 2-form η will similarly be seen to be a rule assigning a real number to each surface $S \subset K$; and a 3-form v as a rule assigning a real number to each elementary subregion of K. The rules for associating real numbers to curves, surfaces, and regions are completely contained in the formal expressions we have defined.

EXAMPLE 5. Let $\omega = xy\,dx + y^2\,dy + dz$ be a 1-form on $\mathbb{R}^3$ and let C be the oriented simple curve in $\mathbb{R}^3$ described by the parametrization $\boldsymbol{\sigma}(t) = (t^2, t^3, 1)$, $0 \leq t \leq 1$. C is oriented by choosing the positive direction of C to be the direction in which $\boldsymbol{\sigma}(t)$ traverses C as t goes from 0 to 1. Then by formula (1)

$$\int_C \omega = \int_0^1 [t^5(2t) + t^6(3t^2) + 0]\,dt = \int_0^1 [2t^6 + 3t^8]\,dt = \tfrac{13}{21}$$

Thus this 1-form ω assigns to each oriented simple curve and each oriented simple closed curve C in $\mathbb{R}^3$ the number $\int_C \omega$.

A 2-form η on an open set $K \subset \mathbb{R}^3$ may similarly be interpreted as a function that associates with each oriented surface $S \subset K$ a real number. This is accomplished by means of the notion of integration of 2-form over surfaces. Let

$$\eta = F(x, y, z)\,dx\,dy + G(x, y, z)\,dy\,dz + H(x, y, z)\,dz\,dx$$

be a 2-form on K, and let $S \subset K$ be an oriented surface parametrized by a function $\boldsymbol{\Phi}: D \to \mathbb{R}^3$, $D \subset \mathbb{R}^2$, $\boldsymbol{\Phi}(u, v) = (x(u, v), y(u, v), z(u, v))$ (see Section 6.3).

DEFINITION. *If S is such a surface and η is a 2-form on K we define $\int_S \eta$ by the formula*

$$\begin{aligned}\int_S \eta &= \int_S F\,dx\,dy + G\,dy\,dz + H\,dz\,dx \\ &= \int_D \left[F(x(u, v), y(u, v), z(u, v)) \cdot \frac{\partial(x, y)}{\partial(u, v)} \right. \\ &\quad + G(x(u, v), y(u, v), z(u, v)) \cdot \frac{\partial(y, z)}{\partial(u, v)} \\ &\quad \left. + H(x(u, v), y(u, v), z(u, v)) \cdot \frac{\partial(z, x)}{\partial(u, v)} \right] du\,dv \end{aligned} \tag{2}$$

OPTIONAL (*Continued*)

where

$$\frac{\partial(x, y)}{\partial(u, v)} = \begin{vmatrix} \dfrac{\partial x}{\partial u} & \dfrac{\partial x}{\partial v} \\ \dfrac{\partial y}{\partial u} & \dfrac{\partial y}{\partial v} \end{vmatrix}, \qquad \frac{\partial(y, z)}{\partial(u, v)} = \begin{vmatrix} \dfrac{\partial y}{\partial u} & \dfrac{\partial y}{\partial v} \\ \dfrac{\partial z}{\partial u} & \dfrac{\partial z}{\partial v} \end{vmatrix}, \qquad \frac{\partial(z, x)}{\partial(u, v)} = \begin{vmatrix} \dfrac{\partial z}{\partial u} & \dfrac{\partial z}{\partial v} \\ \dfrac{\partial x}{\partial u} & \dfrac{\partial x}{\partial v} \end{vmatrix}.$$

If S is composed of several pieces S_i, $i = 1, \ldots, k$, as in Figure 7.4.4, each with its own parametrization $\mathbf{\Phi}_i$, we define

$$\int_S \eta = \sum_{i=1}^{k} \int_{S_i} \eta$$

One must verify that $\int_S \eta$ does not depend on the choice of parametrization $\mathbf{\Phi}$. This result is essentially (but not obviously) contained in Theorem 4, Section 6.6.

EXAMPLE 6. Let $\eta = z^2\, dx\, dy$ be a 2-form on $\mathbb{R}^3$, and let S be the upper unit hemisphere in $\mathbb{R}^3$. Find $\int_S \eta$.

Let us parametrize S by

$$\mathbf{\Phi}(u, v) = (\sin u \cos v, \sin u \sin v, \cos u),$$

where $(u, v) \in D = [0, \pi/2] \times [0, 2\pi]$. So by formula (2)

$$\int_S \eta = \int_D \cos^2 u \left[\frac{\partial(x, y)}{\partial(u, v)}\right] du\, dv$$

where

$$\frac{\partial(x, y)}{\partial(u, v)} = \begin{vmatrix} \cos u \cos v & -\sin u \sin v \\ \cos u \sin v & \sin u \cos v \end{vmatrix}$$

$$= \sin u \cos u \cos^2 v + \cos u \sin u \sin^2 v = \sin u \cos u$$

Therefore

$$\int_S \eta = \int_D \cos^2 u \cos u \sin u\, du\, dv$$

$$= \int_0^{2\pi} \int_0^{\pi/2} \cos^3 u \sin u\, du\, dv = \int_0^{2\pi} \left[-\frac{\cos^4 u}{4}\right]_0^{\pi/2} dv = \frac{\pi}{2}$$

EXAMPLE 7. Evaluate $\int_S x\, dy\, dz + y\, dx\, dy$, where S is the oriented surface described by the parametrization $x = u + v$, $y = u^2 - v^2$, $z = uv$, $(u, v) \in D = [0, 1] \times [0, 1]$.

OPTIONAL (*Continued*)

By definition we have

$$\frac{\partial(y, z)}{\partial(u, v)} = \begin{vmatrix} 2u & -2v \\ v & u \end{vmatrix} = 2(u^2 + v^2)$$

$$\frac{\partial(x, y)}{\partial(u, v)} = \begin{vmatrix} 1 & 1 \\ 2u & -2v \end{vmatrix} = -2(u + v)$$

Consequently

$$\begin{aligned}
\int_S x\, dy\, dz + y\, dx\, dy \\
&= \int_D [(u + v)(2)(u^2 + v^2) + (u^2 - v^2)(-2)(u + v)]\, du\, dv \\
&= 4 \int_D [v^3 + uv^2]\, du\, dv = 4 \int_0^1 \int_0^1 [v^3 + uv^2]\, du\, dv \\
&= 4 \int_0^1 \left[uv^3 + \frac{u^2 v^2}{2} \right]_0^1 dv = 4 \int_0^1 \left[v^3 + \frac{v^2}{2} \right] dv \\
&= \left[v^4 + \frac{2v^3}{3} \right]_0^1 = 1 + \frac{2}{3} = \frac{5}{3}
\end{aligned}$$

Finally, we must interpret 3-forms as functions on the elementary subregions (of types 1 to 4) of K. Let $\nu = f(x, y, z)\, dx\, dy\, dz$ be a 3-form and let $R \subset K$ be an elementary subregion of K. Then to each such $R \subset K$ we assign the number

$$\int_R \nu = \int_R f(x, y, z)\, dx\, dy\, dz \tag{3}$$

which is just the ordinary triple integral of f over R, as described in Section 5.6.

EXAMPLE 8. Suppose $\nu = (x + z)\, dx\, dy\, dz$ and $R = [0, 1] \times [0, 1] \times [0, 1]$. Then

$$\begin{aligned}
\int_R \nu = \int_R (x + z)\, dx\, dy\, dz &= \int_0^1 \int_0^1 \int_0^1 (x + z)\, dx\, dy\, dz \\
&= \int_0^1 \int_0^1 \left[\frac{x^2}{2} + zx \right]_0^1 dy\, dz = \int_0^1 \int_0^1 \left[\frac{1}{2} + z \right] dy\, dz = \int_0^1 \left[\frac{1}{2} + z \right] dz \\
&= \left[\frac{z}{2} + \frac{z^2}{2} \right]_0^1 = 1
\end{aligned}$$

OPTIONAL (*Continued*)

We must now discuss the algebra (or rules of multiplication) of forms, which, together with differentiation of forms, will enable us to state Green's, Stokes', and Gauss' theorems in terms of differential forms.

If ω is a k-form and η is an l-form on K, $0 \le k + l \le 3$ there is a product called the *wedge product* $\omega \wedge \eta$ of ω and η, which is a $k + l$ form on K. The wedge product satisfies the following laws:

(*i*) For each k there is a zero k-form 0 with the property that $0 + \omega = \omega$ for all k-forms ω, and $0 \wedge \eta = 0$ for all l-forms η if $0 \le k + l \le 3$.

(*ii*) (*distributivity*) If f is a 0-form, then

$$(f\omega_1 + \omega_2) \wedge \eta = f(\omega_1 \wedge \eta) + (\omega_2 \wedge \eta)$$

(*iii*) (*anticommutativity*) $\omega \wedge \eta = (-1)^{kl}(\eta \wedge \omega)$.

(*iv*) (*associativity*) If $\omega_1, \omega_2, \omega_3$ are k_1, k_2, k_3 forms, respectively, with $k_1 + k_2 + k_3 \le 3$, then

$$\omega_1 \wedge (\omega_2 \wedge \omega_3) = (\omega_1 \wedge \omega_2) \wedge \omega_3$$

(*v*) (*homogeneity with respect to functions*) If f is a 0-form, then

$$\omega \wedge (f\eta) = (f\omega) \wedge \eta = f(\omega \wedge \eta)$$

Notice that (*ii*) and (*iii*) actually imply (*v*).

(*vi*) The following multiplication rules for 1-forms hold

$$dx \wedge dy = dx\, dy$$

$$dy \wedge dx = -dx\, dy = (-1)(dx \wedge dy)$$

$$dy \wedge dz = dy\, dz = (-1)(dz \wedge dy)$$

$$dz \wedge dx = dz\, dx = (-1)(dx \wedge dz)$$

$$dx \wedge dx = 0, \qquad dy \wedge dy = 0, \qquad dz \wedge dz = 0$$

$$dx \wedge (dy \wedge dz) = (dx \wedge dy) \wedge dz = dx\, dy\, dz$$

(*vii*) If f is a 0-form and ω is any k-form, then $f \wedge \omega = f\omega$.

Using laws (*i*) to (*vii*), we can now find a unique product of any l-form η and any k-form ω, if $0 \le k + l \le 3$.

EXAMPLE 9. Show that $dx \wedge dy\, dz = dx\, dy\, dz$.

By (*vi*), $dy\, dz = dy \wedge dz$. Therefore

$$dx \wedge dy\, dz = dx \wedge (dy \wedge dz) = dx\, dy\, dz$$

OPTIONAL (*Continued*)

EXAMPLE 10. If $\omega = x\,dx + y\,dy$ and $\eta = zy\,dx + xz\,dy + xy\,dz$, find $\omega \wedge \eta$.

Computing $\omega \wedge \eta$, we get

$$\begin{aligned}
\omega \wedge \eta &= (x\,dx + y\,dy) \wedge (zy\,dx + xz\,dy + xy\,dz) \\
&= [(x\,dx + y\,dy) \wedge (zy\,dx)] + [(x\,dx + y\,dy) \wedge (xz\,dy)] \\
&\quad + [(x\,dx + y\,dy) \wedge (xy\,dz)] \\
&= xyz(dx \wedge dx) + zy^2(dy \wedge dx) + x^2z(dx \wedge dy) + xyz(dy \wedge dy) \\
&\quad + x^2y(dx \wedge dz) + xy^2(dy \wedge dz) \\
&= -zy^2\,dx\,dy + x^2z\,dx\,dy - x^2y\,dz\,dx + xy^2\,dy\,dz \\
&= (x^2z - y^2z)\,dx\,dy - x^2y\,dz\,dx + xy^2\,dy\,dz
\end{aligned}$$

EXAMPLE 11. If $\omega = x\,dx - y\,dy$ and $\eta = x\,dy\,dz + z\,dx\,dy$, find $\omega \wedge \eta$.

$$\begin{aligned}
\omega \wedge \eta &= (x\,dx - y\,dy) \wedge (x\,dy\,dz + z\,dx\,dy) \\
&= [(x\,dx - y\,dy) \wedge (x\,dy\,dz)] + [(x\,dx - y\,dy) \wedge (z\,dx\,dy)] \\
&= (x^2\,dx \wedge dy\,dz) - (xy\,dy \wedge dy\,dz) \\
&\quad + (xz\,dx \wedge dx\,dy) - (yz\,dy \wedge dx\,dy) \\
&= [x^2\,dx \wedge (dy \wedge dz)] - [xy\,dy \wedge (dy \wedge dz)] \\
&\quad + [xz\,dx \wedge (dx \wedge dy)] - [yz\,dy \wedge (dx \wedge dy)] \\
&= x^2\,dx\,dy\,dz - [xy(dy \wedge dy) \wedge dz] \\
&\quad + [xz(dx \wedge dx) \wedge dy] - [yz(dy \wedge dx) \wedge dy] \\
&= x^2\,dx\,dy\,dz - xy(0 \wedge dz) + xz(0 \wedge dy) + [yz(dy \wedge dy) \wedge dx] \\
&= x^2\,dx\,dy\,dz
\end{aligned}$$

The last major step in the development of this theory is to show how to differentiate forms. The derivative of a k-form is a $(k + 1)$-form if $k < 3$, and the derivative of a 3-form is always zero. If ω is a k-form we shall denote the derivative of ω by $d\omega$. The operation d has the following properties:

(1) If $f\colon K \to \mathbb{R}$ is a 0-form, then

$$df = \frac{\partial f}{\partial x}\,dx + \frac{\partial f}{\partial y}\,dy + \frac{\partial f}{\partial z}\,dz$$

(2) (*linearity*) If ω_1 and ω_2 are k-forms, then

$$d(\omega_1 + \omega_2) = d\omega_1 + d\omega_2$$

OPTIONAL (*Continued*)

(3) If ω is a k-form and η is an l-form

$$d(\omega \wedge \eta) = (d\omega \wedge \eta) + (-1)^k(\omega \wedge d\eta)$$

(4) $d(d\omega) = 0$ and $d(dx) = d(dy) = d(dz) = 0$ or, simply, $d^2 = 0$.

Properties (1) to (4) provide enough information to allow us to uniquely differentiate any form.

EXAMPLE 12. Let $\omega = P(x, y, z)\,dx + Q(x, y, z)\,dy$ be a 1-form on some open set $K \subset \mathbb{R}^3$. Find $d\omega$.

$$\begin{aligned}
&d[P(x, y, z)\,dx + Q(x, y, z)\,dy] \\
&\quad = d[P(x, y, z) \wedge dx] + d[Q(x, y, z) \wedge dy] && \text{(using 2)} \\
&\quad = [dP \wedge dx] + [P \wedge d(dx)] + [dQ \wedge dy] + [Q \wedge d(dy)] && \text{(using 3)} \\
&\quad = (dP \wedge dx) + (dQ \wedge dy) \\
&\quad = \left[\frac{\partial P}{\partial x}\,dx + \frac{\partial P}{\partial y}\,dy + \frac{\partial P}{\partial z}\,dz\right] \wedge dx + \left[\frac{\partial Q}{\partial x}\,dx + \frac{\partial Q}{\partial y}\,dy + \frac{\partial Q}{\partial z}\,dz\right] \wedge dy \\
& && \text{(using 1)} \\
&\quad = \left(\frac{\partial P}{\partial x}\,dx \wedge dx\right) + \left(\frac{\partial P}{\partial y}\,dy \wedge dx\right) + \left(\frac{\partial P}{\partial z}\,dz \wedge dx\right) \\
&\qquad + \left(\frac{\partial Q}{\partial x}\,dx \wedge dy\right) + \left(\frac{\partial Q}{\partial y}\,dy \wedge dy\right) + \left(\frac{\partial Q}{\partial z}\,dz \wedge dy\right) \\
&\quad = -\frac{\partial P}{\partial y}\,dx\,dy + \frac{\partial P}{\partial z}\,dz\,dx + \frac{\partial Q}{\partial x}\,dx\,dy - \frac{\partial Q}{\partial z}\,dy\,dz \\
&\quad = \left(\frac{\partial Q}{\partial x} - \frac{\partial P}{\partial y}\right)dx\,dy + \frac{\partial P}{\partial z}\,dz\,dx - \frac{\partial Q}{\partial z}\,dy\,dz
\end{aligned}$$

EXAMPLE 13. Let f be a 0-form. Using only differentiation rules (1) to (3) and the fact that $d(dx) = d(dy) = d(dz) = 0$, show that $d(df) = 0$.

By (1)

$$df = \frac{\partial f}{\partial x}\,dx + \frac{\partial f}{\partial y}\,dy + \frac{\partial f}{\partial z}\,dz$$

$$d(df) = d\left(\frac{\partial f}{\partial x}\,dx\right) + d\left(\frac{\partial f}{\partial y}\,dy\right) + d\left(\frac{\partial f}{\partial z}\,dz\right)$$

OPTIONAL (*Continued*)

Working only with the first term, using (3) we get

$$d\left(\frac{\partial f}{\partial x}\,dx\right) = d\left(\frac{\partial f}{\partial x} \wedge dx\right) = d\left(\frac{\partial f}{\partial x}\right) \wedge dx + \frac{\partial f}{\partial x} \wedge d(dx)$$

$$= \left(\frac{\partial^2 f}{\partial x^2}\,dx + \frac{\partial^2 f}{\partial y\,\partial x}\,dy + \frac{\partial^2 f}{\partial z\,\partial x}\,dz\right) \wedge dx + 0$$

$$= \frac{\partial^2 f}{\partial y\,\partial x}\,dy \wedge dx + \frac{\partial^2 f}{\partial z\,\partial x}\,dz \wedge dx$$

$$= -\frac{\partial^2 f}{\partial y\,\partial x}\,dx\,dy + \frac{\partial^2 f}{\partial z\,\partial x}\,dz\,dx$$

Similarly, we find that

$$d\left(\frac{\partial f}{\partial y}\,dy\right) = \frac{\partial^2 f}{\partial x\,\partial y}\,dx\,dy - \frac{\partial^2 f}{\partial z\,\partial y}\,dy\,dz$$

and

$$d\left(\frac{\partial f}{\partial z}\,dz\right) = -\frac{\partial^2 f}{\partial x\,\partial z}\,dz\,dx + \frac{\partial^2 f}{\partial y\,\partial z}\,dy\,dz$$

Adding these up, we get $d(df) = 0$ by the equality of mixed partials.

EXAMPLE 14. Show that $d(dx\,dy)$, $d(dy\,dz)$, and $d(dz\,dx)$ are all zero.

To prove the first case, we use property (3):

$$d(dx\,dy) = d(dx \wedge dy) = [d(dx) \wedge dy - dx \wedge d(dy)] = 0$$

The other cases are similar.

EXAMPLE 15. If $\eta = F(x, y, z)\,dx\,dy + G(x, y, z)\,dy\,dz + H(x, y, z)\,dz\,dx$, find $d\eta$.

By property (2)

$$d\eta = d(F\,dx\,dy) + d(G\,dy\,dz) + d(H\,dz\,dx)$$

We shall compute $d(F\,dx\,dy)$. Using (3) again, we get

$$d(F\,dx\,dy) = d(F \wedge dx\,dy) = dF \wedge (dx\,dy) + F \wedge d(dx\,dy)$$

By Example 14, $d(dx\,dy) = 0$, so we are left with

$$dF \wedge (dx\,dy) = \left(\frac{\partial F}{\partial x}\,dx + \frac{\partial F}{\partial y}\,dy + \frac{\partial F}{\partial z}\,dz\right) \wedge (dx \wedge dy)$$

$$= \left[\frac{\partial F}{\partial x}\,dx \wedge (dx \wedge dy)\right] + \left[\frac{\partial F}{\partial y}\,dy \wedge (dx \wedge dy)\right] + \left[\frac{\partial F}{\partial z}\,dz \wedge (dx \wedge dy)\right]$$

OPTIONAL (*Continued*)

Now

$$dx \wedge (dx \wedge dy) = (dx \wedge dx) \wedge dy = 0 \wedge dy = 0$$

$$\begin{aligned} dy \wedge (dx \wedge dy) &= -dy \wedge (dy \wedge dx) \\ &= -(dy \wedge dy) \wedge dx = 0 \wedge dx = 0 \end{aligned}$$

and

$$dz \wedge (dx \wedge dy) = (-1)^2 (dx \wedge dy) \wedge dz = dx\, dy\, dz$$

Consequently

$$d(F\, dx\, dy) = \frac{\partial F}{\partial z}\, dx\, dy\, dz$$

Analogously, we get that

$$d(G\, dy\, dz) = \frac{\partial G}{\partial x}\, dx\, dy\, dz \quad \text{and} \quad d(H\, dz\, dx) = \frac{\partial H}{\partial y}\, dx\, dy\, dz$$

Therefore

$$d\eta = \left(\frac{\partial F}{\partial z} + \frac{\partial G}{\partial x} + \frac{\partial H}{\partial y} \right) dx\, dy\, dz$$

We have now developed all the concepts needed to reformulate Green's, Stokes', and Gauss' theorems in the language of forms.

THEOREM 13 (GREEN'S THEOREM). *Let D be an elementary region in the xy-plane, with ∂D given the counterclockwise orientation. Suppose $\omega = P(x, y)\, dx + Q(x, y)\, dy$ is a 1-form on some open set K in $\mathbb{R}^3$ that contains D. Then*

$$\int_{\partial D} \omega = \int_D d\omega$$

Here $d\omega$ is a 2-form on K and D is in fact a surface in $\mathbb{R}^3$ parametrized by $\mathbf{\Phi}\colon D \to \mathbb{R}^3$, $\mathbf{\Phi}(x, y) = (x, y, 0)$. Since P and Q are explicitly *not* functions of z, then $\partial P/\partial z$ and $\partial Q/\partial z = 0$, and by Example 12, $d\omega = (\partial Q/\partial x - \partial P/\partial y)\, dx\, dy$. Consequently, Theorem 13 means nothing more than

$$\int_{\partial D} P\, dx + Q\, dy = \int_D \left(\frac{\partial Q}{\partial x} - \frac{\partial P}{\partial y} \right) dx\, dy$$

which is precisely Green's Theorem of Section 7.1. Hence Theorem 13 holds. Likewise, we have the following theorems.

THEOREM 14 (STOKES' THEOREM). *Let S be an oriented surface in $\mathbb{R}^3$ with a boundary consisting of a simple closed surve ∂S (Figure 7.6.3) oriented as the*

OPTIONAL (*Continued*)

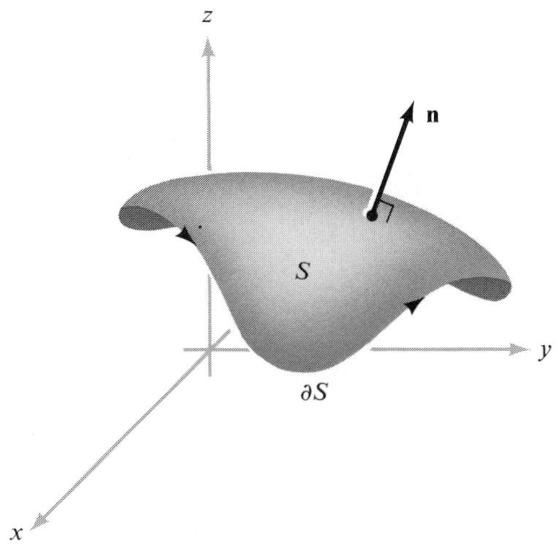

FIGURE 7.6.3
An oriented surface to which Stokes' Theorem applies.
ich the theory of forms applies.

boundary of S (see Figure 7.2.1). Suppose that ω is a 1-form on some open set K that contains S. Then

$$\int_{\partial S} \omega = \int_S d\omega$$

THEOREM 15 (GAUSS' THEOREM). *Let $\Omega \subset \mathbb{R}^3$ be an elementary region with $\partial\Omega$ given the outward orientation (see Section 7.4). If η is a 2-form on some region K containing Ω then*

$$\int_{\partial\Omega} \eta = \int_\Omega d\eta$$

The reader has probably noticed the strong similarity in the statements of these theorems. In the vector field formulations, we used divergence for regions in $\mathbb{R}^3$ (Gauss' Theorem), and the curl for surfaces in $\mathbb{R}^3$ (Stokes' Theorem) and regions in $\mathbb{R}^2$ (Green's Theorem). Here we just use the unified notion of derivative of a differential form for all three theorems; and, in fact, we can state all theorems as one by introducing a little more terminology.

By an *oriented 2-manifold with boundary* in $\mathbb{R}^3$ we mean a surface in $\mathbb{R}^3$ whose boundary is a simple closed curve with orientation as described in Section 7.2. By an *oriented 3-manifold* in $\mathbb{R}^3$ we mean an elementary region

OPTIONAL (*Continued*)

in $\mathbb{R}^3$ (we assume its boundary, which is a surface, is given the outward orientation discussed in Section 7.4). We call the following unified theorem "Stokes' Theorem," according to the current convention.

THEOREM 16 (STOKES' THEOREM). *Let M be an oriented k-manifold in $\mathbb{R}^3$ ($k = 2$ or 3) contained in some open set K. Suppose ω is a $(k-1)$-form on K. Then*

$$\int_{\partial M} \omega = \int_M d\omega$$

EXERCISES

1. Evaluate $\omega \wedge \eta$ if
   (a) $\omega = 2x\,dx + y\,dy$, $\eta = x^3\,dx + y^2\,dy$
   (b) $\omega = x\,dx - y\,dy$, $\eta = y\,dx + x\,dy$
   (c) $\omega = x\,dx + y\,dy + z\,dz$, $\eta = z\,dx\,dy + x\,dy\,dz + y\,dz\,dx$
   (d) $\omega = xy\,dy\,dz + x^2\,dx\,dy$, $\eta = dx + dz$
   (e) $\omega = e^{xyz}\,dx\,dy$, $\eta = e^{-xyz}\,dz$
2. Prove that

$$(a_1\,dx + a_2\,dy + a_3\,dz) \wedge (b_1\,dy\,dz + b_2\,dz\,dx + b_3\,dx\,dy) = \left(\sum_{i=1}^{3} a_ib_i\right) dx\,dy\,dz$$

3. Find $d\omega$ in the following examples.
   (a) $\omega = x^2y + y^3$
   (b) $\omega = y^2 \cos x\,dy + xy\,dx + dz$
   (c) $\omega = xy\,dy + (x + y)^2\,dx$
   (d) $\omega = x\,dx\,dy + z\,dy\,dz + y\,dz\,dx$
   (e) $\omega = (x^2 + y^2)\,dy\,dz$
   (f) $\omega = (x^2 + y^2 + z^2)\,dz$
   (g) $\omega = \dfrac{-x}{x^2 + y^2}\,dx + \dfrac{y}{x^2 + y^2}\,dy$
   (h) $\omega = x^2y\,dy\,dz$
4. Let $\mathbf{V}: K \to \mathbb{R}^3$ be a vector field defined by $\mathbf{V}(x, y, z) = G(x, y, z)\mathbf{i} + H(x, y, z)\mathbf{j} + F(x, y, z)\mathbf{k}$, and let η be the 2-form on K given by

$$\eta = F\,dx\,dy + G\,dy\,dz + H\,dz\,dx$$

   Show that $d\eta = (\text{div}\,\mathbf{V})\,dx\,dy\,dz$.
5. If $\mathbf{V} = A(x, y, z)\mathbf{i} + B(x, y, z)\mathbf{j} + C(x, y, z)\mathbf{k}$ is a vector field on $K \subset \mathbb{R}^3$, define the operation Form_2: Vector Fields $\to$ 2-forms by

$$\text{Form}_2(\mathbf{V}) = A\,dy\,dz + B\,dz\,dx + C\,dx\,dy$$

   (a) Show that $\text{Form}_2(\alpha\mathbf{V}_1 + \mathbf{V}_2) = \alpha\,\text{Form}_2(\mathbf{V}_1) + \text{Form}_2(\mathbf{V}_2)$, where α is a real number.
   (b) Show that $\text{Form}_2(\text{curl}\,\mathbf{V}) = d\omega$, where $\omega = A\,dx + B\,dy + C\,dz$.

OPTIONAL (*Continued*)

6. Using the differential-form version of Stokes' Theorem, prove the vector-field version in Section 7.2. Repeat for Gauss' Theorem.

7. Interpret Theorem 16 in the case $k = 1$.

8. Let $\omega = (x + y)\,dz + (y + z)\,dx + (x + z)\,dy$, and let S be the upper part of the unit sphere; that is, S is the set of (x, y, z) with $x^2 + y^2 + z^2 = 1$ and $z \geq 0$. ∂S is the unit circle in the xy-plane. Evaluate $\int_{\partial S} \omega$ both directly and by Stokes' Theorem.

9. Let T be the triangular solid bounded by the xy-plane, the xz-plane, the yz-plane, and the plane $2x + 3y + 6z = 12$. Compute

$$\int_{\partial T} F_1\,dx\,dy + F_2\,dy\,dz + F_3\,dz\,dx$$

directly and by Gauss' Theorem, if
(a) $F_1 = 3y$, $F_2 = 18z$, $F_3 = -12$; and
(b) $F_1 = z$, $F_2 = x^2$, $F_3 = y$.

10. Evaluate $\int_S \omega$ where $\omega = z\,dx\,dy + x\,dy\,dz + y\,dz\,dx$ and S is the unit sphere, directly and by Gauss' Theorem.

11. Let R be an elementary region in $\mathbb{R}^3$. Show that the volume of R is given by the formula

$$v(R) = \frac{1}{3}\int_{\partial R} x\,dy\,dz + y\,dz\,dx + z\,dx\,dy$$

12. In Section 3.2 we saw that the length $l(\boldsymbol{\sigma})$ of a curve $\boldsymbol{\sigma}(t) = (x(t), y(t), z(t))$, $a \leq t \leq b$ was given by the formula

$$l(\boldsymbol{\sigma}) = \int ds = \int_a^b \left(\frac{ds}{dt}\right) dt$$

where, loosely speaking

$$(ds)^2 = (dx)^2 + (dy)^2 + (dz)^2$$

or

$$\frac{ds}{dt} = \sqrt{\left(\frac{dx}{dt}\right)^2 + \left(\frac{dy}{dt}\right)^2 + \left(\frac{dz}{dt}\right)^2}$$

Now suppose a surface S is given in parametrized form by $\boldsymbol{\Phi}(u, v) = (x(u, v), y(u, v), z(u, v))$, $(u, v) \in D$. Show that the area of S can be expressed as

$$A(S) = \int_D dS$$

where $(dS)^2 = (dx \wedge dy)^2 + (dy \wedge dz)^2 + (dz \wedge dx)^2$.
(HINT:

$$dx = \frac{\partial x}{\partial u}\,du + \frac{\partial x}{\partial v}\,dv$$

and similarly for dy and dz. Use the law of forms for the basic 1-forms du and dv. Then dS turns out to be a function times the basic 2-form $du\,dv$, which we can integrate over D.)

REVIEW EXERCISES FOR CHAPTER 7

1. Use Green's Theorem to find the area of the loop of the curve $x = a \sin \theta \cos \theta$, $y = a \sin^2 \theta$, for $a > 0$ and $0 \le \theta \le \pi$.
2. Let $\mathbf{r}(x, y, z) = (x, y, z), r = \|\mathbf{r}\|$. Show $\nabla^2(\log r) = 1/r^2$ and $\nabla^2(r^n) = n(n + 1)r^{n-2}$.
3. Let $\mathbf{m}$ be a constant vector field and let

$$\phi = \frac{\mathbf{m} \cdot \mathbf{r}}{r^3}, \qquad \mathbf{F} = \frac{\mathbf{m} \times \mathbf{r}}{r^3}$$

Compute $\nabla\phi$, $\nabla^2\phi$, $\nabla \cdot \mathbf{F}$, and $\nabla \times \mathbf{F}$.

4. If $\nabla \times \mathbf{F} = \mathbf{0}$ and $\nabla \times \mathbf{G} = \mathbf{0}$, prove $\nabla \cdot (\mathbf{F} \times \mathbf{G}) = 0$.
5. Prove that if $\nabla \times \mathbf{F} = \nabla \times \mathbf{G}$ then $\mathbf{F} = \mathbf{G} + \nabla f$ for some f.
6. Prove the identity

$$(\mathbf{F} \cdot \nabla)\mathbf{F} = \tfrac{1}{2}\nabla(\mathbf{F} \cdot \mathbf{F}) + (\nabla \times \mathbf{F}) \times \mathbf{F}$$

7. Let $\mathbf{F} = 2yz\mathbf{i} + (-x + 3y + 2)\mathbf{j} + (x^2 + z)\mathbf{k}$. Evaluate $\int_S (\nabla \times \mathbf{F}) \cdot d\mathbf{S}$, where S is the cylinder $x^2 + y^2 = a^2$, $0 \le z \le 1$ (without the top and bottom). What if the top and bottom are included?
8. Let Ω be a region in $\mathbb{R}^3$ with boundary $\partial\Omega$. Prove the identity

$$\int_{\partial\Omega} [\mathbf{F} \times (\nabla \times \mathbf{G})] \cdot d\mathbf{S} = \int_\Omega (\nabla \times \mathbf{F}) \cdot (\nabla \times \mathbf{G})\, dV - \int_\Omega \mathbf{F} \cdot (\nabla \times \nabla \times \mathbf{G})\, dV$$

9. Let $\mathbf{F} = x^2y\mathbf{i} + z^8\mathbf{j} - 2xyz\mathbf{k}$. Evaluate the integral of $\mathbf{F}$ over the surface of the unit cube.
10. Verify Green's Theorem for the line integral

$$\int_C x^2y\, dx + y\, dy$$

when C is the boundary of the region between the curves $y = x$ and $y = x^3$, $0 \le x \le 1$.

11. (a) Show that $\mathbf{F} = (x^3 - 2xy^3)\mathbf{i} - 3x^2y^2\mathbf{j}$ is a gradient vector field.
    (b) Evaluate the integral of $\mathbf{F}$ along the path $x = \cos^3 \theta, y = \sin^3 \theta, 0 \le \theta \le \pi/2$.
12. Can you derive Green's Theorem in the plane from Gauss' Theorem?
13. (a) Show that $\mathbf{F} = 6xy(\cos z)\mathbf{i} + 3x^2(\cos z)\mathbf{j} - 3x^2y(\sin z)\mathbf{k}$ is conservative (see Section 7.3).
    (b) Find f such that $\mathbf{F} = \nabla f$.
    (c) Evaluate the integral of $\mathbf{F}$ along the curve $x = \cos^3 \theta$, $y = \sin^3 \theta$, $z = 0$, $0 \le \theta \le \pi/2$.
14. Suppose $\nabla \cdot \mathbf{F}(x_0, y_0, z_0) > 0$. Show that for a sufficiently small sphere S centered at (x_0, y_0, z_0), the flux of $\mathbf{F}$ out of S is positive.
15. Let the velocity of a fluid be described by $\mathbf{F} = 6xz\mathbf{i} + x^2y\mathbf{j} + yz\mathbf{k}$. Compute the rate at which fluid is leaving the unit cube.
16. Let $\mathbf{F} = x^2\mathbf{i} + (x^2y - 2xy)\mathbf{j} - x^2z\mathbf{k}$. Does there exist a $\mathbf{G}$ such that $\mathbf{F} = \nabla \times \mathbf{G}$?
17. Let $\mathbf{a}$ be a constant vector and $\mathbf{F} = \mathbf{a} \times \mathbf{r}$ [as usual, $\mathbf{r}(x, y, z) = (x, y, z)$]. Is $\mathbf{F}$ conservative? If so, find a potential for it.

*18. Consider the case of incompressible fluid flow with velocity field $\mathbf{F}$ and density ρ.
(a) If ρ is constant for each fixed t, then show that ρ is constant in t as well.
(b) If ρ is constant in t, then show that $\mathbf{F} \cdot \nabla\rho = 0$.

19. (a) Let $f(x, y, z) = 3xye^{z^2}$. Compute ∇f.
(b) Let $\boldsymbol{\sigma}(t) = (3\cos^3 t, \sin^2 t, e^t)$, $0 \le t \le \pi$. Evaluate

$$\int_{\boldsymbol{\sigma}} \nabla f \cdot d\mathbf{s}$$

(c) Verify directly Stokes' Theorem for gradient vector fields $\mathbf{F} = \nabla f$.

20. Using Green's Theorem, or otherwise, evaluate $\int_C x^3\, dy - y^3\, dx$, where C is the unit circle $(x^2 + y^2 = 1)$.

21. Evaluate the integral $\int_S \mathbf{F} \cdot d\mathbf{S}$ where $\mathbf{F} = x\mathbf{i} + y\mathbf{j} + 3\mathbf{k}$ and where S is the surface of the unit sphere $x^2 + y^2 + z^2 = 1$.

22. (a) State Stokes' Theorem for surfaces in $\mathbb{R}^3$.
(b) Let $\mathbf{F}$ be a vector field on $\mathbb{R}^3$ satisfying $\nabla \times \mathbf{F} = \mathbf{0}$. Use Stokes' Theorem to show that $\int_C \mathbf{F} \cdot d\mathbf{s} = 0$ where C is a closed curve.

23. Are the following true or false?
(a) A continuous function is always differentiable.
(b) If the partial derivatives of f exist and are continuous, then f is differentiable.
(c) $\{(x, y) \in \mathbb{R}^2 \,|\, x^2 + y^2 > 0\}$ is an open set.
(d) The magnitude of the curl of a vector field $\mathbf{F}$ represents the amount of circulation per unit area in a plane normal to $\nabla \times \mathbf{F}$.
(e) It is always true that $\mathbf{v} \cdot (\mathbf{v} \times \mathbf{w}) = 0$.
(f) It is always true that $\nabla \cdot (\nabla f) \ge 0$.
(g) $\int_0^1 \int_0^y f(x, y)\, dx\, dy = \int_0^1 \int_0^x f(x, y)\, dy\, dx$ holds for all continuous functions $f(x, y)$.
(h) The area of a region enclosed by a closed curve C is given by $A = \int_C x\, dy$.
(i) If S is a surface and $\mathbf{F}$ is orthogonal to the boundary of S, then it is true that $\int_S (\nabla \times \mathbf{F}) \cdot d\mathbf{S} = 0$.
(j) If the height of a hill is described by a function $h(x, y)$, then a skier who wishes to descend the fastest should ski in a (compass) direction orthogonal to ∇h.
(k) $\nabla \times (\nabla \times \mathbf{F})$ is always equal to $\mathbf{0}$.

24. Evaluate $\int_C yz\, dx + xz\, dy + xy\, dz$ where C is the curve of intersection of the cylinder $x^2 + y^2 = 1$ and the surface $z = y^2$.

25. Evaluate $\int_C (x + y)\, dx + (2x - z)\, dy + (y + z)\, dz$ where C is the perimeter of the triangle connecting $(2, 0, 0)$, $(0, 3, 0)$, and $(0, 0, 6)$ in that order.

26. Which of the following are conservative fields on $\mathbb{R}^3$? For those that are, find a function f such that $\mathbf{F} = \nabla f$.
(a) $\mathbf{F}(x, y, z) = 3x^2y\mathbf{i} + x^3\mathbf{j} + 5\mathbf{k}$
(b) $\mathbf{F}(x, y, z) = (x + z)\mathbf{i} - (y + z)\mathbf{j} + (x - y)\mathbf{k}$
(c) $\mathbf{F}(x, y, z) = 2xy^3\mathbf{i} + x^2z^3\mathbf{j} + 3x^2yz^2\mathbf{k}$

27. Consider the following two vector fields in $\mathbb{R}^3$.
(*i*) $\mathbf{F}(x, y, z) = y^2\mathbf{i} - z^2\mathbf{j} + x^2\mathbf{k}$
(*ii*) $\mathbf{G}(x, y, z) = (x^3 - 3xy^2)\mathbf{i} + (y^3 - 3x^2y)\mathbf{j} + z\mathbf{k}$
(a) Which of these fields (if any) are conservative on $\mathbb{R}^3$ (i.e., which are gradient fields). Give reasons for your answer.
(b) Find potentials for the fields that are conservative.

(c) Let $\boldsymbol{\alpha}$ be the path that goes from (0, 0, 0) to (1, 1, 1) by following edges of the cube $0 \leq x \leq 1, 0 \leq y \leq 1, 0 \leq z \leq 1$ from (0, 0, 0) to (0, 0, 1) to (0, 1, 1) to (1, 1, 1). Let $\boldsymbol{\beta}$ be the path from (0, 0, 0) to (1, 1, 1) directly along the diagonal of the cube. Find the values of the line integrals

$$\int_{\boldsymbol{\alpha}} \mathbf{F} \cdot d\mathbf{s}, \qquad \int_{\boldsymbol{\alpha}} \mathbf{G} \cdot d\mathbf{s}, \qquad \int_{\boldsymbol{\beta}} \mathbf{F} \cdot d\mathbf{s}, \qquad \int_{\boldsymbol{\beta}} \mathbf{G} \cdot d\mathbf{s}$$

28. Consider the *constant* vector field $\mathbf{F}(x, y, z) = \mathbf{i} + 2\mathbf{j} - \mathbf{k}$ in $\mathbb{R}^3$.
(a) Find a scalar field $\phi(x, y, z)$ in $\mathbb{R}^3$ such that $\nabla\phi = \mathbf{F}$ in $\mathbb{R}^3$ and $\phi(0, 0, 0) = 0$.
(b) On the sphere Σ of radius *2* about the origin find all the points at which
(*i*) ϕ is a maximum and
(*ii*) ϕ is a minimum.
(c) Compute the maximum and minimum values of ϕ on Σ.

APPENDICES

Mathematics is difficult, even for mathematicians.

REINHOLD BÖHME

Theorem proofs should be set in large type.

KENNETH ROSS

APPENDIX A

SOME TECHNICAL DIFFERENTIATION THEOREMS

In this appendix we examine the definition of the derivative in further detail and supply proofs omitted from sections 2.2, 2.3, and 2.4.

We shall begin by supplying the proofs of the limit theorems presented in Section 2.2 (The theorem numbering corresponds to that in Chapter 2.) Let us recall the definition of limit.

DEFINITION OF LIMIT. *Let $f: A \subset \mathbb{R}^n \to \mathbb{R}^m$ where A is open. Let $\mathbf{x}_0$ be in A or be a boundary point of A, and let N be a neighborhood of $\mathbf{b} \in \mathbb{R}^m$. We say f is* **eventually in** N **as x approaches** $\mathbf{x}_0$ *if there exists a neighborhood U of $\mathbf{x}_0$ such that $\mathbf{x} \neq \mathbf{x}_0$, $\mathbf{x} \in U$, and $\mathbf{x} \in A$ implies $f(\mathbf{x}) \in N$. We say $f(\mathbf{x})$* **approaches b** *as $\mathbf{x}$ approaches $\mathbf{x}_0$, or, in symbols,*

$$\underset{\mathbf{x}\to\mathbf{x}_0}{\text{limit}}\, f(\mathbf{x}) = \mathbf{b} \quad \text{or} \quad f(\mathbf{x}) \to \mathbf{b} \quad \text{as} \quad \mathbf{x} \to \mathbf{x}_0$$

when, given **any** *neighborhood N of $\mathbf{b}$, f is eventually in N as $\mathbf{x}$ approaches $\mathbf{x}_0$. It may be that as $\mathbf{x}$ approaches $\mathbf{x}_0$ the values $f(\mathbf{x})$ do not get close to any particular number. In this case we say that the* $\underset{\mathbf{x}\to\mathbf{x}_0}{\text{limit}}\, f(\mathbf{x})$ **does not exist**.

Let us first establish that this is equivalent to the ε-δ formulation of limits.

THEOREM 2. *Let $f: A \subset \mathbb{R}^n \to \mathbb{R}^m$ and let $\mathbf{x}_0$ be in A or be a boundary point of A. Then* $\lim_{\mathbf{x}\to\mathbf{x}_0} f(\mathbf{x}) = \mathbf{b}$ *if and only if for every number $\varepsilon > 0$ there is a $\delta > 0$ such that for $\mathbf{x} \in A$ and $0 < \|\mathbf{x} - \mathbf{x}_0\| < \delta$ we have $\|f(\mathbf{x}) - \mathbf{b}\| < \varepsilon$.*

Proof. First let us assume that $\lim_{\mathbf{x}\to\mathbf{x}_0} f(\mathbf{x}) = \mathbf{b}$. Let $\varepsilon > 0$ be given, and consider the ε-neighborhood $N = D_\varepsilon(\mathbf{b})$, the ball or disc of radius ε with center $\mathbf{b}$. By the definition of limit, f is eventually in $D_\varepsilon(\mathbf{b})$ as $\mathbf{x}$ approaches $\mathbf{x}_0$, which means there is a neighborhood U of $\mathbf{x}_0$ such that $f(\mathbf{x}) \in D_\varepsilon(\mathbf{b})$ if $\mathbf{x} \in U$, and $\mathbf{x} \neq \mathbf{x}_0$. Now since U is open and $\mathbf{x}_0 \in U$, there is a $\delta > 0$ such that $D_\delta(\mathbf{x}_0) \subset U$. Consequently, $0 < \|\mathbf{x} - \mathbf{x}_0\| < \delta$ and $\mathbf{x} \in A$ implies $\mathbf{x} \in D_\delta(\mathbf{x}_0) \subset U$. Thus $f(\mathbf{x}) \in D_\varepsilon(\mathbf{b})$, which means that $\|f(\mathbf{x}) - \mathbf{b}\| < \varepsilon$. This is the ε-δ assertion we wanted to prove.

We still have to prove the converse. Assume that for every $\varepsilon > 0$ there is a $\delta > 0$ such that $0 < \|\mathbf{x} - \mathbf{x}_0\| < \delta$ and $\mathbf{x} \in A$ implies $\|f(\mathbf{x}) - \mathbf{b}\| < \varepsilon$. Let N be a neighborhood of $\mathbf{b}$. We have to show that f is eventually in N as $\mathbf{x} \to \mathbf{x}_0$; that is, we must find an open set $U \subset \mathbb{R}^n$ such that $\mathbf{x} \in U$, $\mathbf{x} \in A$, and $\mathbf{x} \neq \mathbf{x}_0$ implies $f(\mathbf{x}) \in N$. Now since N is open, there is an $\varepsilon > 0$ such that $D_\varepsilon(\mathbf{b}) \subset N$. If we choose $U = D_\delta(\mathbf{x})$ (according to our assumption), then $\mathbf{x} \in U$, $\mathbf{x} \in A$, and $\mathbf{x} \neq \mathbf{x}_0$ means $\|f(\mathbf{x}) - \mathbf{b}\| < \varepsilon$, that is $f(\mathbf{x}) \in D_\varepsilon(\mathbf{b}) \subset N$. ■

THEOREM 3 (UNIQUENESS OF LIMITS). *If* $\lim_{\mathbf{x}\to\mathbf{x}_0} f(\mathbf{x}) = \mathbf{b}_1$ and $\lim_{\mathbf{x}\to\mathbf{x}_0} f(\mathbf{x}) = \mathbf{b}_2$, *then* $\mathbf{b}_1 = \mathbf{b}_2$.

Proof. It is convenient to use the ε-δ formulation of Theorem 2. Let us suppose $f(\mathbf{x}) \to \mathbf{b}_1$ and $f(\mathbf{x}) \to \mathbf{b}_2$ as $\mathbf{x} \to \mathbf{x}_0$. Given $\varepsilon > 0$ we can, by assumption, find $\delta_1 > 0$ such that if $\mathbf{x} \in A$ and $0 < \|\mathbf{x} - \mathbf{x}_0\| < \delta_1$, then $\|f(\mathbf{x}) - \mathbf{b}_1\| < \varepsilon$, and similarly, $\delta_2 > 0$ such that $0 < \|\mathbf{x} - \mathbf{x}_0\| < \delta_2$ implies $\|f(\mathbf{x}) - \mathbf{b}_2\| < \varepsilon$. Let δ be the smaller of δ_1 and δ_2. Choose $\mathbf{x}$ such that $0 < \|\mathbf{x} - \mathbf{x}_0\| < \delta$ and $\mathbf{x} \in A$. Such $\mathbf{x}$'s exist, because $\mathbf{x}_0$ is in A or is a boundary point of A. Thus, using the triangle inequality, we have

$$\begin{aligned}\|\mathbf{b}_1 - \mathbf{b}_2\| &= \|(\mathbf{b}_1 - f(\mathbf{x})) + (f(\mathbf{x}) - \mathbf{b}_2)\| \\ &\leq \|\mathbf{b}_1 - f(\mathbf{x})\| + \|f(\mathbf{x}) - \mathbf{b}_2\| < \varepsilon + \varepsilon = 2\varepsilon\end{aligned}$$

Thus for *every* $\varepsilon > 0$ we have $\|\mathbf{b}_1 - \mathbf{b}_2\| < 2\varepsilon$. Hence $\mathbf{b}_1 = \mathbf{b}_2$, for if $\mathbf{b}_1 \neq \mathbf{b}_2$ we could let $\varepsilon = \|\mathbf{b}_1 - \mathbf{b}_2\|/2 > 0$ and we would have $\|\mathbf{b}_1 - \mathbf{b}_2\| < \|\mathbf{b}_1 - \mathbf{b}_2\|$, an impossibility. ■

THEOREM 4. *Let $f: A \subset \mathbb{R}^n \to \mathbb{R}^m$, $g: A \subset \mathbb{R}^n \to \mathbb{R}^m$, $\mathbf{x}_0$ be in A or be a boundary point of A, $\mathbf{b} \in \mathbb{R}^m$, and $c \in \mathbb{R}$; the following assertions hold:*

(*i*) *if* $\lim_{\mathbf{x}\to\mathbf{x}_0} f(\mathbf{x}) = \mathbf{b}$, *then* $\lim_{\mathbf{x}\to\mathbf{x}_0} cf(\mathbf{x}) = c\mathbf{b}$ *where* $cf: A \to \mathbb{R}^m$ *is defined by* $\mathbf{x} \mapsto c(f(\mathbf{x}))$;

(*ii*) *if* $\lim_{\mathbf{x}\to\mathbf{x}_0} f(\mathbf{x}) = \mathbf{b}_1$ *and* $\lim_{\mathbf{x}\to\mathbf{x}_0} g(\mathbf{x}) = \mathbf{b}_2$, *then* $\lim_{\mathbf{x}\to\mathbf{x}_0} (f + g)(\mathbf{x}) = \mathbf{b}_1 + \mathbf{b}_2$ *where* $(f + g): A \to \mathbb{R}^m$ *is defined by* $\mathbf{x} \mapsto f(\mathbf{x}) + g(\mathbf{x})$;

(*iii*) *if* $m = 1$, $\lim_{\mathbf{x}\to\mathbf{x}_0} f(\mathbf{x}) = b_1$, *and* $\lim_{\mathbf{x}\to\mathbf{x}_0} g(\mathbf{x}) = b_2$, *then* $\lim_{\mathbf{x}\to\mathbf{x}_0} (fg)(\mathbf{x}) = b_1 b_2$ *where* $(fg): A \to \mathbb{R}$, $\mathbf{x} \mapsto f(\mathbf{x})g(\mathbf{x})$;

(*iv*) *if* $m = 1$, $\lim_{\mathbf{x}\to\mathbf{x}_0} f(\mathbf{x}) = b \neq 0$, *and* $f(\mathbf{x}) \neq 0$ *for all* $\mathbf{x} \in A$, *then* $\lim_{\mathbf{x}\to\mathbf{x}_0} 1/f = 1/b$ *where* $1/f: A \to \mathbb{R}$, $\mathbf{x} \mapsto 1/f(\mathbf{x})$; *and*

(*v*) *if* $f(\mathbf{x}) = (f_1(\mathbf{x}), \ldots, f_m(\mathbf{x}))$ *where* $f_i: A \to \mathbb{R}$, $i = 1, \ldots, m$, *are the component functions of* f, *then* $\lim_{\mathbf{x}\to\mathbf{x}_0} f(\mathbf{x}) = \mathbf{b} = (b_1, \ldots, b_m)$ *if and only if* $\lim_{\mathbf{x}\to\mathbf{x}_0} f_i(\mathbf{x}) = b_i$ *for each* $i = 1, \ldots, m$.

Proof. We shall illustrate the technique of proof by proving (*i*) and (*ii*). The proofs of the other assertions are a bit more complicated and may be supplied by the reader. In each case, the ε-δ formulation of Theorem 2 is probably the most convenient approach.

To prove (*i*), let $\varepsilon > 0$ be given; we must produce a number $\delta > 0$ such that $\|cf(\mathbf{x}) - c\mathbf{b}\| < \varepsilon$ if $0 < \|\mathbf{x} - \mathbf{x}_0\| < \delta$. If $c = 0$, any δ will do, so we can suppose $c \neq 0$. Let $\varepsilon' = \varepsilon/|c|$; from the definition of limit, there is a δ with the property that $0 < \|\mathbf{x} - \mathbf{x}_0\| < \delta$ implies $\|f(\mathbf{x}) - \mathbf{b}\| < \varepsilon' = \varepsilon/|c|$. Thus $0 < \|\mathbf{x} - \mathbf{x}_0\| < \delta$ implies $\|cf(\mathbf{x}) - c\mathbf{b}\| = |c|\,\|f(\mathbf{x}) - \mathbf{b}\| < \varepsilon$, which proves (*i*).

To prove (*ii*), let $\varepsilon > 0$ be given again. Choose $\delta_1 > 0$ such that $0 < \|\mathbf{x} - \mathbf{x}_0\| < \delta_1$ implies $\|f(\mathbf{x}) - \mathbf{b}_1\| < \varepsilon/2$. Similarly, choose $\delta_2 > 0$ such that $0 < \|\mathbf{x} - \mathbf{x}_0\| < \delta_2$ implies $\|g(\mathbf{x}) - \mathbf{b}_2\| < \varepsilon/2$. Let δ be the lesser of δ_1 and δ_2. Then $0 < \|\mathbf{x} - \mathbf{x}_0\| < \delta$ implies

$$\|f(\mathbf{x}) + g(\mathbf{x}) - \mathbf{b}_1 - \mathbf{b}_2\| \leq \|f(\mathbf{x}) - \mathbf{b}_1\| + \|g(\mathbf{x}) - \mathbf{b}_2\| < \varepsilon/2 + \varepsilon/2 = \varepsilon$$

Thus we have proved that $(f + g)(\mathbf{x}) \to \mathbf{b}_1 + \mathbf{b}_2$ as $\mathbf{x} \to \mathbf{x}_0$. ■

Let us recall the definition of a continuous function.

DEFINITION. *Let* $f: A \subset \mathbb{R}^n \to \mathbb{R}^m$ *be a given function with domain* A. *Let* $\mathbf{x}_0 \in A$. *We say* f *is* **continuous at** $\mathbf{x}_0$ *if and only if*

$$\lim_{\mathbf{x}\to\mathbf{x}_0} f(\mathbf{x}) = f(\mathbf{x}_0)$$

If we just say that f *is* **continuous**, *we shall mean that* f *is continuous at each point* $\mathbf{x}_0$ *of* A.

From Theorem 2, we get Theorem 7: f is continuous at $\mathbf{x}_0 \in A$ if and only if for every number $\varepsilon > 0$ there is a number $\delta > 0$ such that

$$\mathbf{x} \in A \text{ and } \|\mathbf{x} - \mathbf{x}_0\| < \delta \quad \text{implies} \quad \|f(\mathbf{x}) - f(\mathbf{x}_0)\| < \varepsilon$$

One of the properties of continuous functions stated without proof in Section 2.2 was the following:

THEOREM 5. *Let $f: A \subset \mathbb{R}^n \to \mathbb{R}^m$ and let $g: B \subset \mathbb{R}^m \to \mathbb{R}^p$. Suppose $f(A) \subset B$ so that $g \circ f$ is defined on A. If f is continuous at $\mathbf{x}_0 \in A$ and g is continuous at $\mathbf{y}_0 = f(\mathbf{x}_0)$, then $g \circ f$ is continuous at $\mathbf{x}_0$.*

Proof. We shall use the ε-δ criterion for continuity. Thus, given $\varepsilon > 0$, we must find $\delta > 0$ such that for $\mathbf{x} \in A$,

$$\|\mathbf{x} - \mathbf{x}_0\| < \delta \quad \text{implies} \quad \|(g \circ f)(\mathbf{x}) - (g \circ f)(\mathbf{x}_0)\| < \varepsilon$$

Since g is continuous at $f(\mathbf{x}_0) = \mathbf{y}_0 \in B$, there is a $\gamma > 0$ such that for $\mathbf{y} \in B$,

$$\|\mathbf{y} - \mathbf{y}_0\| < \gamma \quad \text{implies} \quad \|g(\mathbf{y}) - g(\mathbf{y}_0)\| < \varepsilon$$

As f is continuous at $\mathbf{x}_0 \in A$, there is, for this γ, a $\delta > 0$ such that for $\mathbf{x} \in A$,

$$\|\mathbf{x} - \mathbf{x}_0\| < \delta \quad \text{implies} \quad \|f(\mathbf{x}) - f(\mathbf{x}_0)\| < \gamma$$

which in turn implies

$$\|g(f(\mathbf{x})) - g(f(\mathbf{x}_0))\| < \varepsilon$$

which is the desired conclusion. ■

To simplify the exposition in Section 2.3 we assumed, as part of the definition of $Df(\mathbf{x}_0)$, that the partial derivatives of f existed. Our next objective is to show that this assumption can be omitted. Let us begin by redefining "differentiable." Theorem 16 will show that this definition is equivalent to the old one.

DEFINITION. *Let U be an open set in $\mathbb{R}^n$ and let $f: U \subset \mathbb{R}^n \to \mathbb{R}^m$ be a given function. We say that f is* **differentiable** *at $\mathbf{x}_0 \in U$ if and only if there exists an $m \times n$ matrix T such that*

$$\operatorname*{limit}_{\mathbf{x} \to \mathbf{x}_0} \frac{\|f(\mathbf{x}) - f(\mathbf{x}_0) - T(\mathbf{x} - \mathbf{x}_0)\|}{\|\mathbf{x} - \mathbf{x}_0\|} = 0 \tag{1}$$

We call T the **derivative** *of f at $\mathbf{x}_0$ and denote it by $Df(\mathbf{x}_0)$. In matrix notation, $T(\mathbf{x} - \mathbf{x}_0)$ stands for (see Section 1.5)*

$$\begin{bmatrix} t_{11} & t_{12} & \cdots & t_{1n} \\ t_{21} & t_{22} & \cdots & t_{2n} \\ \vdots & \vdots & & \vdots \\ t_{m1} & t_{m2} & \cdots & t_{mn} \end{bmatrix} \begin{bmatrix} x_1 - x_{01} \\ \vdots \\ \\ x_n - x_{0n} \end{bmatrix}$$

where $\mathbf{x} = (x_1, \ldots, x_n)$ *and* $\mathbf{x}_0 = (x_{01}, \ldots, x_{0n})$. *Sometimes we write* $T(\mathbf{y})$ *as* $T \cdot \mathbf{y}$ *or just* $T\mathbf{y}$.

Condition (1) can be rewritten as

$$\lim_{\mathbf{h}\to\mathbf{0}} \frac{\|f(\mathbf{x}_0 + \mathbf{h}) - f(\mathbf{x}_0) - T\mathbf{h}\|}{\|\mathbf{h}\|} = 0 \tag{2}$$

by letting $\mathbf{h} = \mathbf{x} - \mathbf{x}_0$. Written in terms of ε-δ notation, (2) says that for every $\varepsilon > 0$ there is a $\delta > 0$ such that $0 < \|\mathbf{h}\| < \delta$ implies

$$\frac{\|f(\mathbf{x}_0 + \mathbf{h}) - f(\mathbf{x}_0) - T\mathbf{h}\|}{\|\mathbf{h}\|} < \varepsilon$$

or in other words,

$$\|f(\mathbf{x}_0 + \mathbf{h}) - f(\mathbf{x}_0) - T\mathbf{h}\| < \varepsilon\|\mathbf{h}\|$$

Notice that because U is open, as long as δ is small enough, $\|\mathbf{h}\| < \delta$ implies $\mathbf{x}_0 + \mathbf{h} \in U$.

Our task is to show that the matrix T is necessarily the matrix of partial derivatives, and hence that this abstract definition agrees with the definition of differentiability given in Section 2.3.

THEOREM 16. *Suppose* $f\colon U \subset \mathbb{R}^n \to \mathbb{R}^m$ *is differentiable at* $\mathbf{x}_0 \in \mathbb{R}^n$. *Then all the partial derivatives of* f *exist at the point* $\mathbf{x}_0$ *and the* $m \times n$ *matrix* T *is given by*

$$(t_{ij}) = \left(\frac{\partial f_i}{\partial x_j}\right)$$

that is,

$$T = Df(\mathbf{x}_0) = \begin{bmatrix} \dfrac{\partial f_1}{\partial x_1} & \cdots & \dfrac{\partial f_1}{\partial x_n} \\ \vdots & & \vdots \\ \dfrac{\partial f_m}{\partial x_1} & \cdots & \dfrac{\partial f_m}{\partial x_n} \end{bmatrix}$$

where $\partial f_i/\partial x_j$ *is evaluated at* $\mathbf{x}_0$. *In particular, this implies that* T *is uniquely determined; there is no other matrix satisfying condition* (1).

Proof. By Theorem 4(v), condition (2) is the same as

$$\lim_{\mathbf{h}\to\mathbf{0}} \frac{|f_i(\mathbf{x}_0 + \mathbf{h}) - f_i(\mathbf{x}_0) - (T\mathbf{h})_i|}{\|\mathbf{h}\|} = 0, \qquad 1 \le i \le m$$

Here $(T\mathbf{h})_i$ stands for the ith component of the column vector $T\mathbf{h}$. Now let $\mathbf{h} = a\mathbf{e}_j = (0, \ldots, a, \ldots, 0)$, the number a in the jth slot and zeros elsewhere.

We get

$$\lim_{a\to 0} \frac{|f_i(\mathbf{x}_0 + a\mathbf{e}_j) - f_i(\mathbf{x}_0) - a(T\mathbf{e}_j)_i|}{|a|} = 0$$

or in other words

$$\lim_{a\to 0} \left| \frac{f_i(\mathbf{x}_0 + a\mathbf{e}_j) - f_i(\mathbf{x}_0)}{a} - (T\mathbf{e}_j)_i \right| = 0$$

so that

$$\lim_{a\to 0} \frac{f_i(\mathbf{x}_0 + a\mathbf{e}_j) - f_i(\mathbf{x}_0)}{a} = (T\mathbf{e}_j)_i$$

But this limit is nothing more than the partial derivative $\partial f_i/\partial x_j$ evaluated at the point $\mathbf{x}_0$. Thus we have proved that $\partial f_i/\partial x_j$ exists and equals $(T\mathbf{e}_j)_i$. But $(T\mathbf{e}_j)_i = t_{ij}$ (see Section 1.5), so our theorem follows. ■

Our next task is to show that differentiability implies continuity.

THEOREM 8. *Let $f: U \subset \mathbb{R}^n \to \mathbb{R}^m$ be differentiable at $\mathbf{x}_0$. Then f is continuous at $\mathbf{x}_0$, and furthermore, $\|f(\mathbf{x}) - f(\mathbf{x}_0)\| < M_1\|\mathbf{x} - \mathbf{x}_0\|$ for some constant M_1 and $\mathbf{x}$ near $\mathbf{x}_0$, $\mathbf{x} \neq \mathbf{x}_0$.*

Proof. We shall need to use the result of Exercise 2 at the end of this section; namely $\|Df(\mathbf{x}_0) \cdot \mathbf{h}\| \leq M\|\mathbf{h}\|$ where M is the square root of the sum of the squares of the matrix elements in $Df(\mathbf{x}_0)$.

Choose $\varepsilon = 1$. Then by the definition of the derivative (see formula (2)) there is a $\delta_1 > 0$ such that $0 < \|\mathbf{h}\| < \delta_1$ implies

$$\|f(\mathbf{x}_0 + \mathbf{h}) - f(\mathbf{x}_0) - Df(\mathbf{x}_0) \cdot \mathbf{h}\| < \varepsilon\|\mathbf{h}\| = \|\mathbf{h}\|$$

Now notice that if $\|\mathbf{h}\| < \delta_1$, then

$$\begin{aligned} \|f(\mathbf{x}_0 + \mathbf{h}) - f(\mathbf{x}_0)\| &= \|f(\mathbf{x}_0 + \mathbf{h}) - f(\mathbf{x}_0) - Df(\mathbf{x}_0) \cdot \mathbf{h} + Df(\mathbf{x}_0) \cdot \mathbf{h}\| \\ &\leq \|f(\mathbf{x}_0 + \mathbf{h}) - f(\mathbf{x}_0) - Df(\mathbf{x}_0) \cdot \mathbf{h}\| + \|Df(\mathbf{x}_0) \cdot \mathbf{h}\| \\ &< \|\mathbf{h}\| + M\|\mathbf{h}\| = (1 + M)\|\mathbf{h}\| \end{aligned}$$

(Note that in this derivation we have used the triangle inequality.) Setting $M_1 = 1 + M$ proves the second assertion of the theorem.

Now let ε' be any positive number, and let δ be the smaller of δ_1 and $\varepsilon'/(1 + M)$. Then $\|\mathbf{h}\| < \delta$ implies

$$\|f(\mathbf{x}_0 + \mathbf{h}) - f(\mathbf{x}_0)\| < (1 + M)\varepsilon'/(1 + M) = \varepsilon'$$

which proves that

$$\operatorname*{limit}_{\mathbf{h}\to\mathbf{0}} f(\mathbf{x}_0 + \mathbf{h}) = f(\mathbf{x}_0)$$

or that f is continuous at $\mathbf{x}_0$. ■

We asserted in Section 2.3 that an important criterion for differentiability is that the partials exist and are continuous. We now prove this.

THEOREM 9. *Let $f\colon U \subset \mathbb{R}^n \to \mathbb{R}^m$. Suppose the partial derivatives $\partial f_i/\partial x_j$ of f all exist and are continuous in a neighborhood of a point $\mathbf{x} \in U$. Then f is differentiable at* $\mathbf{x}$.

Proof. (In this proof we are going to use the Mean Value Theorem from one-variable calculus. See p. 119 above for the statement.)

It suffices to consider the case $m = 1$, for the same reasons as in the proof of Theorem 16, so let us assume $f\colon U \subset \mathbb{R}^n \to \mathbb{R}$. We have to show that

$$\operatorname*{limit}_{\mathbf{h}\to\mathbf{0}} \frac{\left| f(\mathbf{x}+\mathbf{h}) - f(\mathbf{x}) - \sum_{i=1}^{n} \frac{\partial f}{\partial x_i}(\mathbf{x})h_i \right|}{\|\mathbf{h}\|} = 0 .$$

Write

$$\begin{aligned}
&f(x_1 + h_1, \ldots, x_n + h_n) - f(x_1, \ldots, x_n)\\
&= f(x_1 + h_1, \ldots, x_n + h_n) - f(x_1, x_2 + h_2, \ldots, x_n + h_n)\\
&\quad + f(x_1, x_2 + h_2, \ldots, x_n + h_n) - f(x_1, x_2, x_3 + h_3, \ldots, x_n + h\ ) + \cdots\\
&\quad + f(x_1, \ldots, x_{n-1} + h_{n-1}, x_n + h_n) - f(x_1, \ldots, x_{n-1}, x_n + h_n)\\
&\quad + f(x_1, \ldots, x_{n-1}, x_n + h_n) - f(x_1, \ldots, x_n)
\end{aligned}$$

(This is called a "telescoping sum" since each term cancels with the succeeding or preceding one, except the first and last.) By the Mean Value Theorem, this expression may be written as

$$f(\mathbf{x}+\mathbf{h}) - f(\mathbf{x}) = \frac{\partial f}{\partial x_1}(\mathbf{y}_1)h_1 + \frac{\partial f}{\partial x_2}(\mathbf{y}_2)h_2 + \cdots + \frac{\partial f}{\partial x_n}(\mathbf{y}_n)h_n$$

where $\mathbf{y}_1 = (c_1, x_2 + h_2, \ldots, x_n + h_n)$, and c_1 lies between x_1 and $x_1 + h_1$; $\mathbf{y}_2 = (x_1, c_2, x_3 + h_3, \ldots, x_n + h_n)$, and c_2 lies between x_2 and $x_2 + h_2$; and $\mathbf{y}_n = (x_1, \ldots, x_{n-1}, c_n)$ where c_n lies between x_n and $x_n + h_n$. Thus we can write

$$\begin{aligned}
&\left| f(\mathbf{x}+\mathbf{h}) - f(\mathbf{x}) - \sum_{i=1}^{n} \frac{\partial f}{\partial x_i}(\mathbf{x})h_i \right|\\
&\qquad = \left| \left(\frac{\partial f}{\partial x_1}(\mathbf{y}_1) - \frac{\partial f}{\partial x_1}(\mathbf{x})\right)h_1 + \cdots + \left(\frac{\partial f}{\partial x_n}(\mathbf{y}_n) - \frac{\partial f}{\partial x_n}(\mathbf{x})\right)h_n \right|
\end{aligned}$$

By the triangle inequality, this expression is less than or equal to

$$\left|\frac{\partial f}{\partial x_1}(\mathbf{y}_1) - \frac{\partial f}{\partial x_1}(\mathbf{x})\right| |h_1| + \cdots + \left|\frac{\partial f}{\partial x_n}(\mathbf{y}_n) - \frac{\partial f}{\partial x_n}(\mathbf{x})\right| |h_n|$$

$$\leq \left\{\left|\frac{\partial f}{\partial x_1}(\mathbf{y}_1) - \frac{\partial f}{\partial x_1}(\mathbf{x})\right| + \cdots + \left|\frac{\partial f}{\partial x_n}(\mathbf{y}_n) - \frac{\partial f}{\partial x_n}(\mathbf{x})\right|\right\} \|\mathbf{h}\|$$

since $|h_i| \leq \|\mathbf{h}\|$. Thus we have proved that

$$\frac{\left|f(\mathbf{x} + \mathbf{h}) - f(\mathbf{x}) - \sum_{i=1}^{n} \frac{\partial f}{\partial x_i}(\mathbf{x})h_i\right|}{\|\mathbf{h}\|}$$

$$\leq \left|\frac{\partial f}{\partial x_1}(\mathbf{y}_1) - \frac{\partial f}{\partial x_1}(\mathbf{x})\right| + \cdots + \left|\frac{\partial f}{\partial x_n}(\mathbf{y}_n) - \frac{\partial f}{\partial x_n}(\mathbf{x})\right|$$

But since the partials are continuous by assumption, the right side approaches 0 as $\mathbf{h} \to \mathbf{0}$, so that the left side approaches 0 as well. ■

Finally, we shall prove the all-important Chain Rule.

THEOREM 11. *Let $U \subset \mathbb{R}^n$ and $V \subset \mathbb{R}^m$ be open. Let $g: U \subset \mathbb{R}^n \to \mathbb{R}^m$ and $f: V \subset \mathbb{R}^m \to \mathbb{R}^p$ be given functions such that g maps U into V, so that $f \circ g$ is defined. Suppose g is differentiable at $\mathbf{x}_0$ and f is differentiable at $\mathbf{y}_0 = g(\mathbf{x}_0)$. Then $f \circ g$ is differentiable at $\mathbf{x}_0$ and*

$$D(f \circ g)(\mathbf{x}_0) = Df(\mathbf{y}_0)Dg(\mathbf{x}_0)$$

Proof. According to the definition of the derivative, we must verify that

$$\lim_{\mathbf{x}\to\mathbf{x}_0} \frac{\|f(g(\mathbf{x})) - f(g(\mathbf{x}_0)) - Df(\mathbf{y}_0)Dg(\mathbf{x}_0)\cdot(\mathbf{x} - \mathbf{x}_0)\|}{\|\mathbf{x} - \mathbf{x}_0\|} = 0$$

First rewrite the numerator and apply the triangle inequality as follows:

$$\begin{aligned}&\|f(g(\mathbf{x})) - f(g(\mathbf{x}_0)) - Df(\mathbf{y}_0)\cdot(g(\mathbf{x}) - g(\mathbf{x}_0)) \\ &\quad + Df(\mathbf{y}_0)\cdot[g(\mathbf{x}) - g(\mathbf{x}_0) - Dg(\mathbf{x}_0)\cdot(\mathbf{x} + \mathbf{x}_0)]\| \\ &\leq \|f(g(\mathbf{x})) - f(g(\mathbf{x}_0)) - Df(\mathbf{y}_0)\cdot(g(\mathbf{x}) - g(\mathbf{x}_0))\| \\ &\quad + \|Df(\mathbf{y}_0)\cdot[g(\mathbf{x}) - g(\mathbf{x}_0) - Dg(\mathbf{x}_0)\cdot(\mathbf{x} - \mathbf{x}_0)]\|\end{aligned}$$

As in Theorem 8, $\|Df(\mathbf{y}_0)\cdot\mathbf{h}\| \leq M\|\mathbf{h}\|$ for some constant M. Thus the above expression is less than or equal to

$$\begin{aligned}&\|f(g(\mathbf{x})) - f(g(\mathbf{x}_0)) - Df(\mathbf{y}_0)\cdot(g(\mathbf{x}) - g(\mathbf{x}_0))\| \\ &\quad + M\|g(\mathbf{x}) - g(\mathbf{x}_0) - Dg(\mathbf{x}_0)\cdot(\mathbf{x} - \mathbf{x}_0)\| \qquad (3)\end{aligned}$$

Since g is differentiable at $\mathbf{x}_0$, given $\varepsilon > 0$, there is a $\delta_1 > 0$ such that $0 < \|\mathbf{x} - \mathbf{x}_0\| < \delta_1$ implies

$$\frac{\|g(\mathbf{x}) - g(\mathbf{x}_0) - Dg(\mathbf{x}_0)\cdot(\mathbf{x} - \mathbf{x}_0)\|}{\|\mathbf{x} - \mathbf{x}_0\|} < \frac{\varepsilon}{2M}$$

This makes the second term in Equation (3) less than $\varepsilon\|\mathbf{x} - \mathbf{x}_0\|/2$.

Let us turn to the first term. By Theorem 8, $\|g(\mathbf{x}) - g(\mathbf{x}_0)\| < M_1\|\mathbf{x} - \mathbf{x}_0\|$ for a constant M_1 if $\mathbf{x}$ is near $\mathbf{x}_0$, say $0 < \|\mathbf{x} - \mathbf{x}_0\| < \delta_2$. Now choose δ_3 such that $0 < \|\mathbf{y} - \mathbf{y}_0\| < \delta_3$ implies

$$\|f(\mathbf{y}) - f(\mathbf{y}_0) - Df(\mathbf{y}_0)\cdot(\mathbf{y} - \mathbf{y}_0)\| < \varepsilon\|\mathbf{y} - \mathbf{y}_0\|/2M_1$$

Since $\mathbf{y} = g(\mathbf{x})$ and $\mathbf{y}_0 = g(\mathbf{x}_0)$, $\|\mathbf{y} - \mathbf{y}_0\| < \delta_3$ if $\|\mathbf{x} - \mathbf{x}_0\| < \delta_3/M_1$ and $\|\mathbf{x} - \mathbf{x}_0\| < \delta_2$, and so

$$\begin{aligned}\|f(g(\mathbf{x})) - f(g(\mathbf{x}_0)) - Df(\mathbf{y}_0)\cdot(g(\mathbf{x}) - g(\mathbf{x}_0))\|\\ < \varepsilon\|g(\mathbf{x}) - g(\mathbf{x}_0)\|/2M_1 < \varepsilon\|\mathbf{x} - \mathbf{x}_0\|/2\end{aligned}$$

Thus if $\delta = \min(\delta_1, \delta_2, \delta_3/M_1)$, the expression (3) is less than

$$\varepsilon\|\mathbf{x} - \mathbf{x}_0\|/2 + \varepsilon\|\mathbf{x} - \mathbf{x}_0\|/2 = \varepsilon\|\mathbf{x} - \mathbf{x}_0\|$$

and so

$$\frac{\|f(g(\mathbf{x})) - f(g(\mathbf{x}_0)) - Df(\mathbf{y}_0)Dg(\mathbf{x}_0)(\mathbf{x} - \mathbf{x}_0)\|}{\|\mathbf{x} - \mathbf{x}_0\|} < \frac{\varepsilon}{2} + \frac{\varepsilon}{2} = \varepsilon$$

for $0 < \|\mathbf{x} - \mathbf{x}_0\| < \delta$. This proves the theorem. ■

The student has already met with a number of examples illustrating the above theorems. Let us consider two of a more technical nature.

EXAMPLE 1 (Figure A.1). Let

$$f\colon \mathbb{R}^2 \to \mathbb{R},\ f(x, y) = \begin{cases} \dfrac{x^2}{x^2 + y^2} & (x, y) \neq (0, 0) \\ 0 & (x, y) = (0, 0) \end{cases}$$

Is f differentiable?

On $\mathbb{R}^2\backslash(0, 0)$ f is differentiable, since it has continuous partial derivatives on this set. However, f is not differentiable at $(0, 0)$; in fact, f is not even continuous there. To see this, approach $(0, 0)$ along the line $x = y$. Clearly, $f(x, x,) = \frac{1}{2}$; as $x \to 0$ this does not converge to 0 (see Example 15, p. 99).

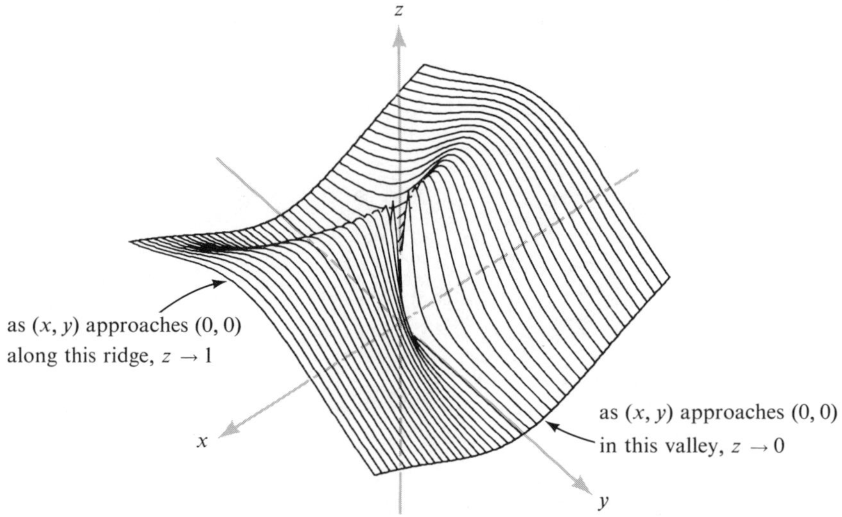

$$z = \frac{x^2}{x^2 + y^2} \qquad -1 \le x \le 1, -1 \le y \le 1$$

FIGURE A.1
This function is not differentiable at (0, 0).

EXAMPLE 2 (Figure A.2). Let

$$f(x, y) = \begin{cases} \dfrac{xy}{\sqrt{x^2 + y^2}} & (x, y) \neq (0, 0) \\ 0 & (x, y) = (0, 0) \end{cases}$$

Is f differentiable at (0, 0)? The answer is not immediately clear. However, we note that

$$\frac{\partial f}{\partial x}(0, 0) = \underset{x \to 0}{\text{limit}} \frac{f(x, 0) - f(0, 0)}{x}$$

$$= \underset{x \to 0}{\text{limit}} \frac{(x \cdot 0)/\sqrt{x^2 + 0} - 0}{x} = \underset{x \to 0}{\text{limit}} \frac{0 - 0}{x} = 0$$

and, similarly, $\partial f/\partial y(0, 0) = 0$. Thus the partials exist at (0, 0). Also, if $(x, y) \neq (0, 0)$, then

$$\frac{\partial f}{\partial x} = \frac{y\sqrt{x^2 + y^2} - 2x(xy)/2\sqrt{x^2 + y^2}}{x^2 + y^2} = \frac{y}{\sqrt{x^2 + y^2}} - \frac{x^2 y}{(x^2 + y^2)^{3/2}}$$

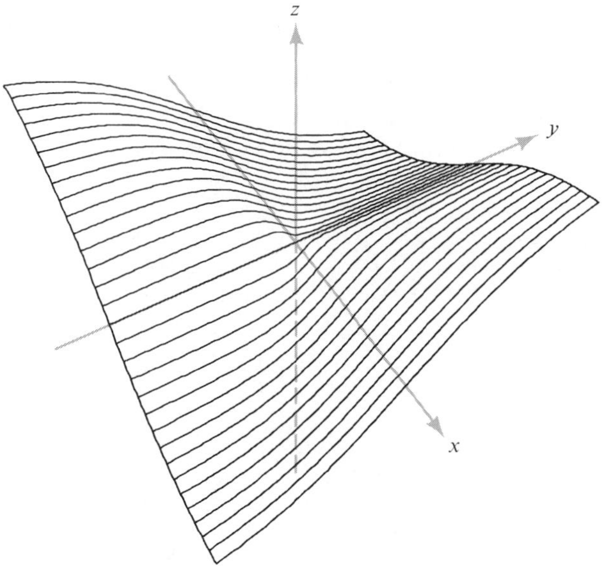

$$z = \frac{xy}{\sqrt{x^2 + y^2}} \qquad -3 \le x \le 3, -3 \le y \le 3$$

FIGURE A.2
This function is not differentiable at (0, 0) *because it is "crinkled."*

which does not have a limit as $(x, y) \to (0, 0)$. Different limits are obtained for different paths of approach, as can be seen by letting $x = My$. Thus the partial are not continuous at (0, 0), so we cannot use Theorem 9.

We might now try to show that f is not differentiable (f itself is continuous, so the method of Example 1 will not work). If $Df(0, 0)$ existed, by Theorem 16 it would have to be the zero matrix, since $\partial f/\partial x$ and $\partial f/\partial y$ are zero at (0, 0). Thus, by the definition of differentiability, for any $\varepsilon > 0$ there would be a $\delta > 0$ such that $0 < \|(h_1, h_2)\| < \delta$ implies

$$\frac{|f(h_1, h_2) - f(0, 0)|}{\|(h_1, h_2)\|} < \varepsilon$$

that is

$$|f(h_1, h_2)| < \varepsilon\|(h_1, h_2)\|$$

or

$$|h_1 h_2| < \varepsilon(h_1^2 + h_2^2)$$

But if we choose $h_1 = h_2$, this reads

$$\tfrac{1}{2} < \varepsilon$$

which is untrue if we choose $\varepsilon \le \frac{1}{2}$. Hence, f is *not* differentiable at (0, 0).

EXERCISES

1. Let $f(x, y, z) = (e^x, \cos y, \sin z)$. Compute Df. In general, when will Df be a diagonal matrix?

2. (a) Let $A: \mathbb{R}^n \to \mathbb{R}^m$ be a linear transformation given by the matrix $A = (a_{ij})$ so that $A\mathbf{x}$ has components $y_i = \sum a_{ij}x_j$. Let $M = (\sum a_{ij}^2)^{1/2}$. Use the Schwarz inequality to prove that $\|A\mathbf{x}\| \le M\|\mathbf{x}\|$.
   (b) Use the inequality derived in (a) to show that a linear transformation T: $\mathbb{R}^n \to \mathbb{R}^m$ given by a matrix $T = (t_{ij})$ is continuous.
   (c) Let $A: \mathbb{R}^n \to \mathbb{R}^m$ be a linear transformation given by a matrix. If

$$\operatorname*{limit}_{\mathbf{x}\to\mathbf{0}} \frac{A\mathbf{x}}{\|\mathbf{x}\|} = \mathbf{0}$$

   show that $A = 0$.

3. Let $f: A \to B$ and $g: B \to C$ be maps between open subsets of Euclidean space, and let $\mathbf{x}_0$ be in A or be a boundary point of A and $\mathbf{y}_0$ be in B or be a boundary point of B.
   (a) If $\operatorname*{limit}_{\mathbf{x}\to\mathbf{x}_0} f(\mathbf{x}) = \mathbf{y}_0$ and $\operatorname*{limit}_{\mathbf{y}\to\mathbf{y}_0} g(\mathbf{y}) = \mathbf{w}$, show that $\operatorname*{limit}_{\mathbf{x}\to\mathbf{x}_0} g(f(\mathbf{x}))$ need not equal $\mathbf{w}$.
   (b) If $\mathbf{y}_0 \in B$ and $\mathbf{w} = g(\mathbf{y}_0)$ show that $\operatorname*{limit}_{\mathbf{x}\to\mathbf{x}_0} g(f(\mathbf{x})) = \mathbf{w}$.

4. A function $f: A \subset \mathbb{R}^n \to \mathbb{R}^m$ is called *uniformly continuous* if for every $\varepsilon > 0$ there is a $\delta > 0$ such that for all $\mathbf{p}$ and $\mathbf{q} \in A$, $\|\mathbf{p} - \mathbf{q}\| < \delta$ implies $\|f(\mathbf{p}) - f(\mathbf{q})\| < \varepsilon$. (Note that a uniformly continuous function is continuous; try to describe explicitly the extra property that a uniformly continuous function has.)
   (a) Prove that a linear map $T: \mathbb{R}^n \to \mathbb{R}^m$ is uniformly continuous. (HINT: Use Exercise 2.)
   (b) Prove $x \mapsto 1/x^2$ on $]0, 1]$ is continuous, but not uniformly continuous.

5. Let $A = (a_{ij})$ be a *symmetric* $n \times n$ matrix (that is, $a_{ij} = a_{ji}$) and define $f(\mathbf{x}) = \mathbf{x} \cdot A\mathbf{x}$ so $f: \mathbb{R}^n \to \mathbb{R}$. Show that $\nabla f(\mathbf{x})$ is the vector $2A\mathbf{x}$.

6. The following function is graphed in Figure A.3:

$$f(x, y) = \begin{cases} \dfrac{2xy^2}{x^2 + y^4} & (x, y) \neq (0, 0) \\ 0 & (x, y) = (0, 0) \end{cases}$$

   Show that $\partial f/\partial x$ and $\partial f/\partial y$ exist everywhere; in fact all directional derivatives exist. But show that f is not continuous at (0, 0). Is f differentiable?

7. Let $f(x, y) = g(x) + h(y)$, and suppose g is differentiable at x_0 and h is differentiable at y_0. Prove f is differentiable at (x_0, y_0).

8. Use the Schwarz inequality to prove that for any vector $\mathbf{v} \in \mathbb{R}^n$,

$$\operatorname*{limit}_{\mathbf{x}\to\mathbf{x}_0} \mathbf{v} \cdot \mathbf{x} = \mathbf{v} \cdot \mathbf{x}_0$$

9. Prove that if $\operatorname*{limit}_{\mathbf{x}\to\mathbf{x}_0} f(\mathbf{x}) = b$ for $f: A \subset \mathbb{R}^n \to \mathbb{R}$ then

$$\operatorname*{limit}_{\mathbf{x}\to\mathbf{x}_0} (f(\mathbf{x}))^2 = b^2 \quad \text{and} \quad \operatorname*{limit}_{\mathbf{x}\to\mathbf{x}_0} \sqrt{|f(\mathbf{x})|} = \sqrt{|b|}$$

   (You may use Exercise 3(b).)

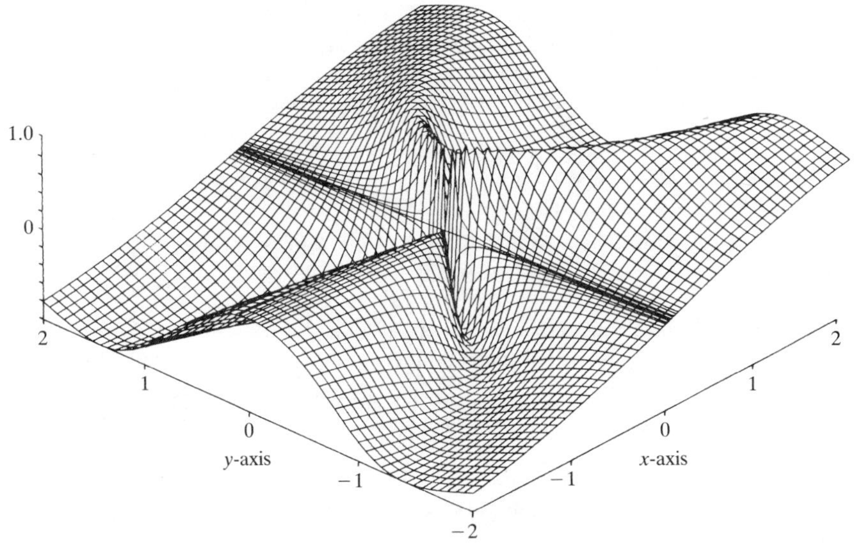

FIGURE A.3
Computer-generated graph of $z = 2xy^2/(x^2 + y^4)$.

10. Show that in Theorem 9 with $m = 1$ it is enough to assume that $n - 1$ partial derivatives are continuous and merely that the other one exists. Does this agree with what you expect when $n = 1$?

11. Define $f\colon \mathbb{R}^2 \to \mathbb{R}$ by

$$f(x, y) = \begin{cases} \dfrac{xy}{(x^2 + y^2)^{1/2}} & (x, y) \neq (0, 0) \\ 0 & (x, y) = (0, 0) \end{cases}$$

Show that f is continuous (see Example 2).

12. (a) Does $\displaystyle\lim_{(x, y) \to (0, 0)} \frac{x}{x^2 + y^2}$ exist?

(b) Does $\displaystyle\lim_{(x, y) \to (0, 0)} \frac{x^3}{x^2 + y^2}$ exist?

13. Find $\displaystyle\lim_{(x, y) \to (0, 0)} \frac{xy^2}{\sqrt{x^2 + y^2}}$.

14. Prove that $s\colon \mathbb{R}^2 \to \mathbb{R}^1$, $(x, y) \mapsto x + y$, is continuous.

15. Using the definition of continuity, prove that f is continuous at $\mathbf{x}$ if and only if

$$\lim_{\mathbf{h} \to \mathbf{0}} f(\mathbf{x} + \mathbf{h}) = f(\mathbf{x})$$

16. (a) A sequence x_n of points in $\mathbb{R}^m$ is said to *converge to* $\mathbf{x}$, written $\mathbf{x}_n \to \mathbf{x}$ as $n \to \infty$, if for any $\varepsilon > 0$ there is an N such that $n \geq N$ implies $\|\mathbf{x} - \mathbf{x}_n\| < \varepsilon$.

Show that $\mathbf{y}$ is a boundary point of an open set A if and only if $\mathbf{y}$ is not in A and there is a sequence of distinct points of A converging to $\mathbf{y}$.

(b) Let $f: A \subset \mathbb{R}^n \to \mathbb{R}^m$ and $\mathbf{y}$ be in A or be a boundary point of A. Then $\operatorname*{limit}_{\mathbf{x}\to\mathbf{y}} f(\mathbf{x}) = \mathbf{b}$ if and only if $f(\mathbf{x}_n) \to \mathbf{b}$ for every sequence $\mathbf{x}_n$ of points in A with $\mathbf{x}_n \to \mathbf{y}$.

(c) If $U \subset \mathbb{R}^m$ is open, show that $f: U \to \mathbb{R}^p$ is continuous if and only if $\mathbf{x}_n \to \mathbf{x} \in U$ implies $f(\mathbf{x}_n) \to f(\mathbf{x})$.

17. If $f(\mathbf{x}) = g(\mathbf{x})$ for all $\mathbf{x} \neq \mathbf{a}$ and if $\operatorname*{limit}_{\mathbf{x}\to\mathbf{a}} f(\mathbf{x}) = \mathbf{b}$, then show that $\operatorname*{limit}_{\mathbf{x}\to\mathbf{a}} g(\mathbf{x}) = \mathbf{b}$ as well.

18. Let $A \subset \mathbb{R}^n$ and let $\mathbf{x}_0$ be a boundary point of A. Let $f: A \to \mathbb{R}$ and $g: A \to \mathbb{R}$ be functions defined on A such that $\operatorname*{limit}_{\mathbf{x}\to\mathbf{x}_0} f(\mathbf{x})$ and $\operatorname*{limit}_{\mathbf{x}\to\mathbf{x}_0} g(\mathbf{x})$ exist, and assume that for all $\mathbf{x}$ in some deleted neighborhood of $\mathbf{x}_0$, $f(\mathbf{x}) \le g(\mathbf{x})$. (A deleted neighborhood of $\mathbf{x}_0$ is any neighborhood of $\mathbf{x}_0$, less $\mathbf{x}_0$ itself.)

(a) Prove that $\operatorname*{limit}_{\mathbf{x}\to\mathbf{x}_0} f(\mathbf{x}) \le \operatorname*{limit}_{\mathbf{x}\to\mathbf{x}_0} g(\mathbf{x})$.

(HINT: Consider the function $\phi(\mathbf{x}) = g(\mathbf{x}) - f(\mathbf{x})$; prove that $\operatorname*{limit}_{\mathbf{x}\to\mathbf{x}_0} \phi(\mathbf{x}) \ge 0$, and then use the fact that the limit of the sum of two functions is the sum of their limits.)

(b) If $f(\mathbf{x}) < g(\mathbf{x})$ do we necessarily have strict inequality of the limits?

19. Given $f: A \subset \mathbb{R}^n \to \mathbb{R}^m$, we say that "$f$ is $o(\mathbf{x})$ as $\mathbf{x} \to \mathbf{0}$" if $\operatorname*{limit}_{\mathbf{x}\to\mathbf{0}} f(\mathbf{x})/\|\mathbf{x}\| = \mathbf{0}$.

(a) If f_1 and f_2 are $o(\mathbf{x})$ as $\mathbf{x} \to \mathbf{0}$, prove that $f_1 + f_2$ is also $o(\mathbf{x})$ as $\mathbf{x} \to \mathbf{0}$ (where $(f_1 + f_2)(\mathbf{x}) = f_1(\mathbf{x}) + f_2(\mathbf{x})$).

(b) Let $g: A \to \mathbb{R}$ be a function with the property that there is a number $c > 0$ such that $|g(\mathbf{x})| \le c$ for all $\mathbf{x}$ in A (g is called *bounded*). If f is $o(\mathbf{x})$ as $\mathbf{x} \to \mathbf{0}$, prove that gf is also $o(\mathbf{x})$ as $\mathbf{x} \to \mathbf{0}$ (where $(gf)(\mathbf{x}) = g(\mathbf{x})f(\mathbf{x})$).

(c) Show that $f(x) = x^2$ is $o(x)$ as $x \to 0$. Is $g(x) = x$ $o(x)$ as $x \to 0$?

APPENDIX B

SOME TECHNICAL INTEGRATION THEOREMS

In this appendix we shall provide the main ideas of the proofs of the existence and additivity of the integral that were stated in Section 5.2. These proofs require more advanced concepts than those needed for the main text of this book. The notions of uniform continuity and the completeness of the real number, both of which are usually treated in a junior level course in mathematical analysis or real variable theory, are called upon here.

DEFINITION. *Let $D \subset \mathbb{R}^n$ and $f: D \to \mathbb{R}$. Recall that f is said to be* **continuous at** $\mathbf{x} \in D$ *provided that for all $\varepsilon > 0$ there is a number $\delta > 0$ such that if $\mathbf{x} \in D$ and $\|\mathbf{x} - \mathbf{x}_0\| < \delta$, then $\|f(\mathbf{x}) - f(\mathbf{x}_0)\| < \varepsilon$. We say f is* **continuous on** *D if it is continuous at each point of D.*

The function f is said to be **uniformly continuous** *on D if for every number $\varepsilon > 0$ there is a $\delta > 0$ such that whenever $\mathbf{x}, \mathbf{y} \in D$ and $\|\mathbf{x} - \mathbf{y}\| < \delta$, then $\|f(\mathbf{x}) - f(\mathbf{y})\| < \varepsilon$.*

If the reader will pause for a few minutes to ponder these definitions, it should become clear that if f is uniformly continuous, it must also be continuous, whereas the converse is not true. Hence uniform continuity appears to be a stronger notion. To show that continuity does not imply uniform continuity and that therefore the latter is indeed stronger, we need only give an example of a function that is continuous but not uniformly continuous. Such a function appears in Exercise 2.

The distinction between the notions of continuity and uniform continuity is simply this: For a function f that is continuous but not uniformly continuous, δ cannot be chosen independently of the point of the domain (the $\mathbf{x}_0$ in the definition). The definition of uniform continuity states explicitly that once you are given an $\varepsilon > 0$, a δ can be found independent of any point of D.

DEFINITION. *A set $D \subset \mathbb{R}^n$ is* ***bounded*** *if there exists a number $M > 0$ such that $\|\mathbf{x}\| < M$ for all $\mathbf{x} \in D$. A set is* ***closed*** *if it contains all its boundary points.*

Thus a set is bounded if it can be strictly contained in some (large) ball. The next theorem states that under some conditions a continuous function is actually uniformly continuous.

THEOREM 7 (THE UNIFORM CONTINUITY PRINCIPLE). *Every function that is continuous on a closed and bounded set D in $\mathbb{R}^n$ is uniformly continuous on D.*

The proof of this theorem will take us too far afield;* however, we can prove a special case of it, which is, in fact, sufficient for many situations relevant to this text.

Proof of a Special Case of Theorem 7. Let us assume that $D = [a, b]$ is a closed interval on the line, that $f: D \to \mathbb{R}$ is continuous, that df/dx exists on the open interval $]a, b[$, and that df/dx is bounded (i.e., there is a constant $C > 0$ such that $|(df/dx)(x)| \leq C$ for all x in $]a, b[$). To show that these conditions imply f is uniformly continuous, we use the Mean Value Theorem as follows: Let $\varepsilon > 0$ be given and let x and y lie in D. Then by the Mean Value Theorem (stated on p. 119),

$$f(x) - f(y) = f'(c)(x - y)$$

for some c between x and y. By the assumed boundedness of the derivative,

$$|f(x) - f(y)| \leq C|x - y|$$

* The proof can be found in any text on mathematical analysis. See, for example, J. Marsden, *Elementary Classical Analysis*, W. H. Freeman and Company, San Francisco, 1974, or W. Rudin, *Principles of Mathematical Analysis*, 3rd ed., McGraw-Hill, New York, 1976.

Let $\delta = \varepsilon/C$. If $|x - y| < \delta$, then

$$|f(x) - f(y)| < C \cdot \frac{\varepsilon}{C} = \varepsilon$$

Thus f is uniformly continuous. (Note that δ depends on neither x nor y, which is a crucial part of the definition.) ■

This proof also works for regions in $\mathbb{R}^n$ that are convex; that is, for any two points $\mathbf{x}, \mathbf{y}$ in D, the line segment $\boldsymbol{\sigma}(t) = t\mathbf{x} + (1 - t)\mathbf{y}, 0 \leq t \leq 1$, joining them also lies in D. We assume f is differentiable (on an open set containing D) and that $\|\nabla f(\mathbf{x})\| \leq C$ for a constant C. Then the Mean Value Theorem applied to the function $h(t) = f(\boldsymbol{\sigma}(t))$ gives

$$h(1) - h(0) = h'(c)$$

or

$$f(\mathbf{x}) - f(\mathbf{y}) = h'(c) = \nabla f(\boldsymbol{\sigma}(c)) \cdot \boldsymbol{\sigma}'(c) = \nabla f(\boldsymbol{\sigma}(c)) \cdot (\mathbf{x} - \mathbf{y})$$

by the Chain Rule. Thus by the Cauchy-Schwarz inequality,

$$|f(\mathbf{x}) - f(\mathbf{y})| \leq \|\nabla f(\boldsymbol{\sigma}(c))\| \, \|\mathbf{x} - \mathbf{y}\| \leq C\|\mathbf{x} - \mathbf{y}\|$$

Then, as above, given $\varepsilon > 0$, we can let $\delta = \varepsilon/C$.

We now move on to the notion of a Cauchy sequence of real numbers. Recall that in the definition of Riemann sums we obtained a sequence of numbers $\{S_n\}$, $n = 1, \ldots$. It would be nice if we could say that this sequence of numbers converges to S (or has a limit S), but how can we obtain such a limit? In the abstract setting we know no more about S_n than that it is the Riemann sum of a (say, continuous) function, and, although this has not yet been proved, it should be enough information to insure its convergence.

Thus we must determine a property for sequences that guarantees their convergence. We shall define a class of sequences called Cauchy sequences, and then take as an axiom of the real number system that all such sequences converge to a limit.* The determination in the nineteenth century that such an axiom was necessary for the foundations of the Calculus was a major breakthrough in the history of mathematics and paved the way for the modern rigorous approach to mathematical analysis. We shall say more about this shortly.

* Texts on mathematical analysis, such as those mentioned in the preceding footnote, sometimes use different axioms, such as the least upper bound property. In such a setting, our completeness axiom becomes a theorem.

DEFINITION. *A sequence of real numbers* $\{S_n\}$, $n = 1, \ldots$ *is said to be* **Cauchy** *if for every* $\varepsilon > 0$ *there exists an* N *such that for all* $m, n \geq N$, $|S_n - S_m| < \varepsilon$.

If a sequence S_n converges to a limit S then S_n is a Cauchy sequence. To see this, recall the definition: For every $\varepsilon > 0$ there is an N such that for all $n \geq N$, $|S_n - S| < \varepsilon$. Given $\varepsilon > 0$, choose N_1 such that for $n \geq N_1$, $|S_n - S| < \varepsilon/2$ (use the definition with $\varepsilon/2$ in place of ε). Then if $n, m \geq N_1$,

$$|S_n - S_n| = |S_n - S + S - S_m| \leq |S_n - S| + |S - S_m| < \varepsilon/2 + \varepsilon/2 = \varepsilon$$

which proves our contention. The completeness axiom asserts that the converse is true as well:

COMPLETENESS AXIOM OF THE REAL NUMBERS. *Every Cauchy sequence* $\{S_n\}$ *converges to some limit* S.

HISTORICAL NOTE

Augustin-Louis Cauchy (1789–1857), one of the greatest mathematicians of all time, defined what we now call Cauchy sequences in his "Cours d'analyse," published in 1821. This book was a basic work on the foundations of analysis, although it would be considered somewhat loosely written by today's standards. Cauchy knew that a convergent sequence was "Cauchy" and remarked that a Cauchy sequence converges. He did not have a proof of this fact, nor could he have had one, as such a proof depends on the rigorous development of the real number system that was achieved only in 1872 by the German mathematician Georg Cantor (1845–1918).

It is now clear what we must do to ensure that the Riemann sums $\{S_n\}$ of, say, a continuous function on a rectangle, converge to some limit S, which would prove that continuous functions on rectangles are integrable; *we must show that* $\{S_n\}$ *is a Cauchy sequence.* In demonstrating this fact we use the uniform continuity principal. The integrability of continuous functions will be a consequence of the following two lemmas.

LEMMA 1. *Let* f *be a continuous function on a rectangle* R *in the plane, and let* $\{S_n\}$ *be a sequence of Riemann sums for* f. *Then* $\{S_n\}$ *converges to some number* S.

Proof. Given a rectangle $R \subset \mathbb{R}^2$, $R = [a, b] \times [c, d]$, we have the regular partition of R, $a = x_0 < x_1 < \cdots < x_n = b$, $c = y_0 < y_1 < \cdots < y_n = d$

discussed in Section 5.2. Recall that

$$\Delta x = x_{j+1} - x_j = \frac{b-a}{n}, \qquad \Delta y = y_{k+1} - y_k = \frac{d-c}{n}$$

and

$$S_n = \sum_{j,k=0}^{n-1} f(\mathbf{c}_{jk})\,\Delta x\,\Delta y$$

where c_{jk} is an arbitrarily-chosen point in $R_{jk} = [x_j, x_{j+1}] \times [y_k, y_{k+1}]$. The sequence $\{S_n\}$ is determined only by the selection of the points c_{jk}.

For the purpose of the proof we shall introduce a slightly more complicated but very precise notation, namely, we set

$$\Delta x^n = \frac{b-a}{n}$$

and

$$\Delta y^n = \frac{d-c}{n}$$

with this notation we have

$$S_n = \sum_{j,k}^{n-1} f(\mathbf{c}_{jk})\,\Delta x^n\,\Delta y^n \tag{1}$$

To show that $\{S_n\}$ is Cauchy we must show that given $\varepsilon > 0$ there exists an N such that for all, $n, m \geq N$, $|S_n - S_m| < \varepsilon$. By the uniform continuity principal, f is uniformly continuous on R. Thus given $\varepsilon > 0$ there exists a $\delta > 0$ such that when $\mathbf{x}, \mathbf{y} \in R$, $\|\mathbf{x} - \mathbf{y}\| < \delta$, then $|f(\mathbf{x}) - f(\mathbf{y})| < \varepsilon/[2\,\mathrm{Area}(R)]$ ($\varepsilon/[2\,\mathrm{Area}(R)]$ is used in place of ε in the definition). Let N be so large that for any $m \geq N$ the diameter of any subrectangle R_{jk} in the mth regular partition of R is less than δ. Thus if $\mathbf{x}, \mathbf{y}$ are points in the same subrectangle we will have $|f(\mathbf{x}) - f(\mathbf{y})| < \varepsilon/[2\,\mathrm{Area}(R)]$.

Fix $m, n \geq N$. We will show that $|S_n - S_m| < \varepsilon$. This shows that $\{S_n\}$ is Cauchy and hence converges. Consider the $(m \cdot n)^{\text{th}} = (m \text{ times } n)^{\text{th}}$ regular partition of R. Then

$$S_{mn} = \sum_{r,t} f(\tilde{\mathbf{c}}_{rt})\,\Delta x^{mn}\,\Delta y^{mn}$$

where $\tilde{\mathbf{c}}_{rt}$ is a point in the $(rt)^{\text{th}}$ subrectangle. Note that each subrectangle of the $(mn)^{\text{th}}$ partition is a subrectangle of both the mth and the nth regular partitions (see Figure B.1).

Let us denote the subrectangles in the mn^{th} subdivision by $\tilde{R}_{rt}$ and those in the nth subdivision by R_{jk}. Thus each $\tilde{R}_{rt} \subset R_{jk}$ for some jk, and hence we can rewrite (1) as

$$S_n = \sum_{j,k}^{n-1} \left(\sum_{\tilde{R}_{rt} \subset R_{jk}} f(\mathbf{c}_{jk})\,\Delta x^{mn}\,\Delta y^{mn} \right) \tag{1'}$$

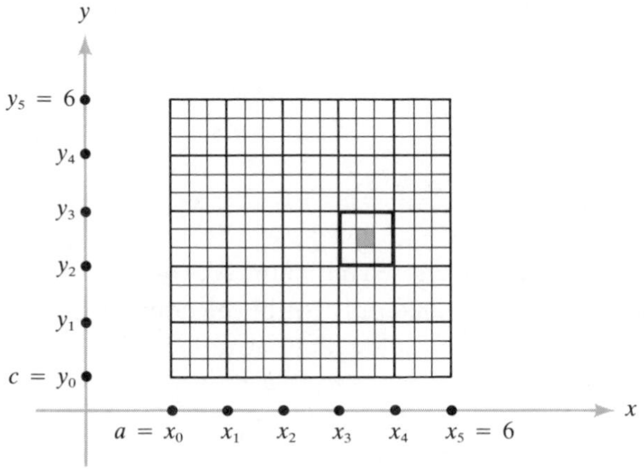

FIGURE B.1

The shaded box shows a subrectangle in the mn^{th} partition and the darkly outlined box a subrectangle in the mth partition.

Here we are using the fact that

$$\sum_{\tilde{R}_{rt} \subset R_{jk}} f(\mathbf{c}_{jk})\, \Delta x^{mn}\, \Delta y^{mn} = f(\mathbf{c}_{jk})\, \Delta x^{n}\, \Delta y^{n}$$

where the sum is taken over all subrectangles in the mn^{th} subdivision contained in a *fixed* subrectangle R_{jk} in the nth subdivision. We also have the identity

$$S_{mn} = \sum_{r,t}^{mn-1} f(\tilde{\mathbf{c}}_{rt})\, \Delta x^{mn}\, \Delta y^{mn} \tag{2}$$

This relation can also be rewritten as

$$S_{mn} = \sum_{j,k} \sum_{\tilde{R}_{rt} \subset R_{jk}} f(\tilde{\mathbf{c}}_{rt})\, \Delta x^{mn}\, \Delta y^{mn} \tag{2$'$}$$

where in (2′) we are first summing over those subrectangles in the mn^{th} partition contained in a fixed R_{jk} and then summing over j, k. Subtracting (2′) from (1′) we get

$$|S_n - S_{mn}| = \left| \sum_{jk} \sum_{\tilde{R}_{rt} \subset R_{jk}} [f(\mathbf{c}_{jk})\, \Delta x^{mn}\, \Delta y^{mn} - f(\tilde{\mathbf{c}}_{rt})\, \Delta x^{mn}\, \Delta y^{mn}] \right|$$
$$\leq \sum_{jk} \sum_{\tilde{R}_{rt} \subset R_{jk}} |f(\mathbf{c}_{jk}) - f(\tilde{\mathbf{c}}_{rt})|\, \Delta x^{mn}\, \Delta y^{mn}$$

By our choice of δ and N, $|f(\mathbf{c}_{jk}) - f(\tilde{\mathbf{c}}_{rt})| < \varepsilon/[2\,\text{Area}(R)]$, and consequently the above inequality becomes

$$|S_n - S_{mn}| \leq \sum_{jk} \sum_{\tilde{R}_{rt} \subset R_{jk}} (\varepsilon/[2\,\text{Area}(R)])\, \Delta x^{mn}\, \Delta^{ymn} = \varepsilon/2$$

Thus $|S_n - S_{mn}| < \varepsilon/2$ and similarly one shows that $|S_m - S_{mn}| < \varepsilon/2$. Since

$$|S_n - S_m| = |S_n - S_{mn} + S_{mn} - S_m| \le |S_n - S_{mn}| + |S_{mn} - S_m| < \varepsilon$$

for $m, n \ge N$, we have shown $\{S_n\}$ is Cauchy, and thus has a limit S. ■

We have already remarked that each Riemann sum depends on the selection of a collection of points $\mathbf{c}_{jk}$. In order to show that a continuous function on a rectangle R is integrable we must further demonstrate that the limit S we obtained in Lemma 1 is independent of the choices of the points $\mathbf{c}_{jk}$.

LEMMA 2. *The limit S in Lemma 1 does not depend on the choice of points $\mathbf{c}_{jk}$.*

Proof. Suppose we have two sequences of Riemann sums $\{S_n\}$ and $\{S_n^*\}$ obtained by selecting two different sets of points, say $\mathbf{c}_{jk}$ and $\mathbf{c}_{jk}^*$ in each nth partition. By Lemma 1 we know that $\{S_n\}$ converges to some number S and $\{S_n^*\}$ must also converge to some number, say S^*. We want to show that $S = S^*$ and shall do this by showing that given any $\varepsilon > 0$, $|S - S^*| < \varepsilon$, which implies that S must be equal to S^* (why?).

To start, we know that f is uniformly continuous on R. Consequently given $\varepsilon > 0$ there exists a δ so that $|f(\mathbf{x}) - f(\mathbf{y})| < \varepsilon/[3 \text{ Area}(R)]$ whenever $\|\mathbf{x} - \mathbf{y}\| < \delta$. We choose N so large that whenever $n \ge N$ the diameter of each subrectangle in the nth regular partition is less than δ. Since $\lim_{n\to\infty} S_n = S$ and $\lim_{n\to\infty} S_n^* = S^*$ we can assume that N has been chosen so large that $n \ge N$ implies that $|S_n - S| < \varepsilon/3$ and $|S_n^* - S^*| < \varepsilon/3$. Also for $n \ge N$ we know by uniform continuity that if $\mathbf{c}_{jk}$ and $\mathbf{c}_{jk}^*$ are points in the same subrectangle R_{jk} of the nth partition then $|f(\mathbf{c}_{jk}) - f(\mathbf{c}_{jk}^*)| < \varepsilon/[3 \text{ Area}(R)]$. Thus

$$\begin{aligned} |S_n - S_n^*| &= \left| \sum_{jk} f(\mathbf{c}_{jk})\, \Delta x^n\, \Delta y^n - \sum_{jk} f(\mathbf{c}_{jk}^*)\, \Delta x^n\, \Delta y^n \right| \\ &\le \sum_{jk} |f(\mathbf{c}_{jk}) - f(\mathbf{c}_{jk}^*)|\, \Delta x^n\, \Delta y^n < \varepsilon/3 \end{aligned}$$

We now write

$$\begin{aligned} |S - S^*| &= |S - S_n + S_n - S_n^* + S_n^* - S^*| \\ &\le |S - S_n| + |S_n - S_n^*| + |S_n^* - S| < \varepsilon \end{aligned}$$

and so the lemma is proved. ■

Putting lemmas 1 and 2 together proves Theorem 1 of Section 5.2:

THEOREM 1. *Any continuous function defined on a rectangle R is integrable.*

HISTORICAL NOTE

Cauchy presented the first published proof of this theorem in his Résumé of 1823, in which he points out the need to prove the existence of the integral as a limit of a sum. In this paper he first treats continuous functions (as we are doing now), but on an interval $[a, b]$. (The proof is essentially the same.) However, his proof was not rigorous, as it lacked the notion of uniform continuity, which was not available at that time.

The notion of a Riemann sum S_n for a function f certainly predates Bernhard Riemann (1826–1866). The sums are probably named after him because he developed a theoretical approach to the study of integration in a fundamental paper on trigonometric series in 1854. His approach, although later generalized by Darboux (1875) and Stieltjes (1894), was to last more than half a century until it was augmented by the theory Lebesgue presented to the mathematical world in 1902. This latter approach to integration theory is generally studied in graduate courses in mathematics.

The proof of Theorem 2 (Section 5.2) is left to the reader as Exercises 4 to 6 at the end of this appendix. The main ideas are essentially contained in the proof of Theorem 1, although the proof of Theorem 2 contains an additional difficulty.

Our next goal will be to present a proof of property (*iv*) of the integral on page 269, namely, its additivity. However, because of some technical difficulties in establishing this result in its full generality, we shall prove it only in the case in which f is continuous.

THEOREM (ADDITIVITY OF THE INTEGRAL). *Let R_1 and R_2 be two disjoint rectangles (i.e., their intersection contains no rectangle) and are such that $Q = R_1 \cup R_2$ is again a rectangle. If f is a function that is continuous over Q and hence over each R_i, then*

$$\int_Q f = \int_{R_1} f + \int_{R_2} f \qquad (3)$$

Proof. The proof depends on the ideas that have already been presented in the proof of Theorem 1.

The fact that f is integrable over Q, R_1, and R_2 follows from Theorem 1. Thus all three integrals in equation (3) above exist, and it is only necessary to establish equality.

Without loss of generality we can assume that $R_1 = [a, b_1] \times [c, d]$ and $R_2 = [b_1, b] \times [c, d]$ (see Figure B.2). Again we must develop some

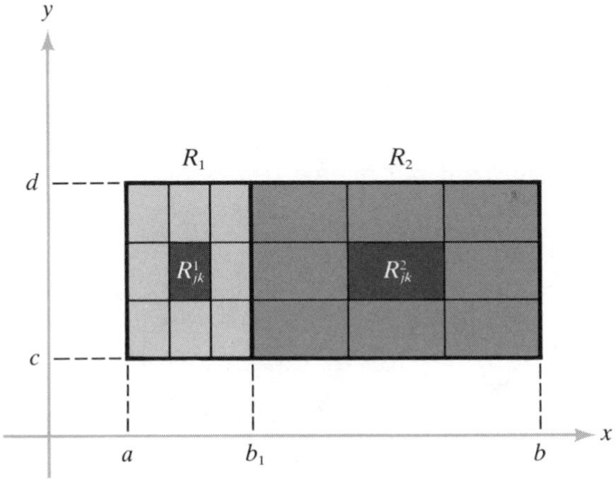

FIGURE B.2
Elements of a regular partition of R_1 and R_2.

notation. Let

$$\Delta x_1^n = \frac{b_1 - a}{n}, \qquad \Delta x_2^n = \frac{b - b_1}{n}, \qquad \Delta x^n = \frac{b - a}{n} \quad \text{and} \quad \Delta y^n = \frac{d - c}{n}.$$

Let

$$S_n^1 = \sum_{j,k} f(\mathbf{c}_{jk}^1)\, \Delta x_1^n\, \Delta y^n \tag{4}$$

$$S_n^2 = \sum_{j,k} f(\mathbf{c}_{jk}^2)\, \Delta x_2^n\, \Delta y^n \tag{5}$$

$$S_n = \sum_{j,k} f(\mathbf{c}_{jk})\, \Delta x^n\, \Delta y^n \tag{6}$$

where $\mathbf{c}_{jk}^1$, $\mathbf{c}_{jk}^2$, and $\mathbf{c}_{jk}$ are points in the jk^{th} subrectangle of the nth regular partition of R_1, R_2, and Q respectively. Let $S^i = \lim_{n\to\infty} S_n^i$, where $i = 1, 2$, and $S = \lim_{n\to\infty} S_n$. It must be shown that $S = S^1 + S^2$, which we will accomplish by showing that for arbitrary $\varepsilon > 0$, $|S - S^1 - S^2| < \varepsilon$.

By the uniform continuity of f on Q we know that given $\varepsilon > 0$ there is a $\delta > 0$ such that whenever $\|\mathbf{x} - \mathbf{y}\| < \delta$, $|f(\mathbf{x}) - f(\mathbf{y})| < \varepsilon$. Let N be so big that for $n \geq N$, $|S_n - S| < \varepsilon/3$, $|S_n^i - S^i| < \varepsilon/3$, $i = 1, 2$, and if $\mathbf{x}$, $\mathbf{y}$ are any two points in any subrectangle of the nth partition of either R_1, R_2, or Q, then $|f(\mathbf{x}) - f(\mathbf{y})| < \varepsilon/[3\,\text{Area}(Q)]$. Let us consider the nth regular partition of R_1, R_2, and Q. These form a collection of rubrectangles that we shall denote by R_{jk}^1, R_{jk}^2, R_{jk} respectively (see Figures B.2 and B.3).

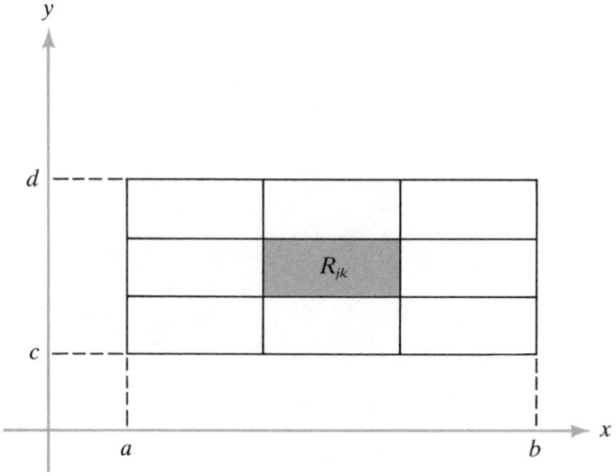

FIGURE B.3
A regular partition of Q.

If we superimpose the subdivision of Q on the nth subdivisions of R_1 and R_2 we get a new collection of rectangles, say $\tilde{R}_{\alpha\beta}$, $\beta = 1, \ldots, n$ and $\alpha = 1, \ldots, m$, $m > n$ (see Figure B.4).

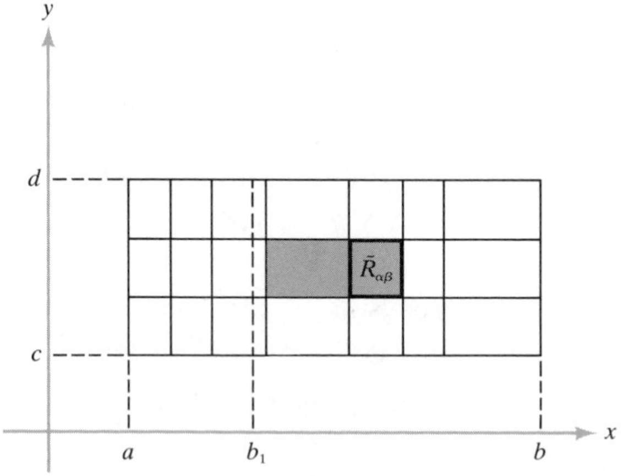

FIGURE B.4
The vertical and horizontal lines of this subdivision are obtained by taking the union of the vertical and horizontal lines of Figures B.2 and B.3.

Each $\tilde{R}_{\alpha\beta}$ is contained in some subrectangle R_{jk} of Q and in some subrectangle of the nth partition of either R_1 or R_2. Consider equalities (4), (5), and (6) above. These can be rewritten as

$$S_n^i = \sum_{j,k} \sum_{\tilde{R}_{\alpha\beta} \subset R_i} f(\mathbf{c}_{jk}^i)\text{Area}(\tilde{R}_{\alpha\beta}) \tag{4'}$$

$$= \sum_{\substack{\alpha\beta \\ \tilde{R}_{\alpha\beta} \subset R_i}} f(\tilde{\mathbf{c}}_{\alpha\beta})\text{Area}(\tilde{R}_{\alpha\beta}) \tag{5'}$$

where $\tilde{\mathbf{c}}_{\alpha\beta} = \mathbf{c}_{jk}^i$ if $\tilde{R}_{\alpha\beta} \subset R_{jk}^i$, $i = 1, 2$ and

$$S_n = \sum_{j,k} \sum_{\tilde{R}_{\alpha\beta} \subset R_{jk}} f(\mathbf{c}_{jk})\text{Area}(\tilde{R}_{\alpha\beta}) = \sum_{\alpha\beta} f(\mathbf{c}_{\alpha\beta}^*)\text{Area}(\tilde{R}_{\alpha\beta}) \tag{6'}$$

where $\mathbf{c}_{\alpha\beta}^* = \mathbf{c}_{jk}$ if $\tilde{R}_{\alpha\beta} \subset R_{jk}$.

For the reader encountering such index notation for the first time, we would like to point out that

$$\sum_{\substack{\alpha\beta \\ \tilde{R}_{\alpha\beta} \subset R_i}}$$

means that the summation is taken over those α's and β's such that the corresponding rectangle $\tilde{R}_{\alpha\beta}$ is contained in the rectangle R_i.

Now the sum for S_n can be split into two parts,

$$S_n = \sum_{\substack{\alpha\beta \\ \tilde{R}_{\alpha\beta} \subset R_1}} f(\mathbf{c}_{\alpha\beta}^*)\,\text{Area}(\tilde{R}_{\alpha\beta}) + \sum_{\substack{\alpha\beta \\ \tilde{R}_{\alpha\beta} \subset R_2}} f(\mathbf{c}_{\alpha\beta}^*)\,\text{Area}(\tilde{R}_{\alpha\beta})$$

From these representations and the triangle inequality, it follows that

$$\begin{aligned} |S_n - S_n^1 - S_n^2| &\le \left| \sum_{\substack{\alpha\beta \\ \tilde{R}_{\alpha\beta} \subset R_1}} (f(\mathbf{c}_{\alpha\beta}^*) - f(\tilde{\mathbf{c}}_{\alpha\beta}))\text{Area}(\tilde{R}_{\alpha\beta}) \right| \\ &\quad + \left| \sum_{\substack{\alpha\beta \\ \tilde{R}_{\alpha\beta} \subset R_2}} (f(\mathbf{c}_{\alpha\beta}^*) - f(\tilde{\mathbf{c}}_{\alpha\beta}))\text{Area}(\tilde{R}_{\alpha\beta}) \right| \\ &\le \frac{\varepsilon}{3\,\text{Area}\,Q} \sum_{\substack{\alpha\beta \\ \tilde{R}_{\alpha\beta} \subset R_1}} \text{Area}(\tilde{R}_{\alpha\beta}) \\ &\quad + \frac{\varepsilon}{3\,\text{Area}\,Q} \sum_{\substack{\alpha\beta \\ \tilde{R}_{\alpha\beta} \subset R_2}} \text{Area}(\tilde{R}_{\alpha\beta}) < \varepsilon/3 \end{aligned}$$

In this step we used the uniform continuity of f. Thus $|S_n - S_n^1 - S_n^2| < \varepsilon/3$ for $n \ge N$. But $|S - S_n| < \varepsilon/3$, $|S_n^1 - S^1| < \varepsilon/3$ and $|S_n^2 - S^2| < \varepsilon/3$. As in Lemma 2 an application of the triangle inequality shows that $|S - S^1 - S^2| < \varepsilon$, which completes the proof. ■

EXERCISES

1. Show that if a and b are two numbers such that for any $\varepsilon > 0$, $|a - b| < \varepsilon$ then $a = b$.
2. (a) Let f be the function on the half open interval $]0, 1]$ defined by $f(x) = 1/x$. Show that f is continuous at every point of $]0, 1]$ but not uniformly continuous.
   (b) Generalize this example to $\mathbb{R}^2$.
3. Let R be the rectangle $[a, b] \times [c, d]$ and f be a bounded function that is integrable over R.
   (a) Show that f is integrable over $[(a + b)/2, b] \times [c, d]$.
   (b) Let N be any positive integer. Show that f is integrable over $[(a + b)/N, b] \times [c, d]$.

Exercises 4 to 6 are intended to give a proof of Theorem 2 of Section 5.2.

4. Let C be the graph of a continuous function $\phi: [a, b] \to R$. Let $\varepsilon > 0$ be any positive number. Show that C can be placed in a finite union of boxes $B_i = [a_i, b_i] \times [c_i, d_i]$ such that C does not contain a boundary point of $\cup B_i$ and such that $\Sigma \text{ Area}(B_i) < \varepsilon$. (Hint: Use the uniform continuity principle presented in this appendix.)
5. Let R and B be rectangles and let $B \subset R$. Consider the nth regular partition of R and let b_n be the sum of the areas of all rectangles in the partition that have a non-empty intersection with B. Show that $\lim_{n\to\infty} b_n = \text{Area}(B)$.
6. Let R be a rectangle and $C \subset R$ the graph of a continuous function ϕ. Suppose that $f: R \to \mathbb{R}$ is bounded and continuous except on C. Use Exercises 4 and 5 above and the techniques used in the proof of Theorem 1 of this appendix to show that f is integrable over R.
7. (a) Use the uniform continuity principle to show that if $\phi: [a, b] \to \mathbb{R}$ is a continuous function then ϕ is bounded.
   (b) Generalize (*a*) to show that the continuous function $f: [a, b] \times [c, d] \to \mathbb{R}$ is bounded.
   (c) Generalize (*b*) still further to show f is bounded where $f: D \to \mathbb{R}$ is a continuous function on a closed and bounded set $D \subset \mathbb{R}^n$.

APPENDIX C

TABLES

Table 1. Trigonometric Functions

Angle					*Angle*				
Degree	*Radian*	*Sine*	*Cosine*	*Tangent*	*Degree*	*Radian*	*Sine*	*Cosine*	*Tangent*
0°	0.000	0.000	1.000	0.000					
1°	.017	.017	1.000	.017	46°	0.803	0.719	0.695	1.036
2°	.035	.035	0.999	.035	47°	.820	.731	.682	1.072
3°	.052	.052	.999	.052	48°	.838	.743	.669	1.111
4°	.070	’070	.998	.070	49°	.855	.755	.656	1.150
5°	.087	.087	.996	.087	50°	.873	.766	.643	1.192
6°	.105	.105	.995	.105	51°	.890	.777	.629	1.235
7°	.122	.122	.993	.123	52°	.908	.788	.616	1.280
8°	.140	.139	.990	.141	53°	.925	.799	.602	1.327
9°	.157	.156	.988	.158	54°	.942	.809	.588	1.376
10°	.175	.174	.985	.176	55°	.960	.819	.574	1.428
11°	.192	.191	.982	.194	56°	.977	.829	.559	1.483
12°	.209	.208	.978	.213	57°	.995	.839	.545	1.540
13°	.227	.225	.974	.231	58°	1.012	.848	.530	1.600
14°	.244	.242	.970	.249	59°	1.030	.857	.515	1.664
15°	.262	.259	.966	.268	60°	1.047	.866	.500	1.732
16°	.279	.276	.961	.287	61°	1.065	.875	.485	1.804
17°	.297	.292	.956	.306	62°	1.082	.883	.470	1.881
18°	.314	.309	.951	.325	63°	1.100	.891	.454	1.963
19°	.332	.326	.946	.344	64°	1.117	.899	.438	2.050
20°	.349	.342	.940	.364	65°	1.134	.906	.423	2.145
21°	.367	.358	.934	.384	66°	1.152	.914	.407	2.246
22°	.384	.375	.927	.404	67°	1.169	.921	.391	2.356
23°	.401	.391	.921	.425	68°	1.187	.927	.375	2.475
24°	.419	.407	.914	.445	69°	1.204	.934	.358	2.605
25°	.436	.423	.906	.466	70°	1.222	.940	.342	2.747
26°	.454	.438	.899	.488	71°	1.239	.946	.326	2.904
27°	.471	.454	.891	.510	72°	1.257	.951	.309	3.078
28°	.489	.470	.883	.532	73°	1.274	.956	.292	3.271
29°	.506	.485	.875	.554	74°	1.292	.961	.276	3.487
30°	.524	.500	.866	.577	75°	1.309	.966	.259	3.732
31°	.541	.515	.857	.601	76°	1.326	.970	.242	4.011
32°	.559	.530	.848	.625	77°	1.344	.974	.225	4.331
33°	.576	.545	.839	.649	78°	1.361	.978	.208	4.705
34°	.593	.559	.829	.675	79°	1.379	.982	.191	5.145
35°	.611	.574	.819	.700	80°	1.396	.985	.174	5.671
36°	.628	.588	.809	.727	81°	1.414	.988	.156	6.314
37°	’646	.602	.799	.754	82°	1.431	.990	.139	7.115
38°	.663	.616	.788	.781	83°	1.449	.993	.122	8.144
39°	.681	.629	.777	.810	84°	1.466	.995	.105	9.514
40°	.698	.643	.766	.839	85°	1.484	.996	.087	11.43
41°	.716	.656	.755	.869	86°	1.501	.998	.070	14.30
42°	.733	.669	.743	.900	87°	1.518	.999	.052	19.08
43°	.750	.682	.731	.933	88°	1.536	.999	.035	28.64
44°	.768	.695	.719	.966	89°	1.553	1.000	.017	57.29
45°	.785	.707	.707	1.000	90°	1.571	1.000	.000	∞

Table 2. Derivatives

1. $\dfrac{dau}{dx} = a\,\dfrac{du}{dx}$

2. $\dfrac{d(u + v)}{dx} = \dfrac{du}{dx} + \dfrac{dv}{dx}$

3. $\dfrac{d(uv)}{dx} = u\,\dfrac{dv}{dx} + v\,\dfrac{du}{dx}$

4. $\dfrac{d(u/v)}{dx} = \dfrac{v(du/dx) - u(dv/dx)}{v^2}$

5. $\dfrac{d(u^n)}{dx} = nu^{n-1}\,\dfrac{du}{dx}$

6. $\dfrac{d(u^v)}{dx} = vu^{v-1}\,\dfrac{du}{dx} + u^v(\log u)\,\dfrac{dv}{dx}$

7. $\dfrac{d(e^u)}{dx} = e^u\,\dfrac{du}{dx}$

8. $\dfrac{d(e^{au})}{dx} = ae^{au}\,\dfrac{du}{dx}$

9. $\dfrac{da^u}{dx} = a^u(\log a)\,\dfrac{du}{dx}$

10. $\dfrac{d(\log u)}{dx} = \dfrac{1}{u}\,\dfrac{du}{dx}$

11. $\dfrac{d(\log_a u)}{dx} = \dfrac{1}{u(\log a)}\,\dfrac{du}{dx}$

12. $\dfrac{d\sin u}{dx} = \cos u\,\dfrac{du}{dx}$

13. $\dfrac{d\cos u}{dx} = -\sin u\,\dfrac{du}{dx}$

14. $\dfrac{d\tan u}{dx} = \sec^2 u\,\dfrac{du}{dx}$

15. $\dfrac{d\cot u}{dx} = -\csc^2 u\,\dfrac{du}{dx}$

16. $\dfrac{d\sec u}{dx} = \tan u \sec u\,\dfrac{du}{dx}$

17. $\dfrac{d\csc u}{dx} = -(\cot u)(\csc u)\,\dfrac{du}{dx}$

18. $\dfrac{d\arcsin u}{dx} = \dfrac{1}{\sqrt{1 - u^2}}\,\dfrac{du}{dx}$

19. $\dfrac{d\arccos u}{dx} = \dfrac{-1}{\sqrt{1 - u^2}}\,\dfrac{du}{dx}$

20. $\dfrac{d\arctan u}{dx} = \dfrac{1}{1 + u^2}\,\dfrac{du}{dx}$

21. $\dfrac{d\,\mathrm{arc\,cot}\, u}{dx} = \dfrac{-1}{1 + u^2}\,\dfrac{du}{dx}$

22. $\dfrac{d\,\mathrm{arc\,sec}\, u}{dx} = \dfrac{1}{u\sqrt{u^2 - 1}}\,\dfrac{du}{dx}$

23. $\dfrac{d\,\mathrm{arc\,csc}\, u}{dx} = \dfrac{-1}{u\sqrt{u^2 - 1}}\,\dfrac{du}{dx}$

24. $\dfrac{d\sinh u}{dx} = \cosh u\,\dfrac{du}{dx}$

25. $\dfrac{d\cosh u}{dx} = \sinh u\,\dfrac{du}{dx}$

26. $\dfrac{d\tanh u}{dx} = \mathrm{sech}^2\, u\,\dfrac{du}{dx}$

27. $\dfrac{d\coth u}{dx} = -(\mathrm{csch}^2\, u)\,\dfrac{du}{dx}$

28. $\dfrac{d\,\mathrm{sech}\, u}{dx} = -(\mathrm{sech}\, u)(\tanh u)\,\dfrac{du}{dx}$

29. $\dfrac{d\,\mathrm{csch}\, u}{dx} = -(\mathrm{csch}\, u)(\coth u)\,\dfrac{du}{dx}$

30. $\dfrac{d\sinh^{-1} u}{dx} = \dfrac{1}{\sqrt{1 + u^2}}\,\dfrac{du}{dx}$

31. $\dfrac{d\cosh^{-1} u}{dx} = \dfrac{1}{\sqrt{u^2 - 1}}\,\dfrac{du}{dx}$

32. $\dfrac{d\tanh^{-1} u}{dx} = \dfrac{1}{1 - u^2}\,\dfrac{du}{dx}$

33. $\dfrac{d\coth^{-1} u}{dx} = \dfrac{1}{u^2 - 1}\,\dfrac{du}{dx}$

34. $\dfrac{d\,\mathrm{sech}^{-1}\, u}{dx} = \dfrac{-1}{u\sqrt{1 - u^2}}\,\dfrac{du}{dx}$

35. $\dfrac{d\,\mathrm{csch}^{-1}\, u}{dx} = \dfrac{-1}{|u|\sqrt{1 + u^2}}\,\dfrac{du}{dx}$

Table 3. Integrals (*An arbitrary constant may be added to each integral.*)

1. $\int x^n \, dx = \frac{1}{n+1} x^{n+1} \quad (n \neq -1)$

2. $\int \frac{1}{x} \, dx = \log|x|$

3. $\int e^x \, dx = e^x$

4. $\int a^x \, dx = \frac{a^x}{\log a}$

5. $\int \sin x \, dx = -\cos x$

6. $\int \cos x \, dx = \sin x$

7. $\int \tan x \, dx = -\log|\cos x|$

8. $\int \cot x \, dx = \log|\sin x|$

9. $\int \sec x \, dx = \log|\sec x + \tan x| = \log|\tan(\frac{1}{2}x + \frac{1}{4}\pi)|$

10. $\int \csc x \, dx = \log|\csc x - \cot x| = \log|\tan \frac{1}{2}x|$

11. $\int \arcsin \frac{x}{a} \, dx = x \arcsin \frac{x}{a} + \sqrt{a^2 - x^2} \quad (a > 0)$

12. $\int \arccos \frac{x}{a} \, dx = x \arccos \frac{x}{a} - \sqrt{a^2 - x^2} \quad (a > 0)$

13. $\int \arctan \frac{x}{a} \, dx = x \arctan \frac{x}{a} - \frac{a}{2} \log(a^2 + x^2) \quad (a > 0)$

14. $\int \sin^2 mx \, dx = \frac{1}{2m}(mx - \sin mx \cos mx)$

15. $\int \cos^2 mx \, dx = \frac{1}{2m}(mx + \sin mx \cos mx)$

16. $\int \sec^2 x \, dx = \tan x$

17. $\int \csc^2 x \, dx = -\cot x$

18. $\int \sin^n x \, dx = -\frac{\sin^{n-1} x \cos x}{n} + \frac{n-1}{n} \int \sin^{n-2} x \, dx$

19. $\int \cos^n x \, dx = \frac{\cos^{n-1} x \sin x}{n} + \frac{n-1}{n} \int \cos^{n-2} x \, dx$

20. $\int \tan^n x \, dx = \frac{\tan^{n-1} x}{n-1} - \int \tan^{n-2} x \, dx \quad (n \neq 1)$

Table 3 *(continued)*

21. $\int \cot^n x\,dx = -\dfrac{\cot^{n-1} x}{n-1} - \int \cot^{n-2} x\,dx \quad (n \neq 1)$

22. $\int \sec^n x\,dx = \dfrac{\tan x \sec^{n-2} x}{n-1} + \dfrac{n-2}{n-1}\int \sec^{n-2} x\,dx \quad (n \neq 1)$

23. $\int \csc^n x\,dx = -\dfrac{\cot x \csc^{n-2} x}{n-1} + \dfrac{n-2}{n-1}\int \csc^{n-2} x\,dx \quad (n \neq 1)$

24. $\int \sinh x\,dx = \cosh x$

25. $\int \cosh x\,dx = \sinh x$

26. $\int \tanh x\,dx = \log|\cosh x|$

27. $\int \coth x\,dx = \log|\sinh x|$

28. $\int \operatorname{sech} x\,dx = \arctan(\sinh x)$

29. $\int \operatorname{csch} x\,dx = \log\left|\tanh \dfrac{x}{2}\right| = -\dfrac{1}{2}\log\dfrac{\cosh x + 1}{\cosh x - 1}$

30. $\int \sinh^2 x\,dx = \frac{1}{4}\sinh 2x - \frac{1}{2}x$

31. $\int \cosh^2 x\,dx = \frac{1}{4}\sinh 2x + \frac{1}{2}x$

32. $\int \operatorname{sech}^2 x\,dx = \tanh x$

33. $\int \sinh^{-1}\dfrac{x}{a}\,dx = x\sinh^{-1}\dfrac{x}{a} - \sqrt{x^2 + a^2} \quad (a > 0)$

34. $\int \cosh^{-1}\dfrac{x}{a}\,dx = \begin{cases} x\cosh^{-1}\dfrac{x}{a} - \sqrt{x^2 - a^2} & \left[\cosh^{-1}\left(\dfrac{x}{a}\right) > 0,\ a > 0\right] \\ x\cosh^{-1}\dfrac{x}{a} + \sqrt{x^2 - a^2} & \left[\cosh^{-1}\left(\dfrac{x}{a}\right) < 0,\ a > 0\right] \end{cases}$

35. $\int \tanh^{-1}\dfrac{x}{a}\,dx = x\tanh^{-1}\dfrac{x}{a} + \dfrac{a}{2}\log|a^2 - x^2|$

36. $\int \dfrac{1}{\sqrt{a^2 + x^2}}\,dx = \log\left(x + \sqrt{a^2 + x^2}\right) = \sinh^{-1}\dfrac{x}{a} \quad (a > 0)$

37. $\int \dfrac{1}{a^2 + x^2}\,dx = \dfrac{1}{a}\arctan\dfrac{x}{a} \quad (a > 0)$

38. $\int \sqrt{a^2 - x^2}\,dx = \dfrac{x}{2}\sqrt{a^2 - x^2} + \dfrac{a^2}{2}\arcsin\dfrac{x}{a} \quad (a > 0)$

39. $\int (a^2 - x^2)^{3/2}\,dx = \dfrac{x}{8}(5a^2 - 2x^2)\sqrt{a^2 - x^2} + \dfrac{3a^4}{8}\arcsin\dfrac{x}{a} \quad (a > 0)$

Table 3 *(continued)*

40. $\displaystyle\int \frac{1}{\sqrt{a^2 - x^2}}\,dx = \arcsin\frac{x}{a} \quad (a > 0)$

41. $\displaystyle\int \frac{1}{a^2 - x^2}\,dx = \frac{1}{2a}\log\left|\frac{a + x}{a - x}\right|$

42. $\displaystyle\int \frac{1}{(a^2 - x^2)^{3/2}}\,dx = \frac{x}{a^2\sqrt{a^2 - x^2}}$

43. $\displaystyle\int \sqrt{x^2 \pm a^2}\,dx = \frac{x}{2}\sqrt{x^2 \pm a^2} \pm \frac{a^2}{2}\log|x + \sqrt{x^2 \pm a^2}|$

44. $\displaystyle\int \frac{1}{\sqrt{x^2 - a^2}}\,dx = \log|x + \sqrt{x^2 - a^2}| = \cosh^{-1}\frac{x}{a} \quad (a > 0)$

45. $\displaystyle\int \frac{1}{x(a + bx)}\,dx = \frac{1}{a}\log\left|\frac{x}{a + bx}\right|$

46. $\displaystyle\int x\sqrt{a + bx}\,dx = \frac{2(3bx - 2a)(a + bx)^{3/2}}{15b^2}$

47. $\displaystyle\int \frac{\sqrt{a + bx}}{x}\,dx = 2\sqrt{a + bx} + a\int \frac{1}{x\sqrt{a + bx}}\,dx$

48. $\displaystyle\int \frac{x}{\sqrt{a + bx}}\,dx = \frac{2(bx - 2a)\sqrt{a + bx}}{3b^2}$

49. $\displaystyle\int \frac{1}{x\sqrt{a + bx}}\,dx = \frac{1}{\sqrt{a}}\log\left|\frac{\sqrt{a + bx} - \sqrt{a}}{\sqrt{a + bx} + \sqrt{a}}\right| \quad (a > 0)$

$\displaystyle\qquad = \frac{2}{\sqrt{-a}}\arctan\sqrt{\frac{a + bx}{-a}} \quad (a < 0)$

50. $\displaystyle\int \frac{\sqrt{a^2 - x^2}}{x}\,dx = \sqrt{a^2 - x^2} - a\log\left|\frac{a + \sqrt{a^2 - x^2}}{x}\right|$

51. $\displaystyle\int x\sqrt{a^2 - x^2}\,dx = -\tfrac{1}{3}(a^2 - x^2)^{3/2}$

52. $\displaystyle\int x^2\sqrt{a^2 - x^2}\,dx = \frac{x}{8}(2x^2 - a^2)\sqrt{a^2 - x^2} + \frac{a^4}{8}\arcsin\frac{x}{a} \quad (a > 0)$

53. $\displaystyle\int \frac{1}{x\sqrt{a^2 - x^2}}\,dx = -\frac{1}{a}\log\left|\frac{a + \sqrt{a^2 - x^2}}{x}\right|$

54. $\displaystyle\int \frac{x}{\sqrt{a^2 - x^2}}\,dx = -\sqrt{a^2 - x^2}$

55. $\displaystyle\int \frac{x^2}{\sqrt{a^2 - x^2}}\,dx = -\frac{x}{2}\sqrt{a^2 - x^2} + \frac{a^2}{2}\arcsin\frac{x}{a} \quad (a > 0)$

Table 3 (*continued*)

56. $\displaystyle\int \frac{\sqrt{x^2 + a^2}}{x}\,dx = \sqrt{x^2 + a^2} - a \log\left|\frac{a + \sqrt{x^2 + a^2}}{x}\right|$

57. $\displaystyle\int \frac{\sqrt{x^2 - a^2}}{x}\,dx = \sqrt{x^2 - a^2} - a \operatorname{arc\,cos} \frac{a}{|x|}$

$$= \sqrt{x^2 - a^2} - a \operatorname{arc\,sec}\left(\frac{x}{a}\right) \quad (a > 0)$$

58. $\displaystyle\int x\sqrt{x^2 \pm a^2}\,dx = \tfrac{1}{3}(x^2 \pm a^2)^{3/2}$

59. $\displaystyle\int \frac{1}{x\sqrt{x^2 + a^2}}\,dx = \frac{1}{a}\log\left|\frac{x}{a + \sqrt{x^2 + a^2}}\right|$

60. $\displaystyle\int \frac{1}{x\sqrt{x^2 - a^2}}\,dx = \frac{1}{a}\operatorname{arc\,cos}\frac{a}{|x|} \quad (a > 0)$

61. $\displaystyle\int \frac{1}{x^2\sqrt{x^2 \pm a^2}}\,dx = \mp\frac{\sqrt{x^2 \pm a^2}}{a^2 x}$

62. $\displaystyle\int \frac{x}{\sqrt{x^2 \pm a^2}}\,dx = \sqrt{x^2 \pm a^2}$

63. $\displaystyle\int \frac{1}{ax^2 + bx + c}\,dx = \frac{1}{\sqrt{b^2 - 4ac}}\log\left|\frac{2ax + b - \sqrt{b^2 - 4ac}}{2ax + b + \sqrt{b^2 - 4ac}}\right| \quad (b^2 > 4ac)$

$$= \frac{2}{\sqrt{4ac - b^2}}\operatorname{arc\,tan}\frac{2ax + b}{\sqrt{4ac - b^2}} \quad (b^2 < 4ac)$$

64. $\displaystyle\int \frac{x}{ax^2 + bx + c}\,dx = \frac{1}{2a}\log|ax^2 + bx + c| - \frac{b}{2a}\int \frac{1}{ax^2 + bx + c}\,dx$

65. $\displaystyle\int \frac{1}{\sqrt{ax^2 + bx + c}}\,dx$

$$= \frac{1}{\sqrt{a}}\log|2ax + b + 2\sqrt{a}\sqrt{ax^2 + bx + c}| \quad (a > 0)$$

$$= \frac{1}{\sqrt{-a}}\operatorname{arc\,sin}\frac{-2ax - b}{\sqrt{b^2 - 4ac}} \quad (a < 0)$$

66. $\displaystyle\int \sqrt{ax^2 + bx + c}\,dx$

$$= \frac{2ax + b}{4a}\sqrt{ax^2 + bx + c} + \frac{4ac - b^2}{8a}\int \frac{1}{\sqrt{ax^2 + b + c}}\,dx$$

67. $\displaystyle\int \frac{x}{\sqrt{ax^2 + bx + c}}\,dx = \frac{\sqrt{ax^2 + bx + c}}{a} - \frac{b}{2a}\int \frac{1}{\sqrt{ax^2 + bx + c}}\,dx$

Table 3 (*continued*)

68. $$\int \frac{1}{x\sqrt{ax^2 + bx + c}}\,dx$$

$$= \frac{-1}{\sqrt{c}} \log\left|\frac{2\sqrt{c}\sqrt{ax^2 + bx + c} + bx + 2c}{x}\right| \quad (c > 0)$$

$$= \frac{1}{\sqrt{-c}} \text{arc sin} \frac{bx + 2c}{|x|\sqrt{b^2 - 4ac}} \quad (c < 0)$$

69. $$\int x^3\sqrt{x^2 + a^2}\,dx = (\tfrac{1}{5}x^2 - \tfrac{2}{15}a^2)\sqrt{(a^2 + x^2)^3}$$

70. $$\int \frac{\sqrt{x^2 \pm a^2}}{x^4}\,dx = \frac{\mp\sqrt{(x^2 \pm a^2)^3}}{3a^2x^3}$$

71. $$\int \sin ax \sin bx\,dx = \frac{\sin(a - b)x}{2(a - b)} - \frac{\sin(a + b)x}{2(a + b)} \quad (a^2 \neq b^2)$$

72. $$\int \sin ax \cos bx\,dx = -\frac{\cos(a - b)x}{2(a - b)} - \frac{\cos(a + b)x}{2(a + b)} \quad (a^2 \neq b^2)$$

73. $$\int \cos ax \cos bx\,dx = \frac{\sin(a - b)x}{2(a - b)} + \frac{\sin(a + b)x}{2(a + b)} \quad (a^2 \neq b^2)$$

74. $$\int \sec x \tan x\,dx = \sec x$$

75. $$\int \csc x \cot x\,dx = -\csc x$$

76. $$\int \cos^m x \sin^n x\,dx = \frac{\cos^{m-1} x \sin^{n+1} x}{m + n} + \frac{m - 1}{m + n}\int \cos^{m-2} x \sin^n x\,dx$$

$$= -\frac{\sin^{n-1} x \cos^{m+1} x}{m + n} + \frac{n - 1}{m + n}\int \cos^m x \sin^{n-2} x\,dx$$

77. $$\int x^n \sin ax\,dx = -\frac{1}{a} x^n \cos ax + \frac{n}{a}\int x^{n-1} \cos ax\,dx$$

78. $$\int x^n \cos ax\,dx = \frac{1}{a} x^n \sin ax - \frac{n}{a}\int x^{n-1} \sin ax\,dx$$

79. $$\int x^n e^{ax}\,dx = \frac{x^n e^{ax}}{a} - \frac{n}{a}\int x^{n-1} e^{ax}\,dx$$

80. $$\int x^n \log ax\,dx = x^{n+1}\left[\frac{\log ax}{n + 1} - \frac{1}{(n + 1)^2}\right]$$

81. $$\int x^n(\log ax)^m\,dx = \frac{x^{n+1}}{n + 1}(\log ax)^m - \frac{m}{n + 1}\int x^n(\log ax)^{m-1}\,dx$$

82. $$\int e^{ax} \sin bx\,dx = \frac{e^{ax}(a \sin bx - b \cos bx)}{a^2 + b^2}$$

Table 3 (*continued*)

83. $\int e^{ax} \cos bx \, dx = \dfrac{e^{ax}(b \sin bx + a \cos bx)}{a^2 + b^2}$

84. $\int \operatorname{sech} x \tanh x \, dx = -\operatorname{sech} x$

85. $\int \operatorname{csch} x \coth x \, dx = -\operatorname{csch} x$

APPENDIX D

ANSWERS TO ODD-NUMBERED EXERCISES

Some solutions requiring proofs are incomplete or are omitted.

SECTION 1.1

1. $x = 0, z = 0, y \in \mathbb{R}; x = 0, y = 0, z \in \mathbb{R}; y = 0, x, z \in \mathbb{R}; x = 0, y, z \in \mathbb{R}$
3. (b) First the similarity of $\Delta((0, 0, 0), (x, 0, 0), (x, y, 0))$ to $\Delta((0, 0, 0), (\alpha x, 0, 0), (\alpha x, \alpha y, 0))$ shows that $(0, 0, 0)$, $(x, y, 0)$, and $(\alpha x, \alpha y, 0)$ are collinear. Second, the similarity of $\Delta((0, 0, 0), (x, y, 0), (x, y, z))$ to $\Delta((0, 0, 0), (\alpha x, \alpha y, 0), (\alpha x, \alpha y, \alpha z))$ shows that $(0, 0, 0)$, (x, y, z), and $(\alpha x, \alpha y, \alpha z)$ are collinear. Thus $(\alpha x, \alpha y, \alpha z)$ is in the proper direction for $\alpha\mathbf{v}$. Finally, use the pythagorean theorem to show that the length is correct.
5. 4; 17
7. $(-104 + 16a, -24 - 4b, -22 + 26c)$
9. $24\mathbf{i} + 0\mathbf{j} + 0\mathbf{k} = 24\mathbf{i}$
11. $\{(2s, 7s + 2t, 7t) \mid s \in \mathbb{R}, t \in \mathbb{R}\}$
13. $\mathbf{l}(t) = (2t - 1)\mathbf{i} - \mathbf{j} + (3t - 1)\mathbf{k}$
15. $\{(x_0, y_0, z_0) + s\mathbf{a} + t\mathbf{b} \mid 0 \leq s \leq 1, 0 \leq t \leq 1\}$
17. If the vertices are at $\mathbf{0}$, $\mathbf{v}$, and $\mathbf{w}$, the midpoints of the sides are at $(\frac{1}{2}\mathbf{v}, \frac{1}{2}\mathbf{w}$, and $\frac{1}{2}(\mathbf{v} + \mathbf{w})$ by Example 7. Check these equations: $(\frac{2}{3})(\frac{1}{2})(\mathbf{v} + \mathbf{w}) = \frac{1}{2}\mathbf{v} + \frac{1}{3}(\mathbf{w} - \frac{1}{2}\mathbf{v}) = \frac{1}{2}\mathbf{w} + \frac{1}{3}(\mathbf{v} - \frac{1}{2}\mathbf{w})$

19. If (x, y, z) lies on the line, then $x = 2 + t$, $y = -2 + t$ and $z = -1 + t$. Therefore $2x - 3y + z - 2 = 4 + 2t + 6 - 3t - 1 + t - 2 = 7$, which is not zero. Hence no (x, y, z) satisfies both conditions.

21. $(1, 0, 0)$ is in the set. If the line is $(1, 0, 0) + t(a, b, c)$, then $(1 + at)^2 - (tb)^2 - (tc)^2 = 1$. That is, $t[2a + (a^2 - b^2 - c^2)t] = 0$. This must hold for every t, so $a = 0$ and $b^2 = c^2$. There are two lines through $(1, 0, 0)$ in the set, namely, $\mathbf{l}_1(t) = (1, 0, 0) + t(0, 1, 1)$ and $\mathbf{l}_2(t) = (1, 0, 0) + t(0, 1, -1)$.

SECTION 1.2

1. (a) Write out the expressions in components and use associative and distributive properties for numbers.
   (b) Use commutativity of multiplication of numbers on the coordinates.
3. $99°$
5. 75.7
7. $\|\mathbf{u}\| = \sqrt{5}$, $\|\mathbf{v}\| = \sqrt{2}$, $\mathbf{u} \cdot \mathbf{v} = -3$
9. $\|\mathbf{u}\| = \sqrt{11}$, $\|\mathbf{v}\| = \sqrt{62}$, $\mathbf{u} \cdot \mathbf{v} = -14$
11. $\|\mathbf{u}\| = \sqrt{14}$, $\|\mathbf{v}\| = \sqrt{26}$, $\mathbf{u} \cdot \mathbf{v} = -17$
13. Any (x, y, z) with $x + y + z = 0$

SECTION 1.3

1. $\begin{vmatrix} 1 & 2 & 1 \\ 3 & 0 & 1 \\ 2 & 0 & 2 \end{vmatrix} = -8; \quad \begin{vmatrix} 3 & 0 & 1 \\ 1 & 2 & 1 \\ 2 & 0 & 2 \end{vmatrix} = 8; \quad \begin{vmatrix} 2 & 1 & 1 \\ 0 & 3 & 1 \\ 0 & 2 & 2 \end{vmatrix} = 8$, etc.
3. $-3\mathbf{i} + \mathbf{j} + 5\mathbf{k}$
5. $\sqrt{35}$
7. $\pm\mathbf{k}$
9. $\pm(113\mathbf{i} + 17\mathbf{j} - 103\mathbf{k})/\sqrt{23667}$
11. $2/\sqrt{338} = \sqrt{2}/13$
13. $\mathbf{u} + \mathbf{v} = 3\mathbf{i} - 3\mathbf{j} + 3\mathbf{k}$; $\mathbf{u} \cdot \mathbf{v} = 6$; $\|\mathbf{u}\| = \sqrt{6}$; $\|\mathbf{v}\| = 3$; $\mathbf{u} \times \mathbf{v} = -3\mathbf{i} + 3\mathbf{k}$
15. (a) $x + y + z - 1 = 0$
    (b) $x + 2y + 3z - 6 = 0$
17. (a) Do the first by working out the coordinates, then use that and $\mathbf{A} \times (\mathbf{B} \times \mathbf{C}) = -(\mathbf{B} \times \mathbf{C}) \times \mathbf{A}$ to get the second.
    (b) Use (a) to write the quantity in terms of inner products.
    (c) Use (a) and collect terms.
19. Compute the results of Cramer's rule and check that they satisfy the equation.
21. $x - 2y + 3z + 12 = 0$

23. $4x - 6y - 10z = 14$

25. $10x - 17y + z + 25 = 0$

SECTION 1.4

1. (a)

Cylindrical r	θ	z	Rectangular x	y	z
1	45°	1	$\sqrt{2}/2$	$\sqrt{2}/2$	1
2	$\pi/2$	-4	0	2	-4
0	45°	10	0	0	10
3	$\pi/6$	4	$3\sqrt{3}/2$	3/2	4

Spherical ρ	θ	ϕ
$\sqrt{2}$	45°	45°
$2\sqrt{5}$	$\pi/2$	π-arccos$(2\sqrt{5}/5)$
10	45°	0
5	$\pi/6$	arccos(4/5)

(b)

Rectangular x	y	z	Spherical ρ	θ	ϕ
2	1	-2	3	arctan 1/2	$\pi/2 +$ arccos $\sqrt{5}/3$
0	3	4	5	$\pi/2$	arcsin 3/5
$\sqrt{2}$	1	1	2	arcsin $\sqrt{3}/3$	$\pi/3$
$-2\sqrt{3}$	-2	3	5	$7\pi/6$	arccos 3/5

Cylindrical r	θ	z
$\sqrt{5}$	arctan 1/2	-2
3	$\pi/2$	4
$\sqrt{3}$	arcsin $\sqrt{3}/3$	1
4	$7\pi/6$	3

3. (a) rotation by π around the z-axis.
   (b) reflection across the xy-plane

5. no; $(r, \theta, \phi) = (-r, \theta + \pi, \pi - \phi)$

7. $\hat{\mathbf{e}}_\rho = (x\mathbf{i} + y\mathbf{j} + z\mathbf{k})/\sqrt{x^2 + y^2 + z^2}$
$\hat{\mathbf{e}}_\theta = (-y\mathbf{i} + x\mathbf{j})/\sqrt{x^2 + y^2}$
$\hat{\mathbf{e}}_\phi = (xz\mathbf{i} - yz\mathbf{j} + (x^2 + y^2)\mathbf{k})/r\rho$
$\hat{\mathbf{e}}_\theta \times \mathbf{j} = -y\mathbf{k}/\sqrt{x^2 + y^2}$, $\hat{\mathbf{e}}_\phi \times \mathbf{j} = (xz/r\rho)\mathbf{k} - (r/\rho)\mathbf{i}$

SECTION 1.5

1. (*ii*) Express in components and use commutativity of multiplication of numbers.
(*iii*) $\mathbf{x} \cdot \mathbf{x}$ is a sum of squares of real numbers.
(*iv*) $\mathbf{x} \cdot \mathbf{x}$ is the sum of the squares of the components of $\mathbf{x}$. This can be 0 only if each component is 0.

3. (a) $|\mathbf{x} \cdot \mathbf{y}| = 10 = \sqrt{5}\sqrt{20} = \|\mathbf{x}\|\,\|\mathbf{y}\|$
$\|\mathbf{x} + \mathbf{y}\| = 3\sqrt{5} = \|\mathbf{x}\| + \|\mathbf{y}\|$
(b) $|\mathbf{x} \cdot \mathbf{y}| = 17 < \sqrt{3690} = \|\mathbf{x}\|\,\|\mathbf{y}\|$
$\|\mathbf{x} + \mathbf{y}\| = \sqrt{165} < 15 < \sqrt{41} + 3\sqrt{10} = \|\mathbf{x}\| + \|\mathbf{y}\|$
(c) $|\mathbf{x} \cdot \mathbf{y}| = 5 < \sqrt{65} = \|\mathbf{x}\|\,\|\mathbf{y}\|$
$\|\mathbf{x} + \mathbf{y}\| = \sqrt{28} < \sqrt{5} + \sqrt{13} = \|\mathbf{x}\| + \|\mathbf{y}\|$

5. $(\det A)(\det A^{-1}) = \det(AA^{-1}) = \det(I) = 1$

7. Compute the matrix product in both orders.

9. HINT: For $k = 2$ use the triangle inequality to show that $\|\mathbf{x}_1 + \mathbf{x}_2\| \le \|\mathbf{x}_1\| + \|\mathbf{x}_2\|$; then for $k = i + 1$ note that $\|\mathbf{x}_1 + \mathbf{x}_2 + \cdots + \mathbf{x}_{i+1}\| \le \|\mathbf{x}_1 + \mathbf{x}_2 + \cdots + \mathbf{x}_i\| + \|\mathbf{x}_{i+1}\|$

11. (a) Check $n = 1$ and $n = 2$ directly. Then reduce an $n \times n$ determinant to a sum of $(n - 1) \times (n - 1)$ determinants and use the induction assumption on these.
(b) The argument is similar to (a). Suppose the first row is multiplied by λ. The first term of the sum will be λa_{11} times an $(n - 1) \times (n - 1)$ determinant with no factors of λ. The other terms obtained (by expanding across the first row) are similar.

13. Not necessarily. Try $A = \begin{bmatrix} 0 & 1 \\ 0 & 0 \end{bmatrix}$ and $B = \begin{bmatrix} 1 & 0 \\ 0 & 0 \end{bmatrix}$.

15. (a) The sum of two continuous functions and a scalar multiple of a continuous function are continuous.
(b) (*i*)
$$(\alpha f + \beta g) \cdot k = \int_0^1 (\alpha f + \beta g)(x)h(x)\,dx$$
$$= \int_0^1 f(x)h(x)\,dx + \beta \int_0^1 g(x)h(x)\,dx$$
$$= \alpha f \cdot h + \beta g \cdot h$$
(*ii*)
$$f \cdot g = \int_0^1 f(x)g(x)\,dx = \int_0^1 g(x)f(x)\,dx = g \cdot f$$

In (*iii*) and (*iv*), the integrand is a perfect square. So the integral is nonnegative and can be 0 only if the integrand is 0 everywhere. If $f(x) \neq 0$ for some x, then it would be positive in a neighborhood of x by continuity, and the integral would be positive.

REVIEW EXERCISES FOR CHAPTER 1

1. $\mathbf{v} + \mathbf{w} = 4\mathbf{i} + 3\mathbf{j} + 6\mathbf{k}$; $3\mathbf{v} = 9\mathbf{i} + 12\mathbf{j} + 15\mathbf{k}$; $6\mathbf{v} + 8\mathbf{w} = 26\mathbf{i} + 16\mathbf{j} + 38\mathbf{k}$; $-2\mathbf{v} = -6\mathbf{i} - 8\mathbf{j} - 10\mathbf{k}$; $\mathbf{v} \cdot \mathbf{w} = 4$; $\mathbf{v} \times \mathbf{w} = 9\mathbf{i} + 2\mathbf{j} - 7\mathbf{k}$

3. $\{st\mathbf{a} + s(1 - t)\mathbf{b} \mid 0 \le t \le 1 \text{ and } 0 \le s \le 1\}$

5. Let $\mathbf{v} = (a_1, a_2, a_3)$, $\mathbf{w} = (b_1, b_2, b_3)$, and apply the CBS inequality.

7. The area is the absolute value of

$$\begin{vmatrix} a_1 & a_2 \\ b_1 & b_2 \end{vmatrix} = \begin{vmatrix} a_1 & a_2 \\ b_1 + \lambda a_1 & b_2 + \lambda a_2 \end{vmatrix}$$

(A multiple of one row of a determinant may be added to another row without changing its value.)

9. The cosines of the two parts of the angle are equal since $\mathbf{a} \cdot \mathbf{v}/\|\mathbf{a}\|\,\|\mathbf{v}\| = (\mathbf{a} \cdot \mathbf{b} + \|\mathbf{a}\|\,\|\mathbf{b}\|)/\|\mathbf{v}\| = \mathbf{b} \cdot \mathbf{v}/\|\mathbf{b}\|\,\|\mathbf{v}\|$.

11. $\mathbf{i} \times \mathbf{j} = \begin{vmatrix} \mathbf{i} & \mathbf{j} & \mathbf{k} \\ 1 & 0 & 0 \\ 0 & 1 & 0 \end{vmatrix} = \mathbf{k}$; etc.

13. (a) (HINT: The length of the projection of the vector between any pair of points, one on each line, onto $(\mathbf{a}_1 \times \mathbf{a}_2)/\|\mathbf{a}_1 \times \mathbf{a}_2\|$ is d.)
    (b) $\sqrt{2}$

15. (a) Note that

$$\frac{1}{2}\begin{vmatrix} 1 & 1 & 1 \\ x_1 & x_2 & x_3 \\ y_1 & y_2 & y_3 \end{vmatrix} = \frac{1}{2}\begin{vmatrix} 1 & 0 & 0 \\ x_1 & x_2 - x_1 & x_3 - x_1 \\ y_1 & y_2 - y_1 & y_3 - y_1 \end{vmatrix} = \frac{1}{2}\begin{vmatrix} x_2 - x_1 & x_3 - x_1 \\ y_2 - y_1 & y_3 - y_1 \end{vmatrix}$$

    (b) $\frac{1}{2}$

17.

Rectangular	Spherical
(a) $(\sqrt{2}/2, \sqrt{2}/2, 1)$	(a) $(\sqrt{2}, \pi/4, \pi/4)$
(b) $(3\sqrt{3}/2, \frac{3}{2}, -4)$	(b) $(5, \pi/6, \arccos(-\frac{4}{5}))$
(c) $(0, 0, 1)$	(c) $(1, \pi/4, 0)$
(d) $(0, -2, 1)$	(d) $(\sqrt{5}, -\pi/2, \arccos(\sqrt{5}/5))$
(e) $(0, 2, 1)$	(e) $(\sqrt{5}, \pi/2, \arccos(\sqrt{5}/5))$

19. $z = r^2 \cos 2\theta$; $\cos \phi = r \sin^2 \phi \cos 2\theta$

21. $|\mathbf{x} \cdot \mathbf{y}| = 6 < \sqrt{98} = \|\mathbf{x}\|\,\|\mathbf{y}\|$; $\|\mathbf{x} + \mathbf{y}\| = \sqrt{33} < \sqrt{14} + \sqrt{7} = \|\mathbf{x}\| + \|\mathbf{y}\|$

23. (a) The associative law for matrix multiplication may be checked by:

$$[(AB)C]_{ij} = \sum_{k=1}^{n} (AB)_{ik}C_{kj} = \sum_{k=1}^{n}\sum_{l=1}^{n} A_{il}B_{lk}C_{kj}$$

$$= \sum_{l=1}^{n} A_{il}(BC)_{lj} = [A(BC)]_{ij}$$

Use this with C taken to be a column vector.

    (b) The matrix for the composition is the product matrix.

25. $\mathbb{R}^n$ is spanned by the vectors $\mathbf{e}_1, \mathbf{e}_2, \ldots, \mathbf{e}_n$. If $\mathbf{v} \in \mathbb{R}^n$, then $A\mathbf{v} = A(\sum_{i=1}^n (\mathbf{v} \cdot \mathbf{e}_i)\mathbf{e}_i) = \sum_{i=1}^n (\mathbf{v} \cdot \mathbf{e}_i) A\mathbf{e}_i$. Let $a_{ij} = (A\mathbf{e}_j \cdot \mathbf{e}_i)$ so that $A\mathbf{e}_j = \sum_{j=1}^n a_{ij}\mathbf{e}_i$. Then $A\mathbf{v} \cdot \mathbf{e}_k = \sum_{i=1}^n (\mathbf{v} \cdot \mathbf{e}_i) a_{ki}$. That is, if

$$\mathbf{v} = \begin{bmatrix} v_1 \\ \vdots \\ v_n \end{bmatrix}, \quad \text{then} \quad A\mathbf{v} = \begin{bmatrix} a_{11} & \cdots & a_{1n} \\ \vdots & & \\ a_{n1} & \cdots & a_{nn} \end{bmatrix} \begin{bmatrix} v_1 \\ \vdots \\ v_n \end{bmatrix}$$

as desired.

SECTION 2.1

1. The level curves and graphs are sketched below.

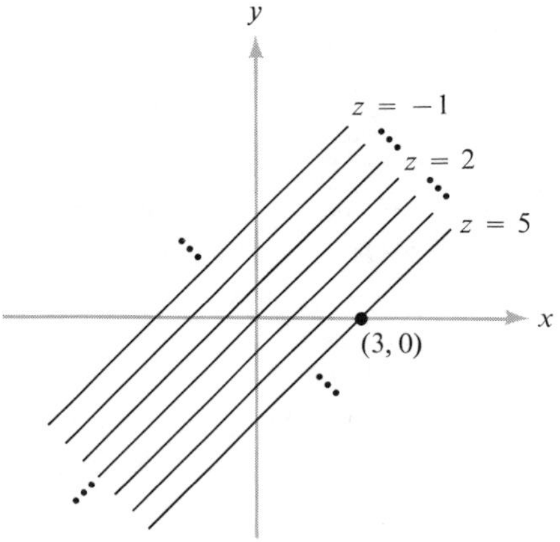

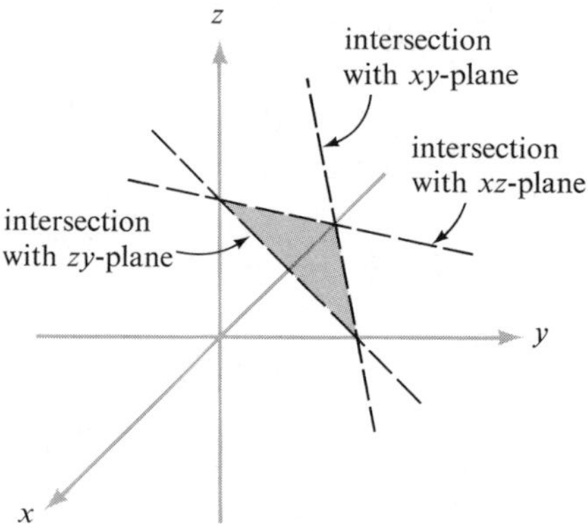

Problem 1(a)

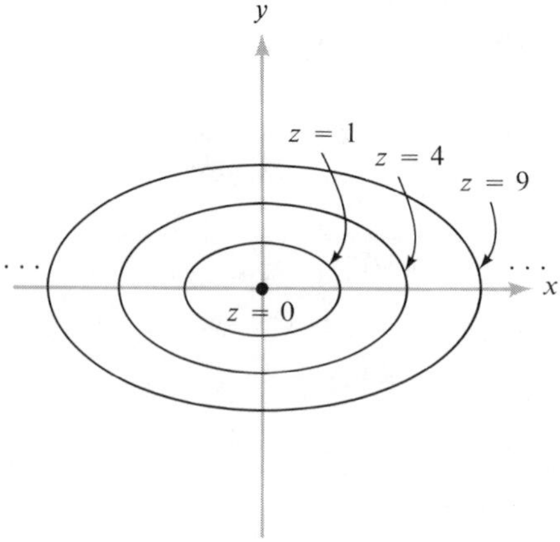

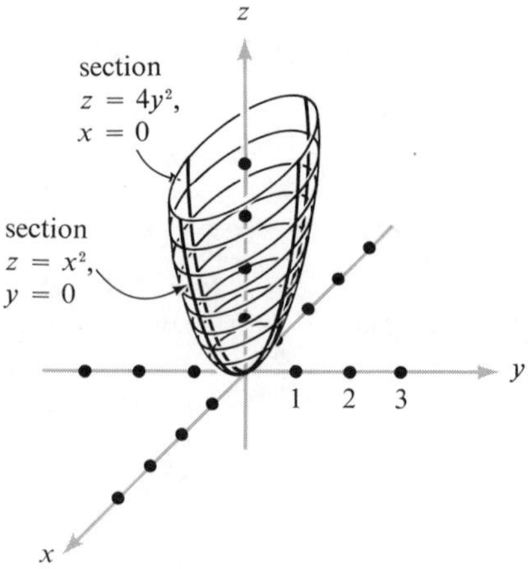

Prob.1(b)

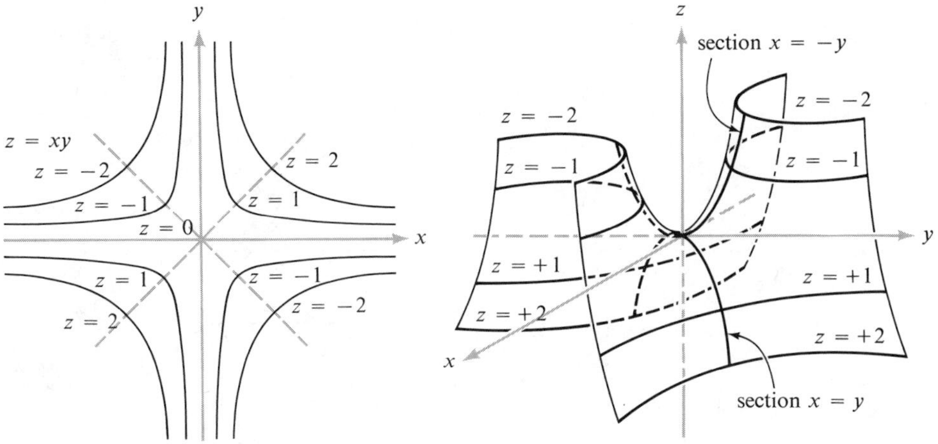

Prob.1(c)

The graph in (c) is a hyperbolic paraboloid like that of Example 4 but rotated 45° and vertically flattened by a factor of $\frac{1}{4}$. To see this, use the variables $u = x + y$ and $v = x - y$. Then $z = (u^2 - v^2)/4$.

3. The level curves are cirlces $x^2 + y^2 = 100 - c^2$ when $c \leq 10$. The graph is the upper hemisphere of $x^2 + y^2 + z^2 = 100$.

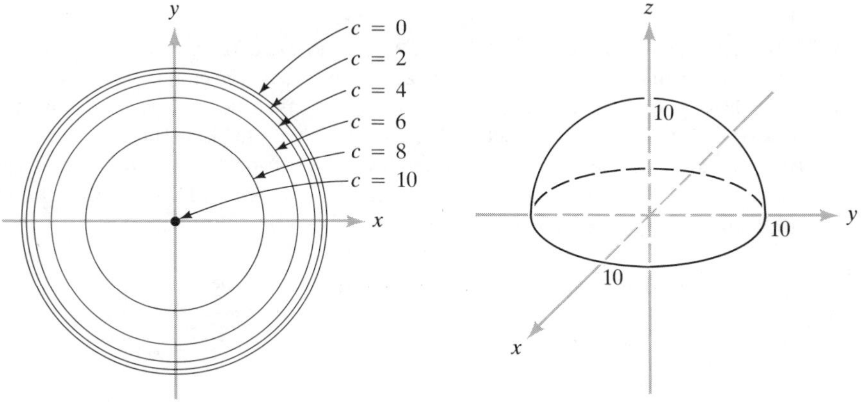

5. The level curves are circles, and the graph is a paraboloid revolution. See Example 3 of this section.

7. If $c = 0$ the level curve is the straight line $y = -x$. If $c \neq 0$, then $y = -x + (c/x)$. The level curve is a hyperbola with the y-axis and the line $y = -x$ as asymptotes.

   The graph is a hyperbolic paraboloid. Sections along the line $y = ax$ are the parabolas $z = (1 + a)x^2 = (1 + a)r^2/(1 + a^2)$.

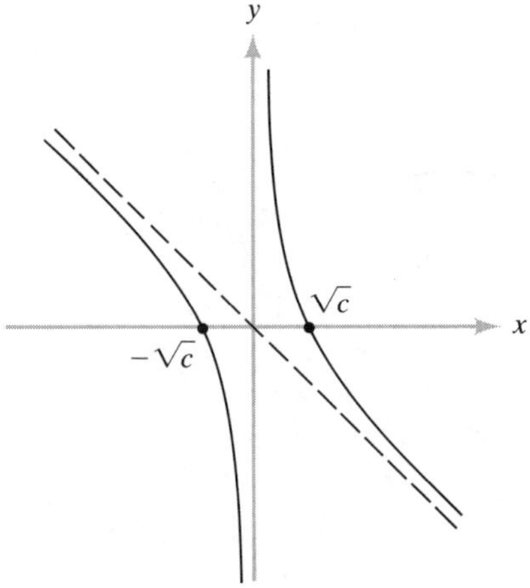

$c < 0$

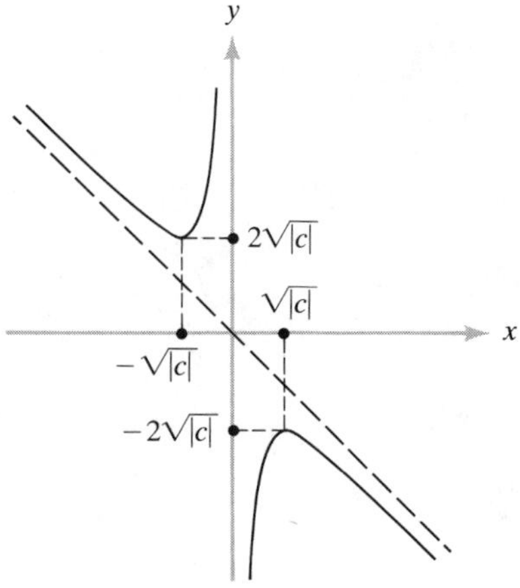

$c < 0$

9. If $c > 0$, the level surface $f(x, y, z) = c$ is empty. If $c = 0$, the level surface is the point $(0, 0, 0)$. If $c < 0$, the level surface is the sphere of radius $\sqrt{c}$ centered at $(0, 0, 0)$. A section of the graph determined by $z = a$ is given by $t = -x^2 - y^2 - a^2$, which is a paraboloid of revolution opening down in (x, y, t)-space.

11. If $c < 0$, the level surface is empty. If $c = 0$ the level surface is the z-axis. If $c > 0$, it is the right circular cylinder $x^2 + y^2 = c$ of radius $\sqrt{c}$ whose axis is the z-axis. A section of the graph determined by $z = a$ is the paraboloid of revolution $t = x^2 + y^2$. A section determined by $x = b$ is a “trough’ with parabolic cross section $t(y, z) = y^2 + b^2$.

13. Setting $u = (x - z)/\sqrt{2}$ and $v = (x + z)/\sqrt{2}$ gives u and v axes rotated 45° around the y-axis from the x and z axes. Since $f = vy\sqrt{2}$, the level surfaces $f = c$ are “cylinders” perpendicular to the vy-plane ($z = -x$) whose cross sections are the hyperbolae $vy = c/\sqrt{2}$. The section $S_{x=a} \cap \text{graph}(f)$ is the hyperbolic paraboloid $t = (z + a)y$ in yzt-space (see Exercise 1(c)). The section $S_{y=b} \cap \text{graph}(f)$ is the plane $t = bx + bz$ in xzt-space. The section $S_{z=b} \cap \text{graph}(f)$ is the hyperbolic paraboloid $t = y(x + b)$ in xyt-space.

15.

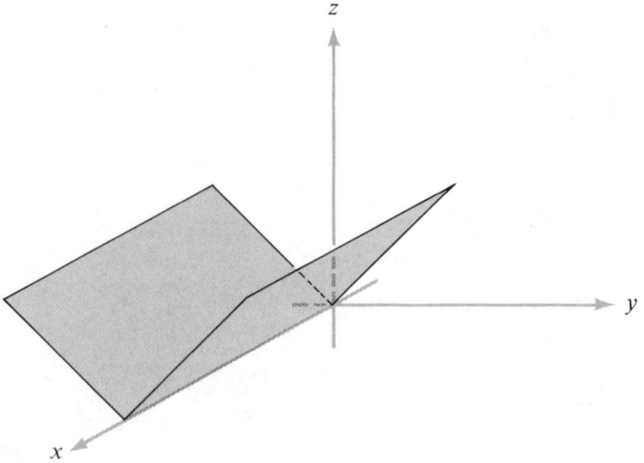

If $c < 0$, the level curve is empty. If $c = 0$, the level curve is the x-axis. If $c > 0$, it is the pair of parallel lines $|y| = c$. The sections of the graph with x-constant are V-shaped curves $x = |y|$ in yz-space. The graph is sketched on p. 536.

17. Complete the square to get $(x + 2)^2 + (y - b/2)^2 + (z + \frac{9}{2})^2 = (b^2 + 4b + 97)/4$. This is an ellipsoid with center at $(-2, b/2, -\frac{9}{2})$ and axes parallel to the coordinate axes.

19. The value of z does not matter, so we get a "cylinder" of elliptic cross section parallel to the z-axis and intersecting the xy-plane in the ellipse $4x^2 + y^2 = 16$.

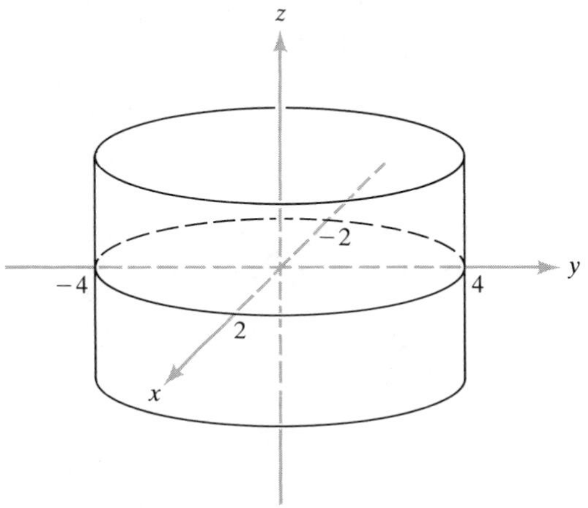

21. The value of x does not matter, so we get a "cylinder" parallel to the x-axis of hyperbolic cross-section intersecting the yz-plane in the hyperbola $z^2 - y^2 = 4$.

23. This is a saddle surface similar to that of Example 4 but the hyperbolae, which are level curves, no longer have perpendicular asymptotes.

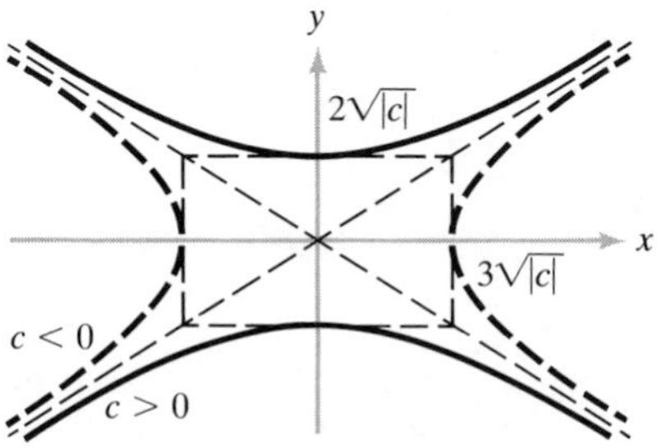

level curves

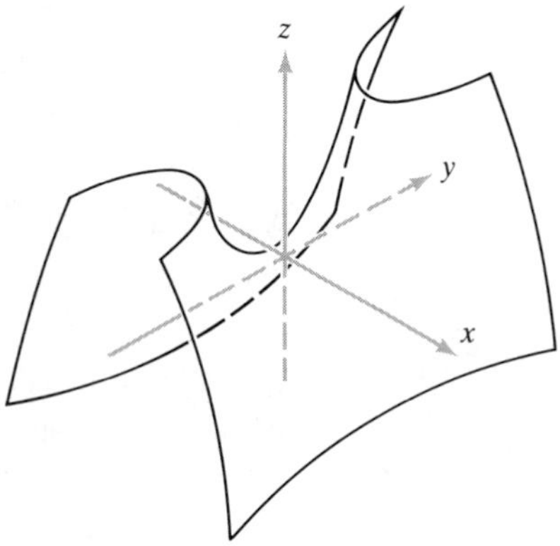

25. A double cone with axis along the y-axis and elliptical cross sections

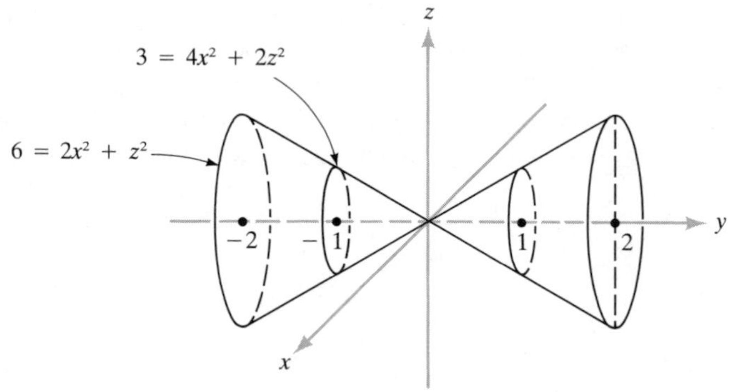

27. An elliptic paraboloid with axis along the x-axis

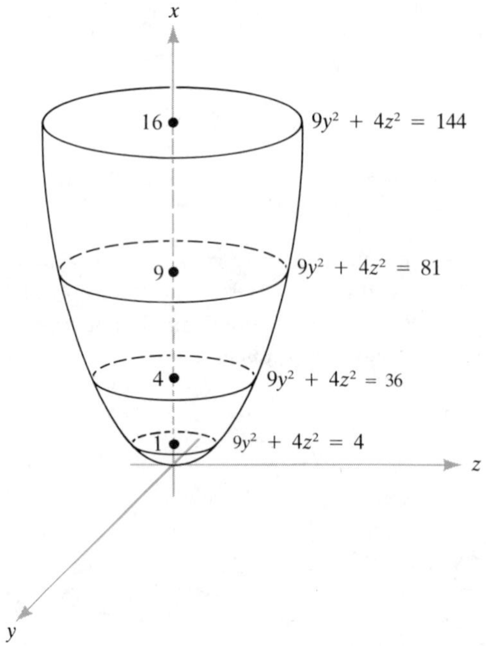

29. Level curves are described by $\cos 2\theta = cr^2$. If $c > 0$, then $-\pi/4 \leq \theta \leq \pi/4$ or $3\pi/4 \leq \theta \leq 5\pi/4$. If $c < 0$, then $\pi/4 \leq \theta \leq 3\pi/4$ or $5\pi/4 \leq \theta \leq 7\pi/4$. In either case you get a figure-8 shape through the origin called a lemniscate. (Such shapes

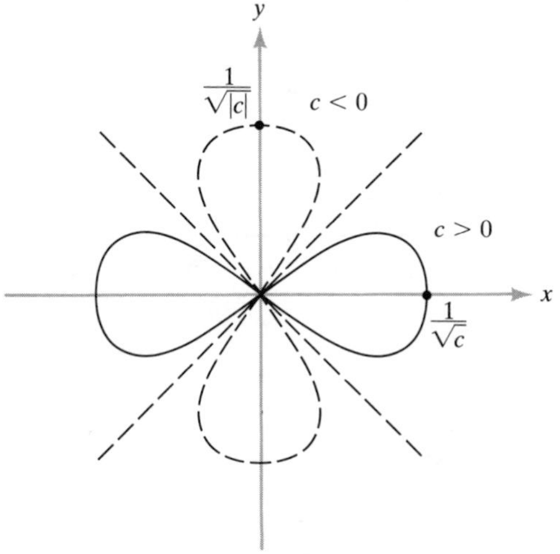

were first studied by Jacques Bernoulli and are sometimes called Bernoulli's leminiscates.)

SECTION 2.2

1. (a) If $(x_0, y_0) \in A$ then $|x_0| < 1$ and $|y_0| < 1$. Let $r < 1 - |x_0|$ and $r < 1 - |y_0|$. Prove that $D_r(x_0, y_0) \subset A$ either analytically or by drawing a figure.
   (b) If $(x_0, y_0) \in B$ and $0 < r < y_0$ (e.g., $r = y_0/2$), then $D_r(x_0, y_0) \subset B$ (prove analytically or by drawing a figure).
   (c) Let $r = \min(4 - \sqrt{x_0^2 + y_0^2}, \sqrt{x_0^2 + y_0^2} - 2)$.
   (d) Let r be the smallest of the three numbers used in (a), (b), and (c).
   (c) Let $r = \min(|x_0|, |y_0|)$.

3. Suppose $\mathbf{y} \in U \cap V$. If s and t are small enough then $D_s(\mathbf{y}) \subset U$ and $D_t(\mathbf{y}) \subset V$ and if $r = \min(s, t)$, then $D_r(\mathbf{y}) \subset U \cap V$. If $\mathbf{z} \in U \cup V$, then some disc around $\mathbf{z}$ is contained in whichever of U or V contains $\mathbf{z}$.

5. For $|x - 2| < \delta = \sqrt{\varepsilon + 4} - 2$, we have $|x^2 - 4| = |x - 2|\,|x + 2| < \delta(\delta + 4) = \varepsilon$. By Theorem 3(*iii*), $\lim_{x\to 2} x^2 = (\lim_{x\to 2} x)^2 = 2^2 = 4$.

7. (a) 1; (b) $\|\mathbf{x}_0\|$; (c) $(1, e)$;
   (d) limit doesn't exist (look at the limits for $x = 0$ and $y = 0$ separately)

9. 0

11. (a) $\lim_{x\to b^+} f(x) = L$ if for every $\varepsilon > 0$ there is a $\delta > 0$ such that $x > b$ and $0 < x - b < \delta$ imply $|f(x) - L| < \varepsilon$.
   (b) $\lim_{x\to 0^-} (1/x) = -\infty$, $\lim_{t\to -\infty} e^t = 0$, so $\lim_{x\to 0^-} e^{1/x} = 0$. Hence $\lim_{x\to 0^-} 1/(1 + e^{1/x}) = 1$. The other limit is 0.

(c)

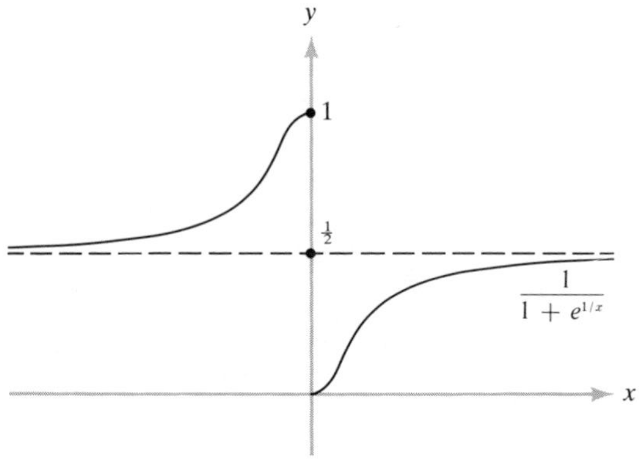

13. (a) 5; (b) 0, (c) $2x$; (c) 1; (e) $-\frac{1}{2}$
15. Compose $f(x, y) = xy$ with $g(t) = (\sin t)/t$ for $t \neq 0$ and $g(0) = 1$.
17. This may be proved using Theorem 7. That is, f is continuous at $\mathbf{x}_0$ if and only if $\lim_{\mathbf{x}\to\mathbf{x}_0} f(\mathbf{x}) = f(\mathbf{x}_0)$. Equivalently, for every $\varepsilon > 0$, there is a $\delta > 0$ such $0 < \|\mathbf{x} - \mathbf{x}_0\| < \delta$ implies $\|f(\mathbf{x}) - f(\mathbf{x}_0)\| < \varepsilon$. Rewrite the last inequality as $|\|f(\mathbf{x}) - f(\mathbf{x}_0)\| - 0| < \varepsilon$, and it says $\lim_{\mathbf{x}\to\mathbf{x}_0} \|f(\mathbf{x}) - f(\mathbf{x}_0)\| = 0$.
19. Use parts (*ii*) and (*iii*) of Theorem 4.
21. Let $r = \|\mathbf{x} - \mathbf{y}\|/2$. If $\|\mathbf{z} - \mathbf{y}\| \leq r$, let $f(\mathbf{z}) = \|\mathbf{z} - \mathbf{y}\|/r$. If $\|\mathbf{z} - \mathbf{y}\| > r$, let $f(\mathbf{z}) = 1$.
23. If $\varepsilon > 0$ and $\mathbf{x}_0$ are given, let $\delta = (\varepsilon/K)^{1/\alpha}$. Then $\|f(\mathbf{x}) - f(\mathbf{x}_0)\| < K\delta^\alpha = \varepsilon$ whenever $\|\mathbf{x} - \mathbf{x}_0\| < \delta$. Notice that the choice of δ does not depend on $\mathbf{x}_0$. This means that f is *uniformly continuous.*

SECTION 2.3

1. (a) $\partial f/\partial x = y$; $\partial f/\partial y = x$;
   (b) $\partial f/\partial x = ye^{xy}$; $\partial f/\partial y = xe^{xy}$
   (c) $\partial f/\partial x = \cos x \cos y - x \sin x \cos y$
   $\partial f/\partial y = -x \cos x \sin y$
   (d) $\partial f/\partial x = 2x(1 + \log(x^2 + y^2))$
   $\partial f/\partial y = 2y(1 + \log(x^2 + y^2))$; $(x, y) \neq (0, 0)$
3. (a) $\partial w/\partial x = (1 + 2x^2)\exp(x^2 + y^2)$; $\partial w/\partial y = 2xy \exp(x^2 + y^2)$
   (b) $\partial w/\partial x = -4xy^2/(x^2 - y^2)^2$; $\partial w/\partial y = 4yx^2/(x^2 - y^2)^2$
   (c) $\partial w/\partial x = ye^{xy} \log(x^2 + y^2) + 2xe^{xy}/(x^2 + y^2)$;
   $\partial w/\partial y = xe^{xy} \log(x^2 + y^2) + 2ye^{xy}/(x^2 + y^2)$
   (d) $\partial w/\partial x = 1/y$; $\partial w/\partial y = -x/y^2$
   (e) $\partial w/\partial x = -y^2e^{xy} \sin ye^{xy} \sin x + \cos ye^{xy} \cos x$;
   $\partial w/\partial y = (xye^{xy} + e^{xy})(-\sin ye^{xy} \sin x)$

5. $z = f(3, 1) + (\partial f/\partial x)(3, 1)(x - 3) + (\partial f/\partial y)(3, 1)(y - 1)$
$= 10 + 6(x - 3) + 3(y - 1) = 6x + 3y - 11$

7. (a) $\begin{bmatrix} 1 & 0 \\ 0 & 1 \end{bmatrix}$; (b) $\begin{bmatrix} e^y & xe^y - \sin y \\ 1 & 0 \\ 1 & e^y \end{bmatrix}$; (c) $\begin{bmatrix} 1 & 1 & e^z \\ 2xy & x^2 & 0 \end{bmatrix}$

9. Both are xye^{xy}

11. (a) $\nabla f = (e^{-x^2-y^2-z^2}(-2x^2 + 1), -2xye^{-x^2-y^2-z^2}, -2xze^{-x^2-y^2-z^2})$
(b) $\nabla f = (x^2 + y^2 + z^2)^{-2}(yz(y^2 + z^2 - x^2), xz(x^2 + z^2 - y^2), xy(x^2 + y^2 - z^2))$

13. They are constant. (HINT: Divide the line segment between x and y into n equal pieces to obtain $\|f(\mathbf{x}) - f(\mathbf{y})\| \le n^{1-\alpha}K\|\mathbf{y} - \mathbf{y}\|^{\alpha}$.) The derivative is the zero matrix.

SECTION 2.4

1. Use parts (*i*), (*ii*), and (*iii*) of Theorem 10. The derivative at $\mathbf{x}$ is $2(f(\mathbf{x}) + 1)Df(\mathbf{x})$.

3. (a) $h(x, y) = f(x, u(x, y)) = f(p(x), u(x, y))$
We introduce p here solely as notation: $p(x) = x$

Written out: $\dfrac{\partial h}{\partial x} = \dfrac{\partial f}{\partial p}\dfrac{dp}{\partial x} + \dfrac{\partial f}{\partial u}\dfrac{\partial u}{\partial x} = \dfrac{\partial f}{\partial p} + \dfrac{\partial f}{\partial u}\dfrac{\partial u}{\partial x}$ since $\dfrac{dp}{dx} = \dfrac{dx}{dx} = 1$

JUSTIFICATION: Call (p, u) the variables of f. In order to use the Chain Rule we must express h as a composition of functions; i.e., first find g such that $h(x, y) = f(g(x, y))$. Let $g(x, y) = (p(x), u(x, y))$. Therefore $Dh = Df \cdot Dg$ Then

$$\begin{bmatrix} \dfrac{\partial h}{\partial x} & \dfrac{\partial h}{\partial y} \end{bmatrix} = \begin{bmatrix} \dfrac{\partial f}{\partial p} & \dfrac{\partial f}{\partial u} \end{bmatrix} \begin{bmatrix} \dfrac{\partial g_1}{\partial x} & \dfrac{\partial g_1}{\partial y} \\ \dfrac{\partial g_2}{\partial x} & \dfrac{\partial g_2}{\partial y} \end{bmatrix} = \begin{bmatrix} \dfrac{\partial f}{\partial p} & \dfrac{\partial f}{\partial u} \end{bmatrix} \begin{bmatrix} 1 & 0 \\ \dfrac{\partial u}{\partial x} & \dfrac{\partial u}{\partial y} \end{bmatrix}$$

$$= \begin{bmatrix} \dfrac{\partial f}{\partial p} + \dfrac{\partial f}{\partial u}\dfrac{\partial u}{\partial x} & \dfrac{\partial f}{\partial u}\dfrac{\partial u}{\partial y} \end{bmatrix}$$

So $\dfrac{\partial h}{\partial x} = \dfrac{\partial f}{\partial p} + \dfrac{\partial f}{\partial u}\dfrac{\partial u}{\partial x}$. You may see $\dfrac{\partial h}{\partial x} = \dfrac{\partial f}{\partial x} + \dfrac{\partial f}{\partial u}\dfrac{\partial u}{\partial x}$ as an answer.

This requires careful interpretation because of possible ambiguity about the meaning of $\partial f/\partial x$, which is why the name p was used.

(b) $\dfrac{\partial h}{\partial x} = \dfrac{\partial f}{\partial x} + \dfrac{\partial f}{\partial u}\dfrac{\partial u}{\partial x} + \dfrac{\partial f}{\partial v}\dfrac{\partial v}{\partial x}$

(c) $\dfrac{\partial h}{\partial x} = \dfrac{\partial f}{\partial u}\dfrac{\partial u}{\partial x} + \dfrac{\partial f}{\partial v}\dfrac{\partial v}{\partial x} + \dfrac{\partial f}{\partial w}\dfrac{\partial w}{\partial x}$

5. Compute each in two ways; the answers are
   (a) $(f \circ \mathbf{c})'(t) = e^t(\cos t - \sin t)$
   (b) $(f \circ \mathbf{c})'(t) = 15t^4 \exp(3t^5)$
   (c) $(f \circ \mathbf{c})'(t) = (e^{2t} - e^{-2t})(1 + \log(e^{2t} + e^{-2t}))$
   (d) $(f \circ \mathbf{c})'(t) = (1 + 4t^2)\exp(2t^2)$

7. Use Theorem 10(*iii*) and replace matrices by vectors.

9. (a) $\partial f/\partial x(0, 0) = \partial f/\partial y(0, 0) = \lim_{t \to 0} ((0/t^2) - 0)/t = 0$
   (b) $(f \circ \mathbf{g})(t) = (ab^2/(a^2 + b^2))t$; $\nabla f(0, 0) \cdot \mathbf{g}'(0) = (0, 0) \cdot (a, b) = 0$

11. (a) $G(x, y(x)) = 0$ so $\dfrac{\partial G}{\partial x} + \dfrac{\partial G}{\partial y}\dfrac{dy}{dx} = 0$

   (b) $$\begin{bmatrix} \dfrac{dy_1}{dx} \\ \dfrac{dy_2}{dx} \end{bmatrix} = -\begin{bmatrix} \dfrac{\partial G_1}{\partial y_1} & \dfrac{\partial G_1}{\partial y_2} \\ \dfrac{\partial G_2}{\partial y_1} & \dfrac{\partial G_2}{\partial y_2} \end{bmatrix}^{-1} \begin{bmatrix} \dfrac{\partial G_1}{\partial x} \\ \dfrac{\partial G_2}{\partial x} \end{bmatrix}$$ where $^{-1}$ means the inverse matrix,

   The first component of this equation reads

$$\frac{dy_1}{dx} = \frac{-\dfrac{\partial G_1}{\partial x}\dfrac{\partial G_2}{\partial y_2} + \dfrac{\partial G_2}{\partial x}\dfrac{\partial G_1}{\partial y_2}}{\dfrac{\partial G_1}{\partial y_1}\dfrac{\partial G_2}{\partial y_2} - \dfrac{\partial G_2}{\partial y_1}\dfrac{\partial G_1}{\partial y_2}}$$

   (c) $\dfrac{dy}{dx} = \dfrac{-2x}{3y^2 + e^y}$

13. Define $R_1(\mathbf{h}) = f(\mathbf{x}_0 + \mathbf{h}) - f(\mathbf{x}_0) - Df(\mathbf{x}_0) \cdot \mathbf{h}$

15. Proof of (*iii*) follows.

$$\frac{|h(\mathbf{x}) - h(\mathbf{x}_0) - (f(\mathbf{x}_0)Dg(\mathbf{x}_0) + g(\mathbf{x}_0)Df(\mathbf{x}_0))(\mathbf{x} - \mathbf{x}_0)|}{\|\mathbf{x} - \mathbf{x}_0\|}$$
$$\le |f(\mathbf{x}_0)|\frac{|g(\mathbf{x}) - g(\mathbf{x}_0) - Dg(\mathbf{x}_0)(\mathbf{x} - \mathbf{x}_0)|}{\|\mathbf{x} - \mathbf{x}_0\|}$$
$$+ |g(\mathbf{x}_0)|\frac{|f(\mathbf{x}) - f(\mathbf{x}_0) - Df(\mathbf{x}_0)(\mathbf{x} - \mathbf{x}_0)|}{\|\mathbf{x} - \mathbf{x}_0\|}$$
$$+ \frac{|f(\mathbf{x}) - f(\mathbf{x}_0)|}{\|\mathbf{x} - \mathbf{x}_0\|}\frac{|g(\mathbf{x}) - g(\mathbf{x}_0)|}{\|\mathbf{x} - \mathbf{x}_0\|}\|\mathbf{x} - \mathbf{x}_0\|$$

As $\mathbf{x} \to \mathbf{x}_0$, the first two terms go to 0 by the differentiability of f and g. The third does so because $|f(\mathbf{x}) - f(\mathbf{x}_0)|/\|\mathbf{x} - \mathbf{x}_0\|$ and $|g(\mathbf{x}) - g(\mathbf{x}_0)|/\|\mathbf{x} - \mathbf{x}_0\|$ are bounded by a constant, say M, on some ball $D_r(\mathbf{x}_0)$. To see this, choose r small enough so that $(f(\mathbf{x}) - f(\mathbf{x}_0))/\|\mathbf{x} - \mathbf{x}_0\|$ is within 1 of $Df(\mathbf{x}_0)(\mathbf{x} - \mathbf{x}_0)/\|\mathbf{x} - \mathbf{x}_0\|$ if $\|\mathbf{x} - \mathbf{x}_0\| < r$. Then we have $|f(\mathbf{x}) - f(\mathbf{x}_0)|/\|\mathbf{x} - \mathbf{x}_0\| \le M_1 + |Df(\mathbf{x}_0)(\mathbf{x} - \mathbf{x}_0)|/\|\mathbf{x} - \mathbf{x}_0\| = M_1 + |\nabla f(\mathbf{x}_0) \cdot (\mathbf{x} - \mathbf{x}_0)|/\|\mathbf{x} - \mathbf{x}_0\| \le M_1 + \|\nabla f(\mathbf{x}_0)\|$ by the Cauchy-Schwarz inequality.

Part (*iv*) follows from (*iii*) and the special case of the Quotient Rule, with f identically 1; that is, $D(1/g)(\mathbf{x}_0) = (-1/g(\mathbf{x}_0)^2)Dg(\mathbf{x}_0)$. To obtain this answer, note that on some small ball $D_r(\mathbf{x}_0)$, $g(\mathbf{x}) > m > 0$. Use the triangle and the Schwarz inequality to show that

$$\left|\frac{1}{g(\mathbf{x})} - \frac{1}{g(\mathbf{x})} + \frac{1}{g(\mathbf{x}_0)^2} Dg(\mathbf{x}_0)(\mathbf{x} - \mathbf{x}_0)\right| \Big/ \|\mathbf{x} - \mathbf{x}_0\|$$

$$\leq \frac{1}{|g(\mathbf{x})|}\frac{1}{|g(\mathbf{x}_0)|}\frac{|g(\mathbf{x}) - g(\mathbf{x}_0) - Dg(\mathbf{x}_0)(\mathbf{x} - \mathbf{x}_0)|}{\|\mathbf{x} - \mathbf{x}_0\|}$$

$$+ \frac{|g(\mathbf{x}) - g(\mathbf{x}_0)|}{|g(\mathbf{x})|g(\mathbf{x}_0)^2}\frac{|Dg(\mathbf{x}_0)(\mathbf{x} - \mathbf{x}_0)|}{\|\mathbf{x} - \mathbf{x}_0\|}$$

$$\leq \frac{1}{m^2}\frac{|g(\mathbf{x}) - g(\mathbf{x}_0) - Dg(\mathbf{x}_0)(\mathbf{x} - \mathbf{x}_0)|}{\|\mathbf{x} - \mathbf{x}_0\|} + \frac{\|\nabla g(\mathbf{x}_0)\|}{m^3}|g(\mathbf{x}) - g(\mathbf{x}_0)|$$

These both go to 0 since g is differentiable and continuous.

17. Apply the Chain Rule to $\partial G/\partial T$ where $G(t(T, P), P(T, P), V(T, P) = P(V - b)e^{a/RVT} - RT$ is identically 0; $t(T, P) = T$; and $p(T, P) = P$.

19. (2, 0)

21. Let g_1 and g_2 be C^1 functions from $\mathbb{R}^3$ to $\mathbb{R}$ such that $g_1(\mathbf{x}) = 1$ for $\|\mathbf{x}\| < \sqrt{2}/3$; $g_1(\mathbf{x}) = 0$ for $\|\mathbf{x}\| > 2\sqrt{2}/3$; $y_2(\mathbf{x}) = 1$ for $\|\mathbf{x} - (1, 1, 0)\| < \sqrt{2}/3$; and $g_2(\mathbf{x}) = 0$ for $\|\mathbf{x} - (1, 1, 0)\| > 2\sqrt{2}/3$. (See Exercise 20.) Let

$$h_1(\mathbf{x}) = \begin{bmatrix} 1 & 0 & 0 \\ 0 & -1 & 0 \\ 0 & 0 & 0 \end{bmatrix}\begin{bmatrix} x_1 \\ x_2 \\ x_3 \end{bmatrix} + \begin{bmatrix} 1 \\ 1 \\ 0 \end{bmatrix} \quad \text{and} \quad h_2(\mathbf{x}) = \begin{bmatrix} 0 & 0 & -1 \\ 0 & 0 & 0 \\ 0 & 0 & 1 \end{bmatrix}\begin{bmatrix} x_1 \\ x_2 \\ x_3 \end{bmatrix}$$

and put $f(\mathbf{x}) = g_1(\mathbf{x})k_1(\mathbf{x}) + g_2(\mathbf{x})k_2(\mathbf{x})$.

23. By Exercise 22 and Theorem 10(*iii*), each component of k is differentiable and $Dk_i(\mathbf{x}_0) = f(\mathbf{x}_0)Dg_i(\mathbf{x}_0) + g_i(\mathbf{x}_0)Df(\mathbf{x})$. Since $Dg_i(\mathbf{x}_0)(\mathbf{y})$ is the ith component of $Dg(\mathbf{x}_0)(\mathbf{y})$ and $Df(\mathbf{x}_0)(\mathbf{y})$ is the number $\nabla f(\mathbf{x}_0) \cdot \mathbf{y}$, we get $Dk(\mathbf{x}_0)(\mathbf{y}) = f(\mathbf{x}_0)Dg(\mathbf{x}_0)(\mathbf{y}) + (Df(\mathbf{x}_0)(\mathbf{y}))g(\mathbf{x}_0) = f(\mathbf{x}_0)Dg(\mathbf{x}_0)(\mathbf{y}) + (\nabla f(\mathbf{x}_0) \cdot \mathbf{y})g(\mathbf{x}_0)$.

SECTION 2.5

1. $\nabla f(1, 1, 2) \cdot \mathbf{v} = (4, 3, 1) \cdot (1/\sqrt{5}, 2/\sqrt{5}, 0) = 2\sqrt{5}$

3. (a) $3x + 8y + 3z = 20$ (b) $4y - 2x = 6$
   (c) $x + y + z = 3$ (d) $6y - 9x - z = 6$
   (e) $y + z = \pi/2$ (f) $z = 1$

5. $\nabla f(x, y, z) = 2(x, y, z)$ is twice the radius vector to $\mathbf{x}$. But $f(\mathbf{x})$ is $\|\mathbf{x}\|^2$ and this increases fastest directly away from $\mathbf{0}$ as Theorem 13 says. The level surfaces are spheres centered at the origin. The radius vector is normal to these spheres and thus to any curve in them, as determined by Theorem 14.

7. The graph of f is the level surface $0 = F(x, y, z) = f(x, y) - z$. So the tangent plane is given by

$$\begin{aligned} 0 &= \nabla F(x_0, y_0, z_0) \cdot (x - x_0, y - y_0, z - z_0) \\ &= \left(\frac{\partial f}{\partial x}(x_0, y_0), \frac{\partial f}{\partial y}(x_0, y_0), -1\right) \cdot (x - x_0, y - y_0, z - z_0) \end{aligned}$$

Since $z_0 = f(x_0, y_0)$, this is $z = f(x_0, y_0) + (\partial f/\partial x)(x_0, y_0)(x - x_0) + (\partial f/\partial y)(x_0, y_0)(y - y_0)$.

9. (a) $\nabla f = (x + y, z + x, x + y)$, $\mathbf{g}'(t) = (e^t, -\sin t, \cos t)$, $(f \circ \mathbf{g})'(1) = 2e \cos 1 + \cos^2 1 - \sin^2 1$
   (b) $\nabla f = (yze^{xyz}, xze^{xyz}, xye^{xyz})$, $\mathbf{g}'(t) = [6, 6t, 3t^2]$, $(f \circ \mathbf{g})'(1) = 108e^{18}$
   (c) $\nabla f = (x, yz)$, $\mathbf{g}'(t) = (e^t, -e^{-t}, 1)$, $(f \circ \mathbf{g})'(1) = e^2 - e^{-2} + 1$

11. Let $f(x, y, z) = 1/r = (x^2 + y^2 + z^2)^{1/2}$; $\mathbf{r} = (x, y, z)$ Then $\nabla f = -(x^2 + y^2 + z^2)^{-3/2}(x, y, z) = -(1/r^3)\mathbf{r}$

13. $\nabla f = (g'(x), 0)$

15. $Df(0, 0, \ldots, 0) = [0, \ldots, 0]$

SECTION 2.6

1. (a) $\dfrac{\partial^2 f}{\partial x^2} = 24\dfrac{x^3y - xy^3}{(x^2 + y^2)^4}, \dfrac{\partial^2 f}{\partial y^2} = 24\dfrac{-x^3y + xy^3}{(x^2 + y^2)^4},$

$$\frac{\partial^2 f}{\partial x\,\partial y} = \frac{\partial^2 f}{\partial y\,\partial x} = \frac{-6x^4 + 36x^2y^2 - 6y^4}{(x^2 + y^2)^4}$$

   (b) $\dfrac{\partial^2 f}{\partial x^2} = \dfrac{2}{x^3}, \dfrac{\partial^2 f}{\partial y^2} = xe^{-y}, \dfrac{\partial^2 f}{\partial x\,\partial y} = \dfrac{\partial^2 f}{\partial y\,\partial x} = -e^{-y}$

   (c) $\dfrac{\partial^2 f}{\partial x^2} = -y^4 \cos(xy^2), \dfrac{\partial^2 f}{\partial y^2} = -2x \sin(xy^2) - 4x^2y^2 \cos(xy^2),$

$$\frac{\partial^2 f}{\partial x\,\partial y} = \frac{\partial^2 f}{\partial y\,\partial x} = -2y \sin(xy^2) - 2xy^3 \cos(xy^2)$$

   (d) $\dfrac{\partial^2 f}{\partial x^2} = y^4e^{-xy^2} + 12x^2y^3, \dfrac{\partial^2 f}{\partial y^2} = -2xe^{-xy^2} + 4x^2y^2e^{-xy^2} + 6yx^4,$

$$\frac{\partial^2 f}{\partial x\,\partial y} = \frac{\partial^2 f}{\partial y\,\partial x} = -2ye^{-xy^2} + 2xy^3e^{-xy^2} + 12x^3y^2$$

   (e) $\dfrac{\partial^2 f}{\partial x^2} = \dfrac{2(\cos^2 x + e^{-y})\cos 2x + 2\sin^2 2x}{(\cos^2 x + e^{-y})^3},$

$$\frac{\partial^2 f}{\partial y^2} = \frac{e^{-y} - \cos^2 x}{e^y(\cos^2 x + e^{-y})^3}$$

$$\frac{\partial^2 f}{\partial x\,\partial y} = \frac{\partial^2 f}{\partial y\,\partial x} = \frac{2\sin 2x}{e^y(\cos^2 x + e^{-y})^3}$$

3. $\frac{\partial^2 f}{\partial x^2}\left(\frac{dx}{dt}\right)^2 + 2\frac{\partial^2 f}{\partial x\,\partial y}\frac{dx}{dt}\frac{dy}{dt} + \frac{\partial^2 f}{\partial y^2}\left(\frac{dy}{dt}\right)^2 + \frac{\partial f}{\partial x}\frac{d^2x}{dt^2} + \frac{\partial f}{\partial y}\frac{d^2y}{dt^2}$,
where $\mathbf{c}(t) = (x(t), y(t))$

5. $2x + 6y - z = 5$

7. (a) $\frac{\partial f}{\partial x} = \arctan\frac{x}{y} + \frac{xy}{x^2 + y^2}$

$\frac{\partial f}{\partial y} = \frac{-x^2}{x^2 + y^2}$

$\frac{\partial^2 f}{\partial x^2} = \frac{y^3}{(x^2 + y^2)^2}, \frac{\partial^2 f}{\partial y^2} = \frac{2x^2 y}{(x^2 + y^2)^2}$

$\frac{\partial^2 f}{\partial x\,\partial y} = \frac{\partial^2 f}{\partial y\,\partial x} = \frac{-2xy^2}{(x^2 + y^2)^2}$

(b) $\frac{\partial f}{\partial x} = \frac{-x\sin(\sqrt{x^2 + y^2})}{\sqrt{x^2 + y^2}}, \frac{\partial f}{\partial y} = \frac{-y\sin(\sqrt{x^2 + y^2}}{\sqrt{x^2 + y^2}}$

$\frac{\partial^2 f}{\partial x^2} = \frac{x^2\sin\sqrt{x^2 + y^2}}{(x^2 + y^2)^{3/2}} - \frac{x^2\cos\sqrt{x^2 + y^2}}{x^2 + y^2} - \frac{\sin\sqrt{x^2 + y^2}}{(x^2 + y^2)^{1/2}}$

$\frac{\partial^2 f}{\partial y^2} = \frac{y^2\sin\sqrt{x^2 + y^2}}{(x^2 + y^2)^{3/2}} - \frac{y^2\cos\sqrt{x^2 + y^2}}{x^2 + y^2} - \frac{\sin\sqrt{x^2 + y^2}}{(x^2 + y^2)^{1/2}}$

$\frac{\partial^2 f}{\partial x\,\partial y} = \frac{\partial^2 f}{\partial y\,\partial x} = xy\left[\frac{\sin\sqrt{x^2 + y^2}}{(x^2 + y^2)^{3/2}} - \frac{\cos\sqrt{x^2 + y^2}}{x^2 + y^2}\right]$

(c) $\frac{\partial f}{\partial x} = -2x\exp(-x^2 - y^2);\quad \frac{\partial f}{\partial y} = -2y\exp(-x^2 - y^2);$

$\frac{\partial^2 f}{\partial x^2} = (4x^2 - 2)\exp(-x^2 - y^2);\quad \frac{\partial^2 f}{\partial y^2} = (4y^2 - 2)\exp(-x^2 - y^2)$

$\frac{\partial^2 f}{\partial x\,\partial y} = \frac{\partial^2 f}{\partial y\,\partial x} = 4xy\exp(-x^2 - y^2)$

9. Since f and $\partial f/\partial z$ are both C^2, we have

$$\frac{\partial^3 f}{\partial x\,\partial y\,\partial z} = \frac{\partial^2}{\partial x\,\partial y}\frac{\partial f}{\partial z} = \frac{\partial^2}{\partial y\,\partial x}\frac{\partial f}{\partial z} = \frac{\partial}{\partial y}\left(\frac{\partial^2 f}{\partial x\,\partial z}\right) = \frac{\partial}{\partial y}\left(\frac{\partial^2 f}{\partial z\,\partial x}\right) = \frac{\partial^3 f}{\partial y\,\partial z\,\partial x}$$

11. $V = -GM/r = -GM(x^2 + y^2 + z^2)^{-1/2}$. Check that

$$\frac{\partial^2 V}{\partial x^2} + \frac{\partial^2 V}{\partial y^2} + \frac{\partial^2 V}{\partial z^2}$$
$$= GM(x^2 + y^2 + z^2)^{-3/2}[3 - 3(x^2 + y^2 + z^2)(x^2 + y^2 + z^2)^{-2/2}] = 0$$

REVIEW EXERCISES FOR CHAPTER 2

1. (a) elliptic paraboloid
   (b) Let $y' = y + 3$ and write $z = xy'$. This is a (shifted) hyperbolic paraboloid.

3. (a) $Df(x, y) = \begin{bmatrix} 2xy & x^2 \\ -ye^{-xy} & -xe^{-xy} \end{bmatrix}$

(b) $Df(x) = \begin{bmatrix} 1 \\ 1 \end{bmatrix}$

(c) $Df(x, y, z) = [e^x \quad e^y \quad e^z]$

(d) $Df(x, y, z) = \begin{bmatrix} 1 & 0 & 0 \\ 0 & 1 & 0 \\ 0 & 0 & 1 \end{bmatrix}$

5. The tangent plane to a sphere at (x_0, y_0, z_0) is normal to the line from the center to (x_0, y_0, z_0).

7. (a) $z = x - y + 2$ (b) $z = 4x - 8y - 8$
(c) $x + y + z + 1 = 0$ (d) $10x + 6y - 4z = 6 - \pi$
(e) $2z = \sqrt{2}x + \sqrt{2}y$ (f) $x + 2y - z = 2$

9. (a) The level curves are hyperbolae $xy = 1/c$.

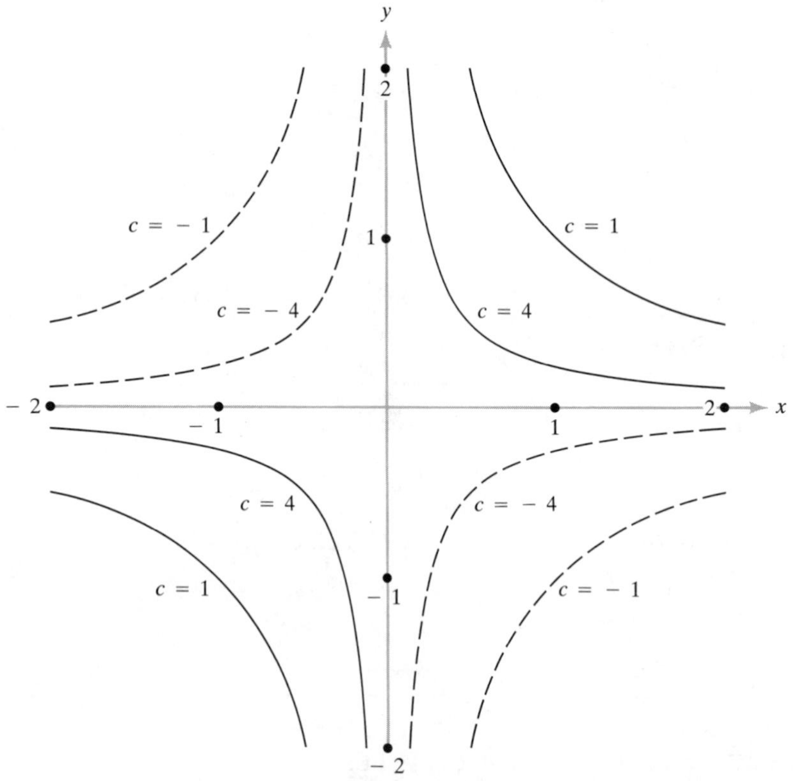

(b) $c = x^2 - xy - y^2 = \left(x - \frac{1 + \sqrt{5}}{2} y\right)\left(x - \frac{1 - \sqrt{5}}{2} y\right)$

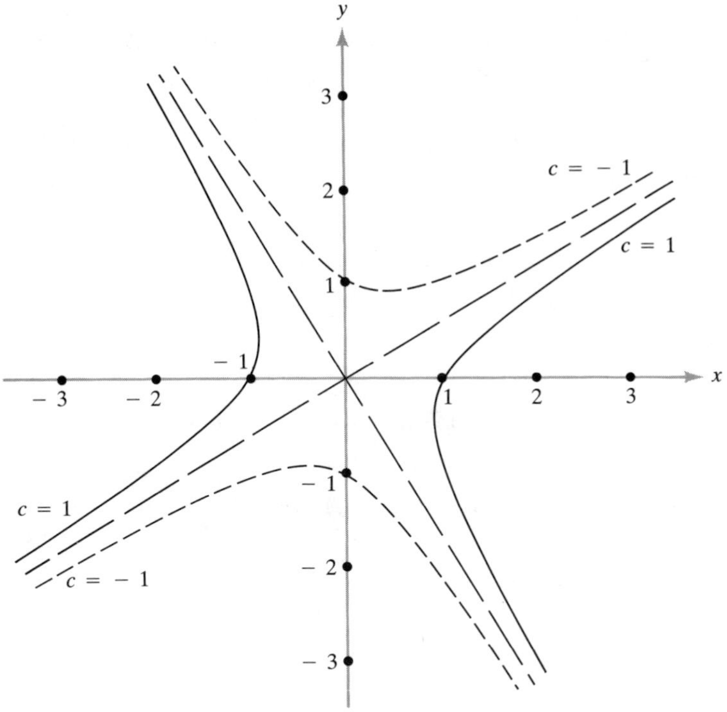

11. (a) 0
    (b) limit does not exist

13. If $\mathbf{F} = \nabla f$, then $\partial F_1/\partial y = \partial^2 f/\partial y\, \partial x$ and $\partial F_2/\partial x = \partial^2 f/\partial x\, \partial y$. Since $\mathbf{F}$ is C^1, the second partials of f are continuous and therefore equal by Theorem 15.

    If $F_1 = y \cos x$ and $F_2 = x \sin y$, then $\partial F_1/\partial y = \cos x$ and $F_2/\partial x = \sin y$. Since these are not equal, $\mathbf{F}$ is not the gradient of anything.

15. (a) The line $\mathbf{L}(t) = (x_0, y_0, f(x_0, y_0)) + t(a, b, c)$ lies in the plane $z = f(x_0, y_0)$ if $c = 0$ and is perpendicular to $\nabla f(x_0, y_0)$ if $a(\partial f/\partial x)(x_0, y_0) + b(\partial b/\partial y)(x_0, y_0) = 0$. (Along $\mathbf{L}$, we have

$$f(x_0, y_0) + \frac{\partial f}{\partial x}(x_0, y_0)(x - x_0) + \frac{\partial f}{\partial y}(x_0, y_0)(y - y_0)$$

$$= f(x_0, y_0) + at\frac{\partial f}{\partial x}(x_0, y_0) + bt\frac{\partial f}{\partial y}(x_0, y_0)$$

$$= f(x_0, y_0) = z$$

Therefore $\mathbf{L}$ lies in the tangent plane. An upward unit normal to the tangent plane is $\mathbf{p} = (1 + \|\nabla f\|^2)^{-1/2}(-(\partial f/\partial x)(x_0, y_0), -(\partial f/\partial y)(x_0, y_0), 1)$. So $\cos\theta = \mathbf{p}\cdot\mathbf{k} = (1 + \|\nabla f\|^2)^{-1/2}$, and $\tan\theta = \sin\theta/\cos\theta = (\|\nabla f\|^2/(1 + \|\nabla f\|^2))^{1/2}/(1 + \|\nabla f\|^2)^{-1/2} = \|\nabla f\|$ as claimed.

(b) The tangent plane contains the horizontal line through (1, 0, 2) perpendicular to $\nabla f(1, 0) = (5, 0)$, that is, parallel to the y-axis. It makes an angle of arctan($\|\nabla f(1, 0)\|$) = arctan 5 ≈ 78.7° with respect to the xy-plane.

17. $(1/\sqrt{2}, 1/\sqrt{2})$ or $(-1/\sqrt{2}, -1/\sqrt{2})$

19. A unit normal is $(\sqrt{2}/10)(3, 5, 4)$. The tangent plane is $3x + 5y + 4z = 18$.

21. Since g is the composition $\lambda \mapsto \lambda x \mapsto f(\lambda x)$, the Chain Rule gives

$$g'(\lambda) = Df(\lambda \mathbf{x}) \begin{bmatrix} x_1 \\ \vdots \\ x_n \end{bmatrix}.$$

Thus

$$g'(1) = Df(\mathbf{x}) \begin{bmatrix} x_1 \\ \vdots \\ x_n \end{bmatrix} = \nabla f(\mathbf{x}) \cdot \mathbf{x}.$$

But also $g(\lambda) = \lambda^p f(\mathbf{x})$, so $g'(\lambda) = p\lambda^{p-1} f(\mathbf{x})$ and $g'(1) = pf(\mathbf{x})$.

23. (a) $\dfrac{\partial^2 f}{\partial x^2} + \dfrac{\partial^2 f}{\partial y^2} = 2\dfrac{y^2 - x^2}{(x^2 + y^2)^2} + 2\dfrac{x^2 - y^2}{(x^2 + y^2)^2} = 0$

(b) $$g_{xx} + g_{yy} + g_{zz} = \frac{y^2 + z^2 - 2x^2}{(x^2 + y^2 + z^2)^{5/2}} + \frac{x^2 - z^2 - 2y^2}{(x^2 + y^2 + z^2)^{5/2}} + \frac{x^2 + y^2 - 2z^2}{(x^2 + y^2 + z^2)^{5/2}} = 0$$

(c) $$h_{xx} + h_{yy} + h_{zz} + h_{ww} = \frac{6x^2 - 2y^2 - 2z^2 - 2w^2}{(x^2 + y^2 + z^2 + w^2)^3} + \frac{6y^2 - 2x^2 - 2z^2 - 2w^2}{(x^2 + y^2 + z^2 + w^2)^3} + \frac{6z^2 - 2x^2 - 2y^2 - 2w^2}{(x^2 + y^2 + z^2 + w^2)^3} + \frac{6w^2 - 2x^2 - 2y^2 - 2z^2}{(x^3 + y^2 + z^2 + w^2)^3} = 0$$

25. $$\frac{\partial^2 w}{\partial u\,\partial v} = \frac{\partial}{\partial u}\left(\frac{\partial f}{\partial x} - \frac{\partial f}{\partial y}\right) = \frac{\partial^2 f}{\partial x^2} + \frac{\partial^2 f}{\partial y\,\partial x} - \frac{\partial^2 f}{\partial x\,\partial y} - \frac{\partial^2 f}{\partial y^2} = \frac{\partial^2 f}{\partial x^2} - \frac{\partial^2 f}{\partial y^2}$$

using the Chain Rule and Theorem 15.

27. (a) If $(x, y) \neq (0, 0)$, then $\partial f/\partial x = (y^3 - yx^2)/(x^2 + y^2)^2$ and $\partial f/\partial y = (x^3 - xy^2)/(x^2 + y^2)^2$. If $x = y = 0$, use the definition directly to find that both partial derivatives are 0.

(b) (*i*) is not continuous at (0, 0); (*ii*) is differentiable but the derivative is not continuous.

29. (a) Use the Chain Rule and assume that f is C^2 so that $\partial^2 f/\partial x\,\partial y = \partial^2 f/\partial y\,\partial x$.

(b) $$\left(\frac{\partial F}{\partial r}\right)^2 + \frac{1}{r^2}\left(\frac{\partial F}{\partial \theta}\right)^2 = \left(\cos\theta\frac{\partial f}{\partial x} + \sin\theta\frac{\partial f}{\partial y}\right)^2 + \frac{1}{r^2}\left(-r\sin\theta\frac{\partial f}{\partial x} + r\cos\theta\frac{\partial f}{\partial y}\right)^2$$
$$= (\cos^2\theta + \sin^2\theta)\left(\frac{\partial f}{\partial x}\right)^2 + (\sin^2\theta + \cos^2\theta)\left(\frac{\partial f}{\partial y}\right)^2$$
$$= \left(\frac{\partial f}{\partial x}\right)^2 + \left(\frac{\partial f}{\partial y}\right)^2 = \|\nabla f\|^2$$

31. $(-4e^{-1}, 0)$

33. (a) See Theorem 11.
(b) $g(u) = (\sin 3u)^2 + \cos 8u$
$g'(u) = 6 \sin 3u \cos 3u - 8 \sin 8u$
$g'(0) = 0$

$\nabla f = (2x, 1)$
$\nabla f(h(0)) = \nabla f(0, 1) = (0, 1)$
$\mathbf{h}'(u) = (3 \cos 3u, -8 \sin 8u)$
$g'(0) = \nabla f(\mathbf{h}(0)) \cdot \mathbf{h}'(0) = (0, 1) \cdot (3, 0) = 0$

SECTION 3.1

1. (a) $\boldsymbol{\sigma}'(t) = (2\pi \cos 2\pi t, -2\pi \sin 2\pi t, 2 - 2t)$, $\boldsymbol{\sigma}'(0) = (2\pi, 0, 2)$
(b) $\boldsymbol{\sigma}'(t) = (e^t, -\sin t, \cos t)$, $\boldsymbol{\sigma}'(0) = (1, 0, 1)$
(c) $\boldsymbol{\sigma}'(t) = (2t, 3t^2 - 4, 0)$, $\boldsymbol{\sigma}'(0) = (0, -4, 0)$
(d) $\boldsymbol{\sigma}'(t) = (2 \cos 2t, 1/(1 + t), 1)$, $\boldsymbol{\sigma}'(0) = (2, 1, 1)$

3. $6m\mathbf{j}$, where m is the particle's mass

5. $\boldsymbol{\sigma}(t) = \left(\dfrac{t^2}{2}, e^t - 6, \dfrac{t^3}{3} + 1\right)$

7. $\mathbf{T} = 5662$ seconds $= 1.57$ hours

9. (a) If $\boldsymbol{\sigma}(t) = (\sigma_1(t), \ldots, \sigma_n(t))$ and $\boldsymbol{\rho}(t) = (\rho_1(t), \ldots, \rho_n(t))$, then

$$\frac{d}{dt}(\boldsymbol{\sigma}(t) \cdot \boldsymbol{\rho}(t)) = \frac{d}{dt}\sum_{i=1}^{n} \sigma_i(t)\rho_i(t) = \sum_{i=1}^{n} \sigma_i'(t)\rho_i(t) + \sum_{i=1}^{n} \sigma_i(t)\rho_i'(t)$$
$$= \boldsymbol{\sigma}'(t) \cdot \boldsymbol{\rho}(t) + \boldsymbol{\sigma}(t) \cdot \boldsymbol{\rho}'(t)$$

(b) $$\frac{d}{dt}(\boldsymbol{\sigma}(t) \times \boldsymbol{\rho}(t)) = \frac{d}{dt}[(\sigma_2\rho_3 - \sigma_3\rho_2)\mathbf{i} - (\sigma_1\rho_3 - \sigma_3\rho_1)\mathbf{j} + (\sigma_1\rho_2 - \sigma_2\rho_1)\mathbf{k}]$$
$$= [\sigma_2'\rho_3 + \sigma_2\rho_3' - \sigma_3'\rho_2 - \sigma_3\rho_2']\mathbf{i} - [\sigma_1'\rho_3 + \sigma_1\rho_3' - \sigma_3'\rho_1 - \sigma_3\rho_1']\mathbf{j}$$
$$+ [\sigma_1'\rho_2 + \sigma_1\rho_2' - \sigma_2'\rho_1 - \sigma_2\rho_1']\mathbf{k}$$
$$= [(\sigma_2'\rho_3 - \sigma_3'\rho_2)\mathbf{i} - (\sigma_1'\rho_3 - \sigma_3'\rho_1)\mathbf{j} + (\sigma_1'\rho_2 - \sigma_2'\rho_1)\mathbf{k}$$
$$+ [(\sigma_2\rho_3' - \sigma_3\rho_2')\mathbf{i} - (\sigma_1\rho_3' - \sigma_3\rho_1')\mathbf{j} + (\sigma_1\rho_2' - \sigma_2\rho_1')\mathbf{k}]$$
$$= (\boldsymbol{\sigma}'(t) \times \boldsymbol{\rho}(t)) + (\boldsymbol{\sigma}(t) \times \boldsymbol{\rho}'(t))$$

(c) Use parts (a) and (b).

11. $\boldsymbol{\sigma}(t) \times \boldsymbol{\sigma}'(t)$ is normal to the plane of the orbit at time t. As in Exercise 10, its derivative is 0, so the orbital plane is constant.

SECTION 3.2

1. (a) 7; (b) $4\sqrt{2} - 2$;
(c) $\frac{1}{2}\pi\sqrt{2 + \pi^2} + \log(\pi + \sqrt{2 + \pi^2}) - \frac{1}{2}\log 2$
(d) $\sqrt{21} + \frac{5}{4}[\log(4 + \sqrt{21}) - \log\sqrt{5}]$
(e) $\frac{1}{2}\sqrt{5} + 2\log(1 + \sqrt{5}) - 2\log 2$
(f) Use the substitution $u = e^t$ to show that the integral is $2(e - e^{-1})$
(g) $\frac{2}{3}[(t_1 + 2)^{3/2} - (t_0 + 2)^{3/2}]$

3. $3 + \log 2$

5. (a) Since α is strictly increasing, it maps $[a, b]$ one-to-one onto $[\alpha(a), \alpha(b)]$. By definition, $\mathbf{v}$ is in the image of $\mathbf{c}$ if and only if there is a t in $[a, b]$ with $\mathbf{c}(t) = \mathbf{v}$. There is one point s in $[\alpha(a), \alpha(b)]$ with $s = \alpha(t)$, so $\mathbf{d}(s) = \mathbf{c}(t) = \mathbf{v}$. Therefore the image of $\mathbf{c}$ is contained in that of $\mathbf{d}$. Use α^{-1} similarly for the opposite inclusion.

(b) $l_{\mathbf{d}} = \int_{\alpha(a)}^{\alpha(b)} \|d'(s)\|\, ds = \int_{s=\alpha(a)}^{s=\alpha(b)} \|\mathbf{d}'(\alpha(t))\|\alpha'(t)\, dt$

$= \int_{t=a}^{t=b} \|\mathbf{d}'(\alpha(t))\alpha'(t)\|\, dt = \int_a^b \|\mathbf{c}'(t)\|\, dt = l_{\mathbf{c}}$

7. (a) $l_{\boldsymbol{\sigma}} = \int_a^b \|\boldsymbol{\sigma}'(s)\|\, ds = \int_a^b ds = b - a$

(b) $\mathbf{T}(s) = \boldsymbol{\sigma}'(s)/\|\boldsymbol{\sigma}'(s)\| = \boldsymbol{\sigma}'(s)$, so $\mathbf{T}'(s) = \boldsymbol{\sigma}''(s)$. Then $k = \|\mathbf{T}'\| = \|\boldsymbol{\sigma}''(s)\|$

(c) Show that if (*i*) $\mathbf{v}$ and $\mathbf{w}$ are in $\mathbb{R}^3$, $\|\mathbf{v} \times \mathbf{w}\| = \|\mathbf{w} - (\mathbf{v} \cdot \mathbf{w}/\|\mathbf{v}\|^2)\mathbf{v}\| \cdot \|\mathbf{v}\|$. Use this to show that if (*ii*) $\boldsymbol{\rho}(t) = (x(t), y(t), z(t))$ is never $(0, 0, 0)$, and $\mathbf{f}(t) = \boldsymbol{\rho}(t)/\|\boldsymbol{\rho}(t)\|$, then

$$\frac{d\mathbf{f}}{dt} = \frac{1}{\|\boldsymbol{\rho}(t)\|}\left(\boldsymbol{\rho}'(t) - \frac{\boldsymbol{\rho}(t) \cdot \boldsymbol{\rho}'(t)}{\|\boldsymbol{\rho}(t)\|^2}\boldsymbol{\rho}(t)\right) \quad \text{and} \quad \left\|\frac{d\mathbf{f}}{dt}\right\| = \frac{\|\boldsymbol{\rho}(t) \times \boldsymbol{\rho}'(t)\|}{\|\boldsymbol{\rho}(t)\|^2}$$

With $\boldsymbol{\rho}(t) = \boldsymbol{\sigma}'(t)$, (*ii*) gives

$$\mathbf{T}'(t) = \frac{\boldsymbol{\sigma}''(t)}{\|\boldsymbol{\sigma}'(t)\|} - \frac{\boldsymbol{\sigma}'(t) \cdot \boldsymbol{\sigma}''(t)}{\|\boldsymbol{\sigma}'(t)\|^3}\boldsymbol{\sigma}'(t) \quad \text{and} \quad \|\mathbf{T}'(t)\| = \frac{\|\boldsymbol{\sigma}'(t) \times \boldsymbol{\sigma}''(t)\|}{\|\boldsymbol{\sigma}'(t)\|^2}$$

If s is the arc length of $\boldsymbol{\sigma}$, $ds/dt = \|\boldsymbol{\sigma}'(t)\|$, and therefore

$$\left\|\frac{d\mathbf{T}}{dt}\right\| = \left\|\frac{d\mathbf{T}}{ds}\frac{ds}{dt}\right\| = k\|\boldsymbol{\sigma}'(t)\|$$

Thus

$$k = \frac{1}{\|\boldsymbol{\sigma}'(t)\|}\left\|\frac{d\mathbf{T}}{dt}\right\| = \frac{\|\boldsymbol{\sigma}'(t) \times \boldsymbol{\sigma}''(t)\|}{\|\boldsymbol{\sigma}'(t)\|^3}$$

(The result is useful in Exercise 9.)

(d) $1/\sqrt{2}$

9. (a) Since $\boldsymbol{\sigma}$ is parametrized by arc length, $\mathbf{T}(s) = \boldsymbol{\sigma}'(s)$, and $\mathbf{N}(s) = \boldsymbol{\sigma}''(s)/\|\boldsymbol{\sigma}''(s)\|$. Use Exercise 9 of Section 3.1 and Exercise 7 to show that

$$\frac{d\mathbf{B}}{ds} = \left(\boldsymbol{\sigma}'' \times \frac{\boldsymbol{\sigma}''}{\|\boldsymbol{\sigma}''\|}\right) + \boldsymbol{\sigma}' \times \left(\frac{\boldsymbol{\sigma}'''}{\|\boldsymbol{\sigma}''\|} - \frac{\boldsymbol{\sigma}'' \cdot \boldsymbol{\sigma}'''}{\|\boldsymbol{\sigma}''\|^3}\boldsymbol{\sigma}''\right)$$

and

$$\tau = -\frac{d\mathbf{B}}{ds} \cdot \mathbf{N} = -\frac{(\boldsymbol{\sigma}' \times \boldsymbol{\sigma}''') \cdot \boldsymbol{\sigma}''}{\|\boldsymbol{\sigma}''\|^2} = \frac{(\boldsymbol{\sigma}' \times \boldsymbol{\sigma}'') \cdot \boldsymbol{\sigma}'''}{\|\boldsymbol{\sigma}''\|^2}$$

(b) Obtain $\mathbf{T}'(t)$ and $\|\mathbf{T}'(t)\|$ as in Exercise 7. $\mathbf{B}$ is a unit vector in the direction of $\boldsymbol{\sigma}' \times \mathbf{T}' = (\boldsymbol{\sigma}' \times \boldsymbol{\sigma}'')/\|\boldsymbol{\sigma}'\|$, so $\mathbf{B} = (\boldsymbol{\sigma}' \times \boldsymbol{\sigma}'')/\|\boldsymbol{\sigma}' \times \boldsymbol{\sigma}''\|$. Use result (*ii*) in the the solution of Exercise 7 with $\boldsymbol{\rho} = \boldsymbol{\sigma}' \times \boldsymbol{\sigma}''$ and Exercise 9 of Section 3.1 to obtain $d\mathbf{B}/dt = (\boldsymbol{\sigma}' \times \boldsymbol{\sigma}''')/\|\boldsymbol{\sigma}' \times \boldsymbol{\sigma}''\| - \{[(\boldsymbol{\sigma}' \times \boldsymbol{\sigma}'') \cdot (\boldsymbol{\sigma}' \times \boldsymbol{\sigma}''')]/\|\boldsymbol{\sigma}' \times \boldsymbol{\sigma}''\|^3\}(\boldsymbol{\sigma}' \times \boldsymbol{\sigma}'')$, and the values of $\mathbf{T}'$ and $\|\mathbf{T}'\|$ to get $\mathbf{N} = (\|\boldsymbol{\sigma}'\|/\|\boldsymbol{\sigma}' \times \boldsymbol{\sigma}''\|)(\boldsymbol{\sigma}'' - (\boldsymbol{\sigma}' \;\; \boldsymbol{\sigma}'')/\|\boldsymbol{\sigma}'\|^2)$. Finally use the Chain Rule and the inner product of these to obtain

$$\tau = -\left[\frac{d\mathbf{B}}{ds}(s(t))\right] \cdot \mathbf{N}(s(t)) = -\frac{1}{|ds/dt|}\frac{d\mathbf{B}}{dt} \cdot \mathbf{N} = \frac{(\boldsymbol{\sigma}' \times \boldsymbol{\sigma}'') \cdot \boldsymbol{\sigma}'''}{\|\boldsymbol{\sigma}' \times \boldsymbol{\sigma}''\|^2}$$

(c) $-\sqrt{2}/2$

11. $\mathbf{N}$ is defined as $\mathbf{T}'/\|\mathbf{T}'\|$, so $\mathbf{T}' = \|\mathbf{T}'\|\mathbf{N} = k\mathbf{N}$. Since $\mathbf{T} \cdot \mathbf{T}' = 0$, $\mathbf{T}$, $\mathbf{N}$ and $\mathbf{B}$ are an orthonormal basis for $\mathbb{R}^3$. Differentiating $\mathbf{B}(s) \cdot \mathbf{B}(s) = 1$ and $\mathbf{B}(s) \cdot \mathbf{T}(s) = 0$ shows that $\mathbf{B}' \cdot \mathbf{B} = 0$ and $\mathbf{B}' \cdot \mathbf{T} + \mathbf{B} \cdot \mathbf{T}' = 0$. But $\mathbf{T}' \cdot \mathbf{B} = \|\mathbf{T}'\|\mathbf{N} \cdot \mathbf{B} = 0$, so $\mathbf{B}' \cdot \mathbf{T} = 0$ also. Thus $\mathbf{B}' = (\mathbf{B}' \cdot \mathbf{T}) + (\mathbf{B}' \cdot \mathbf{N})\mathbf{N} + (\mathbf{B}' \cdot \mathbf{B})\mathbf{B} = (\mathbf{B}' \cdot \mathbf{N})\mathbf{N} = -\tau\mathbf{N}$. $\mathbf{N}' \cdot \mathbf{N} = 0$ since $\mathbf{N} \cdot \mathbf{N} = 1$. Thus $\mathbf{N}' = (\mathbf{N}' \cdot \mathbf{T})\mathbf{T} + (\mathbf{N}' \cdot \mathbf{B})\mathbf{B}$. But differentiating $\mathbf{N} \cdot \mathbf{T} = 0$ and $\mathbf{N} \cdot \mathbf{B} = 0$ gives $\mathbf{N}' \cdot \mathbf{T} = -\mathbf{N} \cdot \mathbf{T}' = -k$ and $\mathbf{N}' \cdot \mathbf{B} = -\mathbf{N} \cdot \mathbf{B}' = \tau$, so the middle equation follows.

SECTION 3.3

1. (a) $\displaystyle \frac{dE}{dt} = \frac{1}{2} m \cdot 2\langle \mathbf{r}'(t), \mathbf{r}''(t)\rangle + \langle \text{grad } V(\mathbf{r}(t)), \mathbf{r}'(t)\rangle$

$$= \langle \mathbf{r}'(t), -\text{grad } V(\mathbf{r}(t))\rangle + \langle \text{grad } V(\mathbf{r}(t)), \mathbf{r}'(t)\rangle = 0$$

(b) Use (a)

3. $\displaystyle \frac{d}{dt} V(\mathbf{c}(t)) = \langle \nabla V(\mathbf{c}(t)), \mathbf{c}'(t)\rangle = -\langle \nabla V(\mathbf{c}(t)), \nabla V(\mathbf{c}(t))\rangle \le 0$

A particle tends to move to a region of lower potential energy. (Water flows downhill.)

5. Use the fact that $-\nabla T$ is perpendicular to the surface $T =$ constant.

7. If $\mathbf{x} = (x_1, x_2, x_3)$, $\phi(\mathbf{x}, t) = (\phi_1, \phi_2, \phi_3)$, and $f = f(x_1, x_2, x_3, t)$, then

$$\frac{d}{dt}(f(\phi(\mathbf{x}, t), t)) = \frac{\partial f}{dt}(\mathbf{x}, t) + \sum_{i=1}^{3} \frac{\partial f}{\partial x_i}(\phi(\mathbf{x}, t), t)\frac{\partial \phi_i}{\partial t}(\mathbf{x}, t)$$

$$= \frac{\partial f}{\partial t}(\mathbf{x}, t) + [\nabla f(\phi(\mathbf{x}, t), t)] \cdot [\mathbf{F}(\phi(\mathbf{x}, t))]$$

SECTION 3.4

1. (a) $\nabla f = \mathbf{r}/\|\mathbf{r}\|$, $\mathbf{r} = (x, y, z)$
   (b) $\nabla f = (y + z, x + z, y + x)$
   (c) $\nabla f = -2\mathbf{r}/\|\mathbf{r}\|^4$, $\mathbf{r} = (x, y, z)$

3. (a) 0; (b) 0; (c) $(10y - 8z, 6z - 10x, 8x - 6y)$

5. $\nabla \cdot \mathbf{F} = (\partial/\partial x)y + (\partial/\partial y)x = 0 + 0 = 0$

7. $\nabla f = (2xy^2, 2x^2y + 2yz^2, 2y^2z)$; $\nabla \times \nabla f = (4yz - 4yz, 0 - 0, 4xy - 4xy)$
   $= (0, 0, 0)$

9. $\nabla \times \mathbf{F} = (0, 0, \sin y - \cos x)$. If $\mathbf{F} = \nabla f$, then, since $\mathbf{F}$ is C^1, f would be C^2, and $\nabla \times \mathbf{F} = \nabla \times \nabla f$ would be 0, but it is not.

SECTION 3.5

1. Only (a)
3. Write out each expression in terms of coordinates.
5. (a) $2xy\mathbf{i} + x^2\mathbf{j}$
   (b) $3y^2zx\mathbf{i} + (4xz - y^3z)\mathbf{j}$
   (c) $4x^2z^2\mathbf{i} + 2y\mathbf{j} + 2y^3z^2x\mathbf{k}$
   (d) $4x^2z^2y + x^2$
   (e) $-y^3zx^3\mathbf{i} + 2x^2y^4z\mathbf{j} + (2x^3z^2 - 2xy)\mathbf{k}$
7. No, consider $\mathbf{F} = x\mathbf{i} + xy\mathbf{j} + \mathbf{k}$; $\nabla \times \mathbf{F} = y\mathbf{k}$, which is not perpendicular to $\mathbf{F}$.
9. (a) $\nabla \cdot \mathbf{F} = (\partial/\partial x, \partial/\partial y, \partial/\partial z) \cdot (F_r\mathbf{e}_r + F_\theta\mathbf{e}_\theta + F_z\mathbf{e}_z)$

$$= \left(\cos\theta\frac{\partial}{\partial r} - \frac{\sin\theta}{r}\frac{\partial}{\partial\theta}, \sin\theta\frac{\partial}{\partial r} + \frac{\cos\theta}{r}\frac{\partial}{\partial\theta}, \frac{\partial}{\partial z}\right)$$

$$\cdot\,(F_r(\cos\theta\mathbf{i} + \sin\theta\mathbf{j}) + F_\theta(-\sin\theta\mathbf{i} + \cos\theta\mathbf{j}) + F_z\mathbf{k})$$

$$= \left(\cos\theta\frac{\partial}{\partial r} - \frac{\sin\theta}{r}\frac{\partial}{\partial\theta}\right)(F_r\cos\theta - F_\theta\sin\theta)$$

$$+ \left(\sin\theta\frac{\partial}{\partial r} + \frac{\cos\theta}{r}\frac{\partial}{\partial\theta}\right)(F_r\sin\theta + F_\theta\cos\theta) + \frac{\partial F_z}{\partial z}$$

$$= \frac{\partial F_r}{\partial r} + \frac{1}{r}F_r + \frac{1}{r}\frac{\partial F_\theta}{\partial\theta} + \frac{\partial F_z}{\partial z} = \frac{1}{r}\left[\frac{\partial}{\partial r}(rF_r) + \frac{\partial F_\theta}{\partial\theta} + \frac{\partial}{\partial z}(rF_z)\right]$$

(You should check the details of this. See Example 4.)

   (b) Make the substitution for $\mathbf{i}, \mathbf{j}, \mathbf{k}, \partial/\partial x, \partial/\partial y, \partial/\partial z$, and F_x, F_y, and F_z found in Example 4 and part (a) and perform elementary column operations to convert the determinant

$$\begin{vmatrix} \mathbf{i} & \mathbf{j} & \mathbf{k} \\ \partial/\partial x & \partial/\partial y & \partial/\partial z \\ F_x & F_y & F_z \end{vmatrix} \quad\text{to}\quad \frac{1}{r}\begin{vmatrix} \mathbf{e}_r & r\mathbf{e}_\theta & \mathbf{e}_z \\ \partial/\partial r & \partial/\partial\theta & \partial/\partial z \\ F_r & F_\theta & F_z \end{vmatrix}$$

(The column operations work since all the operations are linear and we are always adding vectors to vectors, operators to operators, and functions to functions.)

11. Think of polar coordinates as cylindrical coordinates with 0, or missing, z-coordinate to see that the required transformations are those given by Theorem 5. First use part (*i*) to get ∇u, then part (*ii*) to get $\nabla \cdot \nabla u$.

REVIEW EXERCISES FOR CHAPTER 3

1. (a) 2; (b) 0; (c) 14
3. (a) $\nabla f = yz^2\mathbf{i} + xz^2\mathbf{j} + 2xyz\mathbf{k}$
   (b) $\nabla \times F = (x - y)\mathbf{i} - x\mathbf{k}$
   (c) $(2xyz^3 - 3z^2xy^2)\mathbf{i} - (y^2z^3 - 2x^2y^2z)\mathbf{j} + (y^2z^3 - 2x^2yz^2)\mathbf{k}$

5. (a) $z = 0$; (b) $z = 1 + x$; (c) $z = 10x + 10y - 25$

7. 35,880 kilometers

9. $\mathbf{F} = (2m, 0, -m)$

11. $(2\pi, 3\pi^2, -2\pi)$

13. (a) $6/\sqrt{14}$; (b) $\frac{12}{13}$

15. $x(t) = 1/(1-t)$; $y(t) = 0$; $z(t) = e^t/(1-t)$ and $\boldsymbol{\sigma}'(t) = ((1-t)^{-2}, 0, (e^t/(1-t))(1 + 1/(1-t))) = (x(t)^2, 0, z(t)(1 + x(t)) = \mathbf{F}(\boldsymbol{\sigma}(t))$

17. (a) $(3, 0, 0)$;
    (b) The vector $\nabla\sigma$ points in the direction of most rapidly increasing concentration, so $-\nabla\sigma$ is the direction of most rapid decrease.

19. (a) $2x - z = 1$; (b) $8z + 6\sqrt{2}y = 20$

21. (a) $\nabla \cdot \mathbf{F} = 2ye^z + x^2ye^z + 2z$; $\nabla \times \mathbf{F} = \mathbf{0}$
    (b) $f(x, y, z) = x^2ye^z + z^3/3 + C$. Since $\mathbf{F}$ is C^1, an f which works will be C^2, so $\nabla \times \mathbf{F} = \nabla \times \nabla f = 0$. Thus it is necessary that $\nabla \times \mathbf{F} = 0$ for a solution to (b) to exist.

SECTION 4.1

1. $f(h_1, h_2) = h_1^2 + 2h_1h_2 + h_2^2\ (R_2(\mathbf{h}, \mathbf{0}) = 0$ in this case)

3. $f(h_1, h_2) = 1 + h_1 + h_2 + \dfrac{h_1^2}{2} + h_1h_2 + \dfrac{h_2^2}{2} + R_2(\mathbf{h}, \mathbf{0})$

5. $f(h_1, h_2) = 1 + h_1h_2 + R_2(\mathbf{h}, \mathbf{0})$

7. (a) Show that $|R_k(x, a)| \le AB^{k+1}/(k+1)!$ for constants A, B and x in a fixed interval $[a, b]$. Prove $R_k \to 0$ as $k \to \infty$. (Use convergence of the series $\sum c^k/k! = e^c$ and use Taylor's Theorem.)
    (b) The only possible trouble is at $x = 0$. Use L'hopital's rule to show that

$$\operatorname*{limit}_{t\to\infty} p(t)e^t = \infty$$

for every polynomial $p(t)$. Using this, establish that $\operatorname*{limit}_{x\to 0^+} p(x)e^{-1/x} = 0$ for every rational function $p(x)$, and use it to show that $f^{(k)}(0) = 0$ for every k.
    (c) $f\colon \mathbb{R}^n \to \mathbb{R}$ is analytic at $\mathbf{x}_0$ if the series

$$f(\mathbf{x}_0) + \sum_{i=1}^{n} h_i \frac{\partial f}{\partial x_i}(\mathbf{x}_0) + \frac{1}{2}\sum_{i,j=1}^{n} h_ih_j \frac{\partial^2 f}{\partial x_i \partial x_j}(\mathbf{x}_0) + \cdots$$

$$+ \frac{1}{k!}\sum_{i_1,\ldots,i_k=1}^{n} h_{i_1}h_{i_2}\cdots h_{i_k}\frac{\partial^k f}{\partial x_{i_1}\cdots\partial x_{i_k}}(\mathbf{x}_0) + \cdots$$

converges to $f(\mathbf{x}_0 + \mathbf{h})$ for all $\mathbf{h} = (h_1, \ldots, h_n)$ in some sufficiently small disc $\|\mathbf{h}\| < \varepsilon$. The function f is analytic if for every $R > 0$ there is a constant M such that $|\partial^k f/\partial x_{i_1} \cdots \partial x_{i_k}(\mathbf{x})| < M^k$ for each kth order derivative at every $\mathbf{x}$ satisfying $\|\mathbf{x}\| \le R$.

(d) $f(x, y) = 1 + x + y + \frac{1}{2}(x^2 + 2xy + y^2) + \cdots + \frac{1}{k!}\sum_{j=0}^{k}\binom{k}{j}x^j y^{k-j} + \cdots$

$$= \sum_{k=0}^{\infty} \frac{1}{k!}(x + y)^k.$$

SECTION 4.2

1. (0, 0), saddle point
3. The critical points are on the line $y = -x$; they are local minima because $f(x, y) = (x + y)^2 \geq 0$, equaling zero only when $x = -y$.
5. (0, 0), saddle point
7. $(-1/4, -1/4)$, local minimum
9. (0, 0), local maximum (The tests fail, but use the fact that $\cos(z) \leq 1$.)
   $(\sqrt{\pi/2}, \sqrt{\pi/2})$, local minimum
   $(0, \sqrt{\pi})$, local minimum
11. (b) Show that $f(g(t)) = 0$ at $t = 0$, and $f(g(t)) \geq 0$ if $|t| < |b|/3a^2$.
    (c) f is negative on the parabola $y = 2x^2$.
13. The critical points are on the line $y = x$ and they are local minima (see Exercise 3).
15. The only critical point is (0, 0, 0). It is a minimum since

$$f(x, y, z) \geq \frac{x^2 + y^2}{2} + z^2 + xy = \frac{1}{2}(x + y)^2 + z^2 \geq 0$$

17. If $u_n(x, y) = u(x, y) + (1/n)e^x$, then $\nabla^2 u_n = (1/n)e^x > 0$. Thus, u_n is strictly subharmonic and can have its maximum only on ∂D, say at $\mathbf{p}_n = (x_n, y_n)$. If $(x_0, y_0) \in D$, check that this implies $u(x_n, y_n) > u(x_0, y_0) - e/n$. Thus there must be a point $\mathbf{q} = (x_\infty, y_\infty)$ on ∂D such that arbitrarily close to $\mathbf{q}$ we can find an (x_n, y_n) for n as large as we like. Conclude from the continuity of u that $u(x_\infty, y_\infty) \geq u(x_0, y_0)$.

SECTION 4.3

1. maximum at $\sqrt{\frac{2}{3}}(1, -1, 1)$, minimum at $\sqrt{\frac{2}{3}}(-1, 1, -1)$
3. maximum at $(\sqrt{3}, 0)$ minimum at $(-\sqrt{3}, 0)$
5. maximum at $\left(\frac{9}{\sqrt{70}}, \frac{4}{\sqrt{70}}\right)$, minimum at $\left(-\frac{9}{\sqrt{70}}, -\frac{4}{\sqrt{70}}\right)$
7. The minimum value 4 attained at (0, 2). Use a geometric picture rather than Lagrange multipliers.
9. (0, 0, 2) is a minimum of f.
11. The diameter should equal the height, $20/\sqrt[3]{2\pi}$ *cm*.
13. (a) $\nabla f(\mathbf{x}) = A\mathbf{x}$
    (b) S is defined by the constraint function $g(\mathbf{x}) = x_1^2 + x_2^2 + x_3^2 - 1$. Since $\nabla g(\mathbf{x}) = 2\mathbf{x}$ is not $\mathbf{0}$, Theorem 6 applies. At an $\mathbf{x}$ where f is extreme, there is a $\lambda/2$ such that $\nabla f(\mathbf{x}) = (\lambda/2)\nabla g(\mathbf{x})$. That is, $A\mathbf{x} = \lambda\mathbf{x}$.

15. For Exercise 1, the bordered Hessians required are

$$|\bar{H}_2| = \begin{vmatrix} 0 & 2x & 2y \\ 2x & -2\lambda & 0 \\ 2y & 0 & -2\lambda \end{vmatrix} = 8\lambda(x^2 + y^2)$$

$$|\bar{H}_3| = \begin{vmatrix} 0 & 2x & 2y & 2z \\ 2x & -2\lambda & 0 & 0 \\ 2y & 0 & -2\lambda & 0 \\ 2z & 0 & 0 & -2\lambda \end{vmatrix} = -16\lambda(x^2 + y^2 + z^2)$$

At $\sqrt{\frac{2}{3}}(1, -1, 1)$ the Lagrange multiplier is $\lambda = \sqrt{6}/4 > 0$, indicating a maximum at $\sqrt{\frac{2}{3}}(1, -1, 1)$ and $\lambda = -\sqrt{6}/4 < 0$ indicates a minimum at $\sqrt{\frac{2}{3}}(-1, 1, -1)$. In Exercise 5, $|\bar{H}| = 24\lambda(4x^2 + 6y^2)$, so $\lambda = \sqrt{70}/12 > 0$ indicates a maximum at $(9/\sqrt{70}, 4/\sqrt{70})$ and $\lambda = -\sqrt{70}/12 < 0$ indicates a minimum at $(-9/\sqrt{70}, -4/\sqrt{70})$.

17. 11,664 in^3

SECTION 4.4

1. Use Theorem 8 with $n = 1$. (See Example 1). Line (*i*) is given by $0 = (x - x_0, y - y_0) \cdot \nabla F(x_0, y_0) = (x - x_0)(\partial F/\partial x)(x_0, y_0) + (y - y_0)(\partial F/\partial y)(x_0, y_0)$. For line (*ii*), Theorem 8 gives $dy/dx = -(\partial F/\partial x)/(\partial F/\partial y)$, so the lines agree and are given by $y = y_0 - [(\partial F/\partial x)(x_0, y_0)]/[(\partial F/\partial y)(x_0, y_0)](x - x_0)$.

3. (a) If $x < -\frac{1}{4}$ we can solve for y in terms of x using the quadratic formula.
   (b) $\partial F/\partial y = 2y + 1$ is nonzero for $\{y \mid y < -\frac{1}{2}\}$ and $\{y \mid y > -\frac{1}{2}\}$. These regions correspond to the upper and lower halves of a horizontal parabola with vertex at $(-\frac{1}{4}, -\frac{1}{2})$ and to the choice of sign in the quadratic formula. The derivative $dy/dx = -3/(2y + 1)$ is negative on the top half of the parabola, positive on the bottom.

5. With $F_1 = y + x + uv$ and $F_2 = uxy + v$, the determinant in the General Implicit Function Theorem is

$$\begin{vmatrix} \partial F_1/\partial u & \partial F_1/\partial v \\ \partial F_2/\partial u & \partial F_2/\partial v \end{vmatrix} = v - uxy,$$

which is 0 at (0, 0, 0, 0). Thus the Implicit Function Theorem does not apply. If we try directly, we find that $v = -uxy$, so $x + y = u^2xy$. For a particular choice of (x, y) near (0, 0), there are either no solutions for (u, v) or else there are two.

7. No. $f(x, y) = (-1, 0)$ has infinitely many solutions, namely, $(x, y) = (0, y)$ for any y.

9. (a) $x_0^2 + y_0^2 \neq 0$
   (b) $f'(z) = -(x + 2y)/(x^2 + y^2)z$; $\quad g'(z) = (y - 2x)/(x^2 + y^2)z$.

11. Multiply and equate coefficients to get a_0, a_1, and a_2 as functions of r_1, r_2, and r_3. Then compute the Jacobian determinant $\partial(a_0, a_1, a_2)/\partial(r_1, r_2, r_3) = (r_3 - r_2)(r_1 - r_2)(r_1 - r_3)$. This is not zero if the roots are distinct. Thus, the

inverse function theorem shows that the roots may be found as functions of the coefficients in some neighborhood of any point at which the roots are distinct. That is, if the roots r_1, r_2, r_3 of $x^3 + a_2x^2 + a_1x + a_0$ are all different, then there are neighborhoods V of (r_1, r_2, r_3) and W of (a_0, a_1, a_2) such that the roots in V are smooth functions of the coefficients in W.

SECTION 4.5

1. $(-\frac{1}{4}, -\frac{1}{4})$
3. stable equilibrium point $(2 + m^2g^2)^{-1/2}(-1, -1, -mg)$
5. no critical points; no maximum or minimum
7. At the optimum, $qk/\alpha = pL/(1 - \alpha)$.

REVIEW EXERCISES FOR CHAPTER 4

1. (a) saddle point
   (b) if $|C| < 2$, strict minimum; if $|C| > 2$, saddle point; if $C = \pm 2$, minimum
3. (a) $z = 3x - 1$
   (b) $z\sqrt{3} = 2x + 1$
5. (a) 1
   (b) $\sqrt{83}/6$
7. (a) $(2xy^2z^2, 2x^2yz^2, 2x^2y^2z)$
   (b) $(z^2 - y^2)\mathbf{i} - x^2\mathbf{k}$
   (c) $\nabla \cdot \mathbf{F} = 9;\quad \nabla \times \mathbf{F} = -3x^2\mathbf{j}$
   (d) $\nabla \cdot \mathbf{F} = 2x + 2y + 2z;\quad \nabla \times \mathbf{F} = \mathbf{0}$
   (e) $\nabla \cdot \mathbf{F} = 3;\quad \nabla \times \mathbf{F} = e^z\mathbf{i} + \cos x\,\mathbf{j} - 3\mathbf{k}$
9. $z = \frac{1}{4}$
11. $(0, 0, \pm 1)$
13. If $b \geq 2$, the minimum distance is $2\sqrt{b - 1}$; if $b \leq 2$, the minimum distance is $|b|$.
15. not stable
17. $f(-\frac{3}{2}, -\sqrt{3}/2) = 3\sqrt{3}/4$
19. A new orthonormal basis may be found with respect to which the quadratic form given by the matrix

$$A = \begin{bmatrix} a & b \\ b & c \end{bmatrix}$$

takes diagonal form. This change of basis defines new variables ξ and η, which are linear functions of x and y. Manipulations of linear algebra and the Chain Rule show that $Lv = \lambda(\partial^2 v/\partial \xi^2) + \mu(\partial^2 v/\partial \eta^2)$. The numbers λ and μ are the eigenvalues of A and are positive since the quadratic form is positive definite. At a maximum, $\partial v/\partial \xi = \partial v/\partial \eta = 0$. Moreover, $\partial^2 v/\partial \xi^2 \leq 0$ and $\partial^2 v/\partial \eta^2 \leq 0$ since if either were greater than 0 the cross section of the graph in that direction would have a minimum. Then $Lv \leq 0$, thus contradicting strict subharmonicity.

21. Reverse the inequalities in Exercises 19 and 20.

23. The determinant required in the General Implicit Function Theorem is not zero, so we can solve for u and v; $(\partial u/\partial x)(2, -1) = \frac{13}{32}$.

25. $x = (20/3)\sqrt[3]{3};\quad y = 10\sqrt[3]{3};\quad z = 5\sqrt[3]{3}$

27. The equations for a critical point, $\partial s/\partial m = \partial s/\partial b = 0$ when solved for m and b give $m = (y_1 - y_2)/(x_1 - x_2)$ and $b = y_2x_1 - y_1x_2$. The line $y = mx + b$ then goes through (x_1, y_1) and (x_2, y_2).

29. At a minimum of s, we have $0 = \partial s/\partial b = -2\sum_{i=1}^{n}(y_i - mx_i - b)$.

30. $y = \frac{9}{10}x + \frac{6}{5}$

SECTION 5.1

1. (a) $\frac{13}{15}$: (b) $\pi + \frac{1}{2}$; (c) 1; (d) $\log 2 - 1/2$

3. In order to show that the volume of the two cylinders are equal, show that their area functions are equal.

5. (a) $\frac{26}{9}$; (b) $2/\pi$; (c) $(2/\pi)(e^2 + 1)$

7. 196/15

SECTION 5.2

1. (a) $\frac{7}{12}$; (b) $e - 2$; (c) $(1/9)\sin 1$

3. If $f(x_0, y_0) > 0$, use continuity to show there is a small rectangle R_1 containing (x_0, y_0) with $f(x, y) > \frac{1}{2}f(x_0, y_0)$ on R_1. Let $g(x, y)$ be $\frac{1}{2}f(x_0, y_0)$ on R_1 and 0 elsewhere. By Theorem 2, g is integrable. Use properties (*iii*) and (*iv*) of the integral to show this implies that $\int_R f\,dx\,dy > \frac{1}{2}f(x_0, y_0)\text{area}(R_1)$.

5. Use Fubini's Theorem to write

$$\int_R [f(x)g(y)]\,dx\,dy = \int_c^d g(y)\left[\int_a^b f(x)\,dx\right]dy$$

and notice that $\int_a^b f(x)\,dx$ is a constant, and so may be pulled out.

7. $\frac{11}{6}$

9. Since $\int_0^1 dy = \int_0^1 2y\,dy = 1$, we have $\int_0^1 [\int_0^1 f(x, y)\,dy]\,dx = 1$. In any partition of $R = [0, 1] \times [0, 1]$, each rectangle R_{jk} contains points $\mathbf{c}_{jk}^{(1)}$ with x rational and $\mathbf{c}_{jk}^{(2)}$ with x irrational. If in the regular partition of order n we choose $\mathbf{c}_{jk} = \mathbf{c}_{jk}^{(1)}$ in those rectangles with $0 \le y \le \frac{1}{2}$ and $\mathbf{c}_{jk} = \mathbf{c}_{jk}^{(2)}$ when $y > \frac{1}{2}$, the approximating sums are the same as those for

$$g(x, y) = \begin{cases} 1 & 0 \le y \le \frac{1}{2} \\ 2y & \frac{1}{2} < y \le 1 \end{cases}$$

Since g is integrable, the approximating sums must converge to $\int_R g\,dA = \frac{7}{8}$. However, if we had picked all $\mathbf{c}_{ij} = \mathbf{c}_{jk}^{(1)}$, all approximating sums would have the value 1.

11. The function f is not bounded since there must be a volume of -1 over each of the diagonal squares of area $1/[n(n + 1)]^2$.

SECTION 5.3

1. (a) $\frac{1}{3}$, both; (b) $\frac{5}{2}$, both; (c) $(e^2 - 1)/4$, both; (d) $\frac{1}{35}$, both
3. $A = \int_{-r}^{r} \int_{-\sqrt{r^2-x^2}}^{\sqrt{r^2-x^2}} dy\, dz = 2\int_{-r}^{r} \sqrt{r^2 - x^2}\, dx = r^2(\arcsin(1) - \arcsin(-1)) = \pi r^2.$
5. 28,000 ft^3
7. 0
9. type 1; $\pi/2$
11. 50π
13. $\pi/24$
15. Compute the integral with respect to y first. Split that into integrals over $[-\phi(x), 0]$ and $[0, \phi(x)]$ and change variables in the first integral.
17. Let $\{R_{ij}\}$ be a partition of a rectangle R containing D and let f be 1 on D. Thus, f^* is 1 on D and 0 on $R\backslash D$. Let $\mathbf{c}_{jk} \in R\backslash D$ if R_{ij} is not wholly contained in D. The approximating Riemann sum is the sum of the areas of those rectangles of the partition that are contained in D.

SECTION 5.4

1. (a) $\frac{1}{8}$; (b) $\pi/4$; (c) $-\frac{17}{12}$
   (d) $G(b) - G(a)$, where $dG/dy = F(y, y) - F(a, y)$ and $\partial F/\partial x = f(x, y)$
3. Note that the maximum value of f on D is e and the minimum value of f on D is $1/e$. Use the ideas in the proof of Theorem 4 to show that

$$\frac{1}{e} \le \frac{1}{4\pi^2} \int f(x, y)\, dA \le e$$

5. $\frac{4}{3}\pi abc$
7. $\pi(20\sqrt{10} - 52)/3$
9. $\sqrt{3}/4$
11. D looks like a slice of pie.

$$\int_0^1 \left[\int_0^x f(x, y)\, dy\right] dx + \int_1^{\sqrt{2}} \left[\int_0^{\sqrt{2-x^2}} f(x, y)\, dy\right] dx$$

SECTION 5.5

1. (a) 4; (b) $\frac{8}{3}$; (c) $\frac{3}{16}$; (d) $2 - e$
3. Integration of $\iint e^{-xy}\, dx\, dy$ with respect to x first and then y gives log 2. Reversing the order gives the integral on the left side of the equality stated in the exercise.
5. Integrate over $[\varepsilon, 1] \times [\varepsilon, 1]$ and let $\varepsilon \to 0$ to show that the improper integral exists and equals 2 log 2.

7. Use the fact that

$$\frac{\sin^2(x - y)}{\sqrt{1 - x^2 - y^2}} \leq \frac{1}{\sqrt{1 - x^2 - y^2}}$$

9. Use the fact that $e^{x^2+y^2}/(x - y) \geq 1/(x - y)$ on the given region.

SECTION 5.6

1. $\frac{1}{3}$
3. 0
5. $-\frac{1}{6}$
7. $(4\pi/3)(1 - \sqrt{2}/2)$
9. Given $\varepsilon_1 > 0$, continuity of f shows that there is an $\varepsilon_2 > 0$ such that $\varepsilon < \varepsilon_2$ implies $f(x_0, y_0, z_0) - \varepsilon_1 < f(x, y, z) < f(x_0, y_0, z_0) + \varepsilon_1$ whenever $(x, y, z) \in B_\varepsilon$. Integrating gives

$$|B_\varepsilon|(f(x_0, y_0, z_0) - \varepsilon_1) < \int_{B_\varepsilon} f(x, y, z)\, dV < |B_\varepsilon|(f(x_0, y_0, z_0) + \varepsilon_1).$$

Now divide by $|B_\varepsilon|$ and let $\varepsilon \to 0$.

11. The conditions on f and W show that the integral exists. To find its value one can use any approximating Riemann sums. Explain how points may be chosen from boxes of a partition in such a way that contributions from boxes with positive z are cancelled by those from boxes with negative z.
13. $\frac{1}{48}$
15. (a) $a = -1;\ \ b = 1;\ \ \phi_2(x) = -\sqrt{1 - x^2};\ \ \phi_1(x) = \sqrt{1 - x^2};$
$\gamma_2(x.\ y) = \sqrt{x^2 + y^2};\ \ \gamma_1(x, y) = 1$
(b) $a = -\sqrt{3}/2;\ \ b = \sqrt{3}/2;\ \ \phi_2(x) = -\sqrt{\frac{3}{4} - x^2};\ \ \phi_1(x) = \sqrt{\frac{3}{4} - x^2};$
$\gamma_2(x, y) = \frac{1}{2};\ \gamma_1(x, y) = \sqrt{1 - x^2 - y^2}$

SECTION 5.7

1. $S =$ the unit disc minus its center
3. $D = [0, 3] \times [0, 1];$ Yes
5. D is the set of (x, y, z) with $x^2 + y^2 + z^2 \leq 1$ (the unit ball). T is not one-to-one, but is one-to-one on $]0, 1] \times]0, \pi[\times]0, 2\pi]$.
7. Showing that T is onto amounts to showing that the system $ax + by = e$, $cx + dy = f$ can always be solved for x and y, where

$$A = \begin{bmatrix} a & b \\ c & d \end{bmatrix}.$$

When you do this by elimination or by Cramer's rule, the quantity by which you must divide is $\det A$. Thus, if $\det A \neq 0$, the equations can always be solved.

9. Since $\det A \neq 0$, T maps $\mathbb{R}^2$ one-to-one onto $\mathbb{R}^2$. Let T^{-1} be the inverse transformation. Show that T^{-1} has matrix A^{-1}, and $\det(A^{-1}) = 1/\det A \neq 0$. By Exercise 8, $P^* = T^{-1}(P)$ is a parallelogram.

SECTION 5.8

1. $\pi(e - 1)$
3. D is the region $0 \leq x \leq 4, \frac{1}{2}x + 3 \leq y \leq \frac{1}{2}x + 6$; (a) 140; (b) -42
5. D^* is the region $0 \leq u \leq 1, 0 \leq v \leq 2$; $\frac{2}{3}(9 - 2\sqrt{2} - 3\sqrt{3})$
7. π
9. $3\pi/2$
11. $\frac{4}{9}a^{2/3} \int_{D^*} f((au^2)^{1/3}, \ (av^2)^{1/3})u^{-1/3}v^{-1/3}\, du\, dv$
13. $2a^2$
15. π
17. (a) spherical: $b^2c^2 \cos^2\theta \sin^2\phi + a^2c^2 \sin^2\theta \sin^2\phi + a^2b^2 \cos^2\phi = a^2b^2c^2/\rho^2$
    cylindrical: $b^2c^2r^2 \cos^2\theta + a^2c^2r^2 \sin^2\theta + a^2b^2z^2 = a^2b^2c^2$
    (b) spherical: $\cos^2\phi = \sin^2\phi$ or $\phi = \pi/4$ or $3\pi/4$
    cylindrical: $z^2 = r^2$
    (c) spherical: $\theta = \pi/4$, $\cot\phi = \sqrt{2}/2$
    cylindrical: $\theta = \pi/4$, $z = \sqrt{2}r/2$
    (d) spherical: $\rho\cos\phi = \theta$ if $0 \leq \theta < \pi/2$, $= \theta - \pi$ if $\pi/2 < \theta < 3\pi/2$.
    $= \theta - 2\pi$ if $3\pi/2 < \theta < 2\pi$, $\rho^2\sin^2\phi = 1$
    cylindrical: $r = 1$, $z = \theta$
19. $y/x = \tan(\sqrt{x^2 + y^2 + z^2})$ (something like a spiral seashell).
21. $4\pi(\sqrt{3}/2 - \log(1 + \sqrt{3}) + \log\sqrt{2})$
23. $\frac{2}{9}\pi[(1 + a^3)^{3/2} - a^{9/2} - 1]$
25. 4π
27. 24 (try $x = u - v$; $y = u + v$)
29. (a) $\frac{4}{3}\pi abc$; (b) $\frac{4}{5}\pi abc$

REVIEW EXERCISES FOR CHAPTER 5

1. (a) $\frac{81}{2}$; (b) $\frac{29}{70}$; (c) $\frac{1}{4}e^2 - e + \frac{9}{4}$
3. $\frac{1}{3}\pi(4\sqrt{2} - \frac{7}{2})$
5. 2π
7. $4\pi/3$
9. $(5\pi/16)\sqrt{15}$
11. Work the integral with respect to y first on the region $D_{\varepsilon, L} = \{(x, y) | \varepsilon \leq x \leq L, 0 \leq y \leq x\}$ to obtain $I_{\varepsilon, L} = \iint_{D_{\varepsilon,L}} f\, dx\, dy = \int_\varepsilon^L x^{-3/2}(1 - e^{-x})\, dx$. The integrand is positive, so $I_{\varepsilon, L}$ increases as $\varepsilon \to 0$ and $L \to \infty$. Bound $1 - e^{-x}$ above by x for

$0 < x < 1$ and by 1 for $1 < x < \infty$ to see that $I_{\varepsilon, L}$ remains bounded and so must converge. The improper integral does exist.

13. (a) $\frac{1}{6}$; (b) $-\frac{1}{2}$; (c) $\pi^2/8$
15. $abc/6$
17. (a) $6xy$; $(xe^{xy} - z)\mathbf{i} - (ye^{xy})\mathbf{j} + (z - 3x^2)\mathbf{k}$; (b) No. Try $\mathbf{F} = y^2\mathbf{i}$.
19. $16\pi/3$
21. $\frac{1}{2}\pi(1 - 1/e)$; volume under graph of $z = e^{-x^2-y^2}$ over that part of the unit disc with $y < 0$
23. (a) 0; (b) $\pi/24$; (c) 0
25. $\pi/2$
27. $(e - e^{-1})/4$ (Use the change of variables: $u = y - x$, $v = y + x$.)
29. 3 (Use the change of variables: $u = x^2 - y^2$, $v = xy$.)
31. $(9.92 \times 10^6)\pi$ grams
33. (a) 32;
    (b) This occurs at the points of the unit sphere $x^2 + y^2 + z^2 = 1$ inscribed in the cube.
35. (0, 0, 3/8)

SECTION 6.1

1. $\int_{\boldsymbol{\sigma}} f(x, y, z)\, ds = \int_I f(x(t), y(t), z(t)) \cdot |\sigma'(t)|\, dt = \int_0^1 y(t) \cdot 1\, dt = \int_0^1 0 \cdot 1\, dt = 0.$
3. (a) 2
   (b) $52\sqrt{14}$
   (c) $\frac{16}{3} - 2\sqrt{3}$
5. $-\frac{1}{3}(1 + 1/e^2)^{3/2} + \frac{1}{3}(2^{3/2})$
7. (a) The path follows the straight line from (0, 0) to (1, 1) and back to (0, 0) in the xy-plane. Over the path, the graph of f is a straight line from (0, 0, 0) to (1, 1, 1). The integral is the area of the resulting triangle covered twice and equals $\sqrt{2}$.
   (b) $$s(t) = \begin{cases} \sqrt{2}(1 - t^4) & \text{when} \quad -1 \le t \le 0 \\ \sqrt{2}(1 + t^4) & \text{when} \quad 0 < t \le 1 \end{cases}$$
   The path is
   $$\boldsymbol{\sigma}(s) = \begin{cases} (1 - s/\sqrt{2})(1, 1) & \text{when} \quad 0 \le s \le \sqrt{2} \\ (s/(\sqrt{2} - 1))(1, 1) & \text{when} \quad \sqrt{2} \le s \le 2\sqrt{2} \end{cases}$$
   and $\int_{\boldsymbol{\sigma}} f ds = \sqrt{2}$.
9. $2a/\pi$
11. (a) $(2\sqrt{5} + \log(2 + \sqrt{5}))/4$
    (b) $(5\sqrt{5} - 1)/(6\sqrt{5} + 3\log(2 + \sqrt{5}))$
13. The path is a unit circle centered at (0, 0, 0) in the plane $x + y + z = 0$ and so may be parameterized by $\boldsymbol{\sigma}(\theta) = (\cos\theta)\mathbf{v} + (\sin\theta)\mathbf{w}$ where $\mathbf{v}$ and $\mathbf{w}$ are orthogonal

unit vectors lying in that plane. For example, $\mathbf{v} = (1/\sqrt{2})(-1, 0, 1)$ and $\mathbf{w} = (1/\sqrt{6})(1, -2, 1)$ will do. The total mass is $2\pi/3$ grams.

SECTION 6.2

1. (a) $\frac{3}{2}$; (b) 0
3. 9
5. By the Cauchy-Schwarz inequality, $|\mathbf{F}(\boldsymbol{\sigma}(t)) \cdot \boldsymbol{\sigma}'(t)| \le \|\mathbf{F}(\boldsymbol{\sigma}(t))\| \, \|\boldsymbol{\sigma}'(t)\|$ for every t. Thus

$$\left|\int_{\boldsymbol{\sigma}} \mathbf{F} \cdot d\mathbf{s}\right| = \left|\int_a^b \mathbf{F}(\boldsymbol{\sigma}(t)) \cdot \boldsymbol{\sigma}'(t)\, dt\right| \le \int_a^b |\mathbf{F}(\boldsymbol{\sigma}(t)) \cdot \boldsymbol{\sigma}'(t)|\, dt$$

$$\le \int_a^b \|\mathbf{F}(\boldsymbol{\sigma}(t))\| \, \|\boldsymbol{\sigma}'(t)\|\, dt \le M \int_a^b \|\boldsymbol{\sigma}'(t)\|\, dt = Ml$$

7. $\frac{3}{4} - (n-1)/(n+1)$
9. 0
11. the length of $\boldsymbol{\sigma}$
13. If $\boldsymbol{\sigma}'(t)$ is never 0, then the unit vector $\mathbf{T}(t) = \boldsymbol{\sigma}'(t)/\|\boldsymbol{\sigma}'(t)\|$ is a continuous function of t and so is a smoothly-turning tangent to the curve. The answer is no.
15. 7
17. (a) $\|\boldsymbol{\sigma}'(x)\|$
    (b) The inverse exists locally and is differentiable by the Inverse Function Theorem. There is actually an inverse defined for the whole interval at once since $df/dx > 0$, $f(a) = 0$, and $f(b) = L$. This means f is strictly increasing and so maps $[a, b]$ one-to-one onto $[0, L]$.
    (c) $g'(s) = \|\boldsymbol{\sigma}'(g(s))\|^{-1}$
    (d) $\|d\mathbf{p}/ds\| = \|\boldsymbol{\sigma}'(g(s))g'(s)\| = \|\boldsymbol{\sigma}'(g'(s))\| \, |g'(s)| = 1$ using part (c).

SECTION 6.3

1. $z = 2(y-1) + 1$
3. $18(z-1) - 4(y+2) - (x-13) = 0$ or $18z - 4y - x - 13 = 0$.
5. The vector $\mathbf{n} = (\cos v \sin u, \sin v \sin u, \cos u) = (x, y, z)$. The surface is the unit sphere centered at the origin.
7. (a) $x = x_0 + (y - y_0)(\partial h/\partial y)(y_0, z_0) + (z - z_0)(\partial h/\partial z)(y_0, z_0)$ describes the plane tangent to $x = h(y, z)$ at (x_0, y_0, z_0), $x_0 = h(y_0, z_0)$.
    (b) $y = y_0 + (x - x_0)(\partial k/\partial x)(x_0, z_0) + (z - z_0)(\partial k/\partial z)(x_0, z_0)$
9. (a) $u \mapsto u$, $v \mapsto v$, $u \mapsto u^3$, and $v \mapsto v^3$ all map $\mathbb{R}$ onto $\mathbb{R}$.
    (b) $\mathbf{T}_u \times \mathbf{T}_v = (0, 0, 1)$ for $\boldsymbol{\Phi}_1$, and this is never $\mathbf{0}$. $\mathbf{T}_u \times \mathbf{T}_v = 9u^2v^2(0, 0, 1)$ for $\boldsymbol{\Phi}_2$, and this is $\mathbf{0}$ along the u and v axes.
    (c) We want to show that any two parametrizations of a surface that are smooth near a point will give the same tangent plane there. So, suppose $\boldsymbol{\Phi}: D \subset$

$\mathbb{R}^2 \to \mathbb{R}^3$ and $\boldsymbol{\psi}\colon B \subset \mathbb{R}^2 \to \mathbb{R}^3$ are parametrized surfaces such that

(*i*) $$\boldsymbol{\Phi}(u_0, v_0) = (x_0, y_0, z_0) = \boldsymbol{\psi}(x_0, t_0)$$
and
(*ii*) $$(\mathbf{T}_u^{\boldsymbol{\Phi}} \times \mathbf{T}_v^{\boldsymbol{\Phi}})|_{(u_0, v_0)} \neq 0 \text{ and } (\mathbf{T}_s^{\boldsymbol{\psi}} \times \mathbf{T}_t^{\boldsymbol{\psi}})|_{(s_0, t_0)} \neq 0$$

so that $\boldsymbol{\Phi}$ and $\boldsymbol{\psi}$ are smooth and one-to-one in neighborhoods of (u_0, v_0) and (s_0, t_0), which we may as well assume are D and B. Suppose further that they "describe the same surface," that is, $\boldsymbol{\Phi}(D) = \boldsymbol{\Phi}(B)$. To see that they give the same tangent plane at (x_0, y_0, z_0), show that they have parallel normal vectors. To do this, show that there is an open set C with $(u_0, v_0) \in C \subset D$ and a differentiable map $f\colon C \to B$ such that $\boldsymbol{\Phi}(u, v) = \boldsymbol{\psi}(f(u, v))$ for $(u, v) \in C$. Once you have done this, a routine computation shows that the normal vectors are related by $\mathbf{T}_u^{\boldsymbol{\Phi}} \times \mathbf{T}_v^{\boldsymbol{\Phi}} = [\partial(s, t)/\partial(u, v)]\, \mathbf{T}_s^{\boldsymbol{\psi}} \times \mathbf{T}_t^{\boldsymbol{\psi}}$.

To see that there is such an f, notice that since $\mathbf{T}_s^{\boldsymbol{\psi}} \times \mathbf{T}_t^{\boldsymbol{\psi}} \neq 0$, at least one of the 2×2 determinants in the cross product is not zero. Assume, for example, that

$$\begin{vmatrix} \dfrac{\partial x}{\partial s} & \dfrac{\partial y}{\partial s} \\[2ex] \dfrac{\partial x}{\partial t} & \dfrac{\partial y}{\partial t} \end{vmatrix} \neq 0$$

Now use the Inverse Function Theorem to write (s, t) as a differentiable function of (x, y) in some neighborhood of (x_0, y_0).

(d) No.

11. (a) The surface is a helicoid. It looks like a spiral ramp winding around the z-axis. (See Example 2 of Section 6.4.) It winds twice around since θ goes up to 4π.
(b) $\mathbf{n} = \pm(1/\sqrt{1 + r^2})(\sin\theta, -\cos\theta, r)$
(c) $y_0x - x_0y + (x_0^2 + y_0^2)z = (x_0^2 + y_0^2)z_0$
(d) If $(x_0, y_0, z_0) = (r_0 \cos\theta_0, r_0 \sin\theta_0, \theta_0)$, then representing the line segment as $\{(r\cos\theta_0, r\sin\theta_0, \theta_0) | 0 \le r \le 1\}$ shows that the line lies in the surface. Representing the line as $\{(tx_0, ty_0, z_0) | 0 \le t \le 1/(x_0^2 + y_0^2)\}$ and substituting into the result of (c) shows that it lies in the tangent plane at (x_0, y_0, z_0).

13. (a) Using cylindrical coordinates leads to the parametrization $\boldsymbol{\Phi}(z, \theta) = ((1 + z^2)\cos\theta, (1 + z^2)\sin\theta, z)$, $-\infty < z < \infty$, $0 \le \theta \le 2\pi$ as one possible solution.
(b) $\mathbf{n} = (1/\sqrt{1 + 4z^2})(\cos\theta, \sin\theta, -2z)$;
(c) $x_0x + y_0y = 1$;
(d) Substitute the coordinates along these lines into the defining equation of the surface and the result of part (c).

SECTION 6.4

1. 4π
3. $\frac{3}{2}\pi(\sqrt{2} + \log(1 + \sqrt{2}))$

5. $\frac{1}{3}\pi(6\sqrt{6} - 8)$

7. $(2\pi + 4)/(2\pi - 4)$

9. $A(E) = \int_0^{2\pi} \int_0^{\pi} \sqrt{a^2b^2 \sin^2 \phi \cos^2 \phi + b^2c^2 \sin^4 \phi \cos^2 \theta + a^2c^2 \sin^4 \phi \sin^2 \theta}\, d\phi\, d\theta$

11. $(\pi/6)(5\sqrt{5} - 1)$

13. $(\pi/2)\sqrt{6}$

15. $4\sqrt{5}$; For fixed θ, (x, y, z) moves along the horizontal line segment $y = 2x$, $z = \theta$ from the z-axis out to a radius of $\sqrt{5}|\cos \theta|$ into quadrant 1 if $\cos \theta > 0$ and into quadrant 3 if $\cos \theta < 0$.

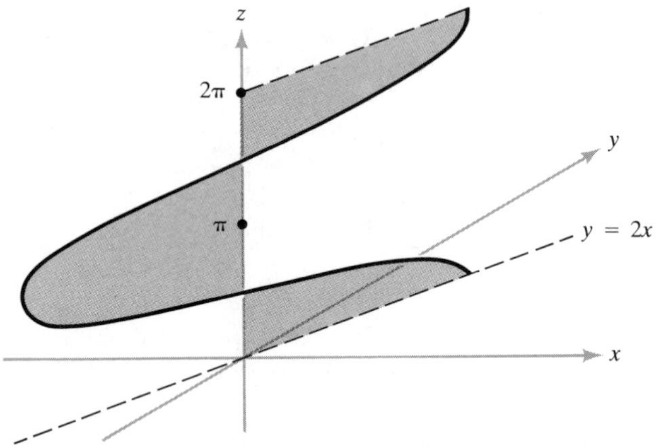

SECTION 6.5

1. $\dfrac{5\sqrt{2} + 3}{24}$

3. πa^3

5. $\sqrt{6}/30$

7. $\dfrac{\pi}{4}\left(\dfrac{5\sqrt{5}}{3} + \dfrac{1}{15}\right)$

9. $16\pi R^3/3$

11. (a) The sphere looks the same from all three axes, so these three integrals should be the same quantity with different labels on the axes.
    (b) $4\pi R^4/3$
    (c) $4\pi R^4/3$

13. $(R/2, R/2, R/2)$

15. Since $\|\partial\mathbf{\Phi}/\partial u\|^2 - 2\|\partial\mathbf{\Phi}/\partial u\|\,\|\partial\mathbf{\Phi}/\partial v\| + \|\partial\mathbf{\Phi}/\partial v\|^2 = (\|\partial\mathbf{\Phi}/\partial u\| - \|\partial\mathbf{\Phi}/\partial v\|)^2 \geq 0$, we have $\|\partial\mathbf{\Phi}/\partial u \times \partial\mathbf{\Phi}/\partial v\| = \|\partial\mathbf{\Phi}/\partial u\|\,\|\partial\mathbf{\Phi}/\partial v\|\,|\sin\theta(u, v)| \leq \frac{1}{2}(\|\partial\mathbf{\Phi}/\partial u\|^2 + \|\partial\mathbf{\Phi}/\partial v\|^2)$ where θ is the angle between $\partial\mathbf{\Phi}/\partial u$ and $\partial\mathbf{\Phi}/\partial v$. This leads to $A(\mathbf{\Phi}) \leq$

$J(\mathbf{\Phi})$. Assuming continuity, equality can hold only if $\|\partial\mathbf{\Phi}/\partial u\|\,\|\partial\mathbf{\Phi}/\partial v\|\,|\sin\theta| = \frac{1}{2}(\|\partial\mathbf{\Phi}/\partial u\|^2 + \|\partial\mathbf{\Phi}/\partial v\|^2) = \frac{1}{2}(\|\partial\mathbf{\Phi}/\partial u\| - \|\partial\mathbf{\Phi}/\partial v\|)^2 + \|\partial\mathbf{\Phi}/\partial u\|\,\|\partial\mathbf{\Phi}/\partial v\|$ at each point. This can occur if $|\sin\theta| = 1$ and if $\|\partial\mathbf{\Phi}/\partial u\| = \|\partial\mathbf{\Phi}/\partial v\|$. Thus $\cos\theta = 0$; that is, $\partial\mathbf{\Phi}/\partial u \cdot \partial\mathbf{\Phi}/\partial v = 0$.

17. (a) $\sqrt{2}\pi r^2/4$; (b) $\pi - 2$
19. $2a^2$

SECTION 6.6

1. $\pm 48\pi$ (The sign depends on orientation.)
3. 4π
5. 2π (or -2π, if you choose a different orientation)
7. With the usual spherical coordinate parametrization, $\mathbf{T}_\theta \times \mathbf{T}_\phi = -\sin\phi\,\mathbf{r}$ (see Example 1). Thus

$$\int_S \mathbf{F}\cdot d\mathbf{S} = \iint \mathbf{F}\cdot(\mathbf{T}_\phi \times \mathbf{T}_\theta)\,d\phi\,d\theta = \iint (\mathbf{F}\cdot\mathbf{r})\sin\phi\,d\phi\,d\theta$$
$$= \int_0^{2\pi}\int_0^{\pi} F_r \sin\phi\,d\phi\,d\theta$$

and

$$\int_S f\,dS = \int_0^{2\pi}\int_0^{\pi} f\sin\phi\,d\phi\,d\theta.$$

9. For a cylinder of radius $R = 1$ and normal component F_r

$$\int_S \mathbf{F}\cdot d\mathbf{S} = \int_a^b\int_0^{2\pi} F_r\,d\theta\,dz$$

11. $2\pi/3$
13. $\frac{2}{3}a^3bc\pi$

REVIEW EXERCISES FOR CHAPTER 6

1. (a) $3\sqrt{2}(1 - e^{6\pi})/13$
   (b) $-\pi\sqrt{2}/2$
   (c) $(236158\sqrt{26} - 8)/35\cdot(25)^3$
   (d) $8\sqrt{2}/189$
3. (a) $2/\pi + 1$
   (b) $-\frac{1}{2}$
5. $2a^3$
7. (a) A sphere of radius 5 centered at (2, 3, 0);
   $\mathbf{\Phi}(\theta,\phi) = (2 + 5\cos\theta\sin\phi, 3 + 5\sin\theta\sin\phi, 5\cos\phi)$;
   $0 \le \theta \le 2\pi; 0 \le \phi \le \pi$
   (b) An ellipsoid with center at (2, 0, 0);
   $\mathbf{\Phi}(\theta,\phi) = (2 + (1/\sqrt{2})3\cos\theta\sin\phi, 3\sin\theta\sin\phi, 3\cos\phi)$;
   $0 \le \theta < 2\pi, 0 \le \phi < \pi$

(c) An elliptic hyperboloid of one sheet;
$\mathbf{\Phi}(\theta, z) = (\frac{1}{2}\sqrt{8 + 2z^2} \cos\theta, \frac{1}{3}\sqrt{8 + 2z^2} \sin\theta, z)$;
$0 \leq \theta < 2\pi, -\infty < z < \infty$

9. $A(\mathbf{\Phi}) = \frac{1}{2}\int_0^{2\pi} \sqrt{3\cos^2\theta + 5}\, d\theta$; $\mathbf{\Phi}$ describes the upper nappe of a cone with elliptical horizontal cross sections.

11. $11\sqrt{3}/6$

13. $\sqrt{2}/3$

15. $5\sqrt{5}/6$

17. (a) $(e^y \cos\pi z, xe^y \cos\pi z, -\pi xe^y \sin\pi z)$
(b) 0

19. $\frac{1}{2}(e^2 + 1)$

21. $\mathbf{n} = (1/\sqrt{5})(-1, 0, 2)$, $2z - x = 1$

23. 0

25. If $\mathbf{F} = \nabla\phi$, then $\nabla \times \mathbf{F} = \mathbf{0}$ (at least if ϕ is C^2: See Theorem 1, Section 3.4). Theorem 3 of Section 6.2 shows that $\int_{\mathbf{c}} \nabla\phi \cdot d\mathbf{s} =$ since $\mathbf{c}$ is a closed curve.

27. (a) 24π (b) 24π (c) 60π

29. (a) $\sqrt{R^2 + p^2}\,(z_0 - z_1)$ (b) $(R^2 + p^2)\sqrt{z_0/2g}$

SECTION 7.1

1. -8

3. (a) 0 (b) $-\pi R^2$ (c) 0

5. $3\pi a^2$

7. $3\pi/2$

9. $3\pi(b^2 - a^2)/2$

11. (a) 2π (b) 0

13. 0

15. πab

17. $-\frac{5}{6}$

19. $9\pi/8$

21. If $\varepsilon > 0$, there is a $\delta > 0$ such that $|u(\mathbf{q}) - u(\mathbf{p})| < \varepsilon$ whenever $\|\mathbf{p} - \mathbf{q}\| = \rho < \delta$. Parametrize $\partial B_\rho(\mathbf{p})$ by $\mathbf{q}(\theta) = \mathbf{p} + \rho(\cos\theta, \sin\theta)$. Then

$$|I(\rho) - 2\pi u(\mathbf{p})| \leq \int_0^{2\pi} |u(\mathbf{q}(\theta)) - u(\mathbf{p})|\, d\theta \leq 2\pi\varepsilon.$$

23. Parametrize $\partial B_\rho(\mathbf{p})$ as in Exercise 21. If $\mathbf{p} = (p_1, p_2)$, then $I(\rho) = \int_0^{2\pi} u(p_1 + \rho\cos\theta, p_2 + \rho\sin\theta)\, d\theta$. Differentiation under the integral sign gives

$$\frac{dI}{d\rho} = \int_0^{2\pi} \nabla u \cdot (\cos\theta, \sin\theta)\, d\theta = \int_0^{2\pi} \nabla u \cdot \mathbf{n}\, d\theta = \frac{1}{\rho}\int_{\partial B_\rho} \frac{\partial u}{\partial \mathbf{n}}\, ds = \frac{1}{\rho}\int_{B_\rho} \nabla^2 u\, dA$$

(the last equality uses Exercise 22).

25. Using Exercise 24,

$$\iint_{B_R} u\,dA = \int_0^R \int_0^{2\pi} u(\mathbf{p} + \rho(\cos\theta, \sin\theta))\rho\, d\theta\, d\rho$$

$$= \int_0^R \left(\int_{\partial B_\rho} u ds\right) d\rho = \int_0^R 2\pi\rho u(\mathbf{p})\, d\rho = \pi R^2 u(\mathbf{p})$$

27. Suppose u is subharmonic. We establish the assertions corresponding to Exercise 26(a) and (b). The argument for superharmonic functions is quite similar with inequalities reversed.

Suppose $\nabla^2 u \geq 0$ and $u(\mathbf{p}) \geq u(\mathbf{q})$ for all $\mathbf{q}$ in $B_R(\mathbf{p})$. By Exercise 23, $I'(\rho) \geq 0$ for $0 < \rho \leq R$, so Exercise 24 shows that $2\pi u(\mathbf{p}) \leq I(\rho) \leq I(R)$ for $0 < \rho \leq R$. If $u(\mathbf{q}) > u(\mathbf{p})$ for some $\mathbf{q} = \mathbf{p} + \rho(\cos\theta_0, \sin\theta_0) \in B_R(\mathbf{p})$, then, by continuity, there is an arc $[\theta_0 - \delta, \theta_0 + \delta]$ on $\partial B_\rho(\mathbf{p})$ where $u < u(\mathbf{p}) - d$ for some $d > 0$. This would mean that

$$2\pi u(\mathbf{p}) \leq I(\rho) = \frac{1}{\rho}\int_0^{2\pi} u(\mathbf{p} + \rho(\cos\theta, \sin\theta))\rho\, d\theta$$

$$\leq (2\pi - 2\delta)u(\mathbf{p}) + 2\delta(u(\mathbf{p}) - d) \leq 2\pi u(\mathbf{p}) - 2\delta d$$

This contradiction shows that we must have $u(\mathbf{q}) = u(\mathbf{p})$ for every $\mathbf{q}$ in $B_R(\mathbf{p})$.

If the maximum at $\mathbf{p}$ is absolute for D, the last paragraph shows that $u(\mathbf{x}) = u(\mathbf{p})$ for all $\mathbf{x}$ in some disc around $\mathbf{p}$. If $\boldsymbol{\sigma}$: $[0, 1[\rightarrow D$ is a path from $\mathbf{p}$ to $\mathbf{q}$, then $u(\boldsymbol{\sigma}(t)) = u(\mathbf{p})$ for all t in some interval $[0, b[$. Let b_0 be the largest $b \in [0, 1]$ such that $u(\boldsymbol{\sigma}(t)) = u(\mathbf{p})$ for all $t \in [0, b[$. (Strictly speaking, this requires the notion of least upper bound from a good Calculus text.) Since u is continuous, $u(\boldsymbol{\sigma}(b_0)) = u(\mathbf{p})$. If $b_0 \neq 1$, then the last paragraph would apply at $\boldsymbol{\sigma}(b_0)$ and u is constantly equal to $u(\mathbf{p})$ on a disc around $\boldsymbol{\sigma}(b_0)$. In particular, there is a $\delta > 0$ such that $u(\boldsymbol{\sigma}(t)) = u(\boldsymbol{\sigma}(b_0)) = u(\mathbf{p})$ on $[0, b_0 + \delta[$. This contradicts the maximality of b_0, so we must have $b_0 = 1$. That is, $\boldsymbol{\sigma}(\mathbf{q}) = \boldsymbol{\sigma}(\mathbf{p})$. Since $\mathbf{q}$ was an arbitrary point in D, u is constant on D.

SECTION 7.2

1. -2π
3. 0
5. 0
7. Using Faraday's Law, $\int_S [\boldsymbol{\nabla} \times \mathbf{E} - \partial\mathbf{H}/\partial t] \cdot d\mathbf{S} = 0$ for any surface S. If the integrand were a nonzero vector at some point, then by continuity the integral over some small disc centered at that point and lying perpendicular to that vector would be nonzero.
9. The orientations of $\partial S_1 = \partial S_2$ must agree.
11. Suppose C is a closed loop on the surface drawn so that it divides the surface into two pieces S_1 and S_2. For the surface of a doughnut (torus) you must use two closed

loops . . . can you see why? Then C bounds both S_1 and S_2 but with positive orientation with respect to one and negative with respect to the other. Therefore

$$\int_S \nabla \times \mathbf{F} \cdot d\mathbf{S} = \int_{S_1} \nabla \times \mathbf{F} \cdot d\mathbf{S} + \int_{S_2} \nabla \times \mathbf{F} \cdot d\mathbf{S} = \int_C \mathbf{F} \cdot d\mathbf{s} - \int_C \mathbf{F} \cdot d\mathbf{s} = 0$$

13. (a) If $C = \partial S$, $\int_C \mathbf{v} \cdot d\mathbf{s} = \int_S \nabla \times \mathbf{v} \cdot d\mathbf{S} = \int_S \mathbf{0} \cdot d\mathbf{s} = 0.$
    (b) $\int_C \mathbf{v} \cdot d\mathbf{s} = \int_a^b \mathbf{v} \cdot \boldsymbol{\sigma}'(t)\, dt = \mathbf{v} \cdot \int_a^b \boldsymbol{\sigma}'(t)\, dt = \mathbf{v} \cdot (\boldsymbol{\sigma}(b) - \boldsymbol{\sigma}(a))$, where $\boldsymbol{\sigma}\colon [a, b] \to \mathbb{R}^3$ is a parametrization of C. (The vector integral is the vector whose components are the integrals of the component functions.) If C is closed, the last expression is 0.

15. Both integrals give $\pi/4$.

17. (a) 0 (b) π (c) π

19. 20π

SECTION 7.3

1. If $\mathbf{F} = \nabla f = \nabla g$, and C is a curve from $\mathbf{v}$ to $\mathbf{w}$, then $(f - g)(\mathbf{w}) - (f - g)(\mathbf{v}) = \int_C \nabla(f - g) \cdot d\mathbf{s} = 0$, so $f - g$ is constant.

3. $x^2yz - \cos x$

5. $-1/\|\mathbf{r}_0\|$

7. No; $\nabla \times \mathbf{F} = (0, 0, x) \neq \mathbf{0}$.

9. $e \sin 1 + \frac{1}{3}e^3 - \frac{1}{3}$

11. 3.54×10^{29} ergs

13. (a) $f = x^2/2 + y^2/2 + C$
    (b) $\mathbf{F}$ is not a gradient field (c) $f = \frac{1}{3}x^3 + xy^2 + C$

15. (a) no
    (b) $(\frac{1}{2}z^2, xy - z, x^2y)$ or $(\frac{1}{2}z^2 - 2xyz - \frac{1}{2}y^2, -x^2z - z, 0)$

17. $-(z \sin y + y \sin x, xz \cos y, 0)$(Other answers are possible.)

19. (a) $\nabla \times \mathbf{F} = (0, 0, 2) \neq \mathbf{0}$
    (b) Let $\boldsymbol{\sigma}(t)$ be the path of an object in the fluid. Then $\mathbf{F}(\boldsymbol{\sigma}(t)) = \boldsymbol{\sigma}'(t)$. Let $\boldsymbol{\sigma}(t) = (x(t), y(t), z(t))$. Then $x' = -y$, $y' = x$, and $z' = 0$, so z is constant and the motion is parallel to the xy-plane. Also, $x'' + x = 0$, $y'' + y = 0$. Thus $x = A \cos t + B \sin t$ and $y = C \cos t + D \sin t$. Substituting these values in $x' = -y$, $y' = x$, we get $C = -B$, $D = A$, so $x^2 + y^2 = A^2 + B^2$ and we have a circle.
    (c) counterclockwise

21. (a) $\mathbf{F} = -\dfrac{GM}{(x^2 + y^2 + z^2)^{3/2}}(x, y, z)$;

$$\nabla \cdot \mathbf{F} = -GM \left\{ \frac{x^2 + y^2 + z^2 - 3x^2}{(x^2 + y^2 + z^2)^{5/2}} + \frac{x^2 + y^2 + z^2 - 3y^2}{(x^2 + y^2 + z^2)^{5/2}} + \frac{x^2 + y^2 + z^2 - 3z^2}{(x^2 + y^2 + z^2)^{5/2}} \right\}$$

$$= 0$$

(b) Let S be the unit sphere, S_1 the upper hemisphere, S_2 the lower hemisphere, and C the unit circle. If $\mathbf{F} = \nabla \times \mathbf{G}$, then

$$\int_S \mathbf{F} \cdot d\mathbf{S} = \int_{S_1} \mathbf{F} \cdot d\mathbf{S} + \int_{S_2} \mathbf{F} \cdot d\mathbf{S} = \int_C \mathbf{G} \cdot d\mathbf{s} - \int_C \mathbf{G} \cdot d\mathbf{s} = 0$$

But $\int_S \mathbf{F} \cdot d\mathbf{S} = -GM \int_S (\mathbf{r}/\|\mathbf{r}\|^3) \cdot \mathbf{n}\, dS = -4\pi GM$ since $\|\mathbf{r}\| = 1$ and $\mathbf{r} = \mathbf{n}$ on S. Thus $\mathbf{F} = \nabla \times \mathbf{G}$ is impossible.

SECTION 7.4

1. $$\int_S \nabla \times \mathbf{F} \cdot d\mathbf{S} = \int_V \nabla \cdot (\nabla \times \mathbf{F}) dV = \int_V \left\{ \left(\frac{\partial^2 F_1}{\partial y\, \partial z} - \frac{\partial^2 F_1}{\partial z\, \partial y} \right) + \left(\frac{\partial^2 F_2}{\partial z\, \partial x} - \frac{\partial^2 F_2}{\partial x \partial z} \right) + \left(\frac{\partial^2 F_3}{\partial x\, \partial y} - \frac{\partial^2 F_3}{\partial y\, \partial z} \right) \right\} dV$$

This is 0 if $\mathbf{F}$ is C^2, because the mixed second derivatives of its component functions are equal.

3. 1

5. (a) 0 (b) $\frac{4}{15}$ (c) $-\frac{4}{15}$

7. If $S = \partial\Omega$, then $\int_S \mathbf{r} \cdot \mathbf{n}\, ds = \int_\Omega \nabla \cdot \mathbf{r}\, dV = \int_\Omega 3\, dV = 3$ volume(Ω).

9. If $\mathbf{F} = \mathbf{r}/r^2$ then $\nabla \cdot \mathbf{F} = 1/r^2$. If $(0, 0, 0) \notin \Omega$, the result follows from Gauss' Theorem. If $(0, 0, 0) \in \Omega$ we compute the integral by deleting a small ball $B_\varepsilon = \{(x, y, z) | (x^2 + y^2 + z^2)^{1/2} < \varepsilon\}$ around the origin and then letting $\varepsilon \to 0$:

$$\int_\Omega \frac{1}{r^2} dV = \lim_{\varepsilon \to 0} \int_{\Omega \setminus B_\varepsilon} \frac{1}{r^2} dV = \lim_{\varepsilon \to 0} \int_{\partial(\Omega \setminus B_\varepsilon)} \frac{\mathbf{r} \cdot \mathbf{n}}{r^2} dS$$

$$= \lim_{\varepsilon \to 0} \left(\int_{\partial\Omega} \frac{\mathbf{r} \cdot \mathbf{n}}{r^2} dS - \int_{\partial B_\varepsilon} \frac{\mathbf{r} \cdot \mathbf{n}}{r^2} dS \right) = \lim_{\varepsilon \to 0} \left(\int_{\partial\Omega} \frac{\mathbf{r} \cdot \mathbf{n}}{r^2} dS - 4\pi\varepsilon \right)$$

$$= \int_{\partial\Omega} \frac{\mathbf{r} \cdot \mathbf{n}}{r^2} dS$$

The integral over ∂B_ε is from Theorem 10 (Gauss' law) since $r = \varepsilon$ everywhere on B_ε.

11. Use Formula 8 of Table 3.1 and the Divergence Theorem for part (a). Use Formula 18 for part (b).

13. If $\phi(\mathbf{p}) = \int_\Omega \rho(\mathbf{q})/(4\pi\|\mathbf{p} - \mathbf{q}\|)\, dV(\mathbf{q})$, then

$$\nabla\phi(\mathbf{p}) = \int_\Omega [\rho(\mathbf{q})/4\pi]\, \nabla_\mathbf{p}(1/\|\mathbf{p} - \mathbf{q}\|)\, dV(\mathbf{q})$$
$$= -\int_\Omega [\rho(\mathbf{q})/4\pi]((\mathbf{p} - \mathbf{q})/\|\mathbf{p} - \mathbf{q}\|^3)\, dV(\mathbf{q})$$

where $\nabla_\mathbf{p}$ means the gradient with respect to the coordinates of $\mathbf{p}$ and the integral is the vector whose components are the three component integrals. If $\mathbf{p}$ varies in $V \cup \partial V$ and $\mathbf{n}$ is the outward unit normal to ∂V, we can take the inner product using these components and collect the pieces as

$$\nabla\phi(\mathbf{p}) \cdot \mathbf{n} = -\int_\Omega \frac{\rho(\mathbf{q})}{4\pi} \frac{1}{\|\mathbf{p} - \mathbf{q}\|^3} (\mathbf{p} - \mathbf{q}) \cdot \mathbf{n}\, dV(\mathbf{q})$$

Thus

$$\int_{\partial V} \nabla\phi(\mathbf{p}) \cdot \mathbf{n}\, dV(\mathbf{p}) = -\int_{\partial V}\left(\int_{\partial\Omega} \frac{\rho(\mathbf{q})}{4\pi} \frac{1}{\|\mathbf{p}-\mathbf{q}\|^3}(\mathbf{p}-\mathbf{q})\cdot\mathbf{n}\, d\mathbf{q}\right) dV(\mathbf{p})$$

There are essentially five variables of integration here, three placing $\mathbf{q}$ in Ω and two placing $\mathbf{p}$ on ∂V. Use Fubini's Theorem to obtain

$$\int_{\partial V} \nabla\phi \cdot \mathbf{n} \cdot d\mathbf{S} = -\int_{\Omega} \frac{\rho(\mathbf{q})}{4\pi}\left(\int_{\partial V} \frac{(\mathbf{p}-\mathbf{q})\cdot\mathbf{n}}{\|\mathbf{p}-\mathbf{q}\|^3}\, dS(\mathbf{p})\right) dV(\mathbf{q})$$

If V is a region of type 4, Theorem 10 says that the inner integral is 4π if $\mathbf{q} \in V$ and 0 if $\mathbf{q} \notin V$. So

$$\int_{\partial V} \nabla\phi \cdot \mathbf{n}\, dS = -\int_{\Omega\cap V} \rho(\mathbf{q})\, dV(\mathbf{q})$$

Since $\rho = 0$ outside Ω,

$$\int_{\partial V} \nabla\phi \cdot \mathbf{n}\, dS = -\int_{V} \rho(\mathbf{q})\, dV(\mathbf{q})$$

If V is not of type 4, subdivide it into a sum of such regions. The equation holds on each piece, and, upon adding them together, the boundary integrals along appropriately oriented interior boundaries cancel, leaving the desired result.

(b) By Theorem 9, $\int_{\partial V} \nabla\phi \cdot d\mathbf{S} = \int_V \nabla^2\phi\, dV$, so $\int_V \nabla^2\phi dV = -\int_V \rho\, dV$. Since both ρ and $\nabla^2\phi$ are continuous and this holds for arbitrarily small regions, we must have $\nabla^2\phi = -\rho$.

15. If the charge Q is spread evenly over the sphere S of radius R centered at the origin, the density of charge per unit area must be $Q/4\pi R$. If $\mathbf{p}$ is a point not on S and $\mathbf{q} \in S$, then the contribution to the electric field at $\mathbf{p}$ due to charge near $\mathbf{q}$ is directed along the vector $\mathbf{p} - \mathbf{q}$. Since the charge is evenly distributed, the tangential component of this contribution will be cancelled by that from a symmetric point on the other side of the sphere at the same distance from $\mathbf{p}$. (Draw the picture.) The total resulting field must be radial. Since S looks the same from any point at a distance $\|\mathbf{p}\|$ from the origin, the field must depend only on radius and be of the form $E = f(r)\mathbf{r}$.

If we look at the sphere Σ of radius $\|\mathbf{p}\|$, we have

$$\begin{aligned}(\text{Charge inside } \Sigma) &= \int_\Sigma \mathbf{E}\cdot d\mathbf{S} = \int_\Sigma f(\|\mathbf{p}\|)\mathbf{r}\cdot\mathbf{n}\, dS \\ &= f(\|\mathbf{p}\|)\|\mathbf{p}\| \text{ area } \Sigma = 4\pi\|\mathbf{p}\|^3 f(\|\mathbf{p}\|)\end{aligned}$$

If $\|\mathbf{p}\| < R$, there is no charge inside Σ; if $\|\mathbf{p}\| > R$, the charge inside Σ is Q, so

$$\mathbf{E}(\mathbf{p}) = \begin{cases} \dfrac{1}{4\pi}\dfrac{Q}{\|\mathbf{p}\|^3}\mathbf{p} & \text{if } \|\mathbf{p}\| > R \\ 0 & \text{if } \|\mathbf{p}\| < R \end{cases}$$

17. By Theorem 10, $\int_{\partial M} \mathbf{F}\cdot d\mathbf{S} = 4\pi$ for any surface enclosing the origin. But, if $\mathbf{F}$ were the curl of some field, then the integral over such a closed surface would have to be 0.

19. If v_i is the ith component of a vector $\mathbf{v}$, then Exercise 18(b) gives

$$\left[\frac{d}{dt}\int_{\Omega_t} f\mathbf{F}\, dx\, dy\, dz\right]_i = \frac{d}{dt}\int_{\Omega_t} (f\mathbf{F})_i\, dx\, dy\, dz = \frac{d}{dt}\int_{\Omega_t} fF_i\, dx\, dy\, dz$$

$$= \int_{\Omega_t}\left(\frac{D(fF_i)}{Dt} + (fF_i)\text{div }\mathbf{F}\right) dx\, dy\, dz$$

$$= \int_{\Omega_t}\left(\frac{\partial}{\partial t}(fF_i) + D_x(fF_i)\cdot\mathbf{F} + (fF_i)\text{div }\mathbf{F}\right) dx\, dy\, dz$$

$$= \int_{\Omega_t}\left(\frac{\partial}{\partial t}(fF_i) + \nabla(fF_i)\cdot\mathbf{F} + (fF_i)\text{div }\mathbf{F}\right) dx\, dy\, dz$$

$$= \int_{\Omega_t}\left[\left(\frac{\partial}{\partial t}(f\mathbf{F})\right) + (D(f\mathbf{F})\mathbf{F})_i + ((f\mathbf{F})\text{div }\mathbf{F})_i\right] dx\, dy\, dt$$

$$= \int_{\Omega_t}\left[\frac{\partial}{\partial t}(f\mathbf{F}) + D(f\mathbf{F})\mathbf{F} + (f\mathbf{F})\text{div }\mathbf{F}\right]_i dx\, dy\, dz$$

$$= \left[\int_{\Omega_t}\frac{\partial}{\partial t}(f\mathbf{F}) + D(f\mathbf{F})\mathbf{F} + (f\mathbf{F})\text{div }\mathbf{F})\, dx\, dy\, dz\right]_i$$

$$= \left[\int_{\Omega_t}\left(\frac{\partial}{\partial t}(f\mathbf{F}) + (\mathbf{F}\cdot\nabla)(f\mathbf{F}) + (f\mathbf{F})\text{div }\mathbf{F}\right) dx\, dy\, dt\right]_i$$

Since each component is equal, the vectors are equal.

SECTION 7.5

1. (a) By Exercise 18, Section 7.4,

$$\frac{d}{dt}\int_{\Omega_t}\rho\, dV = \int_{\Omega_t}\left(\frac{\partial\rho}{\partial t} + \nabla\rho\cdot\mathbf{V} + \rho\text{ div }\mathbf{V}\right) dV$$

By Theorem 11, the conservation law for $\mathbf{V}$ is equivalent to the integrand on the right being identically 0 and so implies the left integral is 0. Conversely, if Ω is a small region, so is Ω_t. If the integrals on the right are 0 for all sufficiently small regions, then the integrand must be 0 (by continuity).

(b) For each time t the change of variables $(u, v, w) = \phi((x, y, z), t)$ gives

$$\int_{\Omega_t}\rho(u, v, w, 0)\, du\, dv\, dw = \int_{\Omega}\rho(x, y, z, t)\frac{\partial(u, v, w)}{\partial(x, y, z)}\, dx\, dy\, dz$$

By (a), the left side is constant in time and is therefore equal to its value at $t = 0$:

$$\int_{\Omega}\rho(x, y, z, 0)\, dx\, dy\, dz = \int_{\Omega}\rho(x, y, z, t)J(x, y, z, t)\, dx\, dy\, dz$$

As this holds for all small regions, the integrands must be equal by continuity.

(c) By Exercise 18, Section 7.4, div $\mathbf{V} = 0$ implies $J(\mathbf{x}, t) = 1$. Applying this to (b), $\rho(\mathbf{x}, t) = \rho(\mathbf{x}, 0)$. The density at each point is constant in time, so $\partial\rho/\partial t = 0$

and the conservation law becomes $\mathbf{V} \cdot \nabla \rho = 0$. The flow is perpendicular to the gradient of ρ so the flow lines lie in surfaces of constant density.

3. (a) Since $\mathbf{V} = \nabla \rho$, $\nabla \times \mathbf{V} = \mathbf{0}$, and therefore $\mathbf{V} \cdot \nabla \mathbf{V} = \frac{1}{2}\nabla(\|\mathbf{V}\|^2)$. Euler's equation becomes

$$-\frac{\nabla \rho}{\rho} = \frac{d\mathbf{V}}{dt} + \frac{1}{2}\nabla(\|\mathbf{V}\|^2) = \nabla\left(\frac{d\phi}{dt} + \frac{1}{2}\|\mathbf{V}\|^2\right)$$

If σ is a path from P_1 to P_2, then

$$-\int_\sigma \frac{1}{\rho}\, dp = -\int \frac{1}{\rho} \nabla p \cdot \boldsymbol{\sigma}'(t)\, dt = \int_\sigma \nabla\left(\frac{d\phi}{dt} + \frac{1}{2}\|\mathbf{V}\|^2\right) \cdot \boldsymbol{\sigma}'(t)\, dt$$

$$= \left(\frac{d\phi}{dt} + \frac{1}{2}\|\mathbf{V}\|^2\right)\Bigg|_{P_1}^{P_2}$$

(b) If $d\mathbf{V}/dt = \mathbf{0}$ and ρ is constant, then $\frac{1}{2}\nabla(\|\mathbf{V}\|^2) = -(\nabla p)/\rho = -\nabla(p/\rho)$ and therefore $\nabla(\frac{1}{2}\|\mathbf{V}\|^2 + p/\rho) = \mathbf{0}$.

5. By Ampere's law, $\nabla \cdot \mathbf{J} = \nabla \cdot (\nabla \times \mathbf{H}) - \nabla \cdot (\partial \mathbf{E}/\partial t) = -\nabla \cdot (\partial \mathbf{E}/\partial t) = -(\partial/\partial t)(\nabla \cdot \mathbf{E})$. By Gauss' law this is $-\partial \rho/\partial t$. Thus $\nabla \cdot \mathbf{J} + \partial \rho/\partial t = 0$.

7. (a) If $\mathbf{x} \in S$, then $r''a/R = r$, so $G = 0$. In general, $r = \|\mathbf{x} - \mathbf{y}\|$ and $r'' = \|\mathbf{x} - \mathbf{y}\|$, and therefore

$$G = \frac{1}{4\pi}\left[\frac{R}{a}\frac{1}{\|\mathbf{x} - \mathbf{y}'\|} - \frac{1}{\|\mathbf{x} - \mathbf{y}\|}\right] \quad \text{and} \quad \nabla_{\mathbf{x}} G = \frac{1}{4\pi}\left[\frac{R}{a}\frac{\mathbf{y}' - \mathbf{x}}{\|\mathbf{x} - \mathbf{y}'\|^3} - \frac{\mathbf{y} - \mathbf{x}}{\|\mathbf{x} - \mathbf{y}\|^3}\right]$$

and $\nabla_{\mathbf{x}}^2 G = 0$ when $\mathbf{x} \neq \mathbf{y}$ just as in the analysis of equation (15) ($\mathbf{x} \neq \mathbf{y}'$ since $\mathbf{x}$ is inside and $\mathbf{y}'$ is outside the sphere). Theorem 10 gives $\nabla^2 G = \delta(\mathbf{x} - \mathbf{y}) - (R/a)\delta(\mathbf{x} - \mathbf{x}')$, but the second term is always 0 since $\mathbf{x}$ is never $\mathbf{y}'$. So $\nabla^2 G = \delta(\mathbf{x} - \mathbf{y})$ for x and y in the sphere.

(b) If $\mathbf{x}$ is on the surface of S, then $\mathbf{n} = \mathbf{x}/R$ is the outward unit normal, and

$$\frac{\partial G}{\partial \mathbf{n}} = \frac{1}{4\pi}\left[\frac{R}{a}\nabla\left(\frac{1}{r''}\right) \cdot \mathbf{n} - \nabla\left(\frac{1}{r}\right) \cdot \mathbf{n}\right] = \frac{1}{4\pi}\left[\frac{R}{a}\frac{\mathbf{y}' - \mathbf{x}}{\|\mathbf{y}' - \mathbf{x}\|^3} - \frac{\mathbf{y} - \mathbf{x}}{\|\mathbf{x} - \mathbf{y}\|^3}\right] \cdot \mathbf{n}$$

If γ is the angle between $\mathbf{x}$ and $\mathbf{y}$, then $\|\mathbf{x} - \mathbf{y}\|^2 = r^2 = R^2 + a^2 - 2aR\cos\gamma$ and $\|\mathbf{x} - \mathbf{y}'\|^2 = r''^2 = R^2 + b^2 - 2bR\cos\gamma = (R^2/a^2)r^2$. So

$$\frac{\partial G}{\partial \mathbf{n}} = \frac{1}{4\pi r^3}\left[\frac{R}{a}\frac{\mathbf{y}' - \mathbf{x}}{(R/a)^3} - (\mathbf{y} - \mathbf{x})\right] \cdot \mathbf{n}$$

But $\mathbf{x} \cdot \mathbf{n} = R$ and $\mathbf{n} = \mathbf{x}/R$, so this becomes

$$\frac{\partial G}{\partial \mathbf{n}} = \frac{1}{4\pi r^3}\left[\frac{a^2}{R^3}\mathbf{y}' \cdot \mathbf{x} - \frac{a^2}{R} - \frac{\mathbf{y} \cdot \mathbf{x}}{R} + R\right]$$

$$= \frac{1}{4\pi r^3 R}\left[\frac{a^2}{R^2}\|\mathbf{y}'\|R\cos\gamma - \|\mathbf{y}\|R\cos\gamma + R^2 - a^2\right]$$

$$= \frac{R^2 - a^2}{4\pi R}\frac{1}{r^3} \quad \text{since} \quad \|\mathbf{y}'\| = R^2/a$$

Integrating over the surface of the sphere,

$$u(\mathbf{y}) = \int_S f\,\frac{\partial G}{\partial \mathbf{n}}\,dS = \int_0^{2\pi}\int_0^{\pi}\left(f(\theta, \phi)\,\frac{R^2 - a^2}{4\pi R}\,\frac{1}{r^3}\,R^2 \sin\phi\right) d\phi\,d\theta$$

$$= \frac{R(R^2 - a^2)}{4\pi}\int_0^{2\pi}\int_0^{\pi}\frac{f(\theta, \phi)\sin\phi\,d\phi\,d\theta}{(R^2 + a^2 - 2aR\cos\gamma)^{3/2}}$$

9. (a) $\dot{u} = d/ds\{u(x(s), t(s))\} = u_x\dot{x} + u_t\dot{t} = u_x f'(u) + u_t = 0$

(b) If the characteristic curve $u(x, t) = c$ (by part (a)), define t implicitly as a function of x; then $u_x + u_t(dt/dx) = 0$. But also $u_t + f(u)_x = 0$; that is, $u_t + f'(u)u_x = 0$. These two equations together give $dt/dx = 1/f'(u) = 1/f'(c)$. So the curve is a straight line with slope $1/f'(c)$.

(c) If $x_1 < x_2$, $u_0(x_1) > u_0(x_2) > 0$, and $f'(u(x_2)) > 0$, then $f'(u_0(x_1)) > f'(u_0(x_2)) > 0$ since $f'' > 0$. The characteristic through $(x_1, 0)$ has slope $1/f'(u_0(x_1))$, which is less than $1/f'(u_0(x_2))$ (that of the characteristic through $(x_2, 0)$). So these lines must cross at a point $P = (\bar{x}, \bar{t})$ with $\bar{t} > 0$ and $\bar{x} > x_2$. The solution must be discontinuous at P as these two crossing lines would give it different values there.

(d) $\bar{t} = (x_2 - x_1)/(f'(u_0(x_1)) - f'(u_0(x_2)))$

11. (a) Since the "rectangle" D does not touch the x-axis and $\phi = 0$ on ∂D and outside D, (25) becomes

(*i*)
$$\iint (u\phi_t + f(u)\phi_x) = 0$$

Since $(u\phi)_t + (f(u)\phi)_x = (u_t + f(u)_x)\phi + (u\phi_t + f(u)\phi_x)$, we have

$$\iint_{D_i} (u\phi_t + f(u)\phi_x) = \iint_{D_i} ((u\phi)_t + (f(u)\phi)_x) - \iint_{D_i} (u_t + f(u)_x)\phi$$

But u is C^1 on the interior of D_i, so Exercise 10(b) says $u_t + f(u)_x = 0$ there. Thus

(*ii*)
$$\iint_{D_i} (u\phi_t + f(u)\phi_x) = \iint_{D_i} ((u\phi)_t + (f(u)\phi)_x)$$

(b) By Green's Theorem,

$$\iint_{D_i} ((f(u)\phi)_x - (-u\phi)_t)\,dx\,dt = \int_{\partial D_i} ((-u\phi)\,dx + f(u)\phi\,dt),$$

so (*ii*) becomes

$$\iint (u\phi_t + f(u)\phi_x) = \int_{\partial D_i} \phi\{-u\,dx + f(u)\,dt\}$$

Adding the above for $i = 1, 2$ and using (*i*) gives

$$0 = \int_{\partial D_1} \phi\{-u\,dx + f(u)\,dt\} + \int_{\partial D_2} \phi\{-u\,dx + f(u)\,dt\}$$

The union of these two boundaries traverses ∂D once and that portion of Γ within D once in each direction, once with the values u_1 and once with the values u_2. Since $\phi = 0$ outside D and on ∂D, this becomes $0 = \int_\Gamma \phi\{[-u]\,dx + [f(u)]\,dt\}$.

(c) Since $\phi = 0$ outside D, the first integral is the same as that of the second conclusion of part (b). The second integral results from parametrizing the portion of Γ by $\alpha(t) = (x(t), t)$, $t_1 \leq t \leq t_2$.

(d) If $[-u]s + [f(u)] = c > 0$ at P, then we can choose a small disc B_ε centered at P contained in D (described above) and such that $[-u](dx/dt) + [f(u)] > c/2$ on the part of Γ inside B_ε. Now take a slightly smaller disc $B_\delta \subset B_\varepsilon$ centered at P and pick ϕ such that $\phi \equiv 1$ on B_δ, $0 \leq \phi \leq 1$ on the annulus $B_\varepsilon \backslash B_\delta$ and $\phi \equiv 0$ outside B_ε. If $\alpha(t_0) = P$ then there are t_3 and t_4 with $t_1 < t_3 < t_0 < t_4 < t_2$ and $\alpha(t) \in B_\delta$ for $t_3 < t < t_4$. But then

$$\int_{t_1}^{t_2} \phi\{[-u]\frac{dx}{dt} + [f(u)]\}\,dt > \frac{c}{2}(t_4 - t_3) > 0$$

contradicting the result of part (c). A similar argument (reversing signs) works if $c < 0$.

13. Setting $P = g(u)\phi$ and $Q = -f(u)\phi$, applying Green's Theorem on rectangular R, and using the function ϕ as in Exercise 10 shows that if u is a solution of $g(u)_t + f(u)_x = 0$, then

$$\iint_{t \geq 0} (g(u)\phi_t + f(u)\phi_x)\,dx\,dt + \int_{t=0} g(u_0(x))\phi(x, 0)\,dx = 0$$

This is the appropriate analogy to (25), defining weak solutions of $g(u)_t + f(u)_x = 0$. Thus we want a u such that

(*i*-weak) $$\iint_{t \geq 0} (u\phi_t + \tfrac{1}{2}u^2\phi_x)\,dx\,dt + \int_{t=0} u_0(x)\phi(x, 0)\,dx = 0$$

holds for all admissible ϕ but such that

(*ii*-weak) $$\iint_{t \geq 0} (\tfrac{1}{2}u^2\phi_t + \tfrac{1}{3}u^3\phi_x)\,dx\,dt + \int_{t=0} \tfrac{1}{2}u_0^2(x)\phi(x, 0)\,dx = 0$$

fails for some admissible ϕ. The method of Exercise 11 produces the jump condition $s[g(u)] = [f(u)]$. For (a), this is $s(u_2 - u_1) = (\frac{1}{2}u_2^2 - \frac{1}{2}u_1^2)$ or

(*i*-jump) $$s = \tfrac{1}{2}(u_2 + u_1)$$

For (b), it is $s(\frac{1}{2}u_2^2 - \frac{1}{2}u_1^2) = (\frac{1}{3}u_2^3 - \frac{1}{3}u_1^3)$ or

(*ii*-jump) $$s = \frac{2}{3}\,\frac{u_2^2 + u_1u_2 + u_1^2}{u_2 + u_1}$$

If we take for $u_0(x)$ a (Heaviside) function defined by $u_0(x) = 0$ for $x < 0$, and $u_0(x) = 1$ for $x > 0$, we are led to consider the function $u(x, t) = 0$ when $t > 2x$ and $u(x, t) = 1$ when $t \leq 2x$. Thus $u_1 = 1$, $u_2 = 0$, and the discontinuity curve Γ is given by $t = 2x$. So the jump condition (*i*-jump) (i.e., $dx/dt = \frac{1}{2}(u_1 + u_2)$) is satisfied.

For any particular ϕ, there are numbers T and a such that $\phi(x, t) = 0$ for $x \geq a$ and $t \geq T$. Letting Ω be the region $0 \leq x \leq a$ and $0 \leq t \leq T$, condition (*i*-weak) becomes

$$\begin{aligned}
0 &= \iint_{\Omega} (\phi_t + \tfrac{1}{2}\phi_x)\, dx\, dt + \int_0^a \phi(x, 0)\, dx \\
&= \int_{\partial\Omega} \left(-\phi\, dx + \frac{\phi}{2}\, dt \right) + \int_0^a \phi(x, 0)\, dx \\
&= -\int_0^a \phi(x, 0)\, dx + \int_0^{T/2} \{-\phi(x, 2x)(-dx) + \tfrac{1}{2}\phi(x, 2x)(-2\, dx)\} \\
&\quad + \int_0^a \phi(x, 0)\, dx \\
&= 0
\end{aligned}$$

Thus (*i*-weak) is satisfied for every ϕ, and u is a weak solution of (a). However, (*ii*-weak) cannot be satisfied for every ϕ, since the jump condition (*ii*-jump) fails. Indeed, if we multiply (*ii*-weak) by 2 and insert u, (*ii*-weak) becomes

$$0 = \iint_{\Omega} (\phi_t + \tfrac{2}{3}\phi_x)\, dx\, dt + \int_0^a \phi(x, 0)\, dx$$

The factor 1/2 has changed to 2/3, and the computation above now becomes

$$0 = -\frac{1}{3} \int_0^{\pi/2} \phi(x, 2x)\, dx$$

which is certainly not satisfied for every admissible ϕ.

SECTION 7.6

1. (a) $(2xy^2 - yx^3)\, dx\, dy$ (b) $(x^2 + y^2)\, dx\, dy$
   (c) $(x^2 + y^2 + z^2)\, dx\, dy\, dz$ (d) $(xy + x^2)\, dx\, dy\, dz$
   (e) $dx\, dy\, dz$

3. (a) $2xy\, dx + (x^2 + 3y^2)\, dy$ (b) $-(x + y^2 \sin x)\, dx\, dy$
   (c) $-(2x + y)\, dx\, dy$ (d) $dx\, dy\, dz$
   (e) $2x\, dx\, dy\, dz$ (f) $2y\, dy\, dz - 2x\, dz\, dx$
   (g) $-\dfrac{4xy}{(x^2 + y^2)^2}\, dx\, dy$ (h) $2xy\, dx\, dy\, dz$

5. (a)
$$\begin{aligned}
\text{Form}_2(\alpha\mathbf{V}_1 + \mathbf{V}_2) &= \text{Form}_2((\alpha A_1 + A_2), \alpha B_1 + B_2, \alpha C_1 + C_2)) \\
&= (\alpha A_1 + A_2)\, dy\, dz + (\alpha B_1 + B_2)\, dz\, dx \\
&\quad + (\alpha C_1 + C_2)\, dx\, dy \\
&= \alpha(A_1\, dy\, dz + B_1\, dz\, dx + C_1\, dx\, dy) \\
&\quad + (A_2\, dy\, dz + B_2\, dz\, dx + C_2\, dx\, dy) \\
&= \alpha\text{Form}_2(\mathbf{V}_1) + \text{Form}_2(\mathbf{V}_2)
\end{aligned}$$

(b) $$d\omega = \left[\frac{\partial A}{\partial x}dx + \frac{\partial A}{\partial y}dy + \frac{\partial A}{\partial z}dz\right] \wedge dx - A(dx)^2$$
$$+ \left[\frac{\partial B}{\partial x}dx + \frac{\partial B}{\partial y}dy + \frac{\partial B}{\partial z}dz\right] \wedge dy - B(dy)^2$$
$$+ \left[\frac{\partial C}{\partial x}dx + \frac{\partial C}{\partial y}dy + \frac{\partial C}{\partial z}dz\right] \wedge dz - C(dz)^2$$

But $(dx)^2 = (dy)^2 = (dz)^2 = dx \wedge dx = dy \wedge dy = dz \wedge dz = 0$, $dy \wedge dx = -dx \wedge dv$, $dz \wedge dy = -dy \wedge dz$, and $dx \wedge dz = -dz \wedge dx$. So

$$d\omega = \left[\frac{\partial C}{\partial y} - \frac{\partial B}{\partial z}\right] dy\, dz + \left[\frac{\partial A}{\partial z} - \frac{\partial C}{\partial x}\right] dz\, dx + \left[\frac{\partial B}{\partial x} - \frac{\partial A}{\partial y}\right] dx\, dy$$
$$= \text{Form}_2(\text{curl } \mathbf{V})$$

7. An oriented 1-manifold is a curve. Its boundary is a pair of points that may be considered a 0-manifold. So ω is a 0-form or function, and $\int_{\partial M} d\omega = \omega(b) - \omega(a)$ if the curve M runs from a to b. Furthermore $d\omega$ is the one form $(\partial\omega/\partial x)\, dx + (\partial\omega/\partial y)\, dy$. So $\int_M d\omega$ is the line integral $\int_M (\partial\omega/\partial x)\, d\omega + (\partial\omega/dy)\, dy = \int_M \nabla\omega \cdot d\mathbf{s}$. Thus we obtain Theorem 3 of Section 6.2, $\int_M \nabla\omega \cdot d\mathbf{s} = \omega(b) - \omega(a)$.

9. Put $\omega = F_1\, dx\, dy + F_2\, dy\, dz + F_3\, dz\, dx$. The integral becomes

$$\int_{\partial T} \omega = \int_T d\omega = \int_T \left[\frac{\partial F_1}{\partial z} + \frac{\partial F_2}{\partial x} + \frac{\partial F_3}{\partial y}\right] dx\, dy\, dz$$

(a) 0 (b) 40

11. Consider $\omega = x\, dy\, dz + y\, dz\, dx + z\, dx\, dy$. Compute that $d\omega = 3\, dx\, dy\, dz$ so that $\frac{1}{3}\int_{\partial R} \omega = \frac{1}{3}\int_R d\omega = \int_R dx\, dy\, dz = v(R)$.

REVIEW EXERCISES FOR CHAPTER 7

1. $\pi a^2\, 4$

3. (a) $\nabla\phi = \dfrac{\mathbf{m}}{r^3} - \dfrac{3\mathbf{m} \cdot \mathbf{r}}{r^5}\mathbf{r}$

(b) $\nabla^2\phi = 0$

(c) $\nabla \cdot \mathbf{F} = 0$

(d) $\nabla \times \mathbf{F} = \dfrac{2}{r^3}\mathbf{m} - \dfrac{3}{r^5}\mathbf{r} \times (\mathbf{m} \times \mathbf{r})$

5. $\nabla \times (\mathbf{F} - \mathbf{G}) = 0$, so $\mathbf{F} - \mathbf{G}$ is a gradient field.

7. (a) $2\pi a^2$ (b) 0

9. 0

11. (a) $f = x^4/4 - x^2y^3$ (b) $-\frac{1}{4}$

13. (a) Check that $\nabla \times \mathbf{F} = \mathbf{0}$

(b) $f = 3x^2y \cos z$; (c) 0

15. $\frac{23}{6}$

17. no; $\nabla \times (\mathbf{a} \times \mathbf{r}) = 2\mathbf{a}$

19. (a) $\nabla f = 3ye^{z^2}\mathbf{i} + 3xe^{z^2}\mathbf{j} + 6xyze^{z^2}\mathbf{k}$
    (b) 0
    (c) both sides are 0

21. $8\pi/3$

23. (a) False (b) True (c) True
    (d) True (e) True (f) False
    (g) False (h) True (i) True
    (j) False (k) False

25. 21

27. (a) $\mathbf{G}$ is conservative; $\mathbf{F}$ is not.
    (b) $\mathbf{G} = \nabla\phi$ if $\phi = (x^4/4) + (y^4/4) - \frac{3}{2}x^2y^2 + \frac{1}{2}z^2 + C$, where C is any constant.
    (c) $\int_\alpha \mathbf{F}\cdot d\mathbf{s} = 0$; $\int_\alpha \mathbf{G}\cdot d\mathbf{s} = -1/2$; $\int_\beta \mathbf{F}\cdot d\mathbf{s} = 1/3$; $\int_\beta \mathbf{G}\cdot d\mathbf{s} = -\frac{1}{2}$

APPENDIX A

1. $$Df(x, y, z) = \begin{bmatrix} e^x & 0 & 0 \\ 0 & -\sin y & 0 \\ 0 & 0 & \cos z \end{bmatrix};$$

   Df is diagonal if each component function f_i depends only on x_i.

3. (a) Let $A = B = C = \mathbb{R}$ with $f(x) = 0$ and $g(x) = 0$ if $x \neq 0$ and $g(0) = 1$. Then $w = 0$ and $g(f(x)) = 1$ for all x.
   (b) If $\varepsilon > 0$, let δ_1 and δ_2 be small enough so that $D_{\delta_1}(\mathbf{y}_0) \subset B$ and $\|g(\mathbf{y}) - \mathbf{w}\| < \varepsilon$ whenever $\mathbf{y} \in B$ and $0 < \|\mathbf{y} - \mathbf{y}_0\| < \delta_2$. Since $g(\mathbf{y}_0) = \mathbf{w}$, the $0 < \|\mathbf{y} - \mathbf{y}_0\|$ restriction may be dropped. Let δ be small enough that $\|f(\mathbf{x}) - \mathbf{y}_0\| < \min(\delta_1, \delta_2)$ whenever $\mathbf{x} \in A$ and $0 < \|\mathbf{x} - \mathbf{x}_0\| < \delta$. Then for such $\mathbf{x}$, $\|f(\mathbf{x}) - \mathbf{y}_0\| < \delta_1$, so $f(\mathbf{x}) \in B$ and $g(f(\mathbf{x}))$ is defined. Also, $\|f(\mathbf{x}) - \mathbf{y}_0\| < \delta_2$, so $\|g(f(\mathbf{x})) - \mathbf{w}\| < \varepsilon$.

5. Let $\mathbf{x} = (x_1, \ldots, x_n)$ and fix an index k. Then

$$f(\mathbf{x}) = a_{kk}x_k^2 + \sum_{\substack{i=1 \\ i\neq k}}^{n} a_{kj}x_kx_j + \sum_{\substack{j=1 \\ j\neq k}}^{n} a_{kj}x_kx_j + [\text{terms not involving } x_k]$$

   and therefore

$$\frac{\partial f}{\partial x_k} = 2a_{kk}x_k + \sum_{i\neq k} a_{ki}x_i + \sum_{i\neq k} a_{ki}x_i = 2\sum_{j=1}^{n} a_{kj}x_j = (2A\mathbf{x})_k.$$

   Since the kth components agree for each k, $\nabla f(\mathbf{x}) = 2A\mathbf{x}$.

7. The matrix T of partial derivatives is formed by placing $Dg(x_0)$ and $Dh(y_0)$ next to each other so that $T(x_0, y_0)(x - x_0, y - y_0) = Dg(x_0)(x - x_0) + Dh(y_0)(y - y_0)$. Now use the triangle inequality and the fact that $\|(x - x_0, y - y_0)\|$ is larger than $|x - x_0|$ and $|y - y_0|$ to show that $\|f(x, y) - f(x_0, y_0) - T(x_0, y_0)(x - x_0, y - y_0)\|/\|(x - x_0, y - y_0)\|$ goes to 0.

9. Use the limit theorems and the fact that the function $g(x) = \sqrt{|x|}$ is continuous. (Prove the last statement.)

11. For continuity at (0, 0), use the fact that

$$\left|\frac{xy}{(x^2 + y^2)^{1/2}}\right| \leq \frac{|xy|}{(x^2)^{1/2}} = |y|$$

or $|xy| \leq (x^2 + y^2)/2$.

13. 0 (See Exercise 11.)

15. Let $\mathbf{x}$ take the role of $\mathbf{x}_0$ and $\mathbf{x} + \mathbf{h}$ that of $\mathbf{x}$ in the definition.

17. The vector $\mathbf{a}$ takes the position of $\mathbf{x}_0$ in the definition of limit or in Theorem 2. In either case, the limit depends only on values of $f(\mathbf{x})$ for $\mathbf{x}$ near $\mathbf{x}_0$, not for $\mathbf{x} = \mathbf{x}_0$. Therefore $f(\mathbf{x}) = g(\mathbf{x})$ for $\mathbf{x} \neq \mathbf{a}$ certainly suffices to make the limits equal.

19. (a) $\operatorname{limit}_{\mathbf{x}\to\mathbf{0}} (f_1 + f_2)(\mathbf{x})/\|\mathbf{x}\| = \operatorname{limit}_{\mathbf{x}\to\mathbf{0}} f_1(\mathbf{x})/\|\mathbf{x}\| + \operatorname{limit}_{\mathbf{x}\to\mathbf{0}} f_2(\mathbf{x})/\|\mathbf{x}\| = 0$.

(b) Let $\varepsilon > 0$. Since f is $o(\mathbf{x})$, there is a $\delta > 0$ such that $\|f(\mathbf{x})/\|\mathbf{x}\|\,\| < \varepsilon/c$ whenever $0 < \|\mathbf{x}\| < \delta$. Then $\|(gf)(\mathbf{x})/\|\mathbf{x}\|\| \leq g(\mathbf{x})|\,\|f(\mathbf{x})/\|\mathbf{x}\|\,\| < \varepsilon$, so $\operatorname{limit}_{\mathbf{x}\to\mathbf{0}} (gf)(\mathbf{x})/\|\mathbf{x}\| = 0$.

(c) $\operatorname{limit}_{x\to 0} f(x)/|x| = \operatorname{limit}_{x\to 0} |x| = 0$, so $f(x)$ is $o(x)$. But $\operatorname{limit}_{x\to 0} g(x)/|x|$ does not exist since $g(x)/|x| = \pm 1$ (as x is positive or negative). Therefore $g(x)$ is not $o(x)$.

APPENDIX B

1. If $a \neq b$, let $\varepsilon = |a - b|/2$.

3. Let $e = 2d - c$, so that $d = (c + e)/2$. Consider the vertical "doubling" of R defined by $Q = R \cup R_1$ where $R_1 = [a, b] \times [d, e]$. If f is extended to Q by letting f be 0 on the added part, then f is integrable over Q by additivity. The nth regular partition of $[(a + b)/2, b] \times [c, d]$ is part of the $2n$th regular partition of Q. For large n, the Riemann sums for that $2n$th partition cannot vary by more than ε as we change the choice of points from the subrectangles, in particular if we change only those in $[(a + b)/2, b] \times [c, d]$. These changes correspond to the possible changes for the Riemann sums for the nth partition of $[(a + b)/2, b] \times [c, d]$. The argument for part (b) is similar.

5. Let $R = [a, b] \times [c, d]$ and $B = [e, f] \times [g, h]$. Since the rectangles of a partition of R intersect only along their edges, their areas can be added, and b_n is the area of the union of all subrectangles of the nth regular partition of R that intersect B. Since B is contained in this union, Area $(B) \leq b_n$. On the other hand, if (x, y) is in the union, then $\varepsilon - (b - a)/n \leq x \leq f + (b - a)/n$ and $g - (d - c)/n \leq y \leq h + (d - c)/n$. This leads to $b_n \leq \text{Area}(B) + 2[(b - a)(h - g) + (d - c)(f - e)]/n + 4(a - b)(d - c)/n^2$. Letting $n \to \infty$ and combining the inequalities proves the assertion.

7. (a) The strategy is to go from point to point within $[a, b]$ by short steps, adding up the changes as you go. Given $\varepsilon > 0$, ϕ is uniformly continuous and therefore there exists a $\delta > 0$ such that $|\phi(x) - \phi(y)| \leq \varepsilon$ whenever $|x - y| < \delta$.

Let $x \in [a, b]$ and introduce intermediate points $a = x_0 < x_1 < \cdots < x_{n-1} < x_n = x$ with $x_{i+1} - x_i < \delta$. This can be done with no more than $[(b - a)/\delta] + 1$ segments. By the triangle inequality,

$$|\phi(x) - \phi(a)| \leq \sum_{i=1}^{n} |\phi(x_i) - \phi(x_{i-1})| \leq \left(\frac{b - a}{\delta + 1}\right) \varepsilon$$

Thus $|\phi(x)| \leq |\phi(a)| + ((b - a)/\delta + 1)\varepsilon$ for every x in $[a, b]$.

(b) Use an argument like that for (a), moving by short steps within the rectangle $[a, b] \times [c, d]$.

(c) This is trickier since D may be composed of many disconnected pieces so that the short steps cannot be taken within D. Nevertheless, given ε there is a δ such that $|f(\mathbf{x}) - f(\mathbf{y})| \leq \varepsilon$ whenever $\mathbf{x}$ and $\mathbf{y}$ are in D and $\|\mathbf{x} - \mathbf{y}\| < \delta$ by the uniform boundedness principle. Since D is bounded, we may find a large "cube" R with sides of length L such that $D \subset R$. Partition R into subcubes by dividing each edge into m parts. The diagonal of each subcube has length $\sqrt{n}L/m$. If we take $m > \sqrt{n}L/\delta$, any two points in the same subcube are less than δ apart, and there are m^n subcubes. If $R_1, \ldots, R_N$ are those that intersect D, choose $\mathbf{x}_i \in D \cap R_i$. For any $\mathbf{x} \in D$, we have $|f(\mathbf{x})| < \varepsilon + \max(|f(\mathbf{x}_1)|, \ldots, |f(\mathbf{x}_n)|)$.

APPENDIX E

A FEW IMPORTANT SYMBOLS

Symbols Are Listed in Order of Their Appearance in the Text

Symbol	**Name**
$\mathbb{R}$	real numbers xv
$\mathbb{Q}$	rational numbers xv
$[a, b]$	closed interval $\{x \mid a \leq x \leq b\}$ xvi
$]a, b[$	open interval $\{x \mid a < x < b\}$ xvi
$[a, b]$	half-open interval $\{x \mid a \leq x < b\}$ xvi
$]a, b]$	half-open interval $\{x \mid a < x \leq b\}$ xvi
$\lvert a \rvert$	absolute value of a xvi
$\mathbb{R}^n$	n-dimensional space 3
$\mathbf{i}, \mathbf{j}, \mathbf{k}$	standard basis in $\mathbb{R}^3$ 9
$\mathbf{v} \cdot \mathbf{w}$	inner product of two vectors 17
$\lVert \cdot \rVert$	norm of a vector 18
$\mathbf{v} \times \mathbf{w}$	cross product 29
(r, θ, z)	cylindrical coordinates 41
(ρ, θ, ϕ)	spherical coordinates 44
$D_r(\mathbf{x}_0)$	disc of radius r about $\mathbf{x}_0$ 81

$\lim_{\mathbf{x}\to\mathbf{b}}$	limit as **x** approaches **b** 87
$\lim_{x\to b^-}$	left-hand limit 102
$\dfrac{\partial f}{\partial x}$	partial derivative 105
$Df(\mathbf{x}_0)$	derivative of f at $\mathbf{x}_0$ 111
∇f	grad f, gradient of f 112
C^1	continuously differentiable 114
C^2	twice continuously differentiable 138
$\boldsymbol{\sigma}$	a path 152
∇	del 178
$\nabla \times \mathbf{F}$	curl **F**, curl 178
$\nabla \cdot \mathbf{F}$	div **F**, divergence 182
∇^2	Laplacian 183
$Hf(x_0)$	Hessian 209
$\int_D f\,dA = \int_D f(x, y)\,dx\,dy$	double integral 255
$\int_W f\,dV = \int_W f(x, y, z)\,dx\,dy\,dz$	triple integral 299
$\dfrac{\partial(x, y)}{\partial(u, v)}$	Jacobian 314
$\int_C f\,ds$	path integral 335
$\int_C \mathbf{F}\cdot d\mathbf{s}$	line integral 342
$\boldsymbol{\sigma}_{\mathrm{op}}$	opposite path 347
$\int_S f\,dS$	scalar surface integral 380
$\int_S \mathbf{F}\cdot d\mathbf{S} = \int_S \mathbf{F}\cdot\mathbf{n}\,ds$	vector surface integral (*flux*) 387

INDEX